Springer Tracts in Advanced Robotics

Volume 92

For further volumes:
http://www.springer.com/series/5208

STAR (Springer Tracts in Advanced Robotics) has been promoted under the auspices of EURON (European Robotics Research Network)

Kazuya Yoshida · Satoshi Tadokoro
Editors

Field and Service Robotics

Results of the 8th International Conference

Springer

Editors
Kazuya Yoshida
Department of Aerospace Engineering
Tohoku University
Sendai
Japan

Satoshi Tadokoro
Graduate School of Information Sciences
Tohoku University
Sendai
Japan

ISSN 1610-7438 ISSN 1610-742X (electronic)
ISBN 978-3-642-40685-0 ISBN 978-3-642-40686-7 (eBook)
DOI 10.1007/978-3-642-40686-7
Springer Heidelberg New York Dordrecht London

Library of Congress Control Number: 2013955063

Printed on acid-free paper

Springer is part of Springer Science+Business Media (www.springer.com)

Foreword

Robotics is undergoing a major transformation in scope and dimension. From a largely dominant industrial focus, robotics is rapidly expanding into human environments and is vigorously engaged in its new challenges. Interacting with, assisting, serving, and exploring with humans, the emerging robots will increasingly touch people and their lives.

Beyond its impact on physical robots, the body of knowledge robotics has produced reveals a much wider range of applications reaching across diverse research areas and scientific disciplines, such as: biomechanics, haptics, neurosciences, virtual simulation, animation, surgery, and sensor networks among others. In return, the challenges of the new emerging areas are proving an abundant source of stimulation and insights for the field of robotics. It is indeed at the intersection of disciplines that the most striking advances happen.

The *Springer Tracts in Advanced Robotics (STAR)* is devoted to bringing to the research community the latest advances in the robotics field on the basis of their significance and quality. Through a wide and timely dissemination of critical research developments in robotics, our objective with this series is to promote more exchanges and collaborations among the researchers in the community and contribute to further advancements in this rapidly growing field.

The Eighth edition of *Field and Service Robotics* edited by Kazuya Yoshida and Satoshi Tadokoro offers in its ten-part volume a collection of a broad range of topics spanning: disaster response, service/entertainment robots, inspection/maintenance robots, mobile robot navigation, agricultural robots, robots for excavation, planetary exploration, large area mapping, SLAM for outdoor robots, and elemental technology for mobile robots. The contents of the 43 contributions represent a cross-section of the current state of robotics research from one particular aspect: field and service applications, and how they reflect on the theoretical basis of subsequent developments. Pursuing technologies aimed at emergency scenarios and realizing robots operating in real and challenging disaster environments is the big challenge running throughout this focused collection.

Rich in topics and authoritative contributors, FSR culminates with this unique reference to the current developments and new directions in field and service robotics. A fine addition to the series!

Naples, Italy, July 2013

Bruno Siciliano

Preface

FSR, the International Conference on Field and Service Robotics, is the leading single track conference of robotics for field and service applications. Its goal is to report and encourage the development and experimental evaluation of field and service robots, and to generate a vibrant exchange and discussion in the community. Field robots are non-factory robots, typically mobile, that operate in complex and dynamic environments: on the Earth, or other planetary surface, under the ground, underwater, in the air, or in space. Service robots are those that work closely and interactively with humans to help them with their lives.

The first FSR conference was held in Canberra, Australia, in 1997. Since then it has been held every two years in rotation in Australia, America, Europe, and Asia, such as Pittsburg, USA (1999), Helsinki, Finland (2001), Mount Fuji, Japan (2003), Port Douglas, Australia (2005), Chamonix, France (2007), and Cambridge, USA (2009). Following this tradition, the eighth FSR conference was originally planned for 2011 with the venue of Matsushima in Tohoku region of Japan. However, on March 11, 2011, a magnitude M9.0 earthquake occurred off the Pacific coast of Tohoku, and a large-scale disaster was caused by the Tsunami which resulted. However, the Matsushima area suffered relatively minor damage, so rather than cancelling the conference, it was decided to postpone it to July 2012.

In fact, this earthquake raised issues concerning the contribution of field and service robotics technology to emergency scenarios. A number of precious lessons were learned from operation of robots in the resulting, very real and challenging, disaster environments. We therefore organized a special session to feature the up-to-date study on disaster response, relief, and recovery on 16 July 2012, the first day of the conference held in Tohoku University. Then, three-day technical sessions were followed in Matsushima on 17–19 July.

This book presents the results of FSR2012, the eighth conference of Field and Service Robotics. There are 43 papers in this volume. Thirty-seven papers were presented in oral sessions and six were presented in an interactive session. The papers cover a broad range of topics including: Disaster Response, Service/ Entertainment Robots, Inspection/Maintenance Robots, Mobile Robot Navigation, Agricultural Robots, Robots for Excavation, Planetary Exploration, Large Area Mapping, SLAM for Outdoor Robots, and Elemental Technology for Mobile Robots.

In addition to the paper presentations, there were eight keynote talks in the conference. We would like to acknowledge these talks with great appreciation and thanks to the speakers:

- "Damage Due to the 2011 Tohoku Earthquake Tsunami and its Reconstruction"
  by Prof. Fumihiko Imamura, Tohoku University, Japan
- "Spatio-Temporal Video Archive of 3.11 Earthquake and Tsunami Disasters and Their Visualization Supported by Computer Vision Techniques"
  by Prof. Koichiro Deguchi, Tohoku University, Japan
- "Human-Robot Interaction Lessons from Unmanned Vehicles at Fukushima and Tohoku Tsunami Response"
  by Prof. Robin R. Murphy, Texas A&M University, U.S.A.
- "Rescue Mobile Robot Quince: Toward Emergency Response to Nuclear Accident at Fukushima Daiichi Nuclear Power Plants on March 2011"
  by Prof. Keiji Nagatani, Tohoku University, Japan
- "Therapeutic Robot, PARO, and Mental Care for Victims of Disaster by Earthquakes in Japan"
  by Dr. Takanori Shibata, National Institute of Advanced Industrial Science and Technology, Japan
- "Autonomous Agile Aerial Robots"
  by Prof. Vijay Kumar, University of Pennsylvania, U.S.A.
- "Design and Navigation of Wheeled, Running and Flying Robots"
  by Prof. Roland Siegwart, ETH Zurich, Switzerland
- "Thoughts on Fully Autonomous Vehicles"
  by Prof. Alberto Broggi, Università di Parma, Italy

The abstract of the talks and the speakers' biographies can be viewed in:
http://www.astro.mech.tohoku.ac.jp/FSR2011/KeynoteSpeakers.html

Kazuya Yoshida
Satoshi Tadokoro

Contents

Utilization of Robot Systems in Disaster Sites of the Great Eastern Japan Earthquake

Fumitoshi Matsuno, Noritaka Sato, Kazuyuki Kon, Hiroki Igarashi, Tetsuya Kimura and Robin Murphy

Abstract In this paper, we report our activities in the real disaster areas damaged by the Great Eastern Japan Earthquake. From March 18–21, 2011, we tried to apply a ground rescue robot to the real disaster sites at Aomori Prefecture and Iwate Prefecture. On March 18, we carried out inspection mission in a damaged gymnasium. From March 19–21, we went to other sites to identify possibility of usage of robots, and we found the potential needs for not only ground robots but also underwater robots. Then, after the first activity we established a joint United States-Japanese team for underwater search. From April 19–23, 2011 the joint team brought four ROVs to Miyagi Prefecture for port inspection and to Iwate Prefecture for searching for submerged bodies. The joint team returned to Miyagi Prefecture October 22–26 with an AUV and two ROVs for cooperative debris mapping needed to assist with resuming fishing. Based on these experiences, we discuss the effectiveness and problems of applying the rescue robot in the real disaster sites.

F. Matsuno (✉) · K. Kon · H. Igarashi
Kyoto University, Kyodai-Katsura, Nishigyo-Ku, Kyoto, Japan
e-mail: matsuno@me.kyoto-u.ac.jp

N. Sato
Nagoya Institute of Technology, Gokiso-cho, Showa-ku, Nagoya, Aichi, Japan
e-mail: sato.noritaka@nitech.ac.jp

T. Kimura
Nagaoka University of Technology, Kamitomioka, Nagaoka, Niigata, Japan
e-mail: kimura@mech.nagaokaut.ac.jp

R. Murphy
Texas A and M University, College Station, TX, USA
e-mail: murphy@cse.tamu.edu

K. Yoshida and S. Tadokoro (eds.), *Field and Service Robotics*,
Springer Tracts in Advanced Robotics 92, DOI: 10.1007/978-3-642-40686-7_1,

1 Introduction

After the Great Eastern Japan Earthquake occurred, we visited the disaster sites with robots. In this paper, we report our activities in the disaster sites.

The damages in the Great Eastern Japan Earthquake were caused not only by the seismic motion but also by the tsunami and the nuclear power plant accident. This made the situation more complicated, and made it difficult to cope with the disaster. One of the roles of robots for the earthquake hazard is inspection in the disaster site such as a damaged building, an underground mall or in/on rubble piles. In order to cope with the earthquake hazard, many robot systems have been developed in Japan since 1995 Hanshin-Awaji Earthquake [1] , and we have also developed robot systems [2–5]. In order to respond to the Great Eastern Japan Earthquake, we took a ground robot named KOHGA3 and used it in the real disaster site. The overview of KOHGA3 and reports of our activities using it are described in Sect. 2.

Although, many robots have been developed for earthquake hazard, there are few robots for the tsunami disaster in Japan. This is because 1995 Hanshin-Awaji Earthquake, which was an inland earthquake and did not have the tsunami, is considered as a kind of benchmark for robot developments.

However, throughout the activities and considerations in real disaster sites of the Great Eastern Japan Earthquake, we found potential needs for the underwater exploration in the harbor areas by the robots. A joint United States-Japanese team was established for recovery mission at ports and sea in stricken areas in cooperation with International Rescue System Institute (IRS) and Center for Robot-Assisted Search and Rescue (CRASAR). In Sect. 3, we report activities in Minami-sanriku, Miyagi Prefecture for inspection of a port and in Rikuzen-takata, Iwate Prefecture for body search by using underwater robots and summarize lessons learned from the missions. The locations of our activities are shown in Fig. 1.

2 Response to Earthquake Hazard: Inspection in the Damaged Gymnasium

The owner of a damaged gymnasium in Hachinohe City, Aomori Prefecture asked us to inspect inside of it. The members of the first activity at Aomori and Iwate Prefecture were Dr. Noritaka Sato, Dr. Kazuyuki Kon, Mr. Hiroki Igarashi and Prof. Fumitsohi Matsuno, Kyoto University.

On March 18, 2011, we arrived at Hachinohe City through Misawa Airport and carried out inspection task with KOHGA3 in the damaged gymnasium. On March 19, we tried to apply our robot to rescue missions at tsunami-hit areas in Kuji City and Noda Village, Iwate Prefecture. On March 20 and 21, we visited Hachinohe Port and the administrative organizations in Hachinohe City to search a task using a robot, and gathered detail information of situation of Hachinohe Port and Misawa Port.

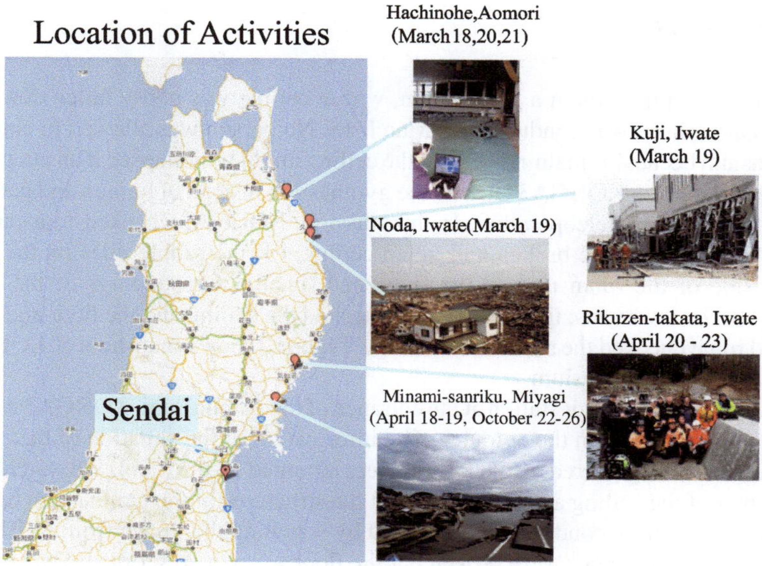

Fig. 1 Locations of our activities

2.1 Specification of KOHGA3

The hardware of KOHGA3 as shown in Fig. 2 consists of one body, four crawler arms which can be independently controlled, and one 4-DOF sensor arm. It is capable of driving on rubble piles by swinging its crawler arms to fit the shape of the ground. A Laser Range Finder (LRF), a posture sensor, a two-way audio (a microphone and a speaker), LED lights, fish-eye cameras, a Pan-Tilt-Zoom (PTZ) camera, a thermal camera, a gas sensor (for flammable gas, CO and H_2S), a wireless LAN device, and batteries are mounted on/in the robot. And the other sensors (a camera, a thermal sensor, a CO_2 sensor) for finding victims are installed at the end of the sensor arm. The size of KOHGA3 is W433 mm × L860 mm × H530 mm, the length of the sensor arm is 950 mm, and the weight of the robot included batteries is 45 kg. It can climb up a 300 mm step and a 43° slope and gather information by using the sensor arm that can reach a target object even if the body of the robot can not approach.

KOHGA3 can be remotely controlled, and it can continuously run for about 1 h with installed batteries. An operator controls the robot using the information sent from the robot via a Graphical User Interface (GUI) as shown in Fig. 3. The maximum communication distance via wireless LAN between the robot and an operator station is about 50 [m] (depends on the condition). It can be teleoperated even in dark environment with installed LED lights.

2.2 Missions

The inspection mission in a gymnasium, whose ceiling was partly fallen down due to the earthquake, was conducted on March 18. No person was allowed to enter the gymnasium because remaining parts of the ceiling might be collapsed if an aftershock occurs. Therefore, KOHGA3 entered the gymnasium instead of human and gathered the information by teleoperation. In this mission, KOHGA3 checked four parts of the gymnasium: (i) the broken and fallen ceiling, (ii) the rubble piles on the floor, (iii) inside of the room next to the stage and (iv) the surroundings of the stage. KOHGA3 had two runs: in the first run it checked the ceiling and the floor and in the second run it checked the room and the stage. Figure 2 shows snapshots of KOHGA3 in the damaged gymnasium.

We set up the operator station at the entrance of the gymnasium. KOHGA3 entered to the gymnasium from the entrance. The information of the situation of the ceiling and the floor was gathered by the PTZ camera mounted on the robot. We watched the fallen part of the ceiling and could find that the structures of the gymnasium has few possibility for the secondary collapse. And by watching the image from KOHGA3, some chairs on the floor were broken but the floor was not so seriously damaged.

We found the door of a room next to the stage was open with about 400 [mm] gap. Since KOHGA3 could not enter through the 400 [mm] opening, we decided to use the sensor arm to check inside of it (Fig. 2(b)). Some pipes were fallen on the floor in the room, but there were no serious damages.

After checking the room, KOHGA3 checked the surroundings of the stage. There were stairs to climb up to the stage, however we were not sure that KOHGA3 could climb the stairs. Therefore we decided to use the sensor arm again. We could find that there were no serious damages around the stage.

(a) (b)

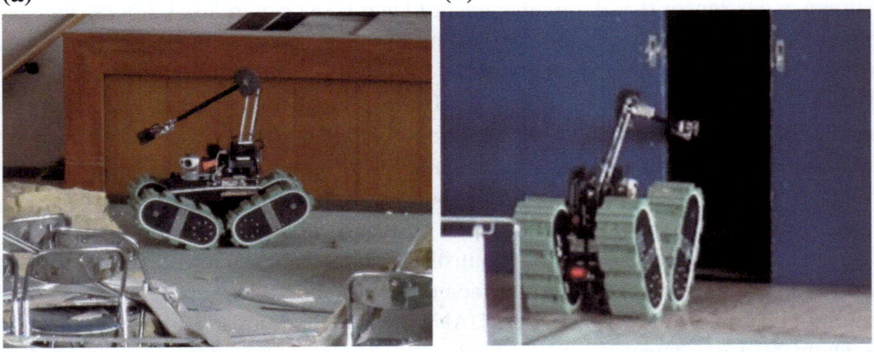

Fig. 2 Snapshots of KOHGA3 in a damaged gymnasium

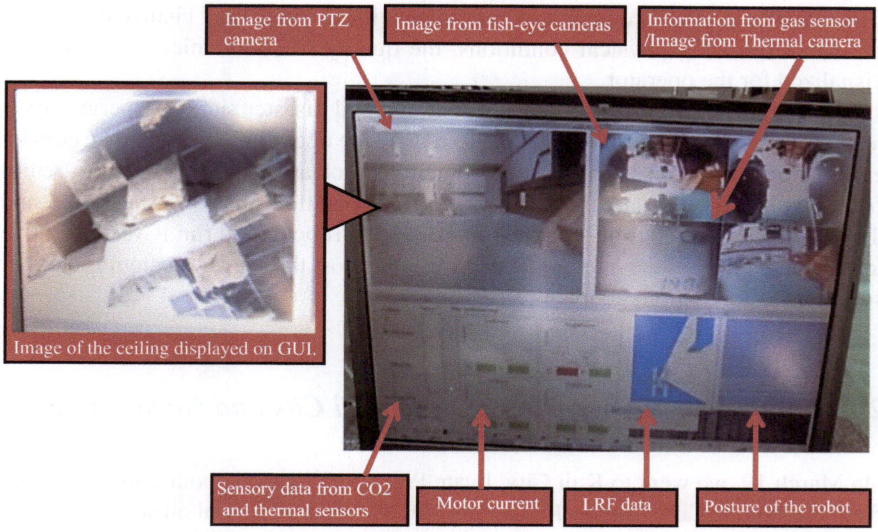

Fig. 3 Graphical user interface for remote control

2.3 Lessons learned

We summarize lessons learned from the inspection mission in the real disaster site by using a ground robot.

We could obtain limited information of the situation in the gymnasium from its entrance because there were many blind areas. We applied a mobile robot to gather information of the blind areas. We think that this is major effectiveness of usage of the robot in the inspection mission.

The PTZ camera was really useful, especially the optical zoom. With its function, we could get a sharp image of a distant target object. If we could not use this function, we might not check the detail of the ceiling because the ceiling of the gymnasium was very high.The sensor arm was also effective in the inspection mission. By using it, we can obtain information of an unaccessible area where a robot can not enter or approach. Fortunately, there were no troubles with KOHGA3 during our activity. Note that our activity by KOHGA3 in the damaged gymnasium is reported in IEEE Spectrum [6].

Throughout the operation in the real disaster situation, we found the following problems of our robot to be solved. The remain electrical power of the batteries must be displayed on the GUI. Decision of changing and charging batteries should be easy. Sensors should be easily changed as necessary. The modularization of the hardware and software might be useful in the real disaster site. The robot should climb steep stairs and slopes in the disaster sites. However, every mobile robots have limitation of the mobility. Therefore, the system should autonomously show the operator whether the robot can climb up the obstacles or not based on information from sensors. As the throughput of communication network may change depending on the surrounding physical conditions, the quality of communication should be visualized for the operator.

For a task by using a teleoperated robot in a real disaster site, the operator feels responsibility. It is a very tough work. To reduce the operator load should be considered. To overcome this problem, some functions such as half-autonomy/auto-return function should be installed. Moreover, for teleoperation mission, calculation and display function of GUI of possible routes that the robot could run is very effective to reduce the load of the operator. Otherwise, low cost throwaway system is another solution for solving this problem.

2.4 Try to find other robotics tasks in Kuji City and Noda Village

On March 19, we went to Kuji City, Iwate Prefecture located about 100 [km] south from Hachinohe City, Aomori Prefecture. Kuji City has a national oil storage station, the ground facilities of the station was completely destroyed by the devastating tsunami as shown in Fig. 4.

In Kuji City, the tasks of inspection in the damaged building or on the rubble piles were almost finished. After discussion with a commander caption in this area, we knew that the inspection task was operated in Noda Village, Iwate Prefecture located about 20 [km] south from Kuji City. And he contacted to the wide-area disaster countermeasures office of Kuji Area and we visited it. In the office, we explained our robot system and the purpose of our activity to the director-general of emergency response headquarters of Kuji Area. Our explanation got a solid understanding, and we could enter Noda Village under leading by a fire-fighting vehicle. After arrival at the disaster site of Noda Village (see Fig. 5), we explained our robot system to the top executives of emergency response headquarters of Noda Area (Fig. 6).

Fig. 4 National oil storage station in Kuji City, Iwate Prefecture

Fig. 5 Tsunami hit Noda Village, Iwate Prefecture

Fig. 6 Negotiation with rescue workers in Noda Village, Iwate Prefecture

They were very interested in our robot and decided to use it for inspection of some damaged buildings. However, in the final adjustment with the mayor in the village office, he wanted to get approval from owners of the buildings. It was very hard to contact them in the confused emergent situation. Then, we should give up the inspection in the building with our robot.

3 Response to Tsunami Hazard using Underwater Robots

The response to the tsunami hazard using underwater robots involved three prefectures: Aomori, Iwate, and Miyagi [7].

On March 20 and 21, we inspected harbor areas of Aomori Prefecture and visited some political administrative organizations to identify possible uses of robots for the tsunami hazard. We visually inspected the damage of Hachinohe Port and Misawa

Port. By the tsunami, many obstacles were scattered on the bottom of the sea. These obstacles were dangerous for ships. Therefore, in order to restore damaged ports, it is necessary to gather information of the situation of the bottom of the sea. As there are many keen debris in the coat of the tsunami-hit areas, works in these areas are very dangerous for divers. Not only the coat but also deep-sea more than 10 [m] depth is very dangerous for divers because of the bends. We considered that the underwarter tasks should be executed by robots instead of the human divers. These tasks are necessary for re-establishment of the port and investigation of missing people. In order to explore needs of usage of robots for recovery mission of ports, we visited some political administrative organizations such as the city office of Hachinohe, Hachinohe Fire Fighting Head Office, and Volunteer Center. Finally, we could reach the office which administrates the ports located in the Hachinohe area. The officer had a need to search the bottom of the sea, and asked us to inspect ports in Hachinohe by using underwater robots immediately. However, we did not have underwater robots available at that time. Then we returned to Kyoto and started to identify new sites and new use for underwater robots. International Rescue System Institute (IRS) reached out to the Center for Robot-Assisted Search and Rescue (CRASAR) in the USA. CRASAR was under travel restrictions stemming from the uncertain state of the Fukushima nuclear plant. Two joint CRASAR-IRS activities in Miyagi and Iwate prefectures one in April and the second in October were carried out.

3.1 Joint United States-Japanese Team Deployments

The two joint CRASAR-IRS teams were headed by Prof. Fumitoshi Matsuno, a vice-president of IRS, and Prof. Robin Murphy, a director of CRASAR. Prof. Tetsuya Kimura took charge of logistics and safety management, while Dr. Eric Steimle coordinated unmanned marine vehicles. The team consisted of UMVs and roboticists, responders and students. The first team, which deployed in April to Minami-sanriku and Rikuzen-takata, brought four different remotely operated vehicles (ROV), which are small tethered underwater robots. The second team, which deployed in October, brought two ROVs and an autonomous underwater vehicle (AUV) and geographical information systems for coordinating robots and distributing information over the internet.

On April 19 and 20, the team conducted inspection missions at a port of Minami-sakriku, Miyagi Prefecture, at the request of the mayor. On April 21-24, the team conducted victim recovery missions along the shore of Rikuzen-takata, Iwate Prefecture, at the request of the disaster countermeasures office of Iwate Prefecture. This deployment illustrates the importance of rescue robots for the recovery phase of a disaster, not just the immediate life-saving response phase.

For Phase 1, the team used tethered remotely operated marine vehicles (ROVs), choosing not to use the other two types of UMVs– autonomous underwater vehicles (AUV) or unmanned surface vehicles (USV). As with any disaster, logistics are important; a UMV must be small and light enough to be transported in cars and handled by one or two people. Furthermore, tsunami disaster response occurs in an extreme environment, where not only must the robot function reliability underwater but the operator control unit must be rugged as it is exposed to salt water spray or rain and snow.

Four commercially-available ROVs were chosen for the Phase 1 deployment based on i) the expectation of critical infrastructure missions which meant that the ROVs had to carry an imaging sonar, not just a video camera, and ii) transportation logistics. AUVs were not considered because they cannot operate near structures as they do not have obstacle avoidance capabilities, cannot be used around unmapped debris, and require a boat in case the AUV gets into trouble and does not return. The three ROVs, SeaBotix SARbot, Seamor, and the Access AC-ROV are shown in Fig. 7 were used, a fourth robot, the SeaBotix LBV-300 was not used as it was essentially the same as the SARbot. All four ROVs were teleoperated. Teleoperation appears to be the industry standard for two reasons. First, navigational autonomy is challenging due to limited sensing (especially as GPS does not work underwater) and, second, mission autonomy (versus navigational autonomy) is hampered by the perceptual complexity of the task and unpredictable attributes of a disaster, thus it requires human interpretation.

For Phase 2, the team was reconfigured to adapt to a different mission: rapid mapping of large scale debris fields. The team brought the SeaBotix SARbot ROV to be in conjunction with a YSI Echomapper AUV as shown in Fig. 8 and the RTV-100 ROV. The intent was for the AUV to map the large portion of Shizugawa Bay and identify potential debris. The team would then deploy on a boat with the ROVs to perform a more accurate identification of the debris and also localize it (the SARbot ROV position could be determined by an external sonar on the boat and then mapped to GPS coordinates). In addition, the data from the robots would be integrated into the General Dynamics GeoSuites system which would allow groups with access to a private web site to view the findings in real-time. However, the AUV experienced a

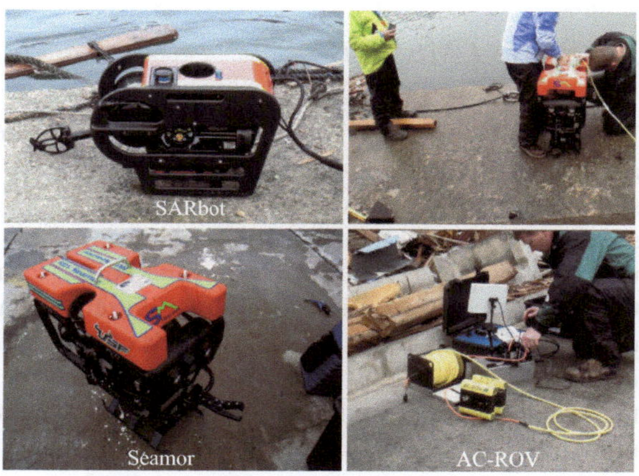

Fig. 7 The ROVs used at the sites for Phase 1

Fig. 8 The AUV and ROV for Phase 2

malfunction and could not be used. Instead, members of the Shizugawa and Utatsu Fishing Cooperatives identified large areas of interest and the ROVs searched there.

3.2 Phase 1 Missions

The CRASAR-IRS team worked at one site at Minami-sanriku, Miyagi Prefecture and five sites at Rikuzen-takata, Iwate Prefecture. We carried out two types of missions. One is port area assessment at Minami-sanriku as shown in Fig. 9 . The team successfully cleared the Minami-sanriku "new port" facility for use. The other is underwater victim search and recovery in Rikuzen-takata in collaboration with Japan Coast Guard as shown in Fig. 10. In Rikuzen-takata, the team searched shallow underwater debris fields and under islands of flotsam for victims in a bay, two small marinas, bridge footings, and a residential coastal area; while no bodies were found, the search was successful in ruling out areas. Table 1 shows number of runs of each robot by site.

Fig. 9 Damaged port in the port area of Minami-sanriku

Fig. 10 Collaboration with Japan Coast Guard

3.2.1 Minami-sanriku

The team used the SARbot and Seamor on April 19-20, 2011, at the "new port" section of the Minami-sanriku fishing port. The objective was to determine if there was at least 5 meters of clearance for fishing boats to enter the port, and if there was rubble within 5 meters of the surface to locate it with GPS. A particular concern was that severed mooring lines floating below the surface could tangle the propellers of fishing boats.

Both the SARbot and Seamor were useful in the highly turbid waters, with the SARbot able to cover an area of interest (inside of thick line in Fig. 11) in 4:45 hours in the water over 1.5 days as shown in Table 1. The transparency of the water in the new port was 30-50 [cm]. It is very hard to identify an object by images from the color video camera. Sonar sensor was very effective for inspection in turbid water. As seen in Fig. 12, the resolution of the DIDSON acoustic camera on the Seamor ROV was

Fig. 11 Covered area of the new port in Minami-sanriku

Table 1 Number of runs of each underwater vehicle by site

Date	Location		Robot	Time in Water
19-Apr-11	Minamisanriku	New port	SARbot	0:16:57
			SARbot	2:02:18
20-Apr-11	Minamisanriku	New port	SARbot	2:25:00
			Seamor	2:58:00
		Bay	SARbot	2:28:31
21-Apr-11	Rikuzentakata	Small marina 1	SARbot	0:44:03
			AC-ROV	0:35:00
		River bridge	SARbot	0:24:18
22-Apr-11	Rikuzentakata		SARbot	1:12:17
		Small marina 2	Seamor	0:42:00
			AC-ROV	0:20:00
23-Apr-11	Rikuzentakata	Residential coastal area	SARbot	0:44:17

sufficiently high to spot cables attached to anchor stones for the moorings. Seamor was in the water for 3 hours. As the activities by a diver in this bad transparency area was very dangerous, ROVs were very useful.

The Minami-sanriku work yielded one device failure, and one instance of tether tangling. The first problem was that a relatively flimsy connection on the Seamor operator control unit immediately broke and had to be repaired in the field; this illustrates the need for reliability and field hardening. The second problem was that the SARbot tether tangled in one of the remaining ship mooring buoys and the lack of situation awareness in the turbid water prevented the ROV operator from being able to untangle it- a nearby leaky rowboat was used to recover the robot.

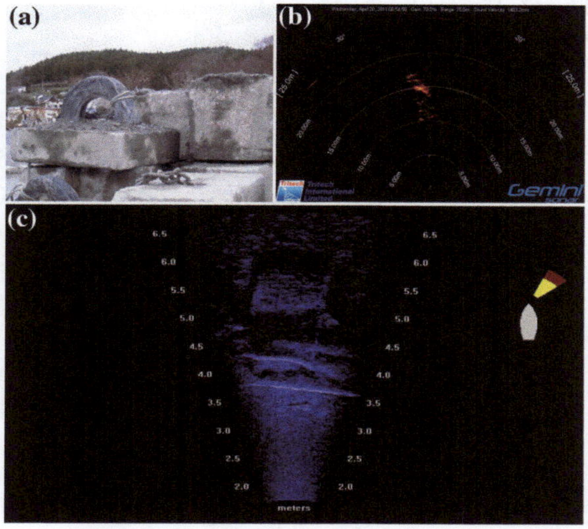

Fig. 12 Example of (b) dual-frequency identification sonar and (c) multi-beam imaging sonar

3.2.2 Rikuzen-takata

The team used the SARbot and Seamor on April 21-23, 2011, at five sites around Rikuzen-takata. The objective was to search for victims underway in shallow littoral areas, where bodies which normally would float to the surface might be trapped on the sea floor under debris or floating under large islands of flotsam.

All three robots were useful in both the clear water in the two small marina sites and the turbid waters in the other three, though no bodies were found. The SARbot was used to explore the shallow debris-filled bay and two floating islands of flotsam. The operators relied on the enhanced video imagery to see despite turbidity within 1.5 m of an object. At a small marina, the SARbot was able to explore under an island of flotsam that included the upper story of a house (Fig. 13).

The Rikuzen-takata work was notable for the mission surprises and the continuing reliability problems with the Seamor. The major surprise was the shift to a victim recovery mission. This had not been considered in the choice of ROVs, which were selected based on the ability to work close to structures. An AUV to provide quick coverage of large areas to find underwater debris piles would have been useful. The SARbot was very useful for working under the islands of flotsam but a surprise was the need to look up into the floating debris or to probe it. A third surprise was the clarity of the water in the coves and shallow regions, which allowed the color video to be used more than had been expected. It should be noted that the SARbot video was superior to the Seamor due to the LYYN software real-time enhancement. The AC-ROV was difficult to control in field, highlighting the need for better cyber-physical controls, and it also lacked the ability to record video, which did not match the mission where stakeholders were not necessarily in the field or dedicated to a device.

Fig. 13 View of house in flotsam

3.3 Phase 2 Missions

The Phase 2 mission was to identify submerged debris that could impact fishing, either through pollution or presenting snags for nets, or to navigation. The Shizugawa Bay is a major food supply for Japan, relying on the pristine water for salmon fishing and aquaculture of oysters and seaweed; thus restoration of fishing has both significant local and national impact. The fishermen were particularly concerned about submerged cars and boats leaking fuel into the pristine waters as well as ghost nets.

The CRASAR-IRS team made four trips with members of the Shizugawa and Utatsu Fishing Cooperatives over Oct. 23-26, 2011, visiting 40 sites with two ROVs. Initially, the team worked from a large fishing boat that was followed with a smaller boat and a diver, who would tie a float to the debris to mark it. However, this practice was quickly abandoned for two reasons. First, the ROV was able to identify debris faster than anticipated and there was more debris than the fishermen expected, thus the diver could not keep up. Indeed, some of the areas that the ROV found submerged cars or fishing boats had been searched by divers and declared clear. Second, the SARbot's GPS location could be inferred from an external sonar, giving the fishermen confidence that they could return to the spot.

The results by the SARbot ROV are shown in Table 2. The ROV was in the water for a total of 371 minutes, finding 104 major pieces of debris marked to be removed and countless other smaller underwater items. The data was taken directly from the GPS trail of the ROV and assumes a 15 m visibility of the sonar. The data for Oct. 26 is an estimate. This means the ROV was able to cover over $80,000m^2$ in slightly more than 6 h of time in the water, averaging $217m^2$ per minute.

The results strongly suggest that ROVs can be of significant value in finding debris, reducing the time divers spend in the water, and reducing the errors. Figure 14 shows divers marking a submerged fishing boat that was found in an area previously searched

Table 2 Data from phase 2 missions. Data in gray is an estimate, not from direct GPS tracking

	#Sites	Time in water (min)	Area covered (m^2)	Distance traveled	Findings
23-Oct-11	6	49	21,524	8,048	5
24-Oct-11	6	105	26,884	6,281	32
25-Oct-11	13	160	51,623	11,480	42
26-Oct-11	7	57	2,000	500	25
totals	**32**	**371**	**80,507**	**26,309**	**104**

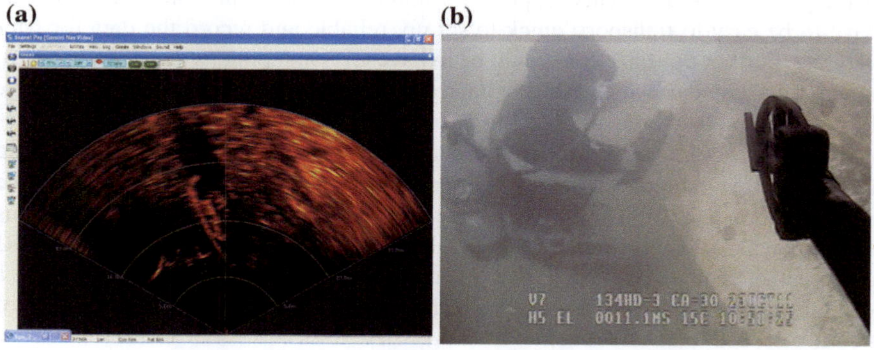

Fig. 14 Submerged fishing boat (a) View from SARbot sonar (b) View of driver attaching a marker float from SARbot

manually. The boat was difficult to find due without sonar due to the turbidity of the water. The diver used the tether of the ROV to quickly go to the boat, and spent only a few minutes in the water. The ROV in theory could have tied off the marker buoy, but mobile manipulation is difficult. The combination of the diver and the ROV was effective and the ROV provided oversight. By working cooperatively, robots and divers can be much more effective.

The use of the GeoSuites tablet-based geographic information system was favorably viewed by the fishermen and local harbormaster. This reinforces the need to develop robot systems with internet connectivity and mi.ddleware.

3.4 Lessons Learned

At least three conclusions can be drawn from the experiences with the recovery operations:

Rescue robots are valuable for both economic and victim recovery, not just response. This deployment extended the scope of rescue robotics from immediate life-saving to the recovery phase. The Minami-sanriku mayor, Hitoshi Sato, emphasized the value of ROVs for the local economy. The mayor is quoted as saying

"The fishery industry is a pillar of our recovery plan. The search result has given us a boost to reopen the fishing port at an early date."

Disaster robots need to be optimized for the unique missions and stakeholder needs. The SARbot, the only ROV optimized for search and recovery, was used 67 % of the total time and the only robot used on the two half-day deployments when time was limited as it took less than 10 min to go from the case to in the water. The AC-ROV used 6 % of the time could be in the water in 3 min. In contrast, the general purpose Seamor often took between 30 minu to an hour to set up and had two equipment failures. This supports the finding that at a minimum disaster robots have to be easy to transport, quick to set up, reliable and record the data for later viewing. Further optimization would be to have GPS mapping (which SARbot has), image enhancement (which SARbot has), and station-keeping abilities (i.e., have the ROV autonomously hold an object in the center of the image).

Human-robot interaction remains a challenge. One example of a HRI challenge is the loss of situation awareness when the tether tangled in turbid water around the mooring buoy line. This highlights the need for increased efforts to design systems and displays that support human perceptual processes. But it should be noted that HRI is more than displays, it is about the larger systems-level design choices that make a system useable for the environment which includes packaging, use, and amount of training needed to be effective.

In the under water tasks a sonar sensor and a camera are only useful to understand surroundings and search a drowned body. Development of underwater sensors is very important, especially for positioning and sensing a body.

4 Conclusion

In this paper, we report our activities in the real disaster areas damaged by the Great Eastern Japan Earthquake. At Hachinohe, Aomori Prefecture, we carried out the inspection mission in the damaged gymnasium by using a ground robot and we identified a potential need for underwater robots. We established a joint United States-Japanese team for underwater search. The joint team visited tsunami-hit areas for inspection of ports in Minami-sanriku, Miyagi Prefecture and search bodies on the shore at Rikuzen-takata, Iwate Prefecture. Based on these experiences, we understand the effectiveness and problems of applying the rescue robots in the real disaster sites.

References

1. S. Tadokoro, *Rescue Robotics: DDT Project on Robots and Systems for Urban Search and Rescue* (Springer, London, 2009)
2. T. Kamegawa, T. Yamasaki, H. Igarashi, F. Matsuno. Development of the snake-like rescue robot KOHGA, in *Proceedings IEEE International Conference on Robotics and Automation*,

pp. 5081–5086 (2004)
3. N. Sato, F. Matsuno, Development of a high mobility wheeled rescue robot with a 1-DOF arm. Int. J. Adv. Mechatron. Sys. **1**(1), 10–23 (2008)
4. T. Kamegawa, N. Sato, M. Hatayama, Y. Uo, F. Matsuno, Design and implementation of grouped rescue robot system using self-deploy networks. J. Field Rob. **28**(6), 813–816 (2008)
5. IEEE Spectrum: Japan Earthquake: More Robots to the Rescue. http://spectrum.ieee.org/automaton/robotics/industrial-robots/japan-earthquake-more-robots-to-the-rescue
6. IEEE Spectrum: Japanese robot surveys damaged gymnasium too dangerous for rescue workers. http://spectrum.ieee.org/automaton/robotics/industrial-robots/japan-earthquake-robot-surveys-damaged-gymnasium
7. R. Murphy, K. Dreger et al., Use of remotely operated marine vehicles at minamisanriku and rikuzentakata japan for disaster recovery. in *Proceedings of The 9th IEEE International Symposium on Safety, Security and Rescue, Robotics*, pp. 19–25 (2011)

Improvements to the Rescue Robot Quince Toward Future Indoor Surveillance Missions in the Fukushima Daiichi Nuclear Power Plant

Tomoaki Yoshida, Keiji Nagatani, Satoshi Tadokoro,
Takeshi Nishimura and Eiji Koyanagi

Abstract On March 11 2011, a huge earthquake and tsunami hit eastern Japan, and four reactors in the Fukushima Daiichi Nuclear Power Plant were seriously damaged. Because of high radiation levels around the damaged reactor buildings, robotic surveillance were demanded to respond to the accident. On June 20, we delivered our rescue robot named Quince which is a tracked vehicle with four sub-tracks, to Tokyo Electric Power Company (TEPCO) for damage inspection missions in the reactor buildings. Quince needed some enhancements such as a dosimeter, additional cameras, and a cable communication system for these missions. Furthermore, stair climbing ability and user interface was implemented for easy operation for novice operators. Quince have conducted six missions in the damaged reactor building. In the sixth mission on October 20, it reached to the topmost floor of the reactor building of unit 2. However, the communication cable was damaged on the way back, and Quince was left on the third floor of the reactor building. Therefore, an alternative Quince is requested recently. In this paper, we report the situation of the missions for Quince, and introduce enhancements of the next Quince for future missions.

1 Introduction

On March 11, 2011, a huge earthquake and tsunami hit eastern Japan, and four reactors in the Fukushima Daiichi Nuclear Power Plant were seriously damaged. There were explosions in three reactor buildings, and a large quantity of radioactive materials was released. The operator of the power plant, Tokyo Electric Power Company (TEPCO), was unable to manage the situation because of the high radiation levels

T. Yoshida (✉) · K. Nagatani · E. Koyanagi
Chiba Institute of Technology, Chiba, Japan
e-mail: yoshida@furo.org

K. Nagatani · S. Tadokoro
Tohoku University, Sendai, Japan

K. Yoshida and S. Tadokoro (eds.), *Field and Service Robotics*,
Springer Tracts in Advanced Robotics 92, DOI: 10.1007/978-3-642-40686-7_2,

measured around the perimeter of the reactor buildings, which prevented people from safely entering the affected areas. They ware unable to plan the best approach for handling the crisis because the situation inside the building was unknown. Therefore, robotic surveillance was required.

The first robot that entered the reactor building was Packbot [1], which was developed by iRobot. On April 17, Packbot opened an airlock gate, which was composed of dual doors, and investigated the situation on the first floor of the reactor buildings [2, 3]. The result of the mission confirmed fears that the radiation dosage in the building was very high humans would have been able remain there for only a short time. Packbot was controlled via a radio communication system, and the radio waves could reach only some parts of the first floor. Furthermore, Packbot could not climb up or down the stairs. Therefore, even after the Packbot mission, TEPCO still needed a robotic surveillance system that could cover extensive areas in the building.

At that time, our joint research group, supported by the New Energy and Industrial Technology Development Organization (NEDO), had been researching and developing tracked robots to assist rescue crews in search and rescue missions in dangerous environments. Quince, the robot developed by us, has high mobility on rough terrains. However, it could not have been mobilized to the reactor building at the time of the crisis because it needed to be reconfigured for the target environment. Therefore, we began to redesign the Quince to resolve the following concerns, which enabled it to be used for surveillance in the reactor building [4].

Communication system At the beginning of the project, we did not know whether the wireless communication system would have been able to control the robot in the reactor building from the outside. We conducted communication tests in the reactor building of the Hamaoka Nuclear Power Plant, which was of the same model as the reactor in the Fukushima Daiichi Power Plant. The test result suggested that a wireless communication link was possible in only a very limited area, so we decided to develop a cable communication system.

Radiation tolerance Our robot was composed of conventional electric devices, and we had very little information about how well it would survive when exposed to gamma rays. Therefore, we conducted experiments to ascertain the radiation tolerance of the components of Quince [5]. The results showed that the essential components of Quince can survive up to 160 Gy of radiation dose.

Reliability We would not have been able to maintain Quince directly after its delivery to the site. Therefore, we focused on improving the reliability of its mechanisms, components, and software by performing numerous tests in a test field.

Operability A good user interface for remote operation was required for Quince to be used by the operators from TEPCO, who had not undergone prior training in its use. This was also significantly improved by performing tests.

Additional sensors In this mission, dosimeter readings were very important. Small dosimeters were supplied by TEPCO and were mounted on Quince. The dosimeters used were the same device used by humans and clipped onto the chest pocket in nuclear power plants. The measured value is displayed on the embedded display of the device, but the device has is unable to communicate with external devices. Therefore, we attached a small camera close to the display for transmitting display images to the operator.

Moreover, at the beginning of May 2011, TEPCO requested two more functions to be added to Quince. At that time, water had been continuously injected into the primary pressure vessels to cool down the fuel core, and the contaminated water had spilled onto the basement floor. It was not possible to measure the depth of water in the basement floor. Thus, functions to install a water level gauge and to sample the contaminated water needed to be added to Quince. Therefore, the following enhancements were added.

Additional functions A simple manipulation mechanism was mounted on Quince. The mechanism included a crane and a winch to install a water level gauge, and a simple mechanism for handling a water-sampling cup.

Mobility improvement For the revised target, the stairs leading down to the basement floor were steeper and narrower than those leading to higher floors. Because of such difficult to maneuver terrain, the robot was required to have a more enhanced stair-climbing ability. Therefore, we optimized its track mechanisms.

On June 20, 2011, the redesigned Quince (Quince 1) was deployed to the Fukushima Daiichi Power Plant. After that, Quince 1 conducted six missions in the damaged reactor buildings. The results of the missions contributed significantly to efforts to restore the plant to a degree of normalcy. However, in the process of doing this, some unforeseen technical issues developed.

The first important issue concerned its communication cable. On October 20, 2011, Quince 1 was left inside the reactor building after its sixth mission because the communication cable failed. The cable rewinding device on Quince 1 was not designed to rewind very long cables. During the mission, Quince 1 had navigated a very long course, and while attempting to rewind the cable, the rewinding device failed and damaged the cable. This limitation and failed operation contributed greatly to the loss of communication.

The second important issue concerned the requirement of additional functions. TEPCO requested a new mission to sample air dust after the first mission by Quince 1 ended. They attached timer-triggered air pumps on Quince 1 to conduct the mission. In this case, the operator was required to direct Quince 1 to the target location within a specific time period, and this was very stressful for the operator. The ability to remotely trigger the air pump would have simplified this process.

Through the six missions, we also identified many others issues that needed improvement, and TEPCO requested an alternative Quince model. Therefore, we developed a revised model of Quince that dealt with the above issues. In addition, to address the communication cable problem, we defined a communication recovery scenario by providing another robot that acts as a communication repeater to re-establish a data link to Quince via a wireless link between the robots.

In this paper, we report the missions performed by Quince 1 in a real disaster site and discuss issues that arose during the missions. Furthermore, we introduce a revised model of Quince that is able to resolve earlier issues and perform a communication recovery scenario.

Table 1 A brief timeline of missions for Quince 1

Date	Event
June 20, 2011	Quince1 was delivered to the site
June 24, 2011	The first mission at unit 2 was conducted. Quince was not able to install a water level gauge
July 8, 2011	The second mission at unit 2 was conducted. Quince was reached the third floor of the building and two sets of air dust samples were successfully captured
July 26, 2011	The third mission at unit 3 was conducted. Target facilities were successfully observed
September 22, 2011	The forth mission at unit 2 was conducted. Target facilities were successfully observed
September 24, 2011	The fifth mission at unit 3 was conducted. Target facilities were successfully observed
October 20, 2011	The sixth mission at unit 2 was conducted. Quince was successfully reached the fifth floor. However it lost a data link at the third floor on its way back

2 Missions of Quince 1

In the damaged reactor buildings in the Fukushima Daiichi Power Plant, a total of six missions were conducted with Quince 1 (Table 1). Furthermore, many practice tests were performed in the reactor building of unit 5, which was not seriously damaged. In this section, we report the missions conducted in the damaged reactor buildings.

2.1 The First Mission on June 24, 2011

The first mission of Quince 1 was conducted on June 24, 2011, at the reactor building of unit 2. The objective of the mission was to install a water level gauge into the contaminated water pool on the basement floor. At that time, water had been continuously injected into the primary pressure vessels to cool down the fuel core, and the injected water had accumulated in the basement floor. An urgent mission was required to survey the status of the contaminated water pool because there was a possibility that the contaminated water could overflow and spill into the sea. To conduct this mission, Quince 1 was fully equipped with a crane, a winch, and a water level gauge.

Quince 1 attempted to descend the staircase and reached the first stair landing. However, the size of the landing was much narrower than what had been communicated to us in advance by TEPCO. In the building, there were multiple staircases to the basement floor, and Quince 1 attempted to descend two of them. However, it was impossible to navigate the landings of the first set of stairs. Ultimately, Quince 1 was unable to accomplish the objective of the first mission.

2.2 The Second Mission on July 8, 2011

The second mission was conducted on July 8, 2011, at the reactor building of unit 2 [3, 6]. The objectives of this mission were to measure the radiation levels of the upper floors and to sample air dust in the building. Quince 1 was equipped with two timer-triggered air pumps for this task. The crane and the winch were removed because they were not required for this mission.

Quince 1 climbed up the staircase and reached the third floor. During this motion, the air dust samplers were activated at the second and third floors. However, on the way back to the entry point, the motor driver boards mounted on Quince 1 encountered problems because the air temperature inside the building was extremely high. The problems were successfully resolved by sending low level commands to the motor driver boards instructing them to reboot. Finally, the robot returned to the entry point, and the second mission, which included the air sampling and dose measurement tasks, was completed. However, it had not been far from being a failure.

2.3 The Third Mission on July 26, 2011

The third mission was conducted on July 26, 2011 [3], at the reactor building of unit 3, which was heavily damaged by an explosion. The objectives of this mission were to investigate the damaged piping of the core spray system and to measure the dose levels around the facility. In the event that the damage was minimal, it may have been possible to re-activate the spray system to cool down the reactor core directly, instead of using the reactor feed water system.

Quince 1 climbed up the staircase and reached the second floor. Then it approached the target piping and captured high-resolution photographs of the target facilities. After that, Quince 1 tried to climb up the staircase to the third floor, but found that the staircase was damaged and blocked with rubble.

Figure 1 shows the high-resolution photographs captured by the wide-angle camera on Quince 1. The left photograph shows the piping of the core spray system, which was located next to the primary containment vessel. The right photograph is of the staircase to the third floor, showing the rubble blocking the staircase.

Based on the exploration results, the restoration process was planned and the core spray system was re-activated on September 1, 2011 [3, 7].

2.4 The Fourth and Fifth Missions on September 22 and 24, 2011

The fourth and fifth missions were conducted on September 22 and 24, 2011, respectively. The objectives of these missions were to inspect the first floor of the reactor buildings of units 2 and 3. The missions were a part of the preparation for the project to investigate inside the primary containment vessel using a borescope. Quince 1 explored the target area, obtained many photographs, and measured the dose rate.

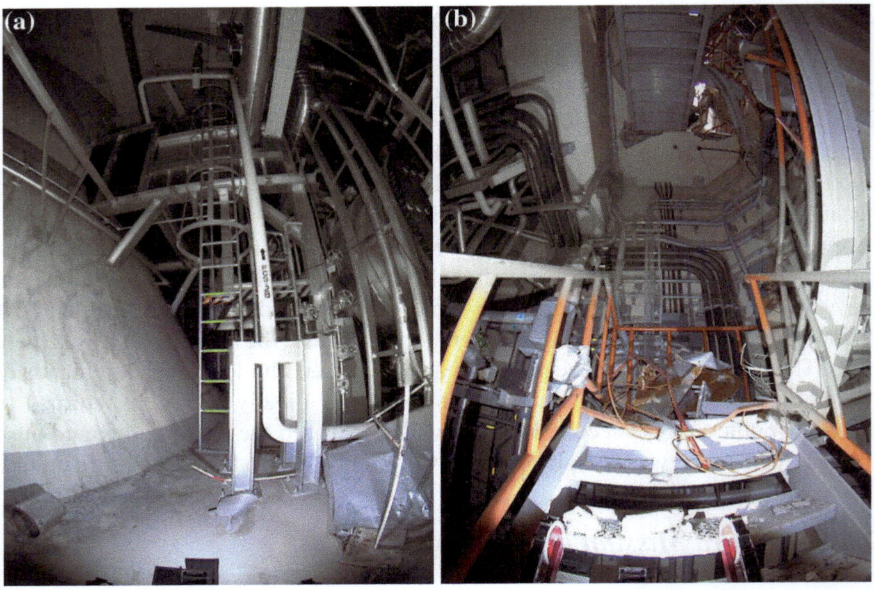

Fig. 1 High-resolution photos captured in the reactor building of unit 3. The *left* shows the piping of the core spray system and the *right* shows the rubble blocking the staircase to the third floor

2.5 The Sixth Mission on October 20, 2011

The sixth mission was conducted on October 20, 2011, at the reactor building of unit 2. The objectives of the mission were to investigate the damage to the the facility on the third floor and to inspect the spent fuel pool on the fifth floor. Air temperature measurements were also a part of tasks performed during this mission, and a conventional thermometer was placed on Quince 1 within the range of vision of a spare camera on Quince 1. The camera image was displayed on the operator console.

Quince 1 climbed up the staircase and reached the third floor. Then, it approached the target facility and captured some photographs. After that, it explored the third floor to measure the radiation level, and climbed up the staircase to the fifth floor. On the fifth floor, it opened a metal accordion curtain with its sub tracks and approached the lid of the primary containment vessel. It measured the radiation level and temperature in the immediate vicinity and took photographs. The dosimeter display showed very high radiation levels (over 200 mSv/h) around the lid.

After inspection of the fifth floor, Quince 1 returned to the entry point. However, the communication cable was caught on piping on the third floor. At the same time, the cable had become jammed in the cable reel because the cable had been continuously rewound from the fifth floor. As a result, the communication cable could not be rewound or be released. Eventually, communication cable was lost and Quince 1 remains there up to the present.

3 Lessons Learned from the Missions of Quince 1

As shown in Sect. 2, the missions completed by Quince 1 contributed significantly to recovery work at the plant. Especially, its mobility and clear camera images were highly evaluated. Quince was the only robot which can climb up the staircases and explore the upper floors. Furthermore, its mobility on rough terrain was very important because there were number of obstacles such as steps, dikes, cables and debris. Still images captured by Quince are very clear and high-resolution (2048 × 1536 pixels). Therefore, Quince was chosen out of other robots for the missions on September 2011, even the targets were on the first floors of the buildings.

Aside from these advantages, the following problems were encountered.

Communication cable The most significant problem was with the communication cable. The cable rewinding device did not function properly toward the end of the sixth mission., and eventually failed causing us to abandon Quince 1 on the site. In our initial implementation of Quince, the ability to make preemptive moves was the first priority. Furthermore, we supposed that the communication cable would have been replaced after each mission, so no rewinding function was installed in the early stages of the project to redesign Quince. However, to enable the switch-back motion of the robot in narrow environments, we added an ad hoc rewinding function, which did not have the ability to wind the cable evenly. Therefore, we specified that it should not be used to rewind cables longer than 20 m. In practice, the device worked much better than we had expected, and at times rewound over 200 m of cable in trial runs. Therefore, during the real missions, longer cable exceeding 20 m was rewound.

Unknown environment In the information received from TEPCO, the the staircase landing down to the basement was reported to 91 cm wide. We built a mock environment in our laboratory using these dimensions, and tested Quince 1 in the mock environment. However, the actual width was 71 cm, and this prevented us from realizing the objectives of the first mission. The data received from TEPCO was based on the construction drawings of the building, but repeated modification to the structure of the building reduced the width of the staircase landing. However, all information pertaining to the re-construction had been washed away by the tsunami.

Carrying method Quince 1 was carried on a stretcher or by manually holding each of the sub tracks. A stretcher could not always be used because of the narrow corners en route to the entry point. After the mission in the reactor building, contaminated dust was stuck on the tracks of Quince 1. When the operators held the sub tracks to transport Quince 1, they became exposed to the radiation source from a very close range.

Additional components The request to sample air particles in the reactor building was requested after Quince 1 had been delivered to the site. The mission was conducted with two timer-driven pumps attached on Quince 1. The timer had to allow a delay time to ensure the robot's arrival at the requested position. This increased the duration of the mission. Furthermore, air temperature measurements were requested for the sixth mission. To do this, a conventional thermometer was attached to Quince 1 within the range of vision of a spare camera. The temperature value was recorded by capturing the screen of the operator console.

These additional functions were not requested until after Quince 1 had been redesigned in our laboratory.

Environment conditions not covered by prior tests Two extreme conditions that were not covered in our laboratory tests were encountered in the missions: strong illumination and high temperature.

The radiation level was displayed on a small screen on the dosimeter, and a small camera captured the screen image and transmitted it to the operator. The screen was a raster scan device. In the reactor building, the only external light was that on Quince 1, and the dosimeter's screen was captured by the camera without any problems. However, in locations that received direct sunlight, e.g., the carry-in-entrance and a location near the blow out panel, the shutter speed of the camera became very fast, and the camera was unable to capture a full frame of the screen on the dosimeter.

In the second mission in the reactor building of unit 2, the air temperature around Quince 1 was very high. The main motor driver overheated and the temperature exceeded its safety limit (50 °C). As a result, the main motor driver was shutdown temporarily. The driver was recovered by remotely sending reset commands, and eventually, Quince 1 returned to the entry point. After the mission, the threshold temperature for shutting down the driver was changed to 80 °C. However, it was difficult to foresee this situation.

4 Enhancements for the New Versions of Quince

The sixth mission was the final mission for Quince 1, as it was left on the third floor of the reactor building unit 2. TEPCO still required a robotic surveillance system that covers extensive areas (particularly upper floors) in the building. Therefore, we prepared two new versions of Quince robots: Quince 2 and Quince 3 (Fig. 2). We redesigned the robot to resolve critical issues encountered in the missions of Quince 1. The enhancements over Quince 1 are as follows:

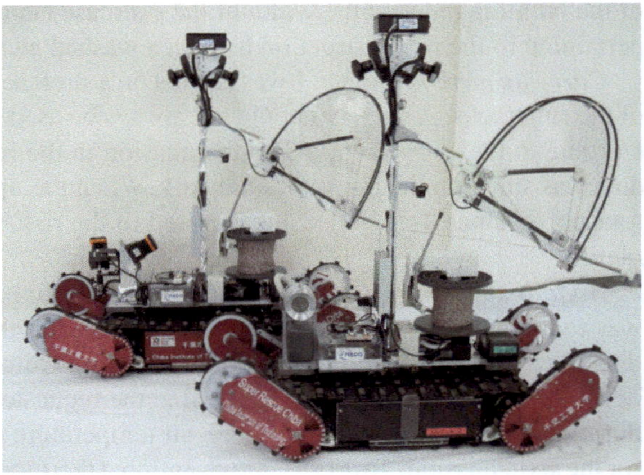

Fig. 2 New versions of Quince. The image on the *right* is Quince 2, which is equipped with an air dust sampler and the image on the *left* is Quince 3, which is equipped with laser scanners

- Detachable carrying handles
  After surveillance missions, the robot may be contaminated by radioactive materials. To carry the robot with minimum radiation exposure to workers, we installed detachable carrying handles to the robot body.
- USB thermometer and hydrogen meter
  To digitally record air temperature and hydrogen concentration values, we mounted a USB thermometer and a hydrogen meter on both of the robots.
- Enclosed dosimeter
  To prevent difficulties in recording reading from dosimeter, we put the dosimeter and a small camera into an opaque box for protection from direct sunlight.
- Remote-controlled air pump
  We mounted a remote-controlled air pump for the air dust sampling mission on Quince 2. This was considered as standard equipment.
- Laser range scanner
  To obtain details of the target environment, we mounted laser range scanners on Quince 3. More information about the scanner is included in previously published papers [4, 8].
- Cable rewinding device
- Backup robot system
- Improvement of the operator console.

The last three enhancements mentioned above are described in detail in the following subsections.

4.1 Cable Rewinding Device

The simple cable rewinding device present on Quince 1 encounters difficulties while rewinding long cables. This is because it does not have the capability to wind the cable evenly. Figure 3 shows the simple rewinding device on Quince 1. The cable is pulled and reeled out passively as Quince 1 moves, and it is rewound by rotating the cable guide by the motor. If the unreeled cable is too long, the rewound cable accumulates on the upper part of the reel. This is because the end of the cable guide arm is aligned at the top of the reel. The robot motion may shake the reel, causing the rewound cable to fall and become jammed. In this case, the reel would no longer be able to release the cable.

The revised Quince models are equipped with a new cable handling device redesigned to avoid the above problem. Features of the device are as follows:

- The cable is rewound evenly on the reel,
- It can handle about 500 m of cable continuously,
- An operator wearing thick gloves can easily change the cable reel,
- The ability to automatically release and rewind is included, and
- It has the mechanical compliance to respond to a sudden change of cable tension.

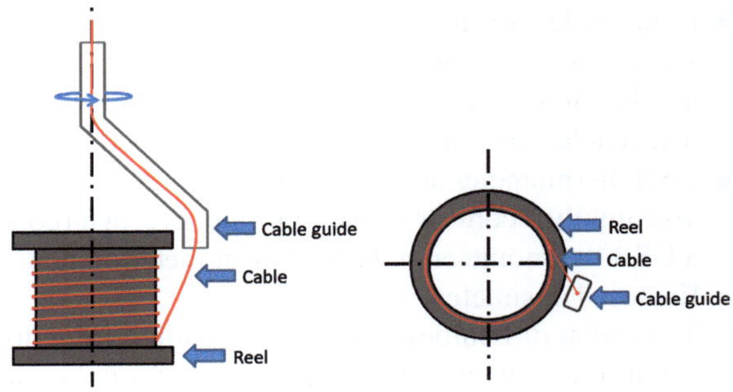

Fig. 3 A cable rewinding device on Quince 1. When the cable is pulled as Quince 1 moves, the cable guide rotates passively and releases the cable. When the operator instructs the device to rewind the cable, the guide is driven by the motor to rotate in the reverse direction, and the cable is rewound

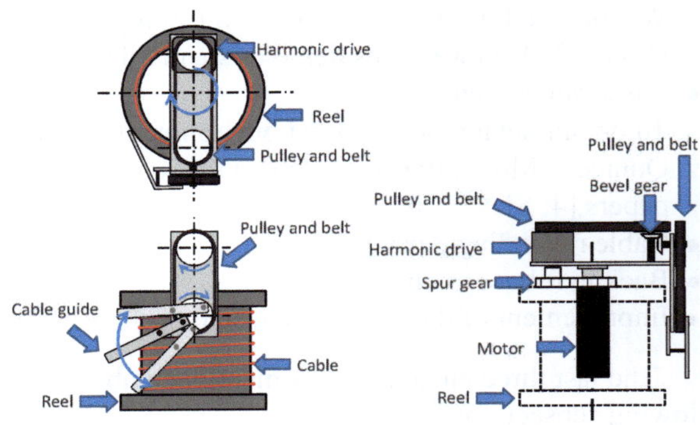

Fig. 4 Cable rewinding module on the reel. The module is driven by a motor located inside the reel. As it rotates around the reel, the cable guide swings up and down to evenly wind the cable

Figure 4 shows the new cable rewinding device. A motor rotates the cable rewinding module, which is shown in the upper part of the right figure, around the reel to wind the cable. While the module rotates around the reel, the cable guide arm is swung up and down to evenly wind the cable around the reel. One of the advantages of the system is that the module can be easily detached, so an operator wearing thick gloves can effortlessly change the cable reel.

To incorporate the automatic release and rewind functions, we implemented a tension control mechanism. Figure 5 shows the mechanism. The tension in the cable pulls down the upper arm that is supported by a spring located at the root. Thus, we can measure the cable tension by measuring the angle θ. By controlling the cable rewinding motor to keep the angle θ constant, the cable will be released or rewound as the robot moves.

To evaluate the new device, we conducted some tests in our mock field, and we identified two problems with the constant-tension control of the cable.

One problem was the difficulty with cable release in case of obstacles. When the cable was obstructed by obstacles, it was difficult for it to be released, because a constant-tension control of the cable was specified. To solve this problem, we installed a "no rewinding mode." In this mode, the tension control mechanism is disabled and the tension in the cable is lowered. This increases the possibility that the cable will be released from the obstacle. In this mode, when the cable is pulled, it releases the cable to prevent damaging it.

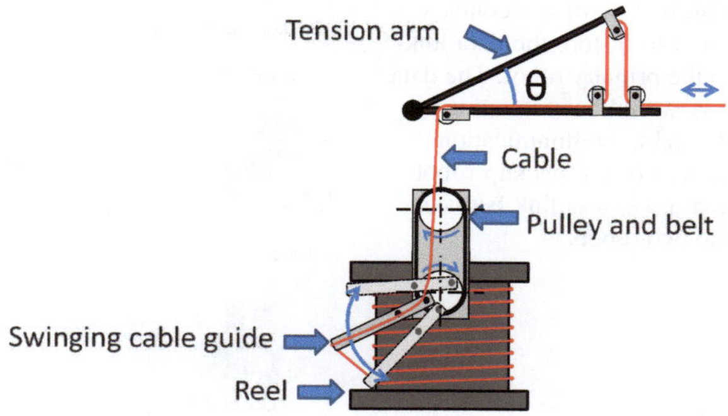

Fig. 5 Redesigned cable handling device. Tension in the cable can be measured by measuring the angle θ, and, by controlling the angle constant, the cable will be either released or rewound

The other problem was due to loosening of the cable after it is caught by obstacles and then released quickly. In such cases, excessive amount of the cable was loosened and lay on the ground for a while. The loosened cable may be run over by the robot while moving backwards. To avoid this situation, the robot should wait until the proper cable tension has been recovered. Therefore, we implemented an alert icon system and tension meter display on the console, which gives a warning to the operator about any abnormal status of the cable.

4.2 Backup Robot System

With Quince 1, there was no communication redundancy, and only one cable failure resulted in loss of the robot. Therefore, to handle such situations involving cable failure, we implemented a backup communication function on a secondary robot to restore the data link of the primary robot using a wireless communication link between the primary and secondary robots.

Figure 6 shows the structure of the backup robot system. Both robots possess a VDSL cable communication system and a 802.11 g wireless LAN system. The primary robot conducts the mission using the cable communication system. If the cable fails, the backup robot is directed to the location, and it approaches the primary robot until a wireless connection is established between the two robots. Then, the data link to the primary robot is restored via the cable communication system of the second robot and the wireless connection between the robots.

For this system, the requirement is simply to mount a wireless transceiver device on both robots. While this was easily implemented, it significantly improved our ability to help stuck robots. Additionally, Quince 2 can function as both a primary and a backup robot, depending to the mission.

The Quince robot has four small cameras and two wide-angle view IP cameras. Images from these cameras are encoded in the MJPEG format and transmitted to the operator console. The VDSL modem on Quince (ABiLINX 1511) has a bandwidth of around 25 Mbps with a 500 m long thin cable. Since multiple video streams are

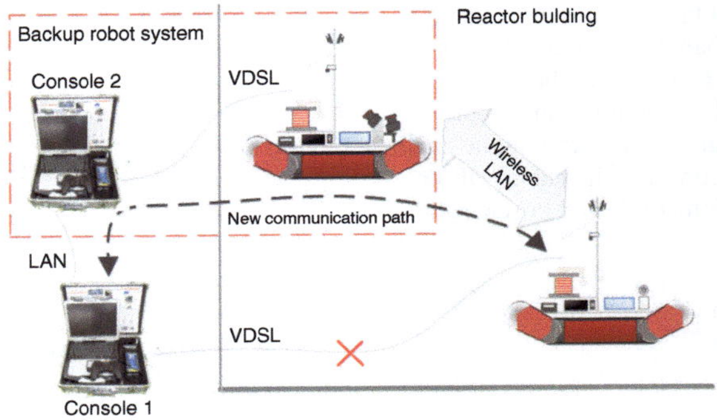

Fig. 6 Use of a secondary robot to restore the data link to the primary robot. The data link would be restored using the cable communication system of the backup robot and a wireless link between the two robots

transmitted over the data link with limited bandwidth, we allocated an optimal bandwidth for each video stream depending on its contents. However, when the cable of the primary robot dies and the backup robot restores the data link, the required bandwidth of the data stream doubles. and the actual data flow may exceed the bandwidth capability. Therefore, we setup two communication modes: the primary mission mode, in which each robot can use a bandwidth of up to 20 Mbps, and the emergency communication mode, in which each robot can use a bandwidth of up to 8 Mbps.

Switching between these communication modes is done by just a few clicks. Therefore, switching from the primary mission mode to the emergency communication mode can be done easily in a real mission.

4.3 Improvement of the Operator Console

The operator console screen of Quince 1 displayed camera images, the robot's posture, battery voltage, and motor driver temperatures. The operators used a gamepad to assign the speeds for the motors of the main and sub tracks. Furthermore, there were two buttons on the console screen that could be used to instruct the sub tracks to take a predefined postures.

The new operator console was improved to consider the feedback given by the operators of the missions conducted by Quince 1. Figure 7 shows the new console screen. The major improvements in response to the feedback were as follows:

The amount of cable remaining should be displayed.

A reel counter is implemented and the value is displayed on the screen.

Abnormal conditions should be signaled.

In the case where abnormal conditions are detected, flashing alert icons will be displayed over the camera image. The conditions include low battery, short

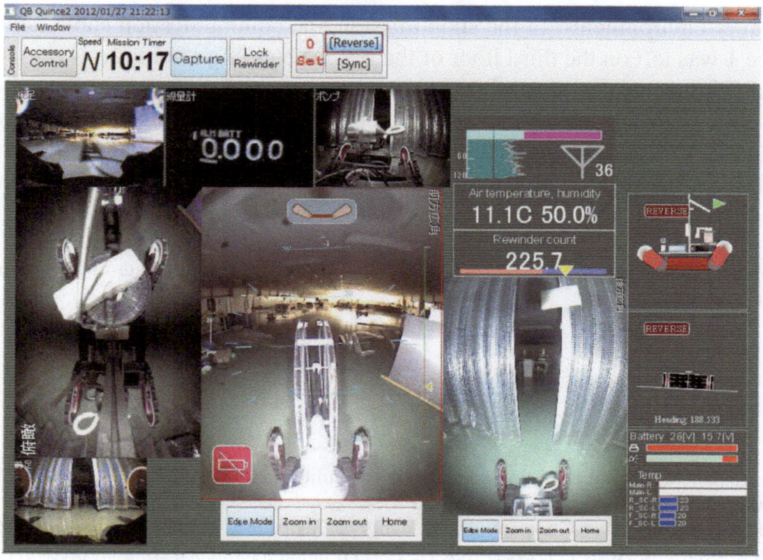

Fig. 7 Improved operator console screen for Quince 2. Status indicators for information such as the amount of cable remaining, temperature, and wireless signal strength, and some alert signals were added

length of cable remaining, low cable tension, and abnormal inclination of the robot.

Commanding sub tracks to take predefined posture should easily be activated

This function was assigned to the gamepad buttons for quick activation.

In addition to the above improvements, we added indicators displaying readings from the newly added components such as sensors for temperature, humidity, wireless signal strength, and cable tension.

5 Summary

Quince 1 was a rescue robot developed to perform surveillance missions in the damaged reactor buildings in the Fukushima Daiichi Reactor Power Plant, and it conducted six missions. The results of the missions significantly contributed to the restoration of the site. However, several issues were identified during the missions. These included the reliability of the communication cable handling device, the method for transporting the contaminated robot, and other problems encountered in the extreme environmental conditions that were not covered by tests in our laboratory environments. In the sixth mission, the communication cable failed and Quince 1 was left on the third floor of the reactor building of unit 2.

TEPCO has requested modified Quince robots, so we upgraded the robots to resolve the critical issues. The upgraded robots were recently completed and delivered to TEPCO in February 2012. We hope that these robots will contribute to the project aimed at recovering from the Fukushima Daiichi Nuclear Power Plant disaster.

Acknowledgments We would like to thank NEDO and the Chiba Institute of Technology for financial support.

References

1. B. Yamauchi, PackBot: a versatile platform for military robotics. Proc. SPIE **5422**, 228–237 (2004)
2. Tokyo Electric Power Company: Plant Status of Fukushima Daiichi Nuclear Power Station (as of 6:00 pm, April 17). Press Release (2011). http://www.tepco.co.jp/en/press/corp-com/release/11041710-e.html
3. Tokyo Electric Power Company: Status of TEPCO's Facilities and its services after the Tohoku-Chihou-Taiheiyou-Oki Earthquake. Press Release (2011). http://www.tepco.co.jp/en/nu/fukushima-np/past-progress/index-e.html
4. K. Nagatani, S. Kiribayashi, Y. Okada, S. Tadokoro, T. Nishimura, T. Yoshida, E. Koyanagi, Y. Hada, Redesign of rescue mobile robot Quince Toward emergency response to the nuclear accident at Fukushima Daiichi Nuclear Power Station on March 2011, in *The 2011 IEEE International Symposium on Safety, Security, and Rescue Robotics (SSRR2011)* (2011)
5. K. Nagatani, S. Kiribayashi, Y. Okada, K. Otake, K. Yoshida, S. Tadokoro, T. Nishimura, T. Yoshida, E. Koyanagi, M. Fukushima, S. Kawatsuma, Gamma-ray irradiation test of Electric components of rescue mobile robot Quince Toward emergency response to nuclear accident at Fukushima Daiichi Nuclear Power Station on March 2011, in *IEEE International Symposium on Safety, Security, and Rescue, Robotics (SSRR2011)* (2011)
6. Tokyo Electric Power Company: Plant Status of Fukushima Daiichi Nuclear Power Station (as of 3:00 pm, July 8). Press Release (2011). http://www.tepco.co.jp/en/press/corp-com/release/11070808-e.html
7. Tokyo Electric Power Company: Plant Status of Fukushima Daiichi Nuclear Power Station (as of 3:00 pm, Sep. 1). Press Release (2011). http://www.tepco.co.jp/en/press/corp-com/release/11090104-e.html
8. T. Yoshida, K. Irie, E. Koyanagi, T. Masahiro, 3D laser scanner with gazing ability, in *IEEE International Conference on Robotics and Automation* (2011)

Collaborative Mapping of an Earthquake Damaged Building via Ground and Aerial Robots

Nathan Michael, Shaojie Shen, Kartik Mohta, Vijay Kumar, Keiji Nagatani, Yoshito Okada, Seiga Kiribayashi, Kazuki Otake, Kazuya Yoshida, Kazunori Ohno, Eijiro Takeuchi and Satoshi Tadokoro

N. Michael (✉)
Robotics Institute, Carnegie Mellon University, Pittsburgh, PA, USA
e-mail: nmichael@cmu.edu

S. Shen · K. Mohta · V. Kumar
GRASP Laboratory, University of Pennsylvania, Philadelphia, PA, USA
e-mail: shaojie@grasp.upenn.edu

K. Mohta
e-mail: kmohta@grasp.upenn.edu

V. Kumar
e-mail: kumar@grasp.upenn.edu

K. Nagatani · Y. Okada · S. Kiribayashi · K. Otake · K. Yoshida
Department of Aerospace Engineering, Tohoku University, Sendai, Japan
e-mail: keiji@astro.mech.tohoku.ac.jp

Y. Okada
e-mail: okada@astro.mech.tohoku.ac.jp

S. Kiribayashi
e-mail: seiga@astro.mech.tohoku.ac.jp

K. Otake
e-mail: otake@astro.mech.tohoku.ac.jp

K. Yoshida
e-mail: yoshida@astro.mech.tohoku.ac.jp

K. Ohno · E. Takeuchi · S. Tadokoro
Graduate School of Information Sciences, Tohoku University, Sendai, Japan
e-mail: kazunori@rm.is.tohoku.ac.jp

E. Takeuchi
e-mail: takeuchi@rm.is.tohoku.ac.jp

S. Tadokoro
e-mail: tadokoro@rm.is.tohoku.ac.jp

K. Yoshida and S. Tadokoro (eds.), *Field and Service Robotics*,
Springer Tracts in Advanced Robotics 92, DOI: 10.1007/978-3-642-40686-7_3,

Abstract We report recent results from field experiments conducted with a team of ground and aerial robots toward the collaborative mapping of an earthquake damaged building. The goal of the experimental exercise is the generation of 3D maps that capture the layout of the environment and provide insight to the degree of damage inside the building. The experiments take place in the top three floors of a structurally compromised engineering building at Tohoku University in Sendai, Japan that was damaged during the 2011 Tohoku earthquake. We provide details of the approach to the collaborative mapping and report results from the experiments in the form of maps generated by the individual robots and as a team. We conclude by discussing observations from the experiments and future research topics.

1 Introduction

In this work we report recent results from field experiments conducted with a team of ground and aerial robots toward the mapping of an earthquake damaged building. We focus on the investigation of the feasibility of deploying aerial robots, specifically a quadrotor, into disaster scenarios where the building is critically damaged but still accessible to robots and humans for experimental purposes. The experimental environment covered the top three floors of an engineering building on the campus of Tohoku University in Sendai, Japan during the first week of August, 2011. Representative images of the interior and exterior of the building are shown in Figs. 1, 2.

On March 11, 2011, a 9.0-magnitude earthquake (on the moment magnitude scale) occurred off the coast of Japan, approximately 130 km from Sendai [1]. The consequences of the earthquake were devastating with significant loss of human life and damage to the environment. Resulting tsunami waves generated further damage and instigated a meltdown at a nuclear power plant near Fukushima, Japan [2].

Fig. 1 Panoramic images depicting the interior of the building. These images are representative of the clutter found throughout the experimental areas

Fig. 2 The building suffered significant structural damage due to the earthquake

Several robotics research groups and companies responded to this natural and nuclear plant disaster [3, 4]. Ground robots with onboard sensing enabled environmental observation of the compromised nuclear power plants in regions inaccessible to humans due to high levels of radioactivity. The ground robots were equipped with long-range cable tethers to enable remote communication, tele-operation, and the transmission of sensor data. These ground robots proved capable in maneuvering through the cluttered environments [5].

We are interested in exploring the possibility of leveraging an autonomous quadrotor in such environments through field experiments that focus on cooperative mapping using both ground and aerial robots. Aerial robots offer several advantages over ground robots including the ability to maneuver through complex three-dimensional environments and gather data from vantages inaccessible to ground robots. Further, quadrotors are able to hover in place, making them well-suited for observation and human-guided or autonomous inspection. However, aerial robots also suffer from several limitations that reduce their applicability in disaster scenarios such as the need for wireless communication and a limited onboard power supply which restricts the platform's payload capacity and flying time.

Given the prior experience of using ground robots at the nuclear power plant disaster site, we designed the experimental scenario based on conditions consistent with those found at the disaster site. Consider an earthquake damaged building with multiple floors that are generally accessible to ground robots. However, various locations in the environment are inaccessible to the ground robots due to debris or clutter. The goal of the experimental exercise is the generation of 3D maps that capture the layout of the environment and provide insight to the degree of damage inside the building. Additionally, there may be specific regions of interest that require attention from operators during the mapping. Throughout the experiments, remote operators must be able to maintain control of the robotics platforms on the ground and in the air.

The experiment design highlights the need for heterogeneity. Ground robots do not suffer as greatly from the same payload limitations as quadrotors and are therefore able to carry larger sensor payloads, maintain tethered communication links, and operate for longer periods of time. However, quadrotors provide mobility and observational capabilities unavailable to ground robots. Hence, to build a rich 3D representation of the environment, we leverage the advantages of each platform and in doing so, mitigate the platform limitations.

The problems of localization and mapping in 3D environments are well-studied for both ground and aerial robots and many methodologies exist to address these problems. In this work, we focus primarily on the integration of our prior work in the areas of localization and mapping for ground and aerial robots. However, there are several examples of prior works employing similar methodologies to our own approach for either ground or aerial platforms [6, 7] including cooperative mapping with ground and aerial platforms [8, 9]. Researchers have also pursued the mapping of complex environments for applications such as search and rescue via ground and aerial platforms [10–12]. Therefore, the contributions of this work are three-fold. First, it experimentally supports the argument that the mapping of complex multi-story environments with ground and aerial robots in disaster scenarios is viable (or nearly viable) given the current state-of-the-art in vehicle design, sensors, computation, and algorithms. Second, it supports the statement that the strengths and weaknesses of individual robot platforms may be overcome by employing heterogeneity in system design. Third, it provides insight into the gap between the current technological capabilities and the remaining challenges we must overcome toward application in true disaster scenarios.

2 Experiment Design and Methodology

To address the requirements of the experimental scenario, we use three different research platforms. The first platform is a ground robot equipped with an onboard sensing suite that enables the generation of dense 3D maps. The vehicle is tele-operated through the multi-floor environment while simultaneously collecting sensor data. After the operators identify locations in the environment that are inaccessible to the ground platform, a second ground platform equipped with an automated helipad is tele-operated to these locations and carries a quadrotor robot equipped with onboard sensing that is able to autonomously open and close the helipad and take-off and land from the helipad (Fig. 3). The aerial robot is physically transported by the ground robot to each location of interest where it autonomously takes-off before an operator is able to guide the robot to map or observe these inaccessible regions. Upon completion of the mapping and observation phase, the aerial robot is remotely signaled to autonomously land and close the helipad. The quadrotor is then guided to the next location of interest via the tele-operated ground robot.

The experiment primarily focuses on the problems of localization and cooperative mapping in 3D environments with ground and aerial robots. In this work, we do not emphasize vehicle autonomy as the experiments required that the operators tele-operate the vehicles. We discuss this requirement further in Sect. 4. During the experiments, tele-operation is conducted over wireless communication. However, we assume that in a disaster scenario, the ground vehicles will communicate with an external operator via a tether as currently employed at the Fukushima site [4].

Fig. 3 The Quince ground platform carries the Pelican aerial robot via a landing pad. The aerial robot opens and closes the landing pad via a wireless interface during autonomous take-off and landing

Communications with the aerial robot are via a local access point carried by the ground robot.

In this work, we leverage our previous efforts in the areas of ground robot design [13], sensor design for 3D map building [14], and ground robot tele-operation [15] toward mapping with ground robots [16–18]. Additionally, we build upon prior work towards autonomous navigation and 3D mapping with an aerial robot [19, 20].

2.1 Robot Platforms

As previously discussed, we employ three robot platforms for this work: two tracked ground platforms (Kenaf and Quince) and a quadrotor (Pelican). We now briefly detail each platform.

2.1.1 Ground Robots

The Kenaf is a tracked ground platform with an onboard rotating laser-scanner that provides feature-rich 3D point clouds of the environment (Fig. 4). The laser scanner

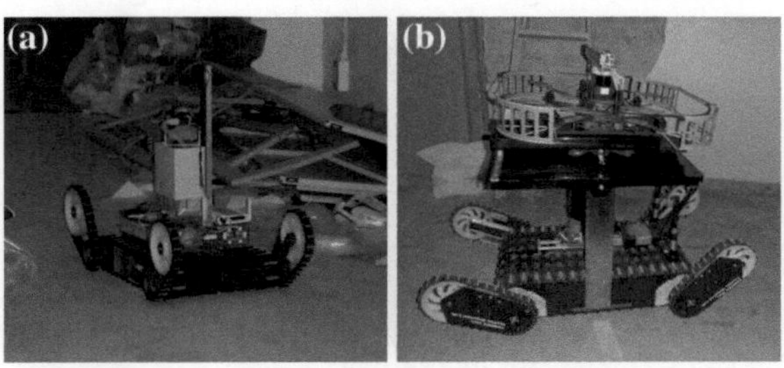

Fig. 4 The three robots used in the experiments include the Kenaf (Fig. 4a) and Quince (Fig. 4b) tracked ground robots. Here we see the Quince transporting the Pelican between discrete sites of interest via the landing pad

on the Kenaf operates at 40 Hz and rotates about the vehicle body-frame at 0.2 Hz. All laser scans from one revolution are assembled into a 3D point-cloud aligned with the robot body-frame origin. Further details of the platform and 3D laser scanner are available in [14, 17] and [18], respectively. The Quince platform (detailed in [13], Fig. 4b) shares a similar tracked design. Both platforms provide odometry and IMU information and are equipped with stabilizing tracked arms that permit climbing stairs and navigating clutter- or debris-filled environments.

The Kenaf and Quince provide visual information for the tele-operation of the vehicle including camera imagery of the surrounding environment during operation. We process any additional sensory information from the Kenaf and Quince off-board.

For this work, we equipped the Quince with a landing pad that opens and closes via a remote signal transmitted over an 802.15.4 wireless interface. The landing pad is equipped with an actuated pair of foam-lined arms that open and close via this remote signal. During the experiments, the aerial robot signals the opening and closing of the arms. The arms grip the base of the aerial robot via the compression of the dense foam lining, keeping the aerial robot firmly in position when it is transported by the Quince. Additionally, the foam helps absorb vibrational or impulse forces on the vehicle due to the Quince going over rough terrain and steps (Fig. 3).

2.1.2 Aerial Robot

The Pelican quadrotor robot platform is sold by Ascending Technologies, GmbH [21] and is equipped with an IMU (accelerometer, gyroscope, magnetometer) and pressure sensor. We developed custom firmware to run at the embedded level to address feedback control and estimation requirements. The other computation unit onboard is a 1.6 GHz Atom processor with 1 GB of RAM. The sensors on the robot include a Hokuyo UTM-30LX (laser), and a Microsoft Kinect (RGB-D sensor). A custom 3D printed mount is attached to the laser that houses mirrors pointing upward and downward. Communication with the robot for monitoring experiment progress and remote tele-operation is via an 802.11n access point mounted on the Quince.

Unlike the ground robots, the aerial robot requires some degree of onboard autonomy to permit autonomous navigation, take-off, and landing. Therefore, the vehicle

Fig. 5 The aerial robot flies through cluttered regions of the environment that are inaccessible to the ground robot and builds a 3D map that will be merged with the maps made by the ground robot

must be able to localize its position based on the current environment map and address the planning and control considerations required to permit autonomous navigation, take-off, and landing during experimentation (Fig. 5). The details of the algorithms employed to enable these capabilities are provided in [19, 20]. Figure 8 depicts a representative 3D map generated online during the experiments that is transmitted to the operator and used for autonomous navigation.

For this work, we require some degree of operator control to permit tele-operation of the vehicle. However, the complexity of the environment and the fact that the operator frequently did not have line-of-sight vision of the vehicle prevented full manual control of the vehicle. Therefore, we provided a “semi-autonomous” mode which permitted the operator to control the vehicle as a kinematic point-model agent or via waypoint control in the current map. Hence, at any moment, the operator could transition between full-autonomy and semi-autonomy to permit closer inspection of a location of interest or override the current behavior of the vehicle.

For this work, the autonomous take-off and landing is based on the originating position of the aerial robot in the current map. Therefore, we required that the Quince not move while the Pelican was flying. Although the autonomous landing maneuver was feed-forward in the sense that it did not observe the platform while landing, we found that the vehicle was able to land without issue in general. However, the autonomous landing maneuver also included a recovery phase should the vehicle detect that it did not successfully land on the platform. This lack of additional feedback information was primarily due to the short time-frame in which these experiments needed to be conducted prior to the experimental site becoming unavailable.

2.2 Map Generation and Merging

We now briefly describe the methods used to generate the 3D maps during the experiment. The experiment consisted of two phases. During the first phase, we tele-operated the Kenaf across the three stories of the building and collected sensor data for 3D map generation. We also identified locations inaccessible to the vehicle (six in total). After completing the first mapping phase, the Quince carried the Pelican to the six locations across the three stories of the building to further extend the map. The maps are generated using a sparse 3D voxel grid representation with a 10 cm resolution [22].

2.2.1 Kenaf

We used two methods to generate 3D maps via the Kenaf sensor data. The first approach uses a 3D iterative closest point (ICP) algorithm to determine incremental body-frame transformations. Details of map generation via this method are discussed in [17]. However, as noted in our prior work, 3D ICP can converge to poor alignment

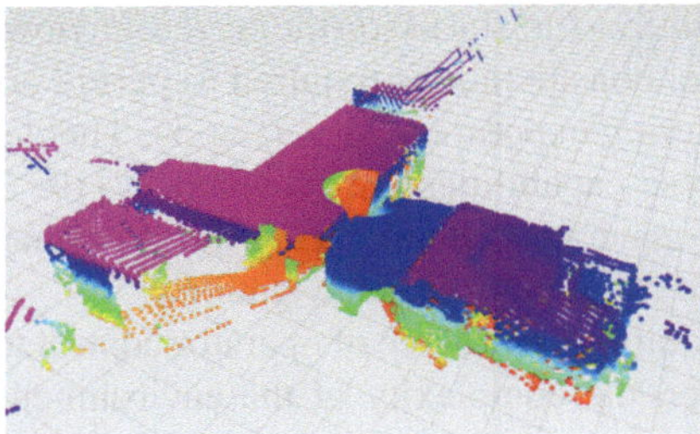

Fig. 6 The 3D rotating laser scanner on the Kenaf generates feature-rich 3D point clouds. Here we show the full output from a single revolution of the scanner

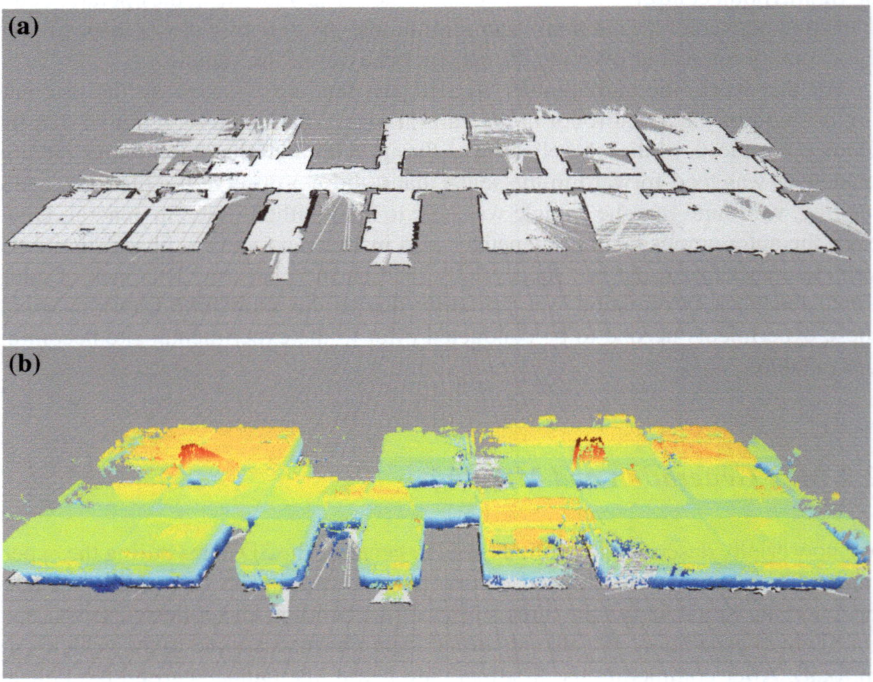

Fig. 7 The 2D occupancy grid map (Fig. 7a) and 3D point-cloud map (Fig. 7b) of the 7th floor generated via the Kenaf sensor data

solutions. We found that when the vehicle was operating on a level z-plane (i.e. not in a stairwell), we could yield a more robust mapping solution by employing the methods discussed in [19] which requires the assumption that the environment is generally described by flat planes and vertical walls (the 2.5D assumption).

For this approach, map corrections are done on a per-revolution basis with the assumption that the odometry error within one revolution is sufficiently small and the assembled point-cloud is accurate. Error in yaw is also corrected using IMU information. Figure 6 shows a typical point-cloud output from one revolution. The point-cloud is down-sampled via a voxel grid filter, from which we generate a 2D point-cloud by choosing all samples at a fixed z-height. We compute SLAM corrections from this 2D point-cloud and odometry data via the methods detailed in [19] to yield corrected robot poses. These corrected poses are used with the 3D point-clouds to generate globally consistent 3D maps of the environment (Fig. 7a) along with 2D occupancy grid maps resulting from the 2.5D assumption (Fig. 7b). In general, we applied the second method when operating on level terrain and only reached for 3D ICP-based SLAM methods when operating in the stairwell regions.

2.2.2 Pelican

As previously noted the Pelican generates a 3D map online during autonomous flight following the methodology detailed in [19]. Unlike the Kenaf, the Pelican collects data at discrete locations in the environment with the origin associated with the take-off location as visited by the Quince. In a manner similar to above, we generate a 3D point-cloud and a 2D occupancy grid map associated with each take-off location. Figure 8 depicts a representative visualization of the sensor data and generated maps. These maps are merged with the Kenaf maps from the previous section to form a complete 3D representation of the environment.

2.2.3 Merging Ground and Aerial Robot Maps

We begin by registering the two types of maps (the Kenaf and Pelican maps) via an initialization point near the known take-off location of the Pelican, as the Quince visits locations defined in the Kenaf map. Further refinement between the two maps is accomplished via ICP [17, 18]. This approach is applied for each of the rooms visited by the Pelican (Fig. 9).

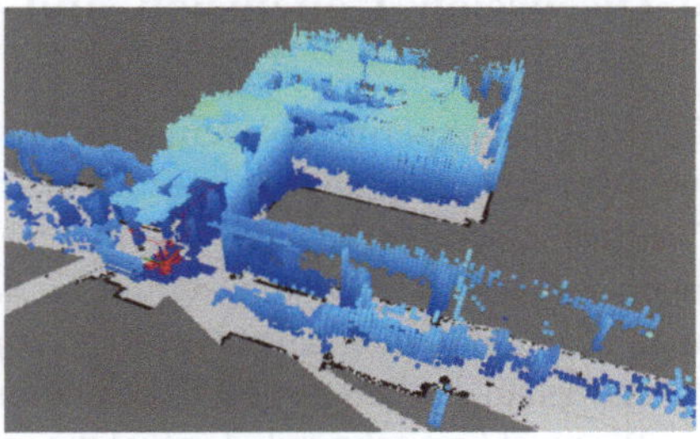

Fig. 8 A representative 3D map generated by the aerial vehicle during flight. A 2D occupancy grid map is also generated at all times. The vehicle and its path are shown as a red mesh and line, respectively

Fig. 9 Merging the Kenaf and Pelican maps. The map generated by the Kenaf is shown on the left while the extended map via the Pelican observations along with the Quince and Pelican trajectories (red and black, respectively) are shown on the right

3 Results

As previously noted, the goal of this work is the generation of 3D maps that capture the layout of the environment and provide insight to the degree of damage inside a multi-story building. In Figs. 10, 11 we provide full 2D and 3D maps of the 7th–9th floors of the building. We can clearly see features in the environment, such as the structural braces placed on the 8th floor (Fig. 11d) to prevent further structural collapse and the locations on the 9th floor (Fig. 10c) where the walls caved out of the building, leaving large openings. In Fig. 12, we show the 3D map for the stairwell between the 7th and 8th floors at various z-height levels.

The experiment lasted a total of 2.5 hours with the Kenaf first generating a 3D map via tele-operation followed by the Quince carrying the Pelican to discrete locations. It is worth noting that while the flight-time of the Pelican in confined environments can be as low as 5 min, we only needed to replace the battery in the vehicle twice due to our use of the aerial robot only when necessary for map extension. Although our Pelican can traverse hallways and stairwells autonomously (as shown in [19]), we conserved the battery power whenever possible by employing the Quince.

4 Discussion, Conclusion, and Areas for Future Work

The original experiments were intended to occur over several days but we found that we were able to complete the full exercise in one afternoon without any failures. While the fact that we were able to map a multi-story building with a heterogeneous team of robots without any significant issues or failures is an encouraging argument that the technological level is close to applicable in real scenarios, there are still some fundamental challenges left to be addressed.

We must first acknowledge that the environment was modified prior to our entry in that it was cleaned of any hazardous materials and structural reinforcements were in place to prevent further building collapse. For this reason, one should be cautious to state that our experiments are completely representative of an earthquake damaged

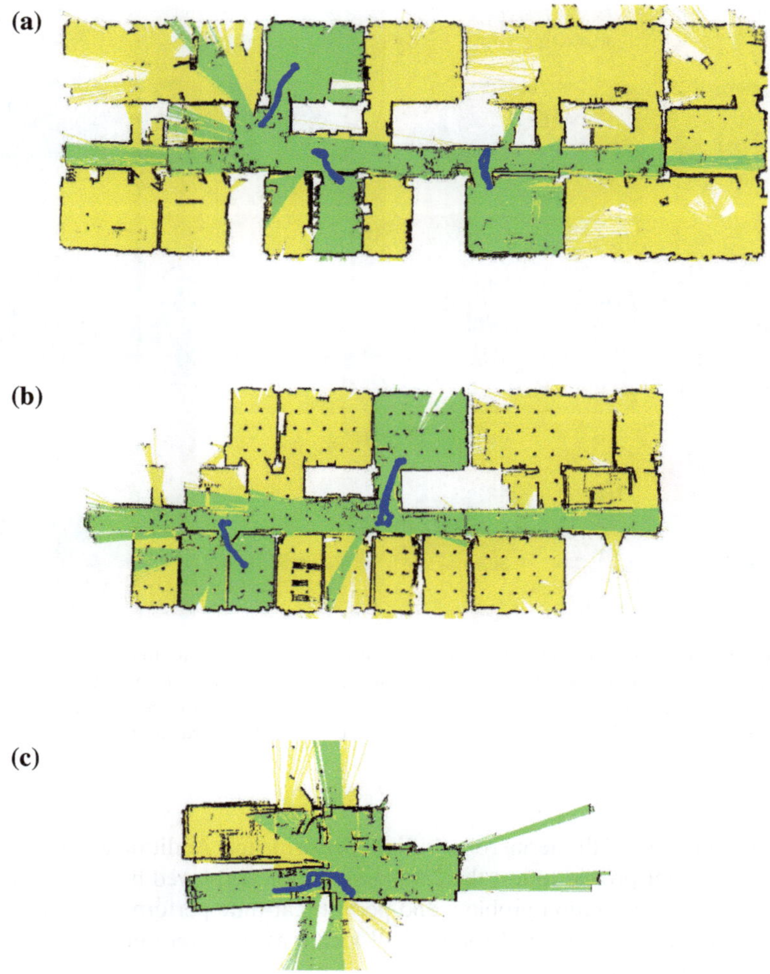

Fig. 10 The 2D occupancy grid maps generated during the experiment of the 7th, 8th, and 9th stories of the Tohoku University Electrical Engineering Building. The contributions to the map made by the Kenaf are shown in yellow with an overlay of the contributions made by the Pelican in green. The path of the Quince is shown in red while the trajectory followed by the Pelican is depicted in blue. The path of the Kenaf is not shown, **a** 7th Floor **b** 8th floor **c** 9th Floor

building. However, the environment still possessed similar attributes to what one would expect: fallen beams, dust and debris throughout the interior, water pools where rain freely entered the building, wires hanging from the ceiling, and personal affects and furniture in disarray. Indeed, loose wall and ceiling materials were of concern for both the ground and aerial robots due to the possibility of breaking the vehicles. Many of the windows and walls were compromised, yielding inconsistent

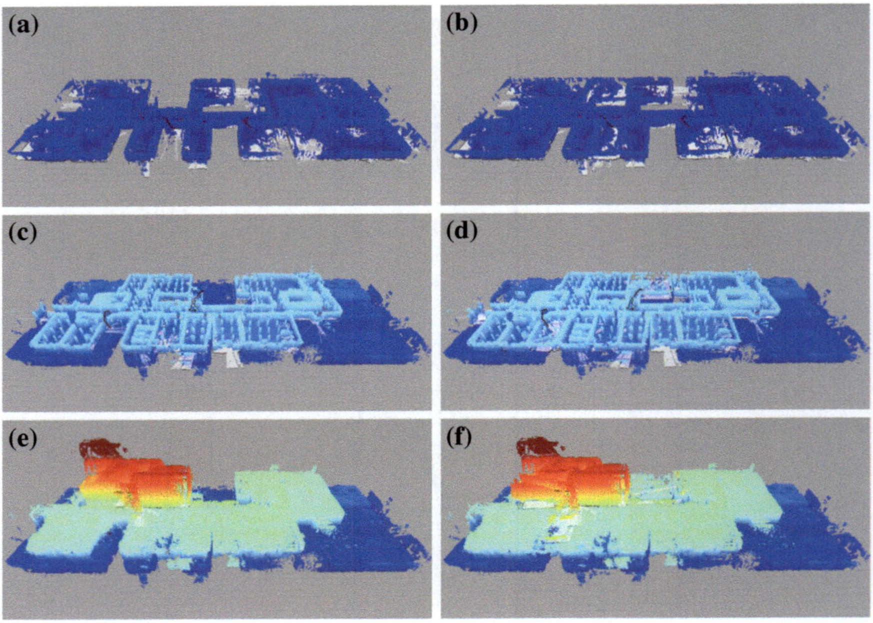

Fig. 11 The 3D voxel grid maps generated during the experiment. The map resulting from the Kenaf sensor data is shown on the left while the merged maps resulting from both the Kenaf and Pelican sensor data are shown on the right, **a** 7th floor - Kenaf **b** 7th floor - Kenaf and Pelican, **c** 7–8th floor - Kenaf, **d** 7–8th floor - Kenaf and Pelican, **e** 7–9th floor - Kenaf, **f** 7–9th Floor - Kenaf and Pelican

air flow that impacted the aerial robot's flight performance. Additionally, some of the debris and clutter proved to test the 2.5D assumption employed by the aerial robot to simplify the localization problem and permit real-time performance. Hence, we were not able to use the aerial robot in all locations that were inaccessible to the ground robot (Fig. 9).

The 3D voxel grid-based maps resulting from this work provide insight into the building layout and structural information but may be too coarse to be of practical use in real search and rescue applications. Recent progress in the area of dense 3D mapping suggests that high fidelity maps suitable for real search and rescue applications are achievable using laser and RGB-D sensors [23, 24]. Further, these richer 3D maps can be generated offline or on an external system as they are not required to enable the individual vehicle autonomy. Thus, the major constraint is the communication bandwidth required to transfer the data between aerial and ground vehicles and the base station. In this work, we consider a tethered ground robot with an aerial vehicle operating in close proximity communicating via 802.11n. Therefore, we believe that given the proposed heterogeneous team, such rich 3D maps are feasible following the implementation methods proposed in [23, 24].

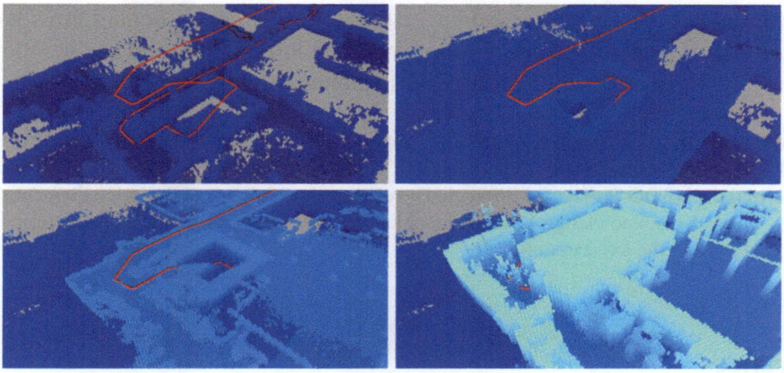

Fig. 12 The 3D map generated for the stairwell traversed by the vehicles between the 7th and 8th floors showing various z-height levels of detail along with the trajectory followed by the Quince robot

In addition to the platform or algorithmic limitations, an interesting consideration that arose in this work is the role of autonomy for aerial robots in search and rescue. We found that tele-operation of an aerial robot can be quite challenging in complex and confined environments, particularly when the operator does not have direct line-of-sight and debris is interacting with the vehicle. An autonomous vehicle may be able to sense and locally avoid those external interactions and preserve stable flight while a tele-operated system may not yield the same result. We found this to be the case at several points during our experimentation when the operator failed to navigate the vehicle through tightly confined spaces but the fully autonomous vehicle was able to find a path and autonomously navigate through the confined space.

From these statements, one may conclude that the areas that require the greatest attention in the future do not lie at the core problems of localization and mapping but more at the boundaries of these problems including the interfaces between the operators and the vehicles and vehicles and the environment. We require a better understanding of the appropriate methods to permit operators to interact with aerial robots in complex and confined environments such as those found in this work. Additionally, we must design aerial vehicles to be more robust to debris in the environment. In this work, we did not notice a significant impact on sensor performance (laser and RGB-D) due to dust or other airborne obscurants. However, this fact is likely due to the nature of the building airflow conditions rather than the sensor robustness, as this is a known concern [25], and should therefore also be considered when determining platform suitability for real applications.

While there are still issues that must be addressed in the algorithms, these problems are primarily of pragmatic concern. At present, we require the 2.5D assumption on the aerial vehicle due to constrained onboard CPU capabilities. As CPUs becoming increasingly capable, we will continue to incorporate more sensor information and eliminate the need for the 2.5D assumption. We are particularly interested in eliminating this assumption in the near future as it is a major algorithmic limitation for the aerial platform. We are also interested in further experimentation with cooperative teams of ground and aerial robots but with multiple ground and aerial robots operating concurrently as opposed to the sequential phases in this work.

Acknowledgments The authors gratefully acknowledge partial support from NSF Grant CNS-1138110, ONR Grant N00014-08-1-0696, ARL Grant W911NF-08-2-0004, the JST J-RAPID program, and the NEDO Project for Strategic Development of Advanced Robotics Elemental Technologies. They also acknowledge Yash Mulgaonkar for the development and manufacturing of the landing pad used in the experiments.

References

1. Magnitude 9.0 - near the east coast of Honshu, Japan, in: United States Geological Survey (March 11, 2011). Source: http://earthquake.usgs.gov/earthquakes/eqinthenews/2011/usc0001xgp
2. H. Tabushi, Company believes 3 reactors melted down in Japan. In: The New York Times (May 24, 2011). Source: http://www.nytimes.com/2011/05/25/world/asia/25nuclear.html
3. E. Ackerman, Japan Earthquake: iRobot Sending Packbots and Warriors to Fukushima Dai-1 Nuclear Plant, in: IEEE Spectrum (March 18, 2011). Source: http://spectrum.ieee.org/automaton/robotics/industrial-robots/irobot-sending-packbots-and-warriors-to-fukushima
4. K. Nagatani, S. Kiribayashi, Y. Okada, S. Tadokoro, T. Nishimura, T. Yoshida, E. Koyanagi, Y. Hada, Redesign of rescue mobile robot Quince—toward emergency response to the nuclear accident at Fukushima Daiichi nuclear power station on March 2011, in *Proceedings of IEEE International Symposium on Safety, Security, and Rescue Robotics* (Kyoto, Japan, 2011) pp. 13–18
5. E. Guizzo, Fukushima robot operator writes tell-all blog, in: IEEE Spectrum (August 23, 2011). Source: http://spectrum.ieee.org/automaton/robotics/industrial-robots/fukushima-robot-operator-diaries
6. A. Bachrach, S. Prentice, R. He, N. Roy, RANGE - robust autonomous navigation in GPS-denied environments. J. Field Robot. **28**(5), 644–666 (2011)
7. J. Pellenz, D. Lang, F. Neuhaus, D. Paulus, Real-time 3D mapping of rough terrain: A field report from Disaster City, in *Proceedings of the IEEE International Workshop on Safety, Security, and Rescue Robot* (Bremen, Germany, 2010)
8. J.P. How, C. Fraser, K.C. Kulling, L.F. Bertuccelli, O. Toupet, L. Brunet, A. Bachrach, N. Roy, Increasing autonomy of UAVs. IEEE Robot. Autom. Mag. **16**(2), 43–51 (2009)
9. B. Kim, M. Kaess, L. Fletcher, J. Leonard, A. Bachrach, N. Roy, S. Teller, Multiple relative pose graphs for robust cooperative mapping, in *Proceedings of the IEEE International Conference on Robot and Autom* (Anchorage, AK, 2010), pp. 3185–3192
10. F. Gonzalez, M.P.G. Castro, P. Narayan, R. Walker, L. Zeller, Development of an autonomous unmanned aerial system to collect time-stamped samples from the atmosphere and localize potential pathogen sources. J. Field Robot. **28**(6), 961–976 (2011)
11. R. Murphy, J. Kravitz, S. Stover, R. Shoureshi, Mobile robot in mine rescue and recovery. IEEE Robot. Autom. Mag. **16**(2), 91–103 (2009)
12. K.S. Pratt, R.R. Murphy, J.L. Burke, J. Craighead, C. Griffin, S. Stover, Use of tethered small unmanned aerial system at Berkman Plaza II collapse, in *Proceedings of the IEEE International Workshop on Safety, Security, and Rescue Robot* (Sendai, Japan, 2008)
13. E. Rohmer, T. Yoshida, K. Ohno, K. Nagatani, S. Tadokoro, E. Koyanagi, Quince: a collaborative mobile robotic platform for rescue robots research and development, in *International Conference on Advanced Mechatronic* (Osaka, Japan, 2010) pp. 225–230

14. K. Ohno, T. Kawahara, S. Tadokoro, Development of a 3D laser scanner for measuring uniform and dense 3D shapes of static objects in dynamic environments, in *Proceedings of IEEE International Conference on Robot and Biomimetics* (Bangkok, Thailand, 2008) pp. 2161–2167
15. Y. Okada, K. Nagatani, K. Yoshida, S. Tadokoro, T. Yoshida, E. Koyanagi, Shared autonomy system for tracked vehicles on rough terrain based on continuous three-dimensional terrain scanning. J. Field Robot. **28**(6), 875–893 (2011)
16. K. Nagatani, N. Tokunaga, Y, Okada, K. Yoshida, Continuous acquisition of three-dimensional environment information for tracked vehicles on uneven terrain, in *Proceedings of the IEEE International Workshop on Safety, Security, and Rescue Robot* (Sendai, Japan, 2008)
17. K. Ohno, S. Tadokoro, K. Nagatani, E. Koyanagi, T. Yoshida, 3-D mapping of an underground mall using a tracked vehicle with four sub-tracks, in *IEEE International Workshop on Safety, Security, and Rescue Robotics* (Denver, CO 2009)
18. K. Ohno, S. Tadokoro, K. Nagatani, E. Koyanagi, T. Yoshida, Trials of 3-D map construction using the tele-operated tracked vehicle Kenaf at Disaster City, in *Proceedings of the IEEE International Conference on Robot and Automation* (Anchorage, AK, 2010) pp. 2864–2870
19. S. Shen, N. Michael, V. Kumar, Autonomous multi-floor indoor navigation with a computationally constrained MAV, in *Proceedings of the IEEE International Conference on Robotics and Automation*, (Shanghai, China, 2011) pp. 20–25
20. S. Shen, N. Michael, V. Kumar, Autonomous indoor 3D exploration with a micro-aerial vehicle, in *Proceedings of the IEEE International Conference on Robot. and Autom* (Saint Paul, MN, 2012) pp. 9–15
21. Ascending Technologies, GmbH. http://www.asctec.de/
22. I. Dryanovski., W. Morris, X. Jizhong, Multi-volume occupancy grids: an efficient probabilistic 3D mapping model for micro aerial vehicles, in *Proceedings of the IEEE/RSJ International Conference on Intelligent Robots and Systems*, (Taipei, Taiwan, 2010) pp. 1553–1559
23. L.P. Ellekilde, S. Huang, J.V. Miro, G. Dissanayake, Dense 3D map construction for indoor search and rescue. J. Field Robot. **24**(1–2), 71–89 (2007)
24. A.S. Huang, A. Bachrach, P. Henry, M. Krainin, D. Maturana, D. Fox, N.Roy, Visual odometry and mapping for autonomous flight using an RGB-D camera, in *Proceedings of the International Symposium of Robot Research* (Flagstaff, AZ, 2011)
25. K.W. Sevcik, N. Kuntz, P.Y. Oh, Exploring the effect of obscurants on safe landing zone indentification. J. Intell. Robotic Syst. **57**(1–4), 281–295 (2010)

Three-Dimensional Thermography Mapping for Mobile Rescue Robots

Keiji Nagatani, Kazuki Otake and Kazuya Yoshida

Abstract In urban search and rescue situations, a 3D map obtained using a 3D range sensor mounted on a rescue robot is very useful in determining a rescue crew's strategy. Furthermore, thermal images captured by an infrared camera enable rescue workers to effectively locate victims. The objective of this study is to develop a 3D thermography mapping system using a 3D map and thermal images; this system is to be mounted on a tele-operated (or autonomous) mobile rescue robot. The proposed system enables the operator to understand the shape and temperature of the disaster environment at a glance. To realize the proposed system, we developed a 3D laser scanner comprising a 2D laser scanner, DC motor, and rotary electrical connector. We used a conventional infrared camera to capture thermal images. To develop a 3D thermography map, we integrated the thermal images and the 3D range data using a geometric method. Furthermore, to enable fast exploration, we propose a method for continuous thermography mapping while the robot is in motion. This method can be realized by synchronizing the robot's position and orientation with the obtained sensing data. The performance of the system was experimentally evaluated in real-world conditions. In addition, we extended the proposed method by introducing an improved iterative closest point (ICP) scan matching algorithm called thermo-ICP, which uses temperature information. In this paper, we report development of (1) a 3D thermography mapping system and (2) a scan matching method using temperature information.

1 Introduction

Recently, there has been a growing demand for preliminary robotic investigation in urban search and rescue missions conducted at sites affected by disasters such as earthquakes or terror attacks. For this purpose, search and rescue robots are being developed by a number of institutes [1], particularly in Japan [2, 3].

K. Nagatani · K. Otake · K. Yoshida (✉)
Tohoku University, Sendai, Japan
e-mail: yoshida@astro.mech.tohoku.ac.jp

K. Yoshida and S. Tadokoro (eds.), *Field and Service Robotics*,
Springer Tracts in Advanced Robotics 92, DOI: 10.1007/978-3-642-40686-7_4,

The important functions of rescue robots are to search for victims and to gather environmental information. for planning rescue operations. Useful information for the former function constitutes biological signals such as body temperature and carbon dioxide, whereas that for the latter function continues images, three-dimensional (3D) shapes [4], temperature, and gas concentration in the target environment. In both cases, infrared cameras are indispensable, particularly in dark conditions. In the 2011 RoboCupRescue competition [5], such cameras mounted on rescue robots were widely used for identifying simulated victims which have heat sources. However, the two-dimensional (2D) information captured by the camera makes it difficult for rescue crews to deduce 3D structures.

Therefore, in this study, we propose a sensor fusion method to integrate a 3D-map captured by a 3D range scanner with infrared images captured by an infrared camera, to construct a 3D thermography map. The integration of a 3D map and color images has been investigated in several studies [6–8]. Recently, some studies on 3D thermography mapping were conducted [9, 10]. However, to the best of our knowledge, this technology has not been applied to rescue robots thus far. The proposed method enables rapid mapping for rescue scenarios. To realize map building, the robot acquires environmental information while it is in motion. We installed the proposed system on a mobile rescue robot testbed called Kenaf in order to verify the validity of the method.

2 3D Range Scanner

Conventional commercial 3D range scanners are expensive and bulky. Therefore, we decided to develop a small, light weight 3D range scanner for this study. In general, a 3D range scanner is realized by rotating a conventional 2D laser range scanner. We adopted this method, as described in the following section.

2.1 Hardware

The 3D range scanner that we developed consists of a 2D laser range scanner (UTM-30LX, HOKUYO Co., Ltd.), DC motor (RE25, Maxon Motor), and rotary connector (Model630, Mercotac). A schematic of the scanner is shown in Fig. 1 (left).

The 2D scanner is attached to a rotating table at a certain angle, β. Thus, the scanning surface of the 2D scanner is inclined to the rotating surface of the table, so that 3D environment information is obtained by rotating the table [11]. At the rotation axis of the table, we installed a rotary electrical connector that having a mercury contact. Therefore, the table with the 2D scanner can rotate freely without causing the cables to twist. The rotation speed of the table can be increased up to 180 rpm, and frequency 3D range scanning can be performed using this device.

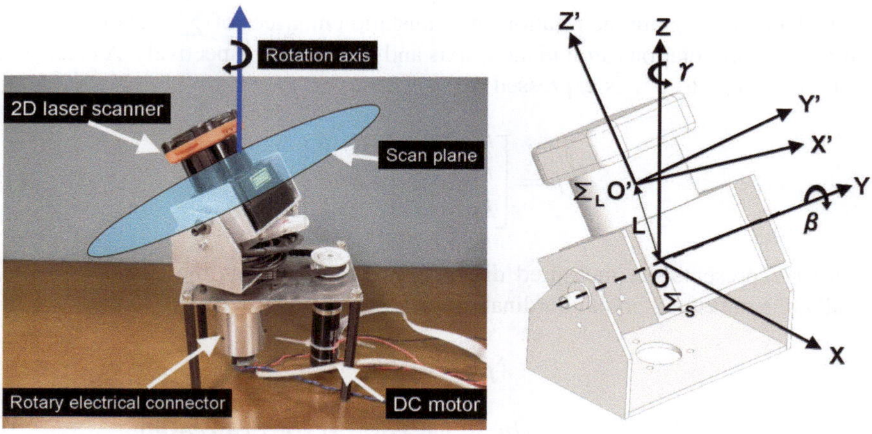

Fig. 1 3D scanner (*left*) and coordinate system (*right*)

2.2 Geometric Model

Here, we introduce the geometric model of the 3D range scanner, as shown in Fig. 1 (right). Let us assume that the coordination system $\sum_L$ is attached to the 2D scanner. The origin O' of the coordinate is located at the center of the scanning surface, the x-axis X' corresponds to the front of the scanner, the z-axis Z' is perpendicular to the scanning surface, and the y-axis Y' is defined to satisfy the right-handed coordinate system. The coordination system $\sum_S$ is attached to the 3D range scanner, and the origin O of the coordinate is located at a certain distance above the center of the rotation table. The x-axis X corresponds to the front of the 3D range scanner, the z-axis Z corresponds to the rotational axis of the table, and the y-axis Y is defined to satisfy the right-handed coordinate system. The mounting angle of the 2D scanner is β, and rotational speed of the table is $\dot{\gamma}$. The homogeneous transform matrix ${}^S\boldsymbol{T}_L$ between $\sum_L$ and $\sum_S$ is expressed by:

$$ {}^S\boldsymbol{T}_L = \begin{bmatrix} {}^S\boldsymbol{R}_L & {}^S\boldsymbol{t}_L \\ 0 & 1 \end{bmatrix} \tag{1} $$

$$ {}^S\boldsymbol{R}_L = \begin{bmatrix} cos\gamma & -sin\gamma & 0 \\ sin\gamma & cos\gamma & 0 \\ 0 & 0 & 1 \end{bmatrix} \begin{bmatrix} cos\beta & 0 & sin\beta \\ 0 & 1 & 0 \\ -sin\beta & 0 & cos\beta \end{bmatrix} \tag{2} $$

$$ {}^S\boldsymbol{t}_L = {}^S\boldsymbol{R}_L \begin{bmatrix} 0 \\ 0 \\ L \end{bmatrix} \tag{3} $$

where ${}^S\boldsymbol{R}_L$ and ${}^S\boldsymbol{t}_L$ are the rotation and translation matrices of $\sum_L$ relative to $\sum_S$, and γ and β are rotations around the Z-axis and the Y-axis, respectively. A measured point ${}^L\boldsymbol{p}$ relative to $\sum_L$ is expressed by:

$$ {}^L\boldsymbol{p} = \begin{bmatrix} {}^Lp_x \\ {}^Lp_y \\ {}^Lp_z \end{bmatrix} = \begin{bmatrix} rcos\theta \\ rsin\theta \\ 0 \end{bmatrix} \tag{4} $$

where r represents the measured distance, and θ represents the scanning angle. Finally, the measured point coordinate ${}^S\boldsymbol{p}$ relative to $\sum_S$ is

$$ {}^S\boldsymbol{p} = {}^S\boldsymbol{T}_L{}^L\boldsymbol{p} \tag{5} $$

$$ \begin{bmatrix} {}^Sp_x \\ {}^Sp_y \\ {}^Sp_z \end{bmatrix} = \begin{bmatrix} cos\gamma & -sin\gamma & 0 \\ sin\gamma & cos\gamma & 0 \\ 0 & 0 & 1 \end{bmatrix} \begin{bmatrix} cos\beta & 0 & sin\beta \\ 0 & 1 & 0 \\ -sin\beta & 0 & cos\beta \end{bmatrix} \begin{bmatrix} rcos\theta \\ rsin\theta \\ L \end{bmatrix}. \tag{6} $$

2.3 Control Architecture

To obtain accurate 3D Cartesian coordinates of sensing points ${}^S\boldsymbol{p}$, synchronization between the scan data and the rotation angle of the table γ is very important. In our control architecture, we use a synchronizing signal from the 2D scanner, a signal of the encoder attached to the motor rotating the table, H8S micro control unit (MCU), and a host PC. Figure 2 shows a conceptual diagram of the synchronization. The 2D scanner sends 1,081 points of measurement data ${}^L\boldsymbol{p}$ every 25 ms to the host PC. The synchronizing digital signal is also generated by the scanner, and it becomes a trigger for reading the encoder data. We assume that the rotation speed $\dot{\gamma}$ is constant, so that linear interpolation is conducted to obtain γ for each sensory reading.

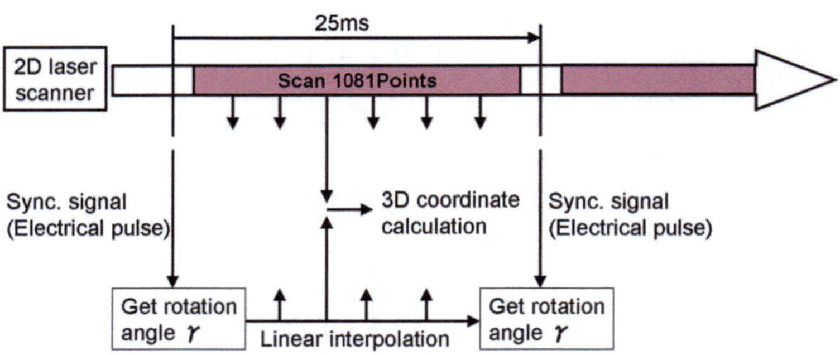

Fig. 2 Conceptual diagram of synchronization

3 3D Thermography Mapping System

To construct a 3D thermography map, we need a wide-view infrared image. However, the view angle of a conventional infrared camera is typically small. For the purpose of 3D thermography mapping, the camera should be mounted on a rotating table. Furthermore, we conducted a calibration of the camera's distortion in order to integrate a 3D range map and infrared images. Finally, we extended the method to provide a moving measurement system for mobile robots, as described in this section.

3.1 Hardware

Ideally, the focal position of the infrared camera and the center of the 3D range scanner should be the same. Practically, this is impossible, and we should ensure that they do not occlude each other. Therefore, we place the infrared camera and the 3D range scanner at different heights, as shown in Fig. 3.

We chose infrared camera module HX0830M2, manufactured by NEC Avio infrared technologies Co., Ltd. as our target infrared camera. The pixel number of the camera is 320 (H) × 240 (V), and the field of view is 50 × 37.5°. To rotate to which the camera is attached, we chose a smart motor, Dynamixel DX-117, manufactured by ROBOTIS Inc. By using the camera module and rotating the table from −90 to 90°, we can obtain a wide infrared image: 50° in the vertical field of view and 217.5° in the horizontal.

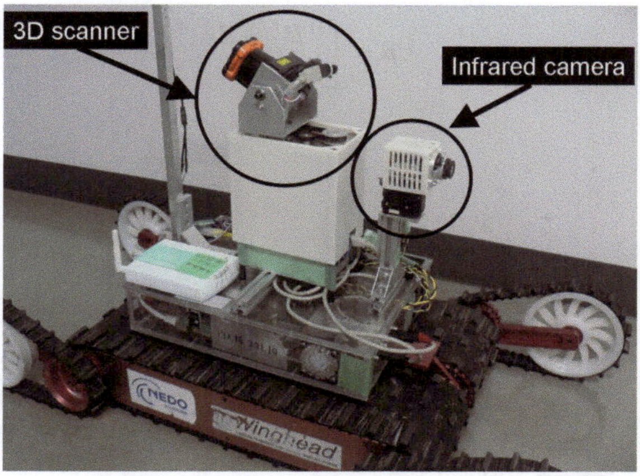

Fig. 3 Kenaf equipped with 3D scanner and infrared camera

3.2 Geometric Model

Here, we introduce a method to match a scan point ${}^{S}\boldsymbol{p}$ with a pixel in an infrared image. The coordination system of the scanner system is shown in Fig. 4 (left). Let us assume that the coordination system $\sum_C$ is attached to a camera. The origin O'' of the coordinate is located at the focal position of the camera, the x-axis X'' corresponds to the optical axis, the z-axis Z'' corresponds to the vertical coordinate of the camera, and the y-axis Y'' is defined to satisfy the right-handed coordinate system. A scan point ${}^{C}\boldsymbol{p}$ represented by $\sum_C$ is calculated by

$$ {}^{C}\boldsymbol{p} = {}^{C}\boldsymbol{T}_S {}^{S}\boldsymbol{p}, \tag{7} $$

where ${}^{C}\boldsymbol{T}_S$ is the homogeneous transform between $\sum_S$ and $\sum_C$. In fact, the positional relationship ${}^{W}\boldsymbol{T}_S$ between $\sum_S$ and $\sum_W$ (the world coordinate system) is known, and ${}^{W}\boldsymbol{T}_C$ between $\sum_C$ and $\sum_W$ is calculated by camera position and camera rotating angle. Therefore, ${}^{C}\boldsymbol{T}_S$ is obtained by the equation

$$ {}^{C}\boldsymbol{T}_S = {}^{C}\boldsymbol{T}_W {}^{W}\boldsymbol{T}_S \tag{8} $$

$$ = {}^{W}\boldsymbol{T}_C^{-1} {}^{W}\boldsymbol{T}_S \tag{9} $$

$$ = \begin{bmatrix} {}^{W}\boldsymbol{R}_C^T & -{}^{W}\boldsymbol{R}_C^T {}^{W}\boldsymbol{T}_C \\ 0 & 1 \end{bmatrix} \begin{bmatrix} {}^{W}\boldsymbol{R}_S & {}^{W}\boldsymbol{T}_S \\ 0 & 1 \end{bmatrix} \tag{10} $$

$$ = \begin{bmatrix} {}^{W}\boldsymbol{R}_C^T {}^{W}\boldsymbol{R}_S & {}^{W}\boldsymbol{R}_C^T \left({}^{W}\boldsymbol{T}_S - {}^{W}\boldsymbol{T}_C\right) \\ 0 & 1 \end{bmatrix}. \tag{11} $$

Finally, Eq. (7) is represented by

$$ {}^{C}\boldsymbol{p} = \begin{bmatrix} {}^{W}\boldsymbol{R}_C^T {}^{W}\boldsymbol{R}_S & {}^{W}\boldsymbol{R}_C^T \left({}^{W}\boldsymbol{T}_S - {}^{W}\boldsymbol{T}_C\right) \\ 0 & 1 \end{bmatrix} {}^{S}\boldsymbol{p}. \tag{12} $$

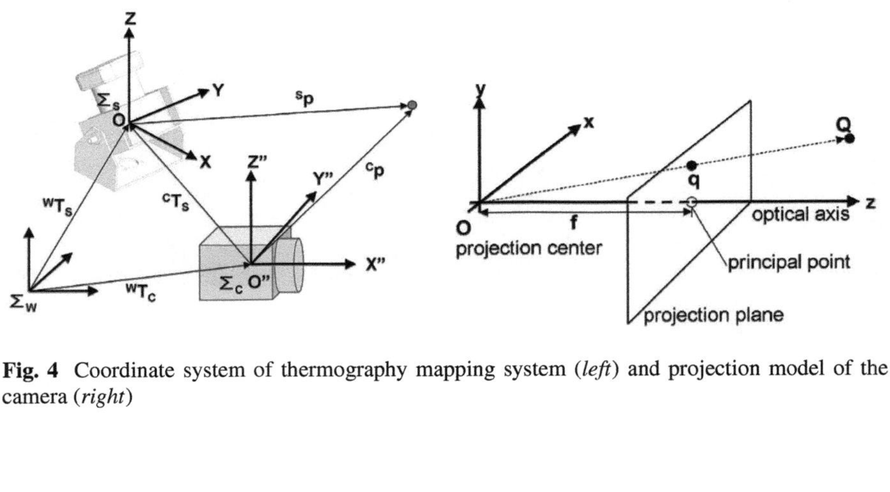

Fig. 4 Coordinate system of thermography mapping system (*left*) and projection model of the camera (*right*)

In case of exchanging from $\sum_W$ to the robot coordinate system $\sum_{RS}$, Eq. (12) represents a scan point when the robot is motionless. In Sect. 3.4, it will be extended to the scan point model when the robot is moving.

The next step is to obtain the scan point coordinate on the infrared image. Figure 4 (right) shows the projection model of the camera. The transform from 3D point $\boldsymbol{Q} = (X, Y, Z)$ to an image point $\boldsymbol{q} = (x, y)$ is conducted by the homogeneous coordination $\hat{q} = (x, y, 1)^T$ and the equation

$$\omega \hat{\boldsymbol{q}} = \boldsymbol{M} \boldsymbol{Q}. \tag{13}$$

$$\boldsymbol{M} = \begin{bmatrix} f_x & 0 & c_x \\ 0 & f_y & c_y \\ 0 & 0 & 1 \end{bmatrix} \tag{14}$$

where M is the internal parameter matrix of the camera, f_x and f_y represent the focal lengths, and c_x and c_y are the offsets of the projection center.

Equation (13) is satisfied only when there is no lens distortion. Practically, distortion calibration of the infrared camera is required, as shown in the next subsection.

3.3 Distortion Calibration of Infrared Camera

To obtain the internal and distortion parameters of the infrared camera, we applied Zhang's method [12]. This method requires images of the calibration pattern to be obtained at different locations. In this study, we used an infrared camera; thus, an ingenious method was required to obtain clear calibration patterns. Therefore, we developed a calibration board composed of high-infrared-reflectance aluminum and low-infrared-reflectance masking tape [13], as shown in Fig. 5 (left). The size of the board was 300 mm × 400 mm, and its grid size was 50 mm. Based on the above method, we were able to obtain clear images of the board (see Fig. 5 (right)) when the surface of the board reflected the sky outside.

Fig. 5 Developed calibration board (*left*) and obtained image (*right*)

3.4 Extension to Moving Measurement System

To enable rapid map building, we extended the thermography mapping method to a moving measurement system for mobile robots. The method is based on a gyroscope-based 3D odometry system [14].

Let us assume that $\sum_{RS}$ is the robot coordinate system when the scanner captures data, and $\sum_{RC}$ is the robot coordinate system when the camera captures an image. The homogeneous transform matrix ${}^{C}\boldsymbol{T}_S$ (to represent $\sum_S$ relative to $\sum_C$) is decomposed as

$$ {}^{C}\boldsymbol{T}_S = {}^{C}\boldsymbol{T}_{RC}\,{}^{RC}\boldsymbol{T}_W\,{}^{W}\boldsymbol{T}_{RS}\,{}^{RS}\boldsymbol{T}_S, \tag{15} $$

where ${}^{RS}\boldsymbol{T}_S$ is the homogeneous transform matrix that represents the scanner coordinate relative to the robot coordinate, and ${}^{RC}\boldsymbol{T}_C$ is the homogeneous transform matrix that represents the camera coordinate relative to the robot coordinate. ${}^{W}\boldsymbol{T}_{RS}$ and ${}^{W}\boldsymbol{T}_{RC}$ are obtained by the 3D odometry system, ${}^{C}\boldsymbol{T}_{RC}$ and ${}^{RC}\boldsymbol{T}_W$ can be calculated from ${}^{RC}\boldsymbol{T}_C$ and ${}^{W}\boldsymbol{T}_{RC}$. Finally, ${}^{C}\boldsymbol{p}$ is obtained by Eq. (7).

3.5 Synchronization Between the Scan Data and the Odometry Data

Because the sampling time of the odometer is different from that of the scanner, we applied time stamps to synchronize the scan data and the odometry data. To achieve precise synchronization, we applied a linear interpolation of the odometry data. The details are explained in [14].

Fig. 6 Setup of the basic experiment (*left*) and the mapping result (*right*)

3.6 Initial Experiment

To evaluate our 3D thermography mapping system, we conducted an initial mapping experiment, as shown in Fig. 6. The translation speed of the robot was typically about 4.3 cm/s, and it surmounted a concrete block while it mapped the environment three-dimensionally. The distance from the robot's path to the target board was 1.2 m. The mounting angle of the 2D scanner, β, was set at 60°, and the rotation speed of the table $\dot{\gamma}$ was set at 6 rpm.

Figure 6 (right) shows a thermography mapping result. In this figure, the block and the target calibration board can be seen: the image does not collapse because of the robot's motion.

Figure 7 shows the binarized result of one scan (10 s) while the robot surmounted the blocks. During the motion, the transition vector of the robot is $(x, y, z, \phi x, \phi y, \phi z) = (+164\,\text{mm}, +4\,\text{mm}, +15\,\text{mm}, -0.4°, +16.7°, +0.5°)$. In this figure, the white frame represents the ideal boundaries of the lattice, red dots represent high-temperature points, blue dots represent low temperature points, and green dots represent excluded measurement points. The total number of dots in the board was 1,733, and the number of dots assigned correctly was 1,492. Thus, the proportion of dots assigned correctly was 86.1 %. According to the above results, the thermography mapping of the range data was conducted precisely, even if the robot surmounted the block. The errors were attributed to the laser beam being slightly conical in shape, so that the reflection close to the edge of the board returned.

3.7 Experiments in Wide Area

To evaluate our 3D thermography mapping system, we conducted a wide-area experiment, shown in Fig. 8. The target environment is the fourth floor of the building

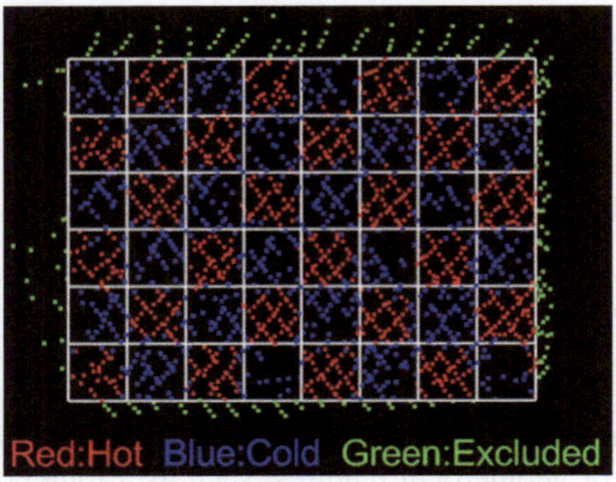

Fig. 7 Binarized thermography map of the target board

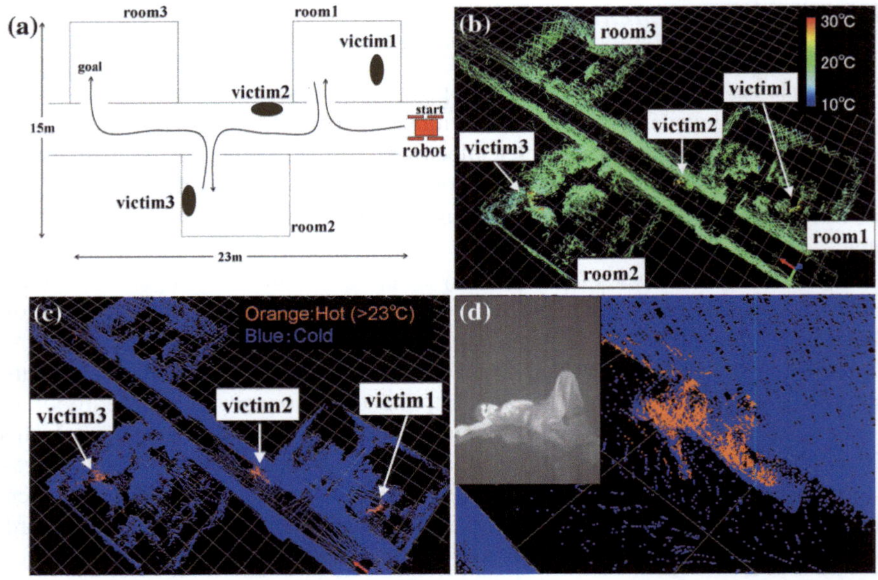

Fig. 8 Experiments in wide area. **a** setup of the experiment, **b** mapping result, **c** binarized mapping result, **d** zoomed-in view of victim 2

that houses our laboratories', which includes three rooms connected by a straight corridor, as shown in Fig. 8a. We located three dummy victims. Figure 8b shows a thermography mapping result, and Fig. 8c shows a binarized thermography mapping result. Figure 8d shows a zoomed-in view of victim 2 and a raw infrared image.

Based on the results, the target environment was measured with no distortion, even if it was measured by only a gyroscope-based odometry system, because the robot navigated on a flat surface. Furthermore, we understood that, since it focuses on the detected victims, a binarized map is an effective aid for human operators, as shown in Fig. 8c.

4 Thermo-ICP (Thermography-Based Iterative Closest Point) Matching Method

As an application of the 3D thermography mapping system, we introduce a thermography-based iterative closest point (ICP) matching method, called Thermo-ICP. A conventional ICP-matching method [15] relies on the shape of the environment's features. Therefore, it is difficult to apply the method to an environment that has a small number of geometrical features, such as a long corridor. Some methods use color information for ICP matching; however, such methods are sensitive to lighting conditions. Therefore, in this study, we used temperature information as a feature for ICP matching.

4.1 Temperature-Shift According to Measurement Position

To use temperature data for ICP matching, the target temperature should be the same, regardless of the position from which it is measured.

First, we consider a temperature measurement at a glancing angle θ. Infrared rays emitted from a point source decrease according to the glancing angle θ (Lambert's cosine law) as:

$$I_\theta = I_0 \times cos\theta. \tag{16}$$

On the other hand, the target size D_θ increases according to the glancing angle θ:

$$D_\theta = D_0 \times \frac{1}{cos\theta}. \tag{17}$$

Thus, the total amount of infrared rays for each pixel of a camera $I_\theta D_\theta$ is equal to $I_0 D_0$.

Practically, an object that has low emissivity of infrared rays has angular dependency. Therefore, it is said that a reasonable measurement angle θ is within $\pm 45°$.

Second, we consider the distance from the camera to the measured object. The emitted power of infrared rays is inversely proportionally with the square of the distance from a heat source. On the other hand, the target size increases proportional to the square of the distance from the heat source. Thus, the total amount of infrared rays for each pixel of a camera is independent of its distance from the heat source. However, in practice, the effect of water vapor or flue dust increases with the distance between the camera and the object.

Accordingly, we used temperature data when the measurement angle θ was within $\pm 45°$, and the distance was not too great for thermo-ICP.

4.2 ICP matching Method

ICP matching is a popular method for fitting two sets of shape data based on geometric features. Mobile robots that can reconstruct environmental information and conduct SLAM (simultaneous localization and mapping) are very useful. In this subsection, we would like to introduce a brief description of conventional ICP matching.

In this method, two sets of sensing points are registered in Cartesian coordinates. In each iteration step, the algorithm selects the closest points as correspondences and calculates the rotation matrix $\boldsymbol{R}$ and the translation matrix $\boldsymbol{t}$ to minimize the equation

$$E(\boldsymbol{R}, \boldsymbol{t}) = \sum_{i=1}^{N_m} \sum_{j=1}^{N_d} \omega_{i,j} \left\| \boldsymbol{m}_i - (\boldsymbol{R}\boldsymbol{d}_j + \boldsymbol{t}) \right\|^2, \tag{18}$$

where N_m and N_d are the number of points in the reference data set $\boldsymbol{M}$ and the matching data set $\boldsymbol{D}$, respectively. $\omega_{i,j} = 1$ when $\boldsymbol{m}_i$ in $\boldsymbol{M}$ is the closest point to $\boldsymbol{d}_j$ in $\boldsymbol{D}$, and $\omega_{i,j} = 0$ otherwise. Newton's method is typically used for calculating $\boldsymbol{R}$ and $\boldsymbol{t}$ in the evaluation function.

4.3 Thermo-ICP Matching

Our Thermo-ICP matching uses not only Euclidean distances but also temperature difference to search for corresponding points. The evaluation function of the algorithm uses not only the squared sum of the closest distance, but also the squared difference of thermo-values. Thus, the evaluation function for thermo-ICP is

$$E_T(\boldsymbol{R}, \boldsymbol{t}) = \sum_{i=1}^{N_m} \sum_{j=1}^{N_d} \hat{\omega}_{i,j} \left(\left\| \boldsymbol{m}_i - (\boldsymbol{R}\boldsymbol{d}_j + \boldsymbol{t}) \right\|^2 + K |h_{mi} - h_{dj}|^2 \right), \tag{19}$$

where K represents a weighting factor of the temperature term. Currently, we don not have any method to determine the value K, instead we set the parameter empirically. h_{mi} is the temperature value of $\boldsymbol{m}_i$, and h_{dj} is the temperature value of $\boldsymbol{d}_j$. $\hat{\omega}_{i,j} = 1$ when the following function e is minimized,

$$e = \left\| \boldsymbol{m}_i - (\boldsymbol{R}\boldsymbol{d}_j + \boldsymbol{t}) \right\|^2 + K |h_{mi} - h_{dk}|^2, \tag{20}$$

and $\omega_{i,j} = 0$ otherwise.

The above calculation is repeated until the value of $E_T(\boldsymbol{R}, \boldsymbol{t})$ converges.

4.4 Simple Experiment

To evaluate the proposed thermo ICP matching, we conducted a simple experiment in a corridor environment that has a few geometric features, and compared the results with those of conventional ICP matching. Figure 9 (left) shows the experimental setup. There is a heat source, a disposable body warmer, attached to the corridor wall. In this environment, the robot captured two scan sets. The first scan was conducted at the origin of the world coordinates, and the second scan was conducted at $(x, y, z, yaw) = (-200, 0, 0\,\text{mm}, 10°)$. Figure 9 (right) shows the motion of the robot. Figure 10 shows top view of the results before ICP matching. It can be seen that the two scans do not overlap at all.

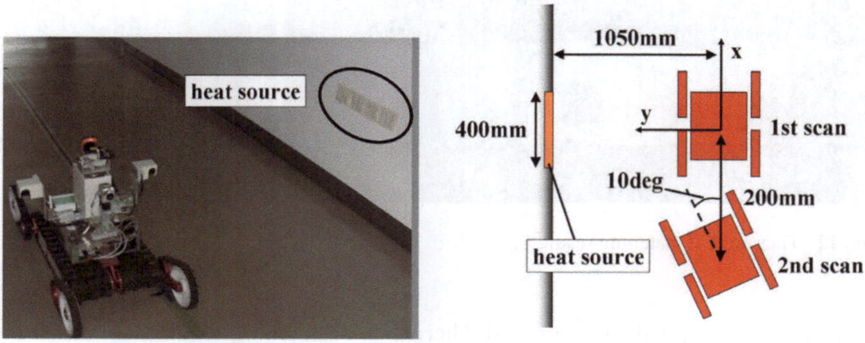

Fig. 9 Experimental scene (*left*) and parameters of motion (*right*)

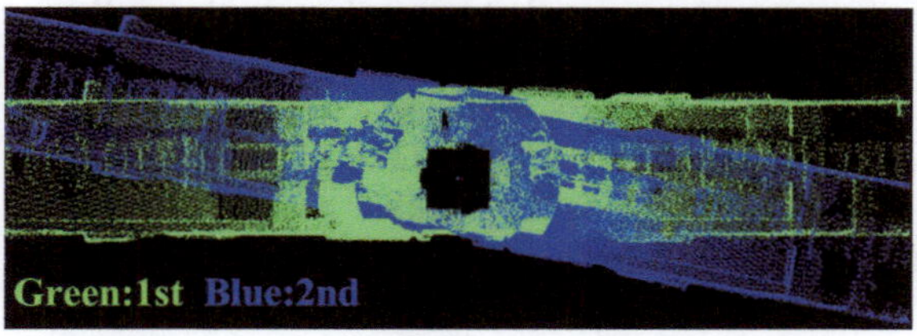

Fig. 10 Before scan matching

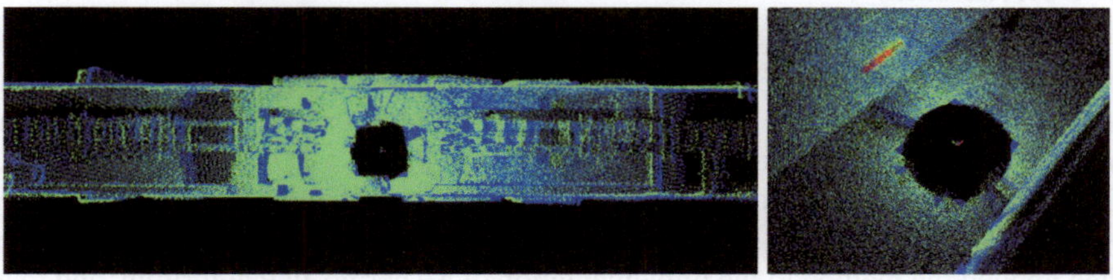

Fig. 11 ICP matching result

First, we applied conventional ICP matching to the data based on Eq. (18), and obtained $\boldsymbol{R}$ and $\boldsymbol{t}$. The result in the top view is shown in Fig. 11 (left), and its zoomed birds-eye-view is shown in Fig. 11 (right). The method seems to have succeeded in matching two scanned data, in the left figure. However in the right figure, you can see that two colors on the wall are not coincident, and the black circles that represent occluded unscanned areas are completely matched even if the scanned position of the robot was different. The calculated translation vector was $(x, y, z, yaw) = (-13.5, 0.0, -8.2\,\text{mm}, 10.7°)$, and the value of the x-axis was far from the correct value, $-200\,\text{mm}$. This is understandable because there were very few features along the x-axis in the corridor.

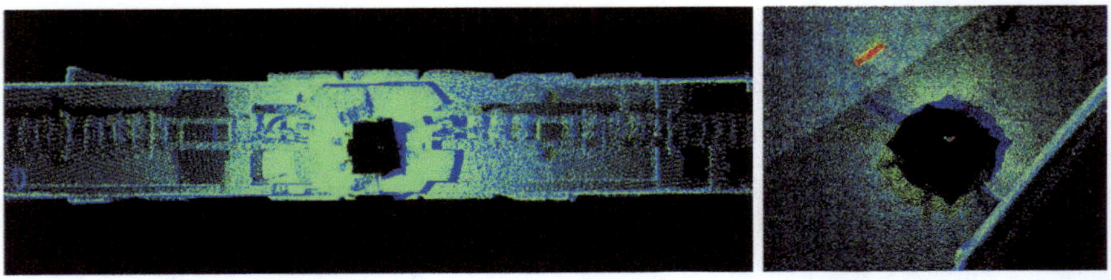

Fig. 12 Thermo ICP matching result

Second, we applied the proposed Thermo-ICP matching to the data based on Eq. (19). The result in the top view is shown in Fig. 12 (left), and its zoomed birds-eye-view is shown in Fig. 12 (right). The method succeeded in matching two scanned data. In the right figure, you can see that two colors on the wall are coincident, and the black circles that represent occluded unscanned areas are out of alignment in the amount of robot's translation. The calculated translation vector was $(x, y, z, yaw) = (-181.6, -7.3, -12.0\,\text{mm}, 11.3°)$. The value of the x-axis was much closer to the correct value: -200 mm. This is a very simple but clear example that proves the effectiveness of the proposed thermo ICP.

5 Conclusions

In this paper, we introduced a 3D range scanner to obtain a 3D range map and an infrared camera to obtain temperature images; in addition, we proposed a fusion method to integrate the data from both deveices to obtain a 3D thermography map. Our system was applied to a simulated disaster environment, and it captured the victims' data successfully. The experiment proved the validity of this method. We then applied the method to the implementation of a Thermo-ICP algorithm, and to a simple experiment. It was a very ad hoc experiment, and we did not apply it in a real environment. However, the result showed the effectiveness of the 3D thermography data for improving matching results. In the future, plan to perform more realistic experiments to confirm the validity of the method.

References

1. Binoy Shah, Howie Choset, Survey on urban search and rescue robots. J. Robot. Soc. Jap. **22**, 582–586 (2004)
2. Special issue of advanced robotics, Advanced research and development of robotics for search and rescue. J. Adv. Robot. **19**(3), 219–347 (2005)
3. F. Matsuno, S. Tadokoro, Rescue robots and systems in japan. in *Proceedings of 2004 IEEE International Conference on Robotics and Biomimetics* (2004), pp. 12–20

4. K. Nagatani, Y. Okada, N. Tokunaga, Kazuya Yoshida, S. Kiribayashi, K. Ohno, E. Takeuchi, S. Tadokoro, H. Akiyama, I. Noda, T. Yoshida, E. Koyanagi, Multirobot exploration for search and rescue missions: a report on map building in robocuprescue 2009. J. Field Robot. **28**(3), 373–387 (2011)
5. H. Kitano, S. Tadokoro, I. Noda, H. Matsubara, T. Takahashi, A. Shinjou, S. Shimada, Robocup rescue: search and rescue in large-scale disasters as a domain for autonomous agents research. in *IEEE International Conference on Systems Man and Cybernetics 1999*, vol. 6. IEEE (2002), pp. 739–743
6. S. Fleck, F. Busch, P. Biber, W. Strasser, H. Andreasson, Omnidirectional 3d modeling on a mobile robot using graph cuts. in *IEEE International Conference on Robotics and Automation* (2005), pp. 1748–1754
7. Y. Bok, Y. Hwang, S. Kweon, Accurate motion estimation and high-precision 3d reconstruction by sensor fusion. in *IEEE International Conference on Robotics and Automation* (2007), pp. 4721–4726
8. K. Nagatani, T. Matsuzawa, K. Yoshida, Scan-point planning and 3-d map building for a 3-d laser range scanner in an outdoor environment. in *Field and Service Robotics* (2010), pp. 207–217
9. I. Grubisic, L. Gjenero, T. Lipic, I. Sovic, T. Skala, Active 3d scanning based 3d thermography system and medical applications. in *MIPRO 2011 Proceedings of the 34th International Convention* (2011), pp. 269–273
10. S. Laguela, J. Martinez, J. Armesto, P. Arias, Energy efficiency studies through 3d laser scanning and thermographic technologies. J. Energy Buildings **43**, 1216–1221 (2011)
11. K. Ohno, T. Kawahara, S. Tadokoro, Development of 3d laser scanner for measuring uniform and dense 3d shapes of static objects in dynamic environment. in *Proceedings of 2008 IEEE International Conference on Robotics and Biomimetics* (2008)
12. Z. Zhang, A flexible new technique for camera calibration. IEEE Transa. Pattern Analy. Mach. Intell. **22**(11), 1330–1334 (2000)
13. IHI Corp, Calibration method and instrument for infrared cameras. Japan patent. 2010–48724 (2010)
14. K. Nagatani, N. Tokunaga, Y. Okada, K. Yoshida, Continuous acquisition of three-dimensional environment information for tracked vehicles on uneven terrain. in *Proceedings of the 2008 IEEE International Workshop on Safety Security and Rescue Robotics* (2008), pp. 25–30
15. Paul J. Besl, Neil D. McKay, A method for registration of 3-d shapes. IEEE Trans. Pattern Analy. Mach. Intell. **14**(2), 239–256 (1992)

Creating Multi-Viewpoint Panoramas of Streets with Sparsely Located Buildings

Takayuki Okatani, Jun Yanagisawa, Daiki Tetsuka, Ken Sakurada and Koichiro Deguchi

Abstract This paper presents a method for creating multi-viewpoint panoramas that is particularly targeted at streets with sparsely located buildings. As is known in the literature, it is impossible to create panoramas of such scenes having a wide range of depths in a distortion-free manner. To overcome this difficulty, our method renders sharp images only for the facades of buildings and the ground surface (e.g., vacant lands and sidewalks) along the target streets; it renders blurry images for other objects in the scene to make their geometric distortion less noticeable while maintaining their presence. To perform these, our method first estimates the three-dimensional structures of the target scenes using the results obtained by SfM (structure from motion), identifies to which category (i.e., the facade surface, the ground surface, or other objects) each scene point belongs based on MRF (Markov Random Field) optimization, and creates panoramic images of the scene by mosaicing the images of the three categories. The blurry images of objects are generated by a similar technique to digital refocus of the light field photography. We present several panoramic images created by our method for streets in the tsunami-devastated areas in the north-eastern Japan coastline because of the Great East Japan Earthquake of March 11, 2011.

1 Introduction

This paper presents a method for creating multi-viewpoint panoramas of long scenes such as city streets. Focusing on urban or residential streets along which buildings are only sparsely located, the method creates their panoramic images that depicts not only building facades but vacant lands and sidewalks along the street within a single image.

This study is one of our series of studies associated with the tsunami devastation due to the Japan earthquake of March 11, 2011. A large number of architectures are

---

T. Okatani (✉) · J. Yanagisawa · D. Tetsuka · K. Sakurada · K. Deguchi
Tohoku University, Sendai 980-8579, Japan
e-mail: okatani@fractal.is.tohoku.ac.jp

K. Yoshida and S. Tadokoro (eds.), *Field and Service Robotics*,
Springer Tracts in Advanced Robotics 92, DOI: 10.1007/978-3-642-40686-7_5,

Fig. 1 Examples of omni-directional images from our image archives of the tsunami-devastated areas in the north-eastern Japan coastline. Displayed as (*single-viewpoint*) cylindrical panoramas

damaged in many urban and residential areas in the coastlines of Iwate, Miyagi, and Fukushima prefectures located in the north-eastern part of Japan. The main purpose of our studies is to develop a method that can visualize these damages as well as the recoveries/reconstructions of these areas that take place thereafter.

We formed a joint team with researchers from Tokyo University and went into these areas about one month after the earthquake; Since then, we have been recording the damages and recoveries mainly using a vehicle-mounted omni-directional camera. In the first one month, our activity mostly covered the entire devastated areas across the three prefectures, and the resulting image archives amount to about 10 terabytes. Examples are shown in Fig. 1. Similar activities are conducted by other parties. For example, Google inc. started collecting similar image archives about three months after us and have made them publicly available on their Street View site.[1]

While omni-directional imagery can create a sensation of immersion into the scenes, they give merely single-viewpoint images. As in Street View of Google inc.,

[1] As our archives contain the images of the devastated areas comparatively right after the disaster, we have not make them publicly available due to privacy consideration.

the user has to change viewpoints to explore a wider space, which is more or less time-consuming. As in the aforementioned tsunami-devastated areas, the damages are often widespread throughout whole cities, an image-presentation method by using which the user can grasp the situation of damages over a large area *at a glance* will be helpful for educational purposes as well as for the purpose of civil engineering aiming at the recovery/reconstruction of the damaged areas. Note that satellite or aerial imagery provides a useful measure in this regard but does not fully replace the ground-level imagery.

In this paper, based on these background, we present a method for rendering streets in a wide panoramic format such as those shown in Fig. 12. There is a long history of the studies in the creation of such multi-viewpoint panoramas [1–8]. While our study is based on these works, it differs from them in the nature of the target scenes. Our target is typically streets along which vacant lands are more dominant than buildings. In a large number of streets in the tsunami-devastated areas, many buildings are swept away by the tsunami or removed in the recovery processes, and thus only a few buildings survived and are located sparsely along the street, as shown in Fig. 1. This nature of the scenes makes it particularly difficult to create their panoramic images, as will be discussed in the Sect. 2

The paper is organized as follows. In the next section, we first discuss the fundamental limitation of multi-viewpoint panoramas that is the most critical for the scenes of our interest, and then present our approach to overcome this limitation. Section 3 presents our model of the three-dimensional structures of scenes and a method to estimate them from the three-dimensional reconstruction obtained by SfM (structure from motion). Section 4 shows a method of creating panoramic images based on the estimated scene structures. Section 5 gives some results and Sect. 6 concludes this study.

2 Previous Works and Our Approach

In general, there is a fundamental limitation in panoramic images of long scenes, which is that objects at different depths cannot be imaged without any distortion in a single image; closer objects have to become narrower and distant objects become wider. This distortion in aspect ratio is less noticeable for scenes of nearly constant depth, whereas it is significant for scenes having a large range of depth. As it is impossible in principle to thoroughly solve this problem, previous studies have been

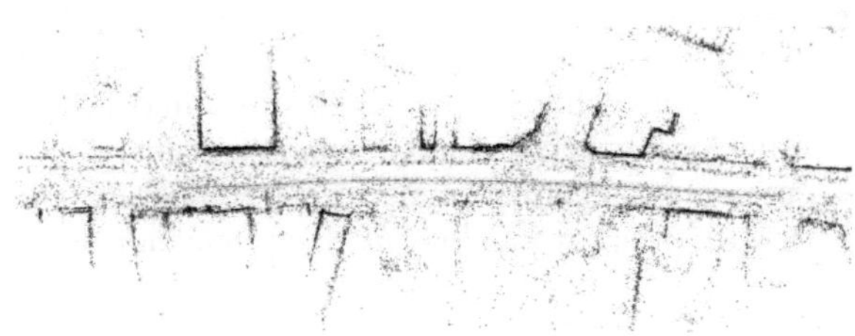

Fig. 2 Example of point cloud of a scene and a trajectory of the camera (about 70 poses) which are obtained by SfM (structure from motion) using hundreds of images

focusing on how to synthesize panoramic images whose distortion is as least noticeable as possible. For example, a semi-automatic method based on cross slits [2] is shown in [5]; it enables the user to design panoramic images based on his/her intention by manually specifying multiple slits. In [6], a method is presented that explicitly estimates the dense depth map of the target scene and creates a panoramic image based on it that minimizes the aspect distortion. These studies, including others such as Street Slide of Microsoft [7], assume more or less that only a few *local* parts of the target scenes have a wide range of depths. For example, scenes of city streets usually have densely located buildings along them and have large depth variation only at intersections.

Such a nature does not hold true for the target scenes of this study. They have sparsely located buildings alongside the streets and these buildings are mixed with far away objects (other buildings, mountains, sky etc.); more importantly, significant portions of the scenes are often filled with such far away objects. This can be thought of as the most difficult type of scenes to create panoramic images with small amount of distortion.

Our approach to overcome difficulty is as follows:

- Only the facades of the buildings alongside the target street and the ground surface (e.g. vacant lands, sidewalks etc.) are chosen as the targets of precise rendering. This choice reflects the fact that these two are important from an application point of view. For this purpose, our method estimates where the building facades and the ground surface are in the three-dimensional space by using the point cloud and the camera trajectory obtained by SfM (structure from motion).
- Objects other than these two are rendered in a blurry manner simulating defocus blurs. As their details are lost in the image, it makes their distortions hardly noticeable while somehow retaining their presence in the scene. The image blur is generated in a similar manner to *digital refocus* of the light field photography [9, 10]. Although the relation between the creation of multi-viewpoint panorama and the light field photography is pointed out by several researchers (e.g., [3, 5]), it has not been intentionally used to design the method for creating panoramas.
- A pushbroom imaging (see [11] for details) is employed for projection model; the projection center is a smooth curve approximating the trajectory of the moving camera. In addition to that the building facades can be rendered free of distortion, the ground surface is rendered in such a way that its image will not provoke sensation of incorrect aspect ratio, owing to the fact that the ground surface extends along the viewing direction and thus the aspect ratio of its texture is not defined.
- A panoramic image is created by mosaicing the textures of the building facades, that of the ground surface, and the blurry images of other objects. To perform this mosaicing, each scene point is classified into these three categories. This is done by using local image similarity across the input images of different viewpoints, while enforcing categorical similarity of neighboring points. The problem is formulated as MRF (Markov Random Field) optimization, which is efficiently solved by graph cuts.

Our aim is to make compatible geometrically accurate rendering of important objects in the scenes and the creation of immersive sensation in a single panoramic image.

3 Model of Scenes

We model the three-dimensional structure of the target scene in such a way that it consists of (i) a surface approximating the facades of buildings alongside the street, (ii) a surface approximating the ground surface, and (iii) other objects. In the simplest case, the two surfaces might be modeled as planes. However, building facades do not generally lie on a single plane. Although we exclude here the streets having steep turns, we still wish to be able to deal with streets that are moderately curved. Therefore, we generate curved surfaces by sweeping a line segment in space to model the surfaces (i) and (ii).

3.1 Camera Trajectory

These surfaces representing the building facades and the ground surface are generated based on the camera trajectory estimated by SfM (Fig. 2). The original images are captured by a vehicle-mounted omni-directional camera; a few examples are shown in Fig. 1. The camera acquires an image at about every 2 m, while the vehicle moves along streets. Although our system is equipped with several GPS sensors and an odometer, they are not used in the three-dimensional reconstruction by SfM. Thus, the images from Google Street View may be used for our purpose here.

The camera trajectory is given as a discrete set of camera poses at which the images are captured, and it could be not smooth due to occasional pose changes of the vehicle. We fit a parametric curve to them to obtain a continuous and smooth approximation of the trajectory. The fitted curve will be used as the projection center of the pushbroom imaging explained later.

Consider the world coordinate system fixed to the scene and the camera coordinate system defined locally on the camera which moves with the camera. Let $\mathbf{X}$ and $\mathbf{X}'$ be the world and camera coordinates of the same scene point, respectively. They are transformed by a Euclidean transform $\mathbf{X}' = \mathtt{T}\mathbf{X}$ with a rotation matrix $\mathtt{R}$ and a translational vector $\mathbf{t}$. Using the indexes $k = 1, 2, \ldots$ to indicate each camera pose and the associated image, we denote the camera coordinates of a point by $\mathbf{X}'_k$ and the pose parameters by $\mathtt{T}_k$ or $\mathtt{R}_k$ and $\mathbf{t}_k$.

When we use K images, the camera trajectory is given as $\{\mathtt{T}_1, \ldots, \mathtt{T}_K\}$. The k-th pose $\mathtt{T}_k$ can be expressed by a 6-vector $\mathbf{c}_k = [\mathbf{t}_k^\top, \omega_k^\top]$, where ω_k is the angle-axis representation of the rotation $\mathtt{R}_k$, i.e., the 3-vector obtained by Rodrigues' formula. We use the following polynomial function to approximate the camera trajectory:

$$\mathbf{c}(s) = \sum_{l=0}^{L} \mathbf{a}_l s^l, \tag{1}$$

where $s \in [0 : 1]$. We set $L = 4$ in our experiments. The coefficients $\mathbf{a}_l$'s are determined by the least squares fitting:

$$\min_{\mathbf{a}_0,\ldots,\mathbf{a}_L} \sum_{k=1}^{K} \|\mathbf{c}(s_k) - \mathbf{c}_k\|^2. \tag{2}$$

We perform the above least squares fitting by setting $s_k = (k-1)/(m-1)$, followed by reparametrizing $s \leftarrow \tilde{s}(s)$ so that s is proportional to the arc length of the curve in space.

3.2 Building Facades and Ground Surface

The surface approximating the building facades, which we will call the facade surface from now on, is generated as follows. We first specify the direction $\mathbf{v}'$ in the camera coordinate system for which we wish to create a panoramic image; for example, if we want to create a panorama of the scene on the right side of the moving vehicle, $\mathbf{v}'$ orients at 90° right relative to the direction of the camera/vehicle motion. It is also necessary to specify the zenith direction $\mathbf{z}'$ in the camera coordinate system. Then we define $\mathbf{n}'$ to be the vector perpendicular to $\mathbf{v}'$ and lying on the plane spanned by $\mathbf{v}'$ and $\mathbf{z}'$, i.e., $\mathbf{n}' \propto (\mathtt{I} - \mathbf{v}'\mathbf{v}'^{\top})\mathbf{z}'$. Note that these vector are all locally defined in the camera coordinate system.

Then, the line segment generating the facade surface is defined to be the one located at a distance d along $\mathbf{v}'$ and is parallel to $\mathbf{n}'$, as shown in Fig. 3a. The parameter d represents the distance from the camera to the building facades, which is also estimated for each s from the point cloud obtained by SfM. To do this, we extract the points belonging to the building facades by the following simple thresholding. Defining for each camera pose k a local coordinate system xyz whose z axis coincides with $\mathbf{v}'$ and y with $\mathbf{n}'$, we denote the coordinates of each point of the SfM point cloud by $\{x_j, y_j, z_j\}$. We then extract points that satisfy

$$\begin{cases} |x_j| < \alpha \\ \beta_1 < y_j < \beta_2 \\ |z_j - d_0| < \gamma d_0 \end{cases} \tag{3}$$

where α, β_1, β_2 and γ are thresholding constants and d_0 is a (rough) estimate of d_k that we manually specify, which is constant for all k's. We use the average of z_j's for the extracted points as d_k, the depth estimate. The thresholding values α, β_1, β_2 and γ are determined as follows: α is determined from the neighboring camera pose

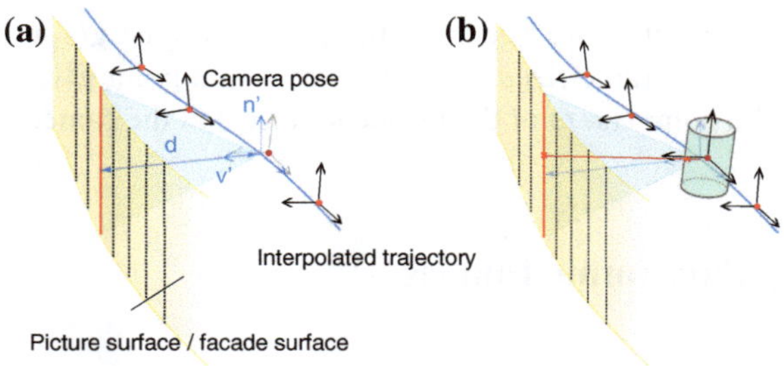

Fig. 3 Geometry of the trajectory of the camera and the surface of building facades. **a** The interpolated camera trajectory and the picture surface of panoramic images. **b** The imaging geometry of the picture surface and an (*omni-directional*) image captured by the camera

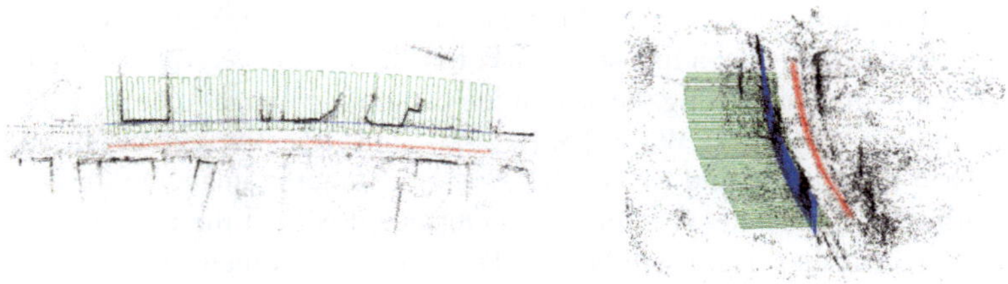

Fig. 4 Estimated surfaces for the three-dimensional reconstruction of Fig. 2. The *blue line* indicates the estimated facade surface and the *green line* the estimated ground surface

distance; β_1 and β_2 are determined from the estimate of the camera height from the ground surface and the vertical field of view of the final panoramic image; γ is set to 0.2. As in the case of the camera trajectory $\mathbf{c}(s)$, we fit a polynomial function to $d_1, \ldots, d_K$ to obtain a continuous representation $d(s)$.

The two functions $\mathbf{c}(s)$ and $d(s)$ specify a single spatial line for each s. As shown in Fig. 3a, we limit the range of the line to define a line segment in such a way that the mid point of the segment will lie on the ray parallel to $\mathbf{v}'$ emanating from $\mathbf{c}(s)$ and moreover its length will be $2d(s)\tan(\phi/2)$, where ϕ is the vertical field of view of the panoramic image. The line segment thus determined is then parametrized by $t \in [0:1]$; $t = 0$ indicates the highest point of the segment and $t = 1$ the lowest point; t is also proportional to the line length. Sweeping this line segment along the fitted camera trajectory $\mathbf{c}(s)$ yields the facade surface, as shown in Fig. 3a, which is parametrized by (s, t) and we denote the surface by $\mathbf{f}(s, t)$. Figure 4 shows the result for the three-dimensional reconstruction of Fig. 2.

The ground surface is determined in a similar way. Suppose a line segment that is perpendicular to $\mathbf{n}'$ and is located by a distance $h(s)$ below the camera. As in the case of $d(s)$, the function $h(s)$ is obtained by first extracting points belonging to the ground surface based on simple thresholding as in (3), then estimating for each k the distance h_k from the camera to the extracted point set, and finally fitting a polynomial function to $h_1, \ldots, h_K$. Sweeping the line segment thus obtained along the camera trajectory $\mathbf{c}(s)$, we generate the ground surface $\mathbf{g}(s,t)$. In the case of ground surfaces, t is parametrized in such a way that a point $\mathbf{g}(s,t)$ will lie on the ray connecting the point $\mathbf{f}(s,t)$ of the facade surface and the camera trajectory $\mathbf{c}(s)$.

4 Rendering Panoramic Images

4.1 Imaging Geometry

We create panoramic images by a pushbroom imaging whose projection center is given by the fitted camera trajectory $\mathbf{c}(s)$. Then, the facade surface $\mathbf{f}(s,t)$ itself becomes the picture surface onto which each ray emanating from an object is projected. Each generator (i.e. a line segment) of $\mathbf{f}(s,t)$ matches a single column in the final panoramic image. Setting the size of the panoramic image to be $W \times H$ pixels, an image point $(u,v) \in [0 : W-1] \times [0 : H-1]$ is transformed to a facade point (s,t) by $s = u/(W-1)$ and $t = v/(H-1)$. Note that the image width W may be chosen arbitrarily, whereas H should be determined so that the aspect ratio of the texture on the facade surface will be 1:1. This imaging geometry determines where on the picture surface each scene point outside the facade surface (e.g. points on the ground surface) is projected.

4.2 Rendering Component Images

Next we describe how the color of each scene point is determined. The real cameras (precisely, the real camera poses at which the images were captured) sparsely distribute near the fitted camera trajectory $\mathbf{c}(s)$. As shown in Fig. 3b, when specifying a scene point, its relative direction from each camera position is immediately determined, from which the color of the scene point viewed from the camera is determined.

We do not know where the scene point corresponding to each image point is located in space, since SfM gives only sparse point cloud and we do not have the dense three-dimensional structure of the scene. Instead, assuming that all the scene point (covered by the image) belong to the facade surface, we can generate a panoramic image for each camera k, as shown in Fig. 5. Similarly, we can generate the images of the ground surface, as shown in Fig. 6. We denote the kth camera image of the facade and ground surfaces by $I_f^{(k)}(u,v)$ and $I_g^{(k)}(u,v)$, respectively.

Consider the projection of the kth camera position onto the picture (facade) surface in such a way that the nearest point on the picture surface to the kth camera position is chosen as its projection. Then, we denote the index of the camera whose projection is the nearest to an image point (u,v) by $k = \tilde{k}(u,v)$. We then define

Fig. 5 Examples of the panoramic image of a facade surface created from each viewpoint image. Selected three out of 70 images

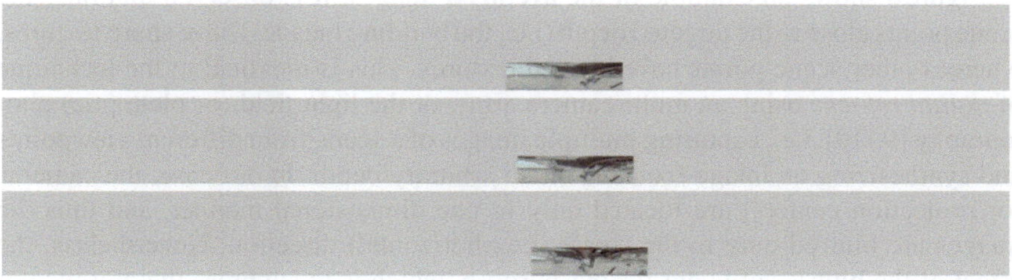

Fig. 6 Examples of the panoramic image of a ground surface for the same viewpoints as Fig. 5

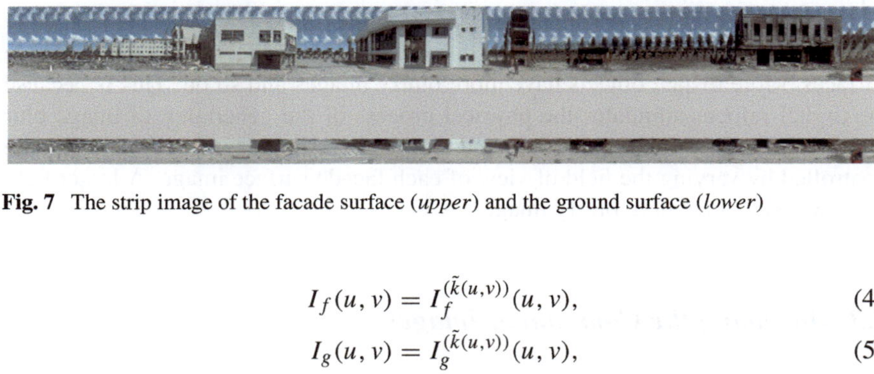

Fig. 7 The strip image of the facade surface (*upper*) and the ground surface (*lower*)

$$I_f(u, v) = I_f^{(\tilde{k}(u,v))}(u, v), \tag{4}$$

$$I_g(u, v) = I_g^{(\tilde{k}(u,v))}(u, v), \tag{5}$$

which are the images generated by copying to the color of each image point from that of the image of the nearest camera k at the same point. As shown in Fig. 7, these images are such that vertical strips cut from the facade/ground surface images of different cameras are orderly placed side by side. We therefore call them strip images. It is observed from Fig. 7 that for the facade-surface strip image, only the textures on the building facades are smoothly stitched; for the ground-surface strip image, only those on the ground surface are smoothly stitched.

Our method renders the objects that are neither close to the facade surface or to the ground surface in a blurry manner. Their blurred images are generated simply by taking the average of the facade surface images over all viewpoints $k = 1, \ldots, K$ as

Fig. 8 The average of all the facade surface images

$$I_{\bar{f}}(u, v) = \frac{1}{K} \sum_{k=1}^{K} I_f^{(k)}(u, v). \tag{6}$$

The trimmed areas of each facade-surface image are excluded in this average.

Figure 8 shows an example of the averaged images. It is observed that only the scene points close to the targeted depth (i.e., the building facades) have sharp textures, whereas other scene points have blurred textures. This is identical to the technique of *digital refocus* using an multi-camera array, or the light field (or plenoptic) photography [9, 10], i.e., capturing multiple images of a scene from different viewpoints and synthesizing an image focusing on an arbitrary depth. In our case, the cameras (or projection centers) are located only in one dimensional manner, and thus the images are blurred only in the single (i.e., horizontal) direction. Nevertheless, the effect of the blur is sufficient for our purpose, which is to suppress the presence of objects outside the target distance range in the panoramic image so that their incorrect aspect ratios will be less noticeable.

It is noteworthy that the image blur of an object thus generated automatically changes its strength depending on the distance from the object to the target facade surfaces. More distant objects have more blurry images and so on. This is because the digital refocus simulates the physical process of the generation of image blur in optical lenses. It should also be noted that the overall strength of the blur can be controlled by varying the field of view of each facade surface image. A larger field of view results in a more blurry image.

4.3 Mosaicing the Component Images

The final image is synthesized by mosaicing the three component images introduced above, the facade-and ground-surface strip images, $I_f(u, v)$ and $I_g(u, v)$, and the averaged facade surface image, $I_{\bar{f}}(u, v)$. For each pixel i whose coordinates are (u_i, v_i), we introduce a variable l_i such that $l_i = 0$ indicates the scene point corresponding to the pixel i belongs to the facade surface, $l_i = 1$ to the ground surface, and $l_i = 2$ to other objects. Then, the final image is mosaiced as follows:

$$I(u_i, v_i) = \begin{cases} I_f(u_i, v_i) & \text{if } l_i = 0, \\ I_g(u_i, v_i) & \text{if } l_i = 1, \\ I_{\bar{f}}(u_i, v_i) & \text{if } l_i = 2. \end{cases} \tag{7}$$

The label l_i of each pixel i is estimated also from the images themselves. We use MRF-based optimization framework, which is widely used in dense stereo matching, to perform this estimation.

As is shown in the facade surface images of different viewpoints (Fig. 5), if a scene point belongs to the facade surface, its imaged position will not move across different camera images, and this holds true for the ground surface images (Fig. 6). Thus, to which of the three scene categories each pixel belongs to can be estimated by checking whether the corresponding scene point moves or not. This can be checked by measuring image similarity between different viewpoint images. Let $\tilde{k}_i = \tilde{k}(u_i, v_i)$ be the index of the camera whose projected position on the image plane is the closest to (u_i, v_i). (This is the same as the one used for generating the strip images.) Employing SSD as a similarity measure, we compute SSD at (u_i, v_i) between the reference image of $\tilde{k}_i$ and a few images around the reference, i.e., $\tilde{k}_i + k$ $(k = \cdots, -2, 1, 1, 2, \cdots)$.

$$E_f(u_i, v_i) = \sum_{k \neq 0} \sum_{(u,v) \in \mathscr{W}(u_i,v_i)} |I_f^{(\tilde{k}_i)}(u, v) - I_f^{(\tilde{k}_i+k)}(u, v)|^2, \tag{8}$$

$$E_g(u_i, v_i) = \sum_{k \neq 0} \sum_{(u,v) \in \mathscr{W}(u_i,v_i)} |I_g^{(\tilde{k}_i)}(u, v) - I_g^{(\tilde{k}_i+k)}(u, v)|^2, \tag{9}$$

where $\mathscr{W}(u_i, v_i)$ is the 5×5 window centered at (u_i, v_i).

It should hold that E_f becomes smaller if (u_i, v_i) is a facade surface point and E_g becomes smaller if it is a ground surface point; they will not be small otherwise. Therefore we define the data term for each pixel as

$$E_{data}(u_i, v_i, l_i) = \begin{cases} E_f(u_i, v_i) & \text{if} \quad l_i = 0 \\ E_g(u_i, v_i) & \text{if} \quad l_i = 1 \\ E_a & \text{if} \quad l_i = 2 \end{cases}, \tag{10}$$

where E_a is a constant representing the cost when the pixel does not belong to the first two categories.

We finally define the total cost for the MRF graph as follows:

$$E(l_1, \ldots, l_N) = \sum_{i=1}^{N} E_{data}(u_i, v_i, l_i) + \beta \sum_{(i,j) \in \mathscr{E}} \delta(l_i, l_j), \tag{11}$$

where N is the number of pixels; $\mathscr{E}$ is the set of the pairs of neighboring pixels in the image; $\delta(l, l')$ is a energy function such that it gives 1 if $l \neq l'$ and 0 if $l = l'$ (i.e., the Potts model). The minimization of this cost is performed by graph cut with alpha expansion. We used the code from GCoptimization version 3.0 [12–14] for this minimization.

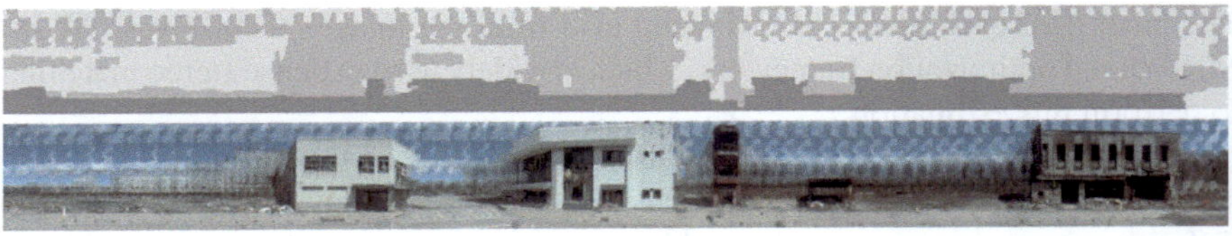

Fig. 9 An example of the optimization performed by graph cuts. *Upper*: Labels for each image pixel. *Lower*: Resulting panoramic image by the image mosaicing based on the labels

Fig. 10 Magnified versions of the strip image (*left*) and our result (*right*) for a certain image region. The strip image could give the viewer an incorrect sensation of the buildings at far away distances. Our image avoids this by blurring those buildings etc

Fig. 11 Magnified versions of the strip image (*left*) and our result (*right*) for a certain image region. Another case

5 Results

Figure 9 shows the result of the optimization described in the previous section for the same data as earlier. As shown in the label image, the building facades, the ground surface, and other objects at far away or close distances are all fairly accurately identified. Only the region of the sky (i.e., the periodic patterns at the image top) is incorrectly identified, which may be due to the presence of the clouds in the sky, whose appearances are obscure and thus difficult to match correctly. However, our purpose is to present the man-made structures on the ground in a geometrically accurate manner, which is achieved in the result. (The sky is not important for our purpose; moreover, if necessary, we may identify the regions of the sky in the input images and eliminate them in advance by color-based image segmentation.)

Fig. 12 **a** A result for a scene (a warehouse in Kamaishi port) with few depth discontinuities . **b** A street in Otsuchi town with sparse buildings. **c** A street in Taro town, Miyako city with a single building. **d** A panoramic image of the city hall of the Rikuzentakata. (It can be confirmed from the broken windows that the tsunami reached as high as its third floor)

Figure 10 shows magnified versions of the strip image shown in Fig. 7 and our result for a certain image region. It is observed that the strip image presents the buildings located at a distance from the our target depth in a corrupted manner. This could give the viewer an erroneous impression of the shapes of these buildings. On the other hand, our result presents these buildings in a blurred manner and thus avoids giving such an incorrect impression.

Figure 11 presents a similar comparison for another result. Besides the incorrect display of the distant building, the strip image also fails to correctly show the road intersecting the street; there appears to be multiple roads, although there is in reality a single road intersecting the street. Our result gives the viewer a correct impression of the geometry of the intersecting road.

Figure 12 presents panoramic images created for several scenes. It can be observed that these results mostly achieves our goal of simultaneously presenting scene structures in a geometrically accurate manner as well as creating immersive sensation in a single image.

6 Summary

We have described a method for creating multi-viewpoint panoramas of streets with sparsely located buildings. To alleviate the difficulty with the creation of panoramas for scenes having a large depth range, our method renders sharp images only for the building facades and the ground surfaces and renders blurry images for other objects. The proposed method satisfies two contradictory demands of the geometrically accurate rendering of important objects existing in the scenes and the creation of immersive sensation in a single image.

References

1. J.Y. Zheng, Digital route panoramas. IEEE Multimedia **10**(3), 57–67 (2003)
2. A. Zomet, D. Feldman, S. Peleg, D. Weinshall, Mosaicing new views: the clossed-slits projection. IEEE PAMI **25**(6), 741–754 (2003)
3. A. Román, G. Garg, M. Levoy, Interactive design of multi-perspective images for visualizing urban landscapes. IEEE Vis. **17**, 537–544 (2004)
4. A. Román, H.P.A. Lensch, Automatic multiperspective images. in *Eurographisc Symposium on Rendering* (2006)
5. A. Agarwala, M. Agrawala, M. Cohen, D. Salesin, R. Szeliski, Photographing long scenes with multi-viewpoint panoramas. ACM Trans. Graph. **25**(3), 853–861 (2006)
6. A. Rav-Acha, G. Engel, S. Peleg, Minimal aspect distortion (MAD) mosaicing of long scenes. Int. J. Comput. Vis. **78**(2–3), 187–206 (2008)
7. J. Kopf, B. Chen, R. Szeliski, M. Cohen, Street slide: Browsing street level imagery. ACM Trans. Graph. (Proc. SIGGRAPH 2010) 29(4), 961–968 (2010)
8. S.M. Seitz, J. Kim, Multiperspective imaging. IEEE Comput. Graph. Appl. **23**(6), 16–19 (2003)
9. T. Adelson, J.Y.A. Wang, Single lens stereo with a plenoptic camera. IEEE PAMI **14**(2), 99–106 (1992)

10. R. Ng, M. Levoy, M. Brédif, G. Duval, M. Horowitz, P. Hanrahan, *Light field photography with a hand-held plenotric camera* (Stanford University Computer Science, Tech. Rep. 2005)
11. R. Guputa, R.I. Hartley, Linear pushbroom cameras. IEEE PAMI **19**(9), 963–975 (1997)
12. Y. Boykov, O. Veksler, R. Zabih, Efficient approximate energy minimization via graph cuts. IEEE PAMI **20**(12), 1222–1239 (2001)
13. V. Kolmogorov, R. Zabih, What energy functions can be minimized via graph cuts? IEEE PAMI **26**(2), 147–159 (2004)
14. Y. Boykov, V. Kolmogorov, An experimental comparison of min-cut/max-flow algorithms for energy minimization in vision. IEEE PAMI **26**(9), 1124–1137 (2004)

Disaster Back-up Support using GIS Contents Composed of Images from Satellite and UAV

Sota Shimizu, Taro Suzuki, Masaya Ogawa, Yoshiyuki Fukazawa, Yuzo Shibayama and Takumi Hashizume

Abstract This manuscript describes a volunteer activity of reconstruction assistance for Great East Japan Earthquake. Concretely, the authors make sand erosion control contents, composed of 3D information from Geospatial Information Authority of Japan, and a combination between wide-area satellite images and a high-resolution mosaic image generated from a movie shot on UAV, i.e., Unmanned Aerial Vehicle, flying in a low-altitude-sky. In addition, we discuss and consider usability of the contents taking into account comments and advices from specialists of geology.

1 Introduction

Great East Japan Earthquake, which occurred on 11th March 2011, has been the most miserable disaster of our country in recent years. Many people over 15,000 died and other people over 3,000 are still missing by the terrible earth-quake and tsunami. Survived people have also been suffered from damages of deep sadness, anxiety of dark prospects, and true crisis of severe winter for their daily lives. For half a year after this tragic disaster, the number of volunteers who help them decreased compared to just after the earthquake in spite of many calls by municipal organizations and volunteer associations. As reported in news, cleaning away rubble is a severe problem.

S. Shimizu (✉) · M. Ogawa
The University of Tokyo, 4-6-1 Komaba, Meguro-ku Tokyo 153-8505, Japan
e-mail: sota@cvl.iis.u-tokyo.jp

T. Suzuki · T. Hashizume
Advanced Research Institute for Science and Engineering,Waseda University, 17 Kikuityou, sinnjyuku-ku Tokyo 162-0044, Japan

Y. Fukazawa
I-NET Corporation, 3-3-1 Minatomirai, Nishi-Ku Yokohama 220-8401, Japan

Y. Shibayama
The University of Tokyo, Bld7, 7-3-1 Hongo, Bunkyo-ku Tokyo 113-8656, Japan

K. Yoshida and S. Tadokoro (eds.), *Field and Service Robotics*, Springer Tracts in Advanced Robotics 92, DOI: 10.1007/978-3-642-40686-7_6,

Many fires were caused by spontaneous ignition of methane gas from a heap of the rubble. We need more years and much more helps continuously in order to revive the disaster areas.

The authors have joined to the reconstruct assistance. The first straw was that there exists a person who worked as a volunteer in the authors' laboratory. Taking into account his experience of volunteer, we knew that it was essential for the survived people to construct temporary housing as soon as possible before severe winter comes. Thus, it is quite important to maintain and control the major arterial roads, e.g., national routes and local railways, in order to carry in heavy machinery and construction materials. Ministry of Land, Infrastructure, Transport and Tourism has planned a mission of "Comb-like road network" for it. We thought that we were also able to assist the mission by surveying and warning dangerous areas where sand erosion, e.g. landslides and rock falls, occurs possibly.

In this paper, the first section described a background of our activity for reconstruct assistance. The second section mentions geographic characteristics of a surveyed area and how and why to choose it. The third section focuses on the following 3 things; (1) Implementing shooting from the sky by a small autonomous flight helicopter, one of UAVs, i.e., Unmanned Aerial Vehicles, (2) Generating a mosaic image from a movie shot in the sky, and making GIS contents of sand erosion database using Google Earth as a platform. The mosaic images and images from a satellite are combined there appropriately and we can observe its geometry as an approximate 3D image by DEM, i.e., Digital Elevation Model, (3) Examining the GIS contents taking into account comments by specialists of geology.

2 Geographic Characteristics of Surveyed Areas

There has been the person, who worked as a volunteer in the early April just after the disaster occurred in the authors' laboratory. He went to the disaster area by train via an active airport in the Tohoku area. He was going to Kesen-numa or Rikuzen-Takada first. But the major railway, Tohoku Shinkansen line, did not work because of comparatively-large earthquake aftershocks. Thus, he had a contact to Iwate Prefectural Council of Social Welfare in Morioka, Iwate. The council asked him to help coastal areas from Aomori prefecture to Iwate prefecture including around Miyako and Kamaishi because the number of volunteers was not enough, although damages in Kesen-numa and Rikuzen-Takada were severe. He reached Miyako along the national route 106 (Route 106 hereafter) by car and worked there. Throughout his volunteer activity, he had an opportunity to look at damages of Miyako, Yamada, O-otsuchi, Kamaishi, and O-ofunato by his own eyes. Significantly, he noticed the following geographic characteristics: (1) Iwate Prefecture has thriving inland cities along the Kitakami River, (2) There exist the major arterial roads and local railways along a river from the inland cities to each community of the coastal cities, where many people live, located in a narrow plain as facing to a cove along a ria shoreline, e.g., the Hei River runs along Route 106 and Yamada line of Japanese Railway (JR)

East Japan company in case of Miyako, (3) Geographically, areas around these major arterial roads and local railways have canyon topography, (4) The coastal cities are connected serially along the national route 45 and Sanriku highways (partly under construction), that is, it is not too much to call this national road "the life line". The major arterial roads from the inland cities have been constructed as being connected like a mesh in order to keep transport on hand with the shortest distance for emergencies such as landslides and rock falls. However, even in case of Route 106, there exist a couple of important spots where it becomes more difficult to carry supplies quickly into the disaster areas when it is decoupled there. These spots in the canyon topographic geography are quite close to the sea. Based on the above information, the authors have decided to implement a geographic survey for maintaining transport between Hanaharashi and Rikuchu-u-Kawai along Route 106 as one of the major arterial roads and JR Yamada line (running almost parallel to Route 106). When we decided the survey area, we avoided the disaster area damaged directly by the tsunami considering pains of the people living there. Figure 1 shows 6 spots for this survey in the Google map [1], chosen exactly as a takeoff and landing points of a small autonomous flight helicopter (mentioned in detail later) for shooting from the sky, by reference to the Sand Erosion GIS contents in Iwate Digital Map [2] in a web page managed by Iwate prefecture. Since this GIS contents in the web page are made by focusing onto communities where people live, we paid different attention mainly to gap areas where people do not live.

3 Generating Mosaic Image from Movie shot in the Sky and Making Sand Erosion GIS Contents

3.1 Implementation of Shooting from the Sky by UAV

The authors implemented shooting from the sky over Route 106 and JR Yamada line along the Hei River in Miyako on the 28th and 29th June, 2011. We chose the small autonomous flight helicopter RMAXG1 (Yamaha motors, Fig. 2) as an UAV for shooting, because it is more suitable for this survey in the canyon topography due to hovering ability, i.e., keeping its posture better than a fixed wing type of UAV we had used before. RMAXG1 can achieve an autonomous round flight to a distant spot (at most 30 km potentially) far away from the takeoff and landing point due to the posture control system and RTK-GPS system. An operator just needs to give it multiple way points in the sky in order to shoot a movie for a long distance. Even when its remote wireless intercommunication is lost, it has a waiting function as hovering over at takeoff point using GPS information. But, in the actual shooting flight, we should manage its safe flight based on rules of JUAV, i.e., Japan UAV association, as keeping its flight distance within a range in which we can confirm it visually and preventing its flight from being beyond houses and Route 106. The flight altitude from the ground was 100 m. We used 3 types of digital cameras,

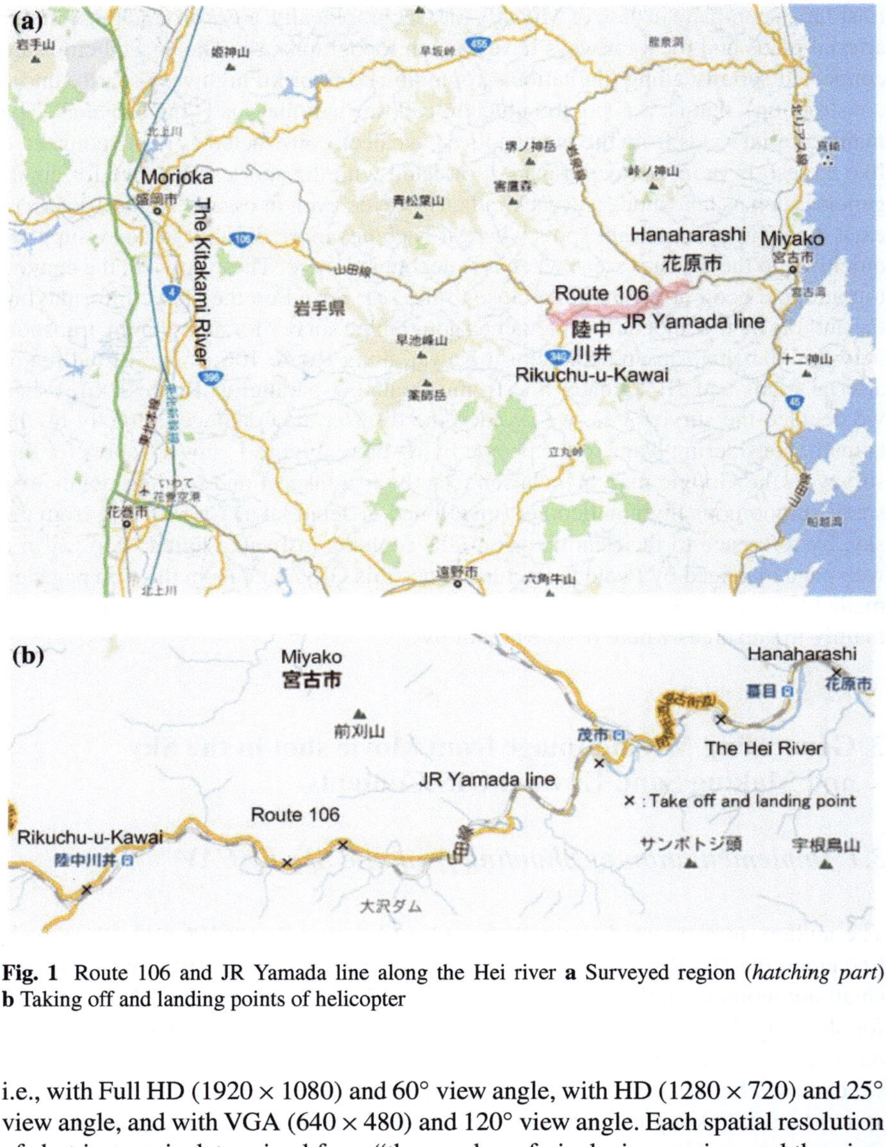

Fig. 1 Route 106 and JR Yamada line along the Hei river **a** Surveyed region (*hatching part*) **b** Taking off and landing points of helicopter

i.e., with Full HD (1920 × 1080) and 60° view angle, with HD (1280 × 720) and 25° view angle, and with VGA (640 × 480) and 120° view angle. Each spatial resolution of shot images is determined from "the number of pixels, image size, and the view angle of lens" of each digital camera.

Fig. 2 Unmanned helicopter RMAX G1

3.2 Generating Mosaic Image and Making Sand Erosion GIS Contents using Google Earth as a Platform

In order to generate a mosaic image (Fig. 3) from still images extracted from a movie shot in the sky with 1[fps] (an extracted still image is shown as in Fig. 4), the authors use not only GPS and IMU information from the sensor unit equipped on the helicopter but also SIFT feature information and digital elevation model (DEM) data provided from Geospatial Information Authority of Japan.

The UAV system as in Fig. 2 consists of a small helicopter type of UAV and a ground control station (GCS).

The UAV comprises a GPS receiver module, an IMU having three gyroscopes and a three-axis accelerometer, and a laser range finder (LRF). In an autonomous flight, the position and attitude data of the UAV calculated by GPS and the IMU

Fig. 3 A merged mosaic image

Fig. 4 A still image extracted from a movie

Fig. 5 Projected images obtained by the UAV using on-board position and attitude data

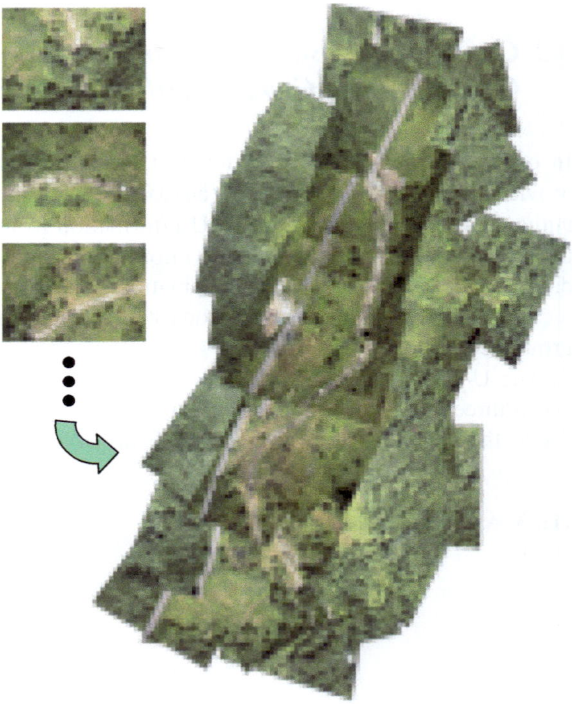

contain errors. This degree of accuracy is short of the accuracy needed to generate well-overlapped mosaic images.

Figure 5 illustrates this problem. The 64 images captured by the digital camera during a flight are transformed and projected onto a 50-meter-mesh digital elevation model (DEM) surface using the position and attitude data in accordance with the method described later. The relative position of these projected images is roughly correct, but the generated image is more distorted than in satellite photography, and multiple images do not accurately overlap one another.

This inaccurate overlap of images is caused by the following reasons:

1. The original position and attitude data of the UAV measured by GPS and the IMU include errors.
2. There is a synchronization error between images data captured by the digital camera and "the position and attitude data".
3. The resolution of DEM is insufficient.

Although these errors may reduce the correlation between the projected images, the authors think that the common underlying cause is the error in the original position and attitude data. The shape of the projected image is susceptible to being affected by small changes in the position and attitude data. For this reason, accurate position and attitude data must in some way be estimated.

This paper presents a technique for the estimation of position and attitude data of the UAV using the integration of the image sensor in order to generate the accurate mosaic image [3, 4]. The procedure of estimating and generating mosaic images is as in the following items.

I. First, the area of the images projected onto the DEM surface is calculated using the original position and attitude data.
II. The overlap ratio is calculated using the areas of the projected images, and each combination of overlapped images is evaluated.
III. Key feature points on the images are then identified and matched to form a combination of the overlapped images.
IV. The position and attitude data are estimated in order to minimize the weighted sum of the squared error cost function, using the image correlation.
V. Finally, the mosaic image is generated using the estimated position and attitude data based on the image projected onto the DEM surface.

Here we describe the method of generating a mosaic image more in detail.

(1) Projection onto a map

First, we project the image onto the DEM surface using the original position and attitude data. It is possible to reduce inaccuracy of distortion using a perspective projection. Figure 6 shows coordinate systems employed in this study.

A 2D image coordinate is denoted by $\mathbf{m} = [u, v]^T$. A 3D camera coordinate is denoted by $\mathbf{M}_C = [X_C, Y_C, Z_C]^T$. The plane that coincides with the DEM surface is denoted by the world coordinate $\mathbf{M}_W = [X_W, Y_W, Z_W]^T$. $\tilde{x}$ denotes an augmented vector obtained by adding 1 as the last element, i.e., $\tilde{\mathbf{m}} = [u, v, l]^T$, $\tilde{\mathbf{M}}_C = [X_C, Y_C, Z_C, 1]^T$ and $\tilde{\mathbf{M}}_W = [X_W, Y_W, Z_W, 1]^T$. The camera is modeled by the usual pinhole; the relationship between a 3D camera coordinate $\tilde{\mathbf{M}}_C$ and its image projection $\tilde{\mathbf{m}}$ (perspective projection) is given by

$$s\tilde{\mathbf{m}} = \mathbf{P}\tilde{\mathbf{M}}_C, \tag{1}$$

where s is an arbitrary scalar factor and $\mathbf{P}$ is a 3×4 projection matrix that contains the camera's intrinsic parameters. These camera intrinsic parameters are assumed to have been calibrated in advance.

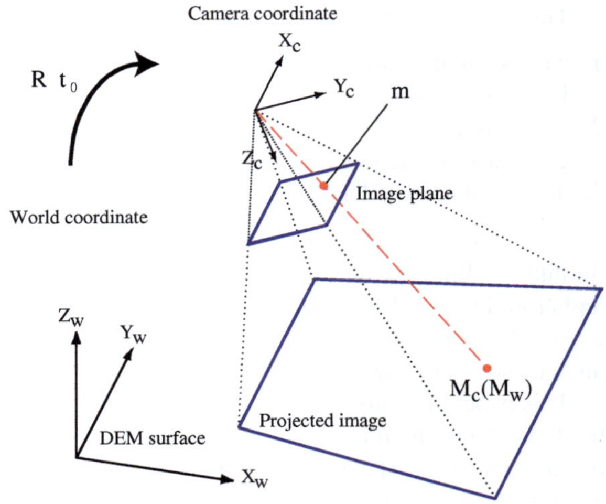

Fig. 6 Coordinates and image projection

The equation expressing the relationship between $\tilde{\mathbf{M}}_C$ and $\tilde{\mathbf{M}}_W$ is defined as

$$\tilde{\mathbf{M}}_C = \mathbf{R}^T(\tilde{\mathbf{M}}_W - \mathbf{t}_0), \tag{2}$$

where $\mathbf{R}$ is a rotation matrix calculated using the UAV attitude estimated by GPS and IMU and $\mathbf{t}_0$ is the UAV position in the world coordinate system. Using Eqs. (1) and (2), the relationship between $\mathbf{m}$ and $\mathbf{M}_W$ is derived, and the image pro-jected onto the DEM surface is determined. This projected image is used to calcu-late overlapped image pairs in the next step.

(2) Image overlap calculation

As shown in Fig. 5, the error in the original position and attitude data was responsible for an overlap error in the mosaic image. Therefore, in order to correct the position and attitude data, the information of feature points that are associated in overlapped images is used. Before this feature extraction and association, the overlapped image pairs need to be determined. Over 100 images are extracted from a movie during a flight, and a flight trajectory of the UAV is a complicated circular form. This renders identification of overlapped areas in all images a difficult task. For this reason, the original position and attitude data are used to determine the overlapped image pairs automatically.

Using Eqs. (1) and (2), the area of each projected image is calculated. Figure 7 illustrates an example of the projected images used to calculate a pair of overlapped images.

In this example, the areas of 4 images are projected onto the DEM surface using the original position and attitude data. In order to determine an overlapped pair, the overlap ratio η is defined. One area of one projected image is denoted as Area1, and another area is denoted as Area2; the overlap ratio η is then defined as follows:

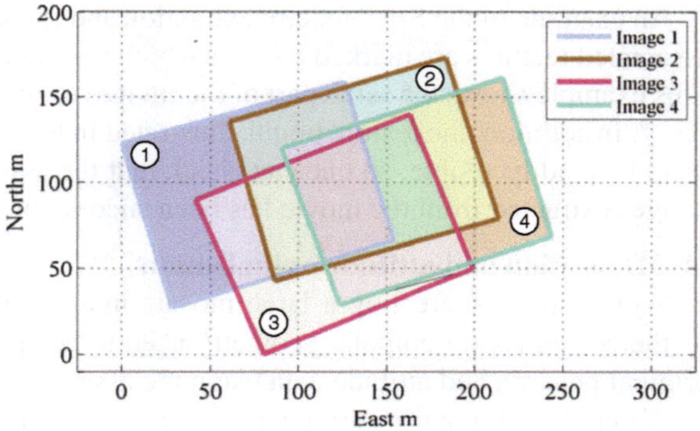

Fig. 7 Overlap detection using the original position and attitude

$$\eta = 2 \times \frac{Area1 \cap Area2}{Area1 + Area2}, \tag{3}$$

The threshold processing of the overlap ratio η is used to define and extract the pair of overlapped images from which we construct whole images. In this study, 0.3 is used as the threshold of overlap ratio η.

Note, however, that even as the original position and attitude data have errors to some degree, the accuracy of the position and attitude data suffice to define a pair of overlapped images.

(3) SIFT feature extraction and association

On this step, in order to correct the position and attitude data using the image correlation, scale invariant feature transform (SIFT) [6] feature points in the overlapped images are extracted and associated. This approach transforms each feature in an image into a 128 dimensional local feature descriptor, i.e., each of which is invariant to image translation, scaling, and rotation, and is partially invariant to illumination changes and projection. In the images extracted from the movie, substantial changes in camera attitude and outdoor illumination occur over the course of a flight circuit. For this reason, the SIFT features make it possible to associate the feature points between a pair of all the overlapped images. To associate the SIFT features, the Euclidean distance D between 128 dimensional local feature descriptor d is used. The feature descriptors in a pair of overlapped images are denoted by d_1 and d_2, and the Euclidean distance D is given as

$$D = \sqrt{\sum_{i=1}^{128} (d_1(i) - d_2(i))^2}. \tag{4}$$

By detecting the minimum distance D using this equation, temporary points of correspondence are defined. These features are matched and associated across multiple frames by using just the image feature descriptor. In order to detect and reject an outlier of the association, the epipolar geometry constraints between the images based on the MAPSAC algorithm [7] are used.

An example of the SIFT feature extraction and association is shown in Fig. 8. The associated features are marked by lines which join their locations in the four images. This example shows the same part of the images used to detect overlapped area as in Fig. 7. In addition, the feature points illustrated in Fig. 8 are reduced till 1/10 to make them be readily visible, so one can check that the association between the multiple images extracted from the movie has been successful.

(4) 3D position and attitude optimization

With respect to creating a large mosaic image, the projection error introduced by the constraints of epipolar geometry need to be optimized in order to correct the original position and attitude data using the associated feature points.

We apply a 3D optimization technique such as the bundle adjustment technique [8] to compute the maximum likely position and attitude data. We denote the sum of the squared error cost function of the projection error. The estimation of the position and attitude data is performed to minimize the cost function. We use the original position and attitude data as initial values for the function. It reduces the likelihood as converging towards a local solution.

The position and attitude data are denoted as $\mathbf{p} = [x, y, z]$ and $\boldsymbol{\Psi} = [\phi, \theta, \varphi]$, respectively. The $[\phi, \theta, \varphi]$ signifies the roll, pitch and yaw angle. The position and attitude in the i-th pair of overlapped images are $\mathbf{p}_{1i}$, $\boldsymbol{\Psi}_{1i}$ and $\mathbf{p}_{2i}$, $\boldsymbol{\Psi}_{2i}$, respectively. The coordinate of the associated feature points in the i-th pair of overlapped images are denoted by $\mathbf{m}_{1ij}$ and $\mathbf{m}_{2ij}$. Here, j stands for the number of associated feature points in each image. The sum of the projection error $\boldsymbol{\Psi}_i(\mathbf{p}, \boldsymbol{\Psi})$ is given by

$$\Phi_i(\mathbf{p}, \Psi) = \sum_j w_{ij} \left| f(\mathbf{m}_{1ij}, \mathbf{p}_{1i}, \Psi_{1i}) - f(\mathbf{m}_{2ij}, \mathbf{p}_{2i}, \Psi_{2i}) \right|, \tag{5}$$

where $f(\mathbf{m}, \mathbf{p}, \Psi)$ is the projection function by which we calculates the coordinate of the feature points projected onto the DEM surface from Eqs. (1) and (2). The w_{ij}

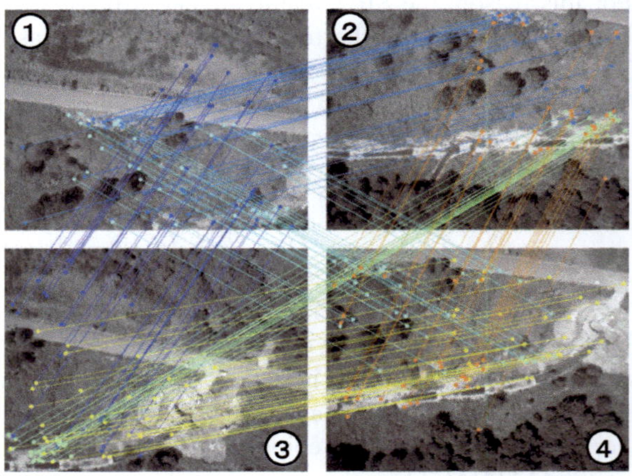

Fig. 8 Example of SIFT feature extraction and association

is a weight of the optimization calculated by an inverse of the Euclid distance of associated feature points D, as calculated from Eq. (4). The cost function E is the sum of the projection error in all pairs of overlapped images. It is given by

$$E = \sum_{j} \Phi_i^2(\mathbf{p}, \Psi), \tag{6}$$

In order to minimize the cost function as in Eq. (6), the UAV position $\mathbf{p}$ and attitude Ψ are estimated. Jacobian matrix $\Delta\mathbf{J}$ is evaluated by computing the Jacobians of Eq. (5). Once $\Delta\mathbf{J}$ is computed for a current iteration of the estimator, the Gauss-Newton iteration step is used to converge toward the optimal state estimate.

A result of the optimization of the position and attitude data is illustrated in Fig. 9. Solid black lines indicate the original UAV position and attitude data, and solid red lines indicate the compensated position and attitude. In this result, 84,351 features in the 64 images have been used, and the algorithm has taken 24 iterations to reach the optimized estimate. The value of the cost function E has decreased from 376, 184 to 363.9. Figure 10a–f shows the error in the on-board position and attitude data as a reference compared to the estimated position and attitude. A green dashed line indicates the mean value and a red line indicates the standard deviation (SD) 1σ.

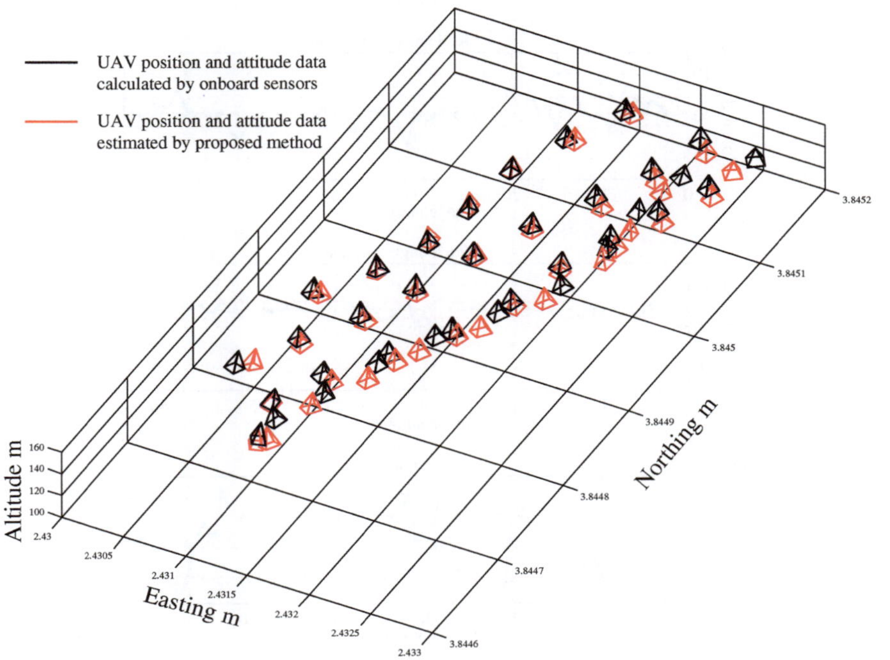

Fig. 9 A result of position and attitude estimation

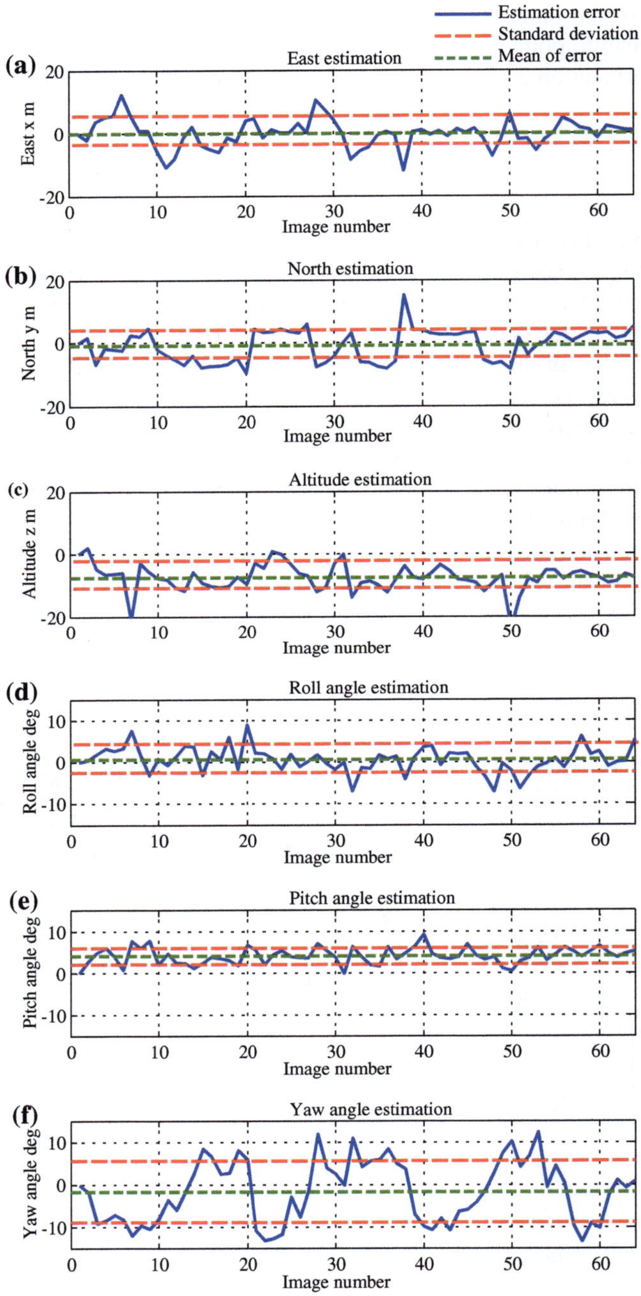

Fig. 10 Error in on-board position and attitude data

The SD of the position $[x, y, z]$ and attitude $[\phi, \theta, \varphi]$ are 4.4, 4.8, 4.2 m, 3.0, 1.9 and 7.2° , respectively. In this experiment, the altitude has the large bias error compared with the other values. Additionally, the SD of the yaw angle indicates a high value. The altitude estimation seems to attribute the error to a characteristic of GPS, i.e., in which the altitude error is generally specified as being 1.5 times larger than the horizontal error. In addition, the roll and pitch angles are observed in the direction of the gravity vector, but the yaw angle is weakly observed, only when the UAV accelerates, using GPS and IMU. It seems that the error of the yaw angle occurs in this way.

(5) Mosaic image generation using estimated state

Using the estimated position and attitude data, the mosaic image is generated based on the method mentioned in Sect. 3.1. Figure 11a shows the mosaic image generated by using the proposed method. Compared to the image shown in Fig. 5, the mosaic image is accurately assembled. In addition, the luminance of this generated mosaic image as in Fig. 11b is adjusted to represent the average luminance of all the images. The problematic correlation of captured images is easily resolved by the proposed method. A large mosaic image is automatically generated with GPS and IMU. In the next section, we evaluate the generated mosaic image in terms of low-cost and flexible vegetation monitoring.

In order to utilize the generated mosaic image and geographic data measured from a laser range finder (LRF) as effective sand erosion GIS information, we make contents which are useful for users intuitively to recognize information they need, e.g., landslides and rock falls, using Google Earth as a platform. The high-resolution mosaic images from the UAV are combined with Google map images shot from satellites in order to improve not only fineness of contents' image but also reliability

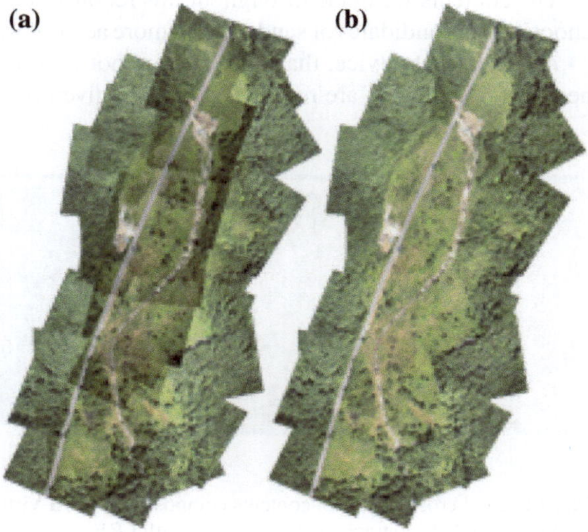

Fig. 11 A result of mosaic image by the proposed method **a** Projected mosaic image **b** Adjusted illumination

because this method can increase frequency to update contents as reducing its cost. When resolution of the image from the helicopter flying at about 100 m height is higher than that of the satellite image, the former is overlapped. On the other hand, when the former is lower than the latter, the latter is survived. In addition, the DEM data are also combined. Figure 12 shows (a) Google Earth displays a combination of the generated mosaic images and the sat-ellite images as a virtual 3D image and (b) LRF geographic data by the vrml form are visualized as changing color graduation from red to green corresponding to the highest height to the lowest one. A part where the color changes from red to green rapidly means to be quite steep (potentially, landslides and rock falls occur more easily). This time, we succeeded in shooting a landslide just after it occurred. Comparing between Figs. 12a, b, we can confirm this land slide occurred in a steep part.

The image from the HD camera at about 100 m height has at least 2 × 2 cm resolution theoretically (Fig. 13). This resolution means that we can not only find empty cans but also discriminate their kinds.

3.3 Examination Taking into Account comments by Geologists

We usually need to go along the following manual in order to make sand erosion GIS contents published as in the web page of Iwate Digital Map:

(1) Choosing candidates from air photos shot by Cessna plane.
(2) Confirming visually by specialists of geology.
(3) Verifying scientifically, e.g., measuring electric resistance, moisture content, and etc.

The contents we made throughout this reconstruction assistance are effective for choosing the candidates of sand erosion more accurately as in (1) of the above manual.

An important advice, that images from not only visible light cameras but also a near infrared camera are required, has been given from a view of geology, because

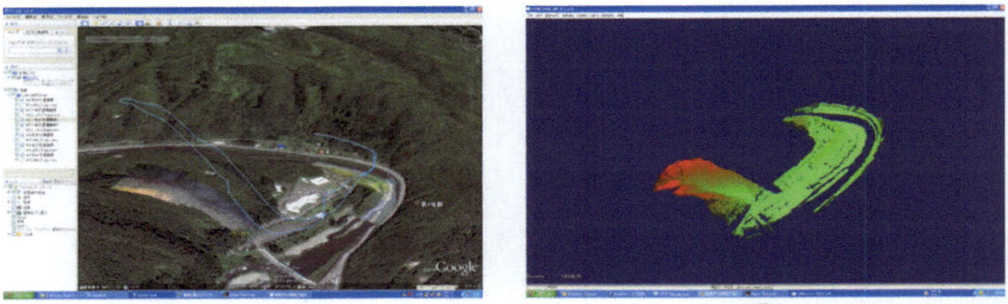

(a) (b)

Fig. 12 Sand erosion control contents composed of 3D. **a** Virtual 3D image of mosaic images on Google earth. **b** Topography data measured with LRF

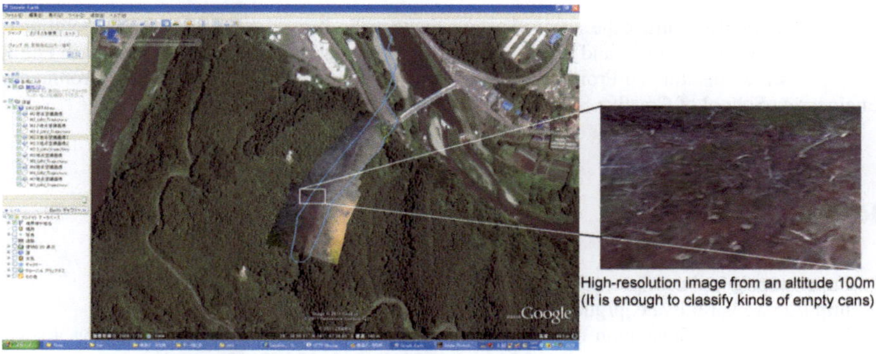

Fig. 13 Quite high resolution due to shooting at 100 m low height

occurrence of landslides, for example, can be presumed from three conditions of geography, type of soil, and moisture content. However, we need to progress an analysis method furthermore as advised, for example, presuming moisture content from vegetation classification using also middle infrared camera, because the ground for measuring moisture content is hidden almost in a rainy season when landslides occur more easily.

4 Conclusion

This paper has reported our reconstruction assistance in Great East Japan Earthquake. The results of this paper are summarized as follows:

1. Geometric characteristics along Route 106 were surveyed for back-up support of reconstruction assistance.
2. Shooting from the sky using UAV was implemented in June.
3. The method of generating a mosaic using DEM data and SIFT has been applied for high-resolution image put on the Google Earth.
4. According to professional comments by the geologists, the sand erosion GIS contents made by our laboratory were examined and discussed as an intuitively-comprehensible form.

Considering practical operation for a long term, one of the most realistic problems seems to be frequency of update images from the UAV and the satellite. It is related deeply to running cost. In addition, we need to progress a method for presuming moisture content much more in order to make a sand erosion GIS contents which is practical for an statistically-irregular weather as a guerrilla heavy rain, i.e., cataract.

Acknowledgments This reconstruction assistance activity has been supported by JKA through its Promotion funds from Keirin RACE. The authors would like sincerely to commemorate many victims, thank and encourage the survived people living in the disaster area for much collaboration in spite of their severe situation, and appreciate objective comments and advices to Mr. Asai in Public Works Research Institute an Prof. Kazuo Kamura in Department of Resources an Environmental Engineering School of Creative Science and Engineering Waseda University.

References

1. http://maps.google.co.jp
2. http://gisweb.pref.iwate.jp/guide/
3. T. Suzuki et al., Generation of large mosaic images for vegetation monitoring using a small unmanned aerial vehicle. J. Robot. Mechatron. **22**(2), 212–220 (2010)
4. T. Suzuki, et. al., SIFT feature based SLAM algorithm using a small UAV for 3 terrain reconstraction. R.S.J. **AC1Q**, 3–5 (2010)
5. S. Shimizu, et al., Bird-Eye View Mapping using wide angle fovea sensor on small UAV for a planned construction site of wind farm. R.S.J. **AC3I**, 1–8 (2010)
6. D.G. Lowe, Distinctive image features from scale-invariant keypoints. Int. J. Comput. Vision **60**, 91–110 (2004)
7. D.G. Lowe, Bayesian model estimation and selection for epipolar geometry and generic manifold fitting. Int. J. Comput. Vision **50**, 35–61 (2002)
8. B. Triggs, P. McLauchlan, R. Hartley, A. Fitzgibbon, Bundle Adjustment—A Modern Synthesis. *Vision Algorithms: Theory and Practice*, LNCS vol 1883 (Springer, Verlag), pp. 298–372

Mine Detecting Robot System

SeungBeum Suh, JunHo Choi, ChangHyun Cho, YeonSub Jin, Seung-Yeup Hyun and Sungchul Kang

Abstract Humanitarian demining, the peaceful and non-explosive de-mining strategies, has been gaining worldwide acceptance lately. As the series of humanitarian demining, tele-operated mine detecting robot system was developed. This paper presents unique demining strategy of the demining robot system. There are two developed system called MIDERS-1 and MIDERS-2. The system is consisted of rough terrain mobile platform, multi degree of freedom manipulator, and the all-in-one mine detecting sensor module between ground penetrating radar and metal detector. We have focused that our cooperative demining procedure between the macroscopic and microscopic demining enhances the conventional human demining. With proposed methodology, the hardware configurations and functions are described.

S. Suh (✉) · J. Choi · Y. Jin · S. -Y. Hyun · S. Kang
KIST, Seongbuk-gu Hawalgok-dong 39-1, Seoul, South Korea
e-mail: keenhurt@gmail.com

J. Choi
e-mail: junhochoi@kist.re.kr

Y. Jin
e-mail: jeyong3@naver.com

S. -Y. Hyun
e-mail: symasi@yahoo.co.kr

S. Kang
e-mail: kasch804@gmail.com

C. Cho
Chosun University, 375 Seosuk-dong, Dong-gu, Kwangju 501-759, Korea
e-mail: chcho@chosun.ac.kr

K. Yoshida and S. Tadokoro (eds.), *Field and Service Robotics*,
Springer Tracts in Advanced Robotics 92, DOI: 10.1007/978-3-642-40686-7_7,

1 Introduction

For many years, the humanitarian demining have been the core motivation for the development of demining robot system [1, 2]. It surely had many challenges and limitations along with the demining robotic researches [3, 4]. Real environments are left unproven for the robotic system to take over human's job in demining [5, 6]. Against the numerous efforts for proving the robotic system can be helpful for demining task, Trevelyan believes that the robotic system is not suitable for this task because it could be too expensive and the sensor technology is still not proficient enough to guarantee the safety [7]. In the sense that the sole robot demining system can be unstable and therefore possibly dangerous, many scientists insisted that the beneficial points would be obtained by the cooperation between human and robotic system [8–10]. In order to accomplish this, the robot system should contain some level of automation or it could be a remote controlled mobile robot system to ensure safety of the operator [11, 12].

There have been many researches regarding tele-operated machines, multi-functional teleoperated robots, and demining survice robots [13–15]. Traditional tele-operated machines include light flail type demining robot system [16, 17]. As tele-operated robotic system enhanced its multi-functional aspects, the system started including other components than demining mechanisms such as mine detecting sensor, laser, vision camera, and etc. [18, 19]. Another advanced type of the demining robot system is demining service robots [20, 21].

In the same motivation under the humanitarian demining service robot development, we suggest brilliant cooperative robotic demining strategy in this project. Suggested strategy of the macroscopic and microscopic demining process would shorten the operating time by utilizing multi-arrayed detecting sensor. It also would enhance the accuracy of mine detection because the mine scanning procedure is repeated numerously in the scanning procedure of MIDERS-1 and MIDERS-2. Section 2 demonstrates overall demining scenario with details on each strategy. Section 3 explains hardware architectures with the functionalities of MIDERS-1 and MIDERS-2.

2 Mine Detection Scenario

In order for a robot system to be used for a dangerous mine detection task, it has to be as efficient as human or even better. Observing current demining process by human, we have concluded that replacing the demining process with a robot system enhances the task performing speed and the safety beneficial for human who operates. In the project, we have developed two mine detection robot systems, macro-scanning-robot and micro-scanning-robot for two phases of a scanning strategy, which are macroscopic scanning process and microscopic scanning process.

Mine detecting strategy with both the macroscopic and the microscopic scanning procedure will be faster and more precise than the conventional human demining strategy. (i.e., overall scenario in Fig. 1). We distinguished the purposes of two different systems as fast and detailed mine detecting robot systems for macro scanning robot and micro scanning robot, respectively. Overall mine detection scheme is described below;

1. Remote operator controls macro-scanning-robot to scan and detect the buried mines on the mine fields with high speed.
2. As mines are detected, macro-scanning-robot marks the detected location with paint (i.e., the paint cannot be sprayed at specific spot).
3. Macro-scanning-robot continues the macroscopic scanning process.
4. Micro-scanning-robot is employed in the vicinity of previously mine detected area by macro-scanning-robot.
5. Micro-scanning-robot detects mine precisely and in great detail with smaller mine detecting sensor.
6. Micro-scanning-robot moves to another locations where macro-scanning-robot suspects that the mine can be buried, and continues the microscopic scanning process.

Cooperation between two different kinds of robot aims to be as efficient as traditional human demining method in addition to guarantee a safe working environment for the operator as well.

2.1 Macroscopic Scanning Strategy

As mentioned above, the macroscopic scanning strategy aims for fast operation. Macro-scanning-robot therefore uses many properties in order to speed up the whole process such as a large size of mine detecting sensor in covering as large as possible area at a single scan or the linear stroke of the manipulator.

Figure 2 demonstrates the sequential macroscopic scanning process. At first step, human develops the safe and mine free region as in Fig. 2a. After it is confirmed that no mines are buried in the scanned area, bush was removed, Fig. 2b–c. Then, macro-scanning-robot operates on the safe zone while it detects mines on the right hand side, because the mine detecting sensor, consisted of three sets of ground penetrating radar (GPR) and metal detector (MD), is installed at right side of macro-scanning-robot as in Fig. 2d. If a buried mine is detected, macro-scanning-robot marks the position with the paint spray for the future detailed mine searching process of micro-scanning-robot, Fig. 2e. Once the mines fields are cleared, the robot continues the macroscopic scanning strategy for the next mine fields, Fig. 2f–g.

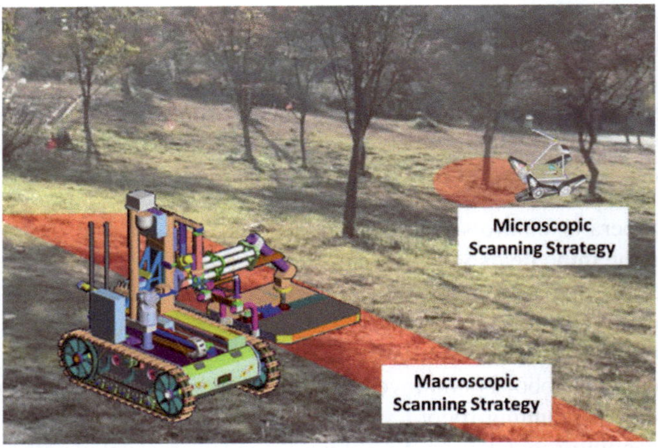

Fig. 1 Overall scenario

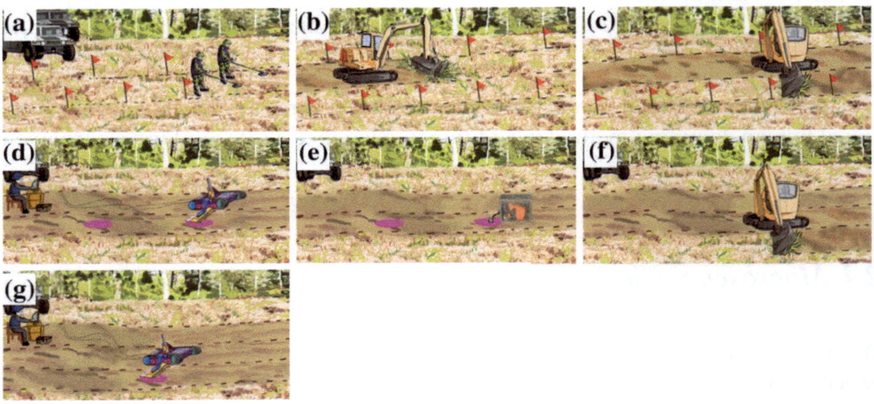

Fig. 2 Macroscopic scanning procedure

2.2 *Microscopic Scanning Strategy*

For the microscopic scanning procedure, the smaller size of mine detecting sensor than macro-scanning-robot was used for relatively slow and detailed detection strategy. The microscopic scanning process is presented in Fig. 3b. Micro-scanning-robot proceeds towards the location where macro-scanning-robot expected the mines are buried as in Fig. 3a. Then, it detects mine carefully and in detail around the designated location with making the semi-circle sensor trajectory. Micro-scanning-robot was equipped with a mine detecting sensor, in which GPR and MD sensor are stacked so that the volume of the sensor becomes smaller. Due to small mine detecting sensor with GPR and MD, it is more suitable for more precise and sophisticated scanning. Once the micro-scanning-robot detects and locates the landmine, it ceases the operation and human removes the mine before micro-scanning-robot continues the operation.

3 Hardware Architecture and Functions

Developed mine detecting robot system is consist of three major parts, mobile platform, manipulator mechanism, and mine detecting sensor unit. This chapter demonstrates the overall hardware compositions, specifications, and functions for hardware architectures of a macro-scanning-robot, which is called "MIDERS-1," and a micro-scanning-robot, which is called "MIDERS-2."

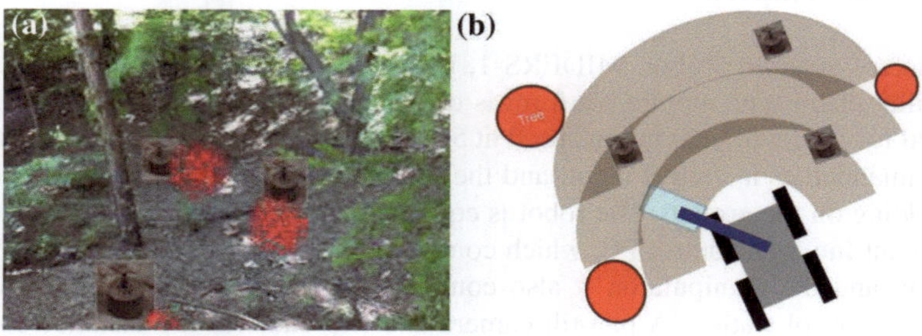

Fig. 3 Macroscopic scanning procedure, **a** Expected mine locations, **b** Schematic of the macroscopic scanning strategy

Fig. 4 MIDERS-1 **a** Integrated system, **b** Mobile platform only

Table 1 Performance specifications

Performance criteria	MIDERS-1	MIDERS-2
Maximum speed	5 (km/h)	12 (km/h)
Maximum obstacle overcome	15 cm	30 cm
Gradability	20°	30°
Size	$1200 \times 780 \times 330\,\text{mm}^3$	$920 \times 620 \times 270\,\text{mm}^3$
Weight	220 kg + payload (100 kg)	85 kg + payload (40 kg)
Operating hour	2 h	3 h

3.1 MIDERS-1

3.1.1 Mobile Platform

The macros-scanning-robot, MIDERS-1, that we operate is shown in Fig. 4. This caterpillar chain type mobile platform is designed for swift mine detection. It can proceed forward with maximum speed at 5 km/h. The velocity and the turn rate can be commanded as a control input, and the mobile platform changes its orientation by skidding on the ground. The robot is equipped with a Compact-Rio by National Instrument Inc. as a control PC, which controls the integrated system of the mobile platform and the manipulator. It also conducts the wireless communication with a remote control station. A pan-tilt camera and two cameras are installed for surveillance at middle, front and back on the platform, respectively. The performance specifications of the mobile platform are shown in Table 1.

3.1.2 Manipulator

The manipulator of MIDERS-1 is designed for fast scan of a mine fields. Figure 5 represents operating strategy of MIDERS-1. MIDERS-1 moves along the safe region which is cleared by human at initial state. A sensor array possessing MD and GPR is located at one side of MIDERS-1. While MIDERS-1 moves along the safe region, the sensor array scans an unrevealed area (i.e., scan region in Fig. 5) for a mine. Detection of a mine is performed with the sequence of move, stop and scan. When a mine is detected, MIDERS-1 marks the location of a mine with paint spray. Note that elimination of a mine is not considered for MIDERS-1.

Since it is planned to perform scanning at the side of MIDERS-1, a manipulator is directed to the side of MIDERS-1. The manipulator of MIDERS-1 is depicted in Fig. 6. The sensor can move in the x, z, roll, and pitch directions. For motions in the x and z directions two linear actuators are equipped, respectively. The stroke in the x direction has value of 1 m and that in the z direction has value of 0.3 m. The sensor array can move with velocity of 0.1 m/s in the x and z directions, respectively. The velocity of 0.1 m/s indicates the scanning speed. The roll and pitch rotations have

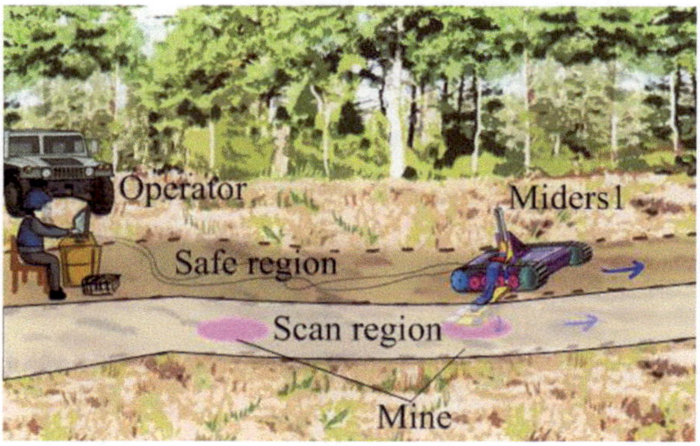

Fig. 5 Operation of MIDERS-1

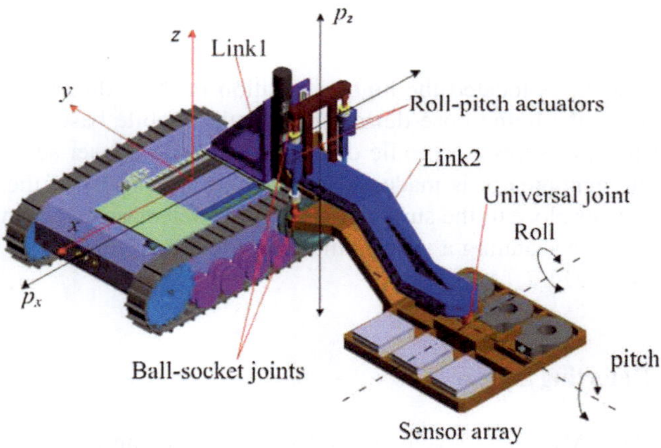

Fig. 6 3D drawing of MIDERS-1

ranges of $\pm 15°$, respectively, and moves with the maximum speed of $30°/s$. It takes around 15 s for a single mine scanning sequence and the covered scanning area at one sequence is $0.82\ m^2$, which makes faster scanning process than the human demining method.

Since the magnetic field of an electric motor can interfere with MD sensors, an electric motor should be located remotely. For MIDERS-1 a parallel mechanism is adopted for the roll-pitch rotations to remotely locate electric motors. Two linear actuators (i.e., roll-pitch actuators in Fig. 6) are equipped at link 2 with ball-socket joints, respectively. The sensor array is attached at link 2 with a universal joint and is connected with two linear motors with ball-socket joints, respectively. When two linear actuators move in the same direction, the roll rotation happens. When one linear actuator moves in the opposite direction to the other actuator, the pitch rotation is made.

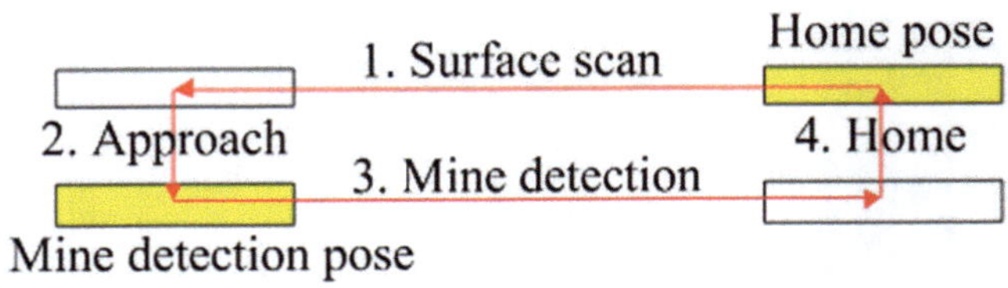

Fig. 7 Motion of the manipulator of MIDERS-1

For precise scan of a mine the manipulator of MIDERS-1 operated the following sequence;

1. Surface scan,
2. Approach,
3. Mine detection,
4. Home.

The sensor array is located the highest position in the z direction at the home pose and stays at the home pose during moving the mobile base. For the surface scan MIDERS-1 measures the profile of the surface with a laser scanner and path planning of the manipulator is made. After scanning the profile of the surface, the sensor array moves close to the surface for the mine detection and scanning a mine is performed. When scanning a mine is finished, the sensor array goes to the home position for move Fig. 7.

3.1.3 Mine Detecting Sensor

The GPR and the MD were developed by Isung Engineering Inc., cooperated with us for the MIDERS-1 project. For the macroscopic scanning strategy, the system should cover as large as possible area at one scan in order to speed up the mine scanning process. Within the purpose, three sets of MD and GPR were sequentially arrayed as in Fig. 8a. There exist space around 15 cm between GPR and MD because MD signal can be interfered by the metal component of GPR. The size of a multiple sensor array is $770 \times 680 \times 46\,\text{mm}^3$ and the total weight of it is 8.1 kg. The frequency bandwidth of GPR is 0.5–1.5 GHz and the detecting frequency is 10 Hz. The anti-personal landmine, so-called KM-16 as in Fig. 8b, was buried under the center of test field. Each set of MD and GPR detects the mine buried under the ground and transmits signal data to remote station. Then three sets of signal data is processed and displayed on PC at the remote station in order for the operator to determine if the data indicates possible mined buried under the ground. Figure 8c shows that the only middle set result detected the landmine. It was possible for MIDER-1 to detect KM-16, buried within 30 cm deep.

However, the gap between GPR and MD, placed for the minimization of signal interference, inhibits the large area scanning comparing to the sensor module size. The need for all-in-one sensor of GPR and MD, thereby, became explicit.

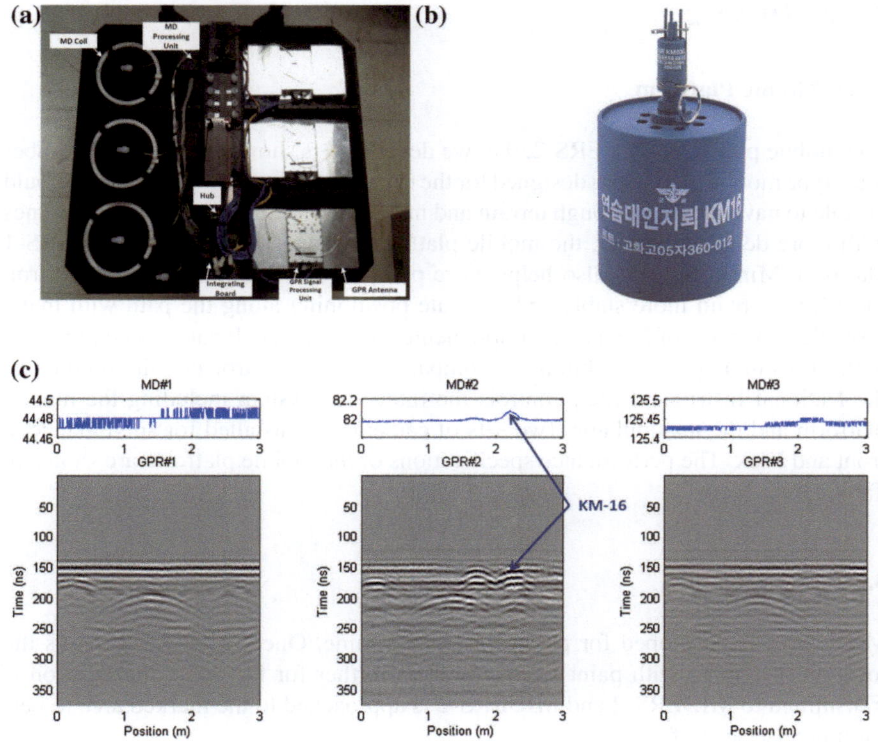

Fig. 8 Mine detecting sensor, **a** Sensor module, **b** KM-16 AP landmine, **c** Signal processed image

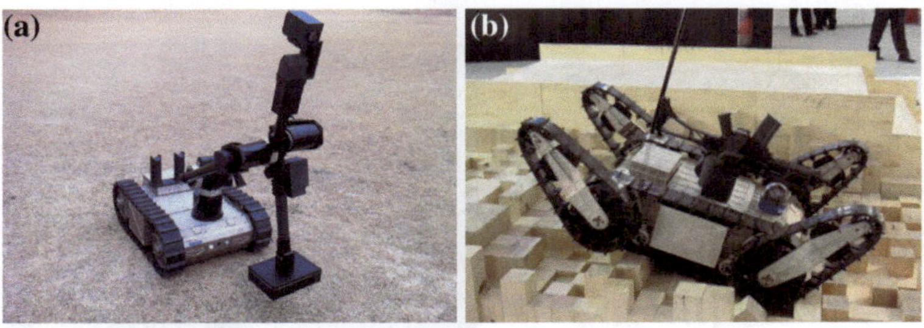

Fig. 9 MIDERS-2, **a** Integrated system, **b** Mobile platform with flippers

3.2 MIDERS-2

3.2.1 Mobile Platform

The mobile platform, MIDERS-2, that we developed is shown in Fig. 9. The rubber track type mobile platform is designed for the rough terrain operations. Since it should be able to navigate on the rough terrain and maintain its stability for detecting mines with more detailed manner, the mobile platform become smaller than MIDERS-1 platform. Minimized size also helps more precise control of the mobile platform. In order to obtain more stable and accurate positioning along the path with many obstacles, four sets of flippers are implemented as in Fig. 9b. It can proceed forward with maximum speed at 3 km/h. A Compact-Rio as a control PC, the product of the National Instrument Inc., controls the integrated system including the mobile platform and the manipulator. Two sets of cameras are installed for surveillance at front and back. The performance specifications of the mobile platform are shown in Table 1.

3.2.2 Manipulator

MIDERS-2 is developed for precise scan of a mine. Once MIDERS-1 marks the location of a mine with paint spray and goes further for fast scan, the location is transmitted to MIDERS-2 and MIDERS-2 is approached to the marked area to perform precise scan of a mine.

The manipulator system of MIDERS-2 is depicted in Fig. 10 and consists of three modules;

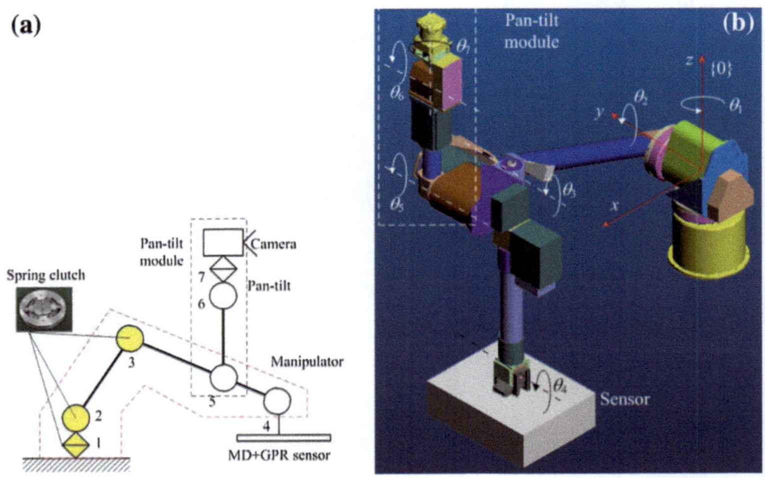

Fig. 10 Manipulator of MIDERS-2, **a** Schematic, **b** 3D drawing

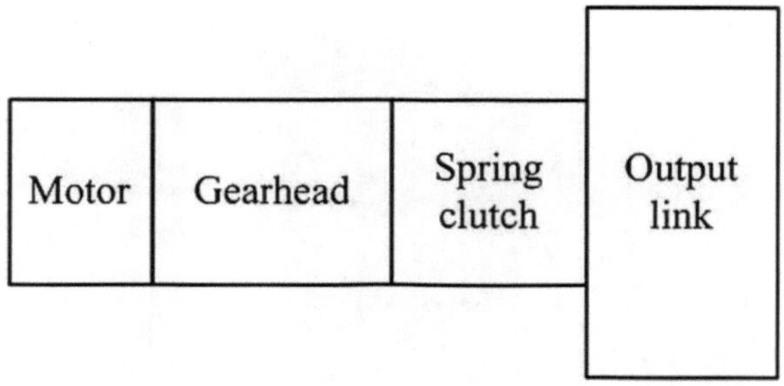

Fig. 11 Joint structure with the spring-clutch

1. A manipulator,
2. A pan-tilt module,
3. Tool (or gripper).

The manipulator and pan-tilt module possess four and three degrees of freedom, respectively. The pan-tilt module can be attached at the distal link or the mobile base. MIDERS-2 can be operated for EOD and surveillance missions by equipping a gripper at the end-effector and attaching the camera module at the distal link. For the mine detection the MD and GPR sensor module is attached at the end-effector and a laser scanner is substituted for a camera. Rotations of each joint are represented in Fig. 10b.

During precise scan contact may happen between the sensor module and the ground. The so-called spring-clutch [22] is adopted for safety reason and equipped at joint 1, joint 2 and joint 3. A joint structure with the spring-clutch is shown in Fig. 11. The spring-clutch is located between the gearhead and the output link. When joint torque exerting at a joint is less than predetermined torque of the spring-clutch, the output link is firmly fixed at the output axis of a gearhead by the spring-clutch. When excessive torque is exerted at the joint by contact, the spring-clutch is unlocked mechanically and the output link can rotate freely. Since no excessive torque can be exerted at the joint, the spring-clutch guarantees to avoid mechanical failures. A joint torque sensor is also equipped at the spring-clutch. External force estimation and force control utilizing the estimated force can be enabled. For details of the spring-clutch refer to [22].

3.2.3 Mine Detecting Sensor

The mine detecting sensor for MIDERS-2 was developed upon the consideration regarding how to diminish or even eliminate the gap between GPR and MD, because the gap caused such a small overlapping detected area at one scan, which eventually retarded the whole process. Isung Engineering Inc. cooperated in this project for developing all-in-one sensor with significantly minimized interference between GPR and MD as in Fig. 12a. In keeping with the purpose of the microscopic scanning strategy, the size of all-in-one mine detecting sensor became much smaller than

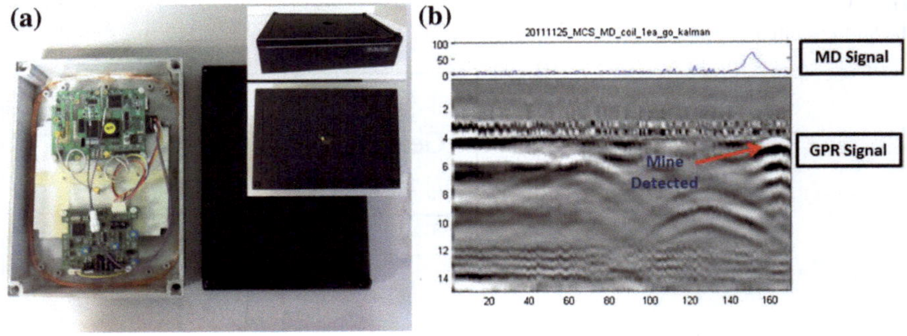

Fig. 12 Mine detecting sensor, **a** All-in-one sensor module, **b** Signal processed image

before in Fig. 9a. The size of it was $260 \times 180 \times 65\,\text{mm}^3$, and the weight was 0.85 kg. MIDERS-2 definitely is to perform detailed and accurate mine detection. It is because the pervious multiple arrayed sensor module shrunk and resulted better ground profile adaptability which enables better and more precise mine detection strategy. With the developed sensor, the operator could locate the exact locations of the buried mines as in Fig. 12b.

4 Conclusion

In this project, we have suggested the cooperative demining strategy between macroscopic and microscopic scanning procedure. The overall purpose was to enhance the traditional human demining in the aspects of the processing time and the accuracy. MIDERS-1 was designed for swift demining in a relatively large and uneven ground condition. It was operated as described macroscopic strategy in the test field, the result showed $0.82\,\text{m}^2$ for 15 s which was satisfactory for proposed strategy.

MIDERS-2 was developed as the microscopic demining strategy for more sophisticated mine detection than MIDERS-1. In order for the system to be compact enough for delicate tasks, all-in-one mine detecting sensor and multi degree of freedom manipulator were equipped. Joint can be protected from the excessive torque by unlocking the spring-clutch mechanically. However, we have not performed the field tests regarding the operations of MIDERS-2. It would be our future work to verify the feasibility of the cooperative demining strategy.

References

1. D. Antonic, Z. Ban, M. Zagar, Demining robots-requirements and constraints. Automatika **42**(3–4) (2001)

2. J. Nicoud, M. Habib, The pemex-b autonomous demining robot: perception and navigation strategies, in *Proceedings of Intelligent Robots and Systems*, Pittsburgh, PA (1995)
3. Y. Baudoin, E. Colon, Humanitarian demining and robots, in *Proceedings of the IEEE International Conference on Control Applications*, Trieste, Italy (1998)
4. H. Cruz, J. Lisboa, P. Pedro Santana, R. ARui Maltez, J. Barata, L. Luis Flores, Two sustainable and compliant robots for humanitarian demining, in *Proceedings of the IARP International Workshop on Robotics and Mechanical Assistance in Humanitarian Demining (HUDEM2005)*, June, Tokyo, Japan, pp. 64–69 (2005)
5. P. G. de Santos, E. Garcia, J. Estremera, M. A. Armada, Dylema: using walking robots for landmien detection and location. Int. J. Syst. Sci. **36**(9) (2005)
6. K. Kato, S. Hirose, Develoment of the quadruped walking robot, titan-IX-mechanical design concept and application for humanitarian demining robot. Adv. Robotics. **15** (2001)
7. J. Trevelyan, Robots and landmines. Industr. Robot: Int. J. **24**(2) (1997)
8. T. Fong, C. Thorpe, C. Baur, Collaboration, dialogue, and human-robot interaction, in *Proceedings of the 10th International Symposium of Robotics Research*, Lourne, Victoria, Australia (2001)
9. P. Kidd, Design of human-centered robotic systems, in *Human Robot Interaction*, ed. by M. Rahimi, W. Karwowski (Taylor and Francis, London, 1992)
10. J. Scholtz, S. Bahrami, Human-robot interaction: development of an evaluation methodology for the bystander role of interaction, in *Proceedings of the IEEE International Conference on Systems, Man and Cybernetics*, vol. 4 (2003)
11. R. Parasuraman, T. Sheridan, C. Wickens, A model for types and levels of human interaction with automation. IEEE Trans. Syst. Man Cybern. Part A Syst. Hum. **30**(3), 286–297 (2000)
12. A. Steinfeld, T. Fong, D. Kaber, M. Lewis, J. Scholtz, A. Schultz, M. Goodrich, Common metrics for human-robot interaction, in *Proceedings of the Conference on Human-Robot Interaction (HRI)*, Salt Lake City, UT (2006)
13. P. M. Blagden, Summary of un demining, in *Proceedings of the International Symposium on Anti-personnel Mines*, Montreux (1993)
14. C. Bruschini, K. D. Bruyn, H. Sahli, J. Cornelis, Eudem: the EU in humanitarian demining, Brussels (1999)
15. S. Burke, The U.S. department of defense humanitarian demining research and development program. J. Mine Action **7**(1) (2003)
16. G. Danielson, P. Blachford, DIANA 44T test and evaluation, Swedish armed forces, Swedish EOD and demining centre. Technical report 13 345:60629 (2003)
17. Testing of MV-4 mine clearing machine, Republic of Croatia, Croatian Mine Action Centre (CROMAC) (2002)
18. E. S. Inc, MR-2 demining and hazardous materials handling robot (2005)
19. D.J. Eisenhauer, C.O. Norman, F.K. Kochanski, J.W. Foley, Enhanced teleoperated ordnance disposal system (ETODS) for humanitarian demining, in *Proceedings of the 8th International Meeting of the American Nuclear Society (ANS)*, Pittsburgh, PA (1999)
20. Y. Tojo, P. Debenest, E. Fukushima, S. Hirose, Robotic system for humanitarian demining, in *Proceedings of the International Conference on Robotics and Automation* (2004)
21. K. Nonami, Q. Huang, D. Komizo, Y. Fukao, Y. Asai, Y. Shiraishi, M. Fujimoto, Y. Ikedo, Development and control of mine detection robot COMET-II and COMET-III. Special issue on advances in motion and vibration control technology, vol. 46, no. 3 (2003)
22. W. Lee, J. Choi, S. Kang, Spring-clutch: A safe torque limiter based on a spring and cam mechanism with the ability to reinitialize its position, in *Proceedings of the IEEE/RSJ International Conference on Intelligent Robots and Systems* (2009)

Experience in System Design for Human-Robot Teaming in Urban Search and Rescue

G. J. M. Kruijff, M. Janíček, S. Keshavdas, B. Larochelle, H. Zender, N. J. J. M. Smets, T. Mioch, M. A. Neerincx, J. V. Diggelen, F. Colas, M. Liu, F. Pomerleau, R. Siegwart, V. Hlaváč, T. Svoboda, T. Petříček, M. Reinstein, K. Zimmermann, F. Pirri, M. Gianni, P. Papadakis, A. Sinha, P. Balmer, N. Tomatis, R. Worst, T. Linder, H. Surmann, V. Tretyakov, S. Corrao, S. Pratzler-Wanczura and M. Sulk

Abstract The paper describes experience with applying a user-centric design methodology in developing systems for human-robot teaming in Urban Search and Rescue. A human-robot team consists of several semi-autonomous robots (rovers/UGVs, microcopter/UAVs), several humans at an off-site command post (mission commander, UGV operators) and one on-site human (UAV operator). This system has been developed in close cooperation with several rescue organizations,

---

The paper reports research funded by the EU FP7 ICT program, Cognitive Systems and Robotics unit, under contract 247870, "NIFTi" (http://www.nifti.eu).

G. J. M. Kruijff (✉) · M. Janíček · S. Keshavdas · B. Larochelle · H. Zender
Nuance Communications Deutschland GmbH, R&D—NLP Research, Kackertstrasse 10, Aachen D-52072, Germany
e-mail: gj@nuance.com

N. J. J. M. Smets · T. Mioch · M. A. Neerincx · J. V. Diggelen
TNO, Soesterberg, The Netherlands

F. Colas · M. Liu · F. Pomerleau · R. Siegwart
ETH Zürich, Zürich, Switzerland
e-mail: francis.colas@mavt.ethz.ch

V. Hlaváč · T. Svoboda · T. Petříček · M. Reinstein · K. Zimmermann
CTU in Prague, Prague, Czech Republic

F. Pirri · M. Gianni · P. Papadakis · A. Sinha
ROMA, Rome, Italy

P. Balmer · N. Tomatis
BLUE, Lausanne, Switzerland

R. Worst · T. Linder · H. Surmann · V. Tretyakov
Fraunhofer, St. Augustin, Germany

S. Corrao
VVFF, Rome, Italy

S. Pratzler-Wanczura · M. Sulk
FDDO, Dortmund, Germany

K. Yoshida and S. Tadokoro (eds.), *Field and Service Robotics*,
Springer Tracts in Advanced Robotics 92, DOI: 10.1007/978-3-642-40686-7_8,

and has been deployed in a real-life tunnel accident use case. The human-robot team jointly explores an accident site, communicating using a multi-modal team interface, and spoken dialogue. The paper describes the development of this complex socio-technical system per se, as well as recent experience in evaluating the performance of this system.

1 Introduction

Urban Search and Rescue is a domain where robots have the potential to make a difference [27]. They can go where people cannot: To help assess a situation, determine an approach to deal with it, even before humans have gone in.

To make this possible, we do need more autonomy in the robot [3], in perceiving the environment, in navigating it. However, disaster areas are harsh places. We inevitably experience what Woods et al. [37] termed "(***Robin***) ***Murphy's Law***: *any deployment of robotic systems will fall short of the target level of autonomy, creating or exacerbating a shortfall in mechanisms for coordination with human problem holders*." Adaptive autonomy is one way of trying to address this problem [24, 31], making explicit the inherent interdependence between humans and robots [13].

Still, this is all for naught if the humans in the team do not *accept* a robot's autonomous capabilities and intelligence. Recent experience with synchronous human-robot interaction in Fukushima (S. Tadokoro, p.c.) and in our own end user studies underline this. A robot's abilities, behaviour, and possible achievements need to be transparent to a human operator: Whether the robot is doing something, what it is doing and why, whether it thinks it has achieved a goal (or not). If an operator is unclear about what to expect from the robot, he or she is unlikely to delegate control to the robot. Instead, no matter what the robot is able to do autonomously, the operator will revert to tele-operation.

And that's not quite what anybody wants. We see this as an issue of (lacking) transparency in experience, behavior and intentions [6]. Robot behavior needs to be transparent, to allow for a proper management of user expectations. A gap between these expectations, and what actually happens, can seriously affect the interaction [17, 23]. A lack of transparency reduces acceptability, which might explain why human-robot interaction (HRI) is a bottleneck in USAR [25]. (For similar observations in the field for asynchronous forms of human-robot interaction see [35], and also the projections for synchronous interaction in [25]).

The problem gets exacerbated in the context of USAR. Humans and robots perform under stress, in complex environments. Situations, interactions, plans change. And with that, expectations change. What we are looking at is not characterizing a gap between expectations before and after a human has interacted with a robot, as is typically done in studies on HRI [17, 23]. Instead, we need to address expectation management online. As situations change, affecting the dynamics of the team, the robot needs to adapt its behavior, and the way it presents that behavior to continue to provide adequate and effective transparency; cf. e.g. [28].

Fig. 1 NIFTi tunnel accident use case: **a** Setting. **b** UAV and UGV in action; **c** control center

In this paper, we try to further understand the problem. We do not offer a solution; but we discuss a way in which we believe we can come to *understand* the problem better, and design systems that can eventually address the problem in real-life. We present a user-centric design methodology (Sect. 2) which draws in end users (first responders from several organizations across Europe) and their experience into the entire R&D process. Following this methodology, we discuss how we design our systems (human-centric, Sect. 3), and how we experiment with them and evaluate them (with end users, under real-life circumstances Sect. 4).

2 User-centric Design Methodology

We adopt a user-centric design methodology, in several respects. Firstly, we include users in all the phases of the development cycle: Requirements analysis, component- and system development, and experiments and evaluations. Users are from various rescue services (Fire Department of Dortmund/Germany, Vigili del Fuoco/Italy). Together, we formulate requirements for hard- and software functionality, and develop physically realistic use cases in which we can experiment with and evaluate our approaches. Figure 1 illustrates one such use case, namely a tunnel accident.

Involving users throughout the yearly development cycle does more than just telling us what they need (requirements), and whether our systems do the job (evaluations). Their involvement provides us with a deeper insight into their needs, their procedures, and what happens out in the field. This is another aspect of the human user-centric design approach we follow. We build systems which can assist humans, doing so in ways that mimic human understanding, and operational procedure. The hypothesis being that this makes robot behavior more transparent to the user.

In the system design, the human perspective is pervasive throughout the representations the robot builds, and the way it determines its behavior. The conceptual understanding of the environment provides a human-like view on the environment, and the inference of spatially grounded affordances results in robot behavior that mimics standard procedure. When it comes to human-robot interaction and planning, humans are explicitly modeled as actors, and action and interaction are planned in ways that conform to human operational practice.

3 Socio-Technical System Design

We approach design from a socio-technical perspective. It concerns the entire system of robots, humans, and how they work together. We focus on the four questions listed below.[1]

1. How to model situation awareness which (a) bridges the gap between a robot's quantitative, and a human's qualitative sense of space, (b) facilitates use by a geographically distributed team, and (c) provides the basis for individual or joint action (4)? See Sect. 3.1.
2. How to model the impact of situations in task- and team-work which influence user performance, given that (a) humans typically perform under stress in USAR missions, and (b) stress alters interaction patterns (3)? See Sect. 3.2.
3. How to model user-adaptive human-robot communication, to adjust how, what, and when a robot communicates given an awareness of the current operative situation (1) and its effects on human performance (2)? See Sect. 3.2.
4. How to model flexible temporal planning and execution, to guide how a robot plans and executes its own actions under different conditions (1)? See Sect. 3.3.

3.1 Intelligence in Situation Awareness

A robot builds up a situation awareness which bridges the gap between its own quantitative forms of perception, and a human qualitative understanding of space. The robot builds up a qualitative structure of dynamic space, and can make inferences about possible actions situated in that space. Mapping therefore builds up several layers of abstraction. First we try to build an accurate metric representation of the environment based on the 3D rolling laser sensor mounted on our robot. Based on this metric representation, we then segment the navigable space into coherent areas linked in a navigation graph. Going 3D requires both to have an efficient 3D representation of the environment and to be able to estimate the six degrees-of-freedom pose of our robot. To avoid part of the distortions, the 3D point clouds are registered into the map only when the robot is static. Preliminary results show that in most cases the distortion when the robot is moving is not too large, but localization may jump from local optima and induce point cloud deformation due to the pose interpolation. The 6 DOF pose estimate is based on a robust 2D map when the robot lies in a mostly 2D part of the environment. We rely on fast and efficient 3D registration implementation (i.e., libpointmatcher) to handle 3D environments online [33].

For the topological segmentation, we take as input the map of the environment. Previously we performed topological extraction based on spectral clustering and mutual information [22]. To better handle changes in the map, both due to exploration

[1] Relations between questions are indicated in brackets, e.g. (4) means a relation to question 4.

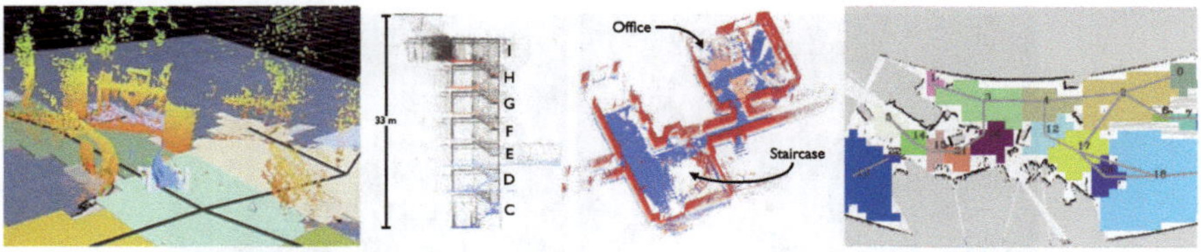

(a) (b) (c)

Fig. 2 3D mapping: **a** 3D point cloud in tunnel, **b** 3D map (indoor environment), **c** Topological segmentation of tunnel, with navigation graph (in *grey*)

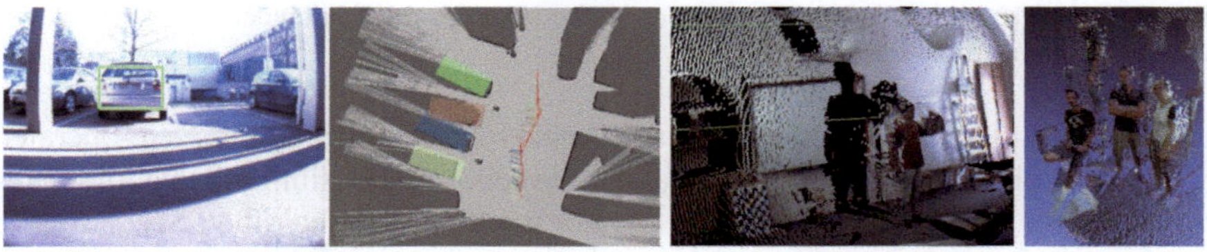

(a) (b) (c)

Fig. 3 Car detection using visual features and 2D mapping (**a**), 3D laser point clouds (**b**) and fusion with visual data from omnicam (**c**)

and due to actual changes, we use incremental topological segmentation. Figure 2c illustrates the result of this new method in the tunnel environment.

In parallel to constructing 3D spatial structure, we combine 3D point clouds and robot positioning to improve vision, and ground observations of objects in the scene. Image-based detection of rear parts of cars in a tunnel accident works relatively well [38], see Fig. 3. Estimating the 3D positions of cars proved more difficult, especially the orientation. To deal with 3D instability we associate 2D features with the 3D metric maps. Figure 3b, c shows an example of assigning image colors to the 3D point clouds. The 2D object detector creates a probabilistic map over the image, and attributes this to points in a 3D point cloud. The 3D information provides an absolute scale, which allows for discarding many false positives.

We use these object observations to perform Functional Mapping, a form of spatial inference [14]. Given an object, and an action to be performed, functional mapping infers areas around the object, where the action can be performed relative to the object. This combines logical inference over associated ontologies for objects and their internal structure, and for actions; and geometric inference. In the tunnel accident, functional mapping infers that being in a particular position relative to a car window facilitates looking into that car. The projection of the areas into 3D space is based on real-time map data and the observed 3D pose of the object. Functional mapping thus combines top-down inferencing, from a priori knowledge of expected objects, and bottom-up inferencing from real-time observations.

Inferring functional areas serves several purposes. First of all, when fire-fighters explore a disaster site, they themselves move between functional areas to make their

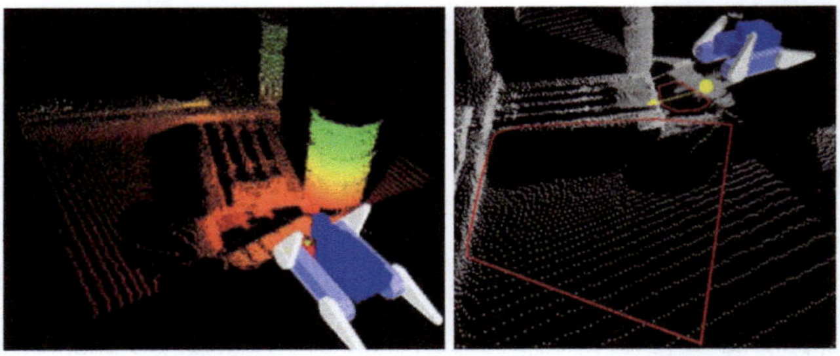

(a) (b)

Fig. 4 Gap detection and analysis; **a** *Top view* of the 3D point cloud and **b** detected gaps together with traversability direction

observations [15]. We observed the same behavior when fire-fighters tele-operated robots to explore an accident [14]. Making the robot follow similar behavior makes that behavior more transparent to an operator working with the robot. Secondly, we use the inference of functional areas to determine optimal vantage points for the robot to perform an observation. Finally, these functional areas serve in maintaining common situation awareness between the robot's metrical environment knowledge and the pilot's qualitative understanding of the environment. Thus, when a human instructs the robot to "go to the car", it goes into a functional area, rather than naively trying to go to (the center of) the car.

Finally, we use map information to perform terrain analysis for traversability, particularly negative obstacle and gap detection. Our approach has two main stages: (i) Application of image morphological and contour detection algorithms and (ii) application of Principal Component Analysis in the orientation domain of the gap contours [29] and extraction of the optimal traversability path. Reasoning with respect to the traversability of the detected gaps is done considering the dimensions and morphological adaptation capabilities of the robot. A representative example of gap detection and analysis is given in Fig. 4.

Adapting the robot's morphology concerns adjusting its articulated components to reduce instabilities that could tip it over [30]. To optimally adapt its morphology with respect to the terrain we consider maximizing the surface contact of the tracks with the ground. (This aims to maximize the traction efficiency of the robot which in parallel results in minimized pressure on the tracks.) Using a set of various terrains classes we first learn the optimal configurations of the robot offline, using a simulation environment (Gazebo) [10], to employ them later on in the real scenario.

3.2 Intelligence in Interaction

HRI is regarded one of the major bottlenecks in rescue robotics [25, 27]. Tele-operating a robot is highly demanding. More autonomy can be a way out of this. But as we already argued, more autonomy requires more transparency, to facilitate common ground and coordination. And that requires communication (Fig. 5).

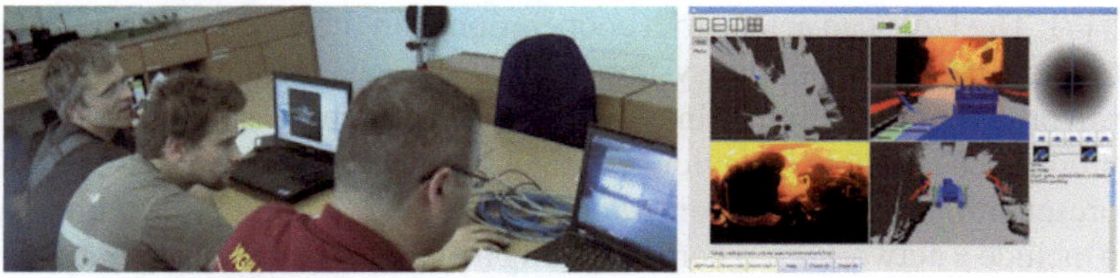

(a) (b)

Fig. 5 Team-based, multi-modal GUI with multiple perspectives (**a**) and multiple info-views (**b**)

Unfortunately, most models of HRI have so far been relatively limited in their use of spoken dialogue, one of the most natural means for humans to interact. Also, these models typically do not ground communication in the social structure, to explain why actors (need to) interact, and what information is to be exchanged. We are working on an approach that takes the social structure and the collaborative ("intentional") context explicitly into account [18–20]. The approach is based in previous collaborative views on dialogue processing [1, 2, 11]. Our approach improves on these by dealing explicitly with uncertain, incomplete information, as is typical for spoken dialogue, and particularly situated dialogue.

We have integrated (limited) spoken dialogue into our multi-modal GUI for human-robot interaction. A user can use dialogue to instruct the robot to move to particular waypoints or landmarks (possibly selected in the GUI), or drive in specific directions [21], similar to [8]. Based on insights in human-human interaction in human-robot teams for USAR (NJEx 2011, Sect. 4), and the recent experience in the end user evaluations at SFO (Sect. 4), we see there is particularly a need for the robot to *produce* contextually appropriate feedback to maintain transparency. (The range of utterances which a robot needs to understand is relatively limited in this domain). Using our recent experimental data, we are investigating the relation between when what is to be communicated by the robot to someone (communication patterns) – and task context, and the user's estimated stress and workload. This should provide an insight in not just *what* to say, but also *how* to say it best such that it is easy to understand by the user under the given circumstances.

3.3 Intelligence in Team Cooperation

Human-robot teams are typically geographically dispersed. For team cooperation this requires the entire system to integrate different views on the environment, (e.g. UAV, UGV, in-field operators), and to facilitate different perspectives and needs [34]. Below we briefly describe the planning approach we use for a robot to share and coordinate control with other team members, to support coordinated execution.

The dynamics of the UGV and UAV can be modeled separately by defining two different temporal declarative models in the Temporal Flexible Situation Calculus

(TFSC) [7]. The UAV can act in strict cooperation with the UGV, so the TFSC model needs to know the states of both system components, via a common model.

The hybrid framework combines temporal constraint reasoning and reasoning about actions. The flexible behaviours of the UAV and UGV are specified in a compact representation by temporal constraint networks T_{UAV} and T_{UGV}, with the possibility to include a network T_{HO} corresponding to an in-field human operator-rescuer. These causal and temporal relations, and their constraints, are learned by continuous interaction with humans, via demonstration and by collected observations of successful processes in controled contexts [15, 32]. The networks are mapped into a structure managing time, resources and actions (model-based control). The model accounts for timelines with time flexibly assigned to each component, to satisfy priorities for both resources and tasks, and which rely on online acquisition of sensor data [9]. The whole set is managed by an execution monitor which continuously checks the environment models $\{M_{HO}, M_{UAV}, M_{UGV}\}$ and the inner states $\{S_{HO}, S_{UAV}, S_{UGV}\}$. The execution loop ensures that the network is kept up to date and consistent. The inner states S_{UAV} and S_{UGV} represent the internal loop which checks on all of the machine components, namely both of the UAV and UGV. The human-robot team shares the information about the environment and the mission, combining together their models of the current percepts. To integrate the different abilities of the UAV, the UGV, and the users, a set of properties $P_1, \ldots, P_n$ is defined on top of the inner states of the team units bridging the different dynamic models (Fig. 6).

4 Field Experiments and Evaluations

During 2011, we performed several experiments with end users operating our systems under realistic circumstances. Almost needless to say, we observed problems along

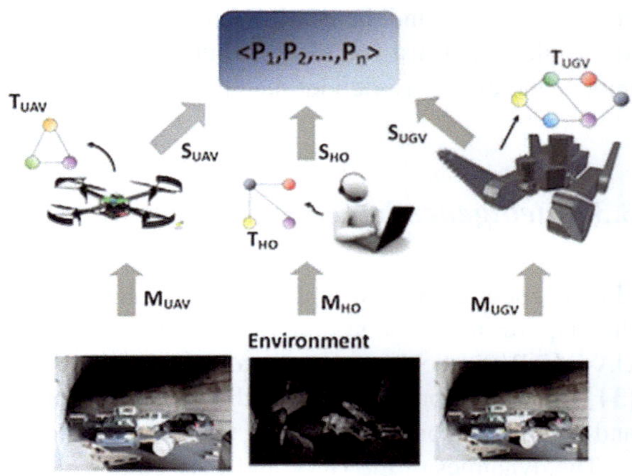

Fig. 6 Properties $P_1, \ldots, P_n$ are defined on inner states S of each team unit, to give a uniform representation of the multi-agent system. M denotes a perceptual model of a unit, T temporal model of unit activities

(a) (b) (c)

Fig. 7 Field experiments at the FDDO training center (**a**; Jan. 2011), control post at joint exercises at FDDO (**b**; Jul. 2011), user evaluations at SFO (**c**; Dec. 2011)

the way that are familiar to anyone operating in the field, see e.g. [5]. Below we focus on the lessons learned with respect to human-robot collaboration.

4.1 January 2011: *Pilot Experiments at FDDO*

The first pilot experiments with users took place at the training center of the Fire Department of Dortmund (FDDO) in January 2011. Users were all professional first responders, and had no prior experience with robots. Each user operated a UGV in a "tunnel accident"—like environment (several crashed cars, a motor bike, debris, set up in a large garage). We ran experiments with two different types of fully tele-operated UGVs: The TNO "Generaal" robot [12], and an ActivMedia P3–AT. The UGVs differed primarily in how the interaction with the robot was supported. The Generaal has a specially designed telepresence control, consisting of a headtracking head-mounted display, whereas the P3–AT was operated via the NIFTi operator control unit (OCU) [21]. The P3–AT was the focus of the pilot experiment, as it used the NIFTi interface and sensory analyses. The users were mostly unfamiliar with robots and and spent 30–60 min training and performing simple tests such as slalom navigation and object detection. Many technical problems arose: signal loss, insufficient battery power, insufficient bandwidth for video-based feedback, and damages due to obstacles and cold temperatures [5]. Out of the three users using the P3-AT to perform the 15-min exploration task, two had to be cut short after 10 and 13 min respectively.

This experiment provided insights similar to what is found in for example [4]. We could observe that users spent about half their time navigating, and about one third of the remaining time trying to find pathways [21]. By observing the paths taken by the users in the scenario, we found out that they were similar to those followed by fire fighters in similar scenarios [15]. A surprising observation was that the users were satisfied with the robot's video feed. The highly-compressed poorly-lit images were typically shown in a low 400×800 resolution. The users indicated that the quality was sufficient, even if the update frequency was too low (< 8Hz) for safe tele-operation. The experiments did reveal that tele-operation increased the cognitive load of the user [21]. This was one reason to develop more autonomous robot behavior (Sect. 3). We were worried that with so much time spent on navigation tasks rather than on observation, the user's situation awareness would be poor. However, sketch maps drawn on white boards during the scenario showed that the users located most objects within one meter of their actual location, and that the relative positioning of the objects with one another was often entirely correct.

It was thus difficult to explain why tele-operation was so difficult if the users were well aware of their surroundings. We have since improved the visualization in the OCU, to make the robot's maps more intuitive to users. For example, we have added a to-scale 3D model of the robot in the metrical map, and provide more and better views of the maps. We also provide means for controlling robot movement through other than manual operation, e.g. by spoken commands.

4.2 *July 2011*: *NIFTi Joint Exercises* (*NJEx*) *at FDDO*

The pilot study focused on single-robot single-operator missions. Given safety ratio and workload considerations (e.g. [26]), we wanted to move to human-robot team setups, and study those in more detail. We organized a more complex joint exercise event at FDDO (NJEx 2011), involving project partners and end users (FDDO, VVFF). During this event, teams of several humans, a NIFTi-specific outdoor UGV (shown in Fig. 1b), and a UAV explored several complex environments. This included a multi-story residential building "on fire". Missions of 45 min were performed by a team consisting of a mission commander, a UGV operator, and a UGV/UAV mission specialist (all in a remote control post), and a UAV operator and two safety directors for the UGV and the UAV (all in-field, line-of-sight). Team members included both first responders, with minimal experience with robots, and scientists, who use the robots on a daily basis.

For NJEx 2011 we focused on human-human interaction within the human-robot team—the robots were fully tele-operated. Interesting observations were that communication primarily concerned the communication of situation awareness ("we see a victim under the shelves in the room at the end of corridor, right"), and goals ("we are going to look under the staircase, at the end of the corridor, left"). Nearly all of the information exchanged was explicitly situated in the environment. The mission commander mostly communicated situation awareness, to maintain common ground within the team, whereas the UGV operator would indicate the next actions of the UGV. Planning exploration tasks was typically done within the control post and the coordination with the in-field team was done through the mission commander. The UAV operator's task was to fly to a particular point, with an explicitly communicated purpose—typically, what kinds of observations the control post would like to make. Video feeds from the UAV were inspected by the UAV mission specialist in the command post, with the mission commander providing feedback to the UAV operator. The two safety directors had the best awareness of the situations around the robots, as they were in line-of-sight.

Analysis so far has yielded that the mission commander and the UGV operator generate the most radio traffic, with one or the other taking on a leading role. In the most effective teams, this was always the mission commander. Furthermore, variations in stress levels could be detected, particularly for team members with high radio traffic, i.e., the mission commander, UGV operator, and the UGV safety commander. We observed that in low stress situations, loosely defined roles and communication protocols can have a slight negative impact on the team performance. However, in more complex situations (e.g., time pressure, high cognitive load, stress), the lack of protocol can break the team cohesion altogether. Face-to-face and radio communication get overloaded, team members get orders from multiple people, situation awareness becomes more local and poorly shared.

It is thus imperative that human team members follow strict rules from the onset of the mission until the end. Adding an autonomous robot to this team thus means that it needs to "fit in". It must be socially accepted by the team. The next experiments studied the introduction of such a robot in a complex USAR mission.

4.3 December 2011: *End User Evaluations at SFO*

This field trial is the third and most recent. The scenario was again a tunnel car accident but this time was located inside a real tunnel, as shown in Fig. 1. The area spanned 25 m into the tunnel by a width of 10 m, filled with debris, pallets, barrels, crashed vehicles, and smoke. Users were professional Vigili del Fuoco firefighters, with no prior experience in operating robots (UGVs). Each user had 40 min to assess the situation with one UGV and one UAV. The human team members were the mission commander and the UGV pilot (in the command post), and the UAV pilot (in-field). The UGV pilots were fire fighters who had minimal experience with robots and received 45 min of training, followed by simple tests like in January 2011.

The team members in the control post had access to a variety of information sources, in a multi-screen multi-modal user interface set-up. The views included robot-specific interfaces, for example one NIFTi OCU per robot (UGV, UAV), and qualitative views for team-level situation awareness (TREX, [36]). Communication between the command post and the in-field UAV operator was via hand-held radio, through the mission commander. The UGV operator communicated with the in-field UGV using the OCU, by touch through spoken dialogue. The UGV was capable of autonomous navigation, and could also use spoken dialogue to provide observations and basic feedback on actions (action-possibility, action-onset).

For these evaluations, we had defined clear roles for the human participants, we had a robust outdoors robot with enough power, space, and bandwidth for the scenario, we had much improved visualization, we had a task planner and a path planner, we had several different levels of autonomy to reduce the amount of tele-operation, and we ensured that the robot could support a basic dialogue with the operator. From our experience in January 2011, we expected that reducing the need to tele-operatethe

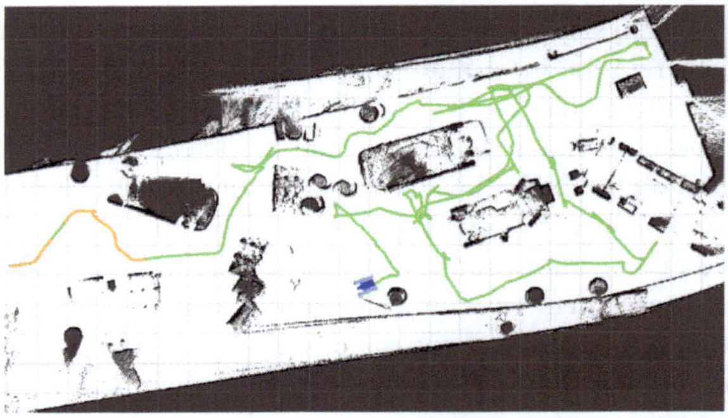

Fig. 8 Path taken in semi-autonomous mode (*orange segments*) versus tele-operation (*green*)

robot to only the situations when autonomous navigation does not work well would free up some time for the user to observe the environment or perform other tasks. However, reality was different.

Figure 8 shows the typical path followed by a user during the 40-min scenario. The first section in orange shows that the user was using spoken commands to control the robot. The green section shows where the robot was tele-operated. Figure 8 illustrates that autonomous behavior of the robot was used less than expected. One explanation might be a lack of transparency of the robot's behavior, and the resulting impact on the user's trust in the robot's capacities. It was often unclear to the user whether a robot had achieved a goal, or why it was failing to do so.

All users started with operating the robot in a high degree of autonomy. However, they all took back control as soon as the path became more difficult to navigate, and many more objects to observe became visible (after about 5 m into the tunnel). This change in sharing control could be grounded in technical reasons (low speed or failures of the autonomous navigation) or on social reasons (lack of trust and understanding about autonomous navigation). While autonomy per se remains crucial to achieving success in robot-assisted USAR, what becomes clear is that we need to find understandable ways to present this autonomy to the users. The robot's state, behavior, and capabilities must be transparent to the human operator. It must be clear what the robot knows, what it is currently doing and why, and what it plans to do. And, perhaps even more importantly from the viewpoint of expectation management, it is crucial that the robot communicates why it *cannot* perform certain actions, or succeed performing them, rather than just failing. If a user is unclear about what to expect from the robot, that user is unlikely to delegate control to the robot. Instead, the operator keeps control, no matter how capable the robot is. In summary, transparency is needed for understanding and trust, and trust is needed for autonomy. Future field experiments will focus on understanding how to achieve transparency, especially in situations of high stress and cognitive load. For example, statements that carry spatial information should convey this spatial part in a multi-modal fashion. Additionally, we need to spend more time on user training, allowing them to adjust their expectations and develop trust in the system.

5 Conclusions

Developing, experimenting, and evaluating USAR robots together with professional users who have much at stake in this domain has turned out to be extremely revealing. In some sense, reality bites. What we believed to be the main issue at stake (autonomy) might well be overshadowed by the problems we are facing in making robot intelligence *acceptable*. Human-robot interaction as "the bottleneck" points into the direction we need to look. We face a socio-technical issue: the entire complex of a robot that can truly behave as a team member in a human-robot team (cf. also [16]). And before we can even talk of common ground, of collaboration, one of the most fundamental lessons we have learnt recently is that this all stands and falls with that robot's autonomous behavior being transparent.

Now that we are slowly beginning to achieve an acceptable level of robot autonomy, it is time to focus our efforts on making this autonomy accepted. In the last field trial, we used a Wizard-of-Oz (WoZ) rather than an automatic speech recognizer (simply to avoid unnecessary complications in the experiment). We plan on pushing this type of setup further, to see how we can control the perception and usage of autonomous behavior. We would like to identify how technical limitations and failures affect the user's perception, compared to how transparency affects the perception of limitations, failures, as well as (situated) capabilities.

References

1. J.F. Allen, C.R. Perrault, Analyzing intention in utterances. Artif. Intell. **15**(3) (1980)
2. J.F. Allen, D.J. Litman. Discourse processing and common sense plans. In P.R. Cohen, J. Morgan, M. Pollack (eds.), *Intentions and Communication*. The MIT Press, 1990
3. A. Birk, S. Carpin, Rescue robotics—a crucial milestone on the road to autonomous systems. Adv. Robot. **20**(5), 595–605 (2006)
4. J.L. Burke, R.R. Murphy, M. Coovert, D. Riddle. Moonlight in Miami: An ethnographic study of human-robot interaction in USAR. Hum. Comput. Interact. **19**(1–2), 85–116 (2004)
5. J. Carlson, R.R. Murphy, How UGVs physically fail in the field. IEEE Trans. Rob. **21**(3), 423–437 (2005)
6. H. Clark, Using Language (Cambridge University Press, Cambridge, 1996)
7. A. Finzi, F. Pirri, Representing flexible temporal behaviors in the situation calculus, in *Proceedings of the 19th International Joint Conference on Artificial Intelligence*, IJCAI'05, (Morgan Kaufmann Publishers Inc, San Francisco, 2005), pp. 436–441
8. T. Fong, C. Thorpe, C. Baur, Multi-robot remote driving with collaborative control. IEEE Trans. Indus. Electron. **54**(4), 699–704 (2003)
9. M. Gianni, P. Papadakis, F. Pirri, M. Liu, F. Pomerleau, F. Colas, K. Zimmerman, T. Svoboda, T. Petricek, G. J. M. Kruijff, H. Zender, Khambhaita, A unified framework for planning and execution-monitoring of mobile robots, in *Proceedings of the AAAI-11 Workshop on Automated Action Planning for Autonomous Mobile Robots*, 2011
10. M. Gianni, P. Papadakis, F. Pirri, M. Pizzoli, Awareness in mixed initiative planning, in *AAAI Fall Symposium Series*, 2011
11. B.J. Grosz, C.L. Sidner, Attention, intention and the structure of discourse. Computat. Linguist. **12**(3), 175–204 (1986)

12. C. Jansen, J.B.F. van Erp. Telepresence control of unmanned systems. in M. Barnes, F. Jentsch, eds. *Human-Robot Interactions in Future Military Operations* (Ashgate Publishing Limited, Hampshire, 2010), pp. 251–270
13. M. Johnson, J.M. Bradshaw, P.J. Feltovich, R.R. Hoffman, C. Jonker, B. van Riemsdijk, M. Sierhuis, Beyond cooperative robotics: the central role of interdependence in coactive design. *IEEE Intelligent Systems*, pp. 81–88, May/June 2011
14. S. Keshavdas, H. Zender, G.J.M Kruijff, M. Liu, F. Colas, Functional mapping: spatial inferencing to aid human-robot rescue efforts in unstructured disaster environments, in *Proceedings of the 2012 AAAI Spring Symposium on Designing Intelligent Robots* (2012)
15. H. Khambhaita, G.J.M. Kruijff, M. Mancas, M. Gianni, P. Papadakis, F. Pirri, M. Pizzoli, Help me to help you: how to learn intentions, actions and plans, in *Proceedings of AAAI Spring Symposium Help Me Help You: Bridging the Gaps in Human-Agent Collaboration*, March 2011
16. G. Klein, D.D. Woods, J.M. Bradshaw, R. Hoffman, P. Feltovich, Ten challenges for making automation a team player in joint human-agent activity. IEEE Intell. Syst. 19(6), 91–95, Nov–Dec 2004
17. T. Komatsu, S. Yamada, Adaptation gap hypothesis: how differences between users' expected and perceived agent functions affect their subjective impression. J. Syst. Cybern. Inform. **9**(1), 67–74 (2011)
18. G.J.M. Kruijff, M. Janíček, Using doctrines for human-robot collaboration to guide ethical behaviour. in Proceedings of the AAAI 2011 Fall Symposium Robot-Human Team-Work in Dynamic Adverse Environments. AAAI, 2011
19. G.J.M. Kruijff, M. Janíček, P. Lison, Continual processing of situated dialogue in human-robot collaborative activities. in Proceedings of the 19th International Symposium on Robot and Human Interactive Communication (RO-MAN 2010). IEEE, 2010
20. G.J.M. Kruijff, M. Janíček, and H. Zender, Situated communication for joint activity in human-robot teams. IEEE Intell. Syst. **27**2, 27–35 (2012) forthcoming
21. B. Larochelle, G.J.M. Kruijff, N. Smets, T. Mioch, P. Groenewegen, Establishing human situation awareness using a multi-modal operator control unit in an urban search and rescue human-robot team. in *Proceedings of the 20th IEEE International Symposium on Robot and Human Interactive Communication*. IEEE, 2011
22. M. Liu, F. Colas, R. Siegwart. Regional topological segmentation based on mutual information graphs. in *Proceedings of the IEEE International Conference on Robotics and Automation (ICRA)*, 2011
23. M. Lohse. The role of expectations and situations in human-robot interaction. In K. Dautenhahn, J. Saunders, eds. *New Frontiers in Human-Robot Interaction* (John Benjamins, Amsterdam, 2011) pp. 35–56
24. C. Miller, R. Parasuraman, Designing for flexible interaction between humans and automation: delegation interfaces for supervisory control. Hum. Factors **49**, 57–75 (2007)
25. R.R. Murphy, Human-robot interaction in rescue robotics. IEEE Trans. Syst. Man Cybern. Part C Appl. Rev. **34**(2), 138–153 (2004)
26. R.R. Murphy, J.L. Burke, The safe human-robot ratio. in M.J. Barnes, F. Jentsch, eds. *Human-Robot Interactions in Future Military Operations, Human Factors in Defence* (Ashgate, Florida, 2010), pp. 31–49
27. R.R. Murphy, S. Tadokoro, D. Nardi, A. Jacoff, P. Fiorini, H. Choset, A.M. Erkmen. Search and rescue robotics. in B. Siciliano, O. Khatib, eds. *Springer Handbook of Robotics* (Springer, Berlin, 2008), Part F, pp. 1151–1173
28. T. Oron-Gilad, P.A. Hancock, Multi-modal information display under stress. in P.A. Hancock, J.L. Szalma, eds. *Performance Under Stress, Human Factors in Defence* (Ashgate, Florida, 2008), pp. 251–270
29. P. Papadakis, I. Pratikakis, S. Perantonis, T. Theoharis, Efficient 3d shape matching and retrieval using a concrete radialized spherical projection representation. Pattern Recognit. **40**(9), 2437–2452 (2007)

30. E.G. Papadopoulos, D.A. Rey. A new measure of tipover stability margin for mobile manipulators. in *Proceedings of the IEEE International Conference on Robotics and Automation*, pp. 3111–3116, 1996
31. R. Parasuraman, M. Barnes, K. Cosenzo, Adaptive automation for human-robot teaming in future command and control systems. Int. J. Command Control **1**(2), 43–68 (2007)
32. F. Pirri, The well-designed logical robot: learning and experience from observations to the situation calculus. Artif. Intell. **175**(1), 378–415 (2011)
33. F. Pomerleau, S. Magnenat, F. Colas, M. Liu, R. Siegwart, Tracking a depth camera: parameter exploration for fast icp. in *Proceedings of the IEEE/RSJ International Conference on Intelligent Robots and Systems (IROS)* (2011)
34. P.M. Salmon, N.A. Stanton, G.H. Walker, D.P. Jenkins, *Distributed Situation Awareness: Theory, Measurement, and Application to Teamwork* (Ashgate, Human Factors in Defence, 2009)
35. K. Stubbs, P.J. Hinds, D. Wettergreen, Autonomy and common ground in human-robot interaction: A field study. IEEE Intell. Syst. **22**, 42–50 (2007) (Special Issue on Interacting with Autonomy)
36. J. van Diggelen, K. van Drimmelen, A. Heuvelink, P.J.M. Kerbusch, M.A. Neerincx, S.M.A van Trijp, E.M. Ubink, and B. van der Vecht. Mutual empowerment in mobile soldier support. Special Issue of the International Journal of Battlefield Technology on Human Factors and Battlefield Technologies. submitted.
37. D.D. Woods, J. Tittle, M. Feil, A. Roesler, Envisioning human-robot coordination for future operations. IEEE Trans. Syst. Man Cybern. Part C Appl. Rev. **34**(2), 210–218 (2004)
38. K. Zimmermann, D. Hurych, T. Svoboda, Improving cascade of classifiers by sliding window alignment in between. in *The 5th International Conference on Automation, Robotics and Applications*. IEEE, 2011

Advancing the State of Urban Search and Rescue Robotics Through the RoboCupRescue Robot League Competition

Raymond Sheh, Adam Jacoff, Ann-Marie Virts, Tetsuya Kimura, Johannes Pellenz, Sören Schwertfeger and Jackrit Suthakorn

Abstract The RoboCupRescue Robot League is an international competition that has grown to be an effective driver for the dissemination of solutions to the challenges posed by Urban Search and Rescue Robotics and accelerated the development of the performance standards that are crucial to widespread effective deployment of robotic systems for these applications. In this paper, we will discuss how this competition has come to be more than simply a venue where teams compete to find a champion and is now "A League of Teams with one goal: to Develop and Demonstrate Advanced Robotic Capabilities for Emergency Responders."

R. Sheh (✉) · A. Jacoff · A.-M. Virts
National Institute of Standards and Technology, Gaithersburg, MD, USA
e-mail: raymond.sheh@robolit.com

A. Jacoff
e-mail: adam.jacoff@nist.gov

A.-M Virts
e-mail: ann.virts@nist.gov

T. Kimura
Nagaoka University of Technology, Niigata, Japan
e-mail: kimura@mech.nagaokaut.ac.jp

J. Pellenz
Bundeswehr Technical Center for Engineer and General Field Equipment, Koblenz, Germany
e-mail: pellenz@uni-koblenz.de

S. Schwertfeger
Jacobs University Bremen, Bremen, Germany
e-mail: s.schwertfeger@jacobs-university.de

J. Suthakorn
Mahidol University, Nakorn Pathom, Thailand
e-mail: egjst@mahidol.ac.th

K. Yoshida and S. Tadokoro (eds.), *Field and Service Robotics*,
Springer Tracts in Advanced Robotics 92, DOI: 10.1007/978-3-642-40686-7_9,

1 Introduction

Competitions have often been used as a way of encouraging collaboration and directing research energy towards particular goals. In this paper, we discuss how the RoboCupRescue Robot League (RoboCup RRL), through competitions and other events, is an effective driver for its three main objectives: increasing awareness of the challenges inherent in Urban Search and Rescue (USAR) Robotics, providing objective performance evaluations for these robots, and promoting collaboration, research, and dissemination of best-in-class solutions to the challenges posed by this domain. In particular, the League has become a competition and community that is geared not to competition against rivals, but rather to moving forward, as a worldwide league of teams, towards solutions to very real challenges. There are many other application domains in the area of field and service robotics where a similar approach may be used to focus development, collaboration, and dissemination among the research community.

The RoboCup RRL was started in response to the 1995 Kobe earthquake and began in association with the broader RoboCup Competitions in 2001. Each year over one hundred teams from academic institutions all over the world compete in regional open competitions that culminate in the International RoboCup RRL Competition. These events are built around the demonstration of advanced robotic capabilities for emergency response applications. These include advanced sensing for situational awareness, mapping and surveying, articulated arms for manipulation and directed perception, mobility platforms for a wide variety of terrains, autonomous behaviours for operator assistance and survivability, algorithms for data interpretation and map building, and human-robot interfaces for highly complex real-time control of robots.

During the course of a week-long competition, 15 to 25 teams, qualified through a competitive regional open and paper submission process, field robots that undertake a series of autonomous and remotely teleoperated search missions to locate

Fig. 1 The RoboCup RRL Arena from the 2011 competition. The tops of the walls are color coded corresponding to the arena areas (the *Red* and *Black* arenas are hidden behind the raised wall sections). A map of this arena appears in Fig. 2

simulated victims in a specially constructed arena, an example of which is shown in Fig. 1. The robots must generate a map with useful information on the state of victims and hazards for emergency responders. This arena consists of a maze of terrains and challenges based on current and emerging standard test method apparatuses for response robots, developed by the US National Institute of Standards and Technology (NIST) through the ASTM International Committee on Homeland Security Operations; Operational Equipment; Robots (E54.08.01). The competition acts as a stepping stone between the laboratory and wider deployment. The apparatuses represent the real world operational requirements gathered from emergency responders, distilled into a form that allows for the repeatable evaluation of robot performance. They reflect the many different challenges that a response robot may encounter. Teams vying for the championship must demonstrate reliable, cutting edge capabilities across the entire arena. Teams may also win Best-in-Class Champion awards for demonstrating Best-in-Class performance in specific problems in USAR robotics.

The RoboCup RRL forms part of the development and dissemination process for standard test methods for response robots, led by NIST [3]. The standards process is driven from the operational requirements of emergency responders who benefit from the development of these tools, such as police, explosive ordinance disposal (EOD) technicians, fire fighters, and rescue and military personnel. At the competition, these test methods are further refined in the presence of the wide variety of robotic solutions to the challenges presented. The competition also helps to disseminate the test methods and apparatuses as teams recreate the arenas in their own labs. Valuable input from researchers, working on the next generation of capabilities, are also gathered and fed back into the standards process. Likewise, exposure to the operationally relevant apparatuses gives teams valuable exposure to real world requirements and highlights the gaps in current capabilities. The competition also gives NIST the opportunity to expose the emergency responders to demonstrations of next generation capabilities. NIST invites teams with best-in-class implementations to response robot evaluation exercises, held at search and rescue training facilities, in order to demonstrate their capabilities to emergency responders, vendors, and officials, with the aim of generating the demand that will shorten the time taken for developments in the lab to reach deployment. As we have focused this paper on how the RoboCup RRL is an effective driver of research and collaboration, we have restricted our reporting of such topics as rules, history, specific test methods, arena construction, and results. For an overview, we invite the reader to refer to the current League Overview [8]. The reader is also invited to refer to the current Rules Outline, Arena Construction Guide, and Community Wiki [4, 6, 9].

2 The RoboCupRescue Robot League

Several unique features of the RoboCup RRL Competition make it particularly effective in promoting research and collaboration. First and foremost, the RoboCup RRL is presented as a competition that does not consider teams as rivals. Rather, the rivalry is between the teams and the application, represented by the arena and the mission specifications. The competition is carefully structured to give teams every chance possible to demonstrate and disseminate their advanced capabilities to their full potential. The RoboCup RRL also specifically encourages teams with well integrated, near-deployable robotic systems, as well as teams with experimental systems that push the boundaries in specific areas and demonstrate Best-in-Class capabilities. It is usually the case that major advances in specialized capabilities come from groups that have specialized and devoted the majority of their energies into that area. Making their inclusion in the competition a priority allows those advances to be demonstrated alongside the more mature, but perhaps less advanced implementations. This makes the competition a unique venue whereby researchers may disseminate their research to others and accelerate the progress of these technologies towards eventual deployment.

The competition starts with a preliminary round where all teams have the opportunity to fully and equally compete in. The aim of this round is to provide all teams with the opportunity to demonstrate their capabilities and obtain statistically significant performance data on their systems. The preliminary round takes up one-half of the week and is held in two half-sized arenas to allow all teams to complete six 15-minutes missions. Team performances in the Preliminary missions decide their qualification for the Championship and Best-in-Class competitions. Teams may compete in all of these competitions or they may decide to focus on one or a few. The Championship competition consists of a semi-final and final round and decides the overall 1st, 2nd, and 3rd place winners. The qualification process, discussed in Sect. 2.3, ensures that teams demonstrating competent, statistically significant performance across the whole arena qualify for the Championship. However, there are also opportunities for more specialized teams to demonstrate their capabilities in the Championship. The Best-in-Class awards, of which there are currently three, are decided on the basis of both the Preliminary missions and a dedicated set of Best-in-Class missions, described in Sect. 2.4. In the rest of this section, we will present in further detail how the different components of the competition work together to foster collaboration among teams with very different capabilities, allow teams to demonstrate their work, and advance the work of the community as a whole. A more general discussion of the competition logistics and the current outline of the rules appears in the current Rules Outline [4].

2.1 Arena

The arena in which the competition is held represents a building, such as a small house or apartment, in various stages of collapse. The arena that was used in the 2011 international competition is shown in Figs. 1 and 2 and consists of current and emerging standard test method apparatuses embedded in a maze. The RoboCup RRL is geared towards showcasing the abilities of robots that are still in the research lab and that are rarely hardened for deployment. Therefore, many of the less practical

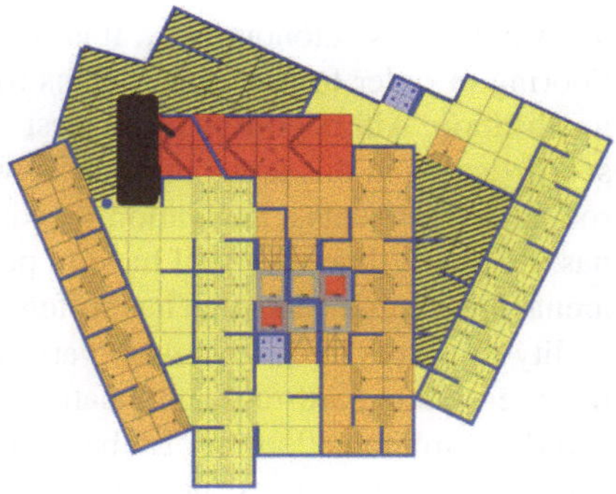

Fig. 2 Map of an example RoboCup RRL Arena showing the *Red*, *Orange*, and *Yellow* arenas, the *Black* arena (car with victims), *Yellow/Black* Radio Dropout Zone, and *Blue* manipulation area. This arena is approximately 15×19 m

or unsafe challenges such as dust, water, mud, fire, chemical hazards, truly unstable terrain, and radio disruptions have been omitted or replaced with more practical substitutes. The simulated victims that teams must locate in order to score points are distributed throughout the arena in such a way as to require them to demonstrate mastery of these apparatuses. In this section we will describe how the design and layout of the arenas and victims were influenced by the objectives of the League. Further details of the makeup of the arena elements and victims may be found in the Arena Construction Guide [9]. As this evolves from year to year, the reader is also invited to refer to the Community Wiki [6]. The arena is divided into three main arenas: Yellow, Orange, and Red, shown in Fig. 3. These form a continuum of increasing difficulty and focus on different current and future challenges faced by Urban Search and Rescue robotics. In addition to the three major arena areas, three specialty arenas exist to encourage developments in specific challenges relevant to USAR Robotics. They are the Blue and Black arenas, shown in Fig. 3, and the Radio Dropout Zone. By dividing the continuous arena into discrete areas, specialist teams are able to focus their energies on the arenas that lie within their area of expertise, where they are able to compete without disadvantage against the Championship teams. It also provides the League with a context in which to introduce new test method apparatuses that reflect the refined operationally relevant requirements gleaned from emergency responders through the ongoing standards development process.

The Yellow arena, in which all missions begin, represents a building to navigate that is largely intact and is designed to showcase teams that are able to navigate and

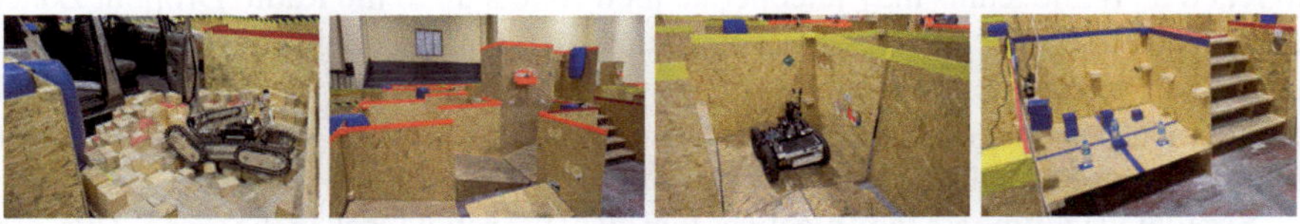

Fig. 3 The *Red* and *Black* arenas, *Orange* arena, *Yellow* arena and *Blue* arena as configured for the Preliminary and Championship missions

identify victims autonomously. It is arranged as a maze with continuous but non-flat flooring in order to encourage teams to implement robust, 3D-aware mapping and localisation. Although it is the easiest of the three arena areas, robots that find and score victims in this arena must do so without any human control. For teleoperated robots, it represents a navigation speed test. The faster they can reach the other arenas, the more time they will have to perform the rest of their mission. The Orange arena represents an area that has suffered from additional collapse and showcases the ability of robots to cover a wide variety of structured terrains and accurately place their sensors in hard-to-reach locations. It is laid out as a maze with multiple levels and discontinuous flooring. Embedded in the Orange arena are structured mobility test method apparatuses such as steep ramps, stairs, steps, and mismatched ramps. All robots may score victims in the Orange arena, however all of these victims are challenging. For instance, they may be placed in difficult locations such as very close to the ground or against steep ramps and only accessible through small holes, or high on walls next to sloped flooring. The Red arena represents complex terrain such as a rubble pile and showcases the abilities of advanced mobility robots and the systems that allow operators to effectively drive these robots in highly unstructured environments. It consists of a stepfield [2], designed to replicate unstructured terrain in a way that is reproducible. This arena tests advanced mobility where the primary challenge is getting to the victim and staying in location while the victim is surveyed—victims are usually easy to view once the robot is co-located. Barriers and victims are placed around the arena so robots must demonstrate well controlled movement. The Black arena, an extension of the Red arena, represents a less standardized part of the arena where robots can demonstrate their abilities in more realistic scenarios. It usually takes the form of victims hidden inside a full sized vehicle.

The Blue arena tests the ability of robots to manipulate objects and carry them as they traverse the terrain in the arena. It consists of shelves containing objects, such as water bottles, radios, and boxes, and grids of holes in which the objects may be placed. In the main competition, teams earn additional points by taking objects from the shelves and placing them with any victims that they find. As mobile manipulation capabilities are still in their infancy among League teams, teams are allowed to start the mission with a single object in the robot's possession. The Radio Dropout Zone, also called the Black/Yellow arena, represents an area of a largely intact building that is experiencing radio interference and encourages teams to incorporate both mobility and autonomy on a single robot. Although it resembles the Yellow arena, it is placed beyond the Orange arena and thus robots require some degree of mobility to reach it. While teams may teleoperate their robots up to the Radio Dropout Zone, once inside the robot must navigate autonomously. On reaching the end of the Zone, teams may resume teleoperation to score victims. As a further incentive, successful autonomous navigation back to the start of the Zone doubles the team's score for finding the zone's sole victim.

Simulated victims are the objectives that all of the robots seek in the arena. They are evenly distributed throughout the three main arenas; an additional victim is assigned to the Radio Dropout Zone. In order to score a victim, the team's robot must be co-located with the victim (usually defined as the centre of the robot's base within

Fig. 4 Examples of simulated victims that appear throughout the arena

1.2 m of the victim) and have line-of-sight to the victim with its sensors. To score well, robots often must deploy their sensor package right up against the victim, requiring slender, dexterous arms and compact sensor packages. Examples of simulated victims are shown in Fig. 4. They consist of several co-located signs of life: a human form, heat, motion, sound, and carbon dioxide. Several of these signs of life taken together are used to identify the state of the simulated victim: unconscious, semi-conscious, or aware. These signs of life have been chosen to be detectable with widely available equipment and yet require sensing and processing that reflects that required to sense real objects of interest. Alongside the simulated victims are test method artifacts that evaluate the ability of the robots to survey the environment, such as hazardous materials labels and eye charts, placed in hard-to-see areas.

Victims are placed in boxes that are either open or accessed through holes of varying sizes, changing the level of difficulty. Victims located in areas that pose greater challenges, such as advanced mobility or full autonomy, are easier to access while victims in areas that have easier terrain are placed in smaller holes that require the robots to be positioned in difficult locations and require novel ways of rapidly and reliably directing perception. Reaching a victim and reporting their condition is only part of the challenge. Teams must also locate the victim on a human-readable georeferenced Geospatial Tag Image File Format (GeoTIFF) map of the arena that their robot autonomously generates as it traverses the arena. Points are scored for the accuracy of the map and its usefulness to an emergency responder who should be able to use it to locate the victim [7].

2.2 Missions

In this section we will first discuss the overall structure of the missions, followed by a brief discussion of the particular features of the missions, which make up the Preliminary and Championship rounds, that help to achieve the goals of the League, as described earlier in this paper. A more complete discussion of the mission and rules appears in the Rules Outline [4]. Each time-limited mission represents a search-and-rescue deployment. Teams deploy their robots to the start point in the arena and their operator interface to the operator station, which is out of sight of the arena. Once set up, the single robot operator uses their single teleoperated robot and multiple autonomous robots to navigate and map the arena, locate the victims, determine their signs of life, and deliver objects. The mission ends when their allotted time has expired, after which they have a short period of time to produce an automatically generated map of the arena with victims identified and localized. The team's score depends on the quality of this map and the extent and accuracy of the information they are able to gather about the victims.

The League uses the scoring process to encourage teams to implement solutions to gaps in current capabilities, as identified through the ongoing standards process. For example, points are awarded for the ability to obtain thermal information about the victim. While teams are able to score some points using non-contact thermometers, more points are available to teams that are able to return a thermal image, resulting in some innovative approaches to performing thermal image surveys of the arena using low cost sensors. Likewise, automatic mapping, 3D perception, and high resolution imaging are encouraged using the scoring metric. As in the rest of the competition, the missions have been designed to facilitate experimentation, allow teams to disseminate their developments to other teams and the general public, collaborate, and assist in the refinement of the standard test methods. The League's objectives are best satisfied if teams are confident that they will be able to showcase their capabilities to their full potential. However, the implementations seen in the competition are rarely deployment-hardened and are highly experimental. The competition has been carefully structured to reduce the risk that luck or random factors result in a team not reaching their full potential. Teams qualify for the Championship rounds based on overall points scored during the preliminaries. To encourage teams to push their robot and to reward teams for bringing innovative, but possibly imperfect, implementations, the worst preliminary mission score for each team is discarded. Furthermore, the qualification cutoff is decided based on the distribution of scores after the preliminary round. There should be a clear performance gap between the best eliminated team and the worst qualified team.

Prior to each mission, the team's operator, or their representative, is asked to walk the arena, identify all of the simulated victims, and check their signs of life. This ensures that bad luck, in the form of the operator randomly choosing a poor path through the maze or a broken simulated victim, plays no part in the ability of the team to showcase their capabilities. While it also means that the operator knows the location and signs of life of each victim, they must still navigate the arena and present convincing evidence to the referee, by way of their user interface, in order to score points. Once the mission has begun, teams are encouraged to have a "spotter" in the arena, who is responsible for ensuring the safety of the robots. Although the mission ends if the spotter touches the robot or interferes with its sensors, it allows teams to experiment and push their robots without fear of damaging them. Teams are also allowed to reset their missions after an intervention, if the robot becomes immobilized, or at any other point of their choosing. This entails moving the robots back to the starting location, resetting their maps and other internal state, repairing, modifying, or replacing the robots as necessary and beginning again. However, while their mission score is set to zero, the clock is kept running so teams effectively have as many missions as they like, within the 15 or 20 minutes they have been allotted. At the end of the time, the mission in which they scored the most points is recorded. In this way, teams may experiment with new settings or approaches, knowing that they will be able to salvage the mission and recover to still earn points. It also reduces the cost of failures early in a mission, before they have had a chance to demonstrate their abilities. During the course of the mission, other teams and members of the public are encouraged to watch the operator and the robots, both of which are projected onto a large screen. This is further facilitated by swapping the order in which teams run between days so that all teams have the opportunity to observe all other teams while not preparing for their own mission. As a further incentive to disseminate their work, a small but significant number of points are given to teams that produce an informative slide and assign a representative to provide a live commentary to the audience during their mission. Following the mission, the maps that each team produces to document their mission is scored by the referees. These are publicly posted, along with their scores and the salient features that either earned points or showed scope for improvement. This allows teams to see how each others' mapping sensors and algorithms have affected their maps and scores and identifies the Best-in-Class implementations to aspire towards.

2.3 Championship

At the start of the semi-final rounds, all team scores are reset to zero. This allows high scoring teams that are confident of qualifying to experiment during the preliminary rounds, where the competition is less fierce, without jeopardising the higher stakes semi-final and final rounds. Depending on the number of qualified teams, the semi-finals are held in either a half-sized or full-sized arena and consume another day of competition. Following the semi-finals, the best performing three or five teams move on to a finals round in the full-sized arena that usually consists of two short missions with scores carrying over from the semi-finals and the final three places decided on points. Teams that present integrated systems that competently perform in all aspects of the arena always qualify for the Championship rounds. Very inexperienced and highly specialized teams are often eliminated. These teams are encouraged to use the remaining one or two days to practice for the Best-in-Class competitions on the final day. Indeed, many specialized teams come to the competition only to compete in the preliminaries and the Best-in-Class competitions with no intention of qualifying for Championship. Yet these teams are not out of the main competition. To further encourage the entry of specialist teams and to foster a spirit of collaboration, each qualified team is strongly encouraged to pair with a team that has been eliminated and progress through to the Championship as a combined team. Each member of the combined team is awarded should the combined team receive the first, second, or third place awards. In recent years, it has been common for teams that focus on mechatronic engineering, and who often perform well across the Orange and Red arenas, to qualify and merge with teams that focus on artificial intelligence, who dominate the Yellow autonomous arena and have excellent mapping but lack the resources to tackle the more physically challenging arenas.

2.4 Best-in-Class

As previously discussed, the field of USAR robotics is so large that many of the breakthroughs in the advancement of capabilities come from specialized research groups. To recognize and reward these groups, who may lack the resources to demonstrate broad competence in enough areas to win the Championship, the League also provides three Best-in-Class awards. Equal in status to the Championship, these awards focus on specific challenges that exist in this field, as identified through consultation with emergency responders as part of the broader standards development process. The current Best-in-Class competitions focus on Mobility, Autonomy, and Mobile Manipulation. Qualification is based on the preliminary missions and performance in these rounds contributes one-half of the Best-in-Class score. The second half comes from a dedicated Best-in-Class round. There is also a separate award, decided by the Organizing Committee, for the team that demonstrates the most remarkable, operationally relevant innovation in user interface development.

The Best-in-Class Mobility competition challenges robots with advanced mobility to demonstrate their ability to rapidly traverse unstructured terrain. Half of a robot's Best-in-Class Mobility score consists of the number of victims that were found by the robot in the Red arena during the preliminary round; teams must find at least one victim in this arena to qualify. The second half of the Best-in-Class Mobility score is based on the standard test method for Advanced Mobility: Stepfields [2]. It takes the form of a race against the clock around the Red arena stepfields. In a fixed time period (usually 10 minutes) teams must drive their robot in a figure-of-eight pattern around the stepfield as many times as possible. While being one of the oldest challenges in the competition, many real world terrains continue to thwart deployed solutions. Through this challenge, teams are encouraged to not only develop highly mobile, robust mobility platforms, but also to develop effective ways of controlling the robots through careful camera placement for situational awareness, reliable low latency communications, and intuitive, low cognitive load user interfaces.

The Best-in-Class Autonomy competition was developed to encourage teams to pursue autonomous victim identification and navigation in continuous but non-flat flooring and in the presence of difficult terrain that must be avoided. Half of the robot's score consists of the number of victims that were found autonomously during the Preliminary missions. The second half of the Best-in-Class Autonomy score is earned during a single Best-in-Class Autonomy mission. The arena is reconfigured so that the Yellow arena becomes one large loop around the whole arena and the mission is for the robot to map as much of the arena as possible within a set time limit. The maps are then evaluated based on coverage and accuracy [7]. There is only a small, albeit growing, subset of teams that are competitive in the Best-in-Class Autonomy competition. However, the technologies required to perform well in this challenge have greater applicability in augmenting existing capabilities. For example, autonomous victim identification drives the development of sensor fusion and object class recognition. This may be equally applied to recognizing not only actual victims but also other objects of interest, such as hazardous material placards or shipping labels across a wide area, something that a robot operator, concentrating on the task at hand, can easily miss. Autonomous navigation can assist the operator by helping to perform long downrange traverses and enhance robot survivability by allowing it to act appropriately should radio communication be impaired. Automatic mapping is also a valuable technology that is only just starting to reach deployment; this competition serves as a valuable proving ground for the test methods that are being developed to evaluate these new capabilities.

The Best-in-Class Mobile Manipulation competition is the newest of the three and encourages the development of dexterous, intuitive mobile manipulation capabilities. Half of the score for this competition comes from the number of objects that were placed with victims during the Preliminary missions. The other half comes from the dedicated Best-in-Class Mobile Manipulation competition, based on the emerging standard test methods for mobile manipulation. Teams use their robots to take as many objects as possible from the shelf and place them in a grid of holes in the Blue Arena within 10 minutes. There is a very real need for dexterous mobile manipulators with intuitive control in the field of USAR robotics. Apart from a few salient examples, robots used in these applications have rudimentary dexterous manipulation capabilities when compared to those in other domains such as telemedicine. Clearly this is a domain where the performance of the bulk of deployable solutions, and to a large part the expectations of the end users, trail the state-of-the-art in the academic community by a significant margin. This competition plays an important part in not only evaluating the performance of the next generation of implementations, but also in helping disseminate their progress to end users and vendors and assisting their rapid integration into fieldable implementations.

One of the main areas of open research in USAR robotics is in the quality of the user experience. This includes all aspects of the robotic system that influence what the user observes through the interface, from the interface itself right through any assistive technologies to the sensor selection, placement, and direction capabilities. It also includes the whole process by which the user controls the robot. To encourage innovations that push the state-of-the-art in this area, the League presents an Innovative User Interfaces award, selected at the end of the competition by the Organizing Committee based on the implementations demonstrated during the week of competition. This award is given to teams that demonstrate implementations that solve an operationally relevant user interface and experience problem in the field of USAR robotics. The League community is made up of groups with a wide variety of approaches to the user interface experience. In the past, this award has been presented to teams for developing wearable interfaces, interfaces based on metaphors such as car driving, and interfaces that use autonomous behaviours to assist the operator.

2.5 Rules and Administration

The RoboCup RRL is primarily administered by NIST as part of its work on standard test methods for response robots. NIST guides the design of the arena and the embedded test method apparatuses, provides the general outline of the competition rules and scoring metrics, defines the Best-in-Class challenges, and manages various competition logistics [4]. Embedded in the arena design, standard test method apparatuses and rule outlines are the operationally relevant requirements that NIST has distilled from its close collaboration with emergency responders and industry. This provides NIST with a powerful way of using its position at the crossroads of researchers, vendors, and emergency responders, to guide the competition towards the challenges that exist in the real world, in a way that is feasible for research-grade implementations to address.

The focus of the RoboCup RRL is not on finding a winner but rather to encourage research and collaboration. To this end, the rules themselves are not set in advance. Rather, based on the outline, the rules are presented, discussed, amended, and agreed upon during meetings at the start of the competition and again at the start of every day with leaders from every competing team. The rules are allowed to evolve from one day to the next to respond to new developments, close loopholes, and adjust scoring to best reward promising implementations. This approach means that teams have little incentive to try to "game" the system—any exploitation of the rules will simply result in that loophole being closed. Instead, teams are encouraged to bring general solutions to the types of challenges that exist in the field of USAR robotics and as emphasised by the published arena design guides and rule outlines. During the competition, much like in a real deployment, teams that do well are the ones that are able to improvize, adapt, and perform rapid development as the conditions, apparatuses, and rules evolve.

3 Dissemination of Solutions

The RoboCup RRL has expanded to host regional open competitions through the year, often serving as qualifying competitions that feed into the international competition. Particularly significant regional opens include those held in Germany, Japan, Iran, and Thailand. The Iranian community, inspired by the 2003 earthquake in Bam, has grown to be one of the major forces in the international competition, fed by a vibrant domestic competition of several hundred teams. The Thai competition is particularly notable, in only a few short years it has rapidly grown to encompass over 100 domestic teams with a championship trophy that is granted and presented by the Thai royal family. As a result of resources and support provided to the domestic Thai RoboCup RRL community, Thai teams dominate the world competition, winning the Championship several years in a row. In the process they have demonstrated many practical advancements that have been disseminated throughout the league, particularly in mobility and directed perception.

It is rare for teams to finish the competition without having made significant improvements to their robot's capabilities in response to its performance in the arenas and the dissemination of best-in-class implementations from other teams. However, it is a high pressure environment, where teams are ultimately concerned with demonstrating their capabilities to their full potential. In order to further encourage collaboration and the dissemination of best-in-class solutions, especially among groups that are considering entering the League, the RoboCup RRL community hosts annual international Rescue Robotics Camps and Summer Schools. Representatives from teams that have exhibited Best-in-Class capabilities in previous competitions are invited to present their work and lead tutorials to help other teams and researchers build on their work. Topics that these events have been particularly effective at disseminating in the past include 2D and 3D mapping, user interface design, and autonomy. These events, attended by around 30 participants and led by around 10 speakers and practical leaders each year, have been instrumental in the development of several new teams who now rank as Best-in-Class in their own right. The League also brings in distinguished speakers in relevant fields to introduce the League community to new developments and encourage research to proceed in the direction of gaps in current capabilities.

Through competition experience, some teams have been able to refine their implementations and move beyond the confines of the relatively safe, structured arenas. Teams that perform well in the competition are invited to demonstrate their implementations at Response Robot Evaluation Exercises, held regularly at fire and rescue training facilities to evaluate the performance of deployable systems. These events represent the next stepping stone between the lab and deployment, with emergency responders, procurement officials, and vendors in attendance. While they are still safe, structured events, the environments are more realistic with real concrete and wood rubble, dirt, dust, and water. Sensors are further challenged with a wider variety of surfaces, longer sight lines, and sunlight, while communications are hampered by steel reinforcing and other structures.

Many teams in the League have implementations that demonstrate capabilities several generations ahead of currently deployed capabilities across the spectrum of user interfaces, manipulation, directed perception, autonomy, mapping, sensing, sensor fusion, and mobility. By demonstrating these capabilities in scenarios that emergency responders are familiar with, the League can spark demand for these technologies and, with vendors also in attendance, hasten their integration into deployable systems. At the same time, researchers are able to see more clearly how their work fits with the overall operational scenarios and identify where and how further improvements may be made. Several teams have already taken advantage of these opportunities, which have proven to be valuable to all parties involved. During subsequent meetings with responders it has been clear that exposure to these cutting-edge implementations has changed their way of thinking about how their problems might be solved. Likewise, researchers exposed to these events have found the experience critical in shaping their research directions to be more application-focused. For procurement officers it has also helped them gain a clearer insight into the capabilities that exist outside of a corporate marketing officer's brief.

Many teams have also formed collaborations with local emergency responder communities and fielded their technologies at Urban Search and Rescue facilities in Australia, Germany, Japan, the United States, and beyond. For example, teams from Japan have long been involved in the domestic Japanese emergency responder community. The design of their robots bear the hallmarks of development in collaboration with emergency responders, with unique capabilities in manipulation and mobility, and unusual and specialized mechanisms such as snake-like robots. German RoboCup RRL teams have also participated in events such as the European Land Robot Trials involving military and civilian applications. Thailand, with its vibrant RoboCup RRL community, has seen teams working with emergency responders to deliver robots for operational purposes. These collaborations with the emergency responder community help to validate both the approaches that the teams have taken, as well as the overall direction of the RoboCup RRL. In an ideal world, the advances made by the teams in the RoboCup RRL would not be required. However, disasters do happen and several teams have already seen their robots deployed in full operational scenarios. The 2011 competition saw the sudden withdrawal of several teams from Japan, when their robots and expertise were required in the aftermath of the 2011 East Japan Earthquake. Robots that were developed for the competition played an instrumental role in the disaster response. This included the Quince robot (CIT, Tohoku University, IRS and NEDO), which was initially designed for traversing the stepfields in the competition. Its proven ability to overcome these terrains resulted in it being chosen for inspecting the buildings housing the reactors at the tsunami-damaged Fukushima Dai-1 nuclear power plant [5].

4 Future Directions

The RoboCup RRL is constantly evolving as the state of the art in implementations and test methods change. In particular, the focus in past years on improving mobility, 2D mapping, and sensing have resulted in most teams achieving a very high level of performance in these areas. The next challenge will be in 3D mapping, manipulation, autonomy, and user interfaces—technologies where a small number of teams have demonstrated mastery. The League organization is also refining the links between these focus areas and the real world challenges through collaboration with emergency responders as part of the standards process and other projects and competitions. As these are reflected in the apparatuses, rules, scores, and Best-in-Class awards, it is anticipated that these focus areas will continue to improve.

New test methods on the horizon are also slated for possible inclusion into the League. In particular, standards for other, related, classes of robots, such as those designed to fight fires, are being considered for adaptation into the League. By carefully matching these new challenges to the skill set already present in the community, the League organization is able to leverage the existing capabilities. This helps to drive innovation in a new, much needed area of research, increases awareness among the research community of a new set of research challenges, and helps to advance a new set of standard test methods by which the performance of future implementations may be evaluated. For the competition as a whole, it also allows new teams a niche in which they can grow and excel and compete with the entrenched players; in the process they benefit the League community by bringing and sharing their unique expertise into the community while themselves learning from the other teams in the League. The explosive ordinance disposal (EOD) community has also been interested in leveraging the RoboCup RRL to further research in their domain. This interest fits well with the increasing focus on manipulation tasks in the competition. The RoboCup RRL is working with the EOD community to ensure the relevance of new competition challenges.

While the RoboCup RRL is focused on robots for disaster response, another league that has its origins in the RoboCup RRL community, RoboCup@Home, focuses on service robots in the home. The two leagues share a common approach to promoting research and collaboration, as well as many common research frontiers. The communities are quite distinct, with the RoboCup RRL community dominated by teams with mechanical and mechatronic backgrounds and the RoboCup@Home community, with its focus on human-robot interactions, heavily focused on artificial intelligence. Going forward, the two leagues plan on working together to encourage the RoboCup RRL community to pursue more complex autonomy in wider, more semantically complex environments, while encouraging the RoboCup@Home community to improve their mechanical platforms to cope with more varied and unstructured terrain. In the process, it is hoped that collaboration and dissemination of technologies between the two communities, and a better understanding of each others' challenges, can be further strengthened.

The RoboCupRescue Virtual Robot League was also formed out of the RoboCup RRL community in order to lower the barrier of entry into this field, especially for teams that are heavily focused on artificial intelligence. This league shares the same challenges as the RoboCup RRL, deployed in a validated physics simulator. Another competition that builds on the RoboCup RRL experience is the DARPA Robotics Challenge [1]. This multi-year competition encourages the development of semi-autonomous robots for disaster response that can operate in a designed-for-human world. It incorporates many of the concepts that make the RoboCup RRL a success and leverages many of the same test methods and processes.

Finally, additional outreach events, such as new summer schools, are planned that aim to bring together not only RoboCup RRL based teams, but also other students and early career researchers to work intensively on focused problems that exist in the league. These include software architectures that improve the ability of teams to share software implementations across different robot platforms and the software modules that will allow teams with little software engineering expertise to begin to compete in this space. Other focus areas for these events include 3D mapping, user interfaces, manipulation, and autonomy. The RoboCup RRL has focused the community on the challenges that matter to the future deployment of Urban Search and Rescue Robotics. Through the combination of a solid grounding in the requirements of emergency responders, integration with the wider standards development process that promotes fair and rigorous evaluation, and an active decision to encourage experimentation, research, collaboration, and dissemination as opposed to rivalry purely for the sake of competition, the RoboCup RRL has succeeded in becoming a "League of Teams with one goal: to Develop and Demonstrate Advanced Robotic Capabilities for Emergency Responders."

Acknowledgments The authors would like to thank past and present members of the RoboCup RRL Organizing Committee, administrators, and teams. The Department of Homeland Security sponsored the production of this material under an Interagency Agreement with the National Institute of Standards and Technology. Certain commercial equipment, instruments, or materials are identified in this document. Such identification does not imply recommendation or endorsement by the National Institute of Standards and Technology, nor does it imply that the products identified are necessarily the best available for the purpose.

References

1. DARPA: DARPA seeks robot enthusiasts (and you) to face off for $2M prize! http://www.darpa.mil/NewsEvents/Releases/2012/04/10.aspx (2012)
2. A. Jacoff, A. Downs, A. Virts, E. Messina, Stepfield pallets: repeatable terrain for evaluating robot mobility, in *Proceedings of the 2008 Performance Metrics for Intelligent Systems Workshop (PerMIS'08)*, NIST Special Publication 1090 (2008)
3. A. Jacoff, H.M. Huang, E. Messina, A. Virts, T. Downs, Comprehensive standard test suites for the performance evaluation of mobile robots, in *Proceedings of the 2010 Performance Metrics for Intelligent Systems Workshop (PerMIS'10)*, NIST Special Publication, p. 1113(2010)
4. A. Jacoff, S. Tadokoro, E. Mihankhah, T. Kimra, J. Pellenz, A. Birk, J. Suthakorn, M. Hofbauer, A.L. Gutierrez, RoboCupRescue Robot League: Rules 2011.2. http://www.nist.gov/el/isd/upload/Robocup_Rules_2011.pdf (2011)
5. K. Nagatani, S. Kiribayashi, Y. Okada, S. Tadokoro, T. Nishimura, T. Yoshida, E. Koyanagi, Y. Hada, Redesign of rescue mobile robot Quince: toward emergency response to the nuclear accident at Fukushima Daiichi nuclear power station on March 2011, in *International Workshop on Safety, Security and Rescue Robotics* (2011)
6. RoboCup Rescue Robot League Organizing Committee: RoboCup Rescue Wiki. http://wiki.robocup.org/wiki/Robot_League (2011)
7. S. Schwertfeger, A. Jacoff, C. Scrapper, J. Pellenz, A. Kleiner, Evaluation of maps using fixed shapes: the fiducial map metric, in *Proceedings of the 2010 Performance Metrics for Intelligent Systems Workshop (PerMIS'10)*, NIST Special Publication, p. 1113, September (2010)
8. R. Sheh, T. Kimura, E. Mihankhah, J. Pellenz, S. Schwertfeger, J. Suthakorn, The RoboCupRescue robot league: guiding robots towards fieldable capabilities, in *International Workshop on Advanced Robotics and Social Impacts* (2011)
9. A. Virts, A. Jacoff, A. Downs, RoboCupRescue arena assembly guide 2011. http://www.nist.gov/el/isd/upload/2011_Assembly_Guide.pdf (2011)

Estimating the 3D Position of Humans Wearing a Reflective Vest Using a Single Camera System

Rafael Mosberger and Henrik Andreasson

Abstract This chapter presents a novel possible solution for people detection and estimation of their 3D position in challenging shared environments. Addressing safety critical applications in industrial environments, we make the basic assumption that people wear reflective vests. In order to detect these vests and to discriminate them from other reflective material, we propose an approach based on a single camera equipped with an IR flash. The camera acquires pairs of images, one with and one without IR flash, in short succession. The images forming a pair are then related to each other through feature tracking, which allows to discard features for which the relative intensity difference is small and which are thus not believed to belong to a reflective vest. Next, the local neighbourhood of the remaining features is further analysed. First, a Random Forest classifier is used to discriminate between features caused by a reflective vest and features caused by some other reflective materials. Second, the distance between the camera and the vest features is estimated using a Random Forest regressor. The proposed system was evaluated in one indoor and two challenging outdoor scenarios. Our results indicate very good classification performance and remarkably accurate distance estimation especially in combination with the SURF descriptor, even under direct exposure to sunlight.

1 Introduction

People detection is an important task in both autonomous machines and human operated vehicles equipped with driver assistant technology. Especially when it comes to applications where machines operate in industrial workspaces shared with humans,

R. Mosberger (✉) · H. Andreasson
AASS Research Centre, School of Science and Technology, Örebro University, Örebro, Sweden
e-mail: rafael.mosberger@oru.se

H. Andreasson
e-mail: henrik.andreasson@oru.se

K. Yoshida and S. Tadokoro (eds.), *Field and Service Robotics*, Springer Tracts in Advanced Robotics 92, DOI: 10.1007/978-3-642-40686-7_10,

it plays a crucial role towards improved safety for the operators. Different sensor modalities have been commonly used in people detection including laser scanners, thermal cameras and vision-based systems. All approaches suffer from drawbacks in safety critical applications. Thermal cameras are expensive and their use depends on the ambient temperature. Laser scanners are also expensive and can fail under extreme conditions such as direct sunshine into the sensor. Vision-based systems offer appealing solutions since they can be inexpensive but require that the ambient illumination is neither too strong nor too weak. Yet, for the application in safety systems for industrial environments, reliable people detection in a variety of different conditions is critical.

In many industrial workplaces such as manufacturing areas, construction sites, warehouses or storage yards the wearing of a reflective safety vest is a legal requirement. In contrast to more general approaches, the work presented in this chapter therefore takes advantage of the enhanced visibility of a person due to the reflective vest to facilitate the detection. Andreasson et al. [1] introduced a people detection system based on a single camera unit which was successfully used to detect humans wearing a reflective vest. The core principle of the detection system is to take two images in short succession, one with and one without IR flash, and to process them as a pair. The processing scheme identifies regions with a significant intensity difference between the two images in order to detect locations where reflective material appears.

The system proposed in this chapter is an extension of the work presented in [1]. The extended system allows not only for detection of people wearing reflective vests but also adds estimation of the 3D position for individual vest features. A machine learning approach is applied in order to estimate the position of a reflective vest based on the description of an image patch extracted from the neighbourhood of the location where the vest was detected in the image.

This chapter is organised as follows. Section 2 briefly discusses related work in the field of vision-based people detection. In Sect. 3, the complete vest detection and position estimation system is described in detail. The section is divided into a part dedicated to the detection of reflective vest features (Sect. 3.1) and a part describing the estimation of a 3D position corresponding to each detected vest feature (Sect. 3.2). In Sect. 4, the performance of the system is evaluated in different environments and conclusions and an outlook on future work are presented in Sect. 5.

2 Related Work

Vision-based people detection for non-stationary environments has been extensively studied for applications in robotic vehicles, (semi-) autonomous cars, driver assistant systems and surveillance. State-of-the-art techniques mainly rely on either the detection of individual body parts or the analysis of templates. Both techniques are commonly used in combination with machine learning techniques. The state-of-the-art method of template based techniques uses the histogram of oriented gradients (HOG) descriptors [2] computed on a dense grid of uniformly spaced cells. The descriptors are fed into a detection system consisting of a binary classifier obtained by supervised learning. Body part based detection systems follow a different approach by representing the body as an ensemble of individual parts pairwise connected with a spring-like link. In [3], the different body parts are represented using a simple appearance model and arranged in a deformable configuration to obtain a pictorial structure which is then matched to the images to be observed.

The performance of vision-based techniques heavily depends on the presence of good visible structures in the images, and thus on a sufficient illumination of the observed scene. Their application is not suitable for dim or completely dark environments. Also, vision-based approaches typically struggle in cases where people have little contrast with the background. For these reasons, existing people detection approaches are not directly applicable in a safety system supposed to operate under challenging conditions, such as rain, snow or direct exposure to sunlight.

The system presented in this chapter focuses on the detection of people wearing a reflective vest using active IR illumination. The detection of retro-reflective material has been successfully applied in motion capture systems where passive markers are used in combination with an array of IR or visible-light LEDs mounted around the lens of one or several cameras in order to detect selected spots on the human body [4]. Yet, to the best of the author's knowledge, there exists no people detection system that makes use of the reflective vest properties in the detection process.

Instead of analysing single images as it is done in most of the related work, our system processes a pair of images one of which is taken with an IR flash and one without. The proposed algorithm exploits the fact that the IR flash is very strongly reflected by the vest reflectors to detect locations in the image where a large intensity difference exists between the two images. Andreasson et al. show in [1] that especially in the mid- and long-range people detection where spatial resolution decreases rapidly in the image, their approach clearly outperforms a state-of-the-art people detection algorithm (Histogram of Oriented Gradient, HOG) applied to a single image.

3 System Description

The reflective vest detection and position estimation system presented in this chapter is described in two parts. Section 3.1 is dedicated to the detection of reflective vest features in the input images while Sect. 3.2 describes the estimation of a 3D position for each detected vest feature. For a schematic overview of the complete system, individual processing steps as well as the data flow in the system see Fig. 1.

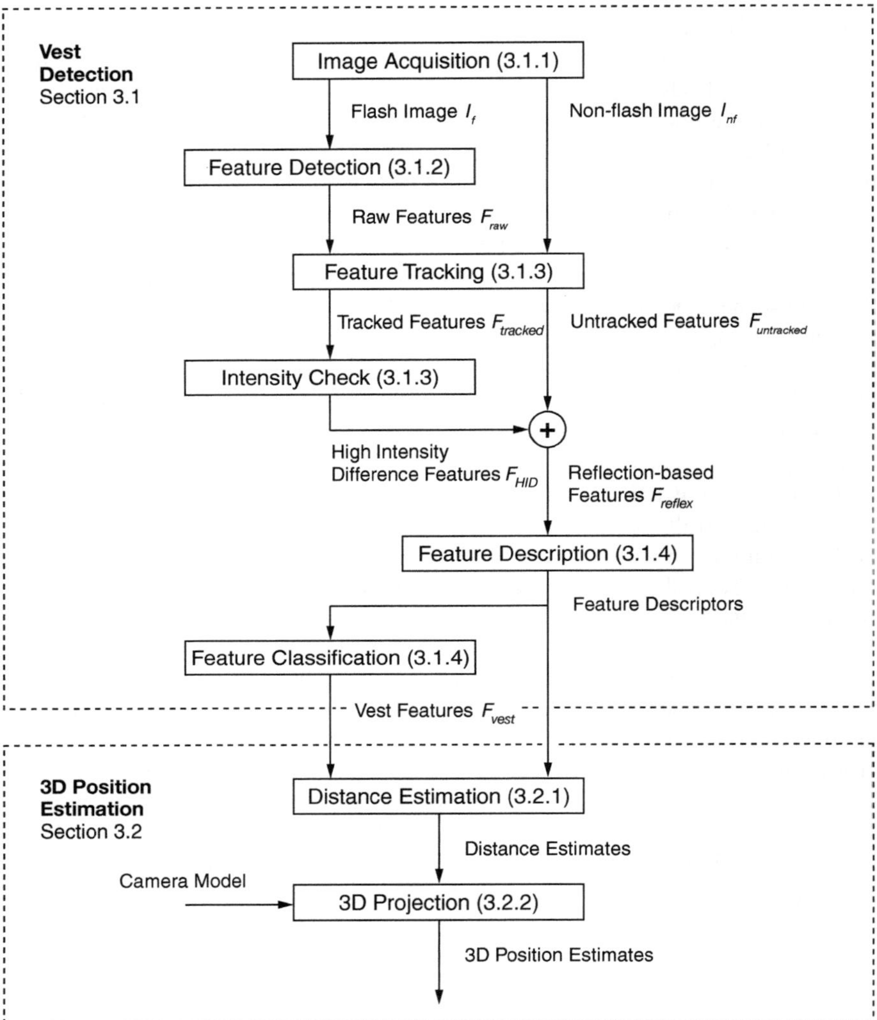

Fig. 1 Overview of the reflective vest detection and position estimation system

3.1 Vest Detection

The upper part of Fig. 1 depicts the detection scheme employed to detect persons wearing reflective vests. The detection system works by comparing two images, one acquired with IR flash, I_f, and one taken without, I_{nf}. A feature detector is used to identify the set F_{raw} of high intensity blob-like interest points in the image I_f. Subsequently, the features detected in I_f are tracked in I_{nf} and, based on the output of the tracker, a subset of features is discarded as not belonging to reflective

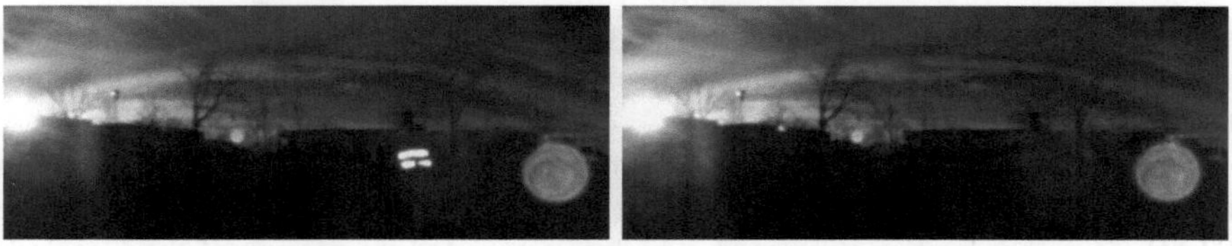

Fig. 2 Example of an image pair taken in short succession. The image on the *left* was taken with IR flash and the image on the *right* without. The images show a panoramic view which is obtained by unwrapping of the raw fish-eye images. The difference in intensity values at locations where the reflective vest appears is clearly visible. The filled *white* circle at the *bottom right* represents a lens artifact originating from direct sunshine into the camera. It may be noted that the overall brightness of the images is very low due to the use of the IR band pass filter in the camera system

material and thus not originating from a reflective vest. Features are discarded if they can be tracked and if the intensity difference between the two images is below a set threshold. This pre-selection process is further described in Sect. 3.1.3. Finally, a binary random forest classifier, trained by supervised learning is used to discriminate vest features from non-vest features.

3.1.1 Hardware and Image Acquisition

The camera unit consists of a standard monochrome CMOS sensor with a resolution of 752×480 pixels and a fish-eye lens with an approximate FOV of 180°. Eight IR LEDs with a wavelength of 850 nm are placed in a ring around the camera. The orientation of the LEDs assures a wide and relatively uniform illumination of the scene in the camera's FOV. A band pass filter with a center wavelength of 852 nm and a full width at half maximum of 10 nm is mounted between the lens and the sensor. The filter corresponds to the dominant IR wavelengths of the IR LEDs.

The image acquisition involves taking a pair of images, one with IR flash and one without. An exemplary pair of panoramic images, obtained from the raw fish-eye images is depicted in Fig. 2. The time increment t_a between the acquisition of the two images is kept as short as possible in order to minimize the difference between the two images due to changes in viewpoint and changes in the scene. The raw fish-eye images are unwrapped to represent a panoramic view containing the area of interest for the reflective vest detection. The unwrapped flash image I_f and the unwrapped non-flash image I_{nf} form the image pair $\boldsymbol{I} = (I_f, I_{nf})$ on which all the post-processing is based.

3.1.2 Feature Detection

The reflection of the IR light by the reflectors of a vest results in high intensity blob-like regions at locations where the vest appears in the image I_f. Shape and size of the high intensity regions depend heavily on the distance between the camera unit and the person wearing the vest as well as on the body pose of the person. Especially at short distances, the reflective markers of a vest appear as elongated regions rather than as circular blobs.

The first step in the vest detection process consists in identifying locations in the image I_f where such high intensity regions appear. It was shown in [1] that the STAR algorithm by Konolige et al. which is a speeded-up version of the CenSurE feature detector [5] yields good results. The STAR detector produces a set of raw features, named F_{raw}, in which every feature is described by the image coordinate pair $\boldsymbol{u} = (u, v)$ indicating the location in the image I_f where it was detected.

An exemplary result of the feature detection is given by the ensemble of crosses in the upper image of Fig. 3. The example illustrates that under the influence of the IR illumination from the flash and the sun, the detected feature set F_{raw} includes many features that do not originate from a reflective vest. Also it is worth mentioning that due to the STAR algorithm's sensitivity to circular shapes, one reflective vest marker can be detected more than once (cf. Fig. 3), especially when its shape appears elongated.

3.1.3 Feature Tracking and Intensity Check

The detected features in the set F_{raw} originate either from a reflective vest or from another bright object in the FOV of the camera. As the images I_f and I_{nf} were taken in short succession, the appearance of non-vest features is assumed to change little from one image to another. In contrast, this *brightness constancy* assumption is not valid for features originating from a reflective vest since the intensity values in the vicinity of a vest feature differ considerably for the image pair $\boldsymbol{I}$. Based on this property, the first processing step to eliminate non-vest features consists in tracking raw features, detected in image I_f, in the corresponding image taken without IR flash, I_{nf}, and perform a check on the intensity difference between the image patches surrounding the detected and the tracked feature locations.

The tracking of the features is performed using a pyramidal implementation of the Lucas-Kanade feature tracker [6]. The tracker is based on the assumption that the temporal increment between two consecutive images is small enough such that the location of a feature changes little from one image to another. As the images I_f and I_{nf} are taken in very short succession, this is the case. Furthermore, the tracker assumes brightness constancy.

In the case of vest features, the tracker is typically unable to find any suitable match in the image I_{nf} because the brightness constancy assumption does not hold true. Thus, features that failed to be tracked are added to the set of reflection based features F_{reflex}. It is worth noting that in contrast to the standard application of a feature tracker, we are not only interested in features that can be successfully tracked. We specifically identify features that cannot be tracked as possible vest features.

In the case of non-vest features, where the brightness constancy assumption holds true, the tracker typically finds the corresponding locationin the non-flash image I_{nf}

Fig. 3 The figure shows an exemplary result of the vest detection process. Blob-like features are detected in the image I_f taken with IR flash (*above*). Detected features are represented by the ensemble of crosses in I_f. The detected features are then tracked in the image I_{nf} taken without IR flash (*below*). The detection area in I_f is restricted to the *white* bounding box to allow features to be tracked even in the case of quick rotational movements. Successfully tracked features are marked as *white* crosses in image I_f and the tracked locations are indicated by *white* 0 crosses in image I_{nf}. All tracked features in the above example show a very low intensity difference and are therefore not considered as vest candidates. Features that failed to be tracked include detections on the reflective vest as well as on the metallic surface of the car standing *right* in *front* of the camera. All the untracked features are considered as vest candidates and classified by the random forest model. A *black* square is drawn around features that are finally classified as vest features

and an intensity difference check within a square window of size w_i surrounding the feature can be performed. If the average difference between the pixels in the window is above a threshold t_i the feature is declared reflection based (and exceptionally tracked) and added to the feature set F_{reflex}. Otherwise, this is not the case and the feature is considered to originate from an area without reflective material and will not be further processed.

3.1.4 Feature Description and Classification

The set F_{reflex} typically contains features that originate from the reflection of the IR light on a reflective material. Yet, other reflective objects than the reflective vest markers can appear in the scene, such as metallic surfaces, windows, mirrors or reflective floor and wall marking tape. An additional processing step therefore aims at classifying the features collected in the set F_{reflex} into a set of vest features F_{vest} and a set of non-vest features $F_{nonvest}$.

The additional processing step is also motivated by the fact that extreme camera movements that cause strong motion blur can result in a high number of detected features that can not be successfully tracked in the image I_{nf}. In such cases, the set F_{reflex}, supposed to contain mainly features representing reflective materials, would contain many other undesired features.

The classification is not performed by directly evaluating the raw intensity values of the image. Instead, a local image descriptor is extracted from the neighborhood of each detected feature in F_{reflex} and serves as input for the classifier. The descriptor is extracted from a square image patch of size w_d centered at the location where a reflection based feature was detected in image I_f. Requirements for an appropriate descriptor include robustness to illumination changes, motion blur and noise as well as computational efficiency of the extraction process. State-of-the-art feature descriptors that were found appropriate include SURF [7] and BRIEF [8].

A random forest classifier [9] is then applied to classify all features in the set F_{reflex}. The forest is an ensemble of n_t binary decision trees with a randomized selection of descriptor variables on which a tree splits. Thus, the classification of a feature descriptor with the random forest classifier provides n_t individual votes, one per each tree in the forest. The probability that a reflection based feature represents a reflective vest can be inferred by dividing the number of trees voting for a reflective vest by the total number of trees n_t in the forest. In a supervised learning task, the random forest classifier is trained on a set of descriptors that are manually labelled with a tag indicating whether the descriptor corresponds to a vest feature or not.

3.2 3D Position Estimation

The lower part of Fig. 1 depicts the two steps the system performs in order to estimate a 3D position for features that were detected in I_f and classified as belonging to the set of vest features F_{vest}. First, the system estimates the distance of a vest feature with a machine learning approach before exploiting the intrinsic camera model together with the distance estimate to produce a 3D position estimate.

3.2.1 Distance Estimation

The same local image descriptors used for the feature classification described in Sec. 3.1.4 are employed to estimate the distance of a reflective vest feature based on machine learning. Using supervised learning we train a random forest regressor on a set of descriptors that are labelled with the ground-truth distance between the camera and the reflective vest that caused the appearance of the corresponding vest feature. The trained model is then applied to obtain a distance estimate d for descriptors of unseen vest features.

A strong emphasis has to be made on the rotation and scale invariance of the feature descriptors adopted in the underlying system. The size of a reflective vest pattern in the image I_f decreases with increasing distance between the vest and the camera. As we aim to estimate the distance to the vest based on the local image descriptor, scale invariance is clearly undesired because it would make it impossible for the regressor to consider the size of the patch. On the other hand, rotation invariance would be beneficial for cases where the regressor has to estimate the distance of an untrained vest feature which is just a rotated version of a trained feature.

In the case of BRIEF the descriptor is neither scale nor rotation invariant but tolerates small amounts of rotation [8]. In contrast, the SURF descriptor is designed to be rotation and scale invariant but this property only holds true if the SURF descriptor is used in combination with the corresponding SURF feature detector which provides a scale and an orientation for every detected feature. The STAR feature detector used in our application does not provide any orientation for the detected features. Thus, we extract the descriptors within a window of fixed size w_d and constant orientation of zero degrees and obtain BRIEF and SURF descriptors that are neither scale nor rotation invariant.

3.2.2 3D Projection

The final step aims at estimating the 3D position relative to the camera for all features in the set F_{vest}. Therefore, an intrinsic camera model of the camera system is obtained by a calibration method dedicated to omni-directional cameras [10]. The method assumes that the image projection function can be described by a Taylor series expansion for which the coefficients are estimated in the calibration process.

Using the obtained camera model and given the image coordinates $\boldsymbol{u} = (u, v)$ of the location at which a feature in F_{vest} was detected, a ray in 3D space can be inferred on which the object that caused the appearance of the feature in the image must lie on. By further taking into account the distance d that was estimated for the corresponding feature, a 3D point on the ray can be located leading to the final position estimate $\boldsymbol{x} = (x, y, z)$ in the coordinate system fixed to the camera.

Table 1 Test scenarios with number of acquired image pairs featured in the system evaluation

Scenario	Environment	Image pairs *I*
1	Indoors, warehouse-like environment	400
2	Outdoors, car parking area, sunny weather conditions	380
3	Outdoors, car parking area, direct sunshine into camera	100

Table 2 Values of the various system parameters used for the evaluation setup

Parameter	Description	Value
f_a	Image pair acquisition rate	~14 Hz
t_a	Time delay between the acquisition of I_f and I_{nf}	~35 ms
w/h	Width and height of the unwrapped input images I_f and I_{nf}	600 x 240 Pixel
b	Feature detection window border size	40 Pixel
w_i	Window size for the intensity difference check	5 Pixel
t_i	Threshold for the intensity difference check	30.0
w_d	Window size for the descriptor extraction	24 Pixel
n_t	Number of trees in the random forest classifier/regressor	20

4 Experiments

The reflective vest detection and position estimation system has been evaluated in three different test scenarios as listed in Table 1. A sensor unit consisting of the camera system and a 2D laser range scanner (SICK LMS-200) both fixed to a solid mechanical frame was used for the data acquisition. An extrinsic calibration was carried out to obtain the position and orientation of the laser range scanner relative to the camera [11]. The sensor unit is mounted at a height of approximately 1.5 m on a mobile platform with four hard rubber wheels.

Several training and validation data sets were acquired for each scenario by simultaneously recording the raw camera images and the 2D laser readings. During the acquisition of all data sets the mobile platform was moving at a speed of approximately 0.5 m/s. One data set per scenario was held back for evaluation purposes while the remaining sets served as training data. Table 2 summarizes the values of the different system parameters used in the evaluation setup.

All the acquired data sets were preprocessed to detect the set of raw features F_{raw} and to extract the corresponding local image descriptors. A BRIEF descriptor of 256 binary variables and a standard SURF descriptor of 64 floating point variables was extracted for every feature. A ground-truth label was manually assigned to each descriptor indicating whether it corresponds to a vest feature (label 1) or not (label 0). Furthermore, the ground-truth distance between the camera and the person wearing the vest was extracted from the laser readings and assigned to the descriptors.

A supervised learning process is applied to obtain the models of the feature classifier and the distance regressor. We trained a random forest on 45 k extracted image descriptors and the corresponding labels to obtain the classifier described in

Table 3 Result of the feature detection process

Scenario	Total detected features	Average features per image I_f	Vest detection rate (%)
1	1,612	4.03	97.50
2	1,540	4.05	97.84
3	4,953	49.53	88.37

Sect. 3.1.4. Likewise, we trained a random forest on 30 k image descriptors labelled as vest features (label 1) and the corresponding ground-truth distance to the person to obtain the regressor described in Sect. 3.2.1.

The evaluation was then performed by processing the validation data set of each scenario and comparing the obtained results with the ground-truth labels assigned during preprocessing. The main processing steps of the system according to Fig. 1 are evaluated individually.

Feature Detection

To evaluate its performance, the feature detector (Sect. 3.1.2) is applied on each image I_f in the validation data sets resulting in a set of raw features F_{raw}. If a reflective vest is identified with at least one feature in F_{raw} the detection process for image I_f is declared successful. Images in which no reflective vest appears are not considered in the evaluation. The vest detection rate is defined as the ratio between images in which the vest is successfully detected and the number of images showing a vest.

Table 3 shows the result of the feature detection evaluation. The data shows that in nearly all images of scenarios 1 and 2 the detector reliably detects at least one raw feature per reflective vest. In scenario 3 the camera faces the sun resulting in lens artifacts frequently appearing in the images. The detector occasionally fails to detect features intersecting with the lens artifacts which leads to a detection rate decreased by approximately 10 %.

Feature Classification

In a second step we evaluate the system's ability to correctly classify a set of detected features F_{raw} into a set of vest features F_{vest} and a set of non-vest features. The evaluation assesses the performance of several processing steps as a group, namely the feature tracking and intensity check (Sect. 3.1.3) as well as the feature description and classification (Sect. 3.1.4). Every set of raw features F_{raw} detected in the series of images I_f is processed to obtain a corresponding set of predicted vest features F_{vest}. The set of predicted non-vest features is defined as $F_{nonvest} = F_{raw} \setminus F_{vest}$. The result of the binary classification into vest and non-vest features is then compared to the ground-truth label manually assigned during preprocessing.

Table 4 The matrices of the binary feature classification into vest- and non-vest features for the case where **a** the classification is based only on the feature tracking and intensity check ($F_{vest} = F_{reflex}$), **b** the feature set F_{vest} is obtained by further classification based on the BRIEF descriptor and **c** by further classification based on the SURF descriptor

Scenario 1

		Predicted			Predicted			Predicted	
		0	1		0	1		0	1
Actual	0	n/a	n/a	0	n/a	n/a	0	n/a	n/a
	1	0.06%	99.94%	1	16.00%	84.00%	1	4.28%	95.72%

Scenario 2

		Predicted			Predicted			Predicted	
		0	1		0	1		0	1
Actual	0	79.07%	20.93%	0	97.94%	2.06%	0	95.09%	4.91%
	1	1.30%	98.70%	1	15.32%	84.68%	1	10.94%	89.06%

Scenario 3

		Predicted			Predicted			Predicted	
		0	1		0	1		0	1
Actual	0	97.45%	2.46%	0	99.29%	0.71%	0	99.51%	0.49%
	1	15.92%	84.08%	1	34.39%	65.61%	1	30.86%	69.14%
		(a)			(b)			(c)	

Table 4 shows the results of the evaluation in form of confusion matrices. Scenario 1 contains images acquired indoors where the only IR source was the flash of the camera system and where no other reflective object than the vest appeared. Thus, true negative and false positive rates are not defined. The results illustrate the effect of the feature description and classification described in Sect. 3.1.4. The false positive rate is decreased by rejecting features of other reflective material than reflective vests. By doing so, the classifier also erroneously discards some actual vest features that look unfamiliar, resulting in an increased false negative rate. Classification based on the SURF descriptor yielded the best trade-off between the two effects.

Distance and Position Estimation

The trained model of the random forest regressor (Sect. 3.2.1) is used to obtain a distance estimate for every predicted vest feature in F_{vest}. The distance estimate is used together with the feature coordinates $\boldsymbol{u} = (u, v)$ and the intrinsic camera model to obtain a corresponding 3D position estimate according to Sect. 3.2.

Figure 4 shows the results of the distance estimation. While the estimations based on the SURF descriptor show a widely stable accuracy over the whole distance range, the BRIEF descriptor only allows a reliable estimation for short range distances up to 7 m. The plots also report sporadic but large outliers indicating a distance estimation

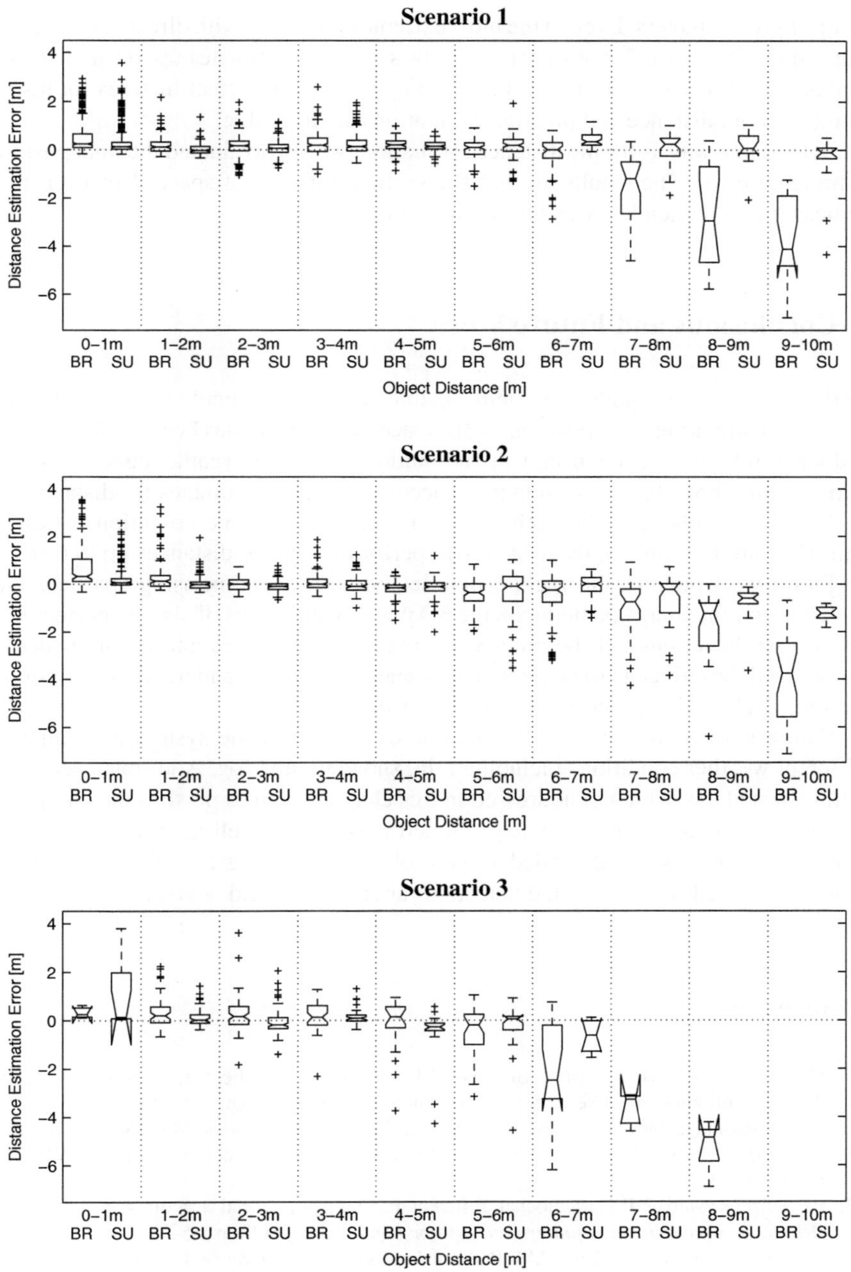

Fig. 4 The box plots show the distance estimation error for the scenarios 1–3. The indications BR (BRIEF) and SU (SURF) specify the image descriptor on which the estimation is based. Missing plots indicate that the vest detection failed and no distance estimation could be performed

error of several meters. Even in the most extreme conditions with direct sunshine into the camera (Scenario 3) the system still gives accurate estimates up to 6 m distance. Under the influence of strong sunlight, the system fails to detect features for higher ranges and no distance and position estimations are available.

In the same way as for the distance estimation, we also evaluated the final position estimation error. The results are not shown here for lack of space. However, they indicate the same tendency as the results shown in Fig. 4.

5 Conclusions and Future Work

In this chapter we presented a system capable of detecting people wearing reflective vests and estimating their position in 3D space. The system has been evaluated in an indoor warehouse-like environment and outdoors in sunny weather conditions. The experiments show that the system gives accurate distance estimates for distances up to 10 m, with only sporadic outliers. Even under the extreme conditions of direct sunshine into the camera, the system still performs well for distances up to 6 m.

Future work includes the tracking of reflective vests over time using a particle filter which is continuously updated with the 3D position estimates of single vest features. Thus, vest detections will be maintained over several frames and the influence of outliers will be reduced. To allow for simultaneous detection and tracking of multiple persons, a clustering process will also be introduced.

Future work also includes a systematic evaluation of the system in a range of different weather conditions including rain, snowfall, and fog. Additional scenarios will be tested that were not addressed in this chapter, such as persons that are partly occluded or lying on the floor (e.g. fainted persons) as well as different types of camera movements. An extended version of the camera system will include more powerful IR LEDs to extend the detection range to 20 m and beyond.

References

1. H. Andreasson, A. Bouguerra, T. Stoyanov, M. Magnusson, A. Lilienthal, Vision-based people detection utilizing reflective vests for autonomous transportation applications. IROS workshop on metrics and methodologies for autonomous robot teams in logistics (MMART-LOG) (2011)
2. N. Dalal, B. Triggs, Histograms of oriented gradients for human detection. In: in CVPR, pp. 886–893 (2005)
3. P.F. Felzenszwalb, D.P. Huttenlocher, Efficient matching of pictorial structures, in *Proceedings IEEE computer vision and pattern recognition conference* (2000), p. 66–73
4. J. Chung, N. Kim, G.J. Kim, C.M. Park, in *International conference on Virtual Systems and MultilMedia*. Postrack: A low cost real-time motion tracking system for vr application (IEEE, 2001), pp. 383–392
5. M. Agrawal, K. Konolige, M.R. Blas, *Censure: Center surround extremas for realtime feature detection and matching*, ed. by D.A. Forsyth, P.H.S. Torr, A. Zisserman ECCV (4), Lecture Notes in Computer Science, vol. 5305 (Springer, Berlin, 2008), p. 102–115

6. J.Y. Bouguet, Pyramidal implementation of the lucas kanade feature tracker description of the algorithm (2000)
7. H. Bay, A. Ess, T. Tuytelaars, L.V. Gool, Surf: Speeded up robust features. Comput. Vis. Image Underst. (CVIU) **110**, 346–359 (2008)
8. M. Calonder, V. Lepetit, C. Strecha, P. Fua, *Brief: Binary robust independent elementary features*, ed. by K. Daniilidis, P. Maragos, N. Paragios ECCV (4), Lecture Notes in Computer Science, vol. 6314 (Springer, Berlin, 2010) p. 778–792
9. L. Breiman, Random forests. Mach Learn **45**(1), 5–32 (2001)
10. D. Scaramuzza, A. Martinelli, R. Siegwart, A flexible technique for accurate omnidirectional camera calibration and structure from motion, in *Proceedings of the IEEE international conference on computer vision systems* (*ICVS*) (2006)
11. Q. Zhang, Extrinsic calibration of a camera and laser range finder, in IEEE International Conference on intelligent robots and systems (IROS), (2004), p. 2301–2306

Impression of Android for Communication Support in Hospitals and Elderly Facilities

Yoshio Matsumoto, Masahiro Yoshikawa, Yujin Wakita, Masahiko Sumitani, Masutomo Miyao and Hiroshi Ishiguro

Abstract In this paper, we report the impression of our android robot obtained in experiments and reactions of people to the android in demonstrations in medical and nursing fields. Our newly developed android robot Actroid-F is utilized for this research. Since the total system for controlling the android is light and compact, it is easy to install in medical and nursing fields. In order to survey the impression of android in the real-world, we conducted a preliminary experiment utilizing the android in a various environments. As a result, it was revealed that most of the subjects have no aversion to the presence of the android, and elderly people tend to have positive impressions of the android. Furthermore we demonstrated the robot in facilities for the elderly, and a school for children with developmental disorders. Findings from the demonstrations together with ideas for the potential application of the android based on the findings are presented.

1 Introduction

In recent years, various kinds of practical assistive robots have been developed for medical and nursing fields, such as exoskeleton power assist robots [1], assistive manipulators [2], robotic beds [3], and intelligent wheelchairs [4]. The main purpose

Y. Matsumoto (✉) · M. Yoshikawa · Y. Wakita
Intelligent Systems Institute, National Institute of Advanced Industrial Science and Technology,Tsukuba, Japan
e-mail: yoshio.matsumoto@aist.go.jp

M. Sumitani
University of Tokyo Hospital, The University of Tokyo, Tokyo, Japan

M. Miyao
Department of Psychosocial Medicine, National Center for Child Health and Development, Tokyo, Japan

H. Ishiguro
Graduate School of Engineering Science, Osaka University, Osaka, Japan

K. Yoshida and S. Tadokoro (eds.), *Field and Service Robotics*,
Springer Tracts in Advanced Robotics 92, DOI: 10.1007/978-3-642-40686-7_11,

of these robots is the direct and physical assistance for either disabled people or their helper. In addition, robots which support psychologically in medical and nursing field have been developed. Paro [5] is a robot which looks like a seal puppy. It gives animal therapy effects to people in environments such as hospitals and extended care facilities where it is difficult to keep living animals. It responds as if it is alive, moving its head and legs, making sounds, and showing preferred behavior through interaction with people. Keepon [6] can express emotion and attention with gestures through its simple yellow puffballs body and eyes. It is used in a therapy for autistic children.

The psychological support robots mentioned above have simple and exaggerative appearances, by extracting essences of shape and function from real animals or human. On the contrary, android robots which have closely resembling appearances with human are developed in last decade. The utilization of the android for psychological support in the medical and nursing fields has just started recently. In Osaka University, an android named ReplieeQ2 was introduced into a medical examination room as a bystander, and synchronized its nodding and smiling motion with patients during the medical treatment. The experimental result revealed that the presence of the bystander android gave positive effects on impressions of a doctor and satisfaction level of medical treatment [7]. This fact suggests that the android robots have a potential to provide new service to medical and nursing fields.

In this paper, we present our newly developed android robot platform, and a result of preliminary field experiment to survey the impressions of people to the android in the real-world. In Sect. 2, the android platform is to be described, and from Sects. 3 to 5, the detail of the result of preliminary experiments and demonstrations are described. Section 6 presents the discussion and Sect. 7 concludes the paper.

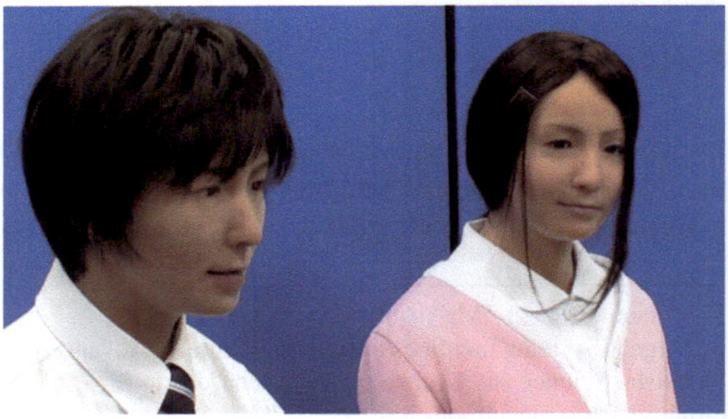

Fig. 1 Actroid-F (male and female versions)

Table 1 Arrangement of pneumatic actuators

Face	7	Eyebrows (up/down)
		Eyebrows (furrow)
		Eyelids (open/close)
		Eyeballs (pan)
		Eyeballs (tilt)
		Mouth (open/close)
		Corners of mouth (up/down)
Neck	3	Head turn (pan)
		Head lean to left (back/front)
		Head lean to right (back/front)
Body	2	Breath
		Waist (bow)

2 Android Platform

2.1 Android Robot Actroid-F

Actroid-F (Fig. 1) is an android robot manufactured by Kokoro Company Ltd. Its appearance highly resembles with human. The face of the android is made of soft silicon rubber by taking a copy of a real human face. The male version, a newly developed android has exactly the same hardware including the skin as the female version. The only differences are the wig, clothes, and makeup.

It is designed to conduct field experiments fields, and has following features:

- It has 12 degrees of freedom in upper body.
- Actuators are driven by a small air compressor.
- It requires standard 100 V power supply and can be connected to household wall socket.
- It has no extra control box with valves. They are installed inside the body.
- Low cost (approx. USD 100,000).

The limbs (i.e. arms and legs) are not movable. All of the degrees of freedom are driven by pneumatic cylinders with air servoing, thus the air compressor is necessary for controlling. The arrangement of pneumatic actuators is shown in Table 1.The face has only seven degrees of freedoms, however as the arrangement was carefully designed to achieve natural communication in medical and nursing fields. As a result, Actroid-F can show more natural facial expressions especially for smile (Fig. 2), which is the most important expression in such situations. Except for the face, the head can pan (shaking motion), tilt (nodding motion), and incline to one side. Furthermore, it has actuators for bending in the waist and for breathing in the chest. The use of the air actuators achieves silent and robust motion without heat problems, and no maintenance is necessary for a year. It also enables the android to have physical compliance and silent motion which are also quite important for field experiments.

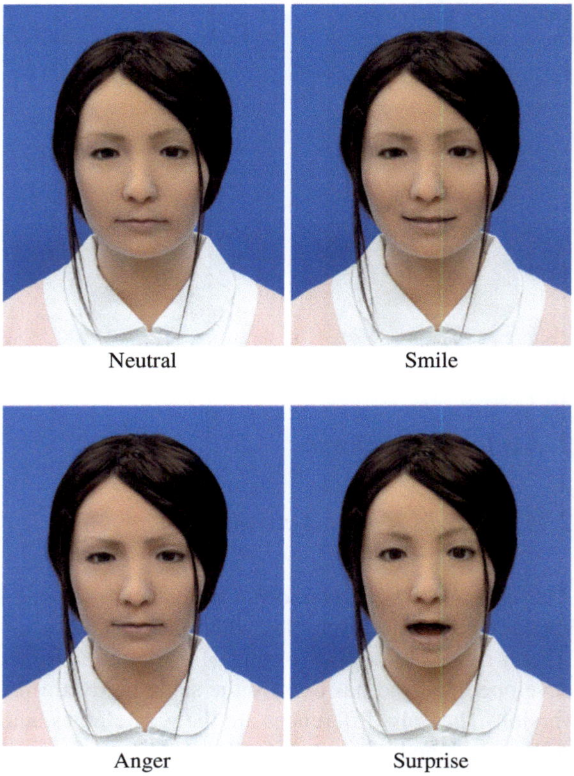

Fig. 2 Facial expression of Actroid-F

2.2 Hardware Configuration

The system of the android is composed as follows: An air compressor and the android are connected using a 20 m air tube. Twelve valves in the body of the android control the quantity of air flow, which drives actuators (i.e. air cylinders). The android is connected to a notebook PC via USB cable, and angles of each joint can be commanded from this PC. A webcam which captures a facial image of a target person for face recognition is also connected to the notebook PC. A microphone is connected to the notebook PC via a voice changer. The voice of the android which is generated from speech synthesizer or the voice changer is output from a speaker behind the android.

2.3 Software Implementation

The graphical interface for the operator has push-buttons which corresponds to pre-determined motions such as "Nodding", "Smile", "Anger", "Surprise", and

"Bowing". All of the 12 degrees of freedom can also be manually controlled by sliders in the interface. In addition, all of the motion generated by the android can be recorded automatically, and replayed afterwards.

Natural human motion such as blinking, gaze movements, head movements, and breathing were implemented on the android. In addition, unconscious head movements of human were recorded beforehand, and were replayed during idle state. Motions of the android can be operated by the motions of the target person using the webcam and faceAPI [8] that recognizes human faces. Roll-pitch-yaw of the head, opening and closing of the mouth, motions of the mouth corner, motions of the eyebrows can follow.

As for speech functions, the system enables the operator to make an utterance using a speech synthesis (text to speech) function. The speech synthesis function is implemented using Speech API (SAPI) of Windows Operating System, and the lip motion synchronizes with the speech.

3 Communication with Patients

3.1 Experimental Setup

We conducted a clinical bystander experiment with android in an examination room of the pain clinic at the outpatient department in The University of Tokyo Hospital for a month and a half. In the clinical examination room, the android was sitting next to the doctor, directing its face to the patient position to make eye-contact with the patient. The air compressor was installed in a distant room in order to prevent disturbing medical examination by noise. Webcams were put on the desk of the doctor, and were connected with the PC for controlling the android. The PC was installed in the neighboring room, and an operator was controlling the android while observing the examination through the webcams (see Fig. 3).

The experimental protocol is as follows: After taking a consent from a subject, the subject enters the clinical examination room. Then the operator makes the android bow in order to attract his/her attention once, after that a normal clinical consultation starts, which takes 10–15 min on average. A single medical doctor performs the clinical consultation throughout the experiment to standardize the experimental condition. Unconscious motion such as blinking, gaze movements, head movements, and breathing are automatically executed during the experiment. The nodding and smiling motion of the android is controlled to synchronize with either the doctor or the patient for each patient, which is not known by the doctor. After the clinical consultation, the subject answers a written questionnaire about their subjective evaluations of the android.

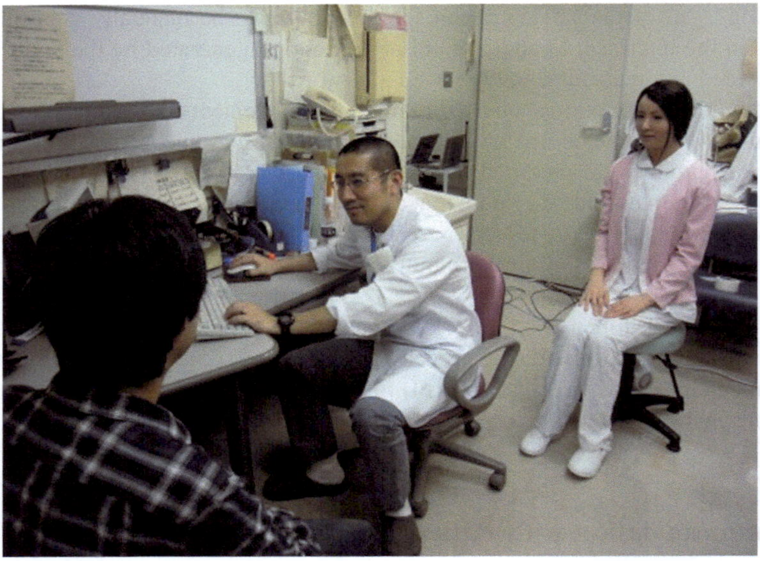

Fig. 3 Experimental scene of bystander experiment in hospital

Table 2 Evaluation about presence of android in examination room

	Under 65	Over 65	Total
Prefer presence	5	18	23
	(7.1)	(25.7)	(32.9)
No difference	18	16	34
	(25.7)	(22.9)	(48.6)
Prefer absence	3	1	4
	(4.3)	(1.4)	(5.7)
N.A	5	4	9
	(7.1)	(5.7)	(12.9)
Total	31	39	70
	(44.3)	(55.7)	(100.0)

Numbers in parentheses are relative frequency (%)

3.2 Experimental Results

Subjects of 70 patients (31 males and 39 females) took part in the experiment. The mean age was 62.1 (±15.4) years. Most of the patients were at their first medical visit to the doctor.

The result of the questionnaire about preference for the presence of the android is shown in Table 2. Approximately half of the patients answered that they did not mind the presence of the android, which was similar to the result shown in Ref. [7]. Since all of the patients visited doctors at the university hospital to take an expert medical

Fig. 4 Comparison of impression on the android between under 65 and over 65 years old

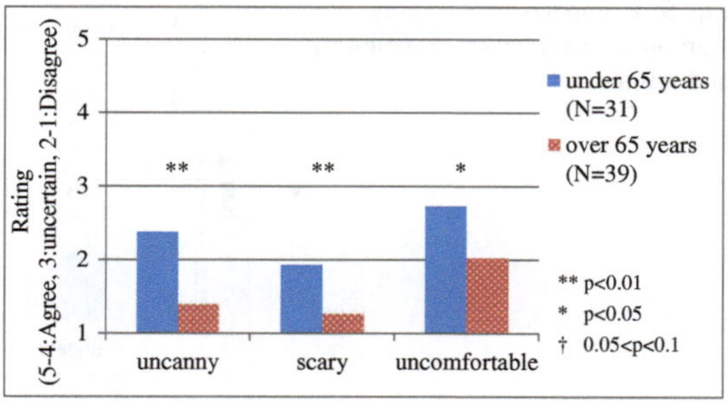

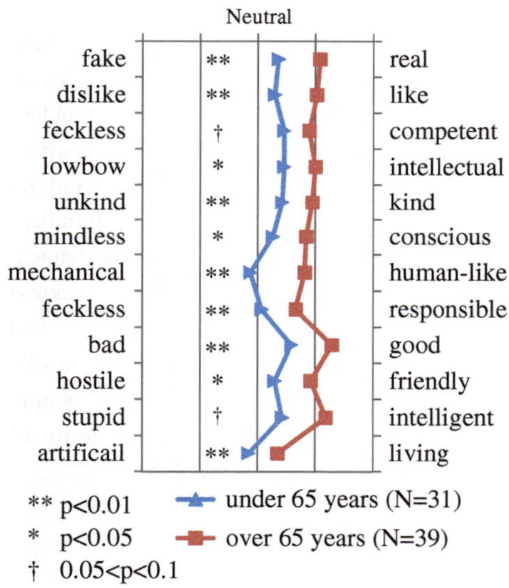

consultation, this is a natural reaction. However one-third of the patients answered that they preferred the presence of the android. The major reasons for the preference were "It was comfortable", "It made me relax", and "I got interested in it". Only 5.7 % answered that they preferred the absence of the android, and the major reason for that was "the android had no relations with the clinical consultation". The total ratio of patients who preferred presence and who did not mind was 81.5 %, thus it was shown that most of the subjects have no aversion to the presence of the android.

Figure 4 shows the relation of impression of the android with the age factor. Patients over 65 years old (N = 39) showed statistically better impression on the android compared with the patients under 65 years old (N = 31) as to the most of adjective words assessed by t-test. Actually, several patients over 65 years old explicitly answered that they enjoyed the experiments and they wanted to talk with the robot more.

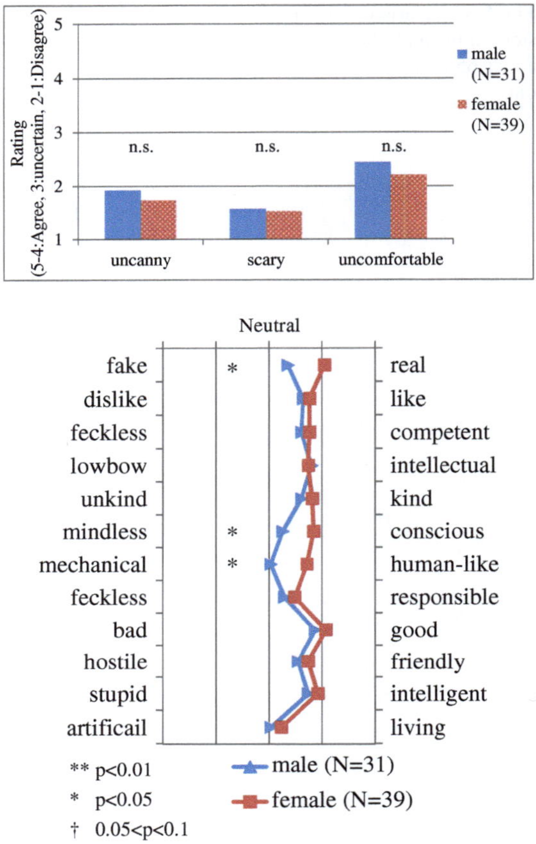

Fig. 5 Comparison of impression on android between male and female

Figure 5 shows the relation of impression of the android with the gender factor. Female patients (N = 39) showed statistically better impression on the android compared with the male patients (N = 31) as to three adjective words "natural", "conscious", and "human-like" assessed by t-test. Female patients also showed better impression as to all other adjective words on average but not statistically.

Before conducting the experiment, we supposed that younger subjects would have positive impressions on the android, since they are familiar with "robots" from comics and movies, and that male subjects would have positive impressions of the android, because the android utilized in this experiment was a young female version. However the experimental results indicated totally opposite tendency. Actually several elderly subjects gave positive comments such as "I enjoyed the experiment and would like to talk with the android more", and some elderly female subject answered "I'm glad to see beautiful thing" and "I prefer a female android than a male android".

4 Communication with Elderly People

4.1 Experimental Setup

The experimental results shown above indicated that most of the subjects have no aversion to the presence of the android in the clinical examination room and elderly subjects had positive impressions of the android. This fact is important to carry on demonstration experiments in the hospital where many of users are elderly people.

We then took the android to facilities for the elderly in order to investigate the effect of the android in the nursing field. Figure 6 shows our androids (both female and male versions) communicating with elderly people in a nursing facility for the elderly. The androids played a role as a photographic model in the "photographic therapy", where elderly people take photographs of human and other subjects, mount them on cardboards, and decorate them by adding words and paper ornaments. Approximately half of the elderly people there suffer from dementia, and the photographic therapy is one of the activities being conducted for preventing dementia.

The androids were tele-operated to show facial expressions such as smiling, and to utter synthesized voice by operators sitting about $5 \sim 10$ m away from the robot. Answers to the frequently asked questions were previously assigned to buttons in the graphical user interface for the operator. Arbitrary sentences can also be uttered by typing it via keyboard utilizing the interface. Captured images from tiny cameras built in the eyeballs of the androids were displayed on the monitor, which were utilized to establish eye contacts with people by the operator.

4.2 Experimental Results

Each android interacted with approximately 50 people. We have not conducted detailed analysis for the experimental results, but a number of interesting acceptance behaviors of the elderly people reacting to the android were observed as follows:

- There were no person who reacted to feel uncanny, scary, or uncomfortable,
- Most of the people enjoyed talking and interacting with the android,
- Many people, especially ladies, were eager to touch and shake hands with the android,
- Once they touch the hand of the android, they seemed to felt closer to the android,
- They seemed to keep interested in the android about half hour or one hour in general,
- A gentleman who usually cannot interact with the staffs of the facility stayed in front of the android for a long time and tried to communicate with it,
- A lady dressed up and came to see the android,
- A lady said that she was more interested in the robotic researchers than the android.

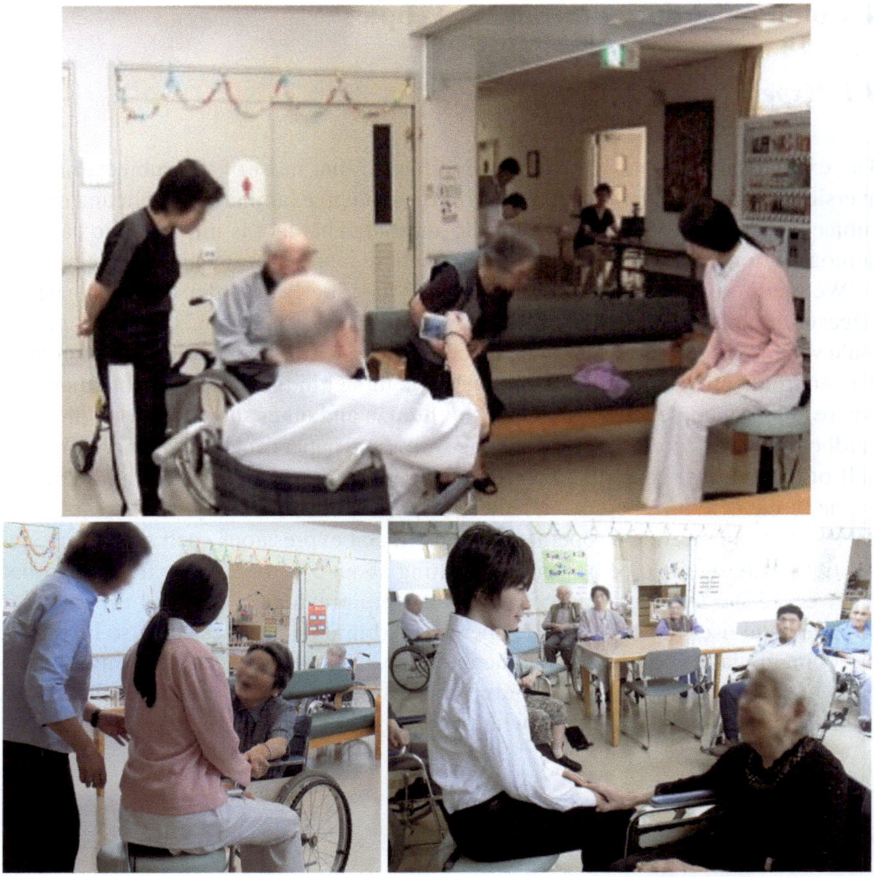

Fig. 6 Experiment at a nursing facility for the elderly

These are very preliminary results of the experiments. We are going perform more experiments to analyze the reactions of elderly people in more detail.

5 Communication with Children

We also took the android to robotic events, where many people of all ages communicated with it. In this section, the reaction of the children to the android is to be described.

Children in elementary age and junior high school age usually regarded the android as a "curious" object, and get close to the robot with touching it. However, as far as we observed, the touching behavior looked not so familiar but rather guarded

Fig. 7 Communication with children

with some caution as shown in the upper figure in Fig. 7. The toughing behavior was totally different from that of elderly people shown in the previous section.

Children in kindergarten age (i.e. from 2 to 5 years old) did not show as curious expression as older children. Many of them even refused to get closer to the robot as shown in the lower figure in Fig. 7. However children in this age are often shy or afraid of strangers in general, and it is still not clear how they regard the android robot.

We also gave a demonstration of the android at a school for children with developmental disorders, especially with Autism and Asperger's syndrome. About 70 students from junior high school to university age joined the demonstration together with five teachers in a large classroom as shown in the upper figure in Fig. 8. They first observed the motion and utterances of the tele-operated android, and then some students voluntarily tried to communicate with the android as shown in the lower figure in Fig. 8. Some students also tried to operate the android by the terminal.

Fig. 8 Demonstration at a school for students with developmental disorder

The response of the children with developmental disorders to the android was mostly positive. They understood that was a tele-operated robot, and most of them commented that they liked the android. Only a few students commented that they feel some fears against the robot, especially looking at the eyes of it. The difficulty of making eye-contact is a typical symptom of autistic patients. The most interesting finding was that a student commented they felt easier to talk to the robot than to the real human, and many students agreed to it. The student explained that he felt easier because he thought the robot would not negatively respond to his words. It seemed that they were afraid of negative response of their family or friends due to their experience, and many of them regarded the android as a conversation partner better than the human.

6 Discussion

6.1 Interpretation of Experimental Results

Although the android has highly real appearance, it might give "unpleasant" or "uncanny" impressions to some people. This phenomenon can be explained by "uncanny valley", [9] and the effect may vary depending on the appearance and the motion of the robot. However as far as we experienced through demonstrations, not many people seemed to have such feelings, and we regard it have overcome the uncanny valley to some extent.

There can be two reasons of the positive impression by elderly people. One is that elderly people tend to have less social interaction with others, since they usually do not work over 65, and the ratio of living alone is high. The other is that as they have lived longer, they feel less fears to any events. We will further investigate these hypothesis through the future experiments.

6.2 Potential Application of Android

We aim to utilize our android platorm in various health-care fields such as hospitals and nursing facilities, in which they need communication support. From our experiments and demonstrations, we came up with some ideas as to how to make use of androids in such situation.

The first usage is utilization of the synchronous android to enhance communication between a medical doctor and a patient as a bystander. As previously mentioned, an android may be able to support medical treatment by decreasing anxiety of patients, or increasing trust of a doctor. Actually, the reason we chose the pain clinic as the first experimental field was that nearly half of the clinical treatments given at the pain clinic is "counseling" rather than "treatment". Since many of the patients are suffering from uncontrolled pains due to unknown physical reasons, the doctor behaves like an counselor, and listens to the patient's talk. Then the doctor encourages the patient to activate the daily life and to control the pain. This kind of treatment is called "pychoeducation" and to be a "good listener" is very important. Therefore we expect that our synchronous android will help the doctor and patients to communicate with each other as a bystander empathetically listening to the patient's talk.

The second usage is a robotic agent at home. An autonomous or semi-autonomous android would be able interact with elderly people as a conversation partner to make them relax and comfortable, to supply useful information from the Internet. It is also important to encourage physical activities and social participation to prevent "disuse syndrome". It should be noticed that elderly people have less "uncanny" feeling toward the android than younger people.

The third usage is a tool for understanding diseases and disabilities, and supporting rehabilitation especially for autistic children. As autistic children commented that they felt easier to talk to the android than to a real human, android might be able to be utilized as a conversation partner for training social skills. This usage will also help other people to understand the state of mind of autistic children, if they can honestly talk what they think to the android. Furthermore, it is thought that some symptoms of autism are caused by disorder of mirror neurons in the brain, which play important roles for imitating and understanding intention of other people. Therefore, it might also be effective for autistic children to try to imitate facial expression of the android which has high reproductivity, for assessment of disorder and training of social skills.

7 Conclusion and Future Works

In this paper, we reported the impression of our android developed to achieve psychological communication support in medical and nursing fields. We conducted a preliminary experiments utilizing the android in various field such as in a medical examination room at a university hospital, at nursing facilities for the elderly, and at a school for children with developmental disorder. As a result, it was revealed that most of the subjects have no aversion to the presence of the android. It was also found that elderly people tend to have positive impressions to the android. We also found that children with autism and Asperger's syndrome can communicate with the robot, and many children even felt it easier to communicate with it than with a real human. As future works, we are planning to conduct further experiments for further investigation of the impression. We are also planning to design the methodology to support communication in medical and nursing fields based on the findings.

Acknowledgments This research was supported by Grant-in-Aid for Scientific Research on Innovative Areas "Human-Robot Symbiosis" (KAKEN No.21118002) from the Ministry of Education, Culture, Sports, Science and Technology, Japan.

References

1. Y. Sankai, HAL: Hybrid Assistive Limb Based on Cybernics, in *Robotics Research:13th International Symposium ISRR*, pp. 25–34, 2010
2. EXACT Dynamics, iARM, http://www.exactdynamics.com/, 2011
3. Panasonic, Robotic bed, http://panasonic.co.jp/ism/robot/robot01.html, 2011
4. H. Uustal, J. Minkel, Study of the independence ibot 3000 mobility system: an innovative power mobility device, during use in community environments. Arch. Phys. Med. Rehabil. **85**(12), 3966–3971 (2004)
5. K. Wada, T. Shibata, Robot therapy in a care house—its sociopsychological and physiological effects on the residents, in *IEEE International Conference on Robotics and Automation* (ICRA 2006), pp. 3966–3971, 2006
6. H. Kozima, M.P. Michalowski, C. Nakagawa, Keepon: a playful robot for research, therapy, and entertainment. Int. J. Soc. Robot. **1**(1), 3–18 (2009)

7. E. Takano, Y. Matsumoto, Y.Nakamura, H. Ishiguro, K. Sugamoto, The psychological effects of an android bystander on human-human communication, in *the 8th IEEE-RAS International Conference on Humanoid Robots* (Humanoids 2008), pp. 25–34, 2008
8. Seeing Machines, faceAPI, http://www.seeingmachines.com/, 2011
9. M. Mori, On the Uncanny Valley, in *Humanoid 2005 workshop: Views of the Uncanny Valley*, 2005

Multi-Robot Formation Control via a Real-Time Drawing Interface

Sandro Hauri, Javier Alonso-Mora, Andreas Breitenmoser, Roland Siegwart and Paul Beardsley

Abstract This paper describes a system that takes real-time user input to direct a robot swarm. The user interface is via drawing, and the user can create a single drawing or an animation to be represented by robots. For example, the drawn input could be a stick figure, with the robots automatically adopting a physical configuration to represent the figure. Or the input could be an animation of a walking stick figure, with the robots moving to represent the dynamic deforming figure. Each robot has a controllable RGB LED so that the swarm can represent color drawings. The computation of robot position, robot motion, and robot color is automatic, including scaling to the available number of robots. The work is in the field of entertainment robotics for play and making robot art, motivated by the fact that a swarm of mobile robots is now affordable as a consumer product. The technical contribution of the paper is three-fold. Firstly the paper presents shaped flocking, a novel algorithm to control multiple robots—this extends existing flocking methods so that robot behavior is driven by both flocking forces and forces arising from a target shape. Secondly the new work is compared with an alternative approach from the existing literature,

S. Hauri
Disney Research, Zurich, Switzerland
e-mail: shauri@student.ethz.ch

J. Alonso-Mora (✉)
ETH and Disney Research, Zurich, Switzerland
e-mail: javiera@disneyresearch.com

A. Breitenmoser
ETH, Zurich, Switzerland
e-mail: andreas.breitenmoser@mavt.ethz.ch

R. Siegwart
ETH, Zurich, Switzerland
e-mail: rsiegwart@ethz.ch

P. Beardsley
Disney Research, Zurich, Switzerland
e-mail: pab@disneyresearch.com

K. Yoshida and S. Tadokoro (eds.), *Field and Service Robotics*,
Springer Tracts in Advanced Robotics 92, DOI: 10.1007/978-3-642-40686-7_12,

and the experimental results include a comparative analysis of both algorithms with metrics to compare performance. Thirdly, the paper describes a working real-time system with results for a physical swarm of 60 differential-drive robots.

1 Introduction

The goal of this work is to allow a non-expert user to represent drawings and animations with a robot swarm. Previous work has demonstrated the artistic potential of robot swarms but the focus has been on abstract visual effects. In contrast, this work describes how to create representational images with robots. A user can make a drawing and the robots will adopt a physical configuration and colors to represent the drawing as shown in Fig. 1. Or the user can create an animation by (a) creating an initial shape and moving it, or (b) creating a sequence of keyframe drawings, and the robots will move to represent the animation.

The motivation for a drawing interface is that it is an intuitive way for a non-expert user to direct a swarm. It is not feasible for a user to explicitly control every robot in a swarm because of the large number of degrees of freedom—three DOF for the location of each robot, two DOF for velocity (on the ground plane), and three DOF for color. Existing flocking algorithms already offer a way for a user to direct a whole swarm, but they do not offer any possibility for representational visual effect. In contrast, our goal with the drawing interface is to allow a user to easily create representational robot art. Our approach has been to build on the power of existing flocking algorithms, and extend them to incorporate shape.

The work belongs in the field of entertainment robotics. The falling price of robots like the ones in Fig. 1 makes for an affordable consumer product. For example, the hardware components of a differential-drive robot with IR communication and RGB

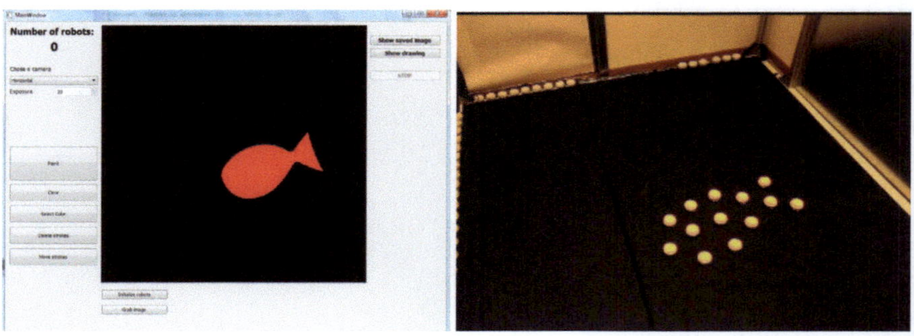

Fig. 1 At *left*, the user interface for creation of drawings and animations. At *right*, the experimental setup and the robot representation of the fish drawing. Robots around the perimeter are docked in battery chargers, and the robots automatically move between the workspace and the chargers as required

LED cost as little as $5, although the robot used in these experiments presents a higher cost. Common portable devices such as smartphones would be suitable to provide the user interface. (The system in this paper uses a fixed overhead camera, but a hand-held device with camera is also a feasible approach for controlling the swarm).

The primary contribution of the paper is shaped flocking, a method for formation control. It combines traditional flocking forces with forces arising from a target shape. It draws on the power of flocking to produce collective motion that is visually compelling, and adds the ability to achieve formations with desired shapes.

Flocking is a non-explicit approach to guiding a multi-robot system, with robot motion being emergent from the flocking forces. A completely alternative approach is to do a geometric analysis of a target shape, and explicitly determine the goal positions of individual robots within the shape. Algorithms for such a goal-directed approach, in particular Voronoi coverage, already exist in the research literature. To provide a comparative analysis, we do a head-to-head comparison of shaped flocking and a goal-directed approach [2] in Sect. 5.

2 Related Work

Outside of robotics, shape constraints have been extensively studied in computer graphics and crowd simulation.

Takahashi et al. [13] achieves smooth transformations between key frames in crowd simulation, using a spectral-based approach, where a transformation is made on the Delauney triangulation from the one key frame to the next. In a next step, a social force field is introduced to avoid collisions. In [5], the complexity of the agent formation is reduced and described only by mean and deviation. It presents a decoupled controller to move and transform a group of robots, but only for simple cases as ellipses and rectangles. In a crowd simulation, Gu [6] samples the outline of the desired shape first and then fills the rest of the shape area using grid points. The agents are assigned to their goal positions by finding correspondences from the current position and the goal deployment. These correspondences are found separately for the outline and the inner part of the shape. The collision avoidance is created by repulsive forces between agents. Rubenstein [10] shows a way to automatically scale the size of the displayed image, so that all robots can fit in the shape. It works without knowing the number of robots in the display, but the robots agree on a factor of the image scale, that is increased, if robots do not fit into the image or reduced, if there is unfilled space around the robot.

A way of controlling coverage in real mobile sensor networks is implemented in [11], where each robot can approximate its weighted Voronoi cell by knowing its local sensor values.

Reynolds famously published a method of simulating the flocking behavior of birds in 1987 [9], using only three behavioral rules to create global bird migration motions. A lot of attention has been given to theoretical and practical research on

Reynolds' rules. Olfati-Saber showed in [8], that all three Reynolds rules can be expressed with a simple control law, that uses attractive and repulsive forces between the robots. Additionally to this, forces of the environment can be added, to affect the behavior of the swarm. The control laws are proven to be stable and collision free for simulated point-agents. Flocking with real non-holonomic robots was achieved by Antonelli [4] by defining a set of rules, how a robot has to react to achieve a specific task, like obstacle avoidance, collision avoidance and approaching the goal. The correct rule is then chosen by a supervisor, which adapts the priority of the rule. In both [3] and [14], shape constraints are created by taking random samples within the wanted shape with a later relaxation to distribute the sample points more evenly. While [14] generally uses a random assignment from agents to those sampling points, [3] minimizes the squared distances of the swarm. In [3], additional constraint at intermediate times can be set and therefore make agents or even the center of the swarm meet exact constraints at specific times. Ho [7] presents a way to control the navigation of a flocking swarm, while navigating through obstacles and along paths with strong curvatures. They assume a flock that already has a given shape, where the connections between the agents is known. The motion of the flock is not rigid, but flexible, which leads to a natural flow of the swarm. Syamsuddin [12] also uses sample points of a skeleton to simulate particles forming a human body. Each particle is attracted by a fixed number of neighboring sample points that are next to each other. Since the particles all start at about the same position and are not real, the assignment and collision avoidance do not matter. The forces are designed in a way, that the particles can follow the sample points without oscillation. A force for obstacle avoidance is also introduced, which allows for simulations including obstacles.

3 Representing Drawings and Animations

This section describes the method for taking a drawing or animation created on the drawing interface and realizing it with robots.

3.1 Representing a Static Drawing

The starting point is the traditional flocking approach defined by Reynolds [9]. Flocking behavior is achieved using three steering rules which define forces between agents:

- Cohesion—steer towards the average position of neighboring agents
- Alignment—steer towards the average heading of neighboring agents
- Separation—steer to avoid close proximity with neighboring agents

The rules can be implemented in different ways, and the formulation here draws on Olfati-Saber [8]. An agent r_i has location q_i and holonomic velocity p_i. It can sense agents within a range r_α. A pair of agents r_i to r_j has a connecting line $\mathbf{n}_{ij}$, and an associated variable a_{ij}:

$$a_{ij} = \begin{cases} 1 & \text{if } \|q_i - q_j\| \leq r_\alpha \\ 0 & \text{otherwise} \end{cases} \tag{1}$$

where $\|q_j - q_i\|$ is the Euclidean distance between agent r_i and r_j.

Olfati-Saber showed that the following formulation for acceleration u_i of agent r_i incorporates all three Reynolds' rules if the robots are sufficiently close:

$$u_i = \sum_{j \in N_i} \phi(\|q_j - q_i\|)\mathbf{n}_{ij} + \sum_{j \in N_i} a_{ij}(p_j - p_i), \tag{2}$$

If the agents are within sensing range, $\phi(\|q_j - q_i\|)$ is a function that drives the two agents to adopt a separation d_α. If the agents are not within sensing range then $\phi(\|q_j - q_i\|)$ is 0. The second term adds a damping force between agents that drives an agent to adopt the same velocity as its neighbors. The result is flocking, an emergent collective motion of all the agents.

Parameter d_α is critical in determining the flocking behavior. Our formulation defines d_α based on characteristics of the target drawing. Assume that the drawing has total area is A and N is the total number of agents. Then:

$$d_\alpha = 2 \cdot c_g \cdot \sqrt{\frac{A}{\pi \cdot N}}, \tag{3}$$

The parameter c_g is related to neighbor connectivity as described in Olfati-Saber.

We augment the traditional steering rules by formulating an additional steering rule called the shape-steering rule, which has the effect of causing the robots to adopt a target shape. The process is:

- The target shape is transformed from screen coordinates of the drawing interface to ground plane coordinates of the physical workspace. The transformation can be done straightforwardly and directly because the aspect ratio of the canvas in the drawing interface is the same as the aspect ratio of the physical workspace.
- For each robot, a circle c_i with radius R is constructed. Subsequent processing for robot r_i depends on whether any of the target shape is contained within c_i.
- If c_i contains none of the target shape, then the nearest boundary point b_i of the target shape is found for r_i, and the steering force is towards b_i.
- If c_i contains some of the target shape, then the center of mass m_i is computed for the parts of the target shape within c_i, and the steering force is towards m_i.

The shape-steering rule causes robots outside the shape to move towards the shape, and it causes robots on the shape perimeter to move within the shape. It has no effect on a robot whose c_i is completely within the shape. As for distribution of robots within the shape, this results from the traditional Reynolds rules because of the choice of d_α in Eq. (3).

The combined steering forces determine an acceleration, and thereby a required velocity p_i, at each robot r_i for the next iteration, suitable for holonomic robots. In fact the robots are non-holonomic, and the velocities p_i are modified to account for this as described in the next section.

If the drawing is formed by more than one shape, each of the robots is assigned to one of the shapes and ignores the other ones. The number of robots assigned to each shape can either be proportional to the area or specified by the user.

3.2 Implementation with Differential-Drive Robots

A holonomic velocity defines a preferred velocity for a non-holonomic agent. The preferred velocity is achieved using the velocity mapping method in [1] as follows.

The aim is for the non-holonomic agent to achieve a trajectory within a bounded error of the trajectory of the holonomic agent. The method has two parameters. Parameter T_0 is the time which determines how quickly the non-holonomic agent should achieve the orientation of the holonomic trajectory. Parameter ε is a a maximum tracking error, which enforces the proximity of the non-holonomic agent to the holonomic trajectory.

These two parameters limit the possible set of allowed holonomic velocities, as illustrated by Fig. 2. If a holonomic velocity lies outside this set, it is optimally

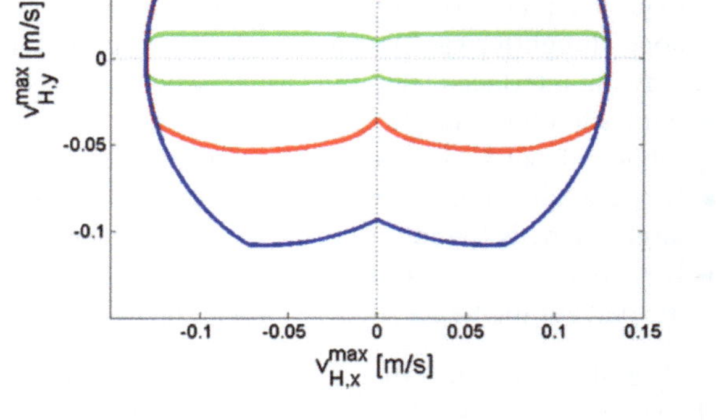

Fig. 2 Set of allowed holonomic velocity (SAHV) for a fixed T_0 and varying ε. A desired holonomic velocity is optimally reduced to the nearest allowed velocity in this set, and then transformed to a non-holonomic velocity and angular velocity

reduced to give the nearest allowed velocity. The allowed holonomic velocity is then transformed to a linear and angular velocity for the non-holonomic agent [1].

There is no explicit handling of collision avoidance in this scheme. The Reynolds steering rule for separation prevents robots coming into close proximity, and collision rarely occurs in the real system. But no guarantee of collision-free trajectories is given.

3.3 Representing an Animation

This section describes how to create a robot animation. A user interacts with the drawing interface in the following way:

- Define the first action by creating one or more shapes on the drawing interface. Then repeat the steps below for each new action.
- Define a new action by adding one or more new shapes.
- Define a new action by transforming an existing shape s to a new shape s' using translation or rotation.
- Define a new action by an unconstrained morphing of an existing shape s to a newly drawn shape s'.

In its most basic form, the animation is thus a sequence of drawings that can be input directly to the algorithm in Sect. 3.1 to drive the robots. But one important modification is needed. In the case that the user creates multiple shape pairs (s_i, s_i') above, it is also possible to specify that the shape pairs maintain an association with the same group of robots g_i. This is useful, for example, for the case of an animated face. The robots for the eyes and mouth should not drift between different facial parts during the animation.

For the case where there are multiple triplets (s_i, s_i', g_i), $i > 1$, the method in Sect. 3.1 is modified as follows. The Reynolds Separation rule is applied across all of the robots, but the Reynolds Cohesion and Alignment rules, and the new shape-steering rule, are applied separately to each triplet (s_i, s_i', g_i). This achieves the desired result—a user can ensure that the robots for individual components of a drawing maintain their identity as they transition through an animation. And it avoids that robots are freely redistributing across the drawing in a confusing and unwanted way.

4 Comparison with a Goal-Directed Approach

The shaped flocking method in Sect. 3 is a non-explicit way to compute robot motion, with the motion being emergent from the flocking rules. A completely alternative approach is to do a geometric analysis of a target shape, and explicitly compute the goal positions for robots within the shape. This goal-directed approach is found in [2] and is not a contribution of this paper, but it does provide a basis to assess this work by a comparative analysis.

The goal-directed approach is described next, followed by a general discussion about both approaches.

4.1 Goal-Directed Approach

Given a target shape, a number of goal positions, equal to the number of robots, is determined by computing the Voronoi coverage of the target shape as described in [2]. The goal positions are fixed, and the robots are advanced to the goal positions over multiple iterations. At each iteration, each robot is associated with a goal position using some proximity criteria. Computed trajectories to reach the goals are modified as required for collision avoidance between robots. Each robot is then advanced for that iteration. The process stops when each robot is within a tolerance of its goal.

4.2 Discussion of Shaped Flocking and Goal-Directed Approach

This section addresses the question of why we are investigating shaped flocking given that the existing goal-directed approach could achieve similar goals. The motivation for our work is that flocking is known to be a powerful method for producing aesthetic effects in multi-agent systems. These effects are both visually compelling and they can be readily varied by changing the steering rules. We seek to build on that, while adding the ability to represent target shapes. In contrast, the goal-directed approach is explicit and rigid, and would require new components in order to vary the computed robot motion.

More specifically, flocking has these interesting properties:

- A change in one part of a shape affects only the local robots in a flocking approach. But it requires a complete recomputation of goal positions in the goal-directed approach.
- It is straightforward to add heterogeneous behavior of agents in a flocking approach. For example, some agents can be more attractive or repulsive. Or agents can be grouped like the shape triplets described in Sect. 3.3.
- Steering parameters in the flocking approach provide an opportunity for automated learning of a range of aesthetic effects.

These arguments suggest that flocking can provide the most flexible basis for producing varied and compelling visual effects.

5 Results

In this section we discuss the execution and the results of the conducted experiments. We first describe the experimental setup and then compare specific measures for the shaped flocking and the goal-directed approach.

5.1 Experimental Setup

The experimental setup is shown in Fig. 3. There is a 2m-by-2m workspace for deployment of the robots. An overhead camera views the workspace, and a central computer provides the drawing interface (see Fig. 1), tracks the robots, computes the required motion of the swarm, and sends wireless commands to direct the robots.

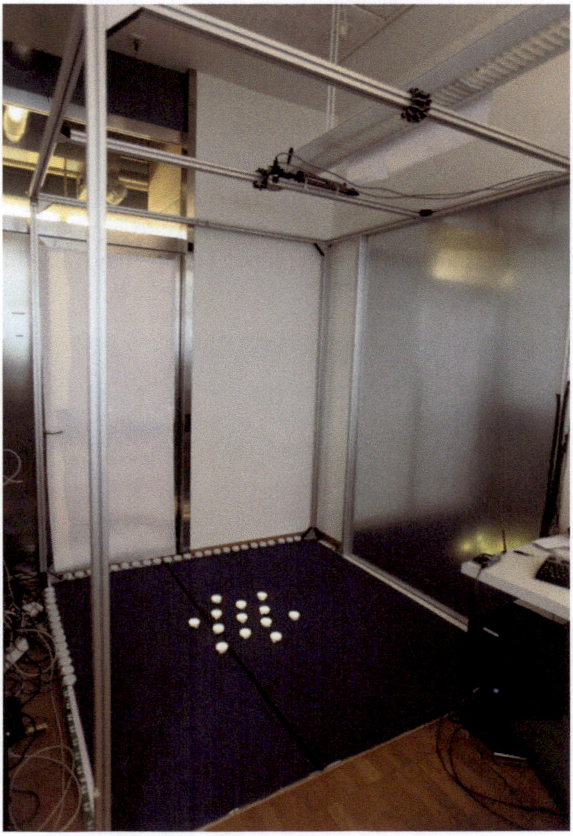

Fig. 3 The experimental setup—robots on the ground plane, overhead camera for tracking, and central computer for processing and sending wireless motion commands to the robots

Battery chargers line the perimeter of the workspace as shown in Fig. 1, and the robots have the ability to automatically dock for recharging.

The overhead camera used for tracking is fixed at a height of 2.3 m above the deployment plane and can detect both visual and infrared light. The camera has 1600×1200 resolution, running at a frame rate of 10 Hz and can localize robots with precision better than 0.01 m.

Figure 4 shows the differential-drive robot used in the experiments. The RGB LED is under a translucent cover. The robot has two wheels driven by geared motors. The motors allow maximum speed of 0.5m/s. The onboard microcontroller runs a fast controller measuring the wheel speed using back-EMF of the motor. Speed commands are transmitted constantly from the host computer through a 2.4 GHz radio with a specific protocol that ensures 10 Hz for up to 100 robots.

Figure 4 also shows a close-up of the charging unit. The robot docks by driving directly into the center channel of the metal component on the green strip, so that it touches the springed recharging contacts that are on either side of the channel.

Figure 5 shows examples of robot faces. Figure 6 shows an animation of a tree and Fig. 7 of a fish.

5.2 Comparative Analysis

Table 1 provides a comparison of the shaped flocking algorithm and the goal-directed algorithm. Both algorithms can apply to a group of robots starting from arbitrary positions, although convergence results may vary. In these experiments the robots

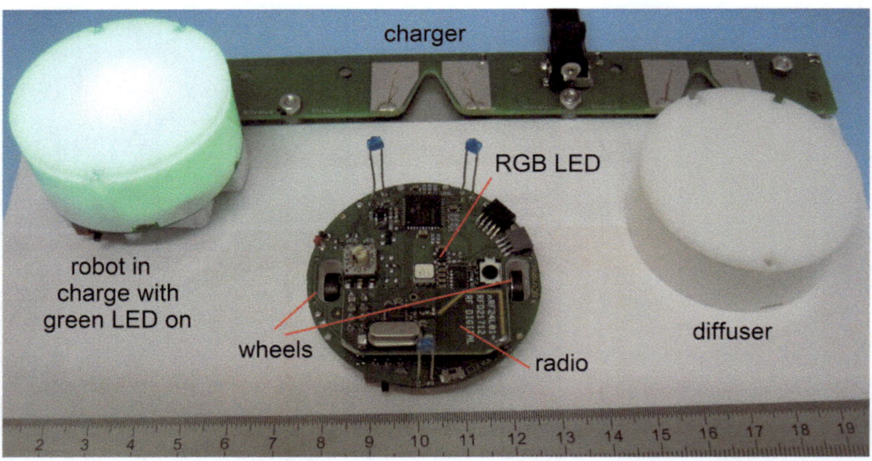

Fig. 4 Differential-drive robot with controllable RGB LED. The robot automatically docks in the charging plate for battery recharging

Fig. 5 Smiley faces—at *top*, the drawings created on the user interface and, at *bottom*, the corresponding robot configurations. Faces make a good choice for animation with wobbling eyebrows and dynamic smiles

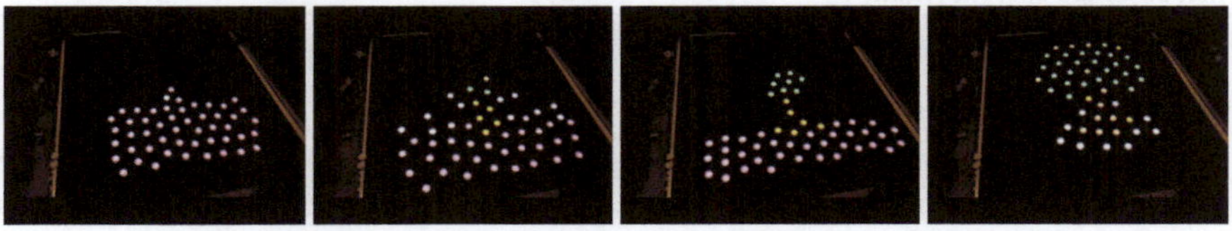

Fig. 6 Tree animation starting with flat earth at *left*, and growing into a full tree at *right*

were randomly distributed in the environment (approximately uniform distribution). The effect of the initial distribution of robots is further described in Sect. 5.3.

The first three rows display the time to converge to a new shape. Shaped flocking needs longer to converge to its final configuration. This is unsurprising because there are no explicit goal positions, and hence no direct motion toward the final configuration, unlike the goal-directed approach.

The coverage cost is a measure of quality of the distribution of robots through the shape and is given by

$$C = \sum_{u \in U} \min_i \|q_u - q_i\| \tag{4}$$

where q_i the location of robot i and U is the set of all pixels u of all patterns, with coordinates q_u.

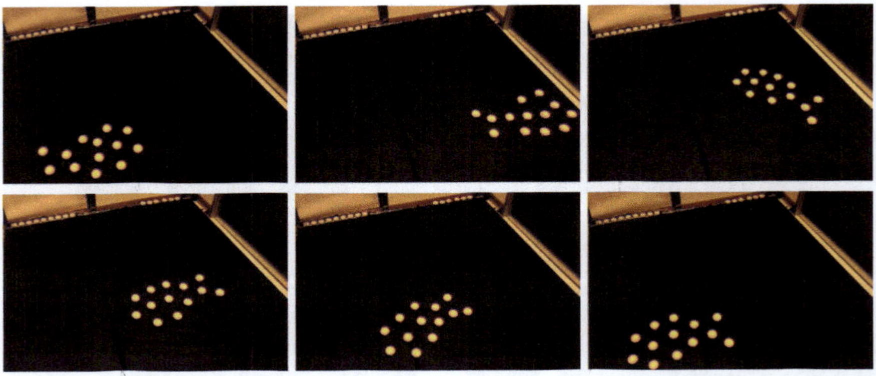

Fig. 7 Fish animation showing motion though user defined keyframes. The image series is arranged from *left* to *right* and *top* to *bottom*

Table 1 Performance metrics for the shaped flocking and goal-directed algorithms

Tree animation	Shaped flocking	Goal-directed
Time to convergence (keyframe 1)	33.1 s	18.1 s
Time to convergence (keyframe 2)	14.5 s	5.1 s
Time to convergence (keyframe 3)	50.0 s	11.1 s
Final coverage cost (keyframe 1)	2647	2486
Final coverage cost (keyframe 2)	2635	2553
Final coverage cost (keyframe 3)	3241	2938

Shaped flocking presents higher coverage cost. Again this is unsurprising because shaped flocking implicitly achieves low coverage costs via steering rules, whereas the goal-directed approach explicitly precomputes goal positions by minimizing the coverage cost.

Figure 8 shows the progression of coverage cost for the animation of the fish in Fig. 7. The shaped flocking method typically shows an increase in cost at the start of a new keyframe, which then reduces as the motion proceeds. As in the discussion above, this is a consequence of the implicit nature of flocking, which is not enforcing low cost configurations of the robots during the motion.

In summary, the goal-directed approach is superior in time to convergence and in coverage cost. However shaped flocking is still the primary focus of our research because

- It is a flexible basis for future work as argued in Sect. 4.2.
- Flocking motion is aesthetically appealing in our experiments, although we have not identified a way to quantify this.

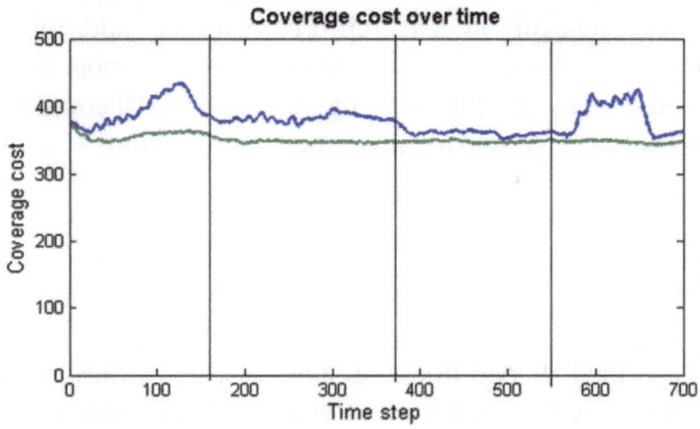

Fig. 8 Coverage cost progression for an animation of a fish. The *blue line* shows the cost of shaped-flocking, the *black line* of goal-driven method. *Vertical lines* indicate the beginning of a new keyframe in the animation

5.3 Bottlenecks and Convergence to Local Minima

The experiments revealed a limitation in the shaped flocking algorithm for certain types of shape and depending on the initial conditions. Figure 9 shows an example with robotic Kanji. Once within the shape, the flocking rules disperse the robots. But for shapes with a pipe-like form, dispersion is slow i.e. it is a bottleneck problem. This is an issue where goal-direction does not encounter the same problem because goal positions are explicitly computed throughout the target shape. We have observed, as expected from the methods, that the goal-directed approach does present similar convergence results to arbitrary shapes independently of the initial conditions, whereas the shaped flocking algorithm strongly depends on the characteristics of the shape and the initial distribution of the robots (convergence to a local minima). For the latter, initial distributions that are homogeneously distributed over an area containing the shape typically present better results.

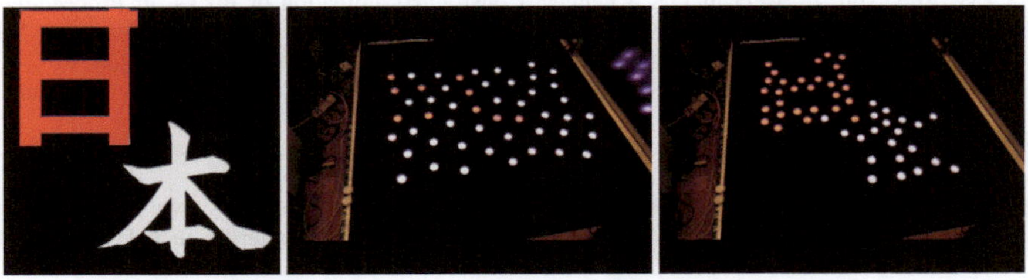

Fig. 9 Illustration of the bottleneck problem. At *left,* the user drawing. At *center,* robot start position. At *right,* the shape is convoluted, and diffusion of robots to achieve the shape is slow

This problem requires further research. For example, additional forces toward unoccupied space in the shape could help to disperse the robots faster and drive the robots through bottlenecks. And the scaling strategy in [10] could be used to define optimal distances between robots to ensure better dispersion.

6 Conclusion

This paper has described shaped flocking, a novel algorithm for representing static and deforming shapes with a multi-agent system. Traditional flocking algorithms produce compelling collective motion in a group of agents. Shaped flocking extends this so that the agents additionally conform to a static or deforming shape. The algorithm was demonstrated in the context of a system that allows a user to do real-time direction of a swarm of robots via a drawing interface, in the field of entertainment robotics.

For comparison, shaped flocking was evaluated against a goal-directed approach. Conceptually the two approaches are different, with shaped flocking relying on emergent behavior to represent a target shape, while goal-direction relies on an explicit geometric analysis of the shape. Both approaches show good representational power and support real-time interaction. Although the goal-based approach shows superior results, we believe that shaped flocking is worth further investigation due to its flexibility and potential for creating robot art.

References

1. J. Alonso-Mora, A. Breitenmoser, M. Rufli, P. Beardsley, R. Siegwart, Optimal reciprocal collision avoidance for multiple non-holonomic robots, in Proceedings of International Symposium on Distributed Autonomous Robotics Systems (2010)
2. J. Alonso-Mora, A. Breitenmoser, M. Rufli, R. Siegwart, P. Beardsley, Image and animation display with multiple mobile robots. Int. J. Robot. Res. **31**, 753–773 (2012)
3. M. Anderson, E. McDaniel, S. Chenney, Constrained animation of flocks, in *Proceedings of the 2003 ACM SIGGRAPH/Eurographics symposium on Computer animation, SCA '03* (Eurographics Association, Aire-la-Ville, Switzerland, Switzerland, 2003), pp. 286–297
4. G. Antonelli, F. Arrichiello, S. Chiaverini, Flocking for multi-robot systems via the null-space-based behavioral control. Swarm Intell. **4**, 37–56 (2010)
5. C. Belta, V. Kumar, Abstraction and control for groups of robots. IEEE Transactions Robot. **20**(5), 865–875 (2004)
6. Q. Gu, Z. Deng, Formation sketching: an approach to stylize groups in crowd simulation, in *Proceedings of Graphics Interface 2011, GI '11* (School of Computer Science, University of Waterloo, Waterloo, Ontario, Canada, Canadian Human-Computer Communications Society, 2011), pp. 1–8
7. C. Ho, Q. Nguyen, Y.S. Ong, X. Chen, Autonomous multi-agents in flexible flock formation, in *Motion in Games, Lecture Notes in Computer Science*, vol. 6459, eds. by R. Boulic, Y. Chrysanthou, T. Komura (Springer, Berlin, 2010), pp. 375–385
8. R. Olfati-Saber, Flocking for multi-agent dynamic systems: algorithms and theory. IEEE Trans. Autom. Control **51**(3), 401–420 (2006)

9. C.W. Reynolds, Flocks, herds and schools: a distributed behavioral model. SIGGRAPH Comput. Graph. **21**, 25–34 (1987)
10. M. Rubenstein, W.M. Shen, Automatic scalable size selection for the shape of a distributed robotic collective, in *Intelligent Robots and Systems (IROS), 2010 IEEE/RSJ International Conference on Intelligent Robots and Systems* pp. 508–513 (2010)
11. M. Schwager, J. Mclurkin, D. Rus, Distributed coverage control with sensory feedback for networked robots, in *Proceedings of Robotics: Science and Systems* (2006)
12. M.R. Syamsuddin, J. Kim, Controllable simulation of particle system, *Proceedings of the 7th International Conference on Advances in Visual Computing - Volume Part II, ISVC'11* (Springer, Berlin, 2011), pp. 715–724
13. S. Takahashi, K. Yoshida, T. Kwon, K.H. Lee, J. Lee, S.Y. Shin, Spectral-based group formation control. Comput. Graphics Forum **28**(2), 639–648 (2009)
14. J. Xu, X. Jin, Y. Yu, T. Shen, M. Zhou, Shape-constrained flock animation. Comput. Animation Virtual Worlds **19**(3–4), 319–330 (2008)

Evaluation and Training System of Muscle Strength for Leg Rehabilitation Utilizing an MR Fluid Active Loading Machine

Hiroshi Nakano and Masami Nakano

Abstract An evaluation and training system of muscle strength for leg rehabilitation has been developed by using a new conceptual loading machine. This loading machine, which is called MR fluid active loading machine, mainly consists of a newly designed magneto-rheological (MR) fluid clutch and a reversible induction motor. The MR fluid clutch produces passively the magnetic field-dependent transmitting torque almost independent of the rotational speed. Because of this feature, the MR fluid clutch will be suitable for the loading machine of a rehabilitation system from a viewpoint of safety and relief. This system can perform the isometric and isokinetic strength evaluations and the isokinetic strength training. And also, the system has applicability to the Range Of Motion training (ROM training). In this paper, the methods of the muscle strength evaluations and training in this system are described, and the performances of the evaluation and training modes are discussed.

1 Introduction

In various rehabilitation machines with passive or active loading devices for muscle strength evaluation and training, it is important to precisely control the load independent of the training inertia and speed, keeping safety. Currently, these machines have been widely using AC, DC servo motor, hydraulic actuator and pneumatic actuator as the loading devices. From the viewpoint of safety and controllability, these

H. Nakano (✉)
Graduate School of Engineering, Tohoku University, 2-1-1, Katahira,
Aoba-ku, Sendai, Miyagi 980-8577, Japan
e-mail: h_nakano@ifc.ifs.tohoku.ac.jp

M. Nakano
Institute of fluid science, Tohoku University, 2-1-1, Katahira,
Aoba-ku, Sendai, Miyagi 980-8577, Japan
e-mail: m-nakano@fmail.ifs.tohoku.ac.jp

K. Yoshida and S. Tadokoro (eds.), *Field and Service Robotics*,
Springer Tracts in Advanced Robotics 92, DOI: 10.1007/978-3-642-40686-7_13,

active actuators are not well suited for rehabilitation machines, because of the lack of intrinsic mechanical protection for human safety and passive controllability such as back-drivability and torque limiting control. The often used muscle strength measurement and training systems such as CYBEX NORM [4] and BIODEX System [12] which do not have Range Of Motion (ROM) training function, are not always safety because of the usage of the active actuator of DC servo motor as a loading device.

The devices using functional fluids such as MR fluids [7, 10, 11] and electro-rheological (ER) fluids, which have variable viscosity depending on the applied magnetic or electric field, have received much attention for use in rehabilitation because of quick response and safety for humans. Mavroidis et al. [3] applied MR and ER fluid brakes to portable rehabilitation devices for elbows and knees of patients. Hakogi et al. [5] applied the combination of motor and MR fluid rotary brake to ROM training robot, which assisted students in the training. Nikitczuk et al. [8] applied an ER fluid brake to knee rehabilitation device. Dong et al. [14] applied an MR fluid damper to knee rehabilitation device. Chen et al. [9] applied the combination of motor and multiple-disc MR fluid clutch to assistive knee braces for disabled people with mobility problem. However, the versatile system, which is possible to perform several functions such as ROM training, muscle strength training and measurement, has not been developed.

Therefore, this study aims at developing a versatile leg rehabilitation system utilizing an active loading machine using the MR fluid. The leg rehabilitation system, which is designed to be unified and compact and can flexibly adjust the load depending on the muscle strength of subjects, has been proposed and developed, featuring higher safety and inward security for human since it measures and trains muscle strengths making good use of the passive load of a developed MR fluid loading machine.

In this paper, the design and performance of the developed MR fluid active loading machine are described, and the evaluation and training system for leg rehabilitation utilizing this machine is presented. And also, the validity of muscle strength measurements and possibility of trainings on this system are discussed, based on the results of isometric and isokinetic strength evaluation, isokinetic training and torque-limiting test of the MR fluid active loading machine for ROM training during knee extension and flexion.

2 MR Fluid Active Loading Machine

The MR fluid, which is one of functional fluids, is changed its rheological properties by applying magnetic field, and behaves like a Bingham fluid having variable yield stress with magnetic field. Because of this inherent feature, the clutch using the MR fluid (MR fluid clutch) generates the magnetic field-dependent transmitting torque almost independent of the rotational speed.

Fig. 1 MR fluid active loading machine

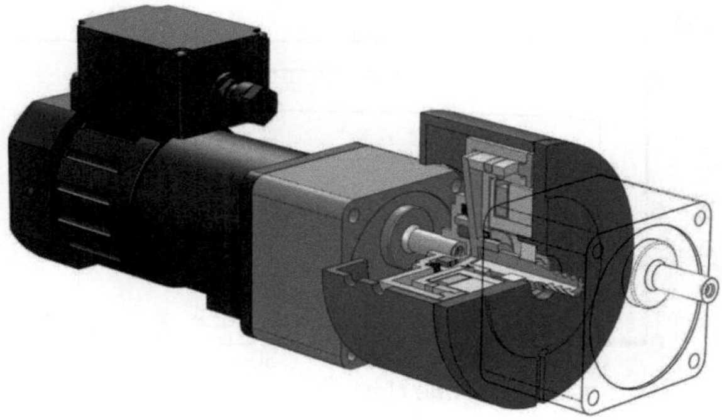

In this study, the MR fluid active loading machine has been developed, which mainly consists of the newly designed MR fluid clutch and a reversible induction motor with an electromagnetic mechanical brake. The MR fluid clutch can transmit the torque below a torque limited by the applied coil current and slip when the torque exceeds the limited value. This loading machine can perform the following two different modes.

Active load mode The load is produced by the induction motor. In this case, since the MR fluid clutch functions as a torque limiter, the loading machine can generate active loads below a limited torque under a safe condition. It is expected to be used to perform the active training such as the ROM training and isokinetic training [2], and the isokinetic strength evaluation.

Passive load mode In the case of fixing the motor shaft on an electromagnetic mechanical brake, the MR fluid clutch functions as an MR fluid brake which works as a passive loading machine. It is expected to be used to perform the muscular evaluation and the isotonic training.

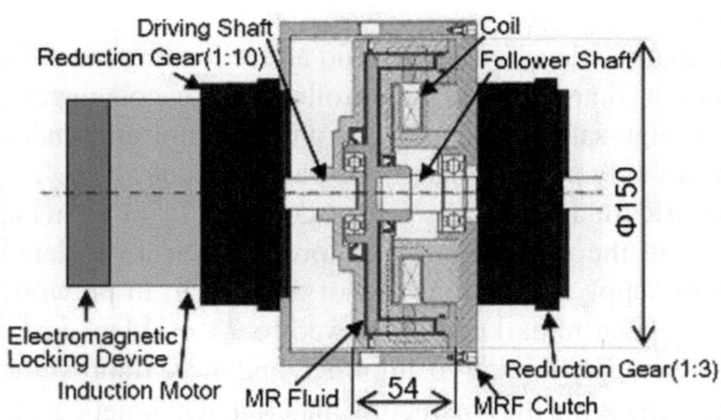

Fig. 2 Cross-sectional view of the MR fluid active loading machine. In order to limit output torque of loading machine, MR fluid clutch is installed to output side of motor

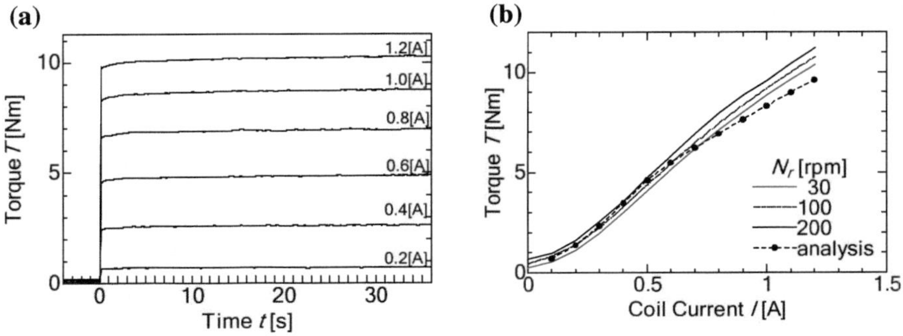

Fig. 3 Steady torque responses of the cup-type MR fluid clutch: **(a)** Time variations of experimental torque for coil current I at rotational speed of motor $N_r = 30$ rpm; **(b)** Comparison between analytical and experimental steady torques versus coil current I

Figures 1 and 2 show the MR fluid active loading machine consisting of the developed cup-type MR fluid clutch, the induction motor and reduction gears. As seen in Fig. 2, the developed clutch has two cup-shaped rotors of the driving and follower shafts, an MR fluid (MRF-132DG, Lord Co.) in the gap of 1 mm between the two rotors and a magnetic coil (wire diameter; 0.4 mm, turns; 675, electric resistance; 23.7 Ω) inside of the cup rotors. The MR fluid clutch was designed based on a steady electromagnetic field analysis using the JMAG software. Analytical and experimental steady torque responses of the clutch are shown in Fig. 3. The clutch can transmit almost constant torque T depending on the coil current I. The transmitting torque increases with increasing the coil current almost independent of the rotational speed as well as analytical one, and generates the torque of at least 8.5 Nm at $I = 1.0$ A. Since the MR fluid clutch has a reduction gear of 1/3 gear ratio connected to the follower shaft, the output torque of the active loading machine is about three times of the clutch torque. And also, the induction motor (rated torque: 1.27 Nm) has an ability to generate the output torque of 12.7 Nm to the MR fluid clutch through a reduction gear of 1/10 connected to the output shaft. Therefore, it can be ensured that the MR fluid active loading machine generates the output torque of at least about 25.5 Nm at $I = 1.0$ A.

It can be said that the developed MR fluid active loading machine has high controllability because its output torque is controlled by only coil current, and shows the specific feature of high safety because it has a torque limiting function in the active mode and generates only passive load in the passive mode.

The cup-type MR fluid clutch has a time constant $T_c = 60$ ms and a pure time delay $L = 25$ ms in the step response of torque when a regulated DC amplifier is used as a power supply for coil current of the clutch. In previous study [6], the step response have been remarkably improved to $T_c = 14$ ms and $L = 11$ ms by changing the power supply to a servo amplifier, and more improvement of magnetic circuit response by changing the conducting material of magnetic isolating parts to an insulator has been demonstrated by an unsteady electromagnetic field analysis. The magnetic isolating parts made of nonmagnetic material (aluminum) in two cylindrical

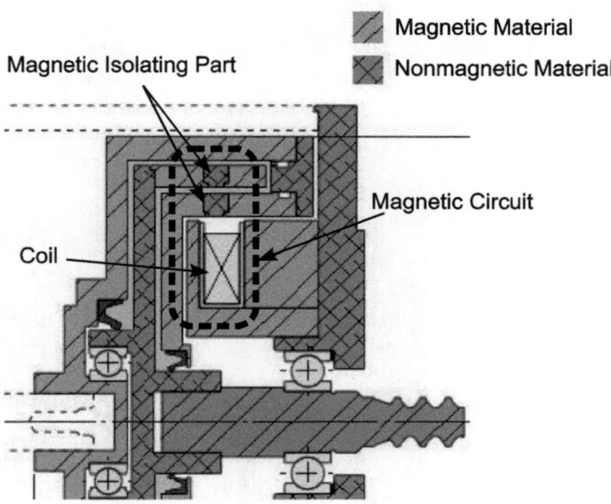

Fig. 4 Cross-sectional view of cup-type MR fluid clutch

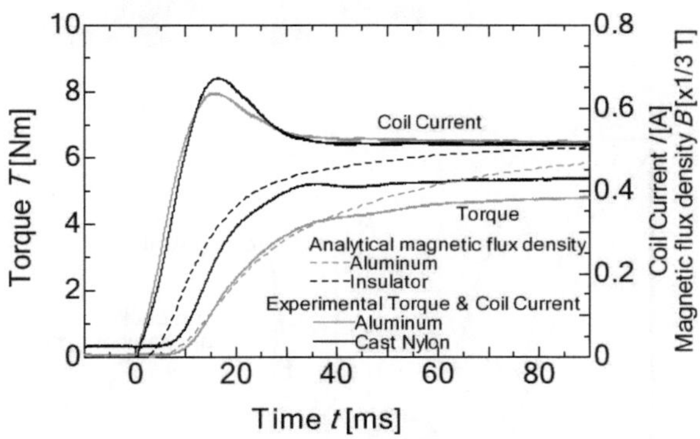

Fig. 5 Step responses of analytical magnetic flux density, experimental torque and coil current before and after changing the material of magnetic isolating part of cup-type MR fluid clutch

rotors are set to form the magnetic circuit through the two rotors, as shown in Fig. 4. The step responses of coil current and torque before and after replacing the magnetic isolating parts of aluminum to that of cast nylon as an insulator are shown in Fig. 5. It is obviously recognized that the change of the materials results in faster step response of torque where $T_c = 9\,\text{ms}$, $L = 9\,\text{ms}$.

3 Evaluation and Training System of Muscle Strength for Leg Rehabilitation Utilizing an MR Fluid Active Loading Machine

Figure 6 shows a schematic view of developed evaluation and training system of muscle strength for leg rehabilitation. This system mainly consists of an adjustable seat, the developed MR fluid active loading machine, a revolving arm which leads to knee extension and flexion to the leg of subject, and the measurements and control

(a)

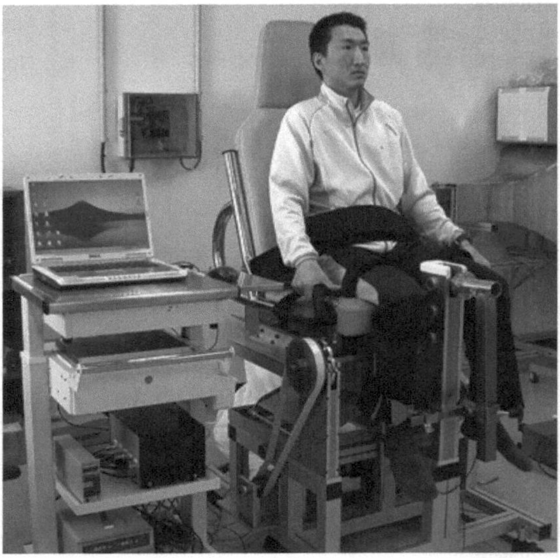

(b)

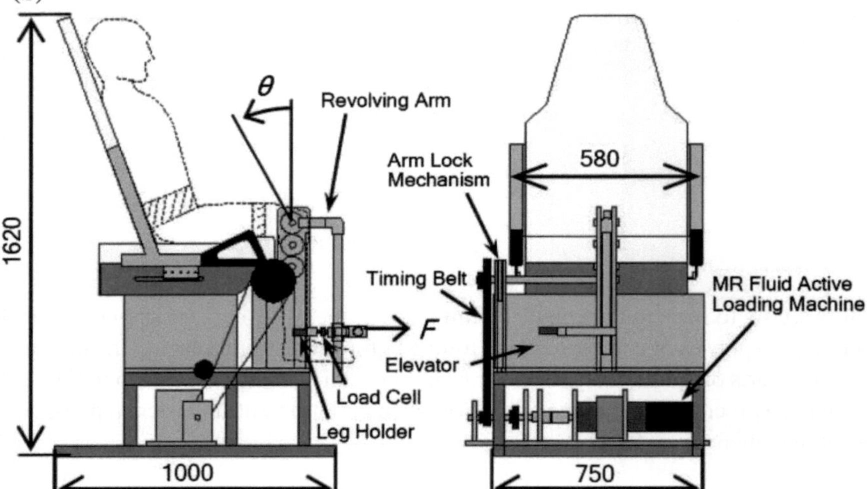

Fig. 6 Overall view **(a)** and front and side views **(b)** of developed evaluation and training system of muscle strength for leg rehabilitation utilizing MR fluid active loading machine. Subject is held on seat by tightening waist and thigh belts, and ankle is held to leg holder of revolving arm by ankle belt

PC unit. The horizontal and vertical position of the seat can be adjusted for the rotational center of the knee of subject to be coincided with that of the revolving arm. The force F acting to the revolving arm from the leg is measured on a load cell. A knee angle θ is measured on a rotary encoder, and is defined as a knee joint

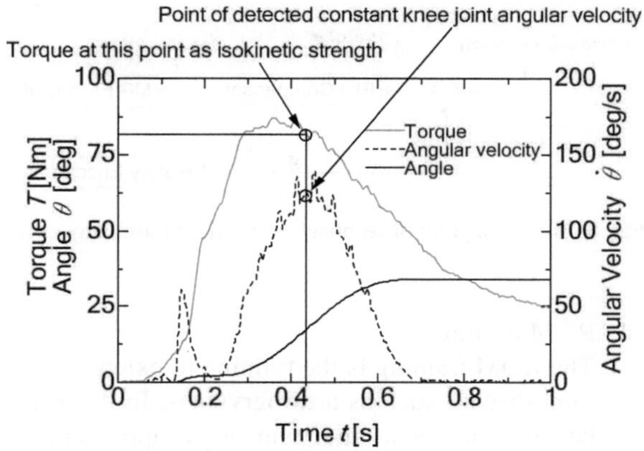

Fig. 7 Measured results and detection of isokinetic strength

angle from vertical direction. The torque of the loading machine is amplified and transmitted to the revolving arm through a reduction gear and a timing belt. The maximum load torque of the revolving arm is about 120 Nm.

This system has four operating modes of evaluations and trainings as follows.

1. Isometric strength evaluation
   The system can lock the revolving arm every 15° of the knee joint angle from 0 to 75° using an arm lock mechanism, like as the CYBEX NORM [4]. Then, the isometric strength of subjects is measured at several knee joint angles.
2. Isokinetic strength evaluation
   In the CYBEX NORM [4], the revolving arm with leg is controlled to actively rotate at constant angular velocity by the servo motor and then the maximum torque acting on the arm from the leg is measured as the isokinetic strength. While, in this system the isokinetic strength is measured by utilizing the passive load mode of the MR fluid active loading machine. In this evaluation, subjects must exert all their strength to kick up the revolving arm under an almost constant load determined by the coil current of the loading machine, and then the torque at constant knee angular velocity detected automatically is measured as the isokinetic strength, as shown in Fig. 7. The measured torque at the constant velocity includes no inertia force of the arm and devices. The measurements are done under several constant loads for the arm by changing the coil current.
3. Isokinetic training
   Isokinetic training is the muscle training of isokinetic motion. In the isokinetic training mode of the CYBEX NORM [4], the revolving arm with leg is controlled to forcibly rotate at constant angular velocity by the servo motor. While, in this system the isokinetic training is performed by utilizing the passive load mode of the MR fluid active loading machine. Figure 8 shows the block diagram of the angular velocity controller in the isokinetic training. During the subject kicks up the revolving arm, the knee joint angular velocity is controlled to be a desired constant speed by controlling the braking torque acting to the revolving arm.

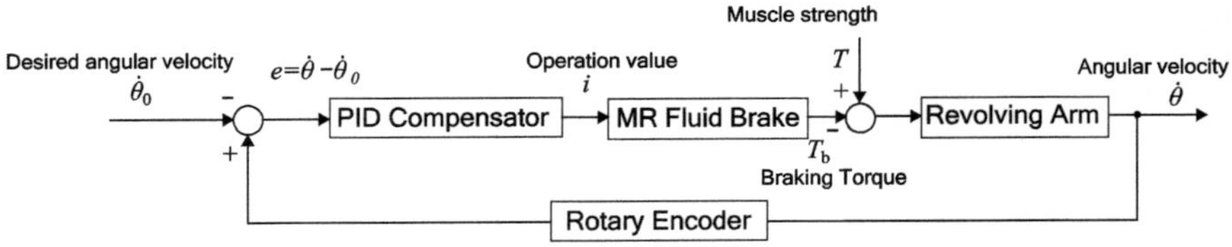

Fig. 8 Block diagram of feedback controller of arm angular velocity

4. ROM training
   The ROM training is the training to extend the range of motion for a patient of joint disease such as arthrogryposis. In the arthrogryposis, the joint resistance becomes larger as the joint angle approaches movable limit. Therefore, the physical therapists try to extend the range of joint motion by adding an effectual load to the joint of angle close to movable limit. In this study, the developed MR active loading machine is considered to utilize to the ROM training. The MR active loading machine has the ability to apply an active load to the joint under safe condition because the limiting load can be set by the MR fluid clutch.

4 Experimental Results

To discuss the validity of strength measurements and the possibility of trainings on this system, the isometric and isokinetic strength evaluations, the isokinetic training and the torque-limiting test for ROM training are demonstrated for healthy men and women in their twenties. The results are shown below. However, the comparisons of the performances between the CYBEX NORM [4] and this developed system are not done here, because the measuring and training methods for muscle strengths in both the two systems are big different each other, as described above.

4.1 Isometric Strength Evaluation

The isometric strengths are measured as a function of knee joint angle θ every 15 ° from 0 to 60 °. As seen in Fig. 9, the measured isometric strengths for five subjects represent the maximum value at the knee joint angle of about 15 ° , and then decreases with increasing θ. These characteristics are in good agreement with that of the existing experimental data of the leg muscle strength [13], showing the validity of this evaluation.

4.2 Isokinetic Strength Evaluation

To discuss the results of this evaluation, the measured isokinetic strengths are fitted by the Hill-based muscle model [1], which is given by the following equation.

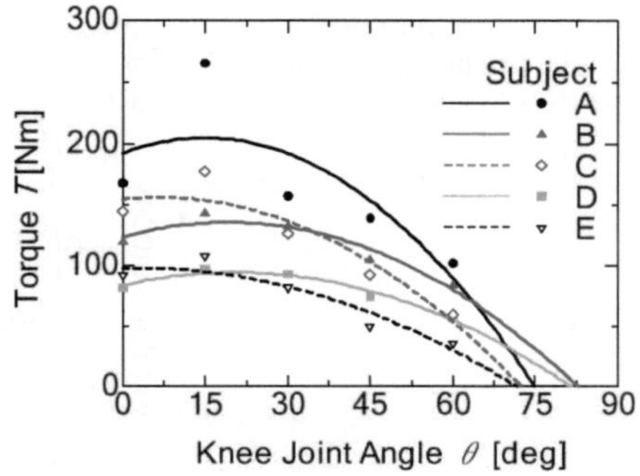

Fig. 9 Measured isometric strengths

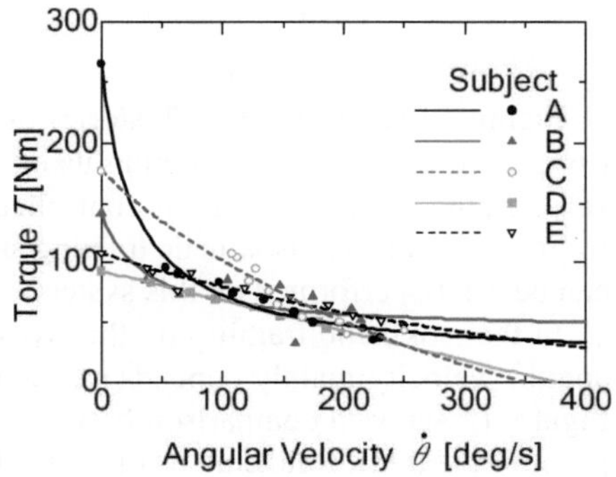

Fig. 10 Measured isokinetic strengths

$$\dot{\theta} \times (F + \alpha) = \beta \times (F_0 - F) \tag{1}$$

where, F is the measured isokinetic strength at angular velocity $\dot{\theta}$, F_0 is the maximum isometric strength, α and β are constants depending on the muscle. Figure 10 shows the results of the isokinetic strength measurement for five subjects. The isokinetic strength decreases with increasing the knee joint angular velocity, and can be approximated well by the Eq. 1. And also, the difference of the isokinetic muscle strength as a function of the angular velocity among five subjects can be obviously compared with each other. Therefore, this evaluation can be said to be valid.

4.3 Isokinetic Training

In the proposed isokinetic training, the revolving arm rotates at a constant angular velocity independent of the muscle strength of subjects, so that the training can be conducted in a full range of joint motion proper to the subject. Therefore, this training also contributes to the keeping and expansion of the range of motion. From the above reasons, this training is considered to be suitable for rehabilitation.

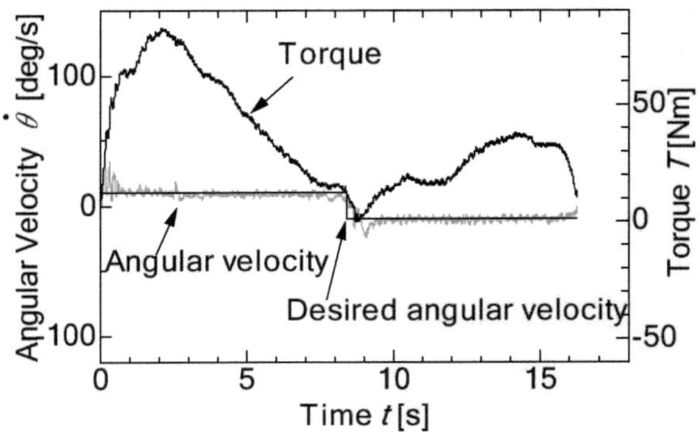

Fig. 11 Time histories of angular velocity and torque in isokinetic training

Figure 11 shows the time histories of knee joint angular velocity $\dot{\theta}$ and strength (torque) in the isokinetic training using this system. In this case of relatively low velocity, the angular velocity is controlled to well follow the desired one. Therefore, it can be said that the isokinetic training at the angular velocity less than $\dot{\theta} = 10\,\text{deg/s}$ can be easily performed on this system.

In the isokinetic training on this system, the following control performance of angular velocity mainly depends on the transient response of the MR fluid clutch. Figure 12 shows a comparison between the isokinetic training results utilizing the pre-improved MR fluid clutch controlled by a DC power amplifier, and the improved fast response MR fluid clutch controlled by a servo amplifier, which are obtained at the desired knee joint angular velocity of 50 deg/s. The result utilizing the fast response MR fluid clutch shown in Fig. 12(b) shows better following control characteristic than the other shown in Fig. 12(a). It exhibits that the improvement of transient torque response in MR fluid clutch results in better following control ability of angular velocity in the isokinetic training.

4.4 Applicability to ROM Training

In order to confirm the applicability of this system to ROM training, the torque limiting function of the MR fluid active loading machine has been investigated to work effectively in the ROM training. In this study, a weight is hanged on the ankle of subject to simulate the arthrogryposis as shown in Fig. 13, and the simulated ROM training is demonstrated experimentally. In the experiments, firstly a constant coil current to the MR fluid clutch is applied to determine the upper limit of the transmitting torque, and then the revolving arm and leg are actively rotated by the motor of the MR fluid active loading machine. Figure 14 shows the time histories of the torque T acting on the leg and the arm angle θ in the experiment where the rotational speeds of the motor and the revolving arm are respectively 100 rpm and 4 deg/s. The torque T and the angle θ increase with time slowly, and then stop

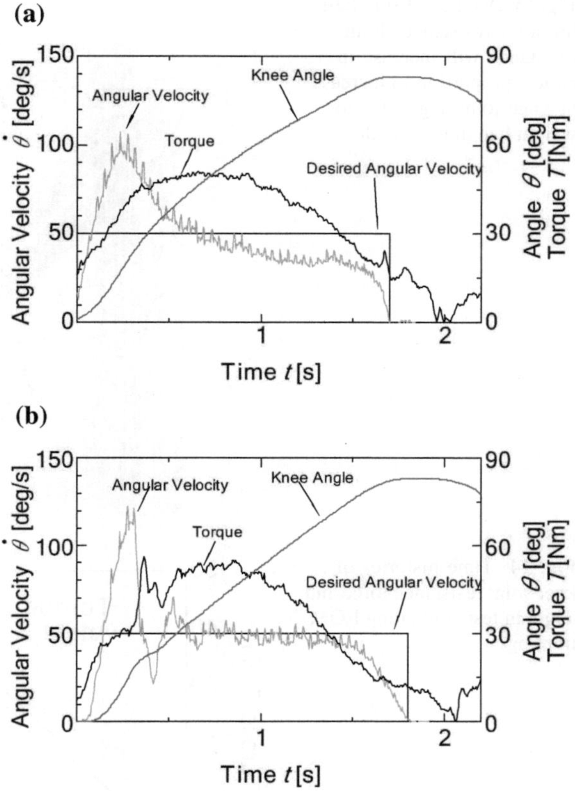

Fig. 12 Isokinetic training results comparison between the system utilizing pre-improved MR fluid clutch (**a**) and improved fast response MR fluid clutch (**b**)

increasing to maintain almost constant values due to the MR clutch slipping at the upper limit torques determined by the applied coil currents I. Thus, this developed system can easily set the upper limiting load appropriate for the subject by the MR fluid clutch, and can avoid overload, showing the applicability to the ROM training.

5 Conclusion

In this paper, we proposed and developed the new conceptual loading machine, which is called the MR fluid active loading machine, and described the design and performance of this loading machine. And also, we proposed and developed the evaluation and training system with safety for leg rehabilitation utilizing this loading machine, and investigated the validity of evaluations such as isometric and isokinetic evaluations, possibility of isokinetic training, and applicability to ROM training. The results are summarized as follows.

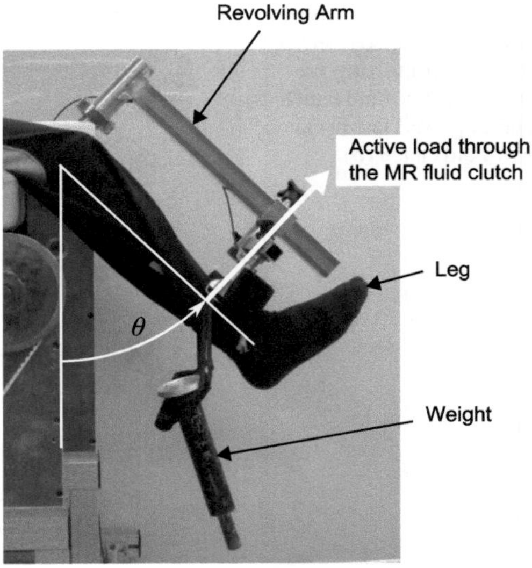

Fig. 13 Method of simulating the joint resistance. Joint resistance will increase in almost proportion to increase of knee joint angle due to weight hanging on ankle

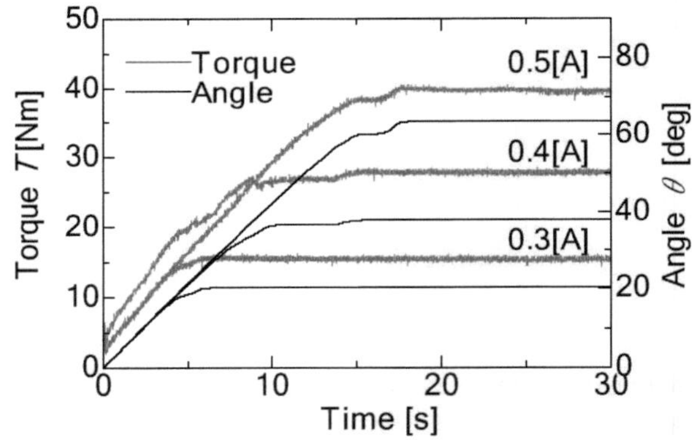

Fig. 14 Time histories of knee joint resistance force and angle in test simulating ROM training

1. It is found from the experiments for steady torque responses of the developed MR fluid clutch that the active loading machine is capable of generating the magnetic field-dependent torque up to about 25.5 Nm for a coil current of 1.0 A. And, the step torque response of the MR fluid clutch is improved to be a first order lag element plus pure time delay system with the time constant of 9 ms and the dead time of 9 ms, by replacing the conducting material (aluminum) of magnetic isolating part of the cylindrical rotor with the non-conducting material (cast nylon).
2. The validity of isometric evaluation is confirmed by obtaining the almost same results as the existing experimental isometric data. And the validity of isokinetic evaluation is confirmed by obtaining the results in good agreement with the Hill-based muscle model and a possibility of comparing with the muscle strengths for each subject.

3. The isokinetic training is well performed in the relatively high speed range up to about 50 deg/s. This system has a high potential for the ROM training, since it can easily set the upper limiting load appropriate for the patient by the MR fluid clutch and can avoid overload.

References

1. A. Nagata, *Biokinetics* (Kyorin Shoin Publication, Tokyo, 1991). (in Japanese)
2. C. Kisner, L.A. Colby, *Therapeutic Exercise* (FA Davis, Philadelphia, 2002)
3. C. Mavroidis et al., Smart portable rehabilitation devices. J. NeuroEng. Rehabilit. **2**, 18 (2005)
4. CYBEX, CSMi. http://www.csmisolutions.com/
5. H. Hakogi, M. Ohaba, N. Kuramochi, H. Yano, Torque control of a rehabilitation teaching robot using magneto-rheological fluid clutches. JSME Int. J. Series B Fluids Therm. Eng. **48**(3), 501–507 (2005)
6. H. Nakano, M. Nakano, Active loading machine using MR fluid clutch for leg rehabilitation system. Int. J. Appl. Electromagnet. Mech. **39**, 463–469 (2012)
7. H. Yamamoto, M. Nakano, Dynamic viscoelasticity and Its mechanical model of an MR suspension in oscillatory slit flow mode. J. Intell. Mater. Syst. Struct. **10**(8), 658–665 (1999)
8. J. Nikitczuk, B. Weinberg, C. Mavroidis, Control of electro-rheological fluid-based torque generation components for use in active rehabilitation devices, in Proceedings of the SPIE, Vol. 6174, 61742C (2006)
9. J.Z. Chen, W.H. Liao, Design, testing and control of a magnetorheological actuator for assistive knee braces. Smart Mater. Struct. **19**, 035029 (2010)
10. M. Nakano et al., Dynamic shear flow behavior of magneto-rheological fluid between two rotating parallel disks under relatively weak magnetic field. JSME Int. J. Series B **48**(3), 494–500 (2005)
11. M. Nakano, H. Yamamoto, M.R. Jolly, Dynamic viscoelasticity of a magnetorheological fluid in oscillatory slit flow. Int. J. Modern Phys. B **13**(14–16), 2068–2076 (1999)
12. N.A.S. Taylor, R.H. Sanders, E.I. Howick, S.N. Stanley, Static and dynamic assessment of the Biodex dynamometer. Eur. J. Appl. Physiol. **62**(3), 180–188 (1991)
13. N. Niwa, M. Ikai, K. Yabe, Relation between joint angle and leg muscle strength. Japan J. Phys. Educ. Health Sport Sci. **9**(1), 173 (1964). (in Japanese)
14. S. Dong, K. Lu, J.Q. Sun, K. Rudolph, A prototype rehabilitation device with variable resistance and joint motion control. Med. Eng. Phys. **28**(4), 348–355 (2006)

Automated and Frequent Calibration of a Robot Manipulator-mounted IR Range Camera for Steel Bridge Maintenance

Andrew Wing Keung To, Gavin Paul, David Rushton-Smith, Dikai Liu and Gamini Dissanayake

Abstract This paper presents an approach to perform frequent hand-eye calibration of an Infrared (IR) range camera mounted to the end-effector of a robot manipulator in a field environment. A set of three reflector discs arranged in a structured pattern is attached to the robot platform to provide high contrast image features with corresponding range readings for accurate calculation of the camera-to-robot base transform. Using this approach the hand-eye transform between the IR range camera and robot end-effector can be determined by considering the robot manipulator model. Experimental results show that a structured lighting-based IR range camera can be reliably hand-eye calibrated to a six DOF robot manipulator using the presented automated approach. Once calibrated, the IR range camera can be positioned with the manipulator so as to generate a high resolution geometric map of the surrounding environment suitable for performing the grit-blasting task.

1 Introduction

In order for a field robot to successfully operate in an unstructured complex environment, a robot must firstly be able to sense and explore the surrounds [3]. For a field operation such as steel bridge maintenance, a robot manipulator can be used to perform the dangerous abrasive blasting task of removing old lead paint and rust [13]. Currently, a mobile manipulator platform (Fig. 1) is being developed for deployment in a steel bridge maintenance field environment to perform grit-blasting. The mobile platform firstly builds a geometric map of the environment using a range camera, and then performs abrasive blasting on the identified steel surfaces. The generated geometric map needs to be accurate and high-resolution for two reasons; so that a safe blasting trajectory can be planned for the manipulator; and so that

A. W. K. To (✉) · G. Paul · D. Rushton-Smith · D. Liu · G. Dissanayake
University of Technology, Sydney, NSW, Australia
e-mail: andrew.w.to@eng.uts.edu.au

K. Yoshida and S. Tadokoro (eds.), *Field and Service Robotics*, 205
Springer Tracts in Advanced Robotics 92, DOI: 10.1007/978-3-642-40686-7_14,

execution of this path produces good abrasive blasting results. However, maintaining accurate mapping is a significant challenge in steel bridge maintenance where the work safety process requires the mobile robot platform to be regularly disassembled to be cleaned and relocated. Additionally, the camera mounting position is prone to change slightly after prolong use due to vibration and blast-hose kinks. Thus, frequent in-situ recalibration is necessary to maintain the required mapping accuracy.

Range camera calibration in robotics is commonly performed to improve both mapping accuracy and multiple scan registration [2, 4]. For example, a checkerboard plane observed from multiple view points [15, 17, 18] is used to calibrate for the range camera's intrinsic and extrinsic parameters (e.g. intrinsic parameters for image projection into 3D coordinates, and extrinsic parameters that describes the position of the range camera relative to a world coordinate). In the case where a range camera is mounted to the end-effector of a robot manipulator [4, 12] extrinsic calibration is also carried out to discover the position of the range camera relative to the end-effector. Generally referred to as hand-eye calibration, the resulting transform is used in geometric map building to register range data taken from different viewpoints into a global world coordinate frame.

Depending on the type of range sensor attached to the robot manipulator, different hand-eye calibration approaches are used. For passive types such as the stereo vision camera, hand-eye calibration is performed using corner and edge features extracted from images of the surrounding environment and/or containing calibration objects [5, 6, 9]. Passive range sensors are limited to a sufficiently illuminated environment to extract features accurately. Therefore, calibration performed in the field environment will require the cumbersome setup of additional lighting infrastructure. Alternatively, active range sensing devices including the laser range scanner and the Infrared (IR) range camera are able to accurately extract scene features without the need for external lighting infrastructure. It has been shown through field trials that the bridge vibration caused by passing traffic affects the scan quality of the laser range scanner. As such, the IR range camera which can instantaneously capture the range data of the whole image frame is preferred in this application. Hand-eye calibration of the IR range camera is performed by extracting 3D edge and plane features from the range data of either the sensed environment and/or a 3D calibration object [10, 14]. However, reliable 3D feature extraction in a dusty and harsh bridge environment is still a challenge that needs addressing to enable the automated and frequent in-situ hand-eye recalibration. This paper proposes an approach that uses image processing to aid the extraction of 3D features for hand-eye calibration of an IR range camera in a bridge. Similar to Gatla et al. [6], which uses a project laser point to generate a high-contrast feature in the image for stereo vision hand-eye calibration, this approach proposes the novel use of reflective material (reflector discs) to create high contrast features in the captured IR image. These reflector discs are arranged in a structured pattern to enable the consistent extraction of 3D points that can be distinctly feature matched for hand-eye calibration.

The paper is structured as follows; Sect. 2 presents the calibration approach which is made up of three main methods: a method for identifying a feature point in 3D

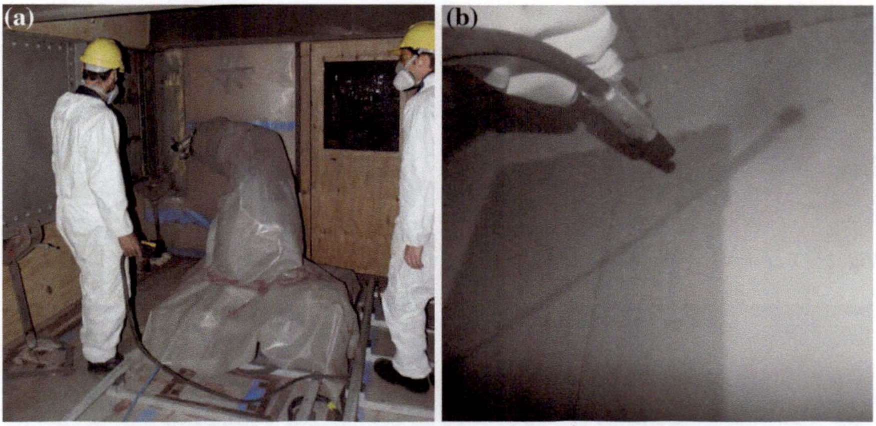

Fig. 1 **a** A mobile robot manipulator in a steel bridge maintenance environment; **b** Grit-blasting work performed using the robot manipulator

from each reflector disc; a method for calculating the camera-to-robot base transform using the identified feature points; and a method for calculating the hand-eye transform between the IR range camera and the end-effector of the robot manipulator. Section 3 presents the experimental results of feature point extraction and the camera calibration. Section 4 presents a discussion of the results along with conclusions and future work.

2 Proposed Approach

2.1 Methodologies

There are three main prerequisites for an automated approach to the calibration of a hand-eye transform ${}^{e}T_{s}$ for an IR range camera mounted to a robot manipulator. Firstly the model of the robot manipulator must be accurate [8, 11]. Secondly, the intrinsic of the IR range camera must be known [9]. Finally, reflector discs must be installed such that the robot base-to-object relationship can be established using the technique [4]. Figure 2 shows the overview of the proposed hand-eye calibration approach. Initially a set of robot manipulator poses are selected so the IR range camera on the end-effector is positioned and orientated to capture IR images containing all the reflector discs. The pixels representing the reflector discs in the IR image are shown to have higher intensity values than the pixels of the surrounding environment. The IR range camera also provides a 3D point cloud generated from the range data and IR camera's intrinsic model. Using the available data, a feature point in 3D can be selected from the centre of each reflector disc using the devised combination of image processing, plane fitting, and circle fitting techniques. Subsequently, the camera-to-robot base transform ${}^{o}T_{s}$ is calculated by matching the extracted feature

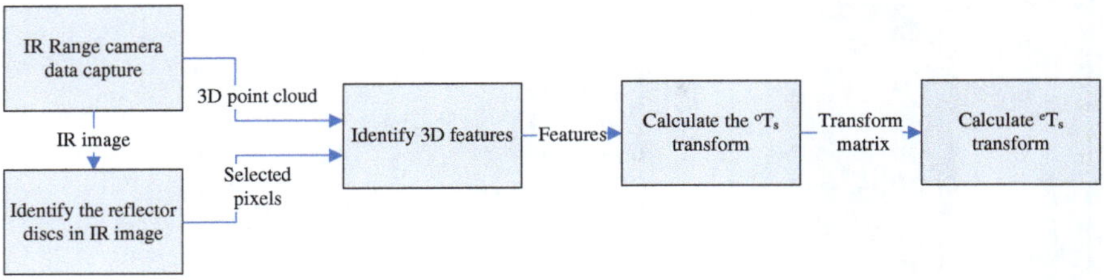

Fig. 2 Overview of the process for hand-eye calibration

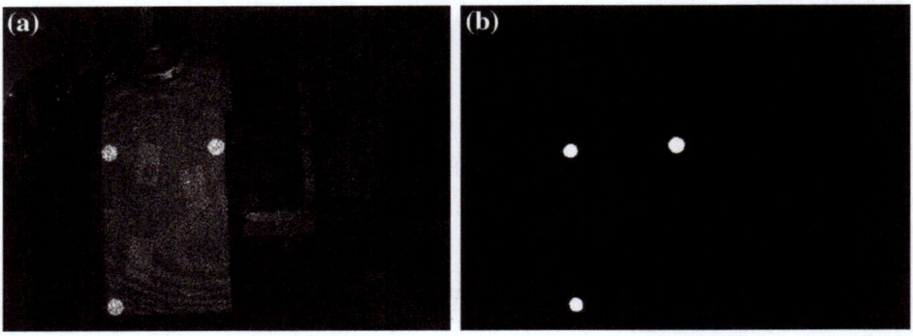

Fig. 3 **a** IR image; **b** Binary image of reflector discs

points (in the camera frame) with the established ground truth location (in the robot base frame) of the feature points. The hand-eye transform ^eT_s is then calculated using the camera-to-robot base transform and the end-effector-to-robot base transform $^oT_e(Q)$ calculated from the model of the robot manipulator.

2.2 Feature Points Identification

Once a set of IR range images is captured, the 3D features corresponding to reflector discs are extracted using the IR image and the 3D point cloud. The captured IR image of the reflector discs (shown in Fig 3a) is processed using a median filter to isolate the reflector disc pixel from the projected IR pattern. An incrementally increasing filter kernel is used to remove the projected IR pattern gradually from the IR image and terminates when only the reflector disc blobs remain in the IR image. Region fill is then applied to the filtered image to remove the holes in the reflector disc blobs and then the image is converted from greyscale to black and white by thresholding the intensity values. The resulting binary image is shown in Fig. 3b. Each of the pixel in the blobs are indexed to the corresponding 3D coordinate in the point cloud to represent the reflector discs location in 3D as shown in Fig. 4a.

Perspective correction is applied to remove the shape distortion caused by a non-perpendicular viewing angle from the IR range camera to the reflector discs. This is achieved by applying the RANSAC plane fitting algorithm to the 3D dataset

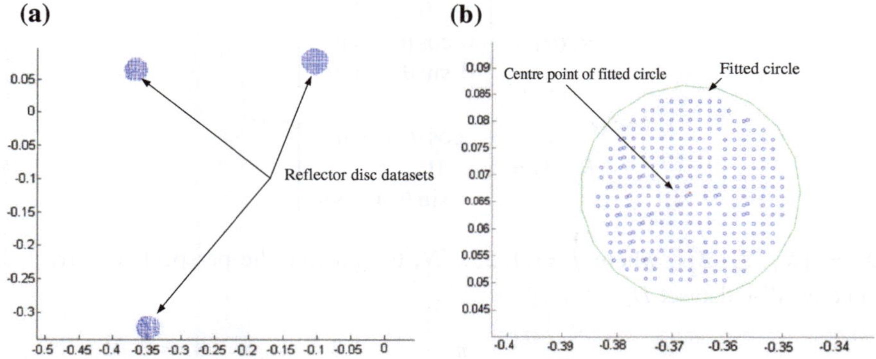

Fig. 4 **a** Datasets of points in 3D representing the reflector discs; **b** Circle fit on a dataset

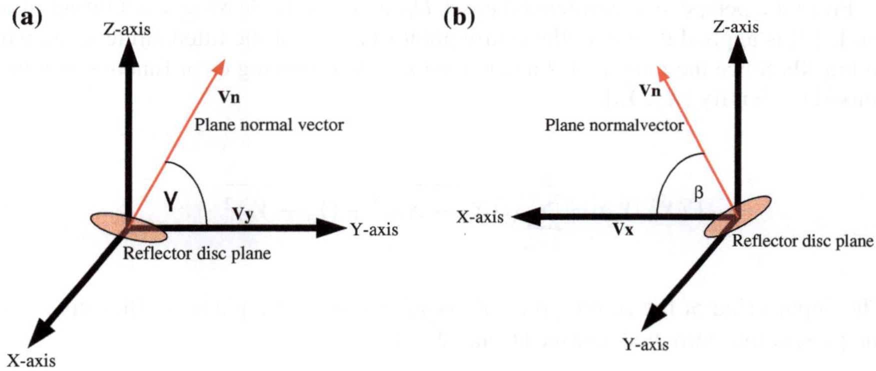

Fig. 5 **a** X-axis rotation; **b** Y-axis rotation

of each reflector disc as in [16], and then identifying the required rotations to align the fitted plane to the image plane of the IR range camera. As illustrated in Fig. 5, it is possible to determine the X and Y axis rotations: β, the angle between the normal vector of the fitted plane V_n and the x-axis of the camera frame V_x, and γ the angle between V_n and the y-axis of the camera frame V_y

$$\beta = \arccos(V_n \cdot V_x) \quad (1)$$

$$\gamma = \arccos(V_n \cdot V_y) \quad (2)$$

The rotations matrices about the x-axis R_x and the y-axis R_y by the angles β and γ, are applied to each reflective disc's 3D point cloud.

$$R_x(\theta) = \begin{bmatrix} 1 & 0 & 0 \\ 0 & \cos\theta & -\sin\theta \\ 0 & \sin\theta & \cos\theta \end{bmatrix} \tag{3}$$

$$R_y(\theta) = \begin{bmatrix} \cos\theta & 0 & \sin\theta \\ 0 & 1 & 0 \\ -\sin\theta & 0 & \cos\theta \end{bmatrix} \tag{4}$$

$D_i = [X_j, Y_j, Z_j]$, where $j = 1, 2, \ldots N$, to generate the perspective corrected reflective disc dataset $\hat{D}_i$,

$$\hat{D}_i = D_i \times R_x(\frac{\pi}{2} - \gamma)R_y(\frac{3}{2}\pi + \beta) \tag{5}$$

From the perspective corrected dataset $\hat{D}_i$, a geometric least-squares fitting of a circle [7] is applied to extract the centre point $[\hat{X}_c, \hat{Y}_c]$, of the fitted circle as shown in Fig. 4b. Since the radius r is known *a priori*, the following error function is minimised to identify $[\hat{X}_c, \hat{Y}_c]$,

$$F(\hat{X}_c, \hat{Y}_c) = \sum_{j=1}^{N} \sqrt{(\hat{X}_j - \hat{X}_c)^2 + (\hat{Y}_j - \hat{Y}_c)^2} - r \tag{6}$$

The depth value of the centre point, $\hat{Z}_c$ is solved using the plane coefficients from the perspective corrected dataset $\hat{D}_j$ and $\hat{X}_c$, $\hat{Y}_c$,

$$\hat{Z}_c = \frac{a\hat{X}_c + b\hat{Y}_c + d}{-c} \tag{7}$$

Finally, $[\hat{X}_c, \hat{Y}_c, \hat{Z}_c]$ is rotated into the IR range camera frame to obtain the 3D feature point $[X_c, Y_c, Z_c]$,

$$[X_c, Y_c, Z_c] = [\hat{X}_c, \hat{Y}_c, \hat{Z}_c] \times R_x(\frac{\pi}{2} - \gamma)^{-1}R_y(\frac{3}{2}\pi + \beta)^{-1} \tag{8}$$

The 3D feature point is calculated for each reflector disc and used to match against a pre-measured ground truth location of the reflector discs relative to the robot base, further discussed in Sect. 2.3.

2.3 Camera-to-Robot Transform Through 3D Feature Matching

In order to calculate the camera-to-robot base transform ${}^{o}T_s$, a feature matching technique is used to relate the IR range camera frame with the robot base frame. The robot base-to-object relationship describing the object in the robot base frame is assumed to be static and in a position and orientation that can be easily measured, and the camera-to-object relationship describing the object in the IR range camera frame is dynamic. As shown in Fig. 6a, the transform ${}^{o}T_s$ describes the translation and rotation required to match the object in the IR range camera frame to the measured location of the object in the robot base frame.

M number of reflector discs placed in a specific configuration on a flat plane are used to determine the correspondence between the two coordinate frames. The location of the reflector discs relative to the robot base is known accurately by moving an end-effector pointer tool to the reflector disc positions as in [4] and the camera-to-object relationship is obtained using the method described in Sect. 2.2. Feature point matching is performed using Singular Value Decomposition (SVD) [1] to find the least-square fit of feature points in the IR range camera frame f_i with the feature points in robot base frame f_i' where $i = 1, 2, \ldots M$, using the equation

$$f_i' = {}^{o}R_s f_i + {}^{o}t_s + N_i \tag{9}$$

where ${}^{o}R_s$ is the rotation matrix, ${}^{o}t_s$ is the translation vector and N_i the noise vector. The least-square solution to ${}^{o}R_s$ and ${}^{o}t_s$ is taken as the camera-to-robot base homogeneous transform ${}^{o}T_s$,

$${}^{o}T_s = \begin{bmatrix} & {}^{o}R_s & & {}^{o}t_s \\ 0 & 0 & 0 & 1 \end{bmatrix} \tag{10}$$

2.4 Hand-Eye Transform and Point Cloud Registration

In order to register a point cloud from the IR range camera frame to the robot base frame, the hand-eye transform ${}^{e}T_s$ is calculated. As shown in Fig. 6b, the hand-eye transform is constant between the camera-to-robot base frame and the end-effector-to-robot base frame for any given robot manipulator pose. Using the camera-to-robot base transform ${}^{o}T_s$ (Sect. 2.3) and the end-effector-to-robot base transform ${}^{o}T_e(Q)$ calculated from the corresponding robot manipulator pose with joint angles, $Q = [q_1, q_2, \ldots q_6]^T$ using the D-H model of the robot manipulator,

$${}^{o}T_e(Q) = \prod_{i=1}^{6} {}^{i-1}T_i(q_i) \tag{11}$$

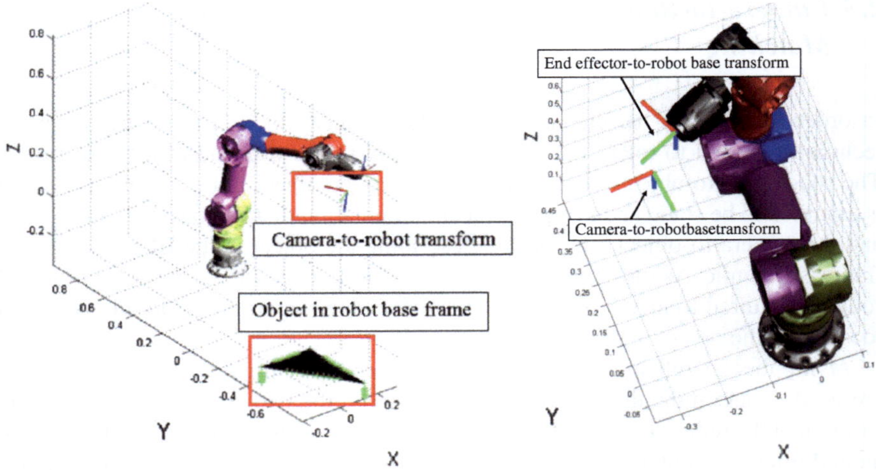

Fig. 6 **a** An IR range camera attached to a robot manipulator observing the calibration plate; **b** Camera-to-robot base frame and end-effector-to-robot base frame

the following expression is solved for the hand-eye transform.

$$^{e}T_{s} = {}^{o}T_{s} \times {}^{o}T_{e}(Q)^{-1} \tag{12}$$

Using the calculated $^{e}T_{s}$, the registration of a point cloud P_k in the IR range camera frame to the robot base frame, P'_k where $k = 1, 2, \dots N$, is achieved by determining the camera-to-robot base transform $^{o}T_{s}(Q)$ for the end effector-to-robot base transform $^{o}T_{e}(Q)$, specific to a robot manipulator pose with joint angles, $Q = [q_1, q_2, \dots q_6]^T$,

$$^{o}T_{s}(Q) = {}^{o}T_{e}(Q) \times {}^{e}T_{s} \tag{13}$$

where

$$[P'_k, 1] = [P_k, 1] \times {}^{o}T_{s}(Q) \tag{14}$$

The hand-eye transform is used to register point clouds, from the IR range camera frame into the robot base frame. A rich geometry map of the environment can be constructed using a single IR range camera by moving the manipulator into different pose [12].

3 Experiments and Results

3.1 Experiment Setup

An IR range camera developed by Microsoft and PrimeSense (Kinect) is used in the following experiments. The sensor interprets 3D scene information through the use of continuous infrared structured light created by an onboard IR projector. Figure 7a shows the mapping experiment setup where the IR range camera is mounted onto the end-effector of a six DOF robot manipulator constructed using Schunk powercube modules. Figure 7b, shows reflector discs setup on the calibration plate used in the experiments, which contains three reflector discs arranged in a right scalene triangle configuration to create a unique feature set. A reflector disc diameter size of 40 mm is selected to provide adequate surface area for depth reading at the tested viewing distance. The reflector discs are arranged such that the disc A-to-B distance is 260 mm, the disc B-to-C distance is 385 mm, and $\angle ABC = 90^\circ$

3.2 Experiment: 3D Feature Point Extraction from Reflector Discs

In this initial experiment, the accuracy of extracting feature points in 3D using the range camera's IR image and range data at increasing distances is evaluated. Figure 8a shows the controlled office environment where the calibration plate is sensed from viewing distances between 500 and 2,000 mm at 100 mm increments. Figure 8b shows the bridge environment setup where a field experiment is conducted to test feature point extraction. The feature points are extracted using the method detailed in Sect. 2.2 at each viewing distance. and the point-to-point distances calculated to evaluate the performance.

Table 1, shows the point-to-point error at each viewing distance for the experiment conducted in the office. The error is calculated by subtracting the sensed point-to-point distance from the known ground truth distance. A positive error value indicates a sensed distance greater than the ground truth and a negative error value indicates a sensed distance shorter than the ground truth. Results shows that the viewing distances between 900 and 1,500 mm have point-to-point distance error values within ± 6 mm for all point-to-point combinations. The maximum error observed in this experiment is at a viewing distance of 1,900 mm with a point A-to-C error of 12.37 mm. Furthermore, Fig. 8c shows the limitations determined from this experiment. Although the IR range camera is capable of capturing range data at distances between 500 and 800 mm. But due to the field-of-view constraints and/or the intensity saturation of the reflector discs resulting in null range returns, feature point extraction was not possible within that range. The outcome of the bridge experiment determined that no significant interference was caused by the addition of field elements (i.e. dust in air, shiny metal, lighting).

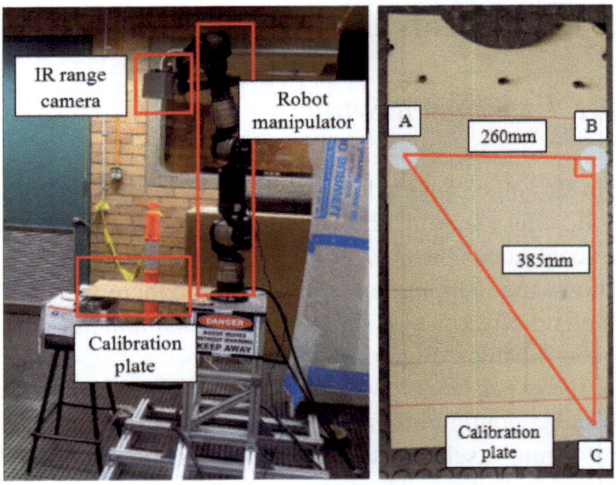

Fig. 7 **a** An IR range camera attached to a robot manipulator observing the calibration plate; **b** The calibration plate with reflector discs

Table 1 Office environment experiment: point to point errors

Viewing distance (mm)	Error A-B (mm)	Error B-C (mm)	Error A-C (mm)
900	2.96	2.22	4.01
1,000	0.16	1.91	0.82
1,100	3.96	1.67	−0.92
1,200	3.06	4.1	3.55
1,300	3.0	3.8	2.03
1,400	1.59	5.24	3.84
1,500	2.13	5.07	4.33
1,600	2.63	6.93	8.2
1,700	4.15	6.74	6.73
1,800	5.72	9.02	11.46
1,900	7.37	9.92	12.37
2,000	9.26	9.86	11.29

3.3 3D Feature Matching Experiment

In this experiment the performance of feature extraction and matching is evaluated with the IR range camera mounted on the end-effector. The configuration of this experiment is shown in Fig. 7a, with the reflector discs positioned statically in a known position relative to the robot base. The robot manipulator is used to position and orientate the IR range camera at 10 different viewpoints to provide various observations of the reflector discs. The method discussed in Sect. 2.3 is used to find a transformation to match the extracted feature points into the measured ground truth location. Performance is based on the distance between the transformed feature point and the corresponding ground truth location of the feature point.

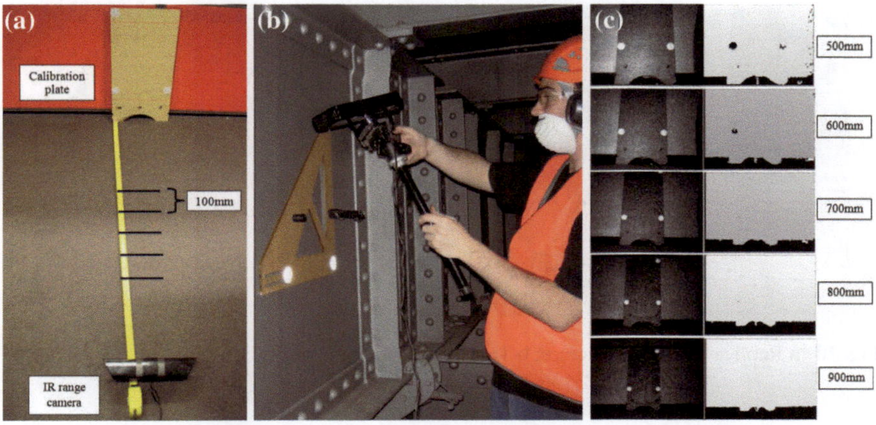

Fig. 8 **a** Office environment experiment setup for 3D feature point extraction; **b** Bridge environment experiment setup for 3D feature point extraction; **c** Example IR image and range image from 500 to 900 mm in controlled environment

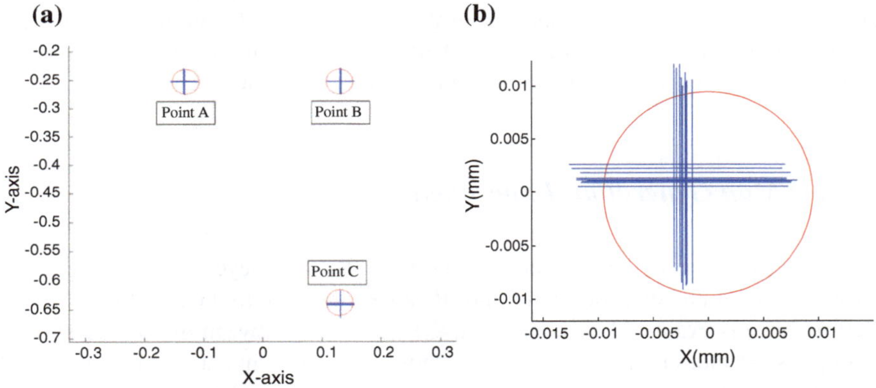

Fig. 9 **a** Feature matching between extracted feature points (+) to measured ground truth location (o) shown in the robot base frame; **b** Expanded view of point A

Table 2 Mean and standard deviation of position error for the 10 manipulator poses

	PointA error (mm)	PointB error (mm)	PointC error (mm)
Mean error	0.2593	0.1758	0.2574
Std Dev	0.09	0.0497	0.09

Figure 6a illustrates the feature matching for a viewpoint provided by a specific robot manipulator pose. The triangle represents the feature points extracted by the IR range camera transformed into the robot base frame using the camera-to-robot base transform. The crosses represent the ground truth location of the feature points.

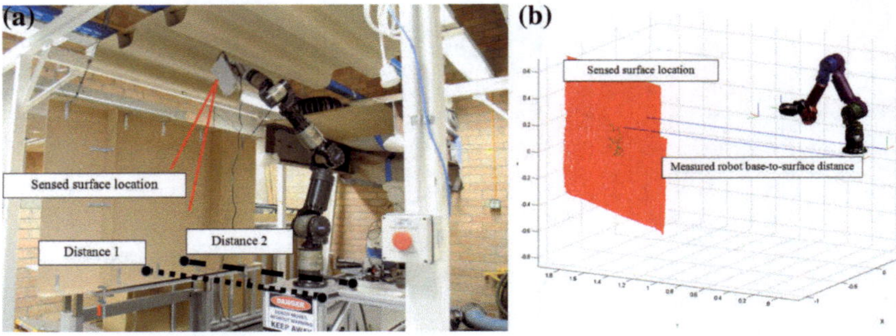

Fig. 10 **a** Replica of bridge structure; **b** Transformed sensed point cloud data with ray-casting

Figure 9a shows an overlay view of the feature points matched from 10 different robot manipulator poses. Figure 9b shows the expanded view of point A, the crosses represent the sensed feature points and the circles represents the ground true location of the reflector discs. Table 2, shows the mean and standard deviation of the position error of the sensed feature points from the ground truth. Execution time for the positioning of the manipulator into the 10 different poses with feature extraction and matching in this experiment required approximately 5 minutes.

3.4 3D Map Generation Experiment

This experiment tests the accuracy of the generated hand-eye transform by registering the point clouds captured by the IR range camera into the robot base frame. Performance is evaluated by comparing the difference between the sensed surface with physical measurements. Figure 10a shows the environment that is a replica of a steel bridge I-beam channel containing a main I-beam web with top and bottom flanges. Ray-casting from the corners of the robot platform is performed to identify the intersection point with the sensed surface as shown in Fig. 10a, b. The distance between the intersection point and the platform corner point is calculated and then compared with physical measurements. The difference between the sensed surface and the physically measured distance is recorded and repeated for a set of 10 different robot manipulator poses that position the IR range camera so as to observe the same side wall surface. Table 3 shows the mean error and standard deviation between the sensed surface distance and measured physical distance.

Table 3 Surface to robot base distance error over the 10 IR camera locations

	Distance 1 error (mm)	Distance 2 error (mm)
Mean error	13.32	16.37
Std Dev	5.04	4.3

3.5 Discussions

The experimental results show that calibration for an IR range camera is possible using the simple reflector discs. Experiment 1 demonstrated that data from an IR range camera can be processed by the presented approach to simultaneously detect a set of reflector discs in the IR image and also to provide range data on the reflector discs for feature extraction. Experiment 2 and 3 demonstrated that the identified feature points are accurate enough to be used in the desired application, and thus can generate the required hand-eye transform such that multiple point clouds can be fused and represented in the robot base frame. These results demonstrate the potential use of reflector discs as a dispensable and effective tool to calibrate an IR range camera in a field environment.

Currently the approach is limited by the viewing distance. Due to intensity saturation at close viewing distances, an accurate range reading of the reflector disc is not possible. Thus, the feature extraction scheme can be modified to select 3D dataset points from the surface surrounding each reflector disc as opposed to the surface on each reflector disc. The range returns for pixels surrounding each reflector disc does not suffer from intensity saturation, therefore feature extraction should be possible for a closer viewing distances than is currently possible. A redesign of the calibration plate will also be necessary to provide the surface plane around each reflector disc. Further onsite experiments will also be conducted to extend the testing of calibration performance in the presence of vibration and dust.

4 Conclusion

This paper has demonstrated an automated approach to hand-eye calibration for use in a harsh field environment. The proposed approach has successfully used an IR range camera to extract feature points from a set of three reflector discs to determine the camera-to-robot manipulator base transform, and subsequently the hand-eye transform by considering the robot manipulator model. Experiment results indicate high performance in feature point extraction using the image processing and 3D manipulation on the IR image and the point cloud. The generated hand-eye transform is able to accurately register point clouds into the robot base frame and verified through physical measurements. Future work will involve extending the range of the feature extraction scheme, the redesign of the calibration plate, and further field testing of the approach.

Acknowledgments This work is supported by the Centre of Excellence for Autonomous Systems (CAS), the Roads and Maritime Services (RMS) and the University of Technology, Sydney.

References

1. K.S. Arun, T.S. Huang, S.D. Blostein, Least-squares fitting of two 3-d point sets, in *IEEE Transactions on Pattern Analysis and Machine Intelligence*, pp. 698–700 (1987)
2. P. Biber, S. Fleck, T. Duckett, 3d modeling of indoor environments for a robotic security guard, in *Proceedings of the IEEE Computer Society Conference on Computer Vision and Pattern Recognition*, pp. 124–124 (2005)
3. Y.C. Chang, Y. Yamamoto, Dynamic decision making of mobile robot under obstructed environment, in *International Conference on Intelligent Robots and Systems*, pp. 4091–4096 (2006)
4. M. Dinham, G. Fang, A low cost hand-eye calibration method for arc welding robots, in *International Conference on Robotics and Biomimetics* (*ROBIO*), pp. 1889–1893 (2009)
5. M. Dinham, G. Fang, Low cost simultaneous calibration of a stereo vision system and a welding robot, in *IEEE International Conference on Robotics and Biomimetics (ROBIO)*, pp. 1452–1456 (2010)
6. C.S. Gatla, R. Lumia, J. Wood, G. Starr, Calibrating pan-tilt cameras in robot hand-eye systems using a single point, in *IEEE International Conference on Robotics and Automation, 2007*, pp. 3186–3191 (2007)
7. W. Gander, G.H. Golub, R. Strebel, Strebel least squares fitting of circles and ellipses. BIT Numer. Math. **34**(4), 558–578
8. J. Ji, L. Sun, L. Yu, A new pose measuring and kinematics calibrating method for manipulators, in *IEEE International Conference on Robotics and Automation, 2007*, pp. 4925–4930 (2007)
9. H. Malm, A. Heyden, Simplified intrinsic camera calibration and hand-eye calibration for robot vision, in *International Conference on Intelligent Robots and Systems* (*IROS*), vol. 1, pp. 1037–1043 (2003)
10. S. May, D. Droeschel, S. Fuchs, D. Holz, A. Nuchter, Robust 3d-mapping with time-of-flight cameras, in *International Conference on Intelligent Robots and Systems* (*IROS*), pp. 1673–1678 (2009)
11. A. Omodei, G. Legnani, R. Adamini, Three methodologies for the calibration of industrial manipulators: experimental results on a scara robot. J. Robot. Syst. **17**(6), 291–307 (2000)
12. G. Paul, N. Kirchner, D. Gamini, D.K. Liu, An effective exploration approach to simultaneous mapping and surface material-type identification of complex three-dimensional environments. J. Field Robot. **26**(11–12), 915–933 (2009)
13. G. Paul, S. Webb, D.K. Liu, G. Dissanayake, A robotic system for steel bridge maintenance: field testing, in *Australasian Conference on Robotics and Automation*, pp. 1–8 (2010)
14. Q. Shi, N. Xi, Y. Chen, W. Sheng, Registration of point clouds for 3d shape inspection, in *International Conference on Intelligent Robots and Systems*, pp. 235–240 (2006)
15. K.H. Strobl, G. Hirzinger, More accurate camera and hand-eye calibrations with unknown grid pattern dimensions, in *IEEE International Conference on Robotics and Automation* (*ICRA*), pp. 1398–1405 (2008)
16. M.Y. Yang, W. Forstner, *Plane detection in point cloud data*. Technical Report, Department of Photogrammetry, pp. 1–16 (2010)
17. Q. Zhang, R. Pless, Extrinsic calibration of a camera and laser range finder (improves camera calibration), in *IEEE/RSJ International Conference on Intelligent Robots and Systems* (*IROS*), vol. 3, pp. 2301–2306 (2004)
18. Z. Zhang, H. Ma, S. Meng, S. Zhang, T. Guo, J. Chen, X. Hu, New calibration method of fringe projection imaging system, in *International Conference on Electronics and Optoelectronics* (*ICEOE*), *2011*, vol. 1, pp. 88–91 (2011)

Vertical Infrastructure Inspection Using a Quadcopter and Shared Autonomy Control

Inkyu Sa and Peter Corke

Abstract This paper presents a shared autonomy control scheme for a quadcopter that is suited for inspection of vertical infrastructure—tall man-made structures such as streetlights, electricity poles or the exterior surfaces of buildings. Current approaches to inspection of such structures is slow, expensive, and potentially hazardous. Low-cost aerial platforms with an ability to hover now have sufficient payload and endurance for this kind of task, but require significant human skill to fly. We develop a control architecture that enables synergy between the ground-based operator and the aerial inspection robot. An unskilled operator is assisted by onboard sensing and partial autonomy to safely fly the robot in close proximity to the structure. The operator uses their domain knowledge and problem solving skills to guide the robot in difficult to reach locations to inspect and assess the condition of the infrastructure. The operator commands the robot in a local task coordinate frame with limited degrees of freedom (DOF). For instance: up/down, left/right, toward/away with respect to the infrastructure. We therefore avoid problems of global mapping and navigation while providing an intuitive interface to the operator. We describe algorithms for pole detection, robot velocity estimation with respect to the pole, and position estimation in 3D space as well as the control algorithms and overall system architecture. We present initial results of shared autonomy of a quadcopter with respect to a vertical pole and robot performance is evaluated by comparing with motion capture data.

I. Sa (✉) · P. Corke
Queensland University of Technology, Brisbane, Australia
e-mail: i.sa@qut.edu.au

P. Corke
e-mail: peter.corke@qut.edu.au

K. Yoshida and S. Tadokoro (eds.), *Field and Service Robotics*,
Springer Tracts in Advanced Robotics 92, DOI: 10.1007/978-3-642-40686-7_15,

1 Introduction

The options for inspecting locations above the ground are quite limited, and all are currently cumbersome. Ladders can be used up to a height of 10–15 m but are quite dangerous: each year 160 people are killed and 170,000 injured in falls from ladders in the United States.[1] A person can be lifted in the basket of a cherry picker up to a height of 15 m but vehicle access is required and the setup time is significant. Beyond that height a person either climbs up the structure or rappels down from the top, both of which are slow and hazardous. Inspection from manned rotorcraft is possible but is expensive and only suitable in non-urban environments. In recent years we have seen significant advances in small VTOL platforms, in particular quadcopters, driven by advances in power electronics, MEMS sensors and microcontrollers. These systems are low-cost and have sufficient payload and endurance for useful inspection missions. They are also low-weight which reduces the hazard due to their deployment.

This paper presents a shared autonomy system for inspection of vertical infrastructure—tall man-made structures such as streetlights, electricity poles or the exterior surfaces of buildings—using a vertical take-off and landing (VTOL) robot platform. Shared autonomy indicates that the major fraction of control is accomplished by the onboard computer. The operator provides "high level" commands in a reduced DOF task space, while the robot is responsible for stable flight, disturbance rejection and collision avoidance. This allows an unskilled operator to easily and safely control a quadcopter to examine locations that are otherwise difficult to reach (Fig. 1).

The presented VTOL flying robot has functionalities of pole detection and task-space operator command input. Implicit in the inspection task is the requirement to fly close to structures with which a collision would significantly damage the vehicle. Air flow around tall structures results in eddies that induce disturbances on

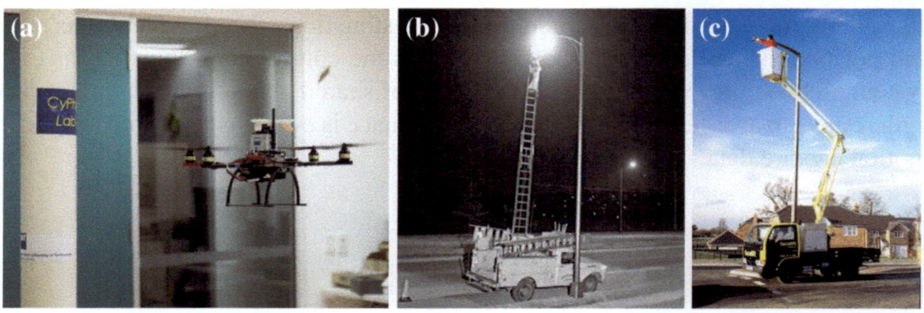

Fig. 1 **a** The Cyphy Lab MikroKopter research platform. The pole can be seen on the left of the image. **b** A dangerous situation to inspect or repair a street light (Baltimore museum of industry; http://www.thebmi.org/). **c** Sufficient space is required for vehicle access and it is a time consuming process to setup operation (Facelift; http://www.facelift.co.uk/)

[1] May 2009 Consumer Reports magazine. http://www.consumerreports.org/cro/magazine-archive/may-2009/may-2009-toc.htm

the vehicle which must be robustly rejected to ensure safety and task performance. This requires accurate and fast velocity and position estimation and an appropriate control methodology.

This paper is organised as follows: Sect. 2 presents relevant research on quadcopter and bio-inspired climbing robots suitable for inspection. Section 3 explains the methodologies: system modeling and identification, velocity estimation and nested controllers, pole detection algorithm, shared control scheme. We present our experimental results in Sect. 4, and important technological trends and conclusions in Sect. 5.

2 Related Work

Robotics and mechatronics researchers have demonstrated a variety of climbing robots for vertical infrastructure inspection. Typically, these robots are inspired by reptiles, mammals and insects and their type of movement varies between sliding, swinging, extension and jumping. The MATS robot has 5 DOF and a symmetrical mechanism that showed good mobility features for travel, however, it requires a special docking stations to hold itself [1]. A bio-mimicking robot, StickyBot, has a hierarchical adhesive structure under its toes to hold itself on any kind of surfaces [2]. RiSE V3, a legged locomotion climbing robot, is designed for high-speed climbing of a uniformly convex cylindrical structure, such as a telephone or electricity pole [3]. A bridge cable inspection robot [4] has wheels held against the cable to create a contact force required to move along the cable. These types of robots could not only replace a worker undertaking risky tasks in a hazardous environment but also increase the efficiency of such tasks. However, they require complex mechanical designs, special materials and complicated dynamics analysis. Their applications are limited to specific type of structures, such as cylindrical-shaped poles. VTOL platforms are a feasible alternative to achieving the same goals as climbing robots and involve a much simpler mechanism. Recently, [5] demonstrates embedded stereo camera based egomotion estimation for structures inspections such as a boiler and general indoor scenarios. Although IMU guided feature matching and stereo based camera pose estimation show impressive real-time achievements, it might need integration of control theory to fly in close quarters.

3 Methodologies

This section describes the key approaches of our system: shared control; modeling and system identification; pole detection; velocity estimation and nested controllers.

3.1 Shared Control and Task Frame

Sheridan [6] introduced a spectrum of approaches for remote operation of a system. At one end is "conventional manual control" (system 1) where the system is fully controlled by a human operator and there is no computer-aided functionalities. At the other end is "fully autonomous system" (system 5) where a human operator can observe but cannot intervene in the process. Our proposed system is modelled on Sheridan's "Supervisory Control" architecture, specifically system 4, in which the control loop is closed through a computer but there are still human interventions. This approach allows the high-bandwidth flight control and obstacle avoidance loops to be closed on board the robot with the "high level" commands from the human being treated as requests that will be implemented if safe to do so (Fig. 2).

A task frame (TF) refers to a coordinate frame that can be attached to an object in the workspace [7]. There is a geometric transformation between the world coordinate and TF. The advantages of a TF is that actions which are difficult to express in the world coordinate can be easily specified in the TF. For an inspection task the TF is associated with the operator's current view of the infrastructure and provides an intuitive control framework to the user in which to express desired motion commands. Fig. 3a shows the world coordinate $\boldsymbol{W}$ and the task frame $\boldsymbol{T}$.

A VTOL platform has four DOF (roll, pitch and yaw angles, and throttle) and significant operator skill is required to control position in 3-dimensional Cartesian space. One aspect of this skill is that the roll and pitch angles induce forces on the vehicle, and with relatively little aerodynamic damping these inputs are effectively Cartesian accelerations. The level of skill required is greatly increased when flying next to a large and unforgiving structure in the presence of wind-induced force disturbance. Manual piloting also requires the vehicle to be in the pilot's visual field of the pilot and sufficiently close that its orientation in space can be determined.

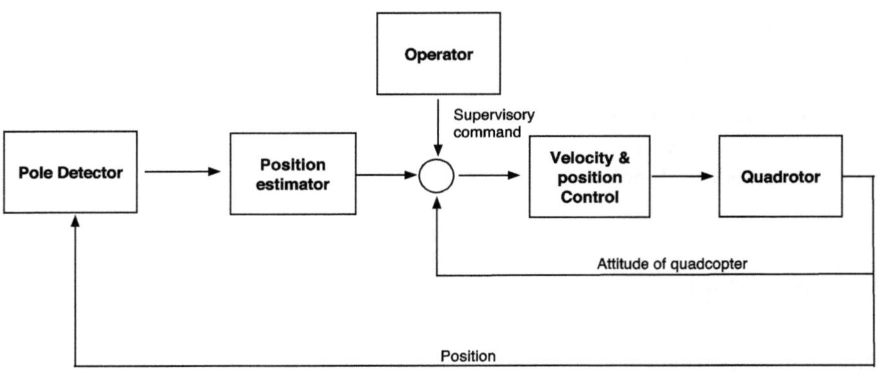

Fig. 2 Hierarchical multi-loop shared control architecture. The inner loop receives a desired goal by the outer loop. Control, Position estimator loops have different update rates for a purpose. *Arrows* indicate data flow directions and specify inputs

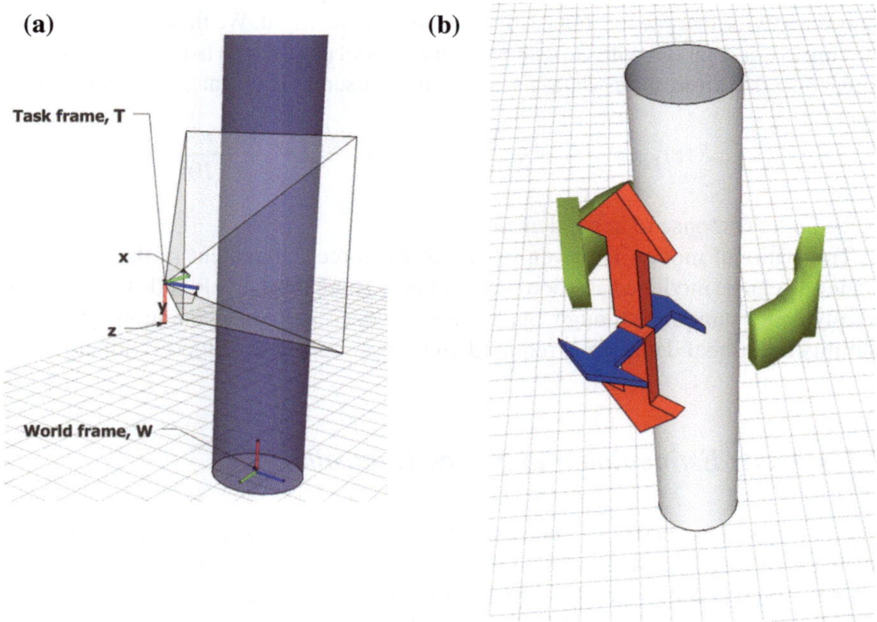

Fig. 3 **a** World frame ***W*** and the task frame ***T***. ***W*** is the centre of the pole and ***T*** denotes a camera coordinate which is equal to a user's point of view. An unskilled operator can easily control the robot because it can localize with respect to the pole. **b** Reduced controllable task degree of freedom (DOF)

For an unskilled operator we need to reduce the number of DOF that must be controlled and make the DOF intuitive and task specific. As shown in Fig. 3b, for a pole inspection task, the operator controls only 2 DOF: distance along the pole and angle around the pole. This is sufficient for inspection of the entire pole area and easy to control.

3.2 Modeling and System Identification

The quadcopter is an under-actuated force-controlled flying vehicle. This force actuation implies that rotational and translational motion can be modeled as a double integrator from command to attitude angle or horizontal position [8, 9]. In our work we use the MikroKopter open-source quadcopter[2] for which there is few engineering documents or published dynamic models [10]. The vehicle has an onboard attitude controller which uses rate and angle feedback from gyroscopes and accelerometers. We identified the dynamics of the closed-loop attitude by recording pilot com-

[2] MikroKopter. http://www.mikrokopter.de/

mands and MikroKopter attitude estimates, for manual flight. We fit an autoregressive moving average model with exogenous inputs model (ARMAX) using recursive least squares to this time series data giving a linear discrete-time (at 50 ms) first-order model

$$F(z)_{pitch} = \frac{0.148}{z - 0.7639}, \quad F(z)_{roll} = \frac{0.145}{z - 0.7704} \tag{1}$$

as the angle response to angle demand.

Translational motion is driven by the thrust force component in the horizontal plane and can be modelled as a double integrator. There is relatively little translational aerodynamic damping, though blade flapping does add some damping [11]. For stability additional damping is required and this necessitates velocity estimation.

3.3 Velocity Estimation and Nested Controllers

The key to stable control of such systems is providing artificial damping through feedback of rotational and translational velocity. In order to introduce damping we require a high quality velocity estimate: smooth, high update rate with low latency. Computing velocity using differentiation of the position from the pole detection and pose estimator results in velocity at 10 Hz with a latency of 100 ms. This significantly limits the gain that can be applied when used for closed-loop velocity control. Instead we use the MikroKopter acceleration measurements (`AccRoll` and `AccNick`) which we read at 20 Hz with low latency and integrate them to create a velocity estimate. We subtract the acceleration due to gravity using the MikroKopter's estimated roll and pitch angles

$$\ddot{x}_Q = \frac{a_x + g \sin\theta}{\cos\theta}, \quad \ddot{y}_Q = \frac{a_y - g \sin\phi}{\cos\phi} \tag{2}$$

where a_x, a_y are the measured acceleration from the flight control board converted to our coordinate system, and θ, ϕ denote the pitch and roll angles respectively. $\{Q\}$ is a coordinate frame centred on the vehicle with axes parallel to the world frame. Acceleration and attitude are returned together in the flight-controller status message at 20 Hz.

As any estimator that relies on integration is subject to substantial errors due to drift, even over quite short time intervals, we therefore fuse these two estimates using a simple discrete-time complementary filter [12] as shown in Fig. 4 and described by

$$\hat{v}_{x_{k+1}} = \hat{v}_{x_k} + (\ddot{x}_{Q_k} + K(v_{x_k} - \hat{v}_{x_k}))\Delta_t \tag{3}$$

where $\hat{v}_{x_k}$ is estimated velocity, v_{x_k} is obtained from differentiation of the laser-based pose estimate and is computed at a slower rate than $\ddot{x}_{Q_k}$ so the filter takes the most recent value, and K is a gain. Complementary filters have been used previously for

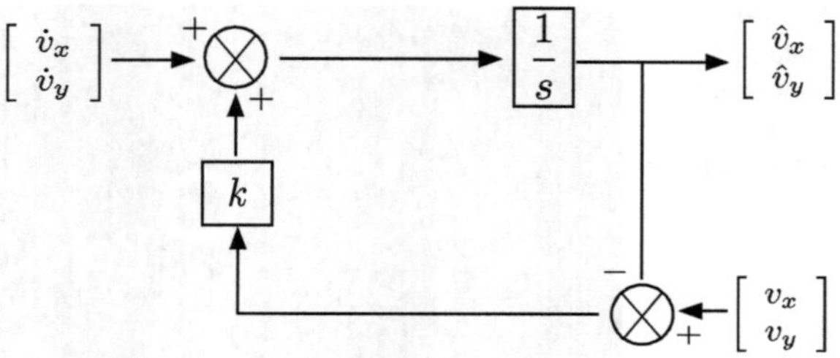

Fig. 4 Complementary filter for velocity estimation. Compared to a Kalman filter the computation is simple, and there is only one tuning parameter, K. $\dot{v}_x$ and $\dot{v}_y$ are obtained from a onboard IMU sensor. v_x and v_y are from a laser range finder

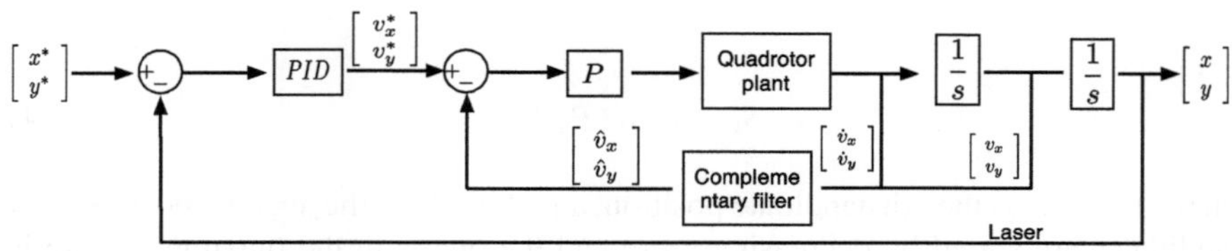

Fig. 5 Velocity estimator and control structure for translational motion. The K_{vel} for the velocity loop is 27 and $K_P = 0.8$, $K_I = 0.1$ and $K_D = 0.7$ for the position PID controller

UAV velocity estimation, such as to fuse velocity from low-rate optical flow with high-rate inertial data [13].

The block diagram of our nested controller is shown in Fig. 5. The inner-loop is a velocity controller with proportional and integral control with feedback of estimated velocity from the complementary filter, Eq. (3). The outer loop is a position controller with proportional control.This structure is equivalent to a proportional-integral-derivative, however the nested structure decouples the different sampling rates of the position sensor and the velocity sensor. The inner-loop runs at 20 Hz and the outer-loop at 10 Hz. As we showed in [10] this simple control architecture gives performance that is comparable with other published results that are using 40 Hz laser scanners and 1 kHz IMU sample rates.

3.4 Pole Detection

We use an Hokuyo model URG-04LX laser range finder (10 Hz and 4 m range) to detect the pole. As shown in Fig. 6 the laser detects the 15 cm radius pole as a straight line rather than a circular arc, and we believe this is an artifact of filtering firmware in the laser range finder. We use a Split-Merge line extraction [14] routine on the raw laser data, followed by target discrimination (see Algorithm 1), tracking and filtering to estimate the range and bearing of the pole with respect to the robot.

We score each candidate using a previous detected averaged position.

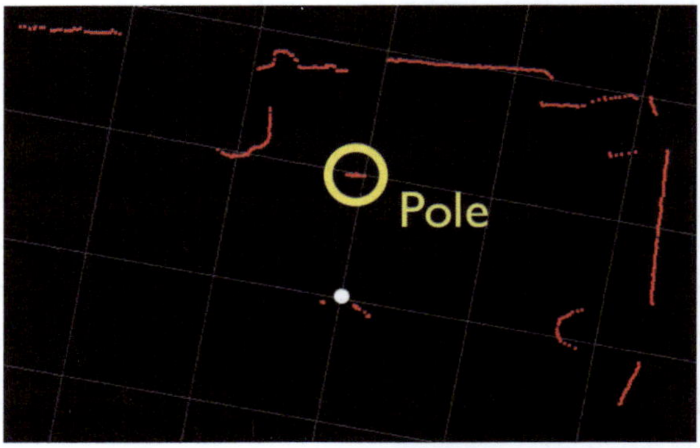

Fig. 6 *Top view*: *Red dots* are raw laser scan data and the *yellow circle* denotes the scan data corresponding to the pole. The *white dot* is the centre of the sensor

$$S_k = dist(\bar{P}, \tilde{P}_k) \tag{4}$$

where $\tilde{P}_k \in \mathbb{R}^2$ is the kth candidate position, and $\bar{P} \in \mathbb{R}^2$ is the average position. All candidates are sorted by decreasing score and the one with the maximum score is selected. For bootstrapping, we assume that a pole, $\mathbf{P}$, is located within discoverable boundary ($\mathbf{P} < \alpha, \beta, \gamma$) at system startup (see Algorithm 1) .

4 Experimental Results

In this section, software and hardware implementation are described in depth. We also present results of estimator performance evaluation while hovering which includes velocity, position and ground-truthed circle trajectory around a pole.

4.1 Software and Hardware Implementation

The ROS framework is used to integrate modules (see Fig. 7), where blue boxes denote the ROS nodes which are individual processes. The onboard Overo Gumstix runs the standard ROS laser scanner node and publishes the topic */scan* over WiFi to the base station every laser scan interval (100 ms). The ROS pole detector subscribes to this topic, and estimates 2D pose (x, y) which it publishes as topic */pole pose2D* . The ROS serial node communicates with the MikroKopter flight control board over the ZigBee link. Every 50 ms it requests a `DebugOut` packet which it receives and the inertial data (converted to SI units) is published as the */mikoImu* topic. This node also subscribes to the */mikoCmd* topic and transmits the command over the ZigBee uplink to the flight controller. Note that the overall software system latency is about

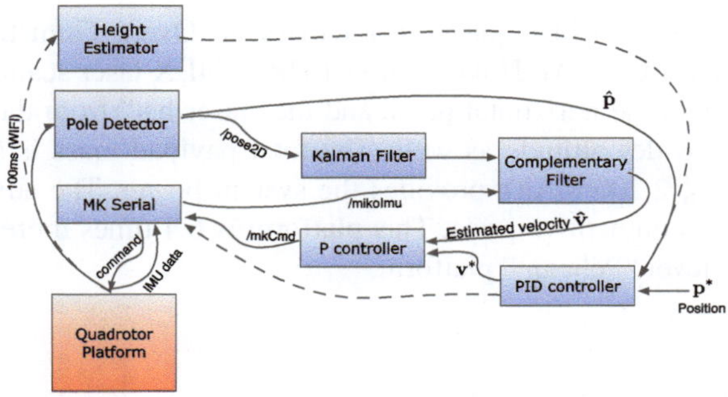

Fig. 7 Software implementation using ROS platform where blue boxes represent ROS nodes running on the ground station in real time and the orange box is the quadcopter platform. The prefix '/' denotes a ROS topic. $\hat{p}$ and p^* are estimated and desired position respectively. v denotes velocity and notation are same as position

Algorithm 1: Pole detection algorithm

while *! (Find a pole)* **do**
  if *(l.length[i] <α) && (l.distance[i] < β)&&(l.angle[i] <γ)* **then**
    $l[i]$ is the pole;
    $Find\ a\ pole =$ TRUE;
  else
    $i =$ i + 1;
  end
end
Continuous : find the best candidate satisfying less strict conditions.
while *! (Find the best candidate)* **do**
  if *(c.length[j] <δ) && (c.distance[j] <ϵ)&&(c.angle[j] <ζ)* **then**
    Put $c[j]$ in the candidate list;
  else
    j = j + 1;
  end
  Calculate scores using $S_k = dist(\bar{P}, \tilde{P_k})$; //Equation 4
  Ascending sorting of the candidate list and pick the best score,c;
  if $c > \xi$ **then**
    pole $= c$;
    $Find\ the\ best\ candidate =$ TRUE;
  end
end
Note that constant parameters $\alpha < \delta$, $\beta < \epsilon$ and $\gamma < \zeta$.
ξ denotes the score threshold.

170 ms and the system response delay is about 200 ms. Technical documentation and this software are available online.[3]

[3] ROS QUT Cyphy wiki page http://www.ros.org/wiki/MikroKopter/Tutorials

Our MikroKopter L4-ME quadcopter carries an Overo Gumstix which runs Ubuntu Linux and ROS.[4] An Hokuyo model URG-04LX laser scanner (10 Hz and 4 m range) scans in the horizontal plane and the "laser hat" from the City College of NewYork[5] provides altitude as well. The total payload mass is 0.18 kg and a Lipo pack (4cells, 2,200 mAh), provides the system power. The advantage of the MikroKopter is a competitive price. This platform is 6.4 times more cost effective than the similar level "Pelican " platform.[6]

4.2 Estimation and Control

The performance evaluation of the velocity estimator is performed by comparing the measured velocities with the ground truth—a sub-millimetre accuracy g-speak/VICON motion capture system.[7] The ground truth velocities are obtained

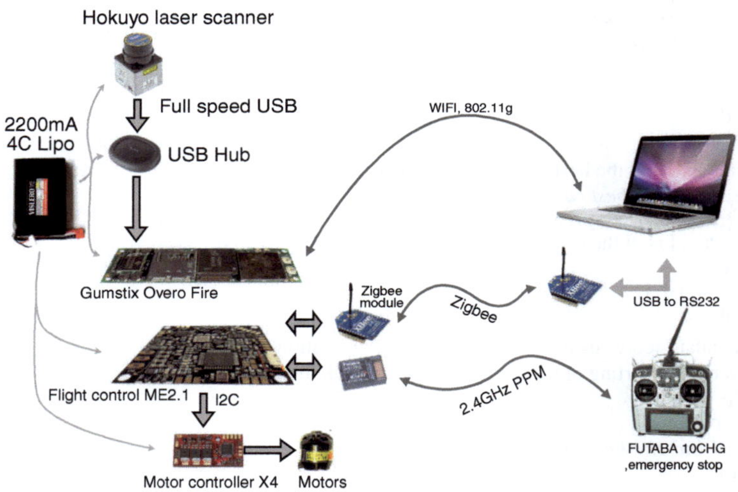

Fig. 8 Hardware integration. The laser scanner is attached to a USB Hub since the Overo Gumstix USB host only supports High Speed USB. The Zigbee module is used to transmit IMU data to the ground station and receive commands. The WiFi connection connects the ROS nodes on the Gumstix to the ground station. For safety a manual pilot transmitter is linked to the quadcopter system

[4] Robot Operating System, http://www.ros.org/wiki/

[5] City College of NewYork Robotics Lab, http://robotics.ccny.cuny.edu/blog/

[6] Ascending Technologies, http://www.asctec.de/

[7] Oblong,g-speak motion capture platform. http://www.oblong.com

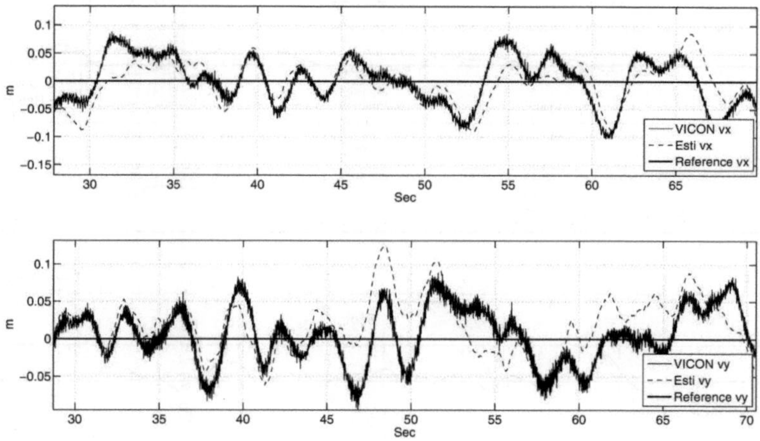

Fig. 9 The lateral velocities estimation results with respect to the pole while hovering. *Solid line* denotes the ground truth and *dash* indicates the complimentary filter velocity estimation output. *Thick solid line* is the reference

by calculating the first derivative of the position and the estimated velocities are generated by the proposed complementary filter, Eq. (3). Note that during takeoff, the quadcopter moves a little horizontally due to poor trim but returns quickly to the desired hovering position (Fig. 8). Figure 9 shows the estimated horizontal velocities compared to the ground truth. The standard deviation values are $\{\sigma_{vx}\ \sigma_{vy}\} = \{0.0495, 0.0375\}$ m/s. Note that these values are calculated over the flight interval between $t = 30$ s (takeoff) and $t = 70$ s (landing).

The vehicle position was estimated using the laser-range-finder, pole detector and Kalman filter and used in a PID controller to maintain the pole at a fixed range and bearing angle—hovering with respect to the pole. Ground truth data obtained from the g-speak system is shown in Fig. 10. The reference position of the vehicle is (0.8, 0, 0.6) m. The standard deviations of the ground truth position are $\{\sigma_x\ \sigma_y, \sigma_z\} = \{0.0483, 0.0455, 0.0609\}$ m. These are again computed over the flight interval.

If we yaw the vehicle while maintaining the pole at a fixed bearing, the result is motion around the pole as shown in Fig. 11. Figure 12 shows the ground truth circle trajectory with the proposed shared control. A current limitation is that yaw angle is estimated from the vertical gyro and drifts with time. A video demonstration is available on our YouTube channel.[8]

[8] YouTube QUT Cyphy channel. http://youtu.be/F1vljjPIglg

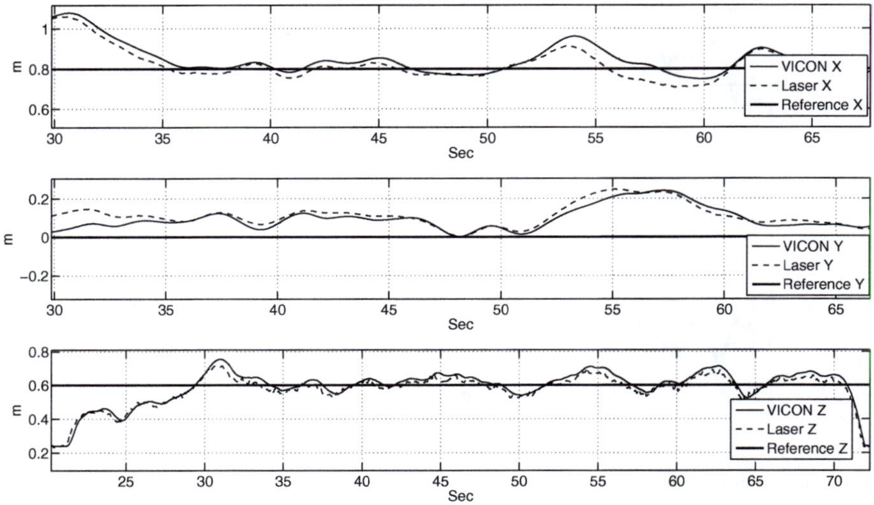

Fig. 10 x, y position estimation with respect to the pole while hovering with the ground truth. *Solid line* denotes the ground truth and dash indicates Kalman filter position estimation. *Thick solid line* is the reference. Median filter is used to estimate z position estimation

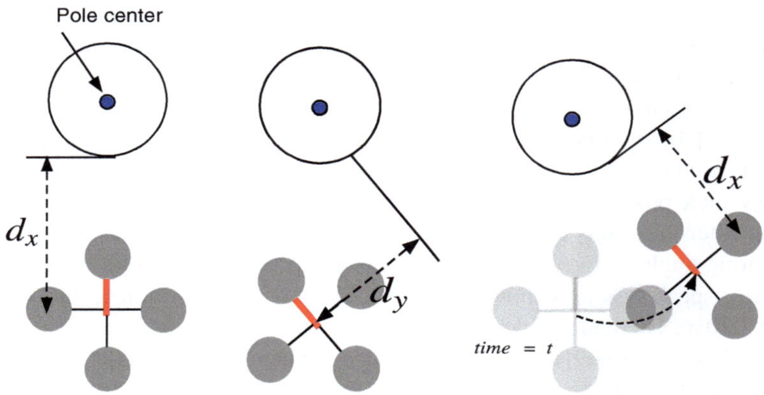

Fig. 11 **a** Changing yaw angle makes the quadcopter circle around the pole (*red bar* indicates the *front* rotor. References for x,y position controllers are d_x and 0 respectively. The robot hovers by keeping d_x distance at $time = t$. **b** An operator sends yaw command and it introduces d_y distance at $time = t + 1$. **c** The robot moves to right to eliminate d_y and keeps d_x distance at $time = t + 2$

5 Conclusion and Future work

We have described our progress toward a shared control scheme that allows an unskilled operator to control a quadcopter easily and safely for a useful class of tasks. Translational velocity estimation is crucially important for quadcopter control and

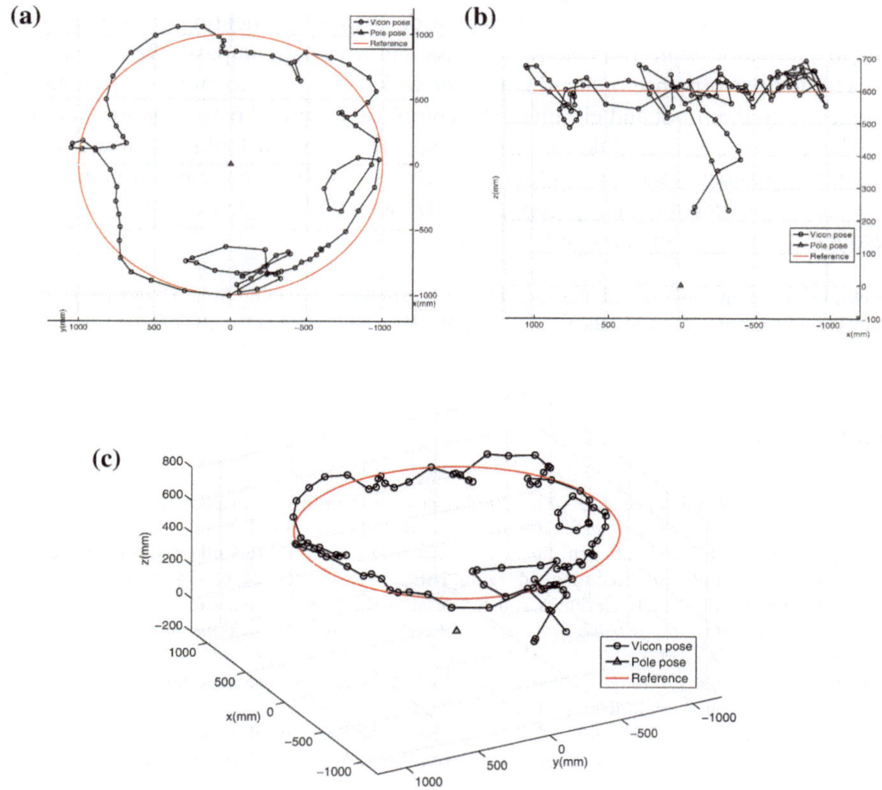

Fig. 12 The ground truth trajectory with shared control. An operator sends only yaw commands using a joystick and the quadcopter keeps the desired distance, $d_x, d_y, d_z = [1, 0, 0.6]$ in metre, with the pole. *Red* denotes the reference. Note that only the ground truth trajectory is presented due to difficulty in estimating yaw angle with a low performance gyroscope

we have presented computationally efficient state estimation and control algorithms which allow for smaller onboard computers. We have demonstrated ground-truthed comparison of lateral velocity, position estimation while hovering and presented circle movement around a pole, done with a platform of less than one fifth the cost and with a laser scanner that scans four times more slowly than other comparable results in the literature.

We used an amateur-class quadcopter, and to achieve a high level of performance required understanding the dynamics of the quadcopter through system identification and reverse engineering. This platform has many advantages such as cost efficiency, high payload, open source firmware and a large user community. Our knowledge about this platform are returned to the community through open documentation and software available online.[9]

[9] ROS QUT Cyphy wiki page http://www.ros.org/wiki/MikroKopter/Tutorials

We have a large program of ongoing work. We are augmenting gyro-based yaw angle estimation with a magnetic compass and a visual compass. We are moving to a higher performance onboard computer which allows us to move computational processes to the robot and eliminate the complexity, limited range and unreliability of the communications link. We are investigating upward looking sensors so the robot can manoeuvre around pole-top structures. Finally, we are investigating high update rate monocular camera (up to 125 Hz) with wide-angle field of view for fast estimation of robot and task-relative state.

Acknowledgments We would like to thank Liz Murphy for providing Split-Merge source code, and Timothy Gurnett for assistance while using the VICON system at the QUT Cube Lab.

References

1. C. Balaguer, A. Gimenez, A. Jardon, Climbing robots' mobility for inspection and maintenance of 3D complex environments. Auton. Robots. **18**, 157–169 (2005)
2. K. Sangbae, M. Spenko, S. Trujillo, B. Heyneman, D. Santos, M.R. Cutkosky, Smooth vertical surface climbing with directional adhesion. IEEE Trans. Robot. **24**, 65–74 (2008)
3. G.C. Haynes, K. Alex, L. Goran, A. Jon, S. Aaron, A.R.D.E.K. Alfred, Rapid pole climbing with a quadrupedal robot, in *IEEE International Conference on Robotics and Automation (ICRA)*, pp. 2767–2772 (2009)
4. F. Xu, X. Wang, L. Wang, Cable inspection robot for cable-stayed bridges: design, analysis, and application. J. Field Robot. **28**, 441–459 (2011)
5. R. Voigt, J. Nikolic, C. Hurzeler, S. Weiss, L. Kneip, R. Siegwart, Robust embedded egomotion estimation, in *International Conference on Intelligent Robots and Systems (IROS)*, pp. 2694–2699 (2011)
6. T.B. Sheridan, *Telerobotics, Automation, and Human Supervisory Control* (The MIT Press, Cambridge, 1992)
7. J. Baeten, H. Bruyninckx, J. De Schutter, Shared control in hybrid vision/force robotic serving using the task frame, in *International Conference on Intelligent Robots and Systems (IROS)*, pp. 2128–2133 (2002)
8. P. Corke, *Robotics, Vision and Control Fundamental Algorithms in MATLAB* (Springer, Berlin, 2011)
9. P. Pounds, R. Mahony, Design principles of large quadrotors for practical applications, in *IEEE International Conference on Robotics and Automation (ICRA)*, pp. 3265–3270 (2009)
10. I. Sa, P. Corke, System identification, estimation and control for a cost effective open-source quadcopter, in *IEEE International Conference on Robotics and Automation (ICRA)* (accepted, 2011)
11. D. Abeywardena, S. Kodagoda, R. Munasinghe, G. Dissanayake, A virtual odometry for a quadrotor micro aerial vehicle, in *Australasian Conference on Robotics and Automation* (2011)
12. M. Roberts, P. Corke, G. Buskey, Low-cost flight control system for a small autonomous helicopter, in *IEEE International Conference on Robotics and Automation (ICRA)*, pp. 546–551 (2003)
13. P. Corke, An inertial and visual sensing system for a small autonomous helicopter. J. Robot. Syst. **21**, 9 (2004)
14. G.A. Borges, M.-J. Aldon, A split-and-merge segmentation algorithm for line extraction in 2D range images, in *International Conference on Pattern Recognition*, pp. 441–444 (2000)

Towards Autonomous Robotic Systems for Remote Gas Leak Detection and Localization in Industrial Environments

Samuel Soldan, Jochen Welle, Thomas Barz, Andreas Kroll and Dirk Schulz

Abstract Detection and localization of escaped hazardous gases is of great industrial and public interest in order to prevent harm to humans, nature and assets or just to prevent financial losses. The development of novel leak-detection technologies will yield better coverage of inspected objects while helping to lower plant operation costs at the same time. Moreover, inspection personnel can be relieved from repetitive work and focus on value-adding supervisory control and optimization tasks. The proposed system consists of autonomous mobile inspection robots that are equipped with several remote gas sensing devices and local intelligence. All-terrain robots with caterpillar tracks are used that can handle slopes, unpaved routes and offer maneuverability in restricted spaces as required for inspecting plants such as petroleum refineries, tank farms or chemical sites as well as sealed landfills. The robots can detect and locate gas leaks autonomously to a great extent using infrared optical spectroscopic and thermal remote sensing techniques and data processing. This article gives an overview of the components of the robotic system prototype, i.e. the robotic platform and the remote sensing and evaluation module.The software

S. Soldan (✉) · A. Kroll
Faculty of Mechanical Engineering, Measurement and Control Department, University of Kassel, Mönchebergstraße 7, 34125 Kassel, Germany
e-mail: samuel.soldan@mrt.uni-kassel.de

A. Kroll
e-mail: andreas.kroll@mrt.uni-kassel.de

J. Welle · T. Barz · D. Schulz
Fraunhofer Institute for Communication, Information Processing and Ergonomics FKIE, Neuenahrer Straße 20, 53343 Wachtberg, Germany
e-mail: jochen.welle@fkie.fraunhofer.de

T. Barz
e-mail: thomas.barz@fkie.fraunhofer.de

D. Schulz
e-mail: dirk.schulz@fkie.fraunhofer.de

K. Yoshida and S. Tadokoro (eds.), *Field and Service Robotics*, Springer Tracts in Advanced Robotics 92, DOI: 10.1007/978-3-642-40686-7_16,

architecture, including the robot middleware and the measurement routines, are described. Results from testing autonomous mobility and object inspection functions in a large test course are presented.

1 Motivation

Detecting gas leaks in pipes and vessels within a plant is of interest for several reasons. At first, escaped hazardous (toxic/explosive) substances have to be detected and removed before people or the environment is harmed. Secondly, pressured equipment is often designed to show “leak-before-break” behavior. This means that a partial failure may result in a detectable leak before a final fracture occurs [1, 2]. Thirdly, leakages are a key maintenance issue: 23.5 % of all damages in a large chemical plant appeared as leak. This made leaks the second most frequent failure mode (valued 1.59 MEUR p. a.); following just 2.2 % behind the #1 symptom (program sequence error) [3].

The use of mobile inspection robots is motivated by various reasons such as better access to remote or tight places, replacing human missions in hazardous environments or to counteract consequences of demographical changes [4]. The development of autonomous robotic systems for remote gas leak detection and localization additionally aims at:

- Quality of process and results: improved inspection due to mobile remote gas sensing technology, central knowledge base increases locally available knowledge, increased coverage of extended inspection areas and increased inspection frequency with available staff
- Operator relief: easier inspection of remote areas by remote sensing technology, automation of repetitive, monotonous routine tasks, automated measurement evaluation and results documentation
- Profitability: more efficient and effective deployment of specialist staff, more time for human operators for activities that create higher value such as planning, supervision, and optimization, efficient deployment of sophisticated measurement devices

In contrast to robots with local (in-situ) gas measuring instruments (see [5] for an overview) remote sensing is contact-free, reactionless and does not disturb the wind field and therefore provides for better leak detection. Also the sensor must not immerge into gas plume/hazardous area and permits measuring of difficult/impossible to access areas.

2 Project Background

The proposed robotic systems are developed in the joint research venture “RoboGas^{Inspector}”[1] with four research, three industrial development and two application partners. The project activities have been grouped to the major working areas

---

[1] See www.robogasinspector.de for further information.

Fig. 1 Testing environments and possible future application areas. *Top* petroleum refinery (© PCK Raffinerie GmbH), *bottom* gas compressor station (© GASCADE Gastransport GmbH)

of 'gas propagation and measurement', 'mobile robots' and 'man-robot-interaction'. The application partners in the project consortium are a petroleum refinery and a gas transportation and distribution provider. They provide industrial environments in which the system could be deployed within in the future (see Fig. 1 for an aerial view of the plants). Therefore these plants are used for requirement elicitation and testing; the target gas species were derived from these applications. In addition a closed and sealed landfill was added as a third application area: in Germany periodical inspections for gas emissions are required by law and their extensiveness makes them a prime target for robotic inspection.

A brief list of general requirements is provided in the following:

- Mobility: The ground mainly consists of stretches of grass, paved and unpaved roads; slopes go up to 40 % (22°). Robust ground propulsion is needed.
- Remote gas sensing: Target objects are up to 20 m elevated above the level of traversable paths/roads, pipe bridges are between 5 m and 15 m high. The material of the components is mainly steel or stainless steel; the surfaces can be bare (pristine or corroded), painted or covered with insulation. Connecting pipes are

usually bundled and can stretch for several hundred meters. The major target gas is methane (CH_4) and in addition other alkanes for the refinery.

- Explosion protection: In some areas of in the plants flammable/explosive gas mixtures may occur. Based on the probability of this occurrence these areas are classified into zones according to IEC 60079.10. The robot should be able to enter a zone 2 area and not serve as an ignition source.
- Industrial application: The overall system has to comply with industrial guidelines. The acceptance of the system from the operator has to be considered during the design phase. System evaluations in industrial plants involving the industrial user group are planned.

3 Robotic Platform and Software Architecture

The prototype of the autonomous mobile inspection robot (see Fig. 2) basically consists of three major components:

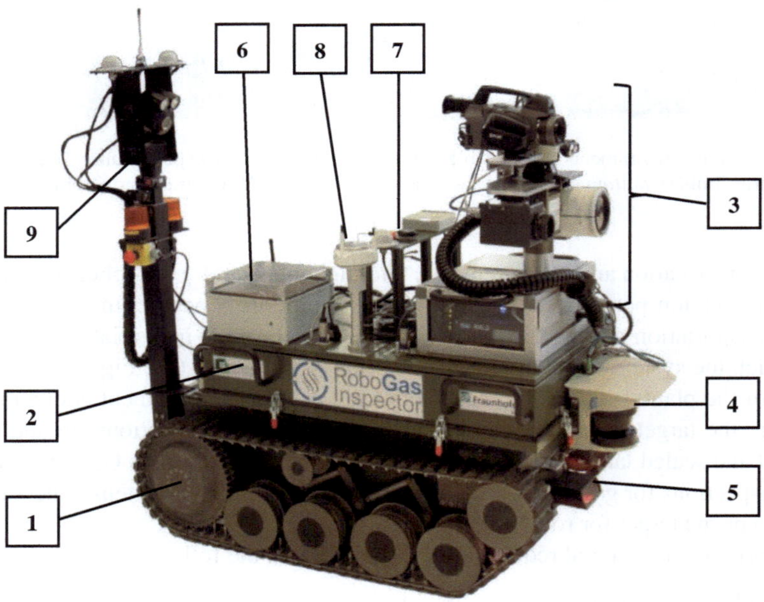

Fig. 2 System overview. *1* telerob tEODor chassis, *2* compartment module, *3* inspection module, *4* front laser scanner, *5* optical odometry, *6* in-situ gas sensor for ex-protection, *7* inertial measurement unit and GPS antennae, *8* weather sensor, *9* extendable rear mast with camera, top mounted communication antennae, emergency stop button and warning lights

- A telerob tEOD or chassis with caterpillar tracks and a rear camera mast. This platform is a modified version of a commercially available component from the field of manual remote explosive ordinance disposal.
- A compartment module, developed by the Unmanned Systems group of the Fraunhofer FKIE, which houses an industrial computer (Spectra PowerBox 1260HP, QX9300 CPU, 4 GB RAM) for local robot control and other non-weatherproof electronics. It serves as baseplate, as well as a central power and information distribution hub for all attached devices and top mounted superstructures.
- The inspection module, developed by the Measurement and Control Department of the University of Kassel, which features remote sensing technology installed on a pan-tilt unit (see Fig. 8).

In its current state the prototype has a maximum speed of 1 m/s, a total weight of approximately 385 kg, an additional payload of up to 340 kg and a drag force of 3,000 N. For communication a Wi-Fi mesh architecture from Proxim is used. To increase communication range the investigation of the usage of the built in video wireless transmitter in conjunction with a separate radio transmitter for control is planned.

Movement is measured by an inertial measurement unit (IMU) that combines local data from acceleration sensors with external positioning information from GPS satellites. The chassis itself provides no odometry information as it was de-signed for manual operation. Therefore, the prototype was enhanced with an experimental optical odometry. The optical odometry uses a mouse sensor with higher power LEDs and a modified lens to measure position changes via relative motion of optical features.

Two planar mounted 2D laser scanners, one at the front and one at the rear, provide a 360° distance measurement for self-localization and collision avoidance. A third laser scanner on the inspection module is used for stationary 3D scanning and additional ground surveillance during movement to detect obstacles not visible in the planar mounted 2D laser scanners. A Vaisala Weather Transmitter WXT520 is used to provide information about wind speed and direction.

Even though other projects have shown the feasibility for mobile robots to get an approval to operate in explosive atmospheres [6] it was decided that for a system prototype this is not necessary. Instead a certified gas warning device (Sewerin EX-TEC HS 680) automatically interrupts the robot's power supply if the local background gas concentration reaches a certain level, well below the lower explosive limit (LEL). Using test gas with only 2.2 vol.% methane it took less than 4 s for the device to sense the concentration and power off the system.

To meet general safety considerations bright warning lights indicate that the robot's drive is in action and increases the visibility to human personnel to a great extent. In addition to an easy to access local emergency stop button the robot can also be stopped and powered off by a certified hardware-based safety remote control system at any time.

3.1 Software Architecture

The system software is based on Fraunhofer FKIE's service-oriented middleware RoSe [7]. This Framework was developed for the use of multi robot systems in rough outdoor environments and especially takes the problem of unreliable wireless communication into account that is often neglected. It has been used in a variety of projects as well as on various robotic platforms of which some have been recently demonstrated in the context of the European Land Robot Trial 2011 [8].

A RoSe service is an independent, multi-threaded stand-alone application with the capability of exchanging messages with other services via the RoSe message system that encapsulates network communication. As an exception, inter-process communication between services on the same host can be handled by shared memory to allow exchange of large data volumes like 3D laser scans.

The RoboGas^{Inspector} software framework consists of a variety of RoSe services that were cooperatively developed by several project partners and:

- *Control Station Service*: Service on the control station that provides an interface for remote controlling the system via graphical user interaction.
- *Robot Control Service*: High-level service on the robot that executes commands from the control station. It controls and supervises other services in order to execute a high-level task, e.g. conducting an inspection.
- *Localization Service*: Responsible for deducing the actual robot position by combining information from different sources, laser scanners, optical odometry, IMU data, GPS and stored map information. The laser-based part of the localization is based on the Monte Carlo localization in [9].
- *Navigation Service*: Handles higher-level navigation points. The system follows GPS-waypoints (outdoors) or 2D map related waypoints (indoors) spaced at intervals of several meters. The navigation service supervises and controls local navigation services that are responsible for local path planning, collision avoidance and issuing actual driving commands to the robot's drive system. At present an 'Expansive Spaces Trees' planner as described in [10] is used.
- *Measurement Control Service*: High-level service that supervises and controls the inspection module and its lower-level support services in order to perform the requested measurement operation. It assesses the gas concentration, informs the control station and robot control system, and initiates measurements for gas leak localization if necessary.
- For each hardware device, such as the robot base platform, laser scanners, pan-tilt unit, gas sensors, cameras and so on, a different service exists, which is responsible for encapsulating the low-level communication and providing an appropriate interface for other higher-level services.

All persistent information gathered or generated by the system is stored in relational databases and can be divided into three categories:

- *Global knowledge* that is needed by the robot and the control station during an inspection, e.g. route information, measurement targets, restricted zones that may not be entered by the robot autonomously. This data has to be synchronized online, when changes occur.

- *Local robot knowledge* that is produced during an ongoing inspection or is only valid in its context. This may be for example extensive amounts of raw measurement data, which can be transferred to the control station offline afterwards, or temporary status information.
- *Local control station knowledge* that has no impact on the robot's immediate inspection duty, e.g. old inspection reports that have to be archived.

As bandwidth of wireless communication is limited, separate databases exist on the robot and on the control station. Only the necessary global knowledge updates or explicitly requested information is transmitted immediately, other data is synchronized later at specific points where a high speed data connection is available.

4 Remote Gas Detection

For the proposed robotic system the remote gas sensing is based on infrared measurements. Three different technologies are used: Tunable diode laser absorption spectroscopy (TDLAS), thermography for gas imaging and thermal imaging for validation of leak hypotheses or detection fluid leakages. The main principles are described in the following section followed by the strategies for detection and localization as well as the practical integration of the instruments into one platform independent module.

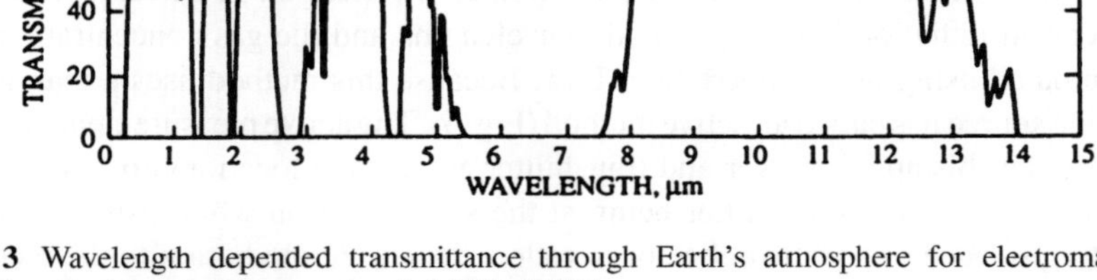

Fig. 3 Wavelength depended transmittance through Earth's atmosphere for electromagnetic radiation (horizontal path at sea level, 1,828 m) [11]

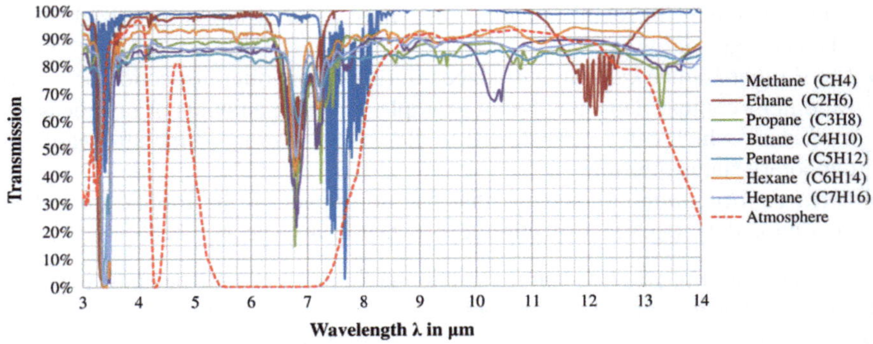

Fig. 4 Spectral transmittance of different hydrocarbons and of atmosphere. Data from [12]

4.1 Remote Gas Sensing

Every object with a temperature above 0 K emits electromagnetic radiation with varying intensity in different wavelengths. For an idealistic object (black body radiator) the amount of radiation of a certain wavelength can be calculated by Planck's law. The transmittance/opacity of the atmospheric gases varies with the wavelength of the electromagnetic radiation and is shown in Fig. 3). The opacity of the atmosphere is caused by the composition of its gases as shown on top of the figure (mainly by oxygen, carbon dioxide and water vapor).

Each gas has a specific absorption behavior, which is caused by different energy levels within the molecules of the gas. Figure 4 depicts the transmission behavior of some gases including the atmosphere (as a gas mixture). It can be seen that in certain wavelengths the atmosphere is approximately transparent while other gas species are opaque (e.g. around 3.3 μm). In TDLAS a laser diode is used to emit electromagnetic radiation in two different wavelengths. The wavelengths are chosen in a region where the atmosphere has a high transmittance and so that the target gas has a high opacity in one wavelength and a low opacity in another wavelength. A detector measures the received radiation intensity in both wavelengths and the gas concentration can be computed using the Lambert-Beer-Law. Because this method uses a man-made radiation source it is called an active method (Fig. 5). The active measurement systems can be either bistatic (receiver and transmitter at separate locations) or monostatic with the receiver and transmitter being at the same location while using a diffuse reflecting surface or a (retro)reflector to reflect the emitted radiation to the receiver (Fig. 5). The measured gas concentration result is an integrated value for the whole measurement path.

In contrast to TDLAS, thermography for gas imaging is deployed as a passive technology with state of the art commercial devices and does not require an artificial radiation source. It is basically a thermography camera with a narrow wavelength band-pass filter that measures the surface temperature of objects. The chosen wavelength region is transparent for the atmosphere but not for the target gas species. This

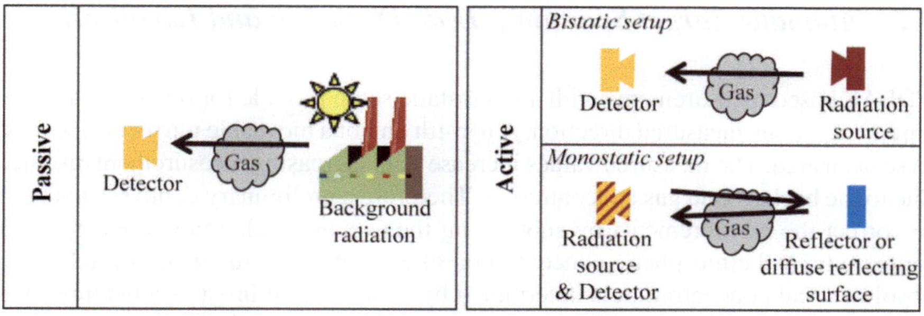

Fig. 5 Measurement arrangement options for passive and active infrared spectroscopy

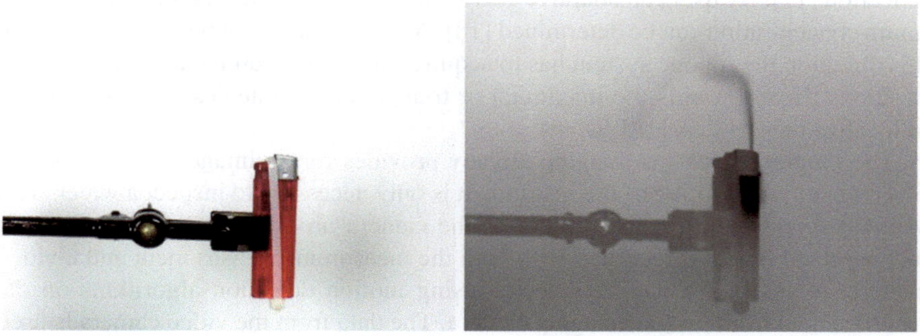

Fig. 6 Lighter with escaping gas seen in video image–wavelength 0.38–0.78 μm (*left*) and thermography for gas imaging—wavelength 3.2–3.4 μm (*right*). Note that the gas plume of the escaping propane/butane is only visible in the right image

makes their temperature visible during measurements (if their temperature is different from the background temperature). The gas plume can be detected as a cloud that occludes the background in an image (Fig. 6). It is therefore only a visualization technique and provides no quantitative reading. Thermography for gas imaging can be used in different wavelength regions and permits detecting multiple gases. However, the sensors do not distinguish single gases but indicate that a gas with absorption in the sensed spectral region is present (e.g. multiple species have absorption characteristics around 3.3 μm in Fig. 4). Similar to the characteristics of gas warning devices this functionality is sufficient for many inspection, monitoring and surveillance tasks.

For more in-depth information about electromagnetic radiation, absorption and thermography Vollmer and Möllmann [13] can be consulted; information about thermography and non-destructive testing can be found in Maldague [14]. Regarding spectroscopic measurements techniques the reader is referred to [15]. An overview of available remote gas sensing systems can be found in [16].

4.2 Automatic Target Scanning, Leak Detection and Localization

TDLAS based measurements with a monostatic setup provide for one integral measurement for one measured direction. A pan-tilt unit or a moveable mirror can be used to scan an area. The measured values increase with increasing measurement distance due to the background gas concentration. Therefore a preliminary depth scan is used to correct the measurements by subtracting the average background concentration (either a typical atmospheric concentration of e.g. 2 ppm for methane is used or the absolute local concentration is determined by an additional in-situ senor) times the measurement path distance. This information is also used to adapt the sensed region e.g. by removing areas that are too far away to be inspected or have no suitable diffuse reflection [17]. By using an adaptive grid scan technique the direction with the maximum concentration can be determined [18]. As the gas leak can be anywhere within this direction the robotic system has to acquire further measurements from different positions. Head-towards-the-maximum or triangulation strategies have been tested to localize the gas leak [18].

Thermography for gas imaging already provides for an image of the inspected object; panning or tilting of the instrument is only necessary to inspect a wider area. Without moving the robotic system or the camera an image series (video) from the inspected object is recorded with both the measurement instrument and a video camera. A gas plume can be detected using motion detection algorithms on the recorded data from the gas imaging camera. The data from the video camera is used to reduce false alarms due to other moving objects that can be seen in the normal visual wavelength region.

Having closed in on a potential leak thermal imaging is used to get a confirmation of the location of the leak hypothesis. Because of expansive cooling, the area around a leak in a gas pipe or similar cools down and this effect can be measured and automatically be detected using pattern matching with known leak patterns (Fig. 7)

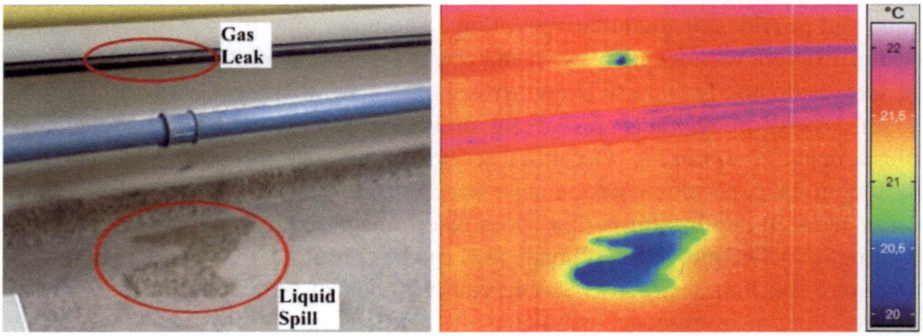

Fig. 7 Water spill on the floor and gas leak in the pipe in video image (*left*) and a corresponding thermogram with temperatures ranging from 20 to 22.1 °C (*right*)

[19, 20]. Additionally, thermal imaging can be used for other inspection tasks like detecting liquid spills (Fig. 7), electric short circuits, broken insulation or bearing wear.

4.3 Inspection Module

For the system prototype three main measurement instruments have been selected and are used:

- TDLAS: Sewerin/Heath RMLD (Remote Methane Leak Detector), wavelength: 1.6 μm, detection distance: up to 30 m, detectable gas: methane.
- Thermography for gas imaging: FLIR GF320, wavelength: 3.2–3.4 μm, detectable gases: hydrocarbons with up to 8 carbon atoms (methane to octane).
- Thermography camera: Infratec VarioCam hr research; wavelength: 7–14 μm, temperature resolution: 30 mK, geometric resolution: 1280 × 960 pixels.

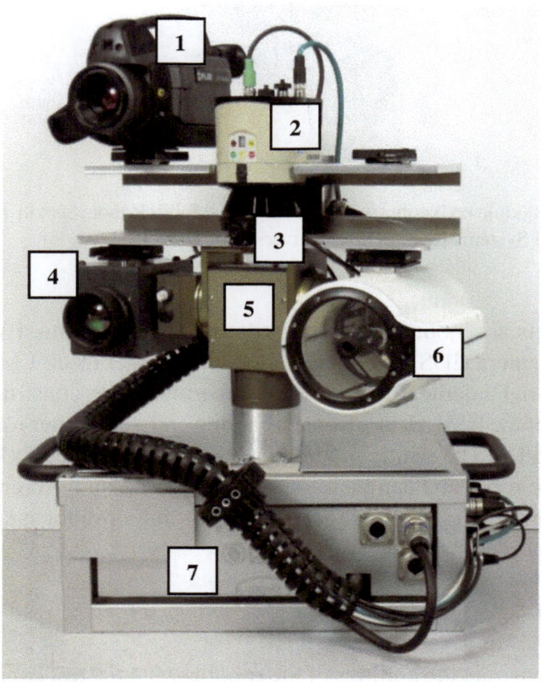

Fig. 8 Inspection module. *1* thermography camera for gas imaging, *2* 2D laser scanner, *3* video camera, *4* thermography camera, *5* pan-tilt-unit, *6* TDALS instrument, *7* compartment with electronics and a control computer

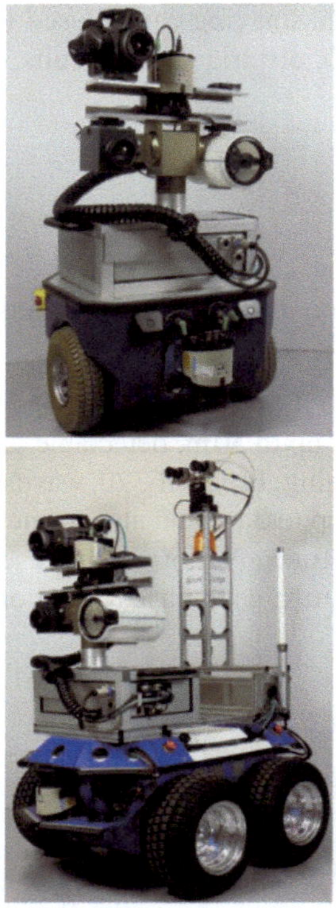

Fig. 9 Inspection module on Neobotix MP-S500 (*top*), MobileRobots Seekur Jr. (*bottom*, courtesy of Human-Machine Systems Engineering, University of Kassel)

These instruments are mounted on a pan-tilt unit (PTU, Schunk PW90) along with a normal USB-camera (IDS uEye) and a 2D laser scanner (Sick LMS 151). In Fig. 8 the setup is shown. Underneath the PTU a compartment was constructed to house the control module of the RMLD, electronics, cabling and a computer for data processing (Intel Core i7 CPU). The whole module only requires one network and one power connection to the robot and can therefore be easily used on different platforms (e.g. lab robots as shown in Fig. 9 or a bigger all-terrain outdoor robot as shown in Fig. 11).

5 Test Results

To test and demonstrate the abilities of the robotic system a test course was designed by the RoboGas^{Inspector} project group with mobility and gas measurement related tasks. An overview of the test field can be seen in Fig. 10). The robot is supposed to complete one round autonomously and thereby pass the following tasks (letters in brackets refer to the area on the test field in Fig. 10):

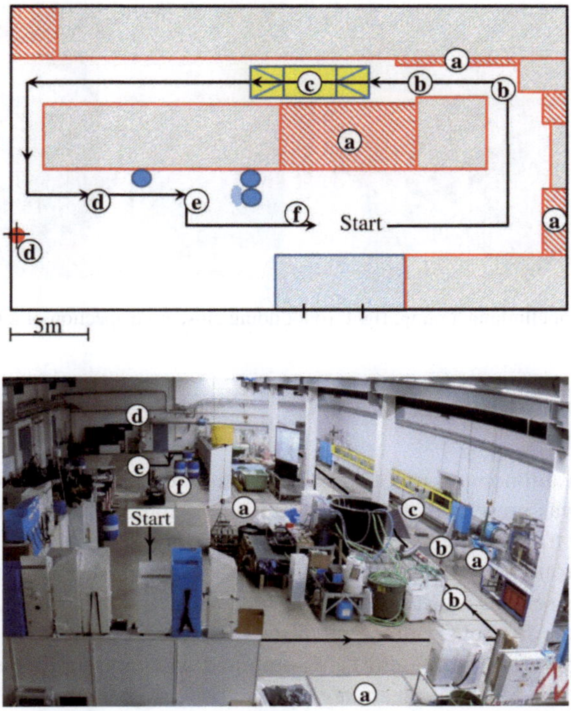

Fig. 10 Overview of test course, *top* schematic, *bottom* photo. Labels are described in the text

- Autonomous driving and obstacle avoidance
- Avoidance of a restricted areas (a)
- Narrow passages (b)
- Ramp climbing and descent (c)—also shown in Fig. 11 (left)
- Scanning of an area and finding of a gas probe (d)—shown in Fig. 11 (right)
- Live stream of video and thermography images
- Handing over to operator for manual control tasks (e)
- Resuming inspection mission after manual control(e)
- Emergency shutdown in case of high background gas concentration (explosion protection) (f)

The test course is approximately 80 m long and only 10 waypoints were provided to the mobile robot. All tasks were successfully completed for 10 rounds, demonstrating the reliability of the system prototype and the reproducibility of the task execution. The robot needed less than five minutes to complete the course (including leak detection). Videos from the robotic system on the test course can be found at www.robogasinspector.de.

Fig. 11 Left: robot climbing a ramp, right: robot doing close up inspection of a flange

6 Summary and Future Works

In this article an autonomous robotic system for remote gas leak detection in industrial environments was presented. The platform is a commercially available EOD robot with caterpillar tracks which has been modified for computer control. Software services enable autonomous path planning and obstacle avoidance. The principles of the used remote gas sensing instruments are based on spectral absorption. Measurement devices based on different technological methods (TDLAS, passive thermography and thermography for gas imaging) are combined on a modular inspection module. Leak detection and localization strategies have been implemented. The system was successfully tested on a test course with challenging tasks like ramp climbing and remote leak detection.

Future activities include testing and improving the strategies in outdoor scenarios in industrial plants.

Acknowledgments The project RoboGas^{Inspector} is funded by the Federal Ministry of Economics and Technology due to a resolution of the German Bundestag. The authors would like to thank the other project partners for their valuable discussions and support as well as Dr.-Ing. R. Hassinger and Dipl-Ing. R. Feldner for very obligingly providing the test hall.

References

1. G. Bartholomé, W. Kastner , E. Keim, Calculation of leakage area and leakage rate for the design of leakage detection systems. Presented at SMIRT 11 Transactions (1991)
2. IAEA (ed.) Applicability of the leak before break concept. IAEA, Vienna, Report No. IAEA-TECDOC-710 (1993)
3. W. Pohrer, Controlling in der Instandhaltung. Presented at Maintenance 2010, Berlin, Germany (2005)
4. A. Kroll, A survey on mobile robots for industrial inspections, in *Proceedings of Intelligent Autonomous Systems* 10: IAS-10 (2008)
5. G. Kowadlo, R.A. Russell, Robot odor localization: a taxonomy and survey. Int. J. Robot. Res. **27**(8), 869–894 (2008)

6. B. Graf, K. Pfeiffer, Mobile robotics for offshore automation, in *Proceedings for Robotics for Risky Interventions and Surveillance of the Environment: IARP/EURON International Workshop* (2008), 2008)
7. A. Tiderko, F. Höller, T. Bachran, A framework for multicast communication over unreliable networks in multi robot systems, in *Proceedings of Towards Autonomous Robotic Systems (TAROS)* (2007)
8. M. Brunner, F. Höller, A. Königs, T. Röhling, F.E. Schneider, A. Tiderko, D. Schulz, D. Wildermuth, The FKIE robot system for the european land robot trial, in *Proceedings of the 5th International IARP Workshop on Robotics for Risky Interventions and Environmental, Surveillance-Maintenance (RISE 2011)*, Leuven, Belgium (2011)
9. S. Thrun, W. Burgard, D. Fox, textitProbabilistic Robotics (MIT Press, Cambridge, 2005)
10. F. Hoeller, D. Schulz, M. Moors, F.E. Schneider, Accompanying persons with a mobile robot using motion prediction and probabilistic roadmaps, in *IEEE/RSJ International Conference on Intelligent Robots and Systems: IROS 2007*, pp. 1260–1265 (2007)
11. G. Gaussorgues, *Infrared Thermography* (Chapman & Hall, London, 1994)
12. National Institute of Standards and Technology (2001) NIST Chemistry WebBook. http://webbook.nist.gov/chemistry. Accessed January 2011
13. M. Vollmer, K.P. Möllmann, *Infrared Thermal Imaging: Fundamentals Research and Applications* (Wiley-VCH, Weinheim, 2010)
14. X.P. Maldague, *Theory and Practice of Infrared Technology for Nondestructive Testing* (John Wiley & Sons, New York, 2001)
15. J.M. Hollas, *Modern Spectroscopy*, 4th edn. (Wiley, Chichester, 2003)
16. G. Bonow, A. Kroll, Zur automatisierten Inspektion von Anlagen mittels Gasfernmesstechnik: Technologien und Geräte, in *Proceedings of AUTOMATION*, Baden-Baden, Germany (2011)
17. S. Soldan, G. Bonow, A. Kroll, RoboGasInspector—a mobile robotic system for remote leak sensing and localization in large industrial environments: overview and first results, in *Accepted for IFAC Workshop-Automatic Control in Offshore Oil and Gas Production*, Trondheim, Norway (2012)
18. A. Kroll, W. Baetz, D. Peretzki, On autonomous detection of pressured air and gas leaks using passive IR-thermography for mobile robot application, in *Proceedings of IEEE International Conference on Robotics and Automation: ICRA 2009*, Kobe, Japan, pp. 921–926 (2009)
19. W. Baetz, A. Kroll, G. Bonow, Mobile robots with active IR-optical sensing for remote gas detection and source localization, in *Proceedings of IEEE International Conference on Robotics and Automation: ICRA 2009*, Kobe, Japan, pp. 2773–2778 (2009)
20. W. Baetz, A. Kroll, S. Soldan, On gas leak detection of pressurised components by using thermograms and pattern recognition algorithms, in *Proceedings of the 8th International Conference on NDE in Relation to Structural Integrity for Nuclear and Pressurised Components*, Berlin, Germany, pp. 503–512 (2010)

To the Bookstore! Autonomous Wheelchair Navigation in an Urban Environment

Corey Montella, Timothy Perkins, John Spletzer and Michael Sands

Abstract In this paper, we demonstrate reliable navigation of a smart wheelchair system (SWS) in an urban environment. Urban environments present unique challenges for service robots. They require localization accuracy at the sidewalk level, but compromise GPS position estimates through significant multi-path effects. However, they are also rich in landmarks that can be leveraged by feature-based localization approaches. To this end, our SWS employed a map-based localization approach. A map of the environment was acquired using a server vehicle, synthesized *a priori*, and made accessible to the SWS. The map embedded not only the locations of landmarks, but also semantic data delineating 7 different landmark classes to facilitate robust data association. Landmark segmentation and tracking by the SWS was then accomplished using both 2D and 3D LIDAR systems. The resulting localization method has demonstrated decimeter level positioning accuracy in a global coordinate frame. The localization package was integrated into a ROS framework with a sample based motion planner and control loop running at 5 Hz to enable autonomous navigation. For validation, the SWS repeatedly navigated autonomously between Lehigh University's Packard Laboratory and the University bookstore, a distance of approximately 1.0 km roundtrip.

1 Introduction and Motivation

In 1997, Professor Illah Nourbakhsh of Carnegie Mellon University established *The Wheelchair Project*. Its goal was to develop an autonomous wheelchair capable of reliable navigation in both indoor and outdoor environments [14]. As an initial

C. Montella (✉) · T. Perkins · J. Spletzer · M. Sands
Lehigh University, 27 Memorial Drive West, Bethlehem, PA18015, USA
e-mail: cmontella@ieee.org

J. Spletzer
e-mail: spletzer@lehigh.edu

K. Yoshida and S. Tadokoro (eds.), *Field and Service Robotics*,
Springer Tracts in Advanced Robotics 92, DOI: 10.1007/978-3-642-40686-7_17,

milestone, Nourbakhsh proposed to demonstrate autonomous outdoor navigation from Smith Hall to the University Center bookstore—a distance of $\approx$ 500 m outdoors. While it does not appear that Nourbakhsh's vision has been realized yet, we draw inspiration from his work to pose a similar challenge to ourselves: demonstrate reliable autonomous navigation of a smart wheelchair system (SWS) from Lehigh University's (LU) Packard Laboratory to the LU Bookstore. Coincidentally, the most direct wheelchair-accessible route (see Fig. 7) for our task is also $\approx$ 500 m. It also required the wheelchair to navigate multiple street crossings, and to maintain a sufficiently accurate localization estimate to realize reliable navigation at the sidewalk scale.

To accomplish this objective, we leveraged technologies unavailable to Nourbakhsh at the time he posed the challenge, the most important of which was a low-cost 3D LIDAR system for ground plane detection, feature detection, and obstacle avoidance. From the development and testing that followed, we provide significant insights into what worked and what did not. By integrating these lessons learned, the SWS was capable of completing our bookstore challenge without incident.

In this chapter, we detail the implementation of a robot platform capable of autonomous navigation in a structured outdoor environment. While the methods in this paper were applied to a SWS, they have broad applicability to general field and service robots operating in urban environments.

2 Related Work

Smart wheelchair systems have been an active research area since the early 1980s. The spectrum of work has ranged from component level safety sensors, to assistive controllers for steering, to completely autonomous indoor solutions. A survey of the field (as of August 2005) can be found in [19]. More recent projects of note include the MIT Intelligent Wheelchair Project [10], the goal of which is to develop a voice-commanded autonomous wheelchair intended for use in indoor environments. The Home, Lift, Position, and Rehabilitation (HLPR) Chair [1] developed by NIST is a special-purpose assistive mobility device to provide independent patient mobility for indoor tasks, such as moving to and placing a person on a chair or bed. HLPR has demonstrated obstacle detection and navigation indoors with promising results. The Personal Mobility and Manipulation Appliance (PerMMA) [2] is being developed at the University of Pittsburgh and Carnegie Mellon University with the objective of combining manipulation and mobility assistance in support of complete independence for its users. The system employs two robotic arms, and has demonstrated object manipulation tasks such as retrieving a drink from a refrigerator. Our own work to date in the smart wheelchair space includes the Automated Transport and Retrieval System (ATRS) [6]. ATRS improves automobile access for power wheelchair users by eliminating the need for an attendant to stow and retrieve the wheelchair.

In contrast to these efforts, the emphasis of our current work is navigation in structured outdoor environments, such as an urban center. Developing robust robotics solutions suitable for use outdoors is a significant challenge compared to indoor environments: the scale is much larger, illumination levels vary from strong sunlight to near complete darkness, the environment is far less structured, environmental conditions can quickly and dramatically change, and simplifying assumptions such as a level ground plane are not reliable. Furthermore, operations at the sidewalk level require localization performance beyond the bounds of what traditional GPS can provide. Other research groups have studied the problem of localization in outdoor and urban environments. These include Georgiev et al., who used a mixture of cameras, GPS, LIDAR, sonar, tilt sensors, and a database of facade models for localization [8]. They demonstrated their approach by traveling a 330 m course and localizing with an error of $\approx$ 1 m. Ramos et al. combined an EKF-SLAM approach with landmark modeling [16]. Driving an automobile over a course of 1.5 km, they were able to localize within an error of 8.6 m.

In this work, we build upon our own previous results in large-scale map based localization. First, our localization approach was improved and extended to enable decimeter level accuracy. Second and more significantly, our SWS was able to employ the localization scheme for reliable outdoor *autonomous* navigation for the first time. This was validated through significant experimental results—including over 10 km of autonomous operations—culminating in the successful completion of our bookstore challenge.

3 Prototype Smart Wheelchair Sytem

The prototype SWS employed in this research was based on an Invacare Pronto Power Wheelchair differential drive platform, pictured in in Fig. 1. The hardware and software architecture of this prototype platform are outlined in this section.

3.1 Development Platform

From our experiences, three-dimensional (3D) perception is a critical enabling technology for autonomous navigation in outdoor environments. Therefore, the primary exteroceptive sensor on the SWS was an IFM O3D200 3D flash LIDAR. New to the robotics arena, the IFM can measure the range to and surface reflectivity of objects in the environment [11]. It is also relatively compact and low cost (<\$1,500 US). More importantly, the IFM performs well in the range of illumination levels encountered outdoors. The trade-offs for this low price point are a relatively low resolution (48×64 pixels), narrow field of view ($30° \times 40°$) and limited effective range ($\approx$ 6–8 m in our application). Despite these limitations, we could identify no sensor on the market that provided reliable 3D measurements outdoors at a comparable price point (note the Microsoft Kinect is not suitable for operations in bright sunlight [12]). In our

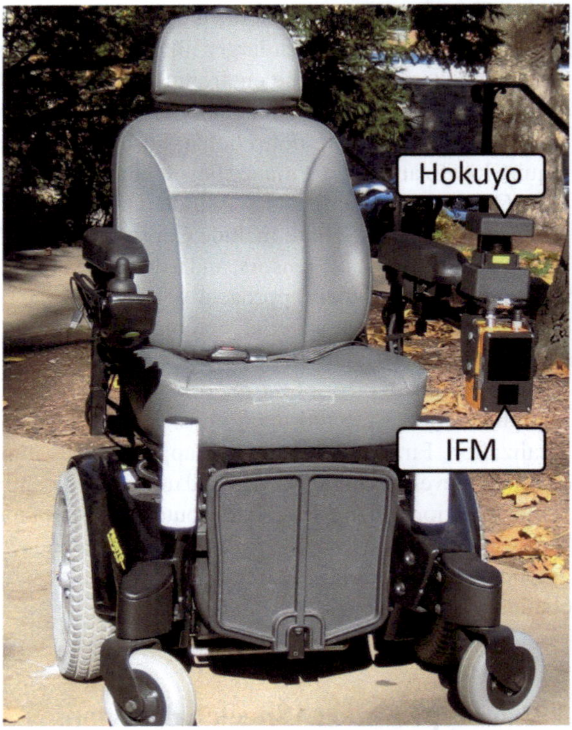

Fig. 1 The SWS integrates both 2D and 3D LIDARs for exteroceptive sensing, high-resolution encoders, and an IMU

application, the IFM established the ground-plane, detected obstacles, and tracked landmark features in 3D at a rate of $\approx 5\,\text{Hz}$.

The SWS prototype also integrated a Hokuyo UTM-30LX 2D LIDAR. The Hokuyo's larger field of view (270°), finer angular resolution (0.25°), higher update rate (40 Hz), and longer effective range (up to 20 m) complemented the limitations of the IFM. Like the IFM, the Hokuyo was mounted as an accessory forward of the left wheelchair armrest and slightly pitched down. It was used for both obstacle detection and tracking landmark features. Motor control for the SWS was accomplished with an on-board embedded computer. To regulate the vehicle's linear and angular velocities, it employed a software-based PID that controlled the individual wheel velocities using feedback from high resolution quadrature encoders (4,096 CPR). A Microstrain 3DM-GX1 IMU mounted to the SWS frame provided gyro corrections for improved odometry performance. All other processing was done by a laptop computer with a 1.6 GHz Intel 720QM processor.

3.2 Software Architecture

The system software architecture was based upon the ROS paradigm of interconnected nodes that communicate via messages [15]. Each node subscribed to topics (data streams) published by other nodes, and published its own messages as it processed data. Nodes also listened to a transform tree, which allowed messages to transform to and from any coordinate frame. For basic algorithms, like RANSAC, pass-through filtering, and nearest neighbor cluster extraction, we leveraged Point Cloud Library (PCL) for processing 2D and 3D LIDAR data [17]. The block diagram in Fig. 2 shows the basic structure of the SWS architecture. Each block represents a major subsystem of the SWS, and the arrows represent inter-process message passing between the systems. These subsystems are detailed in the remainder of the paper.

4 Generating the Global Map

A primary motivation for the map-based localization approach was that by leveraging such maps, robots with lower cost sensor suites gain some of the benefits obtained from the higher fidelity sensors of the mapping platform, but without the cost. This necessitated an *a priori* map construction phase for both localizing landmarks, as well as identifying wheelchair-accessible routes within the global map. These two phases are described in this section.

4.1 The Landmark Map

To obtain the desired sub-meter localization accuracy in the absence of reliable GPS, we used a map-based localization approach, which necessitated the availability of an accurate landmark map. To this end, we leveraged our previous results in large-scale outdoor mapping [7]. In this paradigm, three dimensional map data were acquired by manually driving an automobile (Fig. 3) instrumented with a high precision GPS/INS, in conjunction with LIDAR systems. LIDAR range measurements were subsequently registered to a global (WGS 84) coordinate frame. The resulting map data were then synthesized *a priori* to identify robust, salient features for use as landmarks in localization.

As in [7], the landmarks that were used in this work were limited to "pole-like" features (*e.g.*, lamp posts, trees, parking meters, street signs, *etc.*) that are prevalent in urban landscapes. Each landmark was characterized by 5 parameters: (1) xy position (WGS 84), (2) position uncertainty (covariance) estimate, (3) radius estimate, (4) radius estimate uncertainty, and (5) mean reflectivity. The radius was used to classify the landmarks into one of seven classes, with examples shown in Fig. 4. This

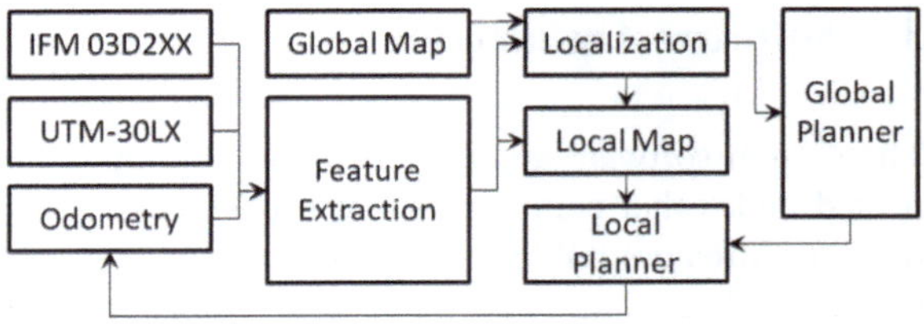

Fig. 2 Major subsystems of the SWS software architecture. Arrows indicate inter-process communication

Fig. 3 The mapping vehicle integrated an OXTS RT-3050 GPS/INS for 6DOF vehicle pose estimation, and a pair of roof mounted Sick LMS291-S14 LIDARs *(circled red)* to capture relative range measurements

classification was used to protect against incorrect landmark data association, and is described in more detail in Sect. 6.2.

We should also note that while the original landmark map of South Bethlehem had been automatically synthesized as outlined in [7], it was augmented for this work "by hand" with additional landmarks. This was necessitated due to the recent razing and reconstruction of a one square block region along our route, and the unavailability of the mapping vehicle during this time frame. The global coordinates of larger features (*e.g.*, lamp posts, trees, *etc.*) were identified from satellite imagery, while smaller features (*e.g.*, parking meters and street signs) were mapped manually using the SWS itself, and subsequently refined during a SLAM phase to improve local consistency. After this refinement phase, their locations and positional covariances were considered fixed within the landmark map.

Fig. 4 Instances of the 7 landmark classes. Landmarks were classified and tracked based upon geometry and reflectivity. Detected landmarks could be reliably associated with their respective classes, which prevented data association errors in our experiments

4.2 The Route Network

The SWS relies upon a route network that exists in the world frame, and serves as the global map for wheelchair accessible paths. The route network was constructed by manually driving the SWS along the desired navigation path while running the localization algorithm described in Sect. 5. Waypoints were then obtained by sampling the SWS pose for every ≈ 4 m traveled. In addition to WGS 84 position information, waypoints were also augmented with semantic information to include labels (*e.g.*, "LU Bookstore"), speed limits, and stop points. Stop points were placed at areas such as cross walks, which currently require human supervision to safely traverse. In our current implementation, when the SWS reaches a designated stop point it will pause until receiving a resume command from the operator.

For global path planning, the route network was represented as a graph $G(V, E)$. Waypoints in the route network corresponded to vertices $v_i \in V$ of G, and the edge set $E \subseteq G$ corresponded to path segments where each $e_{ij} \in E$ connected a pair of waypoints (v_i, v_j). Edges were weighted based upon the expected traversal time as estimated using the edge length and associated waypoints' speed limits. The optimal path to a given destination was then specified via a waypoint sequence as determined from Dijkstra's algorithm.

5 Map-based Localization

Localization of the SWS was based upon a modified version of the FastSLAM 2.0 algorithm [13]. Although FastSLAM is a localization and mapping algorithm, for this work no mapping was conducted during navigation, as the fixed landmark map was provided *a priori*. The algorithm was also adapted to accommodate multiple observations per control, as the control loop (5 Hz) and LIDARs (5 and 40 Hz) operated asynchronously and at different update rates. Each particle was of the form

$$Y_t^{[k]} = \langle x_t^{[k]}, \langle \mu_1^{[k]}, \Sigma_1^{[k]}, s_1^{[k]} \rangle, \ldots, \langle \mu_N^{[k]}, \Sigma_N^{[k]}, s_N^{[k]} \rangle \rangle \tag{1}$$

where $x_t^{[k]}$ was the pose of the kth particle at time t. Each of the N landmarks was parameterized by a mean position $\mu^{[k]}$ and covariance estimate $\Sigma^{[k]}$ which remained fixed. Furthermore, each landmark carried additional semantic data, $s^{[k]}$, used for data association and tracking. In our case, this was the radius of the landmark and reflectivity (in the tracking case). For our tests we used a fixed set of 60 particles. While fewer could be used when the robot was properly localized (≈ 15), we found a large set of particles (> 30) was important to properly initialize the robot.

For the prediction phase of the filter, we sampled from a probabilistic motion model for a differential drive robot where the robot control inputs (v, ω) were corrupted with additive Gaussian noise [20]. Four noise parameters were used: two to indicate control noise for a translational movement ($a_1 = 0.05$ and $a_2 = 0.01$), and

Table 1 Sub decimeter 1D localization accuracy is achieved even when half the landmarks are removed from the map. In each scenario the localized robot pose did not diverge

	100 % Landmarks	75 % Landmarks	50 % Landmarks
Average error (cm)	5.15	5.97	7.18
Standard dev (cm)	4.81	5.20	6.48
Maximum error (cm)	15.71	18.28	23.20
Minimum error (cm)	0.16	0.38	0.31

two to indicate control noise for a rotational movement ($a_3 = 0.001$ and $a_4 = 0.1$). We determined the values for these parameters empirically. Further, we found that with sparsity of landmarks a diverse particle set improves localization performance, thus exaggerated noise in the motion model was preferred.

To perform data association with the landmarks, we used maximum likelihood correspondence (MLC) [20]. Our MLC implementation compared each observation to every landmark in the landmark map, and computed a weight for each association. The weight was approximated by a Gaussian with mean $(z_t - \hat{z_j})$, where z_t is the observation at time t and $\hat{z_j}$ is the predicted observation of landmark j; and covariance $Q_j = H_j \Sigma_j H_j^\top + Q_t$, where H_j is the 3×3 pose Jacobian taken with respect to map features, Σ_j is the covariance of the jth landmark, and Q_t is the linearized vehicle measurement noise. The observation and predicted observation were of the form $z_t = [\rho_t, \phi_t, s_t]^\top$ where ρ, ϕ, and s are respectively the range, bearing, and radius of a feature. The final weight was then approximated by

$$w_j = |2\pi Q_j|^{-\frac{1}{2}} exp\{-\frac{1}{2}(z_t - \hat{z_j})^\top Q_j^{-1}(z_t - \hat{z_j})\} \tag{2}$$

The landmark with the maximum weighted association, w^*, was then used for localization as long as it exceeded a minimum threshold, otherwise it was ignored.

To assess the performance of our localization module, we measured the ground truth distance from reference landmarks and compared this against the localized robot pose. To do this, the robot was manually driven through the course and at 22 reference points we measured the normal distance from the robot base to the nearest landmark. This same distance was calculated through analyzing recorded data on the localized robot pose. We repeated this for three different scenarios: perfect observation, observation of only 75 % landmarks, and observation of only 50 % landmarks. This was done to simulate cases where landmarks may be occluded by dynamic obstacles (*i.e.*, pedestrians). The results of this experiment are summarized in Table 1. While far from exhaustive, these results indicate that while localization accuracy does drop as landmarks are occluded, average 1D accuracy remained sub-decimeter when even half the landmarks were not observed. More significantly, the localization filter did not diverge.

6 Perception

6.1 Ground Plane and Obstacle Detection

Obstacles were detected by both the IFM and Hokuyo LIDARs. The IFM accounted for close range (within 6 m) and low lying obstacles. First, points above the height of the SWS were removed from then scan. Then, a ground plane was established using RANSAC with a planar model [5]. Once a ground plane was established, all points within the noise threshold of the IFM sensor were removed from the raw point cloud data. The 3D points in the remaining point cloud were projected to a 2D xy plane for integration into the local cost map, which is detailed in Sect. 7.1.

The Hokuyo was able to detect obstacles out to 20 m. The raw data was pre-processed by removing max-range scans and outlier points. Since the data were only 2D, no segmentation was needed and the pre-processed scan was integrated directly into the cost map.

6.2 Landmark Segmentation

Landmarks were segmented from both the 2D and 3D LIDAR scans. For the 2D case (Hokuyo LIDAR), each scan was first registered to the fixed global frame to account for any wheelchair motion between scans. Each scan was then decomposed into connected components (CC) with a maximum intra-cluster spacing of 45 cm, and a minimum cluster size of 4 points. Then, a 2D circle was fitted to each CC using RANSAC. Any CC with a fitted circle radius greater than 40 cm was discarded, as this radius was larger than any in our landmark map. Further, any CC that did not fit the circle model well (less than 90 % inliers) was discarded. Each remaining CC was then a potential landmark. Each potential landmark was decomposed into a feature vector for tracking over time, which included the following features: xy coordinate in the global frame, range and bearing to the robot, bounding box dimensions, radius, mean reflectivity, and the number of observations of that potential landmark.

To discriminate between static observations (*e.g.*, trees, parked cars, stationary pedestrians, *etc.*) and dynamic ones (*e.g.*, walking pedestrians, moving cars), each new potential landmark was compared against a list of potential landmarks detected in previous scans. Newly detected potential landmarks that were within the 45 cm intra-cluster threshold and with features (bounding box, radius, and reflectivity) within 10 % of a previously detected potential landmark incremented the observation count of that potential landmark, as well as update all intrinsic parameters to account for the new observation. The range and bearing to a potential landmark was only published when it was observed at least twice.

The 3D case was similar, except given the extra dimension the feature vector for potential landmarks was larger. Since the LIDAR scans were sparse, we concatenated the 5 most recent scans (again registered to the global frame) to create a richer point

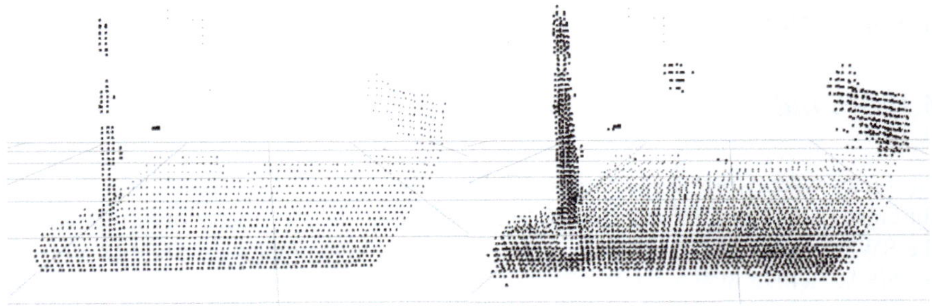

Fig. 5 Scan fidelity was improved by creating composite point clouds from multiple scans. A single scan of a pole (*left*) shows three distinct clusters. The pole in the combined scan (*right*) is contiguous, and will be tracked more reliably due to its higher height/width ratio.

cloud for feature extraction as shown in Fig. 5. We ran this input cloud through a series of filters to remove noise. The ground plane was then removed as described in Sect. 6.1, and the remaining cloud was further processed to segment features. As in the 2D case, the point cloud was broken into connected components of nearest neighbor clusters. A cylindrical model was then fitted to each CC, again using RANSAC, which allowed us to estimate the component's position, orientation, and radius. We necessitated that any potential landmark have a height/width ratio of at least 2, and the lowest point of the feature should be at most 5 cm from the ground plane. Potential landmarks in 3D were tracked just as they were in the 2D case. Again, the range and bearing to a potential landmark was only published as an observation for the localization engine when that potential landmark was observed at least twice.

Fig. 6 Navigation visualization. The *light blue rectangle* is the robot footprint, the *green line* is the desired path, the *yellow line* is the lowest cost trajectory. The *bright red* cells are obstacles of maximum cost. Obstacle cells are inflated with a high cost region in *blue*

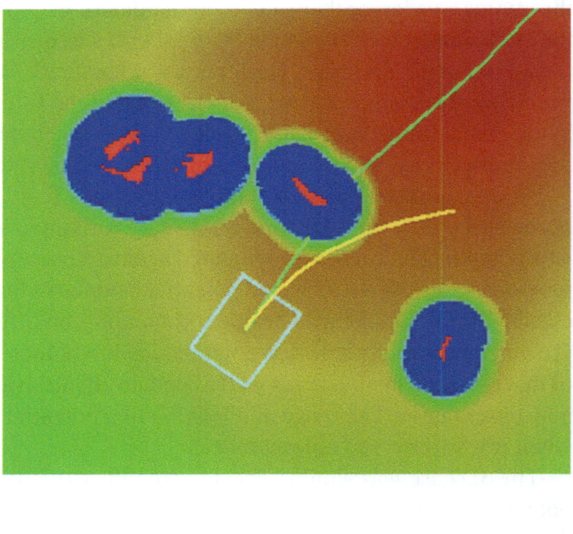

7 Planning and Control

7.1 Generating the Cost Map

As the wheelchair traversed its environment, it maintained a local cost map represented as a 2D occupancy grid [4]. The cost map was 20 m × 20 m in size centered on the robot base, and followed the vehicle in a rolling window fashion. Cell resolution was 5 cm. Each cell was assigned a cost based on three factors: distance to path (C_p), distance to goal (C_g), and occlusion cost (C_o). Cells containing the global path were marked with zero path cost. For all other cells, C_p was proportional to the Euclidean distance from each cell to the closest path cell. A similar process was used to calculate the C_g. Note however that the goal was not the final goal destination, but the last global path point within the dimensions of the rolling window. Any cell that contained an obstacle (as detected by the IFM or Hokuyo) was marked as occupied and given cost C_o which would prohibit traversal. The cost of occupied cells was inflated to the circumscribed radius of the SWS footprint (80 cm) as described in [3]. Occupied cells were only cleared when a ray from both lasers was traced through the occupied cell. The cost map served as input to the local planner for trajectory optimization. A sample costmap is shown in Fig. 6.

7.2 The Local Planner

The SWS employed trajectory rollout for local planning [9]. This is a sample based approach on the input space of the linear and angular control velocities (v, ω). A sample trajectory $T = \{\mathbf{x_0}, v_1, \omega_1, ..., v_k, \omega_k\}$ was specified by the current robot pose $\mathbf{x_0}$ and a sequence of k velocity inputs where k denotes the number of time-steps in the control horizon. The velocities were then integrated forward in time yielding a projected path over the chosen time horizon. An advantage of sampling the control velocities is that we ensure each trajectory is feasible in terms of the wheelchair kinematics. For our implementation, the discretization of v and ω was .06 m/s and .05 rad/s, respectively, and the control horizon $k = 10$ time steps. Each trajectory T_i was then evaluated against the cost map M described in Sect. 7.1 using the cost function $C(T_i, M) = k_o C_o + k_p C_p + k_g C_g$. The optimal trajectory $T^* = \arg\min C(T, M)$ was then selected, and the associated velocity command $(v_1^*, \omega_1^*) \in T^*$ was issued to the wheelchair controller.

8 Experimental Results

Figure 7 provides a satellite view of our bookstore challenge. The green line segments denote the route from Packard Laboratory (right side) to the bookstore (left side). Note that a more direct route was not used, as stairs made it inaccessible

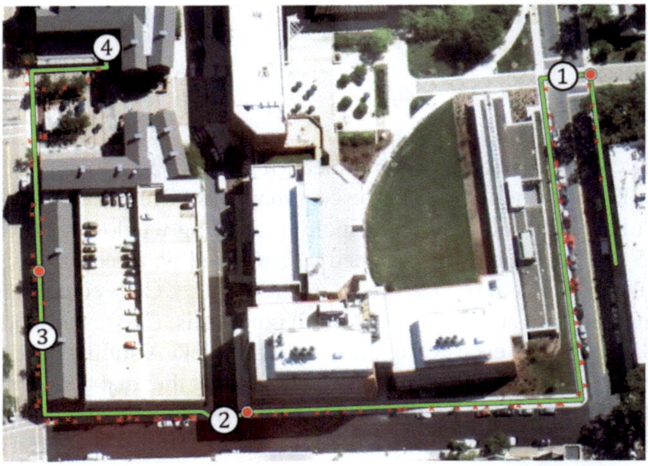

Fig. 7 Satellite view of the route from Packard Laboratory (*right*) to the LU Bookstore (*left*). Stop points are shown as red octagons, and landmarks as red "x" marks. The round-trip distance is 980 m. The four numbered points correspond to locations of interest, as highlighted in Fig. 8

Fig. 8 Ground-level photographs of points of interest along the route from Packard Laboratory to the LU Bookstore. These include: *(1)* a crosswalk on Packard Ave., *(2)* street crossing at Asa Packer Drive with curb cutouts, *(3)* narrow sidewalk Morton St. with significant obstacles, and *(4)* the route destination

for wheelchair users. The red "x" marks denote the locations of landmarks along the route. A maximum of 92 could be observed by the SWS along the route. Red "octagons" denote stop points and are located at the two street crossings as well as the entrance to a parking garage. As noted previously, when the SWS reached a stop point, it paused until manually resumed by the operator (*i.e.*, the user touched the space bar). In a clean run, these and specifying the goal location were the only inputs provided by the user.

The numbered locations in Fig. 7 correspond to the sub-figures in Fig. 8, which provides close-ups of points of interest. These include: (1) the cross-walk on Packard Avenue; (2) the street crossing on Asa Packer Drive; (3) the sidewalk on Morton Avenue, which was only wide enough for single direction traffic; and (4) the LU Bookstore. A successful trip brought the SWS immediately in front of the bookstore doors.

Over the course of a 1 month test period, a total of 11 round-trip missions to/from the bookstore were conducted, for a total distance over 10 km. The nominal SWS speeds were 0.9 m/s (3.1 km/hr) for runs 1–4, 1.0 m/s (3.5 km/hr) for runs 5–8, and 1.2 m/s (4.3 km/hr) for runs 9–11. Of the 11 trips, 3 required user intervention.

The first intervention occurred during the low speed trials, and was attributed to excessive inflation of obstacles in the cost map. This lead the SWS to believe incorrectly that the sidewalk path was blocked. In this case, the SWS attempted to circumvent the perceived obstacle by traveling along the grass to the right. Although this behavior would have been safe, it was obviously incorrect and the run was aborted. A subsequent adjustment to the cost map parameters was made, and this failure mode was not seen subsequently. Success in subsequent trials lead us to increase the SWS speed to 1.0 m/s, and it was this higher speed that exhibited the second failure mode.

At 1.0 m/s, two of the first three trials ended with operator interventions as a result of the SWS leaving the sidewalk path. In one case, a drive wheel migrated onto the grass shoulder and the second time into a mulch bed. Upon review of the log files, we determined that although the SWS perception and planning subsystems were operating properly and sending correct velocity inputs to the motor controller, these were not being actuated in a timely fashion. The cause of failure was identified as the motor controller. The motor controller PID gains had originally been tuned to minimize the steady-state error in the wheel velocities. However, this came at the expense of rise time. As a result, the PID could not achieve the velocity setpoints in a single 200 ms I/O cycle and short duration velocity inputs were lost. This deficiency had been masked at low speeds by the overall system latency, and as a result the local planner could not compensate for the lost actuator inputs. This shortcoming was corrected by re-tuning the motor controller PID using a minimum rise time criterion which was more suitable for real-time control. This allowed setpoints to be approached in a single 200 ms control cycle. An additional benefit from this change was a reduction in the overall actuation latency from approximately 600 to 370 ms as estimated by cross-correlating the velocity I/O response.

After these modifications, an additional trial with the SWS was conducted at the 1.0 m/s velocity. Wheelchair response was noticeably improved, and the trial was completed successfully. As a result of this success, the velocity was further increased to 1.2 m/s, and three trials were completed at this speed without incident. This corresponded to a total distance of over 3.8 km with the new PID tuning, and the SWS behaved predictably at all times. A video of one of these trials can be viewed at http://youtu.be/FUgHMReg4xM. Figure 9 depicts screenshots from this video of the numbered areas in Fig. 7.

9 Conclusions

In this work, we successfully completed our bookstore challenge to demonstrate autonomous navigation of a smart wheelchair system in an urban environment. This was realized by combining 2D and 3D LIDAR sensor systems with a global map to

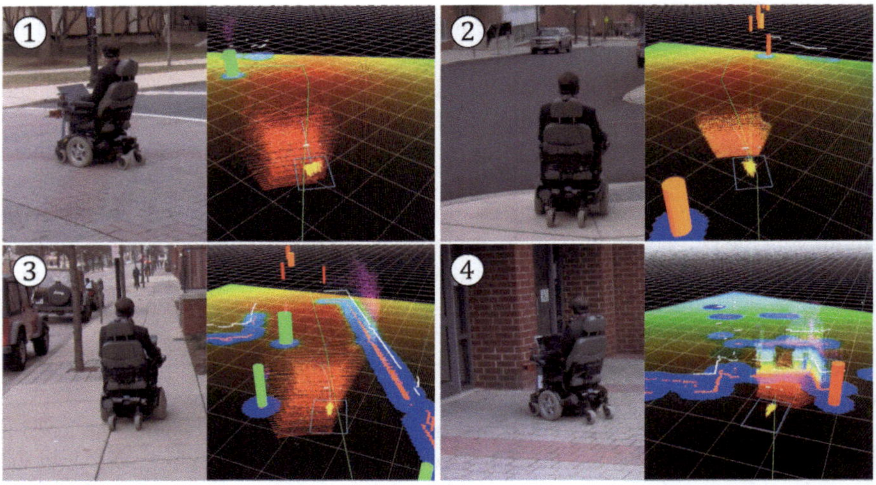

Fig. 9 Screen captures of autonomous navigation. Video capture is on the *left*, visualization of data is on the *right*. In addition to data visualized in Fig. 6, this figure depicts robot pose particles (*yellow arrows*), 2D range data (*white points*), 3D particle cloud (*rainbow points*), and the landmark map (*colored cylinders*). Landmark radii are represented by color and size of the cylinders. Detected landmarks are colored *green*

obtain decimeter level localization accuracy. Open-source software (ROS/PCL) was also leveraged to facilitate platform development.

While we are satisfied with our progress to date, significant work in many aspects of autonomous navigation remains before real-world systems will be available. In the immediate future, we intend to investigate higher speed limits for the SWS, and to grow the global map to investigate longer distance/duration operations. We have no doubt that additional failure modes will manifest as we expand the wheelchair's operational envelope.

In future work, we would also like to address several drawbacks of our current approach. One major disadvtage of the system is that if the localized robot pose diverges, it is very unlikely to ever recover. Therefore, we aim to add new navigation aids like terrain classification so that the wheelchair does not need to be constantly localized in the map, but can navigate the environment and localize only at key points, like when it's nearing an intersection or a destination. The perceptual limitations of the IFM hinder this objective, but we are experimenting with using multiple IFMs to extend the effective perceptual range of the SWS.

A further future emphasis will be navigation in crowds. This will require SWS localization to be robust to occlusion for intermittent periods of time, and likely require the introduction of additional landmark classes into the global map. It will also necessitate reliable people detection and tracking so that appropriate interaction models can be made. We have obtained preliminary results in this area [18], but effectiveness has been hindered by the constrained field-of-view of the IFM. Still, we will continue to make progress, which will inevitably be aided by the future maturation of sensor technologies.

References

1. R. Bostelman, J. Albus, Sensor experiments to facilitate robot use in assistive environemnts, in *Proceedings of the International Conference on Pervasive Technologies Related to Assistive Environments*, Athens, Greece, 2008
2. J. Xu, G.G. Grindle, B. Salatin, D. Ding, R.A. Cooper, Manipulability evaluation of the personal mobility and manipulation appliance (PerMMA), in *International symposium on Quality of Life Technology*, Las Vegas, Nevada, 2010
3. Eitan Marder-Eppstein, "costmap_2d," http://ros.org/wiki/costmap_2d Cited 31 May 2012
4. A. Elfes, Using occupancy grids for mobile robot perception and navigation. IEEE Comput. Mag. **22**, 56–57 (1989)
5. M. Fischler, R. Bolles, Random sample consensus: a paradigm for model fitting with applications to image analysis and automated cartography. Commun ACM **24**, 381–395 (1981)
6. C. Gao, I. Hoffman, T. Miller, T. Panzarella, J. Spletzer, Autonomous docking of a smart wheelchair for the automated transport and retrieval system (ATRS). J. Field Robot. **25**, 203–222 (2008)
7. C. Gao, M. Sands, J. Spletzer, Towards autonomous wheelchair systems in urban environments, in *Proceedings of the International Conference on Field and Service Robotics (FSR)*, Cambridge, Massachusetts, 2009
8. A. Georgiev, P.K. Allen, Localization methods for a mobile robot in urban environments. IEEE Trans. Robot. **20–5**, 851–864 (2004)
9. B.P. Gerkey, K. Konolige, Planning and control in unstructured terrain, in *ICRA Workshop on Path Planning and Costmaps*, Pasadena, California, 2008
10. S. Hemachandra, T. Kollar, N. Roy, S. Teller, Following and interpreting narrated guided tours, in *Proceedings of the International Conference on Robotics and Automation (ICRA)*, Shanghai, China, 2011
11. IFM Effector, "O3D200 3D image sensor product specification." http://www.ifm.com/products/us/ds/O3D200.htm Cited 31 May 2012
12. Microsoft, "Xbox Support - Lighting," http://support.xbox.com/en-US/kinect/setup-and-playspace/lighting Cited 31 May 2012
13. M. Montemerlo, S. Thru, D. Koller, B. Wegbreit, FastSLAM 2.0: an improved particle filtering algorithm for simultaneous localization and mapping that provably converges, in *Proceedings of the International Joint Conference on Artificial Intelligence*, Acapulco, Mexico, 2003
14. I.R. Nourbakhsh, The Wheelchair Project,http://www.cs.cmu.edu/illah/wheelchair.html Cited 31 May 2012
15. M. Quigley, K. Conley, B.P. Gerkey, J. Faust, F. Tully, J. Leibs, R. Wheeler, A.Y. Ng, ROS: an open-source robot operating system, *ICRA Workshop on Open Source Software*, Kobe, Japan, 2009
16. F. Ramos, J. Nieto, H. Durrant-Whyte, Recognising and modelling landmarks to close loops in outdoor SLAM, in *Proceedings of the International conference on Robotics and Automation (ICRA)*, Roma, Italy, 2007
17. R.B. Rusu, S. Cousin , 3D is here: point cloud library (PCL), in *Proceedings of the IEEE International Conference on Robotics and Automation (ICRA)*, Shanghai, China, 2011
18. C. Savtchenko J. Spletzer, in *Sidewalk-leve people tracking with a low-cost 3D LIDAR system*, Lehigh University Technical, Report LU-CSE-11-003, 2011
19. R. Simpson, Smart wheelchairs: a literature review. J. Rehabil. Res. Dev. **42**, 423–436 (2005)
20. S. Thrun, W. Burgard, D. Fox, *Probabilistic Robotics*, (The MIT Press, Cambrige, Massachusetts, 2006)

A Trail-Following Robot Which Uses Appearance and Structural Cues

Christopher Rasmussen, Yan Lu and Mehmet Kocamaz

Abstract We describe a wheeled robotic system which navigates along outdoor "trails" intended for hikers and bikers. Through a combination of appearance and structural cues derived from stereo omnidirectional color cameras and a tiltable laser range-finder, the system is able to detect and track rough paths despite widely varying tread material, border vegetation, and illumination conditions. The approaching trail region is efficiently segmented in a top-down fashion based on color, brightness, and/or height contrast with flanking areas, and a differential motion planner searches for maximally-safe paths within that region according to several criteria. When the trail tracker's confidence drops the robot slows down to allow a more detailed search, and when it senses a dangerous situation due to excessive slope, dense trailside obstacles, or visual trail segmentation failure, it stops entirely to acquire and analyze a ladar-derived point cloud in order to reset the tracker. Our system's ability to negotiate a variety of challenging trail types over long distances is demonstrated through a number of live runs through different terrain and in different weather conditions.

1 Introduction

Roughly linear terrain features such as roads, hiking trails, rivers, powerlines, and pipelines are common in man-made and natural outdoor environments. Such features can be navigationally useful to unmanned ground or aerial vehicles in that they

C. Rasmussen (✉) · Y. Lu · M. Kocamaz
Department of Computer and Information Sciences, University of Delaware,
Newark, NJ, DE 19716, USA
e-mail: cer@cis.udel.edu

Y. Lu
e-mail: yanlu@udel.edu

M. Kocamaz
e-mail: kocamaz@udel.edu

K. Yoshida and S. Tadokoro (eds.), *Field and Service Robotics*,
Springer Tracts in Advanced Robotics 92, DOI: 10.1007/978-3-642-40686-7_18,

both "show the way" and "smooth the way". Finding and keeping to a path by driving along it or flying above it can simplify an autonomous robot's perceptual and motion planning tasks and mitigate hazards which occur in general cross-country navigation. The relative narrowness and continuity of such features implies a certain commonality in the framework of detection, tracking, and control, but each path type has unique appearance and structural characteristics worthy of investigation.

In this chapter we describe a robotic system (shown in Fig. 1a) for following hiking and mountain-biking trails through varied field and forest terrain. We assume that the trail is everywhere traversable with a wheeled vehicle, and also that the trail is non-branching and non-terminating, removing the necessity of intersection or dead-end detection (although our results show that the robot naturally if arbitrarily "chooses" a fork when given a choice). In essence, the task is analogous to "lane keeping" from autonomous road following, involving repeated estimation, or tracking, of the gross shape and appearance attributes of a previously-found trail.

This task echoes the first two DARPA Grand Challenges, which required vehicles to follow rough roads, but there GPS and ladar were sufficient for most successful teams [1, 2]. The DARPA Urban Challenge required more road shape estimation ability, and several teams detailed approaches using primarily vision [3] and rich structural information based on a Velodyne ladar [4]. In the DARPA LAGR program robots had stereo vision instead of ladar and were looking only for open space on their way to a GPS goal, although in constrained areas this was often coincident with path following. Among LAGR-derived work, [5, 6] stand out for explicitly looking for path-like corridors of homogeneous color or texture along the ground.

Fig. 1 **a** Robot in testing area; **b** Sample view from one omnidirectional camera

The European ELROB competitions have also required path-following skills; one robot effectively followed paths by finding "passages" among scattered trees in ladar data [7]. An approach to non-parametric trail detection and tracking using color + intensity saliency maps and agents was discussed in [8].

We reported on earlier versions of our omnidirectional trail-following system in [9, 10]. The former paper discussed a strictly monocular, *appearance*-based approach to discriminating and tracking the oncoming trail region in an image sequence, coupled with differential motion planning within the parametrized trail region while taking into account ladar-detected obstacles. [10] introduced an approach to incorporating stereo-derived scene *structure* estimates as an additional cue at the trail segmentation stage.

Here we present a fully integrated system which uses appearance and structure, not just from stereo but from ladar as well, to find and track the trail in real time. Previous iterations of the system moved at a constant speed regardless of trail or obstacle geometry; now the robot can detect loss-of-trail, excessive slope, or dangerous obstacle events to slow down and even stop in order to more deeply analyze the situation before proceeding. Finally, the differential motion planning system has been updated to lessen the likelihood of collisions while still preserving a basic impetus for forward motion. These changes have yielded vast improvements in the operational performance of the robot in many real-world situations.

2 Methods

As described in [9, 10], the trail region $\mathscr{R}$ immediately in front of the robot is approximated as a constant-width w arc of a circle with curvature κ over a fixed arc range $[d_{\min}, d_{\max}]$. The position of the robot with respect to the trail is given by its lateral offset Δx from the trail centerline and the difference θ between its heading angle and the tangent to the trail arc. Grouping these, we have the current *trail state* $\mathbf{X}$ as the 4-parameter vector $(w, \kappa, \Delta x, \theta)$.

Under the assumption that a unique trail is present in each image, it is segmented in a top-down, maximum likelihood fashion: multiple candidate regions are hypothesized and scored using a *trail likelihood* function L, and the highest-scoring region is the winner. Trail-following entails tracking the trail region over an image sequence, so we use particle filtering [11] to incorporate a prior $p(\mathbf{X}_t|\mathbf{X}_{t-1})$ on the hypotheses which keeps them near the predicted location of the trail in the current frame as derived from the robot's dynamics, as well as setting absolute limits on every state parameter.

2.1 Appearance Likelihood

[9, 12] describe our technique for computing the color appearance likelihood of a candidate region $L_{appear}(\mathscr{R})$ based on the assumption that the trail region has a strong *color* and/or *intensity* contrast with the left and right neighboring regions $\mathscr{R}_L$ and $\mathscr{R}_R$. Briefly, following [5] we compute a small set of exemplar colors for each image using k-means clustering in CIE-Lab space and assign every pixel one of these k labels. A label histogram is computed for each candidate region and its neighbors, and the likelihood is obtained as a weighted combination of *contrast* and *homogeneity* (the entropy of the region color distribution). More details on the approach are given in [9].

In [9, 10] the color contrast is measured by the χ^2 distance between the region and its neighbors, and that measure is used here for some of our results. However, this approach can have some problems with certain scenes such as the one shown in Fig. 8, landmark 3, where several similar shades of grass are found alongside the trail. After k-means clustering the χ^2 metric treats all color clusters as equally dissimilar, meaning that two shades of green are effectively as different as a green and the brown of the actual trail. To avoid accidentally locking onto a marginally distinctive grassy strip beside the trail, we want a measure that preserves some notion of more- and less-similar colors after clustering. The earth mover's distance (EMD) [13, 14] has this property, and so we use this for contrast where noted in Table 1.

Extensive experimentation has shown this approach to trail segmentation to work on a wide range of trail types and illumination conditions without training. Nonetheless, we have found that as a practical matter camera exposure issues can cause serious problems, as with any vision algorithm run outdoors [3]. In particular, bright conditions can be very difficult because of issues with glare (i.e., oversaturation) and deep shadows. These phenomena can obliterate scene colors and make the trail impossible to see in sections, as in Fig. 2. Our cameras (Sect. 3) are in auto-exposure mode by default, but we have found that on sunny days the built-in algorithm frequently gives unsatisfactory results. To mitigate this we implemented our own proportional exposure control method which computes median intensity over a region of interest (ROI) directly in front of the robot and keeps it in a target range by adjusting the shutter speed. This results in much better contrast around the nominal trail region even if other portions of the image are under- or over-exposed.

2.2 Structure Likelihood

The color/intensity contrast between the trail region and neighboring regions depends heavily on the trail material and surrounding terrain and vegetation. While it is sufficient in many situations, when the contrast becomes too low trail tracking may deteriorate or fail entirely. An additional cue which may compensate in these situations is that of scene *structure*. Intuitively the trail region itself is expected to be

Table 1 Testing notes by day

Day	Weather	Key changes	Total km	Number of runs	Mean run (m)	Maximum run (m)
LW	Overcast, few shadows	–	0.77	7	110	310
*S*1	Overcast, few shadows	–	1.57	9	174	410
*S*2	Strong sun, deep shadow	Stop-and-scan for "danger," lowered undersaturation threshold	1.32	18	73	348
*S*3	Strong sun, deep shadow	ROI-based manual exposure control, image capture 7.5 → 10 fps	1.15	19	61	240
*S*4	Overcast, light rain at times	$\chi^2 \rightarrow$ EMD for color contrast, ladar traversability map in likelihood	1.68	7	240	800
*S*5	Scattered clouds, bright sun alternating with gray	Speed $\propto$ trail confidence, max-safety + min-hits motion planning	1.74	4	435	984

relatively smooth while off-trail regions are rougher (i.e., have higher height variance). Moreover, there is often a measurable contrast between the mean height of the trail and the mean height of regions immediately bordering it, whether due to grass, bushes, or rocks that do not exist in the trail or because a "trough" in the soil has been formed from the passage of previous hikers and bikers. More generally, we use the apparent *traversability* of a region as a proxy for the likelihood that the trail goes through it, and then linearly combine this likelihood with the appearance likelihood described above with weighting chosen based on experiments using ground-truth trail segmentations from a separate dataset. There are several sources of scene height information which we exploit:

Stereo A depth map for a subimage of every frame is generated from the robot's stereo omnidirectional cameras. We used the OCamCalib Omnidirectional Camera and Calibration Toolbox for Matlab [15] to obtain intrinsics for the two cameras. Relative extrinsics were initially estimated with manual measurements and then refined with bundle adjustment using *levmar* [16]. Following a common approach

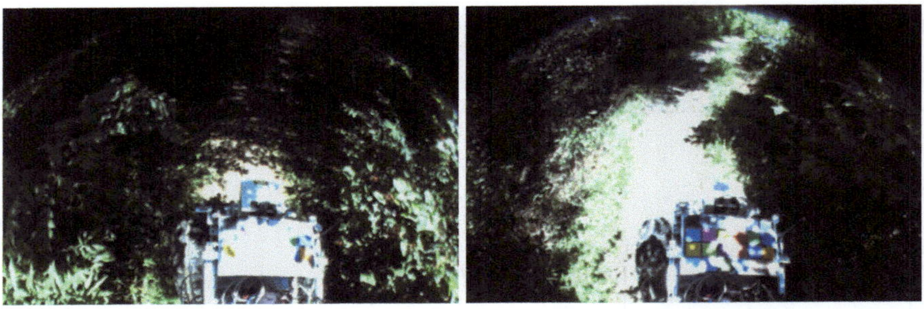

Fig. 2 Problems with oversaturation and shadows. These images are from test *S2* in the mixed section just north of landmark 4

to computing correspondences in omnidirectional imagery [17–19], we rectify the relevant portion of each omnidirectional image into a virtual perspective image such that epipolar lines are image rows; mask out the robot chassis, sensor mast, and peripheral pixels which are not imaged due to the fisheye lens; and then apply a standard pixel correspondence algorithm available in OpenCV, semi-global block matching [20]. The depth map for a sample scene is shown in Fig. 3 at left.

Next, we simplify the approach of [21], which fits planes to robot-sized chunks of a stereo-derived point cloud and combines them into a *traversability map* comprising several hazard-related factors. Full repeated plane-fitting is somewhat expensive, so we approximate it by computing the median absolute deviation (MAD) of the stereo height map over robot-sized bins. If μ_{MAD} is the mean MAD value or "badness" within a hypothesized trail region $\mathscr{R}$, then $L^{stereo}_{structure}(\mathscr{R}) = e^{-\alpha\mu_{MAD}}$. When combined with the appearance likelihood above, this formulation rewards smoother trail region hypotheses and ones which do not contain large step edges (up or down).

Tilting ladar point cloud The exact same MAD approach described for stereo can be used on any point cloud, and the tilting SICK ladar offers very accurate, very detailed point clouds when the robot stops long enough to perform a scan. Because

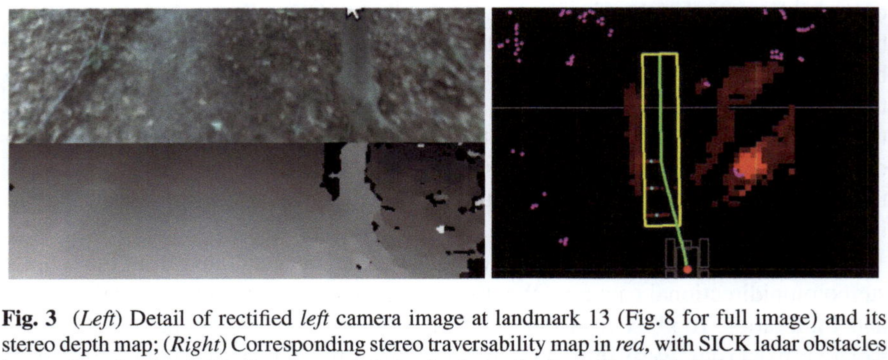

Fig. 3 (*Left*) Detail of rectified *left* camera image at landmark 13 (Fig. 8 for full image) and its stereo depth map; (*Right*) Corresponding stereo traversability map in *red*, with SICK ladar obstacles in *purple* and estimated trail region and planned robot path also indicated (grid is 1 m per square)

such point clouds are not available as the robot is moving, $L^{tilt}_{structure}$ is not part of the normal trail likelihood. When they are gathered, however (see Sect. 2.3 for an explanation of *when* full tilting ladar point clouds are created), they are used as the sole cue to infer the trail location, and the normal visual trail tracker's state is reset to the value indicated by the point cloud search. Some sample point clouds are shown in Fig. 4.

Regular ladar obstacles A traversability map of sorts can be created from the obstacles detected by the SICK ladar in its normal, level configuration as the robot travels. Because obstacles are only detected in one plane, we cannot use height variation as above. Rather, we use the simple criterion of proximity: a bin in the traversability map is incremented if it is within a robot radius of a ladar obstacle. The "badness" of a hypothesized trail region $\mathscr{R}$ is now a sum N of these colliding bins' values, and $L^{ladar}_{structure}(\mathscr{R}) = e^{-\beta N}$. Adding this component to the trail likelihood is extremely helpful because it will push trail hypotheses toward empty or less-dense regions of space even when the robot's visual system is impaired.

2.3 Motion Planning

As described in [9], our motion planner is derived from a Dubins car model [22], which accounts for differential constraints on the robot's motion in the form of a minimum turning radius and rules out reverse motion. The basic Dubins planner, which works for all start and end (x, y, θ) configurations in the absence of obstacles, is used as the kernel of a lookup-table-like approach to planning along the trail in the presence of obstacles. Briefly, given the currently estimated trail region a single *ultimate* goal pose and a set of nearer *candidate* goals are generated and planned for. Each of these plans is evaluated and possibly pruned based on their trajectories colliding with too many obstacles or leaving the trail. From the remaining plans whichever terminates closest to the ultimate goal is selected for execution.

Candidate goal poses are generated in a regular array spanning the trail region laterally in a series of *mini-lanes* and distally from just in front of the robot out to

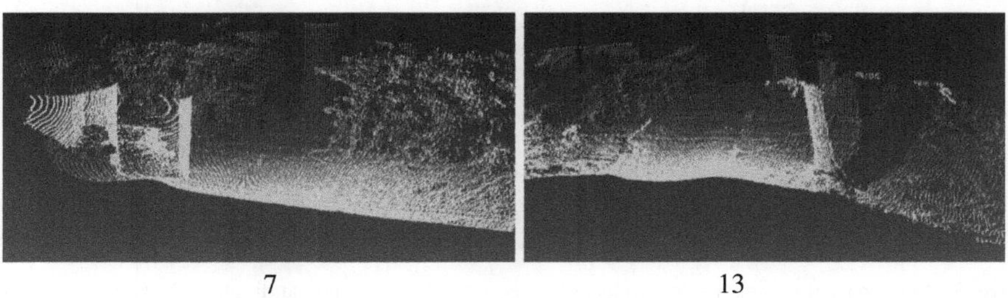

Fig. 4 Tilting ladar point cloud examples for landmarks 7 and 13 (see Fig. 8 for corresponding images). The embankment's drop-off on the *right* is clearly visible in 13

the ultimate goal, all with θ tangent to the trail. A Dubins plan is constructed from the current robot position to each candidate goal pose, and then *extended* along its mini-lane out to the ultimate goal distance. Selecting candidate goals along the same mini-lane but closer to the robot induces more aggressive lane changes in the manner of "swerves" versus "nudges" from [1] or "sharp" versus "smooth" trajectories from [23]. In this work the robot also adaptively inserts lanes which maximize clearance in order to help get through tight spaces. Some examples of plan candidates and their relation to inferred trail regions are shown Fig. 5.

In previous work [9, 10] the criterion for selecting between plans could be termed *min hits*—the path with the fewest collisions was selected. If there was a collision-free path, so much the better, but under the assumption that there is always a way forward, this heuristic would always keep the robot moving with the least amount of obstacle contact—critical for proceeding along trail sections with encroaching foliage such as landmark 5 in Fig. 8. One problem with this approach, however, is that even when there is enough room to stay well away from all obstacles, the robot may pass very close to them because there is no incentive to maximize clearance. Here we implement a two-level path evaluation technique which first ranks plans in terms of *max safety* (the best being a no-collision plan farthest from a collision) and only falls back to the previous min collisions criterion when every plan within the trail region collides.

A further safety and performance improvement can be gained by modulating the robot's speed based on the state's trail likelihood (aka "confidence") and the number of ladar hits anticipated along the robot's planned path. When the trail likelihood is above a certain threshold and the number of expected hits is 0, the robot moves at its maximum speed. As the confidence that it is accurately tracking the trail decreases and/or its expectation that it will be touching or near obstacles goes up, the robot smoothly turns its speed down to a fixed minimum to (a) Allow the trail tracker more time to find or get a better lock on the trail, (b) Allow more time for obstacle avoidance maneuvers to work if the robot is actively turning away from an approaching collision,

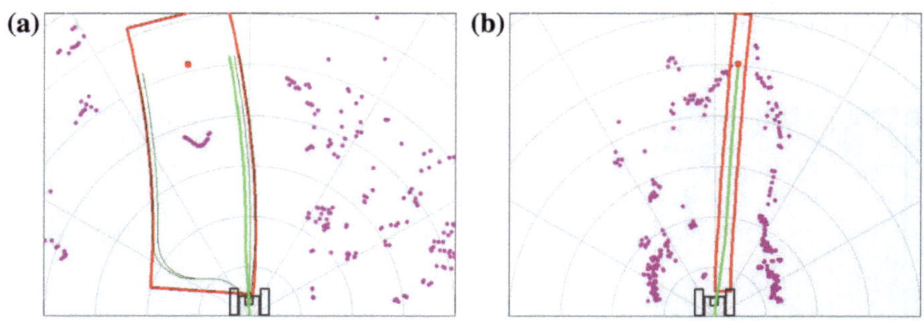

Fig. 5 Motion planning examples: Selected trajectories (*green*) and candidates (*dark green*) within the estimated trail regions. The *red* dot is the ultimate goal pose, while *purple* dots are obstacles detected by the SICK ladar. A *max safety*, no-collision plan is shown in **(a)**, while every plan collides in **(b)**, forcing a *min hits* decision

and (c) Not hit any solid obstacle as hard if a collision is unavoidable. Giving the trail tracker more time is critical at very sharp turns such as landmark 11 where the tracker may lag the angle of the quickly-curving trail, or when the exposure control is working to obtain better contrast in difficult light conditions.

Finally, as a last line of defense the robot will stop completely and perform a full tilt ladar scan when it detects "danger" in form of (a) very large roll or pitch angle which might cause the robot to roll over, or (b) too many collisions in the min collision plan. By performing a search for the trail in the traversability map generated from the full point cloud (as detailed above), the robot can correct mistracking caused by a confused vision system or prevent further travel down a steep slope.

3 Equipment

The primary sensors used for the results in this chapter are two Point Grey Flea2 color cameras and a SICK LMS 291 ladar. Each camera is mounted about 1.15 m off the ground, pointed straight down and rotated so that the longer axis of its CCD is oriented in the direction of vehicle travel. The baseline between them is roughly 0.2 m. The cameras are fitted with omnidirectional Fujinon FE185C046HA-1 lenses which provide a field of view (FOV) of 180° along the vehicle Z axis and 145° along the X axis. In all of these experiments the cameras were set for auto-white balance; where noted they were either in auto-exposure mode or had their exposure manually controlled as described in Sec. 2.1. All images were captured at 640×480 and downsampled as noted for different vision modules.

The SICK ladar is mounted on the robot about 0.6 m off the ground, 0.4 m in front of the robot center, and facing forward with a sweep plane parallel (by default) to the XZ (i.e., ground) plane. Its FOV is 180° and the maximum range is set to 8 m. Its tilt angle is controlled via a Dynamixel EX-106 high-torque servo, enabling the capture of point clouds when the robot is stationary. All point clouds used in this paper were gathered over a tilt range of $[+15, -45]$ degs., with the servo moving at a rate of 20°. / s and the SICK scanning at 50 Hz.

The robot used is a Segway RMP 400 with four-wheel differential steering. The default speed for autonomous trail-following here was 0.75 m/s except where otherwise noted, and the minimum turning radius was limited to 0.75 m. 0.6 m-wide front and rear bumper switches e-stop the motors automatically when pressed with 44.5 N or more of force.

To enable real-time performance, system tasks are distributed over several onboard computers connected via a gigabit Ethernet LAN with IPC message-passing [24]. For all of the experiments described here, the robot's primary computer for image processing, tracking, and motion planning is a Lenovo W520 laptop with an Intel Core i7-2720QM CPU and 8 Gb of RAM. A second computer (a Dell Precision M2400 laptop with an Intel Core Duo T9600 2.80 GHz processor and 4 Gb of RAM) handles and logs all data coming directly from—as well as commands sent to—the front and rear Segway motors, the SICK ladar, and the GPS.

4 Experiments

All autonomous runs documented here were conducted along a combined hiking/mountain-biking trail in a mid-Atlantic US state park which we will term *WCC*. The trail forms a $\approx$ 1.7 km long loop covering varied terrain which on a gross scale can be broken into three largely contiguous types:

(1) Open, grassy fields which are part of a working farm;
(2) A mixture of dense bushes and shorter trees, some overhanging; and
(3) Mature forest, some sections of which have sparse understory foliage and some which are quite dense.

As shorthand, we refer to these categories as *field* (0.6 km long), *mixed* (0.4 km), and *forest* (0.7 km), respectively. The entire loop with the types marked is shown in Fig. 6. A set of notable or difficult locations along the trail are numbered in clockwise order. A short description of each landmark is given in the table next to the trail map, and corresponding images are in Fig. 8.

Testing was conducted on six separate days, with one test occurring in late winter (*LW*) and five spanning the summer months (*S*1–*S*5). Key differences in weather conditions for each day are noted in Table 1, but seasonal variations in vegetation were also important. *LW* presented a challenge with a lack of color contrast (see Fig. 7 for some examples): the grass was dormant, making the fields predominantly brown and yellow, and the trail itself was a wet and muddy brown in many places.

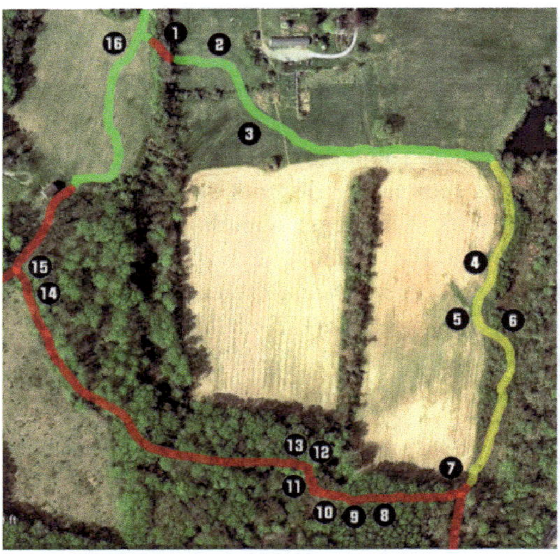

#	Description	#	Description
1	Roots & clustered trees	9	Twin-tree bottleneck
2	Large trailside rocks	10	Narrow section with tall trailside foliage
3	Split for bike-jumping log	11	Sharp left turn before bridge
4	Large isolated trailside rock	12	Ramp to bridge
5	Very narrow section bordered by tall grass	13	Steep embankment after bridge
6	Right trailside sapling	14	Trailside fallen branch
7	Southeast trail junction	15	West trail junction
8	Camouflaged left trailside sapling	16	North trail junction

Fig. 6 Aerial image of $\approx$ 1.7 km *WCC* trail loop. *field* segments are shown in *green*, *mixed* in *yellow*, and *forest* in *red* (the trail location is approximate in the lower forest area). Numbered landmarks referenced in the text are briefly described in accompanying table and pictured in Fig. 8. The area shown is $\approx$ 0.6 km^2

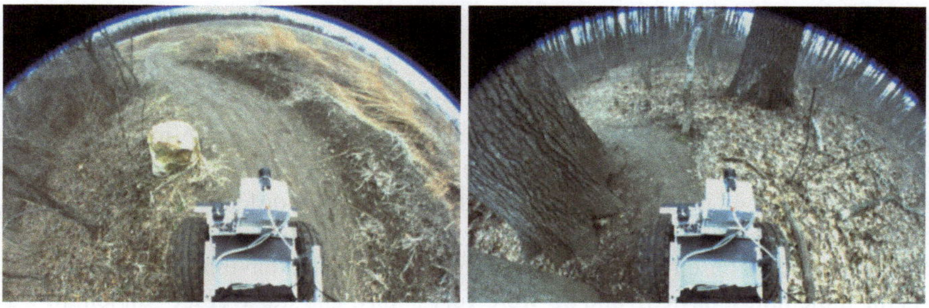

Fig. 7 Low color contrast from late winter (*LW*) at landmarks 4 and 9 on the *WCC* trail. See the summer images of the same landmarks in Fig. 8 for comparison

Trees and shrubs were largely bare and many mixed and forest sections were littered with leaves. By the time of *S*1 and *S*2, spring plant growth made for very strong color contrast for most of the field and mixed sections. This was true for some of the forest as well, but under a dense canopy much of the forest floor remains fairly brown throughout the year. Moreover, the field grass was long, providing height contrast. For *S*3–*S*5, however, the field grass had been cut short for hay.

Two small sections of the loop were not attempted because of terrain characteristics beyond the current perceptual and motion planning abilities of the robot. These were a forest segment in the north (landmark 1 in Figs. 6 and 8) and about 20 m of field just to the east of it (landmark 2). The problematic forest segment has one 2 m section of large, exposed roots and tightly spaced trees that is very difficult to negotiate even manually, and the field segment has a series of large rocks hidden in the grass right alongside the trail. Both are pictured in Fig. 8. Although the robot can track the trail through both of these sections, in the former case it cannot do technical driving that requires reasoning about balance and tire contacts or making zero-radius turns and reversing when necessary. In the latter case grass is growing so close to both sides of the trail that the robot must "brush" past it to proceed (e.g. landmark 5), and it does not detect or reason about the hidden rocks as solid obstacles which must be avoided entirely. Therefore, except for one run on day *S*2 into the rock section (shown in Fig. 8), all runs were started after the rocks.

Each run generally started immediately after the end of the previous run, but in several cases the robot was driven forward manually to get past a difficult spot or to skip a section entirely that it was not deemed ready for. This explains why the total distance traveled on each day varies and why, for example, *LW*'s total distance is not close to the loop length of 1.7 km: the brown grass contrast was so low that no field section was attempted. Time constraints on day *S2* and low motor batteries on *S3* necessitated skipping the last field section.

4.1 Results and Discussion

Top-level results are summarized in Table 1: over the course of six days of testing as the system evolved, 64 separate autonomous runs were attempted. A total of 8.23 km were traveled autonomously by the robot, with a mean run length of 129 m and a median of about 44 m. The five longest individual runs, in order, were 984, 800, 682, 504, and 410 m.

Results varied based on several factors, with the weather having surprising weight. As the table notes, days *S*2 and *S*3 were very sunny, and the average run length was quite low for those days. Of the 33 poor runs which were under 50 m in length, 24 were on these extremely bright days. Even after the ROI-based manual exposure control algorithm described in Sect. 2.1 was implemented for *S*3, the light was still an issue. A major improvement in robustness came with the incorporation of the ladar traversability map from Sect. 2.2 in the trail likelihood with *S4*. This helped the robot navigate areas like landmark 10 in Fig. 8 without being distracted by such bright patches. In general, the modification of the robot's perception and motion

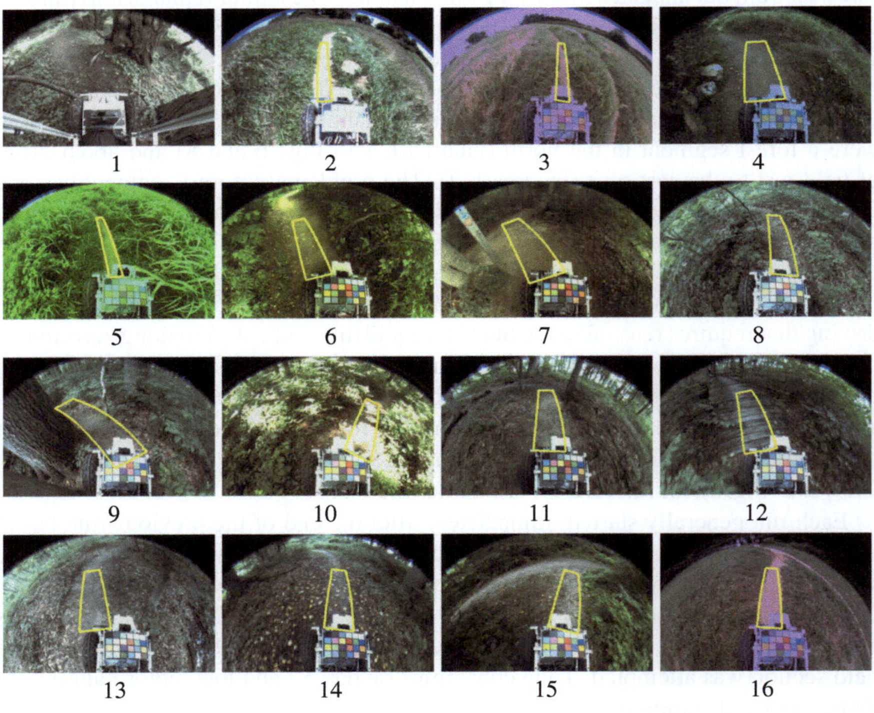

Fig. 8 Robot-view images of trail landmarks during autonomous runs, with the tracked trail region drawn in *yellow*. Landmark 1 was never attempted autonomously and thus no trail region is shown. The images for landmarks 3–16 are all from day S5

planning algorithms over the course of testing significantly improved performance. To underline this, the four longest runs overall were in $S4$ and $S5$. The mean run length for the first two days was 146 m, compared to 311 m for the last two, under similar weather conditions.

On $S4$ the robot completed the loop in seven runs, with its two longest runs that day accounting for 1.3 km or 76 % of the loop. The first of $S4$'s seven runs went from landmark 2 to landmark 6. The run ended when the robot's right front tire rubbed against a small trailside sapling on the way by, causing it to climb slightly and drop suddenly, disconnecting the camera's Firewire cable. The next run ended quickly because a USB cable was still loose. Run 3 was ended manually as the robot approached landmark 7, the trail junction (which it was able to successfully negotiate in $S5$). Run 4 began past landmark 9 and ended at landmark 11 because the turn was too sharp. Run 5 crossed the bridge but was manually terminated because the robot was very close to the embankment on the right (landmark 13). Run 6 quickly missed another sharp left turn, but run 7 took the robot to the end of the loop at landmark 16.

On $S5$ the robot completed the loop in four runs, with its two longest runs totalling 1.66 km or 98 % of the loop. Run 1 went from landmark 2 to landmark 3, where it appeared to mistrack because the trail was obscured by tall grass. Run 2 began at landmark 3, spanned most of the first field section, all of the mixed, and a difficult early portion of the forest before its bumper clipped the sapling in the center-left of the landmark 8 image in Fig. 8. Run 3 ended at landmark 9 because the robot could not maneuver between the two tightly-spaced trees, but on run 4 the robot made it across the bridge, past the embankment, and all the way to the finish line at landmark 16.

By the last two days of testing the trail-tracking system was mature enough that the types of failures observed were primarily ones of motion planning in technical and tight situations, rather than of mistracking the trail. A recurrent problem, echoed in landmarks 2, 6, 8, and 9, stems from a shortcoming in the robot's reasoning about obstacles in the two-level motion planner described in Sect. 2.3. The problem is that when the planner falls down to the *min hits* level, it is basically assuming that the obstacles it will be colliding with are all soft vegetation like grass and twigs. Seeking a least-density path through such obstacles makes sense as a strategy to stay on the trail, but the robot is unable to recognize that some obstacles like saplings and rocks, which may be mixed in with the grass, are solid and *must* be avoided.

5 Conclusion

The trail-following system presented here has been successfully tested over a variety of challenging terrain types, a range of weather conditions and seasons, and at different times of day from mid-morning to late afternoon. In its final form, no parameter changes are necessary for the robot's perceptual component to function in these different situations, nor did the robot have any *a priori* model of the characteristics of its area of operation. Based on previous work using very diverse image data sets [12, 25] and live runs in other locations [10], we believe that the system's perceptual capabilities would transfer quite well to other kinds of trails accessible to wheeled vehicles. In order to increase the overall reliability of the system, however, further improvements in the motion-planning component of the system are being incorporated, including explicit detection of certain classes of obstacles.

Despite these strengths, the possibility of dynamic parameter or sensor changes and exploitation of prior knowledge is attractive. For example, incorporation of trail map information via GPS and visual odometry along GPS-denied trail sections could be quite helpful (1) to let the robot know if it had strayed from the trail, or where the nearest trail was if "lost"; and (2) to allow for intersection anticipation and higher-level route planning over the trail network.

A scenario in which adapation would be desirable is nocturnal trail-following, as neither the color nor stereo structure information derived from the omnidirectional cameras would work in the dark without active lighting. As a dark-capable source of dense structural information for traversability map computations, we have tested a pair of non-overlapping Microsoft Kinect stereo depth cameras and found them highly useful in shady patches and early morning/twilight.

Acknowledgments The authors gratefully acknowledge the support of the National Science Foundation under award 0546410.

References

1. S. Thrun, M. Montemerlo, et al., Stanley the robot that won the DARPA grand challenge. J. Field Robot. **23**(9), 661–692 (2006)
2. C. Urmson et al., A robust approach to high-speed navigation for unrehearsed desert terrain. J. Field Robot. **23**(8), 467–508 (2006)
3. A. Huang, D. Moore, M. Antone, E. Olson, S. Teller, Multi-sensor lane finding in urban road networks, in Robotics: science and systems (Zurich, Switzerland, 2008)
4. C. Urmson et al., Autonomous driving in urban environments: Boss and the urban challenge. J. Field Robot. **25**(1) (2008)
5. M. Blas, M. Agrawal, K. Konolige, S. Aravind, Fast color/texture segmentation for outdoor robots, in *Proceedings of the international conference in intelligent robots and systems* (2008)
6. G. Grudic , J. Mulligan, Outdoor path labeling using polynomial mahalanobis distance, in Robotics: science and systems (2006)
7. C. Armbrust, T. Braun, T. Fohst, M. Proetzsch, A. Renner, B. Schafer, K. Berns, "Ravon—the robust autonomous vehicle for off-road navigation", in *IARP Workshop on Robotics for Risky Interventions & Environmental Surveillance* (2009)
8. P. Santana, N. Alves, L. Correia, and J. Barata, Swarm-based visual saliency for trail detection, in *Proceedings of the international conference in intelligent robots and systems* (2010)
9. C. Rasmussen, Y. Lu, M. Kocamaz, Trail following with omnidirectional vision, in *Proceedings of the international conference in intelligent robots and systems* (2010)
10. C. Rasmussen, Y. Lu, M. Kocamaz, Integrating stereo structure for omnidirectional trail following, in *Proceedings of the international conference in intelligent robots and systems* (2011)
11. A. Blake, M. Isard, *Active Contours* (Springer-Verlag, 1998)
12. C. Rasmussen, Y. Lu, M. Kocamaz, Appearance contrast for fast, robust trail-following, in *Proceedings of the international conference in intelligent robots and systems* (2009)

13. Y. Rubner, C. Tomasi, L. Guibas, A metric for distributions with applications to image databases, in *Proceedings of the international conference in intelligent robots and systems* (1998)
14. G. Mori, Guided model search using segmentation, in *Proceedings of the international conference in computer vision* (2005)
15. D. Scaramuzza, *Omnidirectional vision: from calibration to robot motion estimation* (Ph.D. dissertation, ETH Zurich, Switzerland, 2008)
16. M. Lourakis, levmar: Levenberg-Marquardt nonlinear least squares algorithms in C/C++ (2009), http://www.ics.forth.gr/ lourakis/levmar/. Accessed Nov 2009
17. N. Winters, J. Gaspar, G. Lacey, J. Santos-Victor, Omni-directional vision for robot navigation, in *IEEE Workshop on Omnidirectional Vision* (2000)
18. H. Koyasu, J. Miura, Y. Shirai, Realtime omnidirectional stereo for obstacle detection and tracking in dynamic environments, in *Proceedings of the international conference in intelligent robots and systems* (2001)
19. S. Lin, R. Bajcsy, High resolution catadioptric omni-directional stereo sensor for robot vision, in *Proceedings of the international conference in intelligent robotics and automation* (2003)
20. H. Hirschmuller, Stereo processing by semi-global matching and mutual information, *IEEE Trans. Pattern Anal. Mach. Intell* **25**(2), 328–341 (2008)
21. M. Maimone, C. Leger, J. Biesiadecki, Overview of the mars exploration rovers autonomous mobility and vision capabilities, in *ICRA Space Robotics Workshop* (2007)
22. S. LaValle, *Planning Algorithms* (Cambridge University Press, Cambridge, England, 2006)
23. D. Ferguson, T. Howard, M. Likhachev, Motion planning in urban environments: Part I, in *Proceedings of the international conference in intelligent robots and systems* (2008)
24. R. Simmons, Inter Process Communication (IPC) library (2012), http://www.cs.cmu.edu/~ipc. Accessed Jan 2012
25. C. Rasmussen, Shape-guided superpixel grouping for trail detection and tracking, in *Proceedings of the international conference in intelligent robots and systems* (2008)

Construction of Semantic Maps for Personal Mobility Robots in Dynamic Outdoor Environments

Naotaka Hatao, Satoshi Kagami, Ryo Hanai, Kimitoshi Yamazaki and Masayuki Inaba

Abstract In this paper, a construction system of outdoor semantic maps by personal mobility robots that move in dynamic outdoor environments is proposed. The maps have topological forms based on understanding of road structures. That is, the nodes of maps are intersections, and arcs are roads between each pair of intersections. Topological framework significantly reduces computer resources, and enables consistent map building in environments which include loops. Trajectories of moving objects, landmarks, entrances of buildings, and traffic signs are added along each road. This framework enables personal mobility robots to recognize dangerous points or regions. The proposed system uses two laser range finders (LRFs) and one omnidirectional camera. One LRF is swung by a tilt unit, and reconstruct 3D shapes of obstacles and the ground. The other LRF is fixed on the body of the robot, and is used for moving objects detection and tracking. The camera is used for localization and loop closings. We implemented the proposed system in a personal mobility robot, and demonstrated its effectiveness in outdoor environments.

N. Hatao (✉) · S. Kagami
Digital Human Research Center,
National Institute of AIST, 2-3-26 Aomi, Koto-ku, Tokyo, Japan
e-mail: n.hatao@aist.go.jp; s.kagami@aist.go.jp

R. Hanai
Intelligent Systems Research Institute,
National Institute of AIST, 1-1-1 Umezono, Tsukuba-shi, Ibaraki, Japan
e-mail: ryo.hanai@aist.go.jp

K. Yamazaki
Faculty of Engineering,
Shinshu University, 4-17-1 Wakasato, Nagano-shi, Nagano, Japan
e-mail: kyamazaki@shinshu-u.ac.jp

M. Inaba
Graduate School of Information Science and Technology,
The University of Tokyo, 7-3-1, Hongo, Bukyo-ku, Tokyo, Japan
e-mail: inaba@jsk.t.u-tokyo.ac.jp

K. Yoshida and S. Tadokoro (eds.), *Field and Service Robotics*,
Springer Tracts in Advanced Robotics 92, DOI: 10.1007/978-3-642-40686-7_19,

1 Introduction

Personal mobility robots are required to have ability to move safely in outdoor environments. Outdoor environments are much larger than ranges of sensors and include moving objects faster than robots. Although mapping, localization and path planning methods have been radically improved in recent years, safe navigation in dynamic outdoor environments will not be achieved by just improving their accuracy or reducing their computational costs. To cope with the dangers inherent in dynamic environments, we introduce semantics and reasoning on maps.

Several previous researchers took notice of semantic maps, and proposed their applications. Nüchter et al. developed semantic labeling and object detection framework using a 3D map obtained by a robots [1]. Posnera et al. proposed an online system for semantic labeling of maps in outdoor urban environments [2]. Mozos extracted semantic areas like rooms, corridors, and doors from a 2D map obtained by a robot, and built topological construction using them [3]. This framework enables to decrease computational time required for localization and navigation and to interact with robots using nature language. Galindo et al. improved the performance of their planner using semantic information in the maps [4].

In this paper, we focus on road structures and their semantics, and propose a system of building maps which includes those semantics and a system of navigation using these maps. This paper is organized as follows. We present an overview of the proposed method in Sect. 2. In Sect. 3, we introduce our personal mobility robot and its sensor system for outdoor navigation. In Sect. 4, the geometric semantic map making system performed offline is described. Online navigation system using the semantic maps is presented in Sect. 5, and a conclusion is given in Sect. 6.

2 Concept Description of Outdoor Semantic Maps

This section introduces the advantages of semantic maps for autonomous navigation of mobile robots. Figure 1 shows a simple example of outdoor environment and the presumed path of a person, and Fig. 2 shows the result of shortest path planning using a simple grid map. The path in Fig. 2 would increase a person's chances of being struck by a car.

In this study, we introduce a semantic map to make the mobile robot take a safer path like the one shown in Fig. 1. Semantics useful for outdoor navigation are classified into three categories, "Geometrical Semantics", "Traffic Semantics" and "Symbolic Semantics". Geometrical Semantics indicate geographical properties, such as roads and intersections. Traffic Semantics indicate movement properties of pedestrians and cars on roads, such as which side of the road cars should be on and whether or not cars can enter certain areas. Symbolic Semantics indicate pedestrian crosswalks, traffic signs, etc.

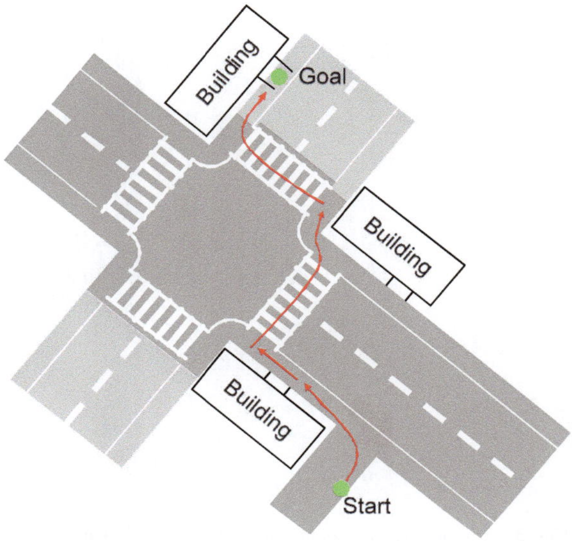

Fig. 1 A simple example of outdoor environments and a person's path

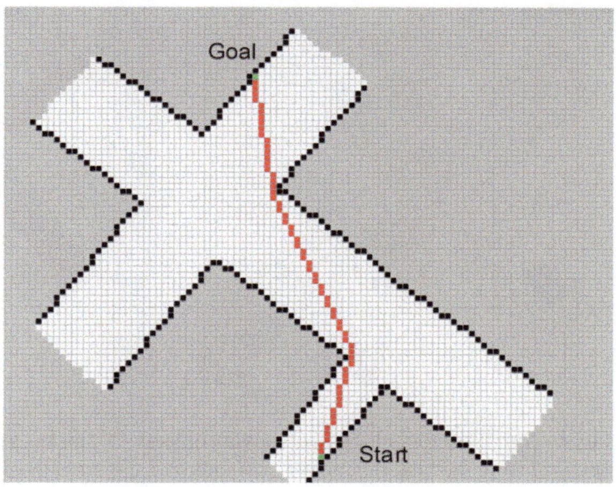

Fig. 2 A path generated from a grid map: it has higher risk of colliding cars

The Geometrical Semantic Maps are constructed as sets of roads. These maps have topological structures that connect the roads with intersections. Even if the map is not geometrically accurate, the positions of robots can be estimated as long as it can recognize intersections or landmarks. Figure 3 shows an example of a topographical semantic map based on road structure.

Other semantics are added in the Geometrical Semantic Maps. The Traffic Semantics can be obtained from the trajectory histories of moving objects (Fig. 4), and the symbolic semantics are obtained from the results of image processing (Fig. 5). The

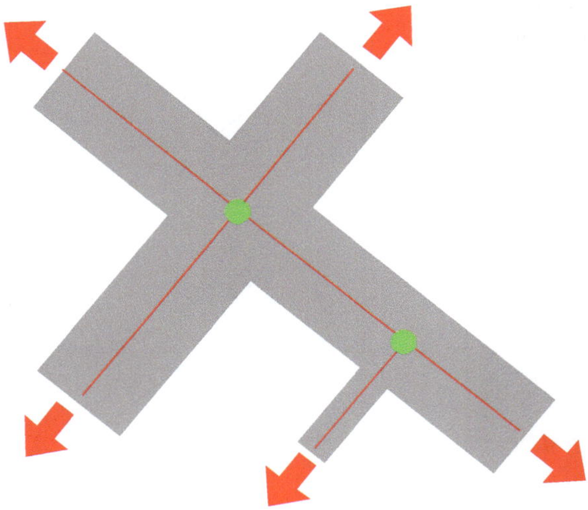

Fig. 3 A geometrical semantic map. *Green circles* mean intersections, and *red arrows* indicates that the roads continue in the indicated directions

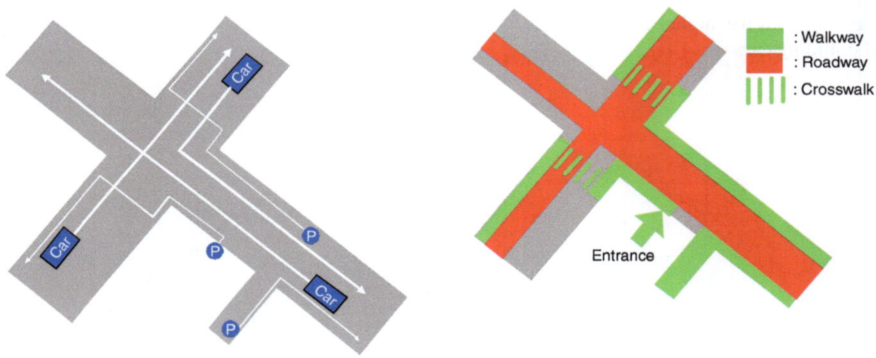

Fig. 4 *Left*: trajectory histories of moving objects ("P" means a pedestrian), *Right*: traffic semantics obtained from the histories. *Green* part indicates the areas in which pedestrians moved around and the *red* part indicates the area cars moved around. The place where a person may appear (written as "entrance") , pedestrian crosswalks, and the directions of cars in each part of the roads can be found

Traffic Semantics in this example are incomplete, but the accuracy can be improved by moving in the same environment repeatedly and updating the trajectory histories.

These semantics makes it possible to plan a path similar to the one shown in Fig. 1 (Fig. 6). Arcs are placed as close as possible to the edges of roads. In addition, reasoning such as "the robot should cross roads at crosswalks", or "the robot should stop temporarily before moving in front of an entrance or entering a crosswalk and confirm that the conditions are safe" will be used to improve safety.

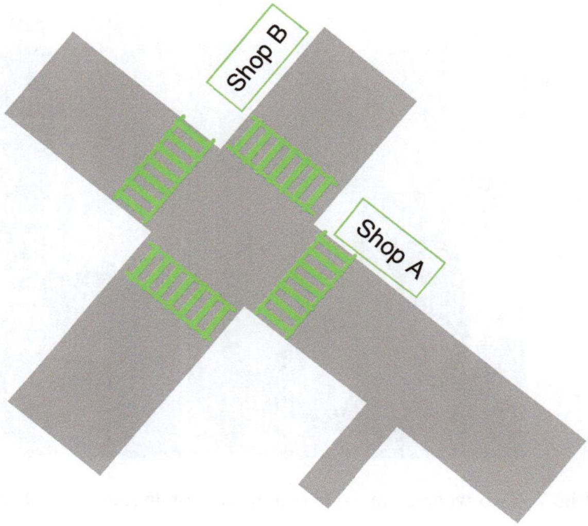

Fig. 5 Symbolic semantics: it is possible to find pedestrian crosswalks and entrances from which pedestrians might appear

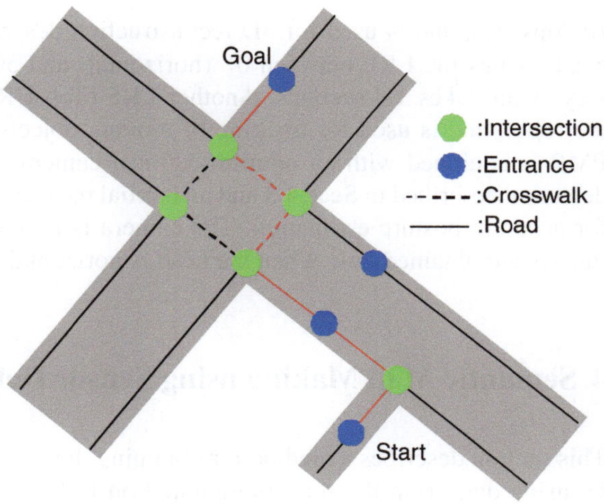

Fig. 6 Path planning result using a semantic map. *Circles* and *lines* indicate nodes and arcs respectively, and *red lines* indicate the path

3 System Configuration of Personal Mobility Robot for Outdoor Environments

Figure 7 shows a mobile robot called "PMR" [5]. It is a single-seat two-wheeled inverted pendulum mobile robot originally developed by Toyota Motor Corporation as "MOBIRO". As PMR can maintain horizontal stance on a slope, it is at little risk of falling over in outdoor environments. Besides, PMR can climb over small bumps up to 50 mm high. The sensor system of PMR is constructed as follows. One LRF (Hokuyo Top-URG 100 Hz version) is mounted on a pan-tilt head (TrackLabs

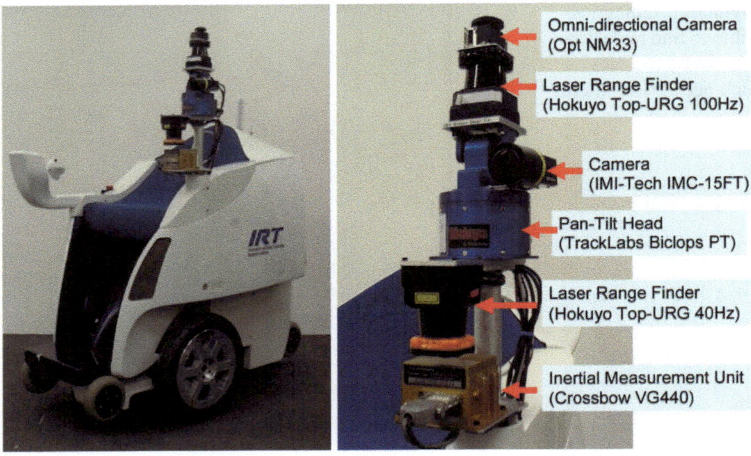

Fig. 7 Two-wheeled inverted pendulum mobile robot "PMR" and its sensor system

Biclops PT), and is used for 3D reconstruction described in Sect. 4.1. The pan-tilt head swings the LRF between 0° (horizontal) and 50° downward, and each one-way swing takes 1.8 seconds. Another LRF (Top-URG 40 Hz version) is fixed on the robot, and is used for tracking of moving objects described in Sect. 4.2. Also, PMR is equipped with an omni-directional camera (Opt NM33) for loop closing detection described in Sect. 4.3 and an inertial measurement unit (Crossbow VG440) for accurate posture estimation. The camera is mounted on the pan-tilt head, and images are obtained only when the head is horizontal.

4 Semantic Map Making using Sensor Data Logs

This section describes a method for obtaining the semantic map described in Sect. 2 by using data from the sensors mounted on PMR. Once PMR runs manually in a targeted outdoor environment, a geometrical semantic map is obtained. Then, PMR can move automatically in the environment, and the map is updated each time it does.

Geometrical Semantic maps are made in the following steps:

1. Local 3D maps are constructed using a every time the swinging LRF reaches the top or bottom of a swing. Local 3D maps are represented by digital elevation maps (DEMs), and each DEM cell is classified as "ground", "obstacle", and so an.
2. "Obstacle" cells corresponding to moving objects are removed using the results of tracking of the moving objects.
3. Local Topological Map Patches are generated from DEMs. Topological Map Patches contain the positions of intersection points and the widths of roads.

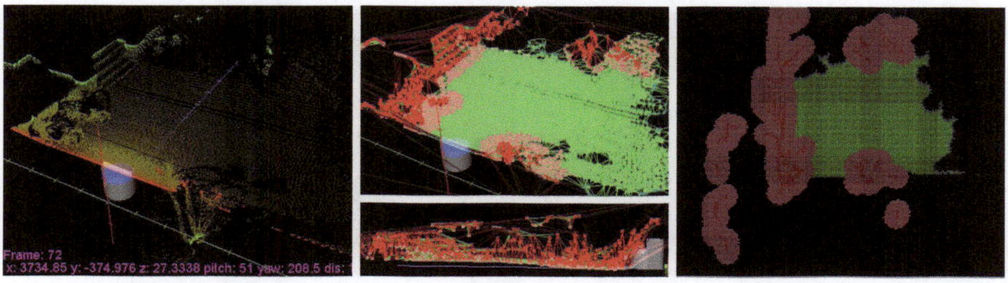

Fig. 8 A result of DEM making on a slope. *Left*: 3D point cloud. *Center*: DEM and Delaunay edges (perspective and side views), *Right*: Interpolated DEM. *Green, red, pink, gray*, cells respectively mean ground, obstacle, near obstacle, and unknown

4. Topological Road Maps are generated from Map Patches. When the robot turns at a intersection, a new road is added to the map.

4.1 Local 3D Map Generation using a Swinging LRF

The proposed method makes local 3D maps by using a swinging LRF to detect static obstacles. It is capable of detecting slopes and small bumps included in outdoor environments. The LRF continues to be swung while the robot is moving. The origin of each 3D map is the point midway between the bottoms of the wheels when the robot start swinging, and the relative coordinates of the robot while moving are estimated using odometry.

As the density of 3D LRF scan points is not uniform, the proposed method uses DEM and Delaunay triangulation. DEM is a kind of 2D grid map in which each grid cell has height information. DEM enables to decrease the number of points in dense areas. If there are several LRF points in one DEM cell, the height of that cell becomes the height of the highest LRF point.

The resolution of LRF scans far away from the robot is sparser than the granularity of DEM, and Delaunay triangulation is used to deal with this disparity. The points for Delaunay triangulation are center points of DEM cells in which there are LRF scan points. The heights of DEM cells without LRF scan points are estimated using Delaunay triangles that include the center points of those cells.

DEM cells are classified as "ground," "obstacle," "near obstacle," and "unknown." It is impossible to classify a cell as "ground" or "obstacle" when using only its height, because outdoor environments include slopes. Therefore, the proposed method uses gradients of Delaunay edges. If the gradient of one of edges connected to a DEM cell is larger than a threshold, that DEM cell is classified as "obstacle". Cells within half the robot's length from "obstacle" cells become "near obstacle" cells, which means the center of the robot cannot enter these cells. Figures 8 and 9 show the results of DEM generations. The size of each DEM cell is 100 mm x 100 mm.

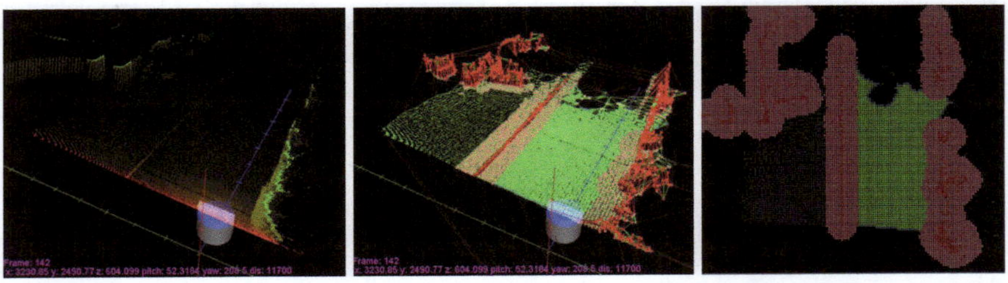

Fig. 9 A result of DEM making. The bump in the center of the DEM is about 80 mm high. The area to the left of the bump is planer, but the robot cannot enter it because of the bump. As a result, most of cells in that area are classified as unknown

4.2 Tracking and Identification of Moving Objects

Outdoor environments include not only static objects but also moving objects. It is difficult to track moving objects by using a swinging LRF because of its slow swing speed. The proposed method therefore uses another LRF fixed to the robot's body to detect and track moving objects.

The tracking algorithm is based on SJPDAF (sample-based joint probabilistic data association filters) [6]. SJPDAF is a kind of multiple hypothesis tracking algorithms [7], and it is robust against false positives and negatives, and makes it possible to flexibly design individual trackers using particle filters.

In the targeted outdoor environments, several kinds of moving objects exist: pedestrians, bicycles, and cars. As the height of the fixed LRF on PMR is about 800 mm, the shapes of LRF scans corresponding to pedestrians are almost the same as those of LRF scans corresponding to bicycles. The, sizes and shapes of LRF scans corresponding to pedestrians and cars, however, are quite different. Besides, as pedestrians often form groups, estimation of the numbers of pedestrians in groups is needed.

The proposed method performs classification of clusters using SVM (Support Vector Machine), and has seven classes: false positive, a car, a pedestrian, and from two to five pedestrians. That is, the method treats estimation of the numbers of pedestrians in a group as a classification problem. As the shapes of LRF scan segments are not stable, the method uses a time-series estimation.

We define the feature vector of LRF scans in a cluster at time t as $z_f(t)$, and a set of feature vectors from time 0 to t as $Z^t_f = \{z_f(0) \cdots z_f(t)\}$. The value we want to estimate is $P(c_n|Z^t_f)$, we obtain

$$P(c_k(t)|Z^t_f) = \alpha \cdot P(z_f(t)|c_k(t)) \cdot P(c_k(t)|Z^{t-1}_f) \tag{1}$$

$$P(c_k(t)|Z^{t-1}_f) = \sum_n [P(c_k(t)|c_k(t-1) = n) \cdot P(c_k(t-1) = n|Z^{t-1}_f)] \tag{2}$$

Also, from Bayes' theorem, we obtain

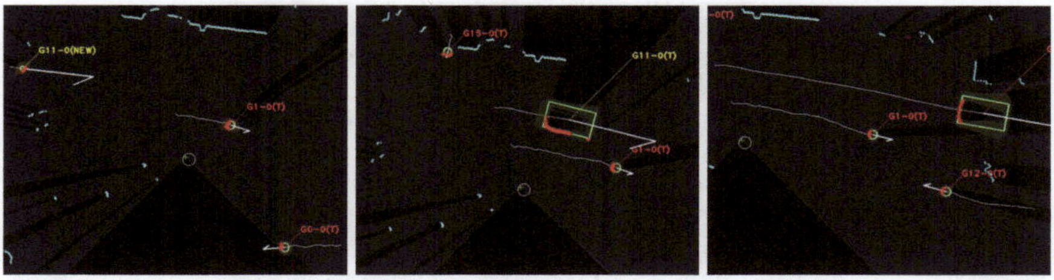

Fig. 10 Result of tracking and identification of moving objects. *Circles* and *squares* respectively indicate pedestrians and cars. Although G11 was initially classified as a pedestrian when it was far away from the robot, it was correctly classified as a car when it approached the robot

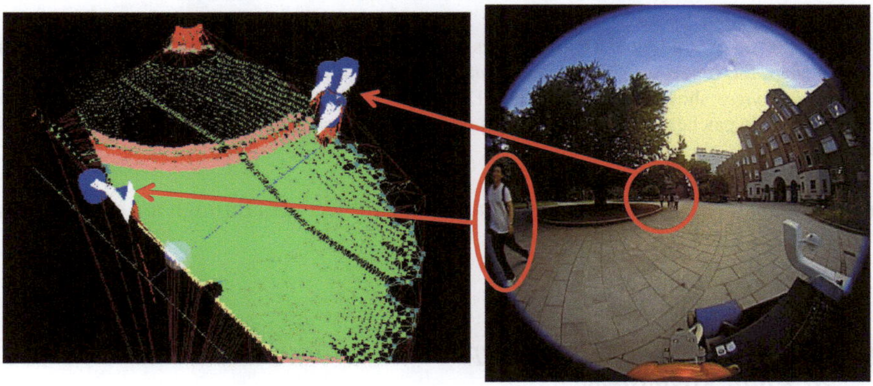

Fig. 11 Trajectory histories in a DEM and an image obtained by the front camera. *White arrows* indicate trajectories of moving objects

$$P(z_f(t)|c_k) = \frac{P(c_k|z_f(t))P(z_f(t))}{P(c_k)} = alpha\,\frac{P(c_k|z_f(t))}{P(c_k)} \qquad (3)$$

$P(c_k|z_f(t))$ can be estimated using SVM, and $P(c_k)$ can be estimated using SVM training sets.

The features for SVM are defined as follows:

z_{f0} : Number of LRF segments
z_{f1} : Sum of lengths of LRF segments
z_{f2} : Average speed
z_{f3} : Difference between angle of directed bounding box and angle of average velocity vector
z_{f4} : Length of long side of directed bounding box
z_{f5} : Length of short side of directed bounding box
z_{f6} : Residual error between directed bounding box and LRF scan points

All features are unaffected by distances and gradients of bounding boxes. Figure 10 shows a result of tracking and identification of moving objects, and Fig. 11 shows

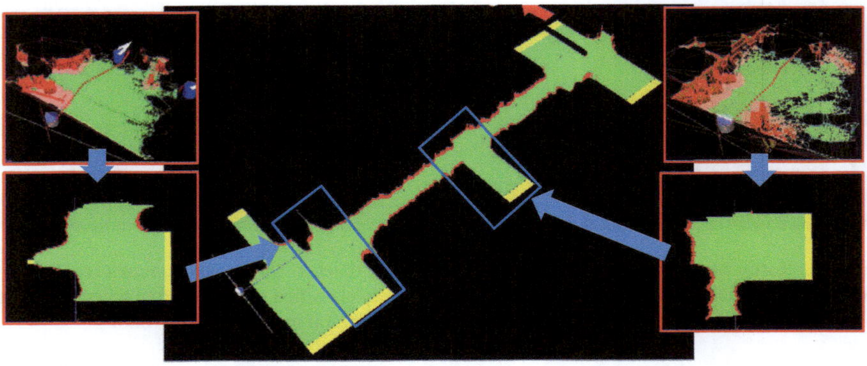

Fig. 12 A result of building of a Road Map using topological map patches. The *center image* shows a Road Map, the *lower side images* show map patches that are parts of that Road Map, and the *upper side images* show the corresponding DEMs before they were converted map patches. *Red* areas in the map patches and the Road Map indicate obstacles and yellow areas indicate open spaces (intersections)

trajectory histories in a DEM. Cells that moving objects got through are classified as "Ground" or "Unknown".

4.3 Making of Topological Map Patches and Reconstruction of Road Structures

The geometrical semantic maps defined in Sect. 2 consist of topological structures. The proposed system integrates local DEMs and builds several Topological Road Maps. Then, topological connections between roads are made using odometry logs and image processing.

Firstly, each DEM is converted to "Topological Map Patch". Topological Map Patches consist of "Raster Segments". Raster Segments are placed every 100 mm (the length of a side of a DEM cell) along x axes, and parallel to y axes. Note that the directions of x axes of DEMs and Topological Map patches are parallel to corresponding roads. The length of a Raster Segment means the width of the road. The left and right edges of each raster can have three kinds of status: "obstacle," "open space," "unknown." "open space" means that there are no obstacles within the range of the swinging LRF, and the system regards "open space" as intersections. Intersections can become nodes that connects two roads.

Next, Road Maps are built using Topological Map Patches. Road Maps have same the structure as Topological Map Patches. Localization of Topological Map Patches is performed using the odometry of the robot, and if overlapped Raster Segments exist, the newer ones are used. Figure 12 shows a result of Road Map Making.

Finally, topological connections between pairs of Road Maps are updated. Because map making is a offline process, the future trajectory of the robot from

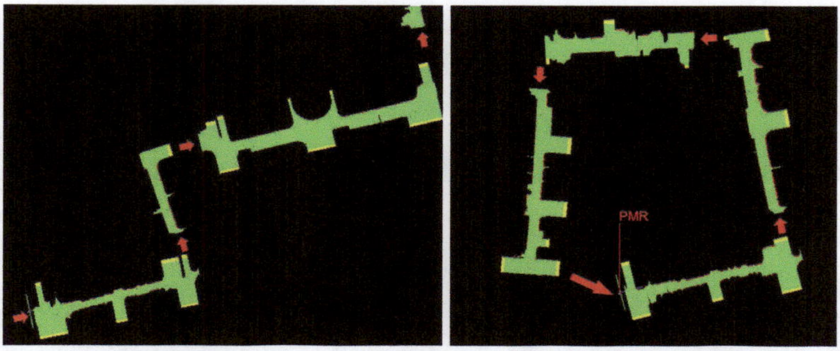

Fig. 13 Results of geometrical semantic map building. *Red arrows* indicate topological connections. The *longest arrow* in the *right-hand image* indicates a connection made by FAB-MAP

the time when the swinging LRF scans were obtained is known. If the future trajectory cross an intersection of the current Road Map, the system considers the robot transitions to another road, and creates a new Road Map when the robot reache the intersection. The two Road Maps are connected with each other by a topological link. The proposed system solves loop problems by using FAB-MAP [8]. FAB-MAP is the method to choose the most similar image with the current one from a sequence of images. Whenever the robot reaches an intersection, the system makes a FAB-MAP by using an omni-directional camera, and checks similarity. If FAB-MAP estimates that the robot has already visited the current intersection, the corresponding topological the connection is created. Figure 13 shows results of semantic map building. Although the shape of the map in the left image is skewed because of inaccurate odometry, the loop is processed appropriately.

5 Outdoor Navigation using a Semantic Map

This section describes outdoor navigation system based on semantic maps. Although the algorithms in the previous section are performed offline, the algorithms in this section are performed online using a laptop computer mounted on PMR.

5.1 Localization Algorithm Using a Geometrical Semantic Map

A geometrical semantic map consists of several Road Maps, and the localization here means to estimate coordinates of robots on a Road Map. It is easy to estimate that which road is the road the robot exists in, because the robot is given the first road and the initial coordinate of the robot in the first road, and the localization enables to detect the intersection where the robot should turn.

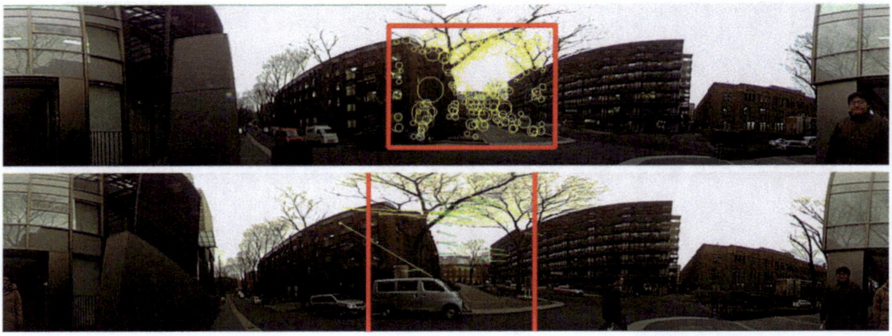

Fig. 14 A result of the angle estimation using FAB-MAP. The *upper image* is the one selected by FAB-MAP as the one most similar to the lower one. *Yellow circles* indicate SURF features used in the angle estimations, and *green lines* indicate the move distances between corresponding points

The localization method in the proposed system uses omni-directional images and topological map patches, and is performed in two stages. Firstly, the angle between the road that the robot is on and the current direction of the robot is estimated using omni-directional images. Secondly, x and y positions are estimated using a topological map patch. Note that x axis of a Road Map is parallel to the road, and the estimations of x positions mean to estimate how far the robot moved along the road.

The angle estimation is based on FAB-MAP, mentioned in Sect. 4.3. Firstly, FAB-MAP selects from the image sequence obtained when the Road Map was constructed the image most similar to the current image. Secondly, the angle estimation is performed using SURF feature points used for FAB-MAP. Only features in the middle 240 pixels (67.5°) are used for the estimation, and it is assumed that the offsets along y direction are much smaller than distances between the robots and SURF feature points. The angle offsets are calculated simply using average horizontal offsets between corresponding points. Outliers are excluded by using Smirnov-Grubbs test. Figure 14 shows a result of the angle estimation method.

The next stage is the estimation of x and y positions in a Road Map by using topological map patches. Before topological map patches are constructed, the result of the angle estimation described above is used to rotate the 3D LRF scan points so that x axis of the patches are parallel to the x axes of the corresponding Road Map.

x positions are estimated using positions of intersections. Intersections in the Road Map corresponding to ones in the current topological map patch are selected, and the x position of the robot is calculated using average values of the positions of lower end of the intersections. If there are no intersections in the current map patch, the x position is updated using the odometry of the robot.

y positions are estimated using the x positions estimated above. First, we define the average values of the y positions of obstacles in the left and right sides of the current map patch as $\overline{y_{l,t}}$ and $\overline{y_{r,t}}$ respectively. Then the Road Map areas corresponding to the current map patch are extracted using the estimated x position of the robot. We

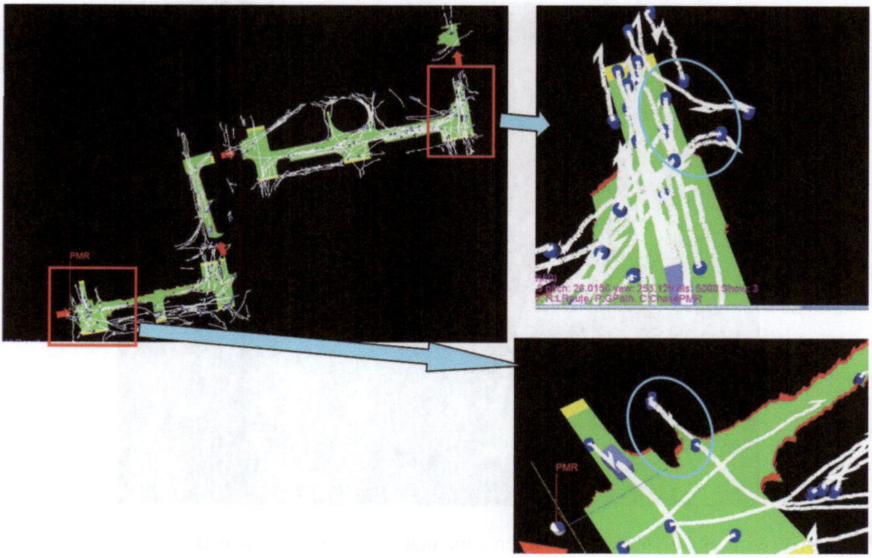

Fig. 15 Traffic semantics: detection of entrances: white lines indicate trajectories of moving objects

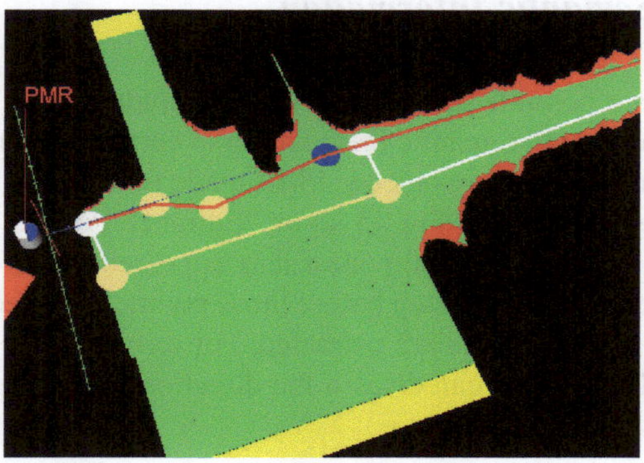

Fig. 16 Nodes in a Road Map: yellow, white, and blue circles respectively indicate Node Type 1, 3, and 4

define the averages values of y positions of obstacles in left and right sides of the extracted part of the Road Map as $\overline{y_{l,g}}$ and $\overline{y_{r,g}}$ respectively. Finally, the estimated y value of the robot $\overline{y_t}$ is calculated as follows:

$$\overline{y_t} = ((\overline{y_{l,t}} - \overline{y_{l,g}}) + (\overline{y_{r,t}} - \overline{y_{r,g}}))/2 \tag{4}$$

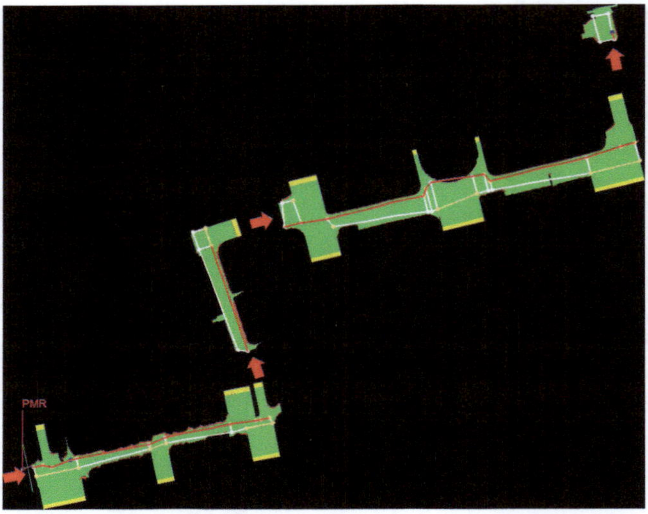

Fig. 17 A result of path planning on a semantic map (*Red lines* are the path)

5.2 Update of Semantic Information

Localization makes it possible for traffic streams or landmarks to be placed into the map. As mentioned above, semantic information is extracted from them. Figure 15 left shows trajectories of moving objects. PMR made four runs in that environment. The images on the right side of Fig. 15 show "Entrance Points." That is, although the robot detected no intersections, pedestrians appeared out of those points. For safe navigation, the robot should stop before those points, and check for pedestrians. Several other kinds of semantics are extracted. For example, the robot can discern road ways and walkways and can discern the directions of road ways by using the trajectory histories of cars.

5.3 Navigation Experiment in an Outdoor Environment

The path planning in the propose system is performed in two stages: global path planning and local path planning.

As mentioned in Sect. 2, Global path planning makes rough routes to the destination by using the semantic map. It decides that that side of each road the robot should move along and where the robot should cross roads. The proposed system places several types of nodes (Fig. 16). All nodes are placed along the right and left sides of each road. The types of nodes are the following:

"Node Type 1" Starting and ending points of intersections.

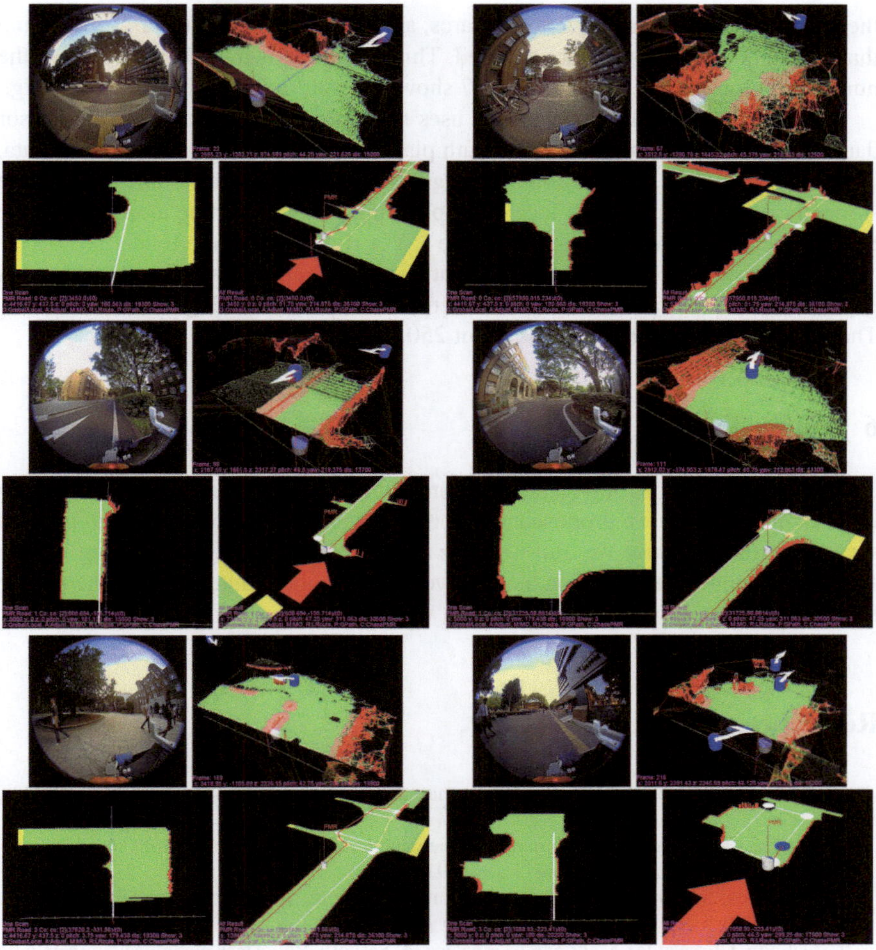

Fig. 18 Navigation experiment results. *Upper left*: front camera images. *Upper right*: DEMs, *Lower left*: local map patches and results of the local path planning. *Lower right*: results of the localization and the global path planning

"Node Type 2" Starting points of roads.
"Node Type 3" Points placed at the opposite sides of "Type 1" nodes. These nodes are not placed if the opposite sides are intersections.
"Node Type 4" "Entrance points" described in Sect. 5.2.

The robot stops temporary in front of "Node Type 1"(if they are starting points) and "Node Type 4" to confirm the safety. Arcs are created as follows. Firstly, adjacent nodes along the x axis of each road are connected. Then "Node Type 3" are connected to corresponding nodes in the opposite side of the road. Finally, "Node Type 2" are connected to "Node Type 1" placed at the corresponding intersections. The costs of the arcs are calculated using length of arcs, and then the costs of the arcs indicating that the robot crosses a road are doubled. This weighting is intended to decrease the number of crossings of roads. Figure 17 shows a result of the global path planning.

Although the global path planning uses semantic maps made using the sensor data obtained in previous runs, local path planning uses the current 3D sensor data. This is because outdoor environments might be changed by quasi-static objects like stopped cars or bicycles. The local path planning generates a line along which the robot can come as close as possible to the left or right edge of the road.

Figure 18 shows the result of the navigation system. The semantic map used in this experiment was the same as that used when the result shown in Fig. 17 was obtained. The moving distance of PMR was about 250 m.

6 Conclusion

To realize autonomous navigation in outdoor environments that include moving objects, in this paper we proposed a system of semantic map making based on road structures. In the future we aim to make a safer navigation system taking human social nature into consideration, and conveying the robot's intentions to the people around it.

References

1. A. Nüchter, J. Hertzberg, Towards semantic maps for mobile robots. Robot. Auton. Syst. **56**(11), 915–926 (2008)
2. I. Posnera, D. Schroeter, P. Newmana, Online generation of scene descriptions in urban environments. Robot. Auton. Syst. **56**(11), 901–914 (2008)
3. O.M. Mozos, Semantic place labeling with mobile robots. Springer Tracts. Adv. Robot. **61**, 15–34 (2010)
4. C. Galindo, J. Fernández-Madrigal, J. González, A. Saffiotti, Robot task planning using semantic maps. Robot. Auton. Syst. **56**(11), 955–966 (2008)
5. N. Hatao, R. Hanai, K. Yamazaki, M. Inaba, Real-time navigation for a personal mobility robot in a environment with pedestrians, in *Proceedings of 18th IEEE International Symposium on Robot and Human Interactive, Communication (RO-MAN'09)*, pp. 619–626 (2009)
6. D. Schulz, W. Burgard, D. Fox, A.B. Cremers, People tracking with a mobile robot using sample-based joint probabilistic data association filters. Int. J. Robot. Res. **22**, 99–117 (2003)
7. Y. Bar-Shalom, Extension of the probabilistic data association filter to multi-target tracking, in *Proceedings of the 5th Symposium on Nonlinear Estimation*, pp. 16–21 (1974)
8. M. Cummins, P. Newman, FAB-MAP: probabilistic localization and mapping in the space of appearance. Int. J. Robot. Res. **27**(6), 647–665 (2008)

Terrain Mapping and Control Optimization for a 6-Wheel Rover with Passive Suspension

Pascal Strupler, Cédric Pradalier and Roland Siegwart

Abstract Rough terrain control optimization for space rovers has become a popular and challenging research field. Improvements can be achieved concerning power consumption, reducing the risk of wheels digging in and increasing ability of overcoming obstacles.In this paper, we propose a terrain profiling and wheel speed adjustment approach based on terrain shape estimation. This terrain estimation is performed using sensor data limited to IMU, motor encoders and suspension bogie angles. Markov Localization was also implemented in order to accurately keep track of the rover position.Tests were conducted in and outdoors in low and high friction environments. Our control approach showed promising results in high friction environment: the profiled terrain was reconstructed well and, due to wheel speed control, wheel slippage could be also decreased. In the low friction sandy test bed however, terrain profiling still worked reasonably well, but uncertainties like wheel slip were too large for a significant control performance improvement.

1 Introduction

Since the first landing of a rover on the moon in 1970 by the Soviet Union, these semi-autonomous, mobile explorers enjoy an increase in popularity. In 1997, the first successful rover named Pathfinder rolled over the Mars surface. On Mars, this is still the only possibility to collect scientific data in such a mobile and interactive manner. Since space rovers are a relatively new way to explore extraterrestrial terrain, mission durations still vary a lot, but the latest missions have been brought to an end due to the rover wheels getting stuck in sand. The two current Mars rovers Spirit and Opportunity were already able to stay operational for more than 5 years, which is 20 times the originally planned mission duration. Nevertheless, they occasionally

P. Strupler · C. Pradalier (✉) · R. Siegwart
Autonomous Systems Lab, ETH Zürich, Switzerland
e-mail: cedric.pradalier@georgiatech-metz.fr

K. Yoshida and S. Tadokoro (eds.), *Field and Service Robotics*, 297
Springer Tracts in Advanced Robotics 92, DOI: 10.1007/978-3-642-40686-7_20,

bogged themselves down in the sand and Spirit was given up and stays immobile because of this issue.

One way to reduce this problem is to minimize wheel slip. During wheel slip the wheels don't move as far as they are supposed to according to their rotational speed. On a sandy surface, this can result in wheels digging themselves in. One of the cause of wheel slippage is often wheels fighting each other because of lack of knowledge about the involved terrain shape. In this paper, we therefore propose a method to adapt the individual wheel speeds of a rover according to the terrain profile. This leads to reduced wheel slippage as well as reduced chances of wheels digging into sandy soil. Because of the complexity of developing new and advanced sensors for space rover wheels, our method is based on sensor input of commonly used and reliable rover sensor technology like inertial measurement unit (IMU), wheel encoders and angle measurements of the bogie suspension system.

1.1 Related Work

Optimizing rough-terrain control for space rovers is a popular field of research. One approach by Iagnemma et al. proposes to estimate force distribution on the wheels by using approximated wheel-ground contact angles [1, 2]. By computing the force distribution of a rover, it is possible to optimize the torques applied on the wheels and therefore reduce wheel slip and power consumption. The estimation of the wheel-ground contact angles is done using simple on-board sensors like IMU inclinometer, joint angle sensors and wheel encoders. Its accuracy strongly depends on dynamic angle measurements and therefore no estimation can be computed when the rover is still. Furthermore, wheel slip and smooth terrain profiles also result in poor wheel contact angle estimation.

Thus Lamon et al. from ETH Zürich developed tactile wheels to measure these wheel-ground contact angles instead of performing an estimation [3, 4]. This method was first implemented on the rovers Octopus [5] and Solero [6]. Later it was also applied to the 6-wheel Crab rover [7]. Although the approach shows promising results [8], embedded wheel sensors are still too complex and unreliable to be used in extraterrestrial environments.

1.2 Goals and Limitations

Our objective is to develop an alternative approach on reducing wheel slip and optimizing control of space rovers in rough terrain. In contrast to the work by Iagnemma et al. mentioned above, our control should also yield good results in smooth terrain. On the other hand, we want to avoid using tactile wheels and other complex sensor systems in order to deliver a realistic approach for current space rovers. Our core idea relies on profiling the terrain shape using commonly used rover sensors such as IMU, wheel encoders and angle measurements of the bogie suspension system. The terrain shape can then be used to achieve wheel speed optimization.

2 Simultaneous Mapping and Control

Our approach proposes a velocity controller based on on-line terrain profiling, called simultaneous mapping and control (SMAC). An overview of this controller is shown in Fig. 1. In the state estimation part, the terrain shape and the rover position are estimated. On one hand, terrain shape estimation highly depends on the rover position, but on the other hand, the rover position estimate can be improved significantly by accounting for the terrain shape in a probabilistic filter. Finally, knowing the terrain and rover position, a wheel controller can be proposed which optimizes the wheel speed to minimize theoretic slippage.

In the following, our implementation is explained using a parallel suspension bogie rover model, as shown in Fig. 1. However it is possible to adapt the controller to other suspension systems by modifying the geometry equations accordingly. Furthermore, we take the following assumptions:

1. We decouple both rover sides from each other and apply our method to each side independently.
2. We do not actually profile the real terrain, but the path traversed by the center of the wheels (see Fig. 1). From now on the term terrain designates this wheel center path. Note that recovering the real terrain shape is not possible due to ambiguities in corners.
3. The rover is assumed to drive straight and does not roll sideways. This allow reducing the profiling problem to 2 dimensions. As another consequence, all the wheel centers will follow the same terrain path which is included in a vertical plane in 3D space (designated as wheel movement plane, also take a look at Fig. 4 in Sect. 3).

In this paper, we focus straight trajectories as a proof of concept for simultaneous mapping and control. This also allows applying the 2 dimensional modeling to each side of the rover independently. However, it is clear that considering curved trajectory would require to consider the full 3D complexity of the problem, for which we cannot propose a solution at this stage.

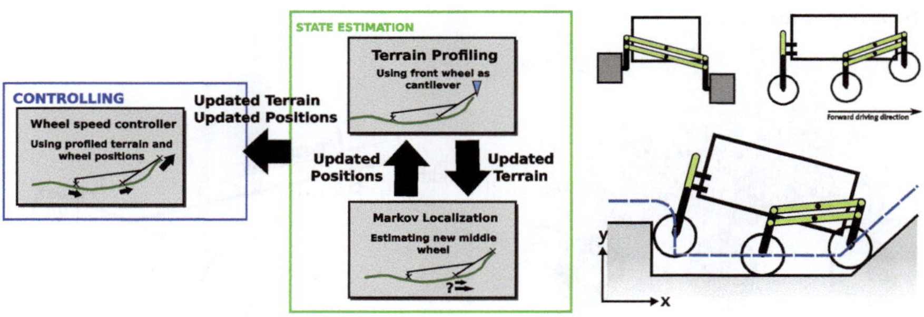

Fig. 1 State estimation and controlling (*left*), rover model (*top right*) and profiled wheel path compared to the real terrain (*bottom right*)

2.1 Terrain Profiling

Terrain profiling allows us to approximate the terrain shape and—knowing the rover position—to optimize the wheel speed. Our objective is to use space-realistic sensors to achieve this goal: IMU, angle measurements of suspension bogies and motor encoder readings. First of all, we assume that, lacking visual or other distant sensing devices, there is no possibility to foreknow the terrain shape. The latter needs to be profiled in the instant the front wheels are traversing it. Therefore, these front wheels can be seen as cantilever-based tactile sensors. They can be used as profiling sensors while the two middle wheels use the former profiled terrain for propagation estimation. During the next iteration, the propagated middle wheels acts as new reference points for the front wheels to profile the next terrain points and so on. Hence, the terrain can be iteratively built up. This procedure is illustrated in the state estimation part of Fig. 1. However, one can easily observe that errors in profiling will accumulate since there are no measurements with absolute reference. To partially mitigate that, the middle wheel position is estimated through a probabilistic filter that reduces the displacement errors along driving direction and thus also improves the quality of the terrain profile. This is described in the next Sect. 2.2.

The illustration of a parallel bogie rover in an arbitrary configuration is shown in Fig. 1. The front and the middle wheels are connected with a parallel bogie (to be called front left/right bogie). In the rear view, one can see that the two back wheels are also connected with a parallel bogie (rear bogie). A simplified model used for the upcoming computations is illustrated in Fig. 2. Note that the parallel bogie-wheel connectors can be disregarded since we only depend on relative wheel positions.

In order to profile the terrain at the front wheel, the position of the middle wheel has to be defined first. Assuming its x-position is iteratively propagated, we can find the y-position by placing the middle wheel on our current terrain profile:

$$x_{MW}(x) = x \tag{1}$$

$$y_{MW}(x) = Terrain(x) \tag{2}$$

The position of the front wheel can then be found using the IMU tilt angle β and the front bogie angle φ:

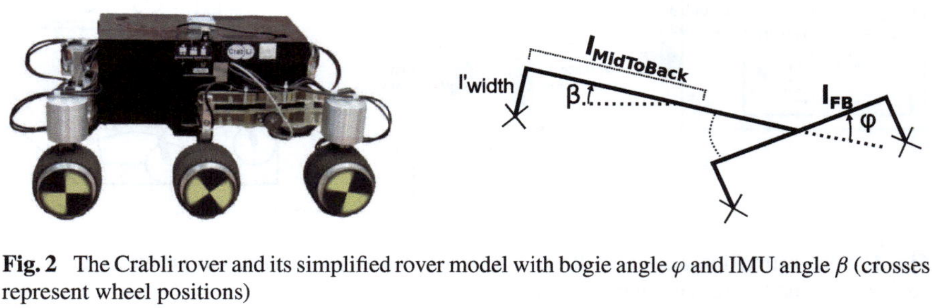

Fig. 2 The Crabli rover and its simplified rover model with bogie angle φ and IMU angle β (crosses represent wheel positions)

$$x_{FW}(x) = x_{MW}(x) + l_{FB}\cos(-\varphi - \beta) \tag{3}$$

$$y_{FW}(x) = y_{MW}(x) + l_{FB}\sin(-\varphi - \beta) \tag{4}$$

where l_{FB} denotes the distance between the middle and the front wheel (length of the front bogie). Hereby, it is possible to profile new terrain points using the current terrain, the middle wheel x-position and the configuration of the rover.

To initialize the system, it is assumed that the rover starts on a flat terrain. Unfortunately, from proprioceptive measurements only, it is not possible to guess the shape of the terrain. As a result, even though the flat terrain assumption is far from perfect it seems to be the only one available. Alternatively, one could use exteroceptive measurements (e.g. from a stereo camera) as a starting point for the terrain profiling. Although very relevant this has not been addressed in the context of this paper.

2.2 Markov Localization

Markov Localization is used to estimate and propagate the middle wheel x-position by fusing the wheel speed information and the rover configuration. The general Markov Localization equation derived from Bayes' law is given by:

$$Bel(x_t|z_t...z_0) = \eta\, p(z_t|x_t) \int p(x_t|u_{t-1}, x_{t-1}) Bel(x_{t-1}) dx_{t-1} \tag{5}$$

where $p(z_t|x_t)$ corresponds to the observation model predicting the observation z_t (see below) given the middle wheel position x_t, and $p(x_t|u_{t-1}, x_{t-1})$ is the rover motion model giving the rover displacement for a control input u_{t-1}.

Belief

The belief $Bel(x_t|z_t...z_0)$ is an arbitrary probability distribution for the middle wheel x position. Since we do not account for the kidnapped rover problem and the updates are quite frequent, the belief does not have to be computed along the whole terrain. We can keep the width of the belief relatively narrow (about 1–2 times the length of the rover) centered around the middle wheel position. This saves a lot of computing power since the belief is involved in the discrete convolution for the action update.

A sequence of beliefs resulting from our test data is shown in Fig. 3 (left). In the beginning, the belief is relatively wide—the position is not known very well yet. As soon as the rover reaches distinctive terrain, the belief gets more narrow (which is due to the sensor model as we can see later). In the second picture of the sequence, the belief gets steeper on the left side. Since the front wheel of the rover is situated higher than the other wheels, it would not make any sense for the rover to be placed more than 0.1 m to the left. In the third illustration, the rover is in a very distinctive

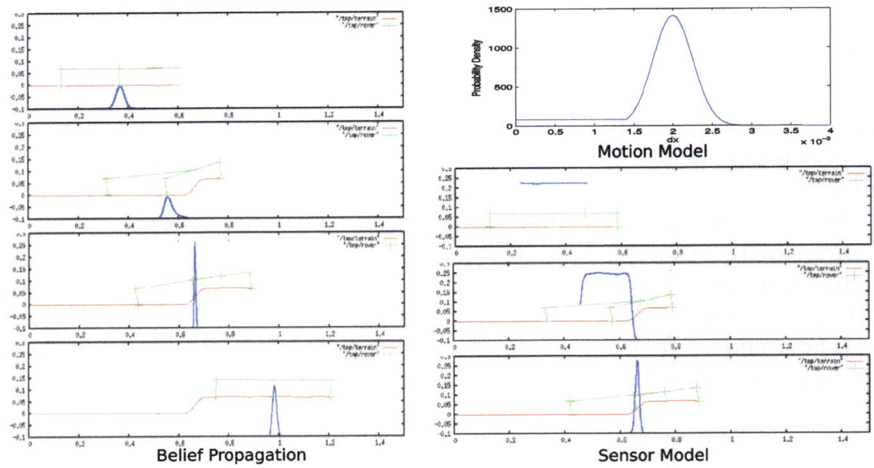

Fig. 3 Markov localization: belief propagation based on sensor model and motion model

configuration. This configuration can only appear if the middle wheel is positioned at the middle of the step slope. Hence the resulting belief is very narrow and precise—the rover is localized. Later, when flat terrain is reached, the belief starts to widen again. This is caused by the convolution with the motion model.

Motion Update

During the motion update, the old belief $Bel(x_{t-1})$ is convolved with the action model $p(x_t|u_{t-1}, x_{t-1})$. The latter defines the expected rover displacement given the motor input. In our case, this is the result of combining two probability distributions:

- A normal distribution centered around the value $u_{t-1}dt$ of the distance traveled since the last update. This distribution models the ordinary motion of the rover without wheel slip.
- A sigmoid distribution (approximation of the uniform distribution) from 0 to the center of the normal distribution and modeling the uncertainty resulting from slippage.

The equation of the motion model is given by:

$$p(x_t|u_{t-1}, x_{t-1}) = \sum_{S_r} p(x_t|u_{t-1}, x_{t-1}, S_r)p(S_r) \tag{6}$$

$$p(x_t|u_{t-1}, x_{t-1}, S_r) = \begin{cases} h(1 - \frac{1}{e^{-x_{end}(x_t-m)}}), & \text{if } S_r = 1 \\ \mathcal{N}\left(x_{t-1} + u_{t-1}dt, \sigma^2\right), & \text{if } S_r = 0 \end{cases} \tag{7}$$

where S_r for wheel slip ($S_r = 1$ if the wheel is slipping, $S_r = 0$ otherwise). Although the slip indicator is binary, the use of a sigmoid distribution means that any movement up to the desired one is considered equally likely in case of slippage, thus covering the cases from the wheel slipping just a bit up to 100 % slippage.

Sensor Update

The sensor model $p(z_t|x_t)$ describes the probability of measuring the observations z_t given x_t. In our case, the measurements are the IMU tilt angle β, the front bogie angle φ and the back bogie angle θ. But in order to reduce computational complexity, we transform these measurements into the y position (height) of the front wheel y_{FW} and the back wheel y_{BW}. Using these transformed measurements, we can write:

$$p(z_t|x_t) = p(y_{FW,t}, y_{BW,t}|x_t) = p(y_{FW,t}|x_t)p(y_{BW,t}|x_t) \tag{8}$$

where we introduce the reasonable approximation of the probability distributions of $y_{FW,t}$ and $y_{BW,t}$ being independent and normally distributed around the position that can be predicted from the terrain shape.

A sequence of sensor models from our test data is shown in Fig. 3 (more precisely, the plots correspond to the posterior localization using the sensor model and a uniform prior). The sensor model is depicted blue, the current rover configuration green and the profiled terrain red. In the first diagram of the sequence, one can see that there is more or less the same probability of this configuration to be placed around the middle wheel position neighborhood. In the next diagram the sensor model probability narrows a little bit due to the front wheel position on the step. Eventually in the third diagram, the configuration appears to be very distinctive and can only be placed in the center of the step. This is also indicated by its narrow sensor model.

2.3 Wheel Speed Controller

Based on the approximated terrain and the current wheel positions, an optimal speed for the front, middle and back wheels can be derived. The principle of our approach is to compute the movement of the front wheel that would result from applying the desired speed for a control interval. From this movement, based on the known length of the suspension element, we can compute the resulting displacement of the middle and rear wheel, and deduce their optimal speed. The challenge is that the terrain is necessarily unknown in front of the front wheel, and consequently, the front wheel movement must be computed by extrapolating the known terrain. Given the low speed we are considering, we currently use a linear extrapolation based on the last measured terrain slope.

3 Simulation and Testing

3.1 *Goals*

To measure a possible performance gain due to the new wheel speed controller compared to the former situation, we intend to evaluate if the controller helps decreasing wheel split and power consumption as well as improving the ability to overcome obstacles.

3.2 *Wheel Tracking System*

In order to measure wheel slip, the difference between actual wheel velocity (ground truth) and commanded velocity needs to be determined. Additionally, we would like to compare the profiled terrain of the controller with the actual path traversed by the wheels. Therefore we decided to implement an external optical wheel tracking system. It is based on a camera recording the test runs from aside the track. Markers placed on the wheels assure reliable detection of the wheel centers. The origin of the 3D coordinate frame is placed below the camera on the ground; y axis aligned with the camera. This setup is illustrated in Fig. 4. For most of the tracking implementation, functionalities from the OpenCV library were utilized. See [9] for more details on the tracking system and its calibration.

3.3 *Testing Environment*

Indoor

In a first approach, tests were performed indoor in the ASL Robolab at ETH Zürich (see Fig. 4). The HD webcam was set at the distance of 1.35m from the track, which appeared to be a reasonable trade-off between track length and resolution of wheel

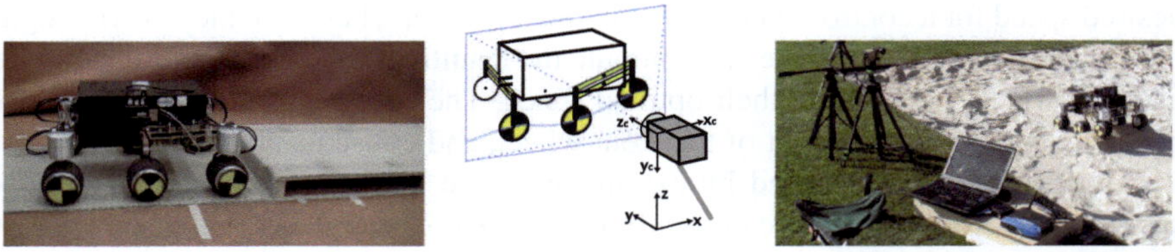

Fig. 4 The indoor test setup (*left*), the wheel tracking setup with HD webcam and markers (*center*) and the outdoor setup (*right*)

tracking. With this distance to the track, a resolution of about 5.75 pixels/cm was achieved.

The track surface consisted of carpet to reduce wheel slip at first. The test runs were performed using 3 different obstacles on the track: a step obstacle (carpet surface), a hill obstacle (wooden surface) and a smooth step obstacle (Wooden surface).

Outdoor

In a second stage, tests were also done outdoors in a sand pool (see Fig. 4). The distances and lengths of the track were similar to the indoor setup. Before each run, the sand surface had to be made smooth and level. Tests with slack sand where also performed to compare the influence of the different types of sand consistencies.

The outdoor obstacles were slightly different from the indoor setup: a step obstacle (bricks), a hill obstacle (wooden surface) and an asymmetric, irregular obstacle (stones).

3.4 Result Discussion

In this subsection we would like to discuss the results of a part of the test runs performed indoors and outdoors. As can be seen later, it is really hard to find a right measure for performance. In our current setup, the performance improvement on wheel slippage due to the controller is small with respect to the measurement noise. Therefore, it has not been possible to end up with quantitative performance metrics describing wheel slip performance or power consumption. In the following, a qualitative analysis of the performance is proposed.

Indoor

The obstacle traversed by the Crabli rover was a 8 cm high step covered by carpet. The situation is shown in Fig. 4.

The resulting terrain shape profiled by the controller is plotted in Fig. 5 as a blue line. The red line originates from the wheel tracking and is seen as ground truth. The most noticeable difference is the slope at the step. The real terrain features a sharp rise followed by an even curve. By contrast, the profiled terrain has an almost constant slope which is more flat. This is actually due to the short wheel slip that occurs when the front wheel touches the step. At that instant, the rover stands still before the front wheel moves up, causing the middle and rear wheels to slip. Our controller is not able to sense this slip though and assumes that the rover is still moving forward. Combined with the new velocity component in vertical direction according to the changing rover configuration, the slope angle appears to be flatter than the one from the ground truth. This effect can also be shown using the diagram

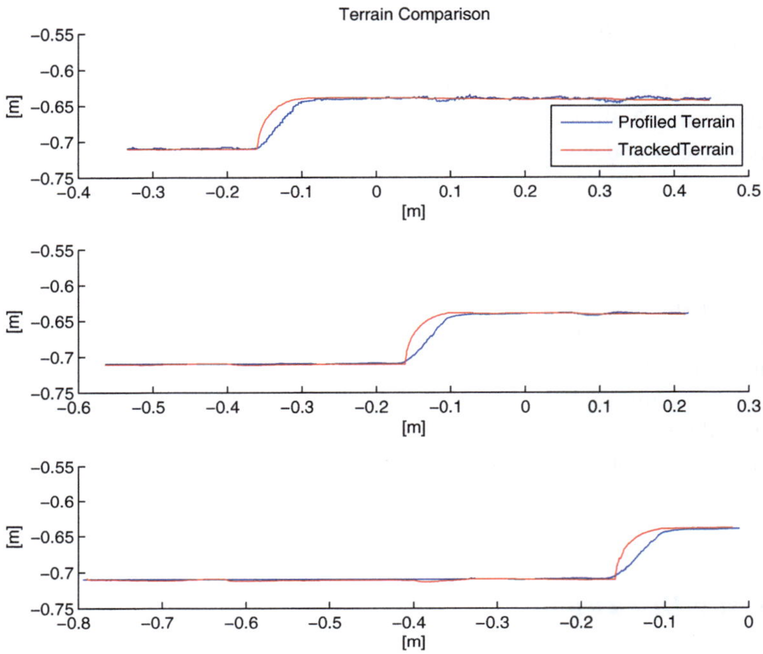

Fig. 5 Profiled terrain compared to tracked terrain (*front*, *middle* and *back wheel*)

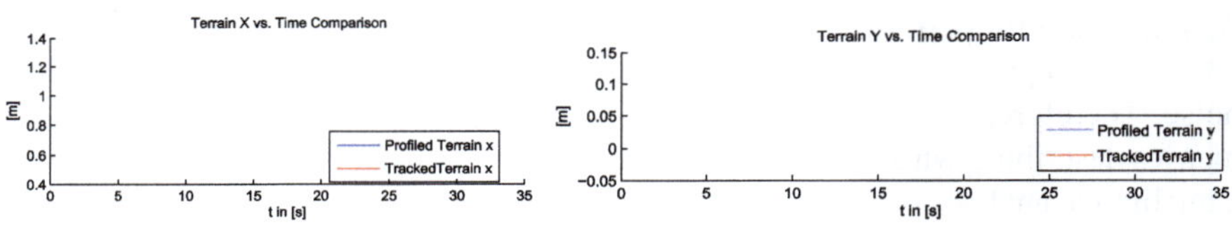

Fig. 6 Profiled terrain compared to tracked terrain, x and y position versus time (*front wheel*)

of the y and x-position over time: As one can see in Fig. 6 the y position of the terrain can be followed quite nicely only having a small lag. Whereas the x position of the profiled terrain deviates from the ground truth in the moment of a wheel reaching the step (Fig. 6). Other than that, the qualitative appearance of the profiled terrain matches the one from tracking.

Incorporating this profiled terrain shape, the controller adapted the wheel speeds accordingly. This can be seen in Fig. 7. There are some key differences between the computed speed by the controller and the real wheel speeds. First of all, it can be observed that the decelerations of the controller are about 30 % lower than in reality. This condition is due to the lower slope gradient of the profiled step. Next the motor speed slow-down lags behind the real wheel deceleration. This is mostly caused by the delay of the front wheel moving up when reaching the step. At that time, the rover stands still, which cannot be sensed by the controller. For the front wheel, in

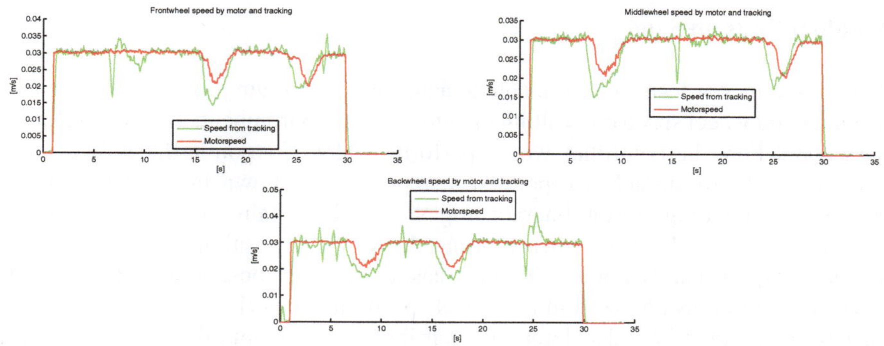

Fig. 7 Front, middle and rear wheel speed

contrast to the middle and back wheel, the controller does not know yet the shape of the terrain ahead. Hence it is not possible to introduce a foresighted controlling for the front wheel. At around 16 s, the middle wheel reaches the terrain. Here the lag of the front and back wheel deceleration appears to be smaller. Finally at 24 s, the back wheel moves up. One can observe that the deceleration lag seems to be larger again. This issue is more complex: It is probably due to a small dent, located at 0.07 to 0.11 m in the profiled terrain. This dent was formed by the profiling front wheel during the middle wheel moving up the step face. In the current situation, the middle wheel is situated at this asperity. At the instant when the back wheel starts to move up, the rover configuration change is interpreted differently by the controller. This configuration change is equivalent to the back wheel remaining level and the middle and front wheel moving down. Whereas the middle wheel moves down anyway due to the dent. At this stage, it is not clear how this perceptual ambiguity can be identified and solved.

Tests run on a smooth step obstacle and a small wooden bump Fig. 8 lead to similar results, with an even better terrain profiling: because there are no sudden change of direction such as on the step obstacle, no wheel slip occur and the terrain profiling performance improves. More details on these test results can be found in [9].

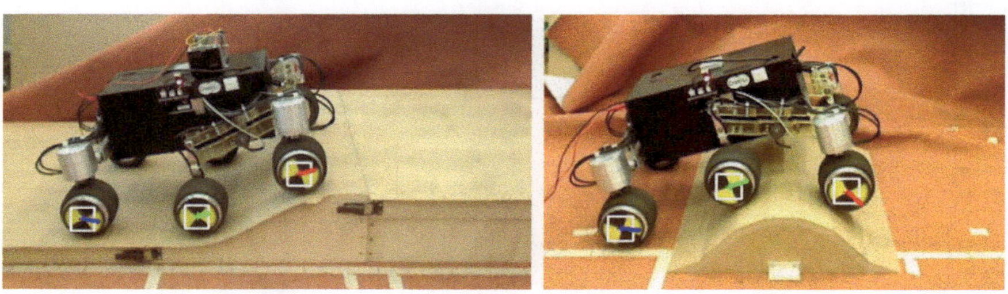

Fig. 8 Smooth step obstacle and small hill

Outdoor Tests on Sand

During the indoor tests, we were able to achieve good terrain profiling and to some extend good wheel speed controlling in high friction environment. The next question was about how the controller would perform on low friction surfaces like sand. One test to be discussed incorporated a step obstacle shown in Fig. 10. Looking at the corresponding terrain diagram (Fig. 9, left) the terrain profiling seems to be reasonably good. However when plotting the y wheel position against time (see Fig. 9, right), lag can be noticed. This increasing lag is responsible for the late wheel speed decelerations observed in the wheel speed diagrams (Fig. 10). One may note that some of the wheel decelerations contributed just about 30 % of what would have been needed to match speeds. This issue can be once again explained by the lower gradient of the step slope in the profiled terrain. As during the indoor test, this happened due to the fact that the rover stood still as the front wheel touched the step face. Now it even takes more time for the slipping middle and rear wheel to build up enough normal force for the front wheel moving up. Additionally the front wheel also slips when moving up—thus using extra time which flattens the slope even more. The high velocity peak at the end is caused by the wheels falling of the second step.

During the outdoor tests, a lot more test runs were performed and analyzed, with similar results: in most cases, the terrain profiling works well, but a significant resulting control performance improvement did not follow. This probably stems out from the lag and strong coupling between the control input and the terrain estimation, as well as out of the impossibility to sense wheel slippage. Should simple sensors such as desktop mouse movement estimator be integrated in future rover, our approach could naturally integrate their input to improve both the terrain profiling and as a result the wheel speed control. More details about this test can also be found in [9].

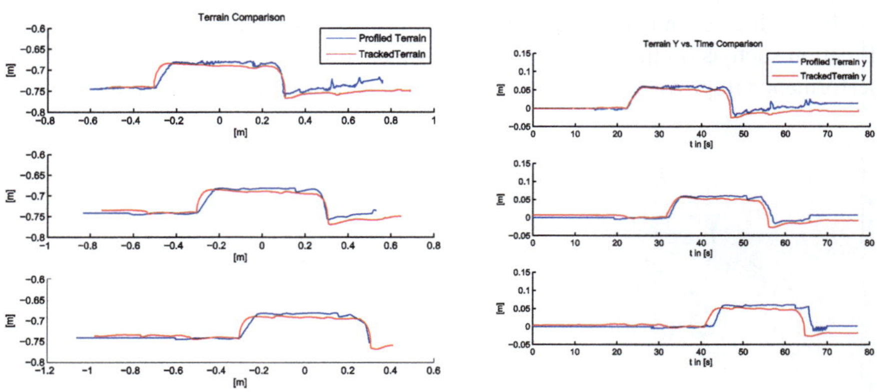

Fig. 9 Profiled terrain compared to tracked terrain (*front*, *middle* and *back wheel*), as a function of distance (*left*) and time (*right*)

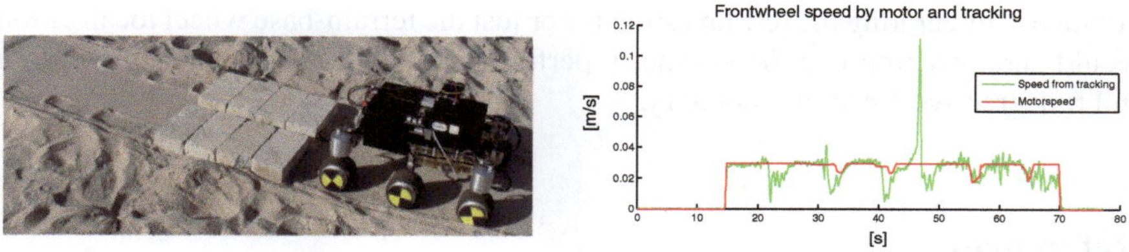

Fig. 10 Illustration of a step obstacle on a sand surface and measured front wheel speed while passing the step. Note in particular the dig-in that occurs each time a wheel climbs the step

4 Conclusion and Outlook

This paper presented an approach for simultaneous terrain profiling and control for 6-wheel rovers with passive suspension. The approach is based on a probabilistic filtering of the vehicle suspension deformation to jointly estimate the rover displacement and the shape of the terrain. Based on this terrain, a wheel speed controller was implemented to minimize the discrepancy between the ideal speed of a wheel following the terrain without slipping and the speed applied by the wheel controller.

Experiments indoor and outdoor have shown that the terrain profiling approach is sound and behave well even in presence of slip. However, at this stage of the implementation, it was not possible to demonstrate a significant control performance improvement resulting from the terrain profiling. The main reason for this lack of performance is mostly the delay introduced in by the terrain estimation, and the absence of wheel-slip sensing leading to noise in the terrain profile.

As for computational load, the SMAC controller was only occupying 10–15 % of the capacity of an Atom-based embedded PC (FitPC2). In cases where the computational load would have to be reduced, there would be enough parameters to influence the controllers computational need. Especially narrowing down of the updated belief yields large differences.

On the hardware side, future work will need to consider integrating ground tracking sensors (e.g. optical mouse sensors) close to the wheels to detect and estimate wheel slippage. On the software side, it would be theoretically feasible to put the terrain estimation in the same Bayesian framework as the current wheel localization. Such a joint estimation would be similar to what is currently implemented in the state of the art of simultaneous localization and mapping (SLAM). Although the computational cost would certainly be higher, a more robust estimation might be possible. Integrating a sensor for wheel sinkage would also help improving the profiling and mitigate the multi-pass effect when 3 wheels drive on the same track.

As a last remark, we would like to point out an alternative application for our SMAC implementation: Since the terrain profiler performs relatively well, one could think of using it for improving odometry or just helping localization on pre-planned path. Let's think of the rover Opportunity on Mars: These rovers plan paths ahead while standing still. Then in a second phase they try to follow this path using mostly odometry. Integrating the terrain estimator or just the terrain-base wheel localization would certainly improve the odometry performance, without having to resort to a full 6-degrees-of-freedom odometry.

References

1. K. Iagnemma, S. Dubowsky, Mobile Robot rough-terrain control (RTC) for planetary exploration. in *Proceedings of the 26th ASME Biennal Mechanisms and Robotics Conference, DETC* (2000)
2. K. Iagnemma, H. Shibley, S. Dubowsky, On-line terrain parameter estimation for planetary rovers. in *IEEE International Conference on Robotics and Automation*, Washington D.C., USA (2002)
3. P. Lamon, A. Krebs, M. Lauria, S. Shooter, R. Siegwart, Wheel torque control for a rough terrain rover, in *IEEE International Conference on Robotics and Automation*, New Orleans, USA (2004)
4. P. Lamon, R. Siegwart, Wheel torque control in rough terrain-Modeling and simulation, in *IEEE International Conference on Robotics and Automation (ICRA)*, Barcelona, Spain (2005)
5. M. Lauria, Y. Piguet, R. Siegwart, Octopus—An autonomous wheeled climbing Robot, in *Proceedings of the Fifth International Conferance on Climbing and Walking Robots* (2002)
6. R. Siegwart, P. Lamon, T. Estier, M. Lauria, R. Piguet, Innovative design for wheeled locomotion in rough terrain. J Robot. Auton. Syst. Elsevier **40**(2–3), 151–162
7. T. Thueer, P. Lamon, A. Krebs, R. Siegwart, CRAB—Exploration rover with advanced obstacle negotiation capabilities. in *9th ESA Workshop on Advanced Space Technologies for Robotics and Automation (ASTRA)*, Noordwijk, The Netherlands (2006)
8. A. Krebs, F. Risch, T. Thueer, J. Maye, C. Pradalier, R. Siegwart, Rover control based on an optimal torque distribution—Application to 6 motorized wheels passive rover. in *IEEE/RSJ International Conference on Intelligent Robots and Systems(IROS)*, Taipei, Taiwan (2010)
9. P. Strupler, *Control Optimization for Space Rovers Using Suspension Angles* (Master Thesis, ETH Zürich, Switzerland, 2010)

Robust Monocular Visual Odometry for a Ground Vehicle in Undulating Terrain

Ji Zhang, Sanjiv Singh and George Kantor

Abstract Here we present a robust method for monocular visual odometry capable of accurate position estimation even when operating in undulating terrain. Our algorithm uses a steering model to separately recover rotation and translation. Robot 3DOF orientation is recovered by minimizing image projection error, while, robot translation is recovered by solving an NP-hard optimization problem through an approximation. The decoupled estimation ensures a low computational cost. The proposed method handles undulating terrain by approximating ground patches as locally flat but not necessarily level, and recovers the inclination angle of the local ground in motion estimation. Also, it can automatically detect when the assumption is violated by analysis of the residuals. If the imaged terrain cannot be sufficiently approximated by locally flat patches, wheel odometry is used to provide robust estimation. Our field experiments show a mean relative error of less than 1 %.

1 Introduction

The task of visual odometry is to estimate motion of a camera, and by association the vehicle it is attached to, using a sequence of camera images. Typically, visual odometry is used in those cases where GPS is not available (eg. in planetary environments), or is too heavy carry (eg. on a small air vehicle), or, is insufficiently accurate at a low cost (eg. in agricultural applications). In ground vehicle applications, visual odometry can provide an alternative or compliment to wheel odometry since it is

J. Zhang (✉) · S. Singh · G. Kantor
The Robotics Institute, Carnegie Mellon University, Pittsburgh, PA, USA
e-mail: zhangji@andrew.cmu.edu

S. Singh
e-mail: ssingh@cmu.edu

G. Kantor
e-mail: kantor@cmu.edu

K. Yoshida and S. Tadokoro (eds.), *Field and Service Robotics*, Springer Tracts in Advanced Robotics 92, DOI: 10.1007/978-3-642-40686-7_21,

not prone to problems such as wheel slippage that can cause serious errors. Recent developments show significant progress in visual odometry and it now possible to estimate 6DOF motion using stereo cameras [1–3]. Stereo cameras help provide scale and some constraints to help recovery of motion but their use comes at a cost. Accuracy is dependant on inter-camera calibration which can be hard to ensure if the cameras are separated significantly. The use of stereo cameras also reduces the field of view because only features that lie in the intersection of the field of view of two cameras can be used. Finally, cost in components, interfacing, synchronization, and computing are higher for stereo cameras compared to a monocular camera.

While it is impossible to recover scale in translation for arbitrary camera motion in 6DOF when using monocular imaging, it is possible to recover scale when some additional information such as the distance and attitude of camera from the ground plane, such as is reasonably constant on a ground vehicle, is available. Recent work shows that under the assumption that the imaged areas are flat and level, it is possible to use visual odometry with monocular imaging [4–6]. This is a significant constraint in that such methods fail if the imaged areas are not guaranteed to be flat.

Here we report on relaxing the constraint such that visual odometry coupled with wheel odometry can be viable in undulating and even in severely 3D settings (Fig. 1) using monocular vision. We do this in two ways. First, our formulation of visual odometry only requires the imaged areas to be locally flat but not necessarily level. Our method recovers the ground inclination angle by finding coplanar features tracked on the ground. Second, the method can automatically determine when the imaged areas are not well approximated by locally flat patches and uses wheel odometry. The result is a monocular system that recovers differential motion with non-holonomic constraint in 3DOF rotation and 1DOF translation. When used on a ground vehicle, our experiments indicate an accuracy comparable to that from state-of-the-art stereo systems even the vehicle is tested in undulating terrain.

To estimate motion from imagery, the standard way is formulating visual odometry into a bundle adjustment problem and solves numerically through iteration.

Fig. 1 **a** An example of the type of terrain over which our ground vehicle based visual odometry is intended to work **b** An example of the type of scene that can be imaged by the visual odometry system. Monocular visual odometry systems that assume a flat environments fail in such a case

Alternatively, by using a steering model, the proposed method decouples the problems of estimating rotation and translation. In the first step, we estimate robot orientation using QR factorization [7] applied to a RANSAC algorithm [8] that minimizes the image reprojection error. In the second step, we use the same set of inlier features found by the RANSAC algorithm and solve an optimization problem that recovers translation together with the ground inclination angles. Since the full blown problem is believed to be NP hard, we utilize an approximation that ensures computational feasibility. The proposed two-step estimation algorithm is able to run with very low computational cost. Further, if the ground patches cannot be approximated as locally flat, the second step estimation becomes inaccurate. Then, wheel odometry is used to compute translation, and visual odometry is only for recovering rotation.

The rest of this paper is organized as follows. In Sect. 2, we present related work. In Sect. 3, we define our problem. The problem is mathematically solved in Sect. 4 with implementation details provided. Experimental results are shown in Sect. 5 and conclusions are made in Sect. 6.

2 Related Work

Today, it is commonly possible to estimate camera motion using visual odometry, that is through the tracking of features in an image sequence. [2, 3]. Typically, the camera motion is assumed to be unconstrained in the 3D space. For stereo systems [9–11], the baseline between the two cameras functions as a reference from which the scale of motion can be recovered. For example Paz, et al's method estimates the motion of stereo hand-hold cameras where scale is solved using features close to the cameras [12]. Konolige, et al's stereo visual odometry recovers 6DOF camera motion from bundle adjustment [1]. The method is integrated with an IMU that handles the orientational drift of visual odometry. It is able to work for lone distance navigation in off-road environments. For monocular systems [13–15], if camera motion is unconstrained, scale ambiguity is unsolvable. Using a monocular camera, Civera, et al formulate the motion estimation and camera calibration into one problem [16]. The approach recovers camera intrinsic parameters and 6DOF motion up to scale.

When a monocular system is used in such a way that the camera motion is constrained to a surface, recovering scale is possible. For example, Kitt, et al's method solves scale ambiguity using Ackermann steering model and assumes the vehicle drives on a planar road surface [5]. Nourani Vatani and Borges use Ackermann steering model along with a downward facing camera to estimate the planar motion of a vehicle [6]. Since the method only recovers the vehicle planar motion, an INS system is used to obtain vehicle pitch and roll angles. Scaramuzza, et al's approach adopts a single omnidirectional camera [4], where Ackermann steering model and steering encoder readings are used as constrains. This approach can recover motion at a low computational cost with a single feature point, and shows significantly improved accuracy compared to unconstrained cases. Scaramuzza also shows that a monocular camera placed with an offset to the vehicle rotation center can recover scale when the vehicle is turning [17]. In straight driving, however, the formulation degenerates and the scale is no longer recoverable.

In [4–6, 17], the methods all assume a planar ground model. However, violation of the assumption can make motion estimation fail. Compared to the existing work, our method does not require the imaged terrain to be flat and level. Our method simultaneously estimates the inclination angle of the ground while recovering motion. Further, our method combines wheel odometry to deal with the case where the system automatically determines if the terrain cannot be well approximated by a local flat patch. Here, we summarize our theoretical analysis of the motion estimation due to space limitations. A more complete analysis will be published in the future.

3 Problem Definition

We assume that the vehicle uses Ackermann steering [18] which limits the steering to be perpendicular to the axles of the robots. We also assume that the camera is well modeled as a pinhole camera [7] in which the intrinsic and extrinsic parameters are calibrated.

3.1 Notations and Coordinate Systems

As a convention in this paper, we use right uppercase superscription to indicate the coordinate systems, and right subscription k, $k \in Z^+$ to indicate the image frames. We use $\mathscr{I}$ to denote the set of feature points in the image frames.

- Camera coordinate system $\{C\}$ is a 3D coordinate system. As shown in Fig. 2, the origin of $\{C\}$ is at the camera optical center with the z-axis coinciding with the camera principal axis. The $x - y$ plane is parallel to the camera image sensor with the x-axis parallel to the horizontal direction of the image pointing to the left. A point i, $i \in \mathscr{I}$, in $\{C_k\}$ is denoted as $\boldsymbol{X}^C_{(k,i)}$.

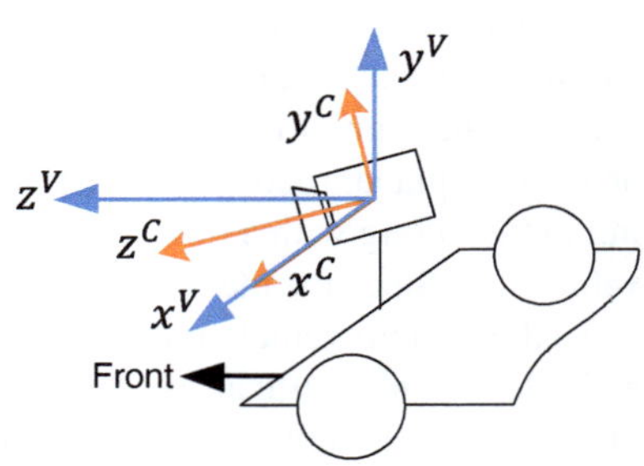

Fig. 2 Illustration of the vehicle coordinate system $\{V\}$ and the camera coordinate system $\{C\}$

- Vehicle coordinate system $\{V\}$ is a 3D coordinate system. The origin of $\{V\}$ is coinciding with the origin of $\{C\}$, the x-axis is parallel to the robot axles pointing to the robot left hand side, the y-axis is pointing upward, and the z-axis is pointing forward. A point i, $i \in \mathscr{I}$, in $\{V_k\}$ is denoted as $\boldsymbol{X}^V_{(k,i)}$.
- Image coordinate system $\{I\}$ is a 2D coordinate system with its origin at the right bottom corner of the image. The u- and v- axes in $\{I\}$ are pointing to the same directions as the x- and y- axes in $\{C\}$. A point i, $i \in \mathscr{I}$, in $\{I_k\}$ is $\boldsymbol{X}^I_{(k,i)}$.

3.2 Problem Description

Since our robot remains on the ground and follows the Ackermann steering model, the translation is limited to the z-direction in $\{V\}$. Let Δz be robot translation between frames $k-1$ and k, Δz is in the $\{V_{k-1}\}$ coordinates. In this paper, we treat the features on the ground in the near front of the robot as coplanar. As shown in Fig. 3, let $\mathbb{W}$ indicate the plane. Let d_0 be the height of the camera above the ground, d_0 is set as a known constant. Let $\mathbb{P}$ be the projection of the camera center. We model $\mathbb{W}$ with 2 rotational DOFs around $\mathbb{P}$. Let $\mathbb{N}$ be the normal of $\mathbb{W}$, and let t_k and r_k be the Euler angles of $\mathbb{N}$ around the x- and z- axes in $\{V_k\}$, respectively. t_k and r_k represent the pitch and roll inclination angles of the ground. Let Δp, Δt, and Δr be robot rotation angles around the y-, x-, and z- axes of $\{V_{k-1}\}$ between frames $k-1$ and k, we have $\Delta t = t_k - t_{k-1}$ and $\Delta r = r_k - r_{k-1}$. In this paper, we want to measure the robot motion between consecutive frames. Our visual odometry problem can be defined as

Problem 1 *Given a set of image frames k, $k \in Z^+$, and the camera height d_0, compute Δp, Δt, Δr, and Δz for each frame k.*

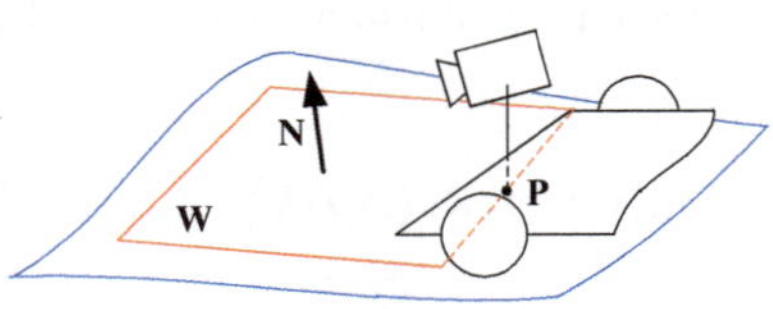

Fig. 3 Modeling the ground. The *blue colored curve* represents the ground, $\mathbb{P}$ is the projection of the camera center, and $\mathbb{W}$ is the plane representing the ground in the near front of the robot. $\mathbb{W}$ has pitch and roll DOFs around $\mathbb{P}$

4 Visual Odometry Algorithm

4.1 Rotation Estimation

In this section, we recover the 3DOF robot orientation. We will show that by using the Ackermann steering model, robot orientation can be recovered regardless of translation. From the pin-hole camera model, we have the following relationship between $\{I\}$ and $\{C\}$,

$$\varsigma \boldsymbol{X}^{I}_{(k,\,i)} = \mathbf{K}\boldsymbol{X}^{C}_{(k,\,i)}, \tag{1}$$

where ς_k is a scale factor, and $\mathbf{K}$ is the camera intrinsic matrix, which is known from the pre-calibration [7].

The relationship between $\{C\}$ and $\{V\}$ is expressed as

$$\boldsymbol{X}^{C}_{(k,\,i)} = \mathbf{R}_z(r_0)\mathbf{R}_x(t_0)\mathbf{R}_y(p_0)\boldsymbol{X}^{V}_{(k,\,i)}, \tag{2}$$

where $\mathbf{R}_x(\cdot)$, $\mathbf{R}_y(\cdot)$, and $\mathbf{R}_z(\cdot)$ are rotation matrices around the x-, y-, and z- axes in $\{V\}$, respectively, and p_0, t_0, and r_0 are corresponding rotation angles from $\{V\}$ to $\{C\}$. Here, note that p_0, t_0, and r_0 are the camera extrinsic parameters, which are known from the pre-calibration [7].

Let $\tilde{\boldsymbol{X}}^{V}_{(k,\,i)}$ be the normalized term of $\boldsymbol{X}^{V}_{(k,\,i)}$, we have

$$\tilde{\boldsymbol{X}}^{V}_{(k,\,i)} = \boldsymbol{X}^{V}_{(k,\,i)}/z^{V}_{(k,\,i)}. \tag{3}$$

where $z^{V}_{(k,\,i)}$ is the 3rd entry of $\boldsymbol{X}^{V}_{(k,\,i)}$. $\tilde{\boldsymbol{X}}^{V}_{(k,\,i)}$ can be computed by substituting (2) into (1) and scaling $\boldsymbol{X}^{V}_{(k,\,i)}$ such that the 3rd entry becomes one.

From the robot motion, we can establish a relationship between $\{V_{k-1}\}$ and $\{V_k\}$ as follows,

$$\boldsymbol{X}^{V}_{(k,\,i)} = \mathbf{R}_z(\Delta r)\mathbf{R}_x(\Delta t)\mathbf{R}_y(\Delta p)\boldsymbol{X}^{V}_{(k-1,\,i)} + [0,\ 0,\ \Delta z]^T\,, \tag{4}$$

where $\mathbf{R}_x(\cdot)$, $\mathbf{R}_y(\cdot)$, and $\mathbf{R}_z(\cdot)$ are the same rotation matrices as in (2).

Substituting (3) into (4) for frame $k-1$ and k, and since Δp, Δt, and Δr are small angles in practice, we perform linearization to obtain the following equations,

$$c_i \tilde{x}^{V}_{(k,\,i)} = \tilde{x}^{V}_{(k-1,\,i)} + \Delta p + \tilde{y}^{V}_{(k-1,\,i)}\Delta r, \tag{5}$$

$$c_i \tilde{y}^{V}_{(k,\,i)} = \tilde{y}^{V}_{(k-1,\,i)} + \Delta t - \tilde{x}^{V}_{(k-1,\,i)}\Delta r, \tag{6}$$

$$c_i = 1 - \tilde{x}^{V}_{(k-1,\,i)}\Delta p - \tilde{y}^{V}_{(k-1,\,i)}\Delta t + \Delta z/z^{V}_{(k-1,\,i)}, \tag{7}$$

where $\tilde{x}^{V}_{(l,\,i)}$ and $\tilde{y}^{V}_{(l,\,i)}$, $l = k-1, k$, are the 1st and the 2nd entries of $\tilde{\boldsymbol{X}}^{V}_{(l,\,i)}$, respectively, $z^{V}_{(l,\,i)}$ is the 3rd entry of $\boldsymbol{X}^{V}_{(l,\,i)}$, and c_i is a scale factor, $c_i = z^{V}_{(k,\,i)}/z^{V}_{(k-1,\,i)}$.

Eqs. (5) and (6) describe a relationship of Δp, Δt, and Δr without interfering with Δz. This indicates that by using the Ackermann steering model, we can decouple the estimation problem and recover Δp, Δt, and Δr separately from Δz. Stacking (5) and (6) for different features, we have

$$\mathbf{A}\boldsymbol{X} = \boldsymbol{b}, \tag{8}$$

where

$$\mathbf{A} = \begin{bmatrix} 1 & 0 & \tilde{y}^V_{(k-1,\,1)} & -\tilde{x}^V_{(k,1)} & 0 & 0 \ldots \\ 0 & 1 & -\tilde{x}^V_{(k-1,\,1)} & -\tilde{y}^V_{(k,1)} & 0 & 0 \ldots \\ 1 & 0 & \tilde{y}^V_{(k-1,\,2)} & 0 & -\tilde{x}^V_{(k,2)} & 0 \ldots \\ 0 & 1 & -\tilde{x}^V_{(k-1,\,2)} & 0 & -\tilde{y}^V_{(k,2)} & 0 \ldots \\ \ldots & \ldots & \ldots & \ldots & \ldots & \ldots \ldots \end{bmatrix},$$

$$\mathbf{b} = -\left[\tilde{x}^V_{(k-1,\,1)},\ \tilde{y}^V_{(k-1,\,1)},\ \tilde{x}^V_{(k-1,\,2)},\ \tilde{y}^V_{(k-1,\,2)},\ \ldots\right]^T,$$

$$\mathbf{X} = [\Delta p,\ \Delta t,\ \Delta r,\ c_1,\ c_2,\ \ldots]^T .$$

Equation (8) can be solved using the QR factorization method. Since $\mathbf{A}$ is a sparse matrix, the QR factorization can be implemented very efficiently. Let $\tilde{x}'^V_{(k-1,\,i)}$ and $\tilde{y}'^V_{(k-1,\,i)}$ be the reprojected coordinates of $\tilde{x}^V_{(k,\,i)}$ and $\tilde{y}^V_{(k,\,i)}$ in $\{V_{k-1}\}$. The QR factorization minimizes the image reprojection error,

$$\min_{\substack{\Delta p, \Delta t, \Delta r,\\ c_i, i \in \mathscr{I}}} \sum_{i \in \mathscr{I}} (\tilde{x}^V_{(k-1,\,i)} - \tilde{x}'^V_{(k-1,\,i)})^2 + (\tilde{y}^V_{(k-1,\,i)} - \tilde{y}'^V_{(k-1,\,i)})^2. \tag{9}$$

With (8) solved, let $e^x_{(k-1,\,i)} = \tilde{x}^V_{(k-1,\,i)} - \tilde{x}'^V_{(k-1,\,i)}$ and $e^y_{(k-1,\,i)} = \tilde{y}^V_{(k-1,\,i)} - \tilde{y}'^V_{(k-1,\,i)}$, $e^x_{(k-1,\,i)}$ and $e^y_{(k-1,\,i)}$ represent the reprojection errors of feature i, $i \in \mathscr{I}$, in $\{V_{k-1}\}$. Using (5) and (6), we can compute $e^x_{(k-1,\,i)}$ and $e^y_{(k-1,\,i)}$ as,

$$e^x_{(k-1,\,i)} = \tilde{x}^V_{(k-1,\,i)} + \Delta p + \tilde{y}^V_{(k-1,\,i)}\Delta r - c_i \tilde{x}^V_{(k,\,i)}, \tag{10}$$

$$e^y_{(k-1,\,i)} = \tilde{y}^V_{(k-1,\,i)} + \Delta t - \tilde{x}^V_{(k-1,\,i)}\Delta r - c_i \tilde{y}^V_{(k,\,i)}. \tag{11}$$

Similarly, let $e^x_{(k,\,i)}$ and $e^y_{(k,\,i)}$ be the reprojection errors in $\{V_k\}$, $e^x_{(k,\,i)}$ and $e^y_{(k,\,i)}$ can be obtained as

$$e^x_{(k,\,i)} = e^x_{(k-1,\,i)}/c_i,\ e^y_{(k,\,i)} = e^y_{(k-1,\,i)}/c_i. \tag{12}$$

Define $\Sigma_{(l,\,i)}$, $l \in \{k-1, k\}$, as a 2×2 matrix,

$$\Sigma_{(l,\,i)} = \text{diag}\left[(e^x_{(l,\,i)})^2, (e^y_{(l,\,i)})^2\right],\ l \in \{k-1,\ k\}. \tag{13}$$

$\Sigma_{(l,\,i)}$ contains the covariance of $\tilde{\boldsymbol{X}}^V_{(l,\,i)}$ measured from the image reprojection error, which will be useful in the following sections.

4.2 Robot Translation

With robot orientation recovered, we derive the expression of translation in this section. The task of recovering translation is formulated into an optimization problem in the next section, and solved in the same section. An shown in Fig. 3, recall that $\mathbb{W}$ is the plane representing the local ground in the near front of the robot, and t_k and r_k are the pitch and roll angles of $\mathbb{W}$. For a feature i, $i \in \mathscr{I}$, on $\mathbb{W}$, the following relationship holds from geometric relationship,

$$-z^V_{(k,\,i)}(\tilde{y}^V_{(k,\,i)} - \tan r_k \tilde{x}^V_{(k,\,i)} + \tan t_k) = d_0, \tag{14}$$

where d_0 is the height of the camera above the ground.

Since t_k and r_k are small angles in practice, we approximate $\tan t_k \approx t_k$ and $\tan r_k \approx r_k$. Then, by substituting (14) into (7) for frames $k-1$ and k, we can derive

$$\alpha t_k + \beta r_k = \gamma, \tag{15}$$

where

$$\begin{aligned}
\alpha &= -\tilde{x}^V_{(k-1,\,i)}\Delta p + \tilde{y}^V_{(k-1,\,i)}\Delta t - c_i + 1,\\
\beta &= (\alpha + 1)\tilde{x}^V_{(k,\,i)} - \tilde{x}^V_{(k-1,\,i)},\\
\gamma &= -(\alpha + 1)(\Delta t + \tilde{x}^V_{(k,\,i)}\Delta r - \tilde{y}^V_{(k,\,i)}) - \tilde{y}^V_{(k-1,\,i)}.
\end{aligned}$$

Equation (15) contains two unknown parameters, t_k and r_k, which indicates that we can solve the function by using two features. Let $(i,\ j)$ be a pair of features, $i,\ j \in \mathscr{I}$, here we use $(i,\ j)$ to solve (15). Then, let $\Delta z_{(i,\,j)}$ be the translation computed from feature pair $(i,\ j)$. From (14), we can derive

$$\Delta z_{(i,\,j)} = \frac{1}{2}(T_{(k,\,i)} + T_{(k,j)} - T_{(k-1,\,i)} - T_{(k-1,\,j)}) \tag{16}$$

where
$$T_{(l,\,h)} = d/(\tilde{y}^V_{(l,\,h)} + \tilde{x}^V_{(l,\,h)} r_k + t_k),\ l \in \{k-1,\ k\},\ h \in \{i,\ j\},$$

Now, let $\sigma_{(i,\,j)}$ be the standard deviation of $\Delta z_{(i,\,j)}$ measured from the image reprojection error, $\sigma_{(i,\,j)}$ will be useful in the next section. From (16), it indicates that $\Delta z_{(i,\,j)}$ is a function of $\tilde{\boldsymbol{X}}^V_{(l,\,h)}$, $l \in \{k-1,\ k\}$, $h \in \{i,\ j\}$. Let $\mathbf{J}_{(l,\,h)}$ be the Jacobian matrix of that function with respect to $\tilde{\boldsymbol{X}}^V_{(l,\,h)}$, $\mathbf{J}_{(l,\,h)} = \partial \Delta z_{(i,\,j)} / \partial \tilde{\boldsymbol{X}}^V_{(l,\,h)}$, we can compute

$$\sigma^2_{(i,\,j)} = \sum_{l \in \{k-1,\,k\}} \sum_{h \in \{i,\,j\}} \mathbf{J}_{(l,\,h)} \Sigma_{(l,\,h)} \mathbf{J}^T_{(l,\,h)}. \tag{17}$$

4.3 Translation Recovery by Optimization

In the above section, we showed that the translation can be recovered using a pair of features. In this section, we want to estimate translation using multiple features, by solving an optimization problem that minimizes the error variance of translation estimation. Suppose we have a total number of n features, $n \in Z^+$, combination of any two features can provide $n(n-1)/2$ feature pairs. Let $\mathscr{J}$ be a set of feature pairs, $1 \leq |\mathscr{J}| \leq n(n-1)/2$. Here, we use the feature pairs in $\mathscr{J}$ to compute the translation Δz. Define Δz as the weighted sum of $\Delta z_{(i,\,j)}$, $(i,\ j) \in \mathscr{J}$,

$$\Delta z = \sum_{(i,\,j) \in \mathscr{J}} w_{(i,\,j)} \Delta z_{(i,\,j)}, \tag{18}$$

where $w_{(i,\,j)}$ is the weight for feature pair $(i,\ j)$, such that

$$\sum_{(i,\,j) \in \mathscr{J}} w_{(i,\,j)} = 1, \text{ and } w_{(i,\,j)} \geq 0,\ (i,\ j) \in \mathscr{J}. \tag{19}$$

Define σ as the standard deviation of Δz measured from the image reprojection error. Here, we want to compute Δz such that σ is minimized. We start with our first question. For a given set of feature pairs $\mathscr{J}$, how to assign the weights $w_{(i,\,j)}$, $(i,\ j) \in \mathscr{J}$, such that σ is the minimum? Mathematically, the problem can be expressed as,

Problem 2 *Given* $\sigma_{(i,\,j)}$, $(i,\ j) \in \mathscr{J}$, *compute*

$$\left\{w_{(i,\,j)}, (i,\ j) \in \mathscr{J}\right\} = \arg\min_{w_{(i,\,j)}} \sigma^2, \tag{20}$$

subject to the constraints in (19).

To solve this problem, we can prove that if each feature $i, i \in \mathscr{I}$ belongs to at most one feature pair in $\mathscr{J}$, then Problem 2 is analytically solvable using the Lagrange multiplier method [19]. However, if a feature exists in multiple feature pairs, the problem becomes a convex optimization problem that has to be solved numerically [20]. Here, we directly give the solution for Problem 2,

$$\min_{w_{(i,\,j)}} \sigma^2 = \sum_{(i,\,j) \in \mathscr{J}} w^2_{(i,\,j)} \sigma^2_{(i,\,j)}, \tag{21}$$

where

Algorithm 1: Feature Pair Selection

input : $\mathscr{I}$ and $\sigma_{(i,\,j)}, i,\ j \in \mathscr{I}$
output : $\mathscr{J}$
begin
    $\mathscr{J} = \emptyset$;
    Sort $\sigma_{(i,\,j)}, i,\ j \in \mathscr{I}$ in increasing order;
    Create a variable σ_i for each $i \in \mathscr{I}$;
    for the decreasing order of $\sigma_{(i,\,j)}, i,\ j \in \mathscr{I}$ **do**
        $\sigma_i = \sigma_{(i,\,j)}, \sigma_j = \sigma_{(i,\,j)}$;
    end
    for each $i,\ j \in \mathscr{I}$ **do**
        if $1/\sigma^2_{(i,\,j)} > 1/\sigma_i^2 + 1/\sigma_j^2$ **then**
            Put $(i,\ j)$ in $\mathscr{J}$, then delete $i,\ j$ from $\mathscr{I}$;
        end
    end
    for the increasing order of $\sigma_{(i,\,j)}, i,\ j \in \mathscr{I}$ **do**
        if $i,\ j \in \mathscr{I}$ **then**
            Put $(i,\ j)$ in $\mathscr{J}$, then delete $i,\ j$ from $\mathscr{I}$;
        end
    end
    Return $\mathscr{J}$.
end

$$w_{(i,\,j)} = \frac{1/\sigma^2_{(i,\,j)}}{\sum_{(p,\,q)\in\mathscr{J}} 1/\sigma^2_{(p,\,q)}},\ (i,\ j) \in \mathscr{J}. \tag{22}$$

With Problem 2 solved, we come to our second question. How to select the feature pairs in $\mathscr{J}$ such that σ is the minimum? Mathematically, the problem is

Problem 3 *Given $\mathscr{I}$ and $\sigma_{(i,\,j)}, i,\ j \in \mathscr{I}$, determine*

$$\left\{\mathscr{J} = \{(i,\ j)\},\ i,\ j \in \mathscr{I}\right\} = \arg\min_{\mathscr{J}} (\min_{w_{(i,\,j)}} \sigma^2), \tag{23}$$

such that each feature $i,\ i \in \mathscr{I}$ belongs to at most one feature pair in $\mathscr{J}$.

Problem 3 can be reformulated into a balanced graph partition problem [21], which is believed to be NP-hard [22]. Here, we focus on an approximation algorithm. The following two inequalities help us to construct the approximation algorithm. First, we find a sufficient condition for selecting the feature pairs. For feature pair $(i,\ j)$, $i,\ j \in \mathscr{I}$, if the following inequality is satisfied, then $(i,\ j) \in \mathscr{J}$,

$$\frac{1}{\sigma^2_{(i,\,j)}} > \frac{1}{\sigma^2_{(i,\,q)}} + \frac{1}{\sigma^2_{(p,j)}},\ \forall p, q \in \mathscr{I},\ p, q \neq i,\ j. \tag{24}$$

Second, we find that if we select the feature pairs $(i,\ j), i,\ j \in \mathscr{I}$ in the increasing order of $\sigma_{(i,\,j)}$, we can obtain a set of feature pairs, let it be $\tilde{\mathscr{J}}$, and let $\tilde{\sigma}$ be the standard deviation of Δz computed using feature pairs in $\tilde{\mathscr{J}}$. Let σ_* be the standard deviation of solving Problem 3 without approximation, we can prove that,

$$\tilde{\sigma}^2 \leq 2\sigma_*^2. \tag{25}$$

Equation (25) indicates that we can solve Problem 3 with an approximation factor of 2. Consequently, the feature pair selection algorithm is shown in Algorithm 1. In Line 5, we first sort the feature pairs in the increasing order of $\sigma_{(i,\,j)}$, $i,\ j \in \mathscr{I}$. Then in Lines 6–14, we go through each feature pair and check if (24) is satisfied. If yes, the feature pair is selected. Then, in Lines 15–19, we select the rest of the feature pairs in the increasing order of $\sigma_{(i,\,j)}$, $i,\ j \in \mathscr{I}$. The algorithm returns $\mathscr{J}$ in Line 20.

4.4 Implementation and Hybrid with Wheel Odometry

To implement the algorithm, we select a number of "good features" with the local maximum eigenvalues using the openCV library, and track the feature points between consecutive frames using the Lucas Kanade Tomasi (LKT) method [23]. To estimate robot rotation, we solve (8) using QR factorization method. The QR factorization is applied to a RANSAC algorithm that iteratively selects a subset of the tracked features as inliers, and uses the inliers to recover the 3DOF rotation, namely Δp, Δt, and Δr. After recovering the rotation, we also obtain the error covariance for each feature point from (13). Using the inliers selected by the RANSAC algorithm and the corresponding error covariance, we can select the feature pairs based on Algorithm 1 and recover robot translation Δz based on (18), (16), and (22).

In the two-step estimation process, the translation estimation requires the ground patches to be locally flat, while the rotation estimation does not rely on such requirement. Therefore, when this requirement is violated, the translation estimation becomes inaccurate. To deal with this case, a checking mechanism is implemented. If the error variance σ^2, the ground inclination angle t_k or r_k is larger than a corresponding threshold, a hybrid odometry system is used. The wheel odometry is used for computing translation, and the visual odometry is for recovering rotation. This strategy allows the system to work robustly even when the camera field of view is blocked by obstacles.

5 Experiments

We conduct experiments using an electrical vehicle as shown in Fig. 4a. The vehicle measures 3.04 m in length and 1.50 m in width. The wheelbase of the vehicle is 2.11 m. The vehicle is embedded with wheel encoders that measures the driving speed. An Imagingsource DFK 21BUC03 camera is attached in front of the vehicle, as shown in Fig. 4b. The camera resolution is set at 640 × 480 pixels and the focal length is

Fig. 4 **a** Our robot, and **b** A monocular camera attached in front of the robot

Table 1 Computation time of the visual odometry algorithm using 300 features

Feature Tracking	State Estimation	Overall
38 ms	5 ms	43 ms

4 mm (horizontal filed of view 64°). The vehicle is equipped with a high accuracy INS/GPS system (Applanix Pos-LV), accurate to better than 10 cm for ground truth acquisition.

5.1 Computation Time

We first show computation time of the proposed visual odometry algorithm. The algorithm is tested on a laptop computer with Quad 2.5GHz CPUs and 6G RAM. We track 300 features at each frame. As shown in Table 1, the feature tracking takes 38 ms and consumes an entire core. The state estimation takes 5 ms and runs on another core. The proposed algorithm is able to run at 26 Hz on average.

Table 2 Accuracy test configuration and relative error computed from 3D coordinates

	Configuration			
Test No.	Driving Distance (m)	Elevation Change (m)	Ground Material	Relative Error (%)
1	903	18	Grass	0.71
2	1117	27	Asphalt	0.87
3	674	15	Grass+Soil	0.74
4	713	17	Concrete	1.13
5	576	21	Asphalt	0.76
6	983	13	Soil	0.81

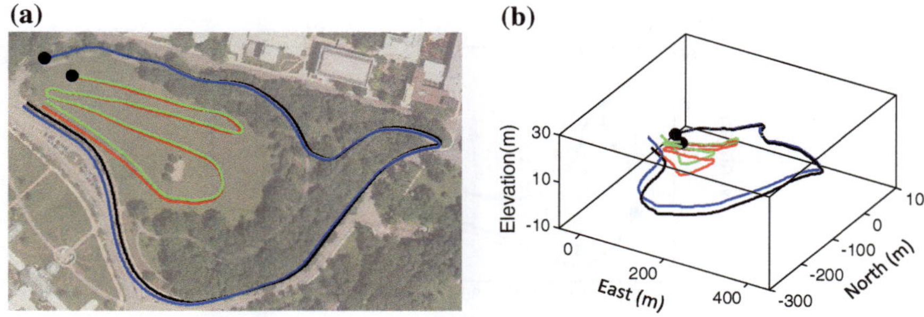

Fig. 5 **a** Planar view and **b** 3D view of the robot trajectories in accuracy test 1–2 (Table 1). The *black colored dots* are the starting points. The *green colored curve* is the visual odometry output for Test 1, and the *red colored curve* is the corresponding ground truth. The *blue colored curve* is the visual odometry output for Test 2, and the *black colored curve* is the ground truth. Ground truth is measured by a high accuracy INS/GPS system

5.2 Accuracy of Test Results

To demonstrate the accuracy of the proposed visual odometry algorithm, we conduct experiments with relatively long driving distance. The test configuration is shown in Table 2. The experiments are conducted with different elevation change and ground material. The overall driving distance for the 6 tests is about 5 km. The mean relative error of the visual odometry is 0.83 %. Specifically, the trajectories of Test 1–2 are presented in Fig. 5.

5.3 Experimental Results

To test the robustness of the proposed method, we conduct experiments with obstacles on the driving path. When the camera field of view is blocked by an obstacle, the requirement on local flatness of the ground pathes is violated. In this case, a hybrid odometry system is used. The translation is measured by wheel odometry and the rotation is estimated by visual odometry. As shown in Table 3, the robustness tests are conducted with different number of obstacles. By using the hybrid odometry system, the relative error is kept much lower than using the visual odometry only. Specifically, the trajectories and obstacles of Test 1 are shown in Fig. 6.

5.4 Analysis of Optimization

Finally, we analyze the effectiveness of the optimization procedure in Sect. 4.3. We compare three different versions of visual odometry algorithms as follows.

1. Visual Odometry (**VO**): The proposed visual odometry algorithm of this paper.

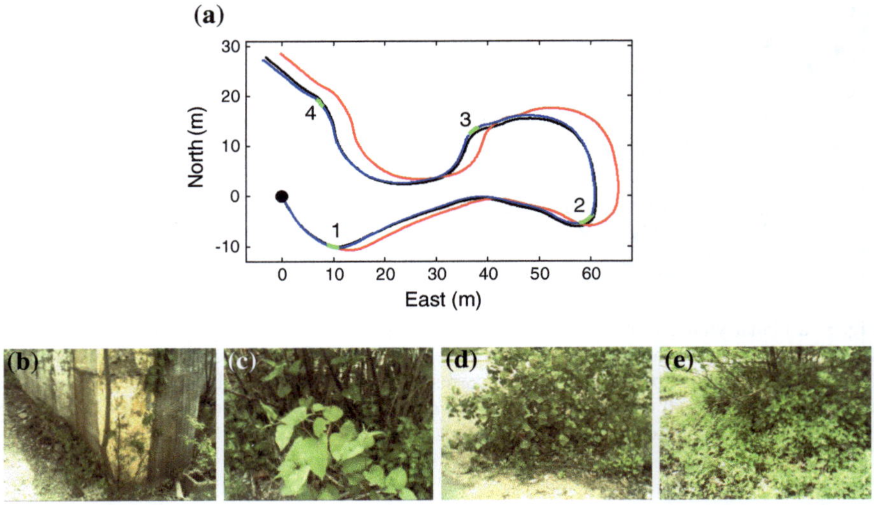

Fig. 6 **a** Robot trajectories for robustness test 1 (Table 2). The test includes 4 obstacles labeled with numbers. The corresponding obstacles are shown in **b–e**. The *black colored dot* is the starting point. The *blue-green colored curve* is measured by the hybrid odometry system, the *blue colored segments* are measured by visual odometry and the *green colored segments* are measured by visual odometry for rotation and wheel odometry for translation, the *red colored curve* is measures by visual odometry only, and the *black colored curve* is the ground truth

2. Visual Odometry Random Pair Selection (**VORPS**): In this version, we turn off the feature pair selection and use randomly selected the feature pairs. By using this algorithm, we can inspect the effect of Problem 3.
3. Visual Odometry Equal Weight (**VOEW**): In this version, we completely turn off the optimization and use equal weights instead of optimized weights in (18). By doing this, we can inspect the effect of Problem 2.

For comparison, we define two evaluation metrics. Let $\bar{\sigma}$ as the mean standard deviation of the one-step translation Δz, and let $\bar{\varepsilon}$ be the mean relative error of

Table 3 Robustness test configuration and relative error computed from 3D coordinates.

Test No.	Configuration Driving Distance (m)	Obstacle No.	Relative Error Visual+Wheel Odometry (%)	Visual Odometry (%)
1	167	4	0.43	1.83
2	124	3	0.39	2.46
3	182	4	0.54	1.54
4	263	6	0.61	4.13
5	106	3	0.47	2.76
6	137	5	0.41	3.81

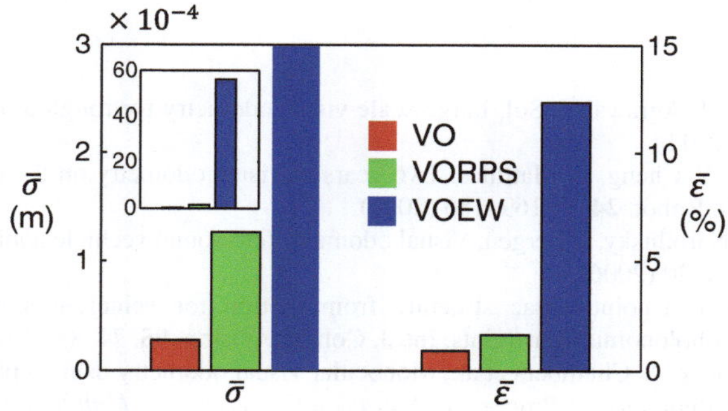

Fig. 7 Comparison of 3 different versions of the visual odometry. VO is the proposed visual odometry algorithm of this paper. VORPS is another version without the feature pair selection, randomly selected feature pairs are used. VOEW uses equal weights instead of optimized weights in (18). $\bar{\sigma}$ is the mean standard deviation of the one-step translation Δz. A full scaled comparison of $\bar{\sigma}$ is shown in the small thumbnail at the left-top corner. $\bar{\varepsilon}$ is the mean relative error of the visual odometry. The results are obtained using combination of the data in Table 2

the visual odometry output, $\bar{\sigma}$ and $\bar{\varepsilon}$ are computed using combination of the data in Table 2. Comparison of the results is presented in Fig. 7. Since $\bar{\sigma}$ of VOEW is significantly larger than that of VO or VORPS, we have to show the full scaled comparison in a small thumbnail at the left-top corner of the figure. From Fig. 7, it is obvious that the errors of VOEW and VORPS are larger then those of VO, especially the errors of VOEW are significantly larger. This result indicates that the optimization functions effectively, while using the optimized weights (Problem 2) plays a more important role than using the selected feature pairs (Problem 3) for reducing the visual odometry error.

6 Conclusion and Future Work

Estimation of camera motion by tracking visual features is difficult because it depends on the shape of the terrain which is generally unknown. The estimation problem is furthermore difficult when a monocular system is used because scale of the translation component cannot be recovered. Our method succeeds in two ways. First, it simultaneously estimates a planar patch in front of the camera along with camera motion, and second recovers scale by taking advantage of the fixed distance from the camera to the ground. In some cases, approximating the terrain in front of the vehicle as a planar patch cannot be justified. Our method automatically detects these cases and uses a hybrid odometry system in which rotation is estimated from visual odometry and translation is recovered by wheel odometry.

Since this paper relies on a kinematical vehicle steering model, lateral wheel slip is not considered. For the future work, we are considering a revision to the vehicle motion model such that the algorithm can handle more complicated ground conditions where lateral wheel slip is noticeable.

References

1. K. Konolige, M. Agrawal, J. Sol, Large-scale visual odometry for rough terrain. Robot. Res. **66**, 201–212 (2011)
2. M. Maimone, Y. Cheng, L. Matthies, Two years of visual odometry on the mars exploration rovers. J. Field Robot. **24**(2), 169–186 (2007)
3. D. Nister, O. Naroditsky, J. Bergen, Visual odometry for ground vechicle applications. J. Field Robot. **23**(1), 3–20 (2006)
4. D. Scaramuzza, 1-point-ransac structure from motion for vehicle-mounted cameras by exploiting non-holonomic constraints. Int. J. Comput. Vision. **95**, 74–85 (2011)
5. B. Kitt, J. Rehder, A. Chambers et al., Monocular visual odometry using a planar road model to solve scale ambiguity, in *Proceeding European Conference on Mobile Robots*, Sept 2011
6. N. Nourani-Vatani, P. Borges, Correlation-based visual odometry for ground vehicles. J. Field Robot. **28**(5), 742–768 (2011)
7. R. Hartley, A. Zisserman, *Multiple View Geometry in Computer Vision* (Cambridge University Press, New York, 2004)
8. M. Fischler, R. Bolles, Random sample consensus: a paradigm for model fitting with applications to image analysis and automated cartography. Commun. ACM **24**(6), 381–395 (1981)
9. A. Howard, Real-time stereo visual odometry for autonomous ground vehicles, in *IEEE International Conference on Intelligent Robots and Systems*, Nice, France, Sept 2008
10. D. Dansereau, I. Mahon, O. Pizarro et al., Plenoptic flow: closed-form visual odometry for light field cameras, in *International Conference on Intelligent Robots and Systems (IROS)*, CA, San Francisco, Sept 2011
11. P. Corke, D. Strelow, S. Singh, Omnidirectional visual odometry for a planetary rover, in *Proceedings of the IEEE/RSJ International Conference on Intelligent Robots and Systems* Sendai, Japan, Sept 2004, pp. 149–171
12. L. Paz, P. Pinies, J. Tardos, Large-scale 6-DOF SLAM with stereo-in-hand. IEEE Trans. Robot. **24**(5), 946–957 (2008)
13. B. Williams, I. Reid, On combining visual slam and visual odometry, in *IEEE International Conference on Robotics and Automation*, Anchorage, Alaska, May 2010
14. M. Wongphati, N. Niparnan, A. Sudsang, Bearing only fast SLAM using vertical line information from an omnidirectional camera, in *Proceedings of the IEEE International Conference on Robotics and Biomimetics*, Bangkok, Thailand, Feb 2009, pp. 494–501
15. A. Pretto, E. Menegatti, M. Bennewitz et al., A visual odometry framework robust to motion blur, in *IEEE International Conference on Robotics and Automation*, Kobe, Japan, May 2009
16. J. Civera, D. Bueno, A. Davison, J. Montiel, Camera self-calibraction for sequential bayesian structure form motion, in *Proceedings of the IEEE International Conference on Robotics and Automation*, Kobe, Japan, May 2009, pp. 130–134
17. D. Scaramuzza, Absolute scale in structure from motion from a single vehicle mounted camera by exploiting nonholonomic constraints, in *IEEE International Conference on Computer Vision*, Kyoto, Japan, Sept 2009
18. T. Gillespie, *Fundamentals of Vehicle Dynamics* (SAE, International, 1992)
19. D. Bertsekas, *Nonlinear Programming* (MA, Cambridge, 1999)
20. J. Zhang, D. Song, On the error analysis of vertical line pair-based monocular visual odometry in urban area, in *Proceedings of the IEEE/RSJ International Conference on Intelligent Robots and Systems*, St. Louis, MO, Oct 2009, pp. 187–191
21. R. Krauthgamer, J. Naory, R. Schwartzz, Partitioning graphs into balanced components, in *The Twentieth Annual ACM-SIAM Symposium on Discrete Algorithms*, NY, New York, Jan 2009
22. K. Andreev, H. Racke, Balanced graph partitioning. Theory Comput. Syst. **39**, 929–939 (2006)
23. B. Lucas, T. Kanade, An iterative image registration technique with an application to stereo vision, in *Proceedings of Imaging Understanding*, Workshop, 1981, pp. 121–130

Lighting-Invariant Visual Odometry using Lidar Intensity Imagery and Pose Interpolation

Hang Dong and Timothy D. Barfoot

Abstract Recent studies have demonstrated that images constructed from lidar reflectance information exhibit superior robustness to lighting changes in outdoor environments in comparison to traditional passive stereo camera imagery. Moreover, for visual navigation methods originally developed using stereo vision, such as visual odometry (VO) and visual teach and repeat (VT&R), scanning lidar can serve as a direct replacement for the passive sensor. This results in systems that retain the efficiency of the sparse, appearance-based techniques while overcoming the dependence on adequate/consistent lighting conditions required by traditional cameras. However, due to the scanning nature of the lidar and assumptions made in previous implementations, data acquired during continuous vehicle motion suffer from geometric motion distortion and can subsequently result in poor metric VO estimates, even over short distances (e.g., 5–10 m). This paper revisits the measurement timing assumption made in previous systems, and proposes a frame-to-frame VO estimation framework based on a novel pose interpolation scheme that explicitly accounts for the exact acquisition time of each feature measurement. In this paper, we present the promising preliminary results of our new method using data generated from a lidar simulator and experimental data collected from a planetary analogue environment with a real scanning laser rangefinder.

1 Introduction

Over the past two decades, advances in computer vision have prompted rapid development of appearance-based estimation techniques for both terrestrial and space applications. For the Mars Exploration Rovers in particular, stereo-vision-based VO

H. Dong (✉) · T. D. Barfoot
University of Toronto Institute for Aerospace Studies, Toronto, Canada
e-mail: hang.dong@utoronto.ca

T. D. Barfoot
e-mail: tim.barfoot@utoronto.ca

K. Yoshida and S. Tadokoro (eds.), *Field and Service Robotics*,
Springer Tracts in Advanced Robotics 92, DOI: 10.1007/978-3-642-40686-7_22,

Fig. 1 Panoramic view of the planetary analogue test site in Sudbury, Ontario, Canada. Located inside a gravel pit, the site was chosen for its lack of vegetation and three-dimensional character. Our field robot can be seen traversing the site here in the daytime, although our test data were gathered at night

provided crucial feedback to operators when wheel odometry could no longer be trusted; the onboard vision system was deemed instrumental in enabling several kilometers of travel for each of the rovers when large error was incurred in wheel odometry due to slip [12].

With greater onboard capability, such as the abundant power available to the Mars Science Laboratory rover, it is desirable to have robust autonomous capabilities that allow a rover to drive in the dark, thereby increasing its daily travel range to maximize scientific return. Other mission scenarios, including exploration of permanently-shadowed regions of the Moon, will require technology that can operate under drastic lighting changes and/or in complete darkness.

Unlike stereo cameras, light detection and ranging (lidar) sensors are active sensors that use one-axis or two-axis scanning lasers to generate 2D or 3D information of the surrounding environment. Due to the dense nature of 3D information, algorithms using two-axis scanning lidar point clouds are typically more computationally intensive than their stereo-vision-based counterparts, which work directly with 2D images.

We have been working toward navigation systems that retain advantages of both cameras and lidar. Specifically, we want to retain the computational advantage of the sparse visual navigation techniques associated with cameras, while gaining the robustness to ambient lighting change for which lidars are known. The key to enabling this vision is the realization that lidar intensity/reflectance information can be processed into a greyscale appearance image of the environment. Moreover, many state-of-the-art computer vision algorithms, such as the SURF feature extractor and descriptor [1], work out of the box on the lidar intensity images [13].

Earlier works on lidar navigation using intensity information revealed promising results. VO estimation performance comparable to that of established stereo-vision-based methods was obtained using a stop-scan-go acquisition method, and the lidar-based approach was shown to be significantly more robust to ambient lighting changes [13]. More recently, appearance-based lidar has also been successfully adapted for use in a VT&R system, enabling long-distance, lighting-invariant autonomous traverse with high accuracy [14].

While stopping for each lidar scan results in high-quality data and has been used in a number of other systems (e.g., [11, 17]), it vastly limits the traversal speed, and in some cases reduces the capability of the autonomous platform. For instance, existing passive camera-based VO allows for continuous vehicle motion, and the velocity derived from pose estimation can be used as feedback in closed-loop speed control; the same cannot be accomplished with lidar in stop-scan-go mode.

Unlike a charge-coupled device (CCD) camera sensor, which captures a complete image at a single instant of time, scanning lidar acquires one time-of-flight (TOF) measurement at a time. Assuming a static scene, any motion experienced by the platform during acquisition translates into a corresponding change in relative displacement between the lidar and the scan subject, resulting in a scan that is geometrically warped in comparison to one acquired while the platform remains stationary. This effect is referred to as *motion distortion*. While McManus et al. [14] achieved accurate autonomous traverse using local sub-maps, the metric accuracy of the estimated rover track in a global frame is in fact very poor primarily due to this motion distortion effect. In addition to warping of the 3D point cloud, motion distortion can also be observed in the appearance image created using intensity information, as shown in Fig. 2.

The extent of motion distortion is related to the vehicle's velocity and scan rate. Off-the-shelf one-axis lidar scanners typically take milliseconds to produce a scan line (e.g., a little as 2 ms on SICK LMS 111), and at a speed of 1 m/s, the vehicle would only have moved few millimeters during a scan. Relatively trivial in magnitude, the motion distortion effect is generally not accounted for when working with a one-axis scanner. In comparison, a two-axis scanner takes significant longer to produce a full scan; depending on the lidar, it can take anywhere from 100 ms to minutes. Hence, the

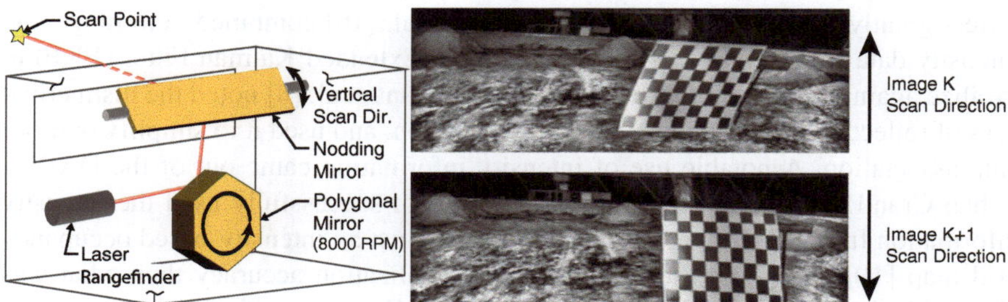

Fig. 2 Simplified illustration of the Autonosys lidar's two-axis scanning mechanism [18] (*left*) and geometric motion distortion effect as seen in intensity images (*right*). Note that in the intensity images, the rover was turning left during both scans. Different skewing effects of the rectangular checkerboard were caused by the nodding behaviour of this lidar. The raw intensity information provided by the lidar is a function of the emitted beam energy, range, target reflectance and its surface orientation with respect to the lidar. See McManus et al. [13] for details on how our intensity images are assembled

effect of motion distortion is no longer inconsequential. For our experiment, we have selected Autonoys' LVC0702 two-axis scanning lidar over the well-known Velodyne HDL-64E for its higher (and adjustable) vertical scan resolution. At its 2 Hz scan setting, the lidar produced intensity images that still allow for reliable SURF feature extraction and matching with a nominal rover speed of 0.5 m/s as demonstrated by McManus et al. [14].

To achieve accurate VO estimates despite lidar motion distortion, this paper proposes an algorithm that explicitly compensates for motion distortion by accounting for the exact measurement time of each lidar return, and still remains computationally tractable using a novel pose interpolation scheme. While many other pose interpolation schemes have been proposed in the past, to our knowledge this is the first one that cleanly handles rotations, and at the same time allows for derivation of analytical Jacobians that are used during a bundle adjustment nonlinear optimization, resulting in a more efficient algorithm than comparable systems using numerical Jacobians [4].

We validate the resulting motion-compensated VO algorithm using data produced by a lidar simulator, and 525 m of real lidar data collected around midnight from a planetary analogue site, shown in Fig. 1, located in Sudbury, Ontario, Canada.

The rest of the paper is organized as follows. Section 2 reviews related navigation systems using lidar intensity data, other lidar-based systems that deal with lidar motion distortion, and our approach at a high level. Section 3 describes our overall VO pipeline, our proposed pose interpolation scheme, and its application to the VO estimation problem. Our hardware configuration as well as simulated and field testing results are documented in Sect. 4, followed by conclusions in Sect. 5 (Figs. 2, and 3).

2 Related Works

The intensity information is often available on laser rangefinders, though its quality differs greatly depending on the model. Neira et al. [16] combined both range and intensity data from a one-axis scanner using an Extended Kalman Filter (EKF) to localize against a known indoor planar map. Guivant et al. [8] noted the distinctiveness of reflective marks in the intensity information, and used it to simplify outdoor data association. A notable use of intensity information came out of the DARPA Urban Grand Challenge; the Stanford racing team successfully used the intensity information from a Velodyne lidar to localize against an intensity-based occupancy grid map [10], providing the vehicle higher localization accuracy than what was obtainable from GPS. Similar technology enabled Google's self-driving car to complete over 300,000 km of autonomous traverse [20]. It is worth noting that the Google system tightly couples intensity-data-based localization with an inertially-aided GPS, and requires a preprocessed map of the environment, neither of which are available for space rovers.

As for lidar motion distortion, thus far there have been three primary approaches to mitigate its impact on estimation accuracy:

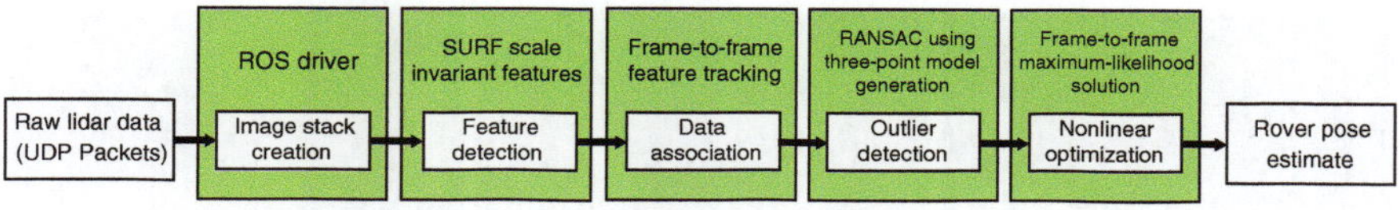

Fig. 3 Major processing blocks of our VO algorithm

1. Reducing acquisition time per scan: given the same platform velocity, spending less time on each scan results in less motion distortion. Most lidars have a fixed data acquisition rate, so this approach typically involves a trade-off between scan rate and scan resolution.
2. Dewarping the point cloud directly using an external motion estimate such as from an IMU.
3. Dewarping the point cloud iteratively by calculating a motion estimate using the Iterative Closest Point (ICP) method and motion-distorted scan, then updating/dewarping the scan using the motion estimate. The process repeats until the motion estimate converges. To speed up the algorithm, Bosse and Zlot [2] preprocessed the dense point cloud into more sparse voxelized version, while Moosmann and Stiller [15] used sub-sampled surface normals for scan matching.

We approach this problem a bit differently. Instead of viewing the entire scan as a unit of acquisition and attempting to correct for motion distortion by dewarping the scan, we view each lidar time-of-flight measurement as our base unit of acquisition. This change of perspective effectively turns a motion-distorted scan (say containing 100,000 points) as viewed from a single pose into 100,000 accurate lidar measurements taken from 100,000 slightly different poses. The large number of poses is non-ideal; each pose in 3D space has six degrees of freedom (DOF). As the number of poses increases, the solution quickly becomes computationally intractable.

Assuming that the platform travels with reasonably constant linear/angular velocity between two consecutive lidar scans, which is typically true for mobile ground robots if the time between scans is short (in our case 0.5 s), it is then possible to represent the large number of poses by interpolating between only two poses.

Faced with similar motion distortion effects in complementary metal-oxide-semiconductor (CMOS) camera sensors and the need to represent large numbers of poses, Forssén and Ringaby [4] choose to use SLERP [19], a quaternion interpolation scheme commonly used in computer graphic animation. Since there exists no simple way to incorporate SLERP analytically into the optimization process, Forssén and Ringaby [4] resorted to using numerical Jacobians for nonlinear optimization.

Another alternative is to parameterize rotations using Euler angles, which are subject to singularities. Moreover, interpolating Euler angles can lead to strange results [7]. As such we set out to find a simple 'linear' interpolation scheme that not only handles rotations cleanly, but also allows for derivation of analytical Jacobians that can be used during the optimization process. Our scheme will be presented in detail in Sect. 3.

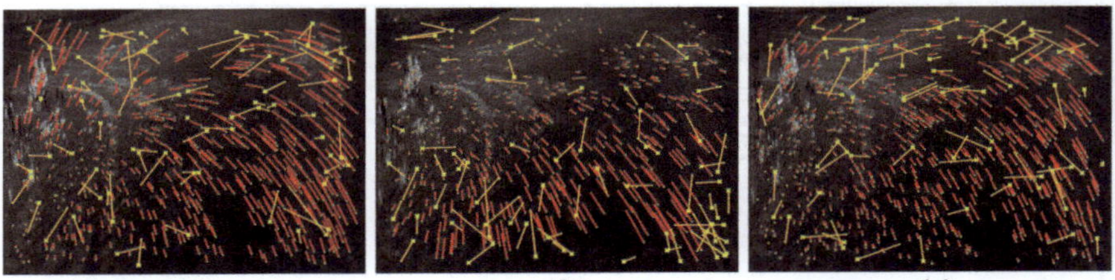

(a) **(b)** **(c)**

Fig. 4 Sample post-RANSAC feature tracks from four consecutive frames (three pairs). The outliers that made it past RANSAC but were detected by the M-estimator are shown in yellow. Note the feature tracks' lengths change between consecutive frame-to-frame matching pairs, as a result of the nodding behaviour of the scanning mechanism. **a** A downward scan matched to an upward scan. **b** An upward scan matched to a downward scan. **c** Repeating pattern in (**a**)

3 System Description

The data flow in our VO algorithm, shown in Fig. 3, is similar to its stereo-vision counterpart, and is nearly identical to the system presented in [13]. We no longer use stereo geometry to extract 3D information as this is directly available in the lidar data. In brief, we extract and match sparse SURF features in consecutive pairs of lidar intensity images, introduce the associated range images to obtain 3D feature locations, run RANdom SAmple Consensus (RANSAC) to reject outliers, and then perform a bundle adjustment nonlinear optimization to determine pose change from frame to frame. We record timestamps for every laser return, which enables pose interpolation inside the frame-to-frame maximum-likelihood solution. Moreover, with motion distortion, we find it necessary to relax the RANSAC matching threshold to avoid throwing away valid feature matches that are temporally far apart. At the same time, this change does allow more outliers to pass RANSAC. We handle these remaining outliers using a Geman-McClure M-estimator during iterative nonlinear optimization. A typical set of post-RANSAC feature tracks is shown in Fig. 4. The most significant change occurs inside the formulation of the maximum-likelihood solution, to which we will devote the remainder of this section.

3.1 Problem Setup

Our VO algorithm is essentially a frame-to-frame bundle adjustment technique that solves for the following incremental variables:

$\mathbf{r}_k^{k+1,k}$: translation of camera pose $k+1$ relative to pose k, expressed in frame k,

$\mathbf{C}_{k+1,k}$: rotation matrix of camera pose $k+1$ (from pose k to pose $k+1$),

$\mathbf{p}_k^{j,k}$: position of landmark j relative to pose k, expressed in frame k, where $j = 1 \ldots J$. After calculating the incremental transforms, we can compose them to obtain metric pose estimates with respect to the initial coordinate frame. Thus from here, we will focus on solving for the incremental variables.

The measurement error term is given by

$$\mathbf{e}_{jl}(\mathbf{r}_k^{l,k}, \mathbf{C}_{l,k}, \mathbf{p}_k^{j,k}) := \mathbf{y}_{jl} - \mathbf{f}\left(\mathbf{C}_{l,k}(\mathbf{p}_k^{j,k} - \mathbf{r}_k^{l,k})\right), \qquad \forall (j,l)$$

where $\mathbf{y}_{jl}$ is the measured quantity (measured at time $t_l \in [t_k, t_{k+1}], l = 1 \ldots L$) and $\mathbf{f}(\cdot)$ is a nonlinear camera model. We seek to find the values of $\mathbf{C}_{k+1,k}$, $\mathbf{r}_k^{k+1,k}$, and $\mathbf{p}_k^{j,k}$ to minimize the following objective function:

$$J(\mathbf{x}) := \frac{1}{2} \sum_{j,l} \mathbf{e}_{jl}(\mathbf{r}_k^{l,k}, \mathbf{C}_{l,k}, \mathbf{p}_k^{j,k})^T \mathbf{R}_{jl}^{-1} \mathbf{e}_{jl}(\mathbf{r}_k^{l,k}, \mathbf{C}_{l,k}, \mathbf{p}_k^{j,k}), \tag{1}$$

where $\mathbf{x}$ is the full state that we wish to estimate (pose and landmarks) and $\mathbf{R}_{jl}$ is the symmetric, positive-definite covariance matrix associated with the (j, l)-th measurement. The usual approach to this problem is to apply the Gauss-Newton method [6]. The added challenge here lies in the fact that our state variables are at times t_k while our measurements are at times t_l, which do not line up. Our approach will be to 'linearly' interpolate poses between the t_k times.

3.2 Interpolation Strategy

In this section we present the necessary formulation to implement our algorithm. Due to the paper length restriction, we refrain from going into detailed background and derivation. First we define the interpolation variable, $\alpha_l \in [0, 1]$, as

$$\alpha_l := \frac{t_l - t_k}{t_{k+1} - t_k}. \tag{2}$$

Recognizing that we can write any rotation matrix using the exponential map,

$$\mathbf{C} = e^{-\phi \mathbf{a}^\times} = e^{-\boldsymbol{\phi}^\times},$$

where ϕ is the angle, $\mathbf{a}$ is the unit-length axis, $\boldsymbol{\phi} := \phi \mathbf{a}$, and $(\cdot)^\times$ is the skew-symmetric operator, we then define our interpolation of rotation variables to be

$$\mathbf{C}_{l,k} := \mathbf{C}_{k+1,k}^{\ \alpha_l} = e^{-\alpha_l \boldsymbol{\phi}_{k+1,k}^\times}. \tag{3}$$

Here we are effectively just scaling the angle of rotation by α_l and leaving the axis untouched. This is not the only way we could define the interpolation, but it is a notationally simple one that avoids singularities. Trivially, we interpolate the translation variable according to

$$\mathbf{r}_k^{l,k} := \alpha_l \, \mathbf{r}_k^{k+1,k}. \tag{4}$$

It is easy to see that $\alpha_l = 0$ implies $\mathbf{C}_{l,k} = \mathbf{1}$ (identity matrix) and $\mathbf{r}_k^{l,k} = \mathbf{0}$, and $\alpha_l = 1$ implies $\mathbf{C}_{l,k} = \mathbf{C}_{k+1,k}$ and $\mathbf{r}_k^{l,k} = \mathbf{r}_k^{k+1,k}$.

3.3 Perturbing Interpolated Poses

In order to linearize our error terms, we perturb the pose variables according to

$$\mathbf{C}_{k+1,k} = e^{\delta\phi_{k+1,k}^{\times}} \bar{\mathbf{C}}_{k+1,k}, \tag{5}$$

$$\mathbf{r}_k^{k+1,k} = \bar{\mathbf{r}}_k^{k+1,k} + \delta\mathbf{r}_k^{k+1,k}, \tag{6}$$

whereupon it can be shown (derivation not included due to space constraints) that the interpolated variables can be perturbed according to

$$\mathbf{C}_{l,k} = \left(\mathbf{1} - \alpha_l \left(\boldsymbol{\Phi}_l \, \delta\phi_{k+1,k}\right)^{\times}\right) \bar{\mathbf{C}}_{l,k}, \tag{7}$$

$$\mathbf{r}_k^{l,k} = \bar{\mathbf{r}}_k^{l,k} + \alpha_l \delta\mathbf{r}_k^{k+1,k}, \tag{8}$$

$$\mathbf{p}_k^{j,k} = \bar{\mathbf{p}}_k^{j,k} + \delta\mathbf{p}_k^{j,k}, \tag{9}$$

where

$$\bar{\mathbf{C}}_{l,k} := \bar{\mathbf{C}}_{k+1,k}{}^{\alpha_l},$$

$$\bar{\mathbf{r}}_k^{l,k} := \alpha_l \bar{\mathbf{r}}_k^{k+1,k},$$

$$\boldsymbol{\Phi}_l := \mathbf{S}\left(\alpha_l \bar{\phi}_{k+1,k}\right) \mathbf{S}\left(\bar{\phi}_{k+1,k}\right)^{-1},$$

$$\mathbf{S}(\phi) := \frac{\sin\phi}{\phi}\mathbf{1} + \left(1 - \frac{\sin\phi}{\phi}\right)\mathbf{a}\mathbf{a}^T - \frac{1-\cos\phi}{\phi}\mathbf{a}^{\times},$$

and $\bar{\phi}_{k+1,k}$ can be determined from $\bar{\mathbf{C}}_{k+1,k}$ exactly and should not be near a singularity so long as the rotation is not large. In the next section we will use these perturbations to linearize our error terms.

After solving for the incremental quantities at each iteration of Gauss-Newton, we will update the mean quantities according to the following update rules:

$$\begin{aligned}\bar{\mathbf{C}}_{k+1,k} &\leftarrow e^{-\delta\boldsymbol{\phi}_{k+1,k}^{\times}}\bar{\mathbf{C}}_{k+1,k},\\ \bar{\mathbf{r}}_k^{k+1,k} &\leftarrow \bar{\mathbf{r}}_k^{k+1,k} + \delta\mathbf{r}_k^{k+1,k},\\ \bar{\mathbf{p}}_k^{j,k} &\leftarrow \bar{\mathbf{p}}_k^{j,k} + \delta\mathbf{p}_k^{j,k}.\end{aligned}$$

3.4 Linearized Error Terms

The last step is to use our perturbed pose expressions to come up with the linearized error terms. Consider just the first nonlinearity before the camera model:

$$\mathbf{p}_l^{j,l} := \mathbf{C}_{l,k}\left(\mathbf{p}_k^{j,k} - \mathbf{r}_k^{l,k}\right).$$

Inserting (7), (8), and (9) and dropping products of small terms we have

$$\begin{aligned}\mathbf{p}_l^{j,l} &\approx \left(\mathbf{1} - \alpha_l\left(\boldsymbol{\Phi}_l\,\delta\boldsymbol{\phi}_{k+1,k}\right)^{\times}\right)\bar{\mathbf{C}}_{l,k}\left(\bar{\mathbf{p}}_k^{j,k} + \delta\mathbf{p}_k^{j,k} - \bar{\mathbf{r}}_k^{l,k} - \alpha_l\delta\mathbf{r}_k^{k+1,k}\right)\\ &\approx \underbrace{\bar{\mathbf{C}}_{l,k}\left(\bar{\mathbf{p}}_k^{j,k} - \bar{\mathbf{r}}_k^{l,k}\right)}_{\bar{\mathbf{p}}_l^{j,l}} + \underbrace{\begin{bmatrix}-\alpha_l\bar{\mathbf{C}}_{l,k} & \alpha_l\bar{\mathbf{p}}_l^{j,l^{\times}}\boldsymbol{\Phi}_l & \bar{\mathbf{C}}_{l,k}\end{bmatrix}}_{=:\mathbf{D}_{jl}}\underbrace{\begin{bmatrix}\delta\mathbf{r}_k^{k+1,k}\\ \delta\boldsymbol{\phi}_{k+1,k}\\ \delta\mathbf{p}_k^{j,k}\end{bmatrix}}_{=:\delta\mathbf{x}_{jl}}\\ &= \bar{\mathbf{p}}_l^{j,l} + \mathbf{D}_{jl}\,\delta\mathbf{x}_{jl}.\end{aligned}$$

Inserting this into the full error expression we have

$$\begin{aligned}\mathbf{e}_{jl}(\bar{\mathbf{x}}_{jl} + \delta\mathbf{x}_{jl}) &\approx \mathbf{y}_{jl} - \mathbf{f}\left(\bar{\mathbf{p}}_l^{j,l} + \mathbf{D}_{jl}\,\delta\mathbf{x}_{jl}\right)\\ &\approx \underbrace{\mathbf{y}_{jl} - \mathbf{f}\left(\bar{\mathbf{p}}_l^{j,l}\right)}_{=:\bar{\mathbf{e}}_{jl}} - \underbrace{\mathbf{F}_{jl}\mathbf{D}_{jl}}_{=:-\mathbf{E}_{jl}}\,\delta\mathbf{x}_{j,l}\\ &= \bar{\mathbf{e}}_{jl} + \mathbf{E}_{jl}\,\delta\mathbf{x}_{jl},\end{aligned}$$

where

$$\mathbf{F}_{jl} := \left.\frac{\partial\mathbf{f}}{\partial\mathbf{p}}\right|_{\bar{\mathbf{p}}_l^{j,l}}.$$

We can then insert this approximation into the objective function in (1), causing it to become quadratic in $\mathbf{x}$, and proceed in the usual Gauss-Newton fashion, being sure to update our rotation variables properly at each iteration.

4 VO Algorithm Test

Testing of the algorithm was conducted in two stages:

1. Using data produced by a custom lidar simulator, which forms a controlled environment and provides 6 DOF groundtruth information.
2. Using real lidar data collected from a planetary analogue site in complete darkness, demonstrating the lighting-invariant aspect of the system, as well as its actual performance in an outdoor unstructured 3D environment. Groundtruth is provided by DGPS.

At each stage we will compare the VO estimates with and without motion compensation against the groundtruth.

4.1 System Testing using Lidar Simulator

The purpose of testing with simulated data was to validate the algorithm as much as possible in a controlled environment. We configured the lidar simulator using parameters from a real scanning laser rangefinder (see details in Sect. 4.2). This included giving it the identical scan resolution of 480×360 pixels and the scanning frequency of 2 Hz, as well as the exact same scanning pattern caused by the nodding mirror inside the real sensor (see Fig. 2).

During the simulation, the rover was given a sinusoidal trajectory with amplitude 0.25 meters in the xy-plane. Yaw heading angle was tangential to the xy planar trajectory.

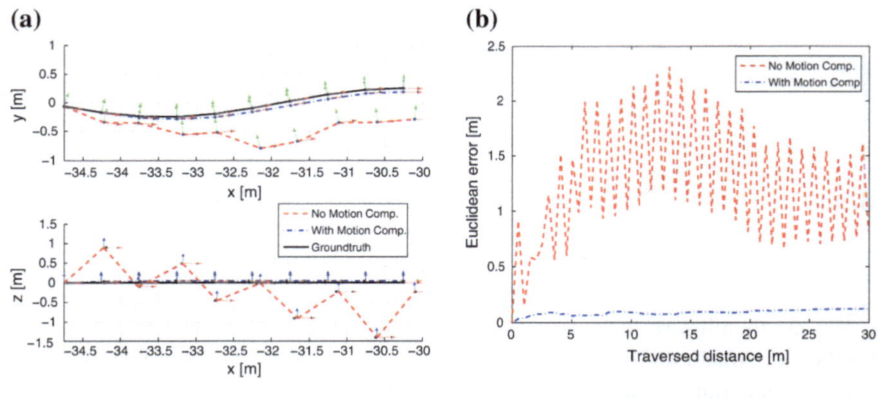

Fig. 5 VO estimates using simulated lidar data. The estimated rover track with motion compensation is smoother and accumulates error at a much lower rate. **a** Groundtruth versus estimates from first 5 s of traverse. **b** Euclidean error over traversed distance

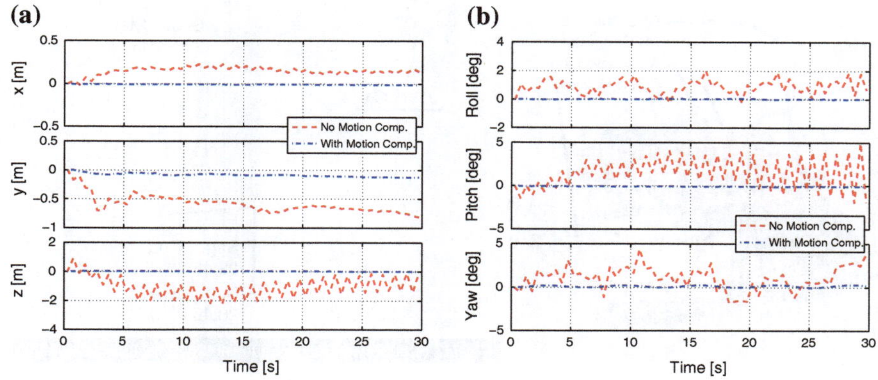

Fig. 6 VO estimation errors in all 6 DOF. Note that during estimation, attitude was represented using a rotation matrix, which is decomposed into roll, pitch, and yaw here for comparison against groundtruth. **a** Position estimation error. **b** Orientation estimation error

As shown in Fig. 5, the VO estimate without motion compensation struggled to match with the groundtruth; a distinctive sawtooth-shaped estimate is clearly visible in the xz-plane as a result of motion distortion. On the other hand, the VO algorithm with motion compensation produced a smooth estimate that closely follows the groundtruth. Figure 6 provides 6 DOF error plots to further demonstrate the contrast in quality between the two estimators. Note the small sinusoidal error in yaw angle even with motion compensation, indicating that the interpolation-based motion compensation is not sufficient to completely capture a nonlinear sinusoidal motion, which is expected.

4.2 Hardware Description

Our field experiment was carried out using a ROC6 skid-steered rover, an Autonosys LVC0702 lidar sensor, and a differential GPS for groundtruth positioning. The configuration is shown in Fig. 7a.

The Autonosys lidar employs a unique two-axis scanning system (see Fig. 2). While the horizontal scanning direction is consistent, a nodding-mirror-based vertical scanning mechanism switches scanning direction after each scan to avoid the need for quick return. As a result, the motion distortion in adjacent scans has completely opposite distortion effects, as shown in Fig. 2. With a field of view (FOV) similar to traditional stereo cameras, this lidar is also referred to as a lidar video camera by the manufacturer, capable of producing 480 × 360 pixel lidar scans at 2 Hz. The lidar sensor has a horizontal FOV of 90 degrees, and vertical FOV of 30 degrees. In order to maximize valid lidar returns, the sensor was aimed 15 degrees down, giving it an effective vertical FOV from −30 to 0 degrees.

Fig. 7 Appearance-based lidar VO experiment setup (*left*) and traverse path groundtruth (*right*) **a** Hardware configuration. **b** GPS tracks of two traverse segments used for VO testing and their associated collection times

Since it was nearly impossible and dangerous for a human operator to pilot the robot in complete darkness, all the data used in this section were collected autonomously; during the daytime, the path was driven once manually, and repeated at night autonomously at a nominal speed of 0.5 m/s using VT&R [14]. The data used in this paper were logged as a byproduct of this unrelated test. The traversed path is shown in Fig. 7b.

Segment 1 is 225 m in length, and contains mostly a long smooth traverse and gradual turns. There is one direction switch in the latter half of the traverse. Segment 2 is 300 m in length, and contains many sharp turns and more elevation changes. There is also a three-point turn in the middle of this traverse.

4.3 Field Testing Results

As shown in Fig. 8, the VO estimator with motion compensation performed better qualitatively in both traverse segments than the estimator that does not address motion distortion. Closer inspection of the Euclidean error plots (see Fig. 9) reveals that the motion-compensated case has much lower error in segment 2 than in segment 1. The Euclidean-error-to-distance-travelled ratio remains under 5 % on segment 2 as compared to 7 % on segment 1, despite the fact that segment 2 was anecdotally a more challenging traverse. From the VO estimate result of segment 1, we can see that while the incremental heading estimate appears to be accurate, there was a gradual accumulation of heading error. As the attitude estimate drifted, the error grew superlinearly, similarly to stereo-camera-based VO results reported by Lambert et al. [9]. The direction switches and heading changes in segment 2 incidentally had a net effect of partially cancelling estimation error in different parts of the traverse, therefore resulting in a better VO estimate.

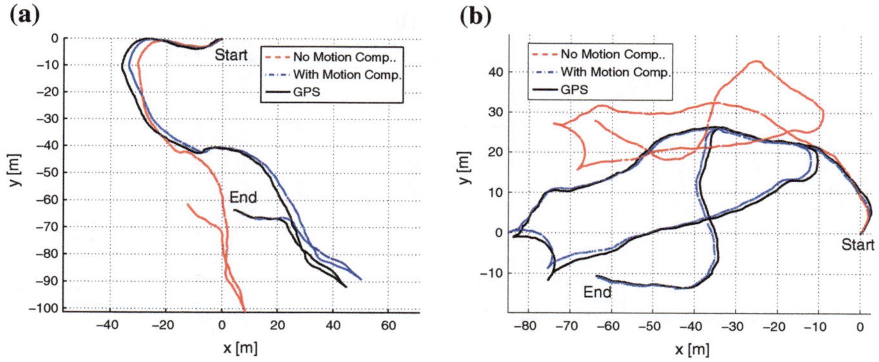

Fig. 8 VO estimates of the two traverse segments in Fig. 7b. The estimator without motion compensation severely underestimates both rotational and translational changes **a** Segment 1. **b** Segment 2

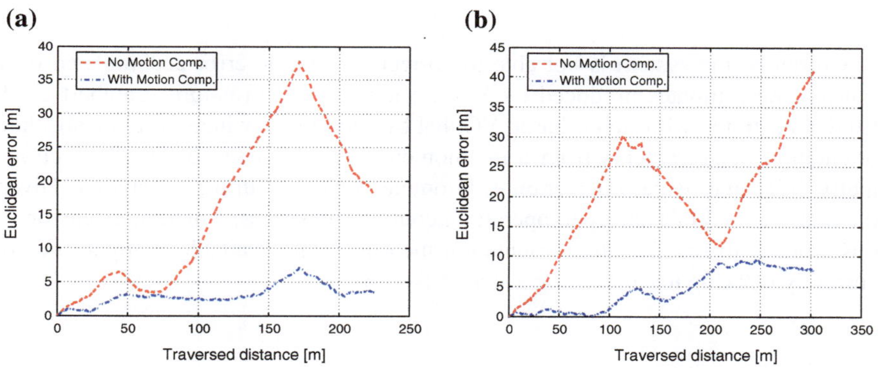

Fig. 9 Euclidean estimation error vs. traverse distance plots, showing the cumulative pose error grows significantly slower with motion compensation. **a** Segment 1. **b** Segment 2

As for the reason behind why VO without motion compensation performed much worse in segment 2, we can explain it by zooming in on short stretches of traverse, shown in Fig. 10. The estimate exhibited a choppy sawtooth shape due to motion distortion, and as a whole underestimated both rotational and translational motion. In contrast, the estimate with motion compensation was smooth and closely followed the groundtruth. This agrees with our earlier observation made using simulated lidar data.

Another source of error is the assumption of constant velocity between poses previously introduced by our pose interpolation scheme. Since our formulation only estimates one pose placed at the center of each scan, it is insufficient at times to fully capture the motion of the rover in a rough terrain (e.g., when driving over a rock). This error can be mitigated by reducing the temporal spacing between poses (similar to higher sampling rate in analogue to digital conversion).

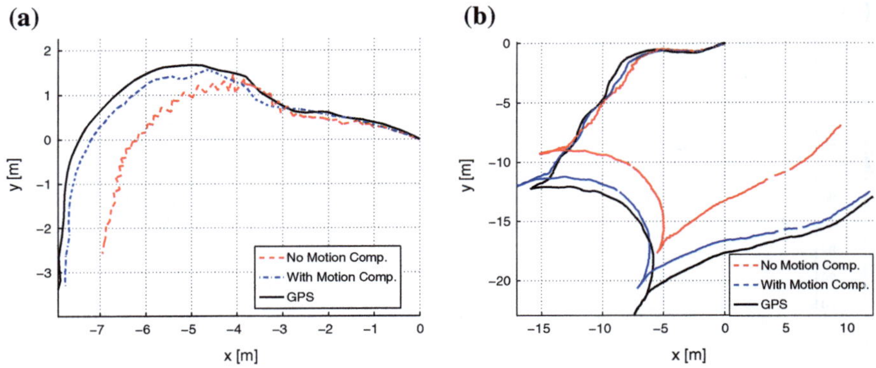

Fig. 10 Close-ups of segment 2. In the estimates without motion compensation, the sawtooth-shaped error previously observed during simulation is clearly visible at this scale. **a** Sharp turn located at (−30 m, 25 m) in segment2. **b** Three-point turn located at (−75 m, −5 m) in segment 2

Currently, the system is not able to correct orientation error accumulated over time as the estimates produced by VO are incremental in nature. Lambert et al. [9] demonstrated with stereo-based VO that by continuously incorporating absolute orientation measurements from a inclinometer and a sun sensor, highly accurate metric VO can be expected over multi-kilometer traverses during the daytime. Given the lidar-based VO's ability to operate in complete darkness, a natural extension of this work is to fuse absolute orientation measurements available at night; we are currently working towards using a star tracker [3].

5 Conclusion

We have presented an improved appearance-based lidar navigation system that exhibits the computational efficiency of sparse visual techniques, typically associated with stereo cameras, while overcoming the lighting dependence of traditional cameras. The contributions of this work include:

1. A novel pose interpolation strategy based on the exponential map that allows for derivation of analytical Jacobians used during a bundle adjustment nonlinear optimization.
2. A VO algorithm based on #1 that compensates for motion distortion in lidar scans acquired during continuous vehicle motion.
3. Testing of #2 using a simulated lidar dataset and over 500 meters of experimental data collected from a planetary analogue environment with a real scanning laser rangefinder in complete darkness.

Our results demonstrate clear improvement in the VO estimate by compensating for the motion distortion effect. We obtained 5–7 % linear error growth in hundred-meter-scale traverses during our field experiment using only lidar data and no other sensor information. This work can be further improved by carrying out global minimization over a small set of scans and/or introducing additional attitude sensors, such as an inclinometer or star tracker into the system. Furthermore, we are interested in moving beyond linear interpolation, as a spline-based interpolation approach may better approximate the nonlinear motion of the robot [5, 21].

Finally, any estimation problem in which there is a large collection of sensor data with distinct measurement times can be solved using far fewer poses/variables by interpolating poses. With new sensors producing measurements at higher and higher rates, and the fact that it may not always be possible to trigger/synchronize different sensors, the proposed interpolation scheme and its associated derivation could be useful for a general set of problems.

Acknowledgments We would like to extend our deepest thanks to the staff of the Ethier Sand and Gravel in Sudbury, Ontario, Canada for allowing us to conduct our field tests on their grounds. We also wish to thank Dr. James O'Neill from Autonosys for his help in preparing the lidar sensor for our field tests. In addition, we would also like to acknowledge (from the Autonomous Space Robotics Laboratory) Colin McManus for being instrumental in gathering the data used in this paper, Andrew Lambert for his help in preparing the GPS payload, Paul Furgale and Chi Hay Tong for their work on the GPU SURF algorithm, Goran Basic for designing and assembling the Autonosys payload mount, and Keith Leung for providing onsite photography for the field tests. Lastly, we also wish to thank the Natural Sciences and Engineering Research Council of Canada and the Canada Foundation for Innovation, Defence R&D Canada at Suffield (particularly Jack Collier), the Canadian Space Agency, and MDA Space Missions (particularly Cameron Ower, Raja Mukherji, and Joseph Bakambu) for providing us with the financial and in-kind support necessary to conduct this research.

References

1. H. Bay, A. Ess, T. Tuytelaars, L. Van Gool, Speeded-up robust features (SURF). Comput. Vis. Image Underst. **110**(3), 346–359 (2008)
2. M. Bosse, R. Zlot, Continuous 3d scan-matching with a spinning 2d laser. in *Robotics and Automation (ICRA), 2009 IEEE International Conference on*, (IEEE, 2009), pp. 4312–4319
3. J. Enright, T.D. Barfoot, M. Soto, Star tracking for planetary rovers. in *Proceedings of the IEEE Aerospace Conference*, (Big Sky, MT, 2012), pp. 1–13
4. P.E. Forssén, E. Ringaby, Rectifying rolling shutter video from hand-held devices. in *IEEE Conference on Computer Vision and Pattern Recognition*, (IEEE Computer Society, IEEE, San Francisco, USA, 2010)
5. P.T. Furgale, T.D. Barfoot, G. Sibley, Continuous-time batch estimation using temporal basis functions. in *Proceedings of the IEEE International Conference on Robotics and Automation (ICRA)*, (St. Paul, USA, 2012), pp. 2088–2095
6. K. Gauss, Theory of the motion of the heavenly bodies moving about the sun in conic sections: Reprint 2004 (1809)
7. F. Grassia, Practical parameterization of rotations using the exponential map. J. Graphics Tools **3**, 29–48 (1998)
8. J. Guivant, E. Nebot, S. Baiker, Localization and map building using laser range sensors in outdoor applications. J. Robot. Syst. **17**(10), 565–583 (2000)

9. A. Lambert, P.T. Furgale, T.D. Barfoot, J. Enright, Field testing of visual odometry aided by a sun sensor and inclinometer. J. Field Robot. (special issue) Space Robot. **29**(3), 426–444 (2012)
10. J. Levinson, Automatic laser calibration, mapping, and location for autonomous vehicles. Ph.D. thesis, Stanford University 2011
11. M. Magnusson, A. Lilienthal, T. Duckett, Scan registration for autonomous mining vehicles using 3d-ndt. J. Field Robot. **24**(10), 803–827 (2007)
12. L. Matthies, M. Maimone, A. Johnson, Y. Cheng, R. Willson, C. Villalpando, S. Goldberg, A. Huertas, A. Stein, A. Angelova, Computer vision on Mars. Int. J. Comput. Vision **75**(1), 67–92 (2007)
13. C. McManus, P.T. Furgale, T.D. Barfoot, Towards appearance-based methods for lidar sensors. in *Robotics and Automation (ICRA), 2011 IEEE International Conference on*, (Shanghai, China, 2011), pp. 1930–1935
14. C. McManus, P.T. Furgale, B.E. Stenning, T.D. Barfoot, Visual teach and repeat using appearance-based lidar. in *Proceedings of the IEEE International Conference on Robotics and Automation (ICRA)*, (St. Paul, USA, 2012), pp. 389–396
15. F. Moosmann, C. Stiller, Velodyne SLAM. in *Intelligent Vehicles Symposium (IV), 2011 IEEE*, (IEEE, 2011), pp. 393–398
16. J. Neira, J. Tardós, J. Horn, G. Schmidt, Fusing range and intensity images for mobile robot localization. IEEE Trans. Robot. Autom. **15**(1), 76–84 (1999)
17. A. Nüchter, K. Lingemann, J. Hertzberg, H. Surmann, 6D SLAM-3D mapping outdoor environments. J. Field Robot. **24**(8–9), 699–722 (2007)
18. J. O'Neill, W. Moore, K. Williams, R. Bruce, Scanning system for lidar (Oct 30 2007), US Patent App. 12/447,937
19. K. Shoemake, Animating rotation with quaternion curves. ACM SIGGRAPH Comput. Graphics **19**(3), 245–254 (1985)
20. S. Thrun, C. Urmson, How Google's self-driving car works. Intelligent Robots and Systems (IROS) (2011). http://spectrum.ieee.org/automaton/robotics/artificial-intelligence/how-google-self-driving-car-works
21. C.H. Tong, P.T. Furgale, T.D. Barfoot, Gaussian process gauss-newton: Non-parametric state estimation. in *9th Canadian Conference on Computer and Robot Vision (CRV)*(to appear), Toronto, Canada, 28–30 May 2012

Modeling and Calibrating Visual Yield Estimates in Vineyards

Stephen Nuske, Kamal Gupta, Srinivasa Narasimhan and Sanjiv Singh

Abstract Accurate yield estimates are of great value to vineyard growers to make informed management decisions such as crop thinning, shoot thinning, irrigation and nutrient delivery, preparing for harvest and planning for market. Current methods are labor intensive because they involve destructive hand sampling and are practically too sparse to capture spatial variability in large vineyard blocks. Here we report on an approach to predict vineyard yield automatically and non-destructively using images collected from vehicles driving along vineyard rows. Computer vision algorithms are applied to detect grape berries in images that have been registered together to generate high-resolution estimates. We propose an underlying model relating image measurements to harvest yield and study practical approaches to calibrate the two. We report on results on datasets of several hundred vines collected both early and in the middle of the growing season. We find that it is possible to estimate yield to within 4 % using calibration data from prior harvest data and 3 % using calibration data from destructive hand samples at the time of imaging.

1 Introduction

Harvest yield prediction is critical to any vineyard grower for deciding when and how to make adjustments to their vines to optimize growth, for preparing a grower for the harvest operation, for shipping their crop, storing their crop and also selling their crop on the market. The typical process of estimating yield is for workers to manually sample a small percentage of the vineyard and extrapolate these measurements to

S. Nuske (✉) · S. Narasimhan · S. Singh
Robotics Institute, Carnegie Mellon University, Pittsburgh, PA15213, USA
e-mail: nuske@cmu.edu

K. Gupta
Indian Institute of Technology Delhi, Delhi, India

K. Yoshida and S. Tadokoro (eds.), *Field and Service Robotics*,
Springer Tracts in Advanced Robotics 92, DOI: 10.1007/978-3-642-40686-7_23,

the entire vineyard. The manual practice is labor intensive, expensive, inaccurate, spatially sparse, destructive and riddled with subjective inputs.

Vineyard managers currently do not have the information they require to make informed decisions on their operations with accuracy and precision. We present technology that can make dense predictions of harvest yield efficiently and automatically using cameras. Here we report results of an approach to automatically detect and count grapes to forecast yield with both precision and accuracy. The approach is to take conventional visible light cameras through a vineyard to image the vines and detect the crop and predict yield, see Fig. 1b.

Visually detecting grape berries is difficult because often there is a lack of color contrast to the background, which is often similarly colored to the grapes, (Fig. 1a). We specifically address the issues of lighting and lack of color contrast, by using shape and texture cues for detection. Also not all berries are visible and therefore the camera makes incomplete measurements that need to be calibrated.

Preliminary results of our approach were reported in Nuske et al. [9], where we presented a method to detect berries and count berries to correlate against yield and we extend our prior work in the following ways:

1. we study the underlying model relating the image measurements of the grape berries to the harvest weight
2. we demonstrate two different approaches to calibrate our image berry measurements to harvest yield
3. we present experimental results with image data collected at various stages during the growing season and in both wine and table-grape vineyards

We deployed our method on three different vine varieties and conducted experiments in which manual per-vine harvest weights were collected and used as ground truth to evaluate our automated yield measurements. The size of the experiment includes 860 individual vines, of three different varieties and vineyards over a linear distance of 2.5 km. Our method predicts weight with approximately 4 % error of the overall actual harvest yield and approximately 8 % error of the harvest weight for individual vineyard rows.

2 Related Work

Current practices to forecast yield are inaccurate because of sampling approaches that tend to adjust towards historical yields and include subjective inputs [2]. The calculation of final cluster weight from weights at véraison use fixed multipliers from historic measurements [12]. Unfortunately, multipliers are biased towards healthier vines thus discriminating against missing or weak vines and multipliers for cluster weights vary widely by vineyard, season and variety.

Sensor-based yield estimation in vineyards has been attempted with trellis tension monitors, multispectral sensors, terahertz-wave imaging and visible-light image processing. A dynamic yield estimation system based on trellis tension monitors

Fig. 1 Photos of the vineyards and equipment from our experiments. Equipment mounted on an aluminum frame fixed on the back tray of a Kawasaki Mule farm utility vehicle. The sensing equipment used is a Nikon D300s color camera facing sideways from the vehicle detecting the fruit, an AlienBees ARB800 ring flash mounted around the lens of the color camera illuminating the scene, a PointGrey BumbleBee2 stereo camera facing back down the row tracking the vehicle motion. Images collected of the vines are processed to detect and measure the vine fruit. **a** Chardonnay vineyard used in experiments. **b** Sensors used in experiments. **c** Raw images. **d** Detected berries

has been demonstrated [1] but it requires permanent infrastructure to be installed. Information obtained from multispectral images has been used to forecast yields with good results but is limited to vineyards with uniformity requirements [8]. A proof of concept study by Federici et al. [5] has shown that terahertz imaging can detect the curved surfaces of grapes and also has the potential to detect these through occluding thin canopy. The challenge for this approach is to achieve fast scan rates to be able to deploy the scanner on a mobile platform.

Small scale yield estimation based on simple image color discrimination has been developed by Dunn and Martin [4]. This approach was attempted on Shiraz post-véraison (i.e. after color development, very close to harvest) in short row segments. The method would not be applicable for the majority of real world examples where the fruit appears over a background of similarly-colored leaves, as is the case in white grape varieties and in all varieties before véraison. Other recent small scale experiments in vineyard work is Dey [3] present a method for classifying plant structures, such as the fruit, leaves, shoots based on 3D reconstructions generated from image sequences which unlike our work is sensitive to slight wind whilst imaging. Other crop detection based on computer vision methods using color pixel classification or shape analysis has been attempted on various fruit types—Jimenez et al. [6] provides a summary of fruit detection work, Singh et al. [10] present a method for detecting and classifying fruit in apple orchards and Swanson et al. [11] use the shading on the curved surfaces of oranges as a cue for detection.

Our prior work, Nuske et al. [9], demonstrated how to detect berries similar in color to the background of leaves using a combination of image processing techniques including a radial symmetry transform and classification of texture properties. In the previous section we list the extensions to our earlier work.

3 Modelling Harvest Yield with Image-Based Measurements

Viticulturalists have long studied the process of predicting the size of the harvest yield and have developed models of the various yield components [2]. In the most basic form, the weight of the harvest (W_h) can be expressed as a product of the number of berries (N_b) and the mean weight of the berries (W_b).

$$W_h = N_b W_b \tag{1}$$

Once berry set has passed the number of berries can be considered constant throughout the remainder of the season, whereas the mean berry weight will substantially increase in size. The standard practice for generating accurate yield predictions is to measure the number of berries early in the season and use historic information of the mean berry weight to project to a harvest estimate. Using historical records of mean berry weight will introduce error into the projection as the mean berry weight at harvest will vary from year to year. However the variation is berry weight is small and is known only to account for 10 % of the year-to-year variations in harvest yield. The berry count contributes to the remaining 90 % [2] of year-to-year yield variations.

Our approach to predict the yield at harvest time is also focussed on measuring the number of berries and we do so using cameras mounted on vehicles driving collecting images of the vines. From a single input image we use our visual berry detection algorithm detailed in Nuske et al. [9] for calculation of the number of berries. Often the canopy of the vines visually similar to the fruit. To detect the fruit we find possible berry locations based on their shape for robustness and classify based on texture and color. The method is to look for radially symmetric points in a fixed, pre-decided radius range followed by color and texture based classification. Fig. 1d presents examples of the berries being detected by our algorithm.

The number of berries found by our visual berry detection algorithm (N_b^d) is the measurement that we pass to a yield forecasting function, $f(\cdot)$, which outputs an estimate ($\hat{N}_b$) of the actual berry count:

$$\hat{N}_b = f(N_b^d) \tag{2}$$

Multiplying our estimated berry count by the expected berry weight at harvest (W_b) gives our yield prediction.

To generate accurate estimates, the function f must model several biases that are inherent to the visual detection process. In Nuske et al. [9] we treated the biases together as a single first order linear factor (k):

$$\hat{N}_b = f(N_b^d) = \frac{1}{k} N_b^d \tag{3}$$

Here in this work we study the individual causes of the bias in an attempt to better understand the system as a whole. First, we introduce three different types of occluders that cause many berries to not be visible to the camera and bias the counts:

1. Self-occlusions (k_s): Berries hidden behind berries within the same grape cluster
2. Cluster-occlusions (k_c): Berries hidden behind other grape clusters
3. Vine-occlusions (k_v): Berries hidden behind the leaves and shoots of the vine

All of these physical biases can be seen in Fig. 1. There are also biases in the visual detection process. In Nuske et al. [9] the performance of the detection algorithm was analyzed to find that a fraction of the berries visible to the camera are not detected by the algorithm and also small fraction of times the algorithm falsely reports a berry where there was not, and these two factors combined introduces a detection bias (k_d). Further, when combining detections from several overlapping images, the system is susceptible to errors where berries are either double-counted or mistakenly not counted, introducing a mis-registration bias, k_r.

The naive approach is to combine these bias terms as linear factors as follows:

$$f(N_b^d) = \frac{N_b^d}{k_s k_c k_v k_d k_r} \tag{4}$$

Later, in the results section, we attempt to isolate the five bias parameters aiming to deepen our understanding of the origins of the error in our visual estimation framework.

3.1 Estimating Self-Occlusions

In our prior work [9] we used the visible berry count as a prediction of the cluster size, assuming the visible berry count is proportional to the total berry count:

$$N_b \propto N_b^d \tag{5}$$

In the results section we study the visible berry count in controlled experiments and also in an attempt to improve the measurement of the occluded berries in a cluster, we propose two potential modifications.

The first alternative measurement we propose is to take the convex hull formed by all the berries in the cluster. Assuming the cluster has uniform density, k, and an average thickness of the grape cluster to be d, we multiply the area A to this fixed cluster depth, and normalize with the average berry radius R_b.

$$N_b \propto d\frac{A}{R_b^2} \tag{6}$$

The second alternative to measuring the size of a cluster is to extend the convex hull model by predicting the berries occluded by the visible layer of berries using an ellipsoid model. A grape cluster's volume can be approximated with an ellipsoid cutting off the image plane as an ellipse. We find the best fit ellipse for the berry center locations with same normalized second central moments. Given the semi-axes of the ellipse in pixels R_1 and R_2, with $R_1 \geq R_2$, volume of the corresponding ellipsoid would be proportional to volume occupied by the berries (B_v) in the cluster. Using the average berry radius (R_b) of the cluster, we can hence calculate the total number of berries occupied by the cluster:

$$\begin{aligned} V_c &\propto \frac{4}{3}\pi r_1 {r_2}^2 \\ N_b &= V_c/(\frac{4}{3}\pi r_b^3) \end{aligned} \tag{7}$$

We study these three approaches to measuring grape cluster size in controlled laboratory tests in the results section. In the vineyard experiments we focus just on using the visible berry count will evaluate the alternatives in future vineyard experiments once we have developed a method to accurately segment and separate neighboring clusters.

4 Calibrating Image Measurements to Harvest Yield

Here we take into consideration different procedures that are possible for calibrating our image measurements. In our prior work [9] we demonstrated that computing a ratio between berries detected and the harvested fruit on one portion of data is sufficient for predicting yield on another portion of data by the applying a linear ratio to the image measurement. In practice, knowing the mean occlusion ratio of a given vineyard would be needed at the time of imaging, because it would defeat the purpose of predicting yield if it were necessary to wait for the harvest data before it were possible to measure the occlusion ratio. We propose two methods for acquiring the calibration ratio at (or before) the time of imaging, well in advance of harvest.

4.1 Calibration of Occlusion Ratio from Destructive Hand Samples at Time of Imaging

One approach takes a small number of destructive hand samples in the vineyard at the time of imaging. The vines are imaged first and then on a small sample of vines the fruit is destructively removed and weighed. The hand fruit weight is projected to harvest using the ratio between current berry weight and expected berry weight at harvest. Taking the hand estimate against the image berry count for these specific vines produces an occlusion ratio that can be estimated well in advance of harvest, and applied to predict yield of the remaining vines that were not destructively sampled.

Figure 2a shows a satellite image of the Chardonnay vineyard, highlighted with red to indicate the six vineyard rows that were imaged in our experiment. On bottom row, purple marks indicate the 15 vines in which the hand samples used for calibration. In Fig. 2b a graph shows the relationship between the hand fruit samples collected the day after imaging and the image berry counts. We derive a calibration function from this relationship and predict the crop weight based on the image berry counts of the remaining vines that were not a part of the destructive hand sample.

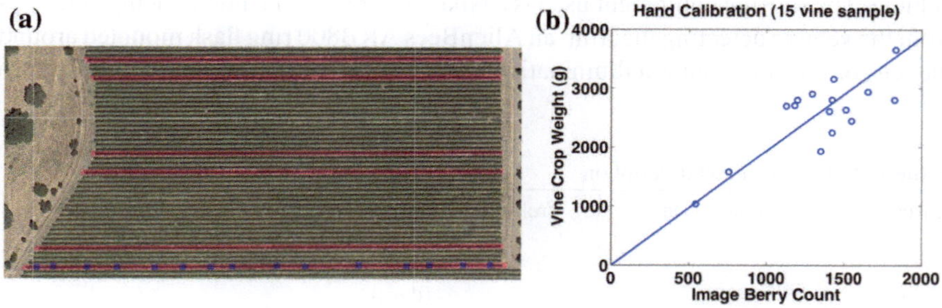

Fig. 2 Satellite image of the Chardonnay vineyard. Highlighted with red to indicate rows that were imaged by our setup and marked with purple on the bottom row to indicate where destructive hand samples were measured the day after imaging. Overall six rows were imaged totaling 665 vines, and 15 vines on the bottom row were destructively hand sampled. Graph showing calibration between the hand sample and the corresponding image berry counts. **a** Satellite image of Chardonnay vineyard. **b** Calibrating with destructive hand sample

4.2 Calibration of Occlusion Ratio from Prior Year Harvest Data

The second method we evaluate for calibrating the image measurements is to use harvest data from prior growing seasons. We have analyzed harvest data from vines trained and prepared in a similar manner from year to year and noticed consistencies. The advantage of calibrating from a prior harvest season is that hand samples are not necessary. The method simply takes a total measurement of the fruit harvested and compared against the berry count detected in the imagery.

We compute calibration functions from datasets collected in 2010 and normalize the calibration based on the mean berry weight at harvest. We then apply the calibration to the image berry counts we collected a year later in 2011 in a vineyard of a different varietal. In the following results section we compare the accuracy of the various approaches to predict harvest yield.

5 Results

5.1 Vineyard Datasets

We deployed our method on three different vine varieties and conducted experiments in which manual per-vine harvest weights were collected and used as ground truth to evaluate our automated yield measurements. The size of the experiment is significant, including 860 individual vines, totaling 2.5 km vines, including following varieties: Traminette, Riesling and Chardonnay. See Table 1 for details of the different datasets, and see Fig. 1b for an example of the equipment we use to image the vines. Equipment mounted on an aluminum frame fixed on the back tray of a Kawasaki Mule farm utility vehicle. The sensing equipment used is a Nikon D300s color camera facing sideways from the vehicle detecting the fruit, an AlienBees ARB800 ring flash mounted around the lens of the color camera illuminating the scene, a PointGrey BumbleBee2 stereo

Table 1 Vineyard dataset description

Variety	Location	Date	Trellis	Time before harvest at imaging (days)	Mean berry weight at imaging (g)	Num. vines
Traminette	Fredonia, NY	Sep. 2010	VSP	10	1.6	88
Riesling	Fredonia, NY	Sep. 2010	VSP	10	1.5	124
Chardonnay	Modesto, CA	June 2011	Semi-VSP	90	0.15	648

camera facing back down the row tracking the vehicle motion and a synchronization box generating pulses to keep the two cameras synchronized. We also collected a small dataset in controlled laboratory for studying self-occlusions, as described in the following section.

To generate a measure for each vine we use the stereo camera using a visual odometry algorithm [7] to track position along the row. We reduce drift in the odometry by detecting the stakes supporting the trellis infrastructure which are fixed at known spacings and provide an extra source of positioning. We detect the stakes by simple image processing searching for near verticle lines in the images. Finding the stakes in neighboring frames enable us to triangulate the stakes location and depth from the camera. We assume the fruiting zone lies in a single plane at the depth of the stakes. Using this assumption we can compute the overlapping content of the images and crop images to avoid double counting the fruit.

5.2 Evaluating Berry Self-Occlusion

First, we evaluate the occlusion of berries within a cluster by the outer layer of clusters (k_b) and study some approaches to potentially improve the estimate of the number of hidden berries. For this specific study, we use a controlled laboratory environment where we collected images individually of 56 grape clusters. We use ripe clusters of the Thompson Seedless variety. For each cluster we collected several images from different orientations, at a fixed distance, and collected a weight and a count of the number of berries. In the laboratory dataset we do not use our automatic detection algorithm and instead hand mark all berries visible within the images to replicate a perfect detection algorithm and remove any bias from errors in the detection algorithm (k_d and k_r). Also, in the laboratory dataset there are no biases from the vine (k_v) or from other clusters (k_c) and hence we can isolate and study the bias from self-occlusions (k_b).

Initially we compare the total berry count (gathered manually) of each cluster against its weight, Table 2. The correlation score for total berry count to weight is $r^2 = 0.95$ and mean squared error from least squares fit of 9.3 %. We consider this an upper bound for the yield predictions as the best yield prediction we could achieve is accurately knowing the berry count.

Table 2 Measurement correlation to fruit weight

Measure-type	r^2 correlation	Mean squared error (%)
Total berry count	0.95	9.3
Visible berry count (Eq. 5)	0.88	15.4
Ellipsoid model (Eq. 7)	0.85	17
Convex hull model (Eq. 6)	0.92	13.7

Next we study different image measurements starting with the visible berry count and present the results in Table 2. The visible berry count correlates with $r^2 = 0.88$ which provides a mean squared error of 15.4 %. The error is just 6 % greater than the total berry count and indicates a similar fraction of visible berries is present for small clusters as with large. The ellipsoidal model has a correlation score of $r^2 = 0.88$ and the lowest mean squared error of 17 %. Even though the ellipsoidal model attempts to predict the occluded berries behind the visible layer of berries, it correlates with a lower score than the visible berry measure. The ellipsoidal model could be less accurate because it violates one of our assumptions; the clusters do not have uniform density or the clusters are not ellipsoidal, or the model could suffer from errors in the designation of the cluster contour.

The final image measurement model we evaluate is the convex hull in Table 2. The correlation measures at $r^2 = 0.92$ which is the best of the three image measurements we study. One possible reason for the high correlation is because it encompasses the entire cluster contour, therefore it includes a measure of the partially visible berries as well as the completely visible berries, thus being more accurate than visible berry count alone. Despite finding that the contour area in the image is more accurate measure other than visible berry count, we do not yet deploy this measure outside the laboratory environment. In the datasets collected in the vineyards, several clusters are visible in each image and we have yet to develop a technique for successful segmenting one cluster from another—a requirement of the ellipsoid and convex hull models. Hence, at present we have only been able to demonstrate precise detection of individual berries, regardless of which cluster they belong, and therefore in the following vineyard results we consider just the visible berry count that our algorithm provides. However, we have indicated with these laboratory tests an avenue for future improvement of our fielded system.

5.3 Biases in Visual Measurement

Here we take data both collected in the vineyard and laboratory we attempt to segregate the biases involved in the visual detection process. Table 3 presents a study of the different visual estimation biases measured in the various datasets. The laboratory dataset was used to estimate the self-occlusion parameter k_s by manually counting visible berries in the images and hand counting the total number of berries physically pulling apart the clusters. The dataset showed that 46 % of the berries are visible to the camera.

The visual detection bias k_d was estimated by manually assessing false positive and false negative detections of berries in images with the marked output of the detection algorithm. The false positive and false negatives combined with the true berry detections gave us the parameter k_d. The detection algorithm is biased towards under-counting the berries which was discovered in our prior work Nuske et. al [9] and here we see between 25 % and 35 % fewer berries than are visible are reported by the algorithm. The algorithm detected about 10 % fewer of the visible berries in

Table 3 Biases in visual yield estimation (Sect. 3). Bias parameters are unit-less and standard deviations where applicable are presented as percentages in brackets. Missing entries for the self-occlusion parameter (k_s) are due to data not collected in a controlled environment in the vineyard datasets, the other missing entries are for the biases not applicable to the laboratory dataset

Dataset name	Mean berry weight at harvest (g)	k_s	k_d	k_r	$k_v * k_c$
Riesling	1.5	-	0.74 (13.4 %)	1.0	0.29
Traminette	1.6	-	0.75 (10.7 %)	1.0	0.24
Chardonnay	0.9	-	0.65 (12.3 %)	1.03 (8.7 %)	0.31
Laboratory	-	0.46 (22 %)	-	-	-

the Chardonnay 2011 dataset which captured around 90 days from harvest when the berries were much smaller. Similarly we take images marked by the detection and image alignment algorithm and manually assess double-counting and mis-counting berries between overlapping images. In the 2011 Chardonnay dataset the automatic aligning algorithm was biased to slightly over-count berries by 3 %. In the 2010 datasets (Riesling and Traminette) we did not deploy our automatic alignment algorithm, and instead manually cropped images to have zero overlap and hence we list a bias of 1.0 here.

Finally, the bias from vine and cluster occlusions we have not experimentally determined, but we combine the other bias factors, assume a similar self-occlusion rate to the laboratory, use the mean berry weight at harvest and deduce the combined effect of the terms k_v and k_c indirectly. We see here that there are significant occlusions from the vine leaves, shoots and clusters occluding clusters with between two thirds and three quarters of the fruit being occluded by these factors.

5.4 Yield Prediction Results

Now we present yield prediction results by applying our calibration approaches described in Sect. 4, including calibration from destructive hand samples at the time of imaging and also calibration from prior harvest data and compare average error for the prediction of the individual vine weights of the Chardonnay dataset. After discovering in the previous section that the visible berry count has the most accurate correlation to yield in the vineyard datasets we use this as our image measure for the following results. We take the Traminette and Riesling datasets, collected in 2010, compute a calibration between image measure and yield, and apply to the Chardonnay dataset from 2011. Figure 3a shows a comparison between the data collected in the two vineyards after normalizing for respective berry weights. The graph shows that

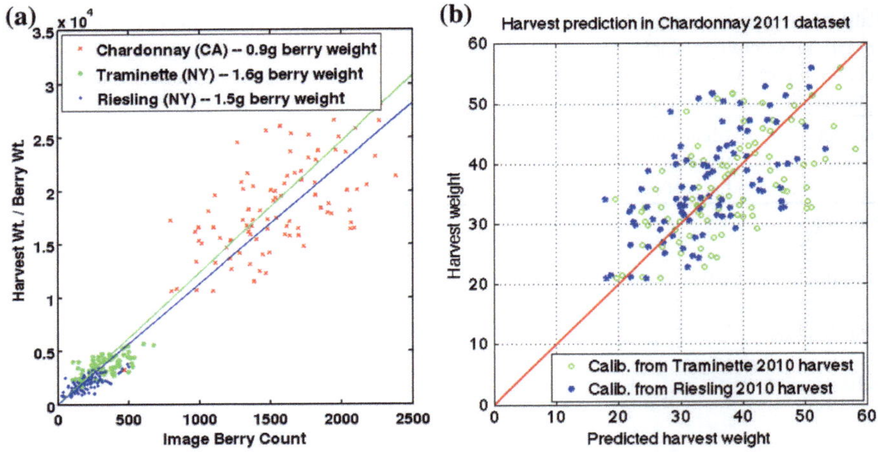

Fig. 3 Graphs demonstrating calibration from prior harvest data. After normalizing for berry weight there is a trend between the 2010 and 2011 datasets. The calibration functions computed from the 2010 datasets applied to the Chardonnay image berry counts produces harvest predictions 12 weeks prior to harvest. **a** Comparison of Traminette and Chardonnay datasets—normalized for berry weight. **b** Predicting Chardonnay 2011 yield from 2010 calibration (lbs)

the Traminette and Riesling vines despite holding much less fruit, do have a trend between the image berry counts in the Chardonnay data. We apply the Traminette and Riesling calibration to the Chardonnay data and show the predicted weight in Fig. 3b. This result demonstrates the prediction of harvest yield 12 weeks out from harvest. We also evaluate calibrating using hand samples collected in the vineyard at the time of imaging, see Fig. 2b.

The results of the calibration on the Chardonnay dataset are presented in Fig. 4a. The errors are between 17 and 19 % for the different calibration approaches, with the calibration from destructive hand samples slightly more accurate. We see that some of the error averages out when comparing yield of entire rows, where the error is now between 7 % and 8 %. In Fig. 4b we present the error for prediction of the entire yield of the vines in the dataset. The hand calibration was most accurate at 3 % error and using calibration from the 2010 Traminette dataset had 4 % error. We see an under-prediction of overall weight by 4.5 % using the 2010 Riesling dataset calibration. It is apparent that despite average absolute per-vine errors of around 18 % for all approaches, the overall error is below 5 % indicating the individual vine errors are well distributed and average out. For comparison in Fig. 4b we present the estimate taken by extrapolating the hand samples alone, which is the traditional industry practice, and was found to be the least accurate estimate with −13 % error.

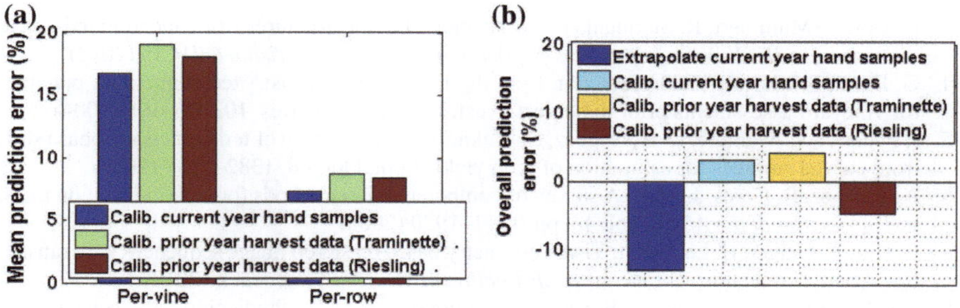

Fig. 4 Results on the prediction of harvest yield in the Chardonnay dataset. We evaluate the prediction accuracy when calibrating using a destructive hand sample at the time of imaging compared to calibrating using prior harvest data. We present two statistics in (**a**); the average absolute error computed for individual vine predictions and the average absolute error for the estimate of row weights. In (**b**) we present the error for prediction of the entire yield of the vines in the dataset. The calibration from hand samples is slightly more accurate than the calibration from prior harvest data. For comparison we present the estimate taken by extrapolating the hand samples alone, with is the least accurate estimate with 13 % error. **a** Vine and Row Mean Error. **b** Overall Prediction Error

6 Conclusion and Future Work

We have demonstrated a method to automatically generate non-destructive, high-resolution, yield predictions vineyards and shown practical approaches to calibrate the measurements. We evaluate the system at both 10 days from harvest and 90 days, finding similar results, with the algorithm detecting only slightly less of the visible berries earlier in the season. Of the two methods of calibrating image measurements—using destructive hand samples at the time of imaging was slightly more accurate at giving predictions within 3 % of harvest yield. Nevertheless calibration from prior harvest data from a different varietal, albeit grown in a similar vine trellis and training structure, gave only 4 % error.

In future work we will look to develop an approach to count grape clusters early in the season, even before berries have formed, to give vineyard managers information with maximum time before harvest to make the necessary adjustments to their vines. We also look at extending the experimentation to more varietals and trellis structures.

Acknowledgments Work funded by the National Grape and Wine Initiative, (info@NGWI.org). Narasimhan was supported partially by NSF awards IIS-0964562 and CAREER IIS-0643628 and an ONR grant N00014-11-1-0295.

References

1. P. Blom, J. Tarara, Trellis tension monitoring improves yield estimation in vineyards. HortScience **44**, 678–685 (2009)
2. P. Clingeleffer, G. Dunn, M. Krstic, S. Martin, Crop development, crop estimation and crop control to secure quality and production of major wine grape varieties: A national approach. Grape and Wine Research and Development Corporation, Australia, Technical report, 2001

3. D. Dey, L. Mummert, R. Sukthankar, Classication of plant structures from uncalibrated image sequences, in *IEEE Workshop on the Applications of Computer Vision (WACV)* (2012)
4. G. Dunn, S. Martin, Yield prediction from digital image analysis:A technique with potential for vineyard assessments prior to harvest. Aust. J. Grape Wine Res. **10**, 196–198 (2004)
5. J. Federici, R. Wample, D. Rodriguez, S. Mukherjee, Application of terahertz gouy phase shift from curved surfaces for estimation of crop yield. Appl. Opt. **48**, 1382–1388 (2009)
6. A. Jimenez, R. Ceres, J. Pons, A survey of computer vision methods for locating fruit on trees, in *Transaction of the ASAE*, vol. 43, pp. 1911–1920 (2000)
7. B. Kitt, A. Geiger, H. Lategahn, Visual odometry based on stereo image sequences with ransac-based outlier rejection scheme, in *IEEE Intelligent Vehicles Symposium* (2010)
8. J.A. Martinez-Casasnovas, X. Bordes, Viticultura de precisiŏn: Predicciŏn decosecha a partir de variables del cultivo e índices de vegetaciŏn. Revista de Teledetecciŏn **24**, 67–71 (2005)
9. S. Nuske, S. Achar, T. Bates, S. Narasimhan, S. Singh, Yield estimation in vineyards by visual grape detection, in *Proceedings of the 2011 IEEE/RSJ International Conference on Intelligent Robots and Systems* (2011)
10. S. Singh, M. Bergerman, J. Cannons, B. Grocholsky, B. Hamner, G. Holguin, L. Hull, V. Jones, G. Kantor, H. Koselka, G. Li, J. Owen, J. Park, W. Shi, J. Teza, Comprehensive automation for specialty crops: Year 1 results and lessons learned. J. Intell. Serv. Robot. Spec. Issue Agric. Robot. **3**(4), 245–262 (2010)
11. M. Swanson, C. Dima, A. Stentz, A multi-modal system for yield prediction in citrus trees, in *ASABE Annual International Meeting*, Pittsburgh, PA (2010)
12. J.A. Wolpert, E.P. Vilas, Estimating vineyard yields: Introduction to a simple, two-step method. Am. J. Enol. Viticulture **43**, 384–388 (1992)

Forest 3D Mapping and Tree Sizes Measurement for Forest Management Based on Sensing Technology for Mobile Robots

Takashi Tsubouchi, Asuka Asano, Toshihiko Mochizuki, Shuhei Kondou, Keiko Shiozawa, Mitsuhiro Matsumoto, Shuhei Tomimura, Shuichi Nakanishi, Akiko Mochizuki, Yukihiro Chiba, Kouji Sasaki and Toru Hayami

Abstract This research work is aimed at application of sensing and mapping technologies that have been developed in mobile robotics, so as to perform equipment measurements of forest trees. This research work utilizes a small sized laser scanner and SLAM (Simultaneous Localization and Mapping) technology for the problem of performing forest mensurements. One of the key pieces of information required for forest management, especially in artificial forests, is accurate records of the tree sizes and the standing timber volume per unit area. The authors have made measurement equipment fore a pre-production trial which consists of small sized laser range scanners with a rotating (scanning) mechanism of them. SLAM and related technologies are applied for the information extraction. In the development of SLAM algorithm for this application, the sparseness of the standing trees and the inclination of the forest floor are considered. After performing the SLAM and obtaining a map based on the data from several measurement points, we can obtain useful information including a map of the standing trees, the diameter at chest height of every tree, and the height at crown base (length of the clear bole). The authors will present the experimental results from the forest including the map and the measured tree sizes.

T. Tsubouchi (✉) · A. Asano
University of Tsukuba, Tsukuba, Japan
e-mail: tsubo@roboken.esys.tsukuba.ac.jp

T. Mochizuki · S. Kondou · K. Shiozawa · K. Sasaki
Adin Research Inc., Tokyo, Japan

M. Matsumoto
Kurume National College of Technology, Kurume, Japan

S. Tomimura · S. Nakanishi · A. Mochizuki · T. Hayami
Forest Revitalization System Corporation, Tokyo, Japan

Y. Chiba
Forestry and Forest Products Research Institute, Tsukuba, Japan

K. Yoshida and S. Tadokoro (eds.), *Field and Service Robotics*,
Springer Tracts in Advanced Robotics 92, DOI: 10.1007/978-3-642-40686-7_24,

1 Introduction

The motivation of this research work arises from the need for a labor-saving alternative that can perform the in forest measurements required for proper forest management. The present authors have experience performing research work in other areas of mobile robotics, and are now applying similar sensing and mapping techniques and technologies into the problem of measurements. Although the field work of the measurements are laborious, manual measurement is still undergoing. The standard method employed for forest survey is the belt transect method (Fig. 1) [1]: e.g., which is 30 m long and 10 m wide with a base line running up and down slopes in a survey plot. Within the belt transect, the distance between every tree and the base line is used to determine the trees' location. Its diameter at the chest height and its height at crown base (length of clear bole) are also measured. Currently, these measurements must be performed manually with measuring tapes by a three person party. The measured parameters are once recorded in a field notes by one person, while the other two people measure the survey items. It takes at least 30 min for a unit area to be measured. After taking measurement, the team inputs the data in an information processing facility.

These data that are collected are processed to obtain the standing timber volume and the growth rate of forest. Such human-intensive manual measurements for forest survey are necessary and common in forest management throughout the world [2]. Even though a remote sensing approach, such as airborne measurement with laser profiler, has been being utilized successfully in some specific cases [3–7], it is still necessary to establish models of correspondence between tree measurement from the ground and measurements from the air [8].

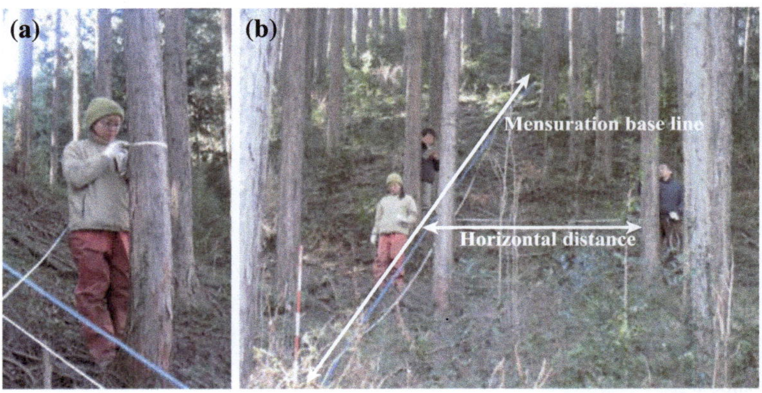

Fig. 1 Manual measurement: **a** Measure the diameter of each tree within a sampling unit area using a tape measure. **b** Measure the horizontal distance from the base linfe and its location on the base line. Three people cooperate to perform the tree location measurements

The main problem with performing measurement manually is that the task is so time-consuming that it is intractable to cover all the targeted forests in Japan. The Japanese government, as a country engaged in the Montreal Process that aims at sustainable forest management, has set the forest monitoring points at 15,000 for the whole of Japan, it is said that the number of the monitoring points is still not enough to obtain sufficient data. Increasing the number of the point is desired, but it is not often possible in reality because of insufficiency of budgets to employ enough workers to perform the required measurements.

The authors consider that replacing such manual measurements with equipment measurements is a good idea to improve the efficiency of the forest measurements. Although an airborne approach (Nelson et al. [3] and enclosed references) is one efficient method of equipment measurement, it can only provide a macroscopic view that will not provide the required detailed observation. Yao et al. [9] already pointed out such a problem and their institution - CSIRO proposes several ideas to gather the forestry information on the ground in the forest. In Yao et al. [9], measurement results by using an Echidna lider are presented. They placed the lider at 5 points in a 100m by 100m area and took point clouds as the primary measurement data. There seems to be differing overlaps of the point clouds depending upon the measurement locations. There are several other reports where a laser profiler is used in a forest (for example Ref. [8]). However, many available profilers have in the past been expensive and heavy. Such expensive and heavy infrastructure will thus not solve the budgetary problem, and will not be portable enough for the required measurement task.

On the other hand, in field robotics, it becomes popular to scan the environment using a small-sized light-weight laser scanner on a mobile robot. Such a laser scanner is less expensive than the laser profiler. Furthermore, some appropriate SLAM (Simultaneous Localization and Mapping) technologies which are detailed in Thrun et al. [10] can be applied to estimate measured locations, to build a map, and to assemble point clouds that are obtained by the laser scanner at several measured locations. The authors considered that such sensing and mapping technologies for mobile robots can be applied to the forest measurement problem. When we scan the forest using the small laser scanner and obtain point clouds in many locations, the assembled point clouds assembled by SLAM will provide shape and arrangement of the trees, and will yield data of tree sizes and standing timber volume.

The authors propose the desired equipment as a rotating laser scanner at the top of a pole. A person can easily carry this equipment in the forest. Once the person stands the pole on the forest floor, they can collect a three dimensional scan there. The person will repeat such a scanning technique at several locations to gather multiple three dimensional point clouds within a sampling unit area in the forest. Since some portions of the point clouds at different standing points will overlap, an ICP scan matching algorithm such as in [11–13] for the registration of overlapped portions can be applied to align all the obtained data to produce a 3D map of the forest. In this paper, the authors propose a step by step matching algorism that takes the sparseness of the standing trees into account before the application of ICP. The proposed scan matching algorithm also takes the steepness of the slope of the forest floor into account. Following the proposed idea, the authors built a scanner device

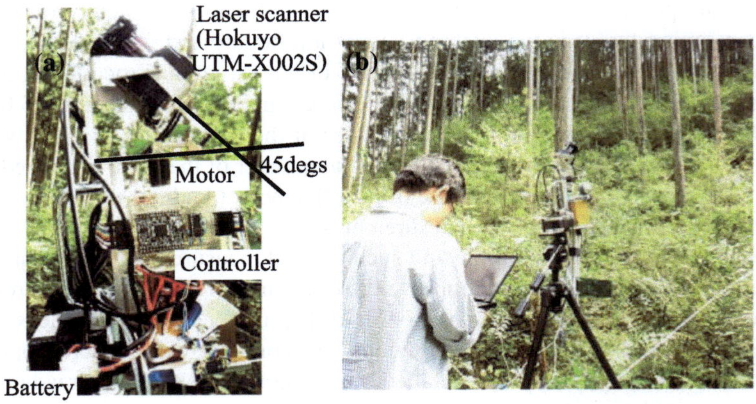

Fig. 2 A scanner device is illustrated in **a** An axle of the motor stands vertically followed by a special crank. An axis coincident with laser rotation axle of the scanner will rotate in precession by the rotation of the motor and crank mechanism. **b** The device is deployed for the field trial measurements

and performed experiments in the forest. This paper reports the result of the 3D mapping from the obtained data of the rotating laser scanner at several points in the forest and the results of diameter measurement of trees in the map.

2 A Scanner and Measurement in the Forest

Figure 2 illustrates the scanner device. A small sized Hokuyo laser scanner UTM-X002S whose planer view angle is 270° (0.6° per steps) is employed. The scanner is placed on a stage driven by a motor and a crank (45°) where the authors made use of a mechanism proposed by [14]. This mechanism performs precession rotation along the center axis which is coincident with the laser rotation axle so that the scanning plane is swayed. In the experiments, the period of stage swaying is 2 seconds. During one period of the swaying, 200 scans of the laser scanner are performed. One scan consists of 440 points of the range data. However, only the 400 points of the range data that corresponds to $\pm 120^\circ$ from the center of the scanner are used. A coordinate system is defined on the scanner device such that origin is at the base of the device, the x axis os toward the front, the z axis coincides with the motor axle and the y axis forms the right hand coordinate system. The authors placed the scanner device whose front, or x axis, looks up 15° from the horizontal level by adjusting the lengths of the legs of the tripod. Therefore, the $x - y$ plane of the scanner device also has an angle of 15° from the ground.

Figure 3 illustrates the location where the scans of the environment are taken and obtained point clouds of surface of the trunks in the forest. In the experiment of this

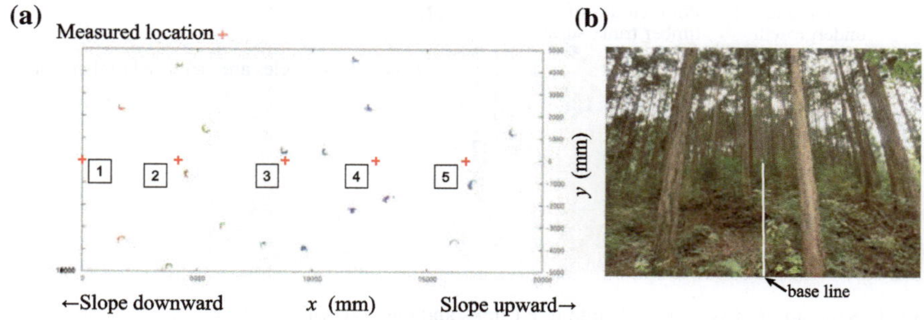

Fig. 3 Measured environment and locations in the sampled slope. Measured locations are numbered from 1 to 5

paper, the authors set the baseline in the direction of the steepest slope at the sampled area as shown in Fig. 3b. The steepness of the slope is 30° on average there. The authors set the device from the lowest to the highest location in approximately at 5 m interval along the base line of the slope. The locations where the device is placed are illustrated with '+' and numbered from 1 to 5 in Fig. 3a. The front of the device is facing in the direction of the base line (Fig. 2b).

3 ICP Considering the Slope and Sparseness of Standing Trees

For the purpose of so called scan matching, ICP [11–13] is a popular method for aligning reference point clouds with current scans which are obtained from two or more locations. After scanning the surroundings by a laser scanner, the authors applied the ICP to the point clouds in planer forest and obtained enough performance for the scan matching [15]. However, the same technique is applied in a forest with a slopped floor, the matching performance in vertical direction deteriorates because trees are still standing vertically and have less features on their trunks. Separation of point clouds into tree trunks and undergrowth is also necessary. Therefore, the scan matching process is designed and proposed as follows:

1. extraction of timbers from the point clouds in a current scan and a reference scan,
2. determine the correspondence between the timbers in the current scan and the reference scan,
3. extraction of the slope ground level, and
4. application of ICP for the scans.

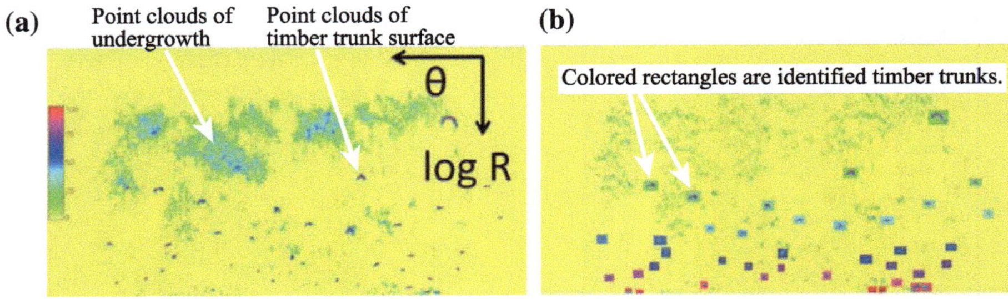

Fig. 4 Separation of point clouds into tree trunks and undergrowth

3.1 Extraction of Timbers

First a point cloud density histogram is obtained from a set of scan data of one location by projecting onto the $x - y$ plane of the scanner device. However, the points are translated onto the $\log R, \theta$ plane, where $R = \sqrt{x^2 + y^2}$, $\theta = atan2(x, y)$ and (x, y) is at the existing cloud point that z ignored. Figure 4a illustrates an example histogram, where color presents the frequency of the cloud points—green presents low frequency, blue mid and red high. When we see the histogram as a density function of $F(\log R(\theta), \theta)$, high frequency is observed at specific θ for the part of undergrowth. On the other hand, it is not observed at specific θ for the part of tree trunk surface. Therefore, a filter is applied to reduce the density if high frequency is observed at specific θ. An example result of the filtering is illustrated in Fig. 4b, then, the tree trunk surfaces are extracted and identified.

A least square circle fitting is applied to each identified surface of the trunk in the $x - y$ coordinate and center location of the timber is obtained. The plot of such points of the timber center on the $x - y$ plane is named "timber center map". The timber center map could be seen as constellations which are used for the next step of determining the correspondence between the constellation of the reference and the current scans.

3.2 Determine the Correspondence Between Trees

When applying ICP, prior knowledge about the correspondence between the same objects in the reference and current scan is helpful to avoid unexpected convergence because of the local minima problem. For this purpose, the correspondence between the trees in the tree trunk centroid map of the reference and current scans must be established [16]. This process has similarities to matching the constellations in two star charts.

The algorithm to make correspondence among the timbers is summarized as follows (Fig. 5) :

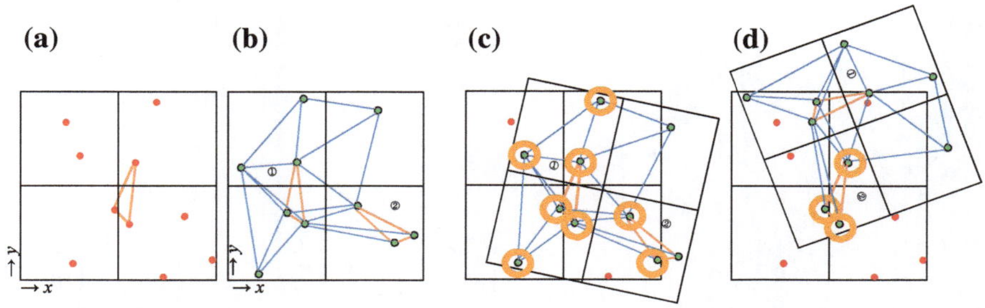

Fig. 5 Determining correspondence between the tree trunk centers in the tree trunk centroid map of the reference and the current scans **a** Timber locations of the reference scan **b** Timber locations of the current scan **c** Sccess match of the two scans **d** Failed match of the two scans

1. Take three points of the timber center randomly in the tree trunk centroid map of the reference scan and form a triangle.
2. Take three points of the trunk center randomly in the tree trunk centroid map of the current scan and form a triangle.
3. Check whether the two triangles are congruent.
4. If the two are congruent, translate and rotate the current map so as to quadrate the triangle in the reference scan with the one in the current scan. Calculate the sum of squared distance between every center point of a trunk in the current scan and a point in the reference scan within 100 mm of the point of the current scan. Count the number of 'paired' points to calculate the sum.
5. Iterate 1. to 4. to find smallest sum and larger the number of the paired points in 4.
6. When found in 4., translation and rotation parameters are obtained and the pairs are recognized as corresponding trunks in the reference and the current scans.
7. Use the translation and rotation parameters obtained in 6. as the initial and find most likelihood parameters for them by means of least square method to minimize the sum of distance between paired corresponding timbers in the tree trunk centroid maps of the reference and the current scans.

The sparseness of the distribution of the timber centers is considered in the process proposed here.

3.3 Extraction of the Forest Ground Level Slope

For the purposes to obtaining good convergence in the z direction, the ground level slope is extracted from the point clouds both in the reference and current scans. The algorithm is summarized as follows:

1. Define a grid in the scanner device $x - y$ plane.

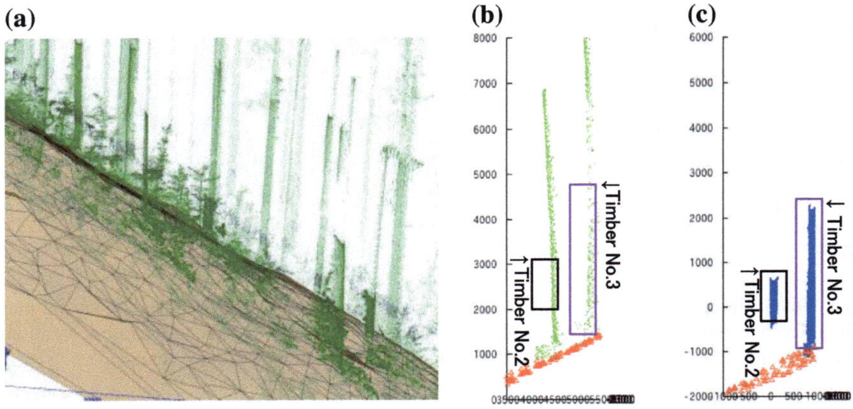

Fig. 6 Extraction of the slope ground level **a** Extracted slope ground level after Delaunay triangulation **b** Timber point clouds measured from the location 1 **c** Timber point clouds measured from the location 2

2. Obtain a cloud point such that z coordinate value is the smallest within every square grid lattice, which is a candidate for the ground and not the undergrowth vegetation.
3. Associate all the points obtained in 2 and construct a surface by means of Delaunay triangulation.

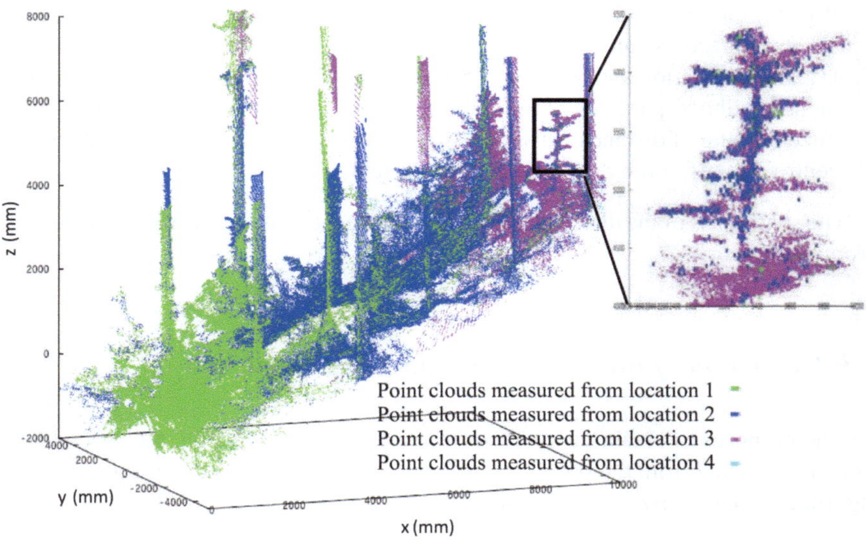

Fig. 7 Example result of ICP scan matching among the 4 locations illustrated in Fig. 3a

Figure 6a illustrates the extracted ground level slope. After the ground level slope is determined, the part of point clouds on the surface of each of the corresponding tree trunks in the different measurement locations is extracted at the same height (Fig. 6b, c). These point clouds are subject to the ICP scan matching technique presented in the next section.

3.4 Application of ICP

For this application the steepest descent ICP algorithm is applied. Translation and rotation parameters that are obtained at the step 6 in Sect. 3.2 are used for initial

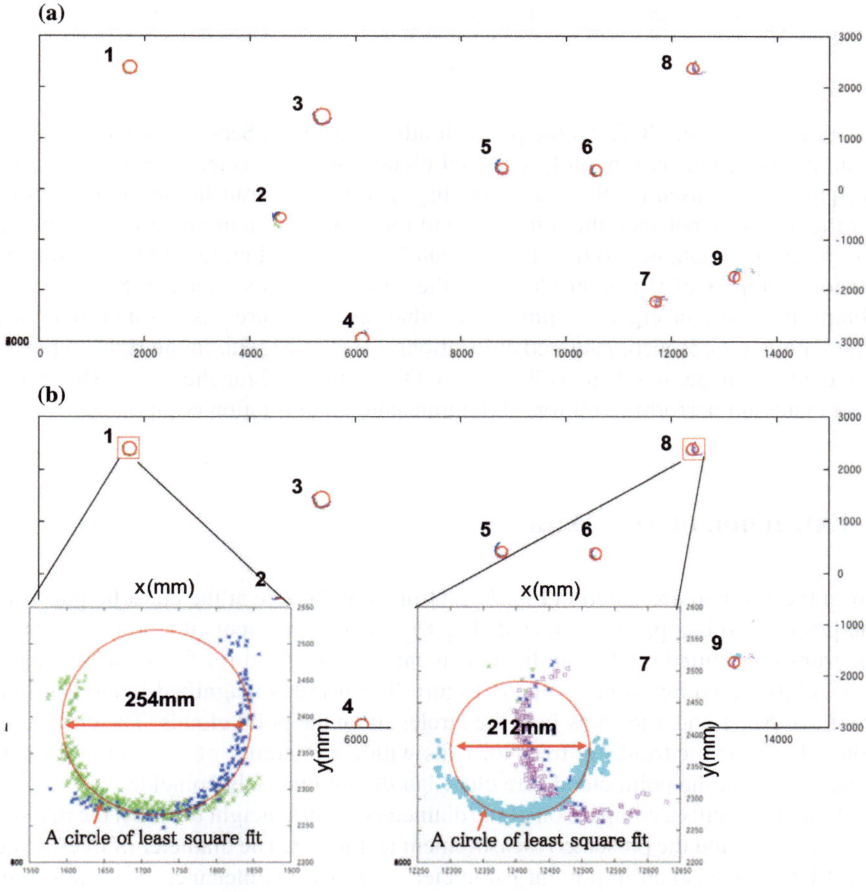

Fig. 8 Example result of ICP scan matching among the 4 locations illustrated in Fig. 3a

Table 1 Comparison between manual measurement and proposed equipment measurement for diameter at the chest height

Timber No.	Diameters at the breast height					
	Manual	Proposed	Errors		Standard deviation of the residue of fitted circle	The number of cloud points
	(cm)	(cm)	(cm)	(%)	(cm)	
1	25.8	25.4	−0.4	1.6	0.94	432
2	22.3	18.2	−4.1	18.4	0.78	10750
3	30.1	30.6	0.5	1.7	0.93	2113
4	23	23.8	0.8	3.5	0.99	369
5	24.5	21.2	−3.3	13.5	0.85	11452
6	21.5	22	0.5	2.3	0.82	1257
7	24.6	21.2	−3.4	13.8	1.52	523
8	21.3	21.2	−0.1	0.5	1.54	737
9	21	20.2	−0.8	3.8	0.99	1492

location to begin the ICP. All the point clouds extracted in Sect. 3.3 in the reference scan are used, but 60 randomly sampled cloud points extracted in Sect. 3.3 in the current scan are used for the scan matching. The x and y translations and rotations for the tree axes between the reference and the current scan are obtained. Then the z translation is obtained to match the ground level slope. Iterate 1000 times for the random samples of the point clouds of the current scan extracted in Sect. 3.3 and obtain the best converged or minimal residue result. Figure 7 is 3D map obtained where point clouds were gathered at locations 1 to 4 in Fig. 3a. In the right of Fig. 7, a magnified image of a shrub is illustrated. Observing the branches of the shrub, the scan matching performs well for all the translation and rotation components.

4 Extraction of Tree Data

From the 3D map presented in Fig. 7, the trunks of the tree at the chest height from the ground level slope are extracted (Fig. 8). Figure 8a presents the point clouds of the trunks and fitted circles on them. This map can be used for forest maintenance to yield the standing timber volume. Figure 8b illustrates magnified images of tree trunks of No. 1 and 8 to show how the circles fit on the point clouds. The circle and point clouds on the trunks of tree No. 1 fits within a tolerance of 5 cm. On the other hand for No. 8, the point clouds are close but do not precisely coincide.

Table 1 presents a comparison of the diameters at chest height between the manual measurement and the presented measurement technique. The diameter of trees at the chest height is also an important parameter for forest maintenance. Six trunks out of nine have error under 2 cm, while the other two have an error for the diameter in the order of 3 cm and the remaining has 4.1 cm error. It is said that the desired performance for the error must not be more than 2 cm. Given this, the proposed equipment measurement achieves the required level of the performance. However, more improvement to reduce the errors is desirable.

5 Conclusions

The authors presented the necessity for equipment-based measurement for the forest measurement and pointed out that it is a good application field for sensing and SLAM technology for mobile robotics. An experimental setup using a for mechanically rotating small-sized laser scanner has been prepared. By operating the setup in the forest, point clouds are obtained at several locations. The authors proposed a step by step scan matching algorithm utilizing ICP taking into account both the existence of the slope and the sparseness of the tree trunks. Such equipment measurement in the forest has been shown to be effective. However more improvements are necessary so as to obtain more accurate data. These improvements are set as future work.

Acknowledgments This study was partly funded by the national program on Research and Development Projects for Application in Promoting New Policy of Agriculture, Forestry and Fisheries provided by the Ministry of Agriculture, Forestry, and Fisheries of Japan.

References

1. J.T. Holloway, W.J. Wendelken, Some unusual problems in sample plot design. New Zealand J. Forest. **7**(4), 77–83 (1957)
2. R.H. Whittaker, F.H. Bormann, G.E. Likens, T.G. Siccama, The hubbard brook ecosystem study: forest biomass and production. Ecol. Monogr. **44**, 233–254 (1974)
3. S. Nelson, W. Krabill, J. Tonelli, Estimating forest biomass and volume using airbone laser data. Remote Sens. Environ. **24**, 247–267 (1988)
4. E. Naesset, Determination of mean tree height of forest stands using airborne laser scanner data. ISPRS J. Photogramm. Remote Sens. **52**, 49–56 (1997)
5. M. Nilsson, Estimation of tree heights and stand volume using an airborne lider system. Remote Sens. Environ. **56**, 1–7 (1996)
6. E. Naesset, T. Gobakken, Estimating forest growth using canopy metrics derived from airborne laser scanner data. Remote Sens. Environ. **96**, 453–465 (2005)
7. A. Kato, L.M. Moskal, P. Schiess, M.E. Swanson, D. Calhoun, W. Stuetzle, Capturing tree crown formation through implicit surface reconstruction using airborne lidar data. Remote Sens. Environ. **113**, 1148–1162 (2009)
8. R. Saito, T. Tanaka, H. Hara, H. Oguma, T. Takamura, H. Kuze, T. Yokota, Aircraft and ground-based observations of boundary layer CO_2 concentration in anticyclonic synoptic condition. Geophys. Res. Lett. **36**, L07807 (2009)
9. T. Yao, X. Yang, F. Zhao et al., Measuring forest structure and biomass in new England forest using echidna ground-based lidar. Remote Sens. Environ. **115**(11), 2965–2974 doi:10.1016/j.rse2010.03.019 (2011)
10. S. Thrun, W. Burgard, D. Fox, *Probab. Robot.* (MIT Press, Cambridge, 2005)
11. Y. Chen, G. Medioni, Object modeling by registration of multiple range images, in *Proceedings of IEEE International Conference on Robotics and Automation*, pp. 2724–2729 (1991)

12. P.J. Besl, N.D. McKay, A method for registration of 3-D shapes. IEEE Trans. PAMI **14**(2), 239–256 (1992)
13. S. Rusinkiewicz, M. Levoy, Efficient variants of the ICP algorithm, in *Proceedings of the 3rd 3-D Diginal Imaging and Modeling 2001*, pp. 145–152 (2001)
14. M. Matsumoto, T. Yoshida, T. Mori, S. Yuta, 3D SOKUIKI sensor module by roundly swinging mechanism and SCIP-3D command system. Trans. Japan Soc. Mech. Eng. C **75**(760), 3314–3323 (2009). (in Japanese)
15. S. Kondo, K. Shiozawa, T. Mochizuki et al., Three-dimensional map building for forest structure analysis applying ICP, in *Proceedings of the 28th Annual Conference of the Robotics Society of Japan, 3I3-8* (in Japanese) (2010)
16. S. Kondo, K. Shiozawa, T. Tsubouchi et al., Three-dimensional map building for forest structure analysis using SOKUIKI sensor, in *Proceedings of the 2010 JSME Conference on Robotics and, Mechatronics, 1A1-D16* (in Japanese) (2010)

Iterative Autonomous Excavation

Guilherme J. Maeda, David C. Rye and Surya P. N. Singh

Abstract This paper introduces a Cartesian impedance control framework in which reaction forces exceeding control authority directly reshape bucket motion during successive excavation passes. This novel approach to excavation results in an iterative process that does not require explicit prediction of terrain forces. This is in contrast to most excavation control approaches that are based on the generation, tracking and re-planning of single-pass tasks where the performance is limited by the accuracy of the prediction. In this view, a final trench profile is achieved iteratively, provided that the forces generated by the excavator are capable of removing some minimum amount of soil, maintaining convergence towards the goal. Field experiments show that a disturbance compensated controller is able to maintain convergence, and that a 2-DOF feedforward controller based on free motion inverse dynamics may not converge due to limited feedback gains.

1 Introduction

Autonomous excavation has the potential to improve the quality and throughput in a variety of field domains. However, it also represents a challenging low-level control problem. Autonomous excavation control attempts date back more than twenty years with very few successful and realistic systems implemented so far. Despite a

S. P. N. Singh is now at the The University of Queensland.

G. J. Maeda (✉) · D. C. Rye · S. P. N. Singh
Australian Centre for Field Robotics, The University of Sydney, Sydney, Australia
e-mail: g.maeda@acfr.usyd.edu.au

D. C. Rye
e-mail: d.rye@acfr.usyd.edu.au

S. P. N. Singh
e-mail: spns@uq.edu.au

K. Yoshida and S. Tadokoro (eds.), *Field and Service Robotics*, Springer Tracts in Advanced Robotics 92, DOI: 10.1007/978-3-642-40686-7_25,

significant history, direct force control remains elusive due to compliance (of both the hydraulic actuation and terrain), coupling, and limited observability of ground reactions. These factors, while complex, are structured (they are not chaotic). Given that the task can be viewed as a multiple-query, successive operation towards a desired profile, an iterative and adaptive control approach is advocated in which the disturbance and estimated reactions are differentially used to reshape (bucket) actuation for subsequent (digging) processes.

The dominant problem in excavation control is that the reaction forces generated through interaction with the environment are difficult to predict, and may equal the force capability of the machine. In the literature, proposed solutions to this problem fall into two broad categories: explicit modelling and reactive strategies.

Explicit modelling. Soil-tool interaction modelling allows for force prediction. In excavation control, prediction is useful for the generation of digging strategies (feasible and optimal scoop trajectories), and anticipative excavation force compensation (computation of feedforward commands that accounts for soil-tool interaction forces). A flat blade is the basic geometry studied in soil mechanics from which the majority of explicit models are derived. A widely accepted model based on flat blade assumptions is known as the fundamental equation of earthmoving (FEE)[19]. Experimental results have shown that flat blade models are helpful in assisting machine design [7] and equipment selection [6]. In regards to excavation, in [23] the author shows that the FEE predicts well when the bucket is not full, however prediction deteriorates as the bucket fills up. In [15] the author adapts the FEE for the excavation case at the cost of global and local optimisation methods for fitting model parameters. The work in [4] is notable for providing a comparison between an analytical and a regression methods and to effectively use their outputs for generation and selection of candidate trajectories.

Beyond the flat blade a variety of 3D models for the excavator bucket address the presence of side walls and surcharge (a review is found in [2]), however those models lack experimental validation even for the simple purpose of force prediction. A step further would require experimental validation of 3D models in terms of low-level control improvements. Alternative methods for modelling include energy methods [24] and exhaustive laboratory experiments [14].

Reactive strategies. In this category control strategies do not make use of model prediction, but instead reactive strategies are used to adjust control actions according to some variable of interest. Since experiments are necessary for the tunning of parameters the literature in reactive approaches is rich in field trials. In general, reactive excavation cannot be achieved by pure position tracking since the commands under feedback can either saturate actuators or generate excessive structural stress. Because no prediction is available, the underlying behaviour (despite different strategies) is that of generating some form of accommodation as reactive forces build. This could be in the form of sensor based active compliance [17, 20] where the forces or trajectories are continuously adjusted. A simple, but experimentally validated, strategy is to simply slow down and decrease the depth of the desired path according to the load conditions of the drives [5]. Artificial intelligence methods have been applied to encode and blend expert operator reactions and other empirical rules [3, 22] in an attempt to address the problem of removing or contouring in the unpredictable presence of large rocks that can constrain the motion. Robust methods [9] have also been applied in excavation, however since the execution is a based on tracking of force or position, the generation of a reference without an explicit model requires restrictive assumptions on terrain forces, usually in the form of an impedance model.

This paper proposes a different solution for the excavation problem. The solution is based on a reactive approach in order to avoid the difficulties imposed by predictive methods; mainly, parameter and structure adaptation, observability, and terrain profile estimation. The solution explores the use of the undesirable compliance of the arm and iteration. Here, "iteration" means making multiple passes with the bucket, where each pass comes closer (iterates) to the desired profile. In principle this approach is orthogonal to the usual idealisation of excavation, where both compliance and iteration are undesirable. The ideal controller would be stiff enough to overcome any reactions, finishing any dig in a single pass. Both compliance and iteration are, however, intrinsic to excavation and thus addressing them is fundamental since:

- Iteration is required because the finite volume capacity of the bucket is usually much smaller than the amount of material to be removed (final profile shown as "target" in Fig. 1). Also, due to the finite force and power that the excavator can apply on the environment the bucket tends to undershoot the desired path, requiring at least one subsequent clean-up pass.
- Compliance in excavation is caused by a lack of control authority. It becomes apparent when forces generated by the controller are lower than the forces required to cut the soil, resulting in position and velocity deviations. Those deviations

Fig. 1 Excavating (**a**) the face of a mine or (**b**) a trench for piping are iterative processes where there is a desired profile to be achieved. The number of scoops and their paths, however, depend on the interaction forces between machine and soil which are difficult to model and to predict. Illustration (**a**) reproduced with permission from P&H. Extracted from: P&H MinePro Services, Peak Performance Practices Excavator Selection, 2006. Illustration (**b**) reproduced from http://www.findfreegraphics.com/image-94/excavator.htm

resemble a situation described as "force in, motion out" in impedance control [11] or, in excavation terms, "reaction in, deviation out". This lack of stiffness can not be avoided since the maximum closed-loop gains are limited by the low bandwidth of the mechanism (around 3 Hz in excavators).

From a perspective of iteration, the problem of robotic excavation is that of maintaining convergence towards a goal that defines the desired trench profile while accounting for unavoidable compliant motion. Notice that compliance and iteration are present in many other situations where motion is dominated by reactions that can be decreased iteratively. This includes tasks as diverse as scooping ice-cream with a plastic spoon or CNC machining a tough material; both are potential candidates for the proposed control strategy.

2 Excavation as Compliant Manipulation

In this paper an excavator arm is viewed as a manipulator where end-effector motion is dominated by large, somewhat unpredictable soil reactions. If the forces required to cut the soil exceed the excavator's control authority, the resulting motion exhibits a compliant characteristic ("reaction in, deviation out" [11]). With a suitable control law, this behaviour can be used naturally to reshape the motion towards areas of less resistance while maintaining attraction towards the goal.

Recently, compliant behaviour in manipulation has received a great deal of attention in control and actuator design. Compliance not only allows manipulation to be safe and to adapt to uncertainties [1] but also increases success rates in tasks where high-gain feedback tracking fails [12]. Cartesian impedance control [11, 18] has been adopted in several of those implementations. The impedance methods used in manipulation have a very intuitive appeal in excavation. In the case where the force generated by the control impedance is larger than the soil resistance, excavation proceeds towards the target by removal of material. When the opposite occurs, the bucket will drift from its desired course while imposing on the environment a recovering response given by the controller impedance. By iterating this control strategy several times, excavation is expected to converge towards the desired dig profile without the need of additional high-level prediction-dependent trajectory planning.

Note that the Cartesian impedance control used in this work [18] differs significantly from previous impedance controllers used in excavation [9, 21]. Those works were based on the idea of generating "target impedances" between a hydraulic cylinder and its load, where the load is the sum of the arm dynamic forces and an assumed linear mass-spring-damper model used to represent terrain forces. The model is used to generate target impedance values which are then tracked by an inner force feedback loop at cylinder level.

3 Low-Level Control

The basic control implementation uses operational-space [13] for feedback control and feedforward joint commands for decoupling and linearisation. End-effector (bucket tip) position is projected into Cartesian space using the excavator forward kinematics. The difference between the bucket and the desired trench positions, multiplied by the proportional feedback gain $\boldsymbol{K}_p$, generates a virtual spring force. Similarly, the difference in velocities multiplied by the derivative feedback gain $\boldsymbol{K}_d$ generates a virtual damping force. The virtual spring-damper 'connects' the bucket tip to the desired trench profile, generating the impedance of the system.

Figure 2 shows a simplified block diagram of the two controllers evaluated during the experiments reported here. The controller at the left, referred as the inverse dynamics controller (ID), is composed of a feedforward compensator and a Cartesian PD feedback law. The controller at the right, termed the ID-VSO controller, is the ID controller augmented with a disturbance estimator in the form of a variable structure observer.

3.1 Cartesian Impedance Control with Feedforward

The ID and ID-VSO controllers use the same gains and are tuned with (1) to the highest possible impedance values by selecting the largest set of gains that do not excite the first resonant mode of the arm.

$$\mathbf{F} = \boldsymbol{K}_p \mathbf{e}_x + \boldsymbol{K}_d \dot{\mathbf{e}}_x \tag{1}$$

The bucket force on the environment is related to the actuator joint torques by projection into the Cartesian space using

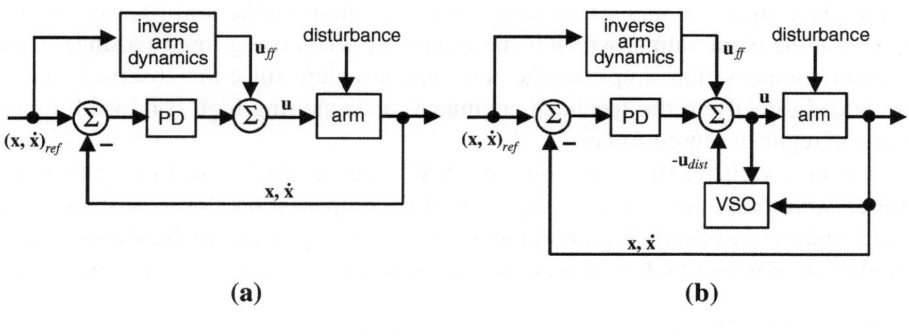

(a) (b)

Fig. 2 The two controllers used to evaluate the iterative approach: (**a**) Inverse dynamics controller and (**b**) inverse dynamics with disturbance observer controller. Here, $\mathbf{u}_{ff}$ feedforward torques, $\mathbf{u}_{dist}$ estimated disturbances, and $\mathbf{x} = [x, y, \theta]$ is the bucket position in Cartesian space. This simplified representation omits the joint/Cartesian space transformations

$$\mathbf{u} = \boldsymbol{J}^T(\mathbf{q})\mathbf{F}\,, \tag{2}$$

where $\mathbf{F}$ is a vector $[F_x, F_y, \tau_z]$ of horizontal and vertical forces at the bucket tip and the torque on the bucket, $\boldsymbol{K}_p$ and $\boldsymbol{K}_d$ are diagonal matrices of proportional and derivative gains for each Cartesian direction, $\mathbf{e}_x$ is the position error in relation to the desired trench, $\mathbf{u}$ are the torques at the joints, $\boldsymbol{J}$ is the Jacobian of the manipulator, and $\mathbf{q}$ are the joint angles.

The original implementation of the operational space control [13] requires an inverse dynamics compensator to achieve linearisation and decoupling. In excavation large modelling errors permit only partial compensation; in [16] this was used in a feedforward scheme to improve performance while avoiding destabilisation. The hydraulic compliance of the experimental platform severely limits the gains of the feedback controller and the feedforward element is essential for position tracking. In [16], feedforward actions were pre-cached by computing values in a forward simulation. In the present work, the 2DOF controller structure in Fig. 2a is used, with the difference being that the pre-cached actions are computed from the inverse arm dynamics instead of from the forward simulation.[1]

3.2 Disturbance Compensation

In the controller shown in Fig. 2a, the only forces that are reactive to disturbances are those given by the feedback actions. As results will show, this controller can not always maintain convergence towards the goal. Forces generated by the impedance controller may be insufficient to cut the soil.

Improving performance in the presence of low feedback impedance is possible by measuring reaction forces and subtracting them from the feedback output, generating compensation. In this work, a disturbance observer is used to generate this compensation, even though some force sensing is available for monitoring purposes. The disturbance values are estimated directly as actuator inputs (that is, disturbances at the plant input) as opposed to external forces acting on the arm (that is, disturbances at the plant output, which is the usual case when using force sensing). This form of compensation simplifies the controller structure since the observed values are added directly to the feedback command, not requiring high bandwidth inner loops to regulate sensed forces.

A robust variable structure observer (VSO) and its dual, a sliding mode controller, were presented in [8] aiming at friction compensation. The robustness of a variable structure observer against model error has been proven suitable for hydraulic manipulators where high seal friction and temperature effects cause parameters to

[1] Forward simulation is used in [16] to pre-cache feedforward commands because it allows the inclusion of soil-tool interaction models in the simulator. Since this work does not make use of a soil-tool model, computation of the inverse dynamics of the arm only is more efficient for obtaining the same required free motion actions.

drift and make identification problematic. However, in this work an attempt to use the original VSO resulted in excessive oscillatory behaviour. The oscillation was a consequence of the observer also compensating the natural mechanical stability due to friction, yielding a system with marginally stable dynamics. Damping those oscillations by high feedback gains amplifies noise that is caused by differentiation of encoder positions. Therefore, the present work proposes friction compensation by feedforward action, avoiding issues introduced by high feedback damping gains. This technique, however, requires a modification of the VSO so that it can be combined with a feedforward command. The following transfer function is proposed:

$$X_1 = \frac{X_2}{s} + \frac{\sigma}{ms} \tag{3}$$

$$X_2 = (-U + U_{dist} + L_1\sigma)\frac{1}{ms+d} \tag{4}$$

$$U_{dist} = \frac{-L_2\sigma}{s} \tag{5}$$

$$\sigma = W \cdot sign\left(e_q\right), \tag{6}$$

where X_1, X_2, and U_{dist} are estimates of position, velocity and disturbance torques; m, L_1, L_2, and W are design parameters and e_q is the error in position estimation (for details on the original observer refer to [8]). The term d is the damping that is added to the observer model, reflected to the joint. The inclusion of damping means that since the observer knows about friction, it does not compensate for it (it has already been compensated by the feedforward action). In this work viscous damping is assumed to be the dominant frictional term and other terms such as stiction and Stribeck effects are unaccounted for, but could be also added to the observer.

Two additional benefits are obtained by including friction in the observer. First, since friction parameter values are found by off-line identification, the observer compensates for its variation and additional modelling errors. Second, feedforward commands do not overlap with compensation commands, thus the observer can be added to an existing controller structure without further modifications.

4 Trajectory Generation

Figure 3 shows an example of a path used to specify a desired dig. In this work, the path design is based on the conclusion in [3] where studies with skilled operators showed that excavation on hard soil requires a penetrate-drag strategy. High angles of attack are used here for the penetration phase in order to generate trenches with close-to-vertical walls.

The bucket is oriented so that the segment A–B, defined as the tangent to the bucket surface that passes through the bucket tip, is made parallel to the path during penetration and dragging (Fig. 3). This condition minimises the force that arises by

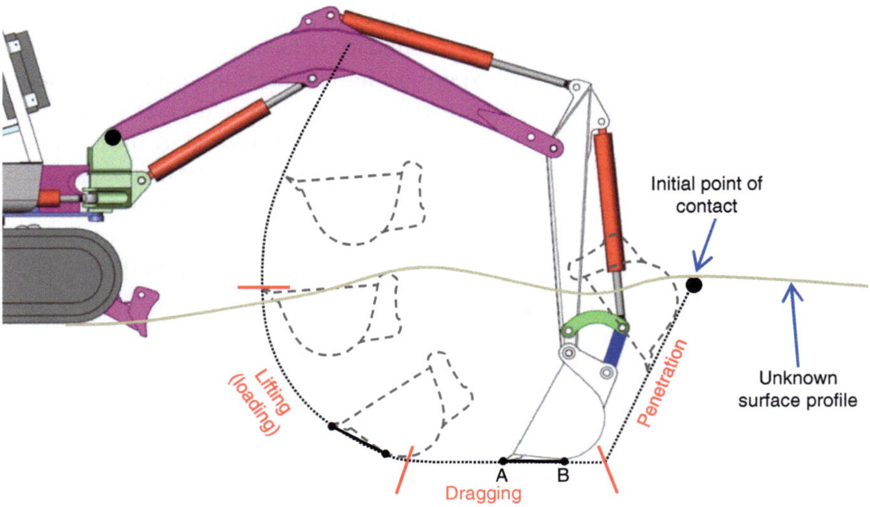

Fig. 3 Example of a path defining a desired dig. The number of passes required is assumed to be unknown, but a function of the impedance of the controller and the reactions of the terrain, and can only be answered after the trench is finished or the convergence stops

compacting the soil in front of the bucket [10]. Intuitively, the bottom surface of the bucket must slide during motion, rather than pushing or compacting the soil. During the lifting phase the bucket orientation gradually changes so that the bucket top becomes horizontal, minimising spillage.

Time along the path is imposed with smooth velocity profiles. The only requirement for trajectory feasibility is that the resulting acceleration does not cause saturation of actuators in free motion.

Notice that saturation *is* allowed during intermediate passes. Assuming that (1) each pass will have a minimum of control authority to overcome reactions, and (2) the "spare" authority is used to capture soil without compacting it, digging resistances will decrease iteratively. Disturbances and saturation will therefore also decrease, ideally to the point where during the last pass disturbances are reduced to sliding friction on the bucket surface because no shearing of soil is required.

Saturation in free motion caused by infeasible accelerations can easily be verified by inverse arm dynamics. The desired trench coordinates are first transformed to joint angles through the inverse kinematics before solving (7):

$$\mathbf{u} = \boldsymbol{M}(\mathbf{q})\ddot{\mathbf{q}} + \mathbf{v}(\mathbf{q}, \dot{\mathbf{q}}) + \mathbf{g}(\mathbf{q}) \,, \tag{7}$$

where $\mathbf{u}$ is the vector of required torques, $\boldsymbol{M}$ is the inertia matrix, $\mathbf{v}$ is the vector of centrifugal and Coriolis forces, and $\mathbf{g}$ is the gravity vector. Figure 4 shows an example of such verification for control actions required for one pass on a trench of 60 cm depth.

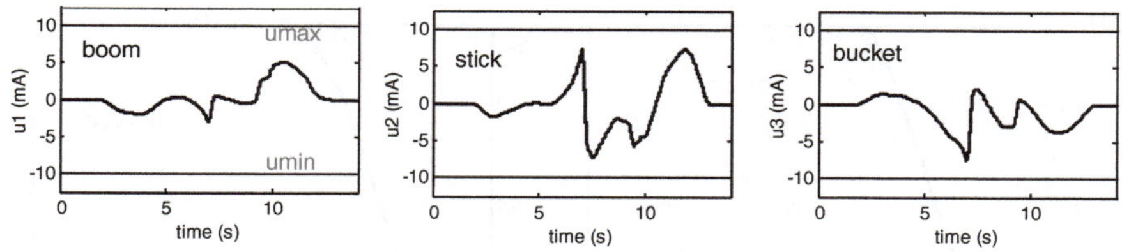

Fig. 4 Control actions required for a single pass on a 60 cm deep trench in free motion. Actions are computed by an inverse arm dynamics model only

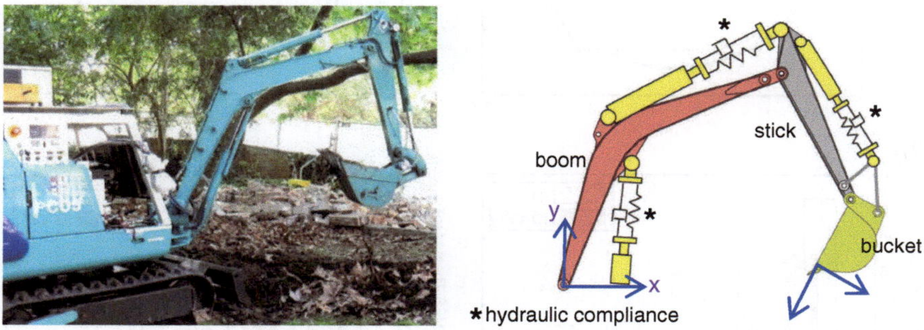

Fig. 5 The experimental platform is a 1.5 tonne excavator with a 110 kg hydraulic arm. The compliance due to flexible hoses is modelled as spring-dampers at the cylinders

5 Experiments

5.1 Experimental Platform

The experimental platform is a 1.5 tonne Komatsu PC05-7 (Fig. 5). The arm links and cylinders weigh a total of 110 kg and the arm reaches 3 m from the boom base. The hydraulic cylinders are flow controlled by servo-valves. All cylinders are supplied from the same accumulator, which is charged to 70 bar by a hydraulic pump driven by a diesel engine. Command signals sent to the servo-valves are spool position references; these are controlled by analog feedback loops internal to the servo-valves. More details on the platform can be found in [9] and issues related to hydraulic compliance and friction are described in [16].

5.2 Results

Figures 6a and 7a show the path described by both the inverse dynamics controller (ID) and the controller with disturbance observer (ID-VSO). In all cases, only the final desired dig profile is given to the controller, shown as the dotted trajectory. In Fig. 6a the reference trajectory depth is of 20 cm and in Fig. 7a the depth is 60 cm.

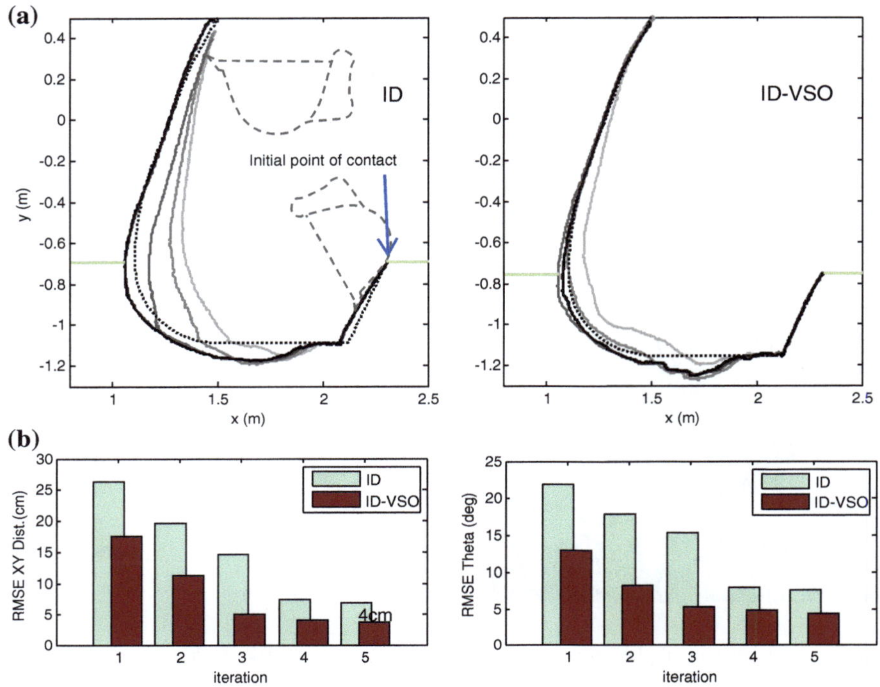

Fig. 6 Iterative excavation aiming for a 20 cm depth trench. (**a**) Workspace motion. The desired trench profile is shown in dotted lines. The first scoop is shown in grey and iterates five times until the final black trajectory. (**b**) Comparison of RMS errors over the five scoops. *Left*: distance in relation to the final trench. Right: Orientation of the bucket in relation to the desired trajectory. The result shows that the controller with VSO always performs better and achieves a final trench accuracy of 4 cm

A characteristic behaviour seen during experiments was that the bucket tended to achieve the best tracking during the beginning of the passes. Apart from the surcharge, this is caused by the progressive loss in the cutting geometry of the tool, which is maximal at the beginning of the scoop when the cutting surface is clear. This could also be an indication that the soil suffered compaction as the tool dragged soil towards the other end. In principle, those effects could be minimised by pulling the bucket out as soon as it captures a desired volume, avoiding unnecessary dragging. One could argue that this form of detection could be achieved by visual feedback. However, visual methods suffer from the dusty environment typical of excavation and the true volume in the bucket is usually partially hidden by the roughness of the trench walls and spillage. Monitoring forces to estimate material weight is effective when the bucket is filled and moving in free motion, however when scooping, estimation lumps soil-soil and soil-tool friction which are not related to the amount of material inside the bucket. For this reason, the experiments were carried out with the

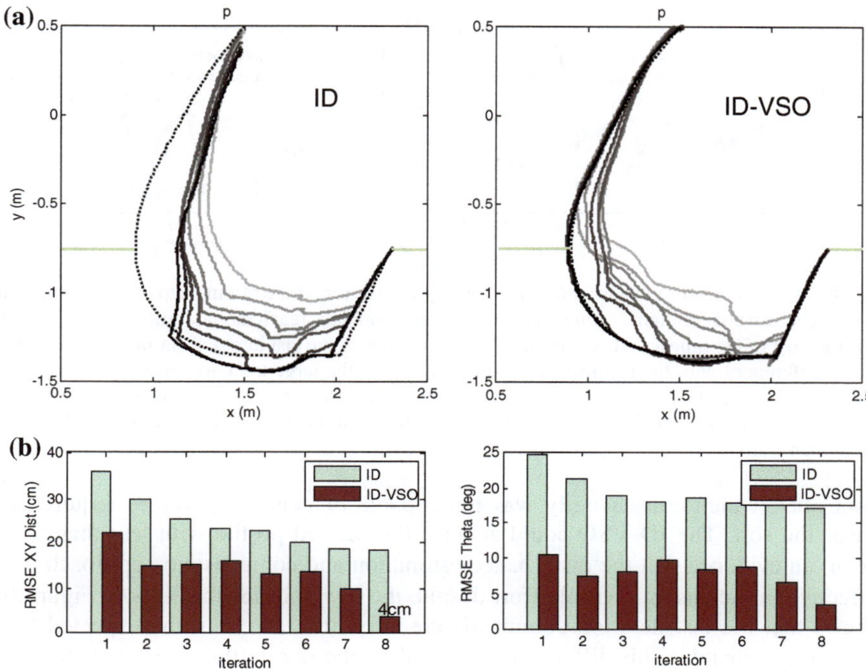

Fig. 7 Iterative excavation towards a 60 cm deep trench. (**a**) For a 60 cm trench the ID controller convergence stops at the 8th iteration while ID-VSO achieves the final desired profile. This is not caused by saturation but is due to low feedback gains. (**b**) Comparison of RMSE over the eight scoops shows that the ID-VSO performs better at every iteration, with final trench profiling error of 4 cm

sub-optimal strategy of repeating a full cycle scoop motion, independent of the amount of material collected in the bucket.

The plots in Figs. 6b and 7b show the RMS error of the distance between the tip of the bucket, where the virtual spring is attached, and the desired trench. The plots also show the RMS error of the orientation of the bucket in relation to the ideal orientation calculated in Sect. 4. The errors were calculated along the whole trajectory of each iteration. In Fig. 6 both controllers have slow convergence after the 5th pass, with the ID-VSO producing roughly half of the error at each iteration in comparison to the ID controller. Despite the larger tracking error, the ID controller was able to achieve the final profile with an RMSE error of 7 cm showing that even with low control authority the iterative method can succeed if some progress is made in each pass.

In Fig. 7 the digging aimed a 60 cm deep profile which could not be achieved by the ID controller. While it could be argued that lack of convergence was a consequence of actuator saturation, Fig. 8a shows that from the 6th pass the actuator was not saturated, and yet the resulting motion was far from the desired trajectory. This shows that the lack of convergence was due to the low Cartesian stiffness of the

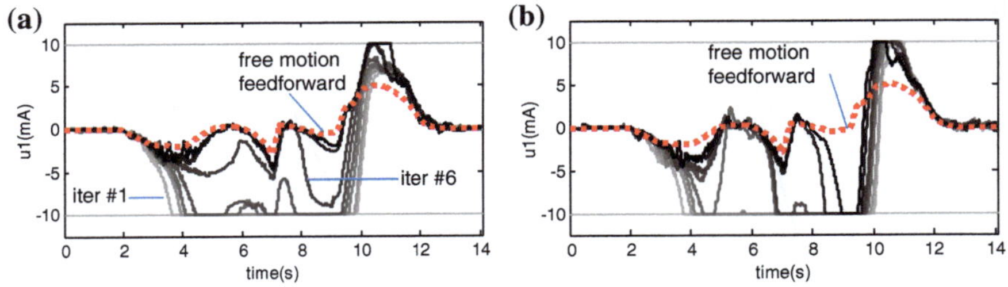

Fig. 8 Boom servo commands (*solid lines*) during excavation of the 60 cm deep trench, compared to the feedforward command (*dotted line*). Note that commands reach the saturation limit of 10 mA during most of the time. (**a**) ID controller. The commands of the last iteration (in *solid black*) show small differences with the feedforward command despite the large tracking error: an evidence of lack in feedback gains. (**b**) ID-VSO controller. Despite small tracking error the final scoop command is very different from the free motion feedforward input due to sliding friction of the bucket with the trench walls

controller, which consequently was not capable of generating forces required to shear the soil. The ID-VSO could achieve the desired profile with less than 5 cm error, an evidence that the disturbance estimation and compensation approach was effective in increasing control effort despite the low gain feedback loop. Figure 8b shows that the last iteration commands are very different from the expected free motion commands. This difference is caused by the (larger than expected) friction between the soil and the tool. The disturbance observer was essential to compensate for this friction.

A load cell was installed at the bucket cylinder for monitoring purposes only. The measured forces required to control bucket orientation exceeded 1.5 tonne during the whole dragging phase. Visual inspection on the trenches (Fig. 9) shows that most of the material below 20–30 cm was clay with scattered pieces of brick and roots. The polished and smooth surfaces at bottom of the trench were caused by the bucket sliding and compacting the clay soil during scooping.

Fig. 9 Visual inspection of the opened trenches shows that except for the initial few centimetres of dry top soil the dominant material was clay. Shearing and dragging a full bucket of this material was enough to generate more than 1.5 tonne of reactions at the cylinders

6 Conclusions

This work presented a low-level control approach for excavation from an iterative perspective. Since forces required to shear soil often surpass control actuation, end-effector motion is dominated by the terrain reactive forces. In this situation the manipulator assumes a compliant behaviour in relation to the environment and Cartesian impedance control was used as a natural approach to address this behaviour.

Experimental results showed that convergence towards the goal is possible if two conditions are satisfied: (a) there is a minimum control authority to counter some amount of reaction, and (b) that this authority is used to capture soil without compacting it. A feedforward controller with bounded gains was not sufficient to satisfy condition (a) requiring the addition of a disturbance observer. Condition (b) was addressed by careful design of the trajectory and the orientation of the bucket.

As shown in Fig. 8, the desired trench was initially unfeasible with respect to required forces. While most of approaches would aim at predicting and avoiding those forces, the combination of impedance and iteration allows feedback to reshape motion as imposed by the terrain, while still achieving the final trench.

Future work will aim at complementing the low-level controller with high-level strategies in two ways. First, actions will be added that go beyond low-level control. For example, consider the case where all areas towards the goal are unfeasible but there may be a route of escape made available by loosening some rocks on the way. While a pure impedance strategy would probably fail, shaking the bucket tip could allow the dig to proceed. Second, concatenation of short trench profiles (used in this paper) will be investigated to achieve realistic longer, wider and deeper trenches.

Acknowledgments The authors thank Pak Hung (Victor) Chan for his support with the experimental platform and Javier Martinez and Seong Ho Lee for their assistance during experiments. This work was supported by the Rio Tinto Centre for Mine Automation and the Australian Centre for Field Robotics funded in part by the New South Wales State Government.

References

1. A. Albu-Schaffer, O. Eiberger, M. Grebenstein, S. Haddadin, C. Ott, T. Wimbock, S. Wolf, G. Hirzinger, Soft robotics. IEEE Robot. Autom. Mag. **15**(3), 20–30 (2008)
2. S. Blouin, A. Hemami, M. Lipsett, Review of resistive force models for earthmoving processes. J. Aerosp. Eng. **14**, 102 (2001)
3. D. Bradley, D. Seward, The development, control and operation of an autonomous robotic excavator. J. Intell. Rob. Syst. **21**(1), 73–97 (1998)
4. H. Cannon, S. Singh, Models for automated earthmoving. in *Lecture Notes in Control and Information Sciences—International Symposium Experimental Robotics (ISER)*, (Springer, 2000), pp. 163–172
5. M. Dunbabin, P. Corke, Autonomous excavation using a rope shovel. J. Field Robot. **23**(6–7), 379–394 (2006)
6. J. Fielke, T. Riley, The universal earthmoving equation applied to chisel plough wings. J. Terrramech. **28**(1), 11–19 (1991)

7. C. Gallo, R. Wilkinson, R. Mueller, J. Schuler, A. Nick, Comparison of ISRU excavation system model blade force methodology and experimental results. in *American Institute of Aeronautics and Astronautics (AIAA)*, (Aerospace Sciences Meetings, 2009)
8. Q. Ha, A. Bonchis, D. Rye, H. Durrant-Whyte, Variable structure systems approach to friction estimation and compensation. in *Proceedings of the IEEE International Conference on Robotics and Automation (ICRA)*, vol. 4 (2000), pp. 3543–3548
9. Q. Ha, M. Santos, Q. Nguyen, D. Rye, H. Durrant-Whyte, Robotic excavation in construction automation. IEEE Robot. Autom. Mag. **9**(1), 20–28 (2002)
10. A. Hemami, Study of bucket trajectory in automatic scooping with load-haul-dump loaders. Trans. Inst. Min. Metall., Sect. A **102**, 37–42 (1993)
11. N. Hogan, Impedance control: an approach to manipulation. in *American Control Conference*, (IEEE, 1984), pp. 304–313
12. M. Kalakrishnan, J. Buchli, P. Pastor, M. Mistry, S. Schaal, Learning, planning, and control for quadruped locomotion over challenging terrain. Int. J. Robot. Res. **30**(2), 236 (2011)
13. O. Khatib, A unified approach for motion and force control of robot manipulators: the operational space formulation. IEEE J. Robot. Autom. **3**(1), 43–53 (1987)
14. J. Kuśmierczyk, J. Szlagowski, Automated excavation process analysis for given trajectory and soil parameters. in *International Symposium on Automation and Robotics in Construction (ISARC)*, (2008), pp. 95–99
15. O. Luengo, S. Singh, H. Cannon, Modeling and identification of soil-tool interaction in automated excavation. in *Proceedings of the IEEE International Conference on Intelligent Robots and Systems (IROS)*, vol. 3 (IEEE, 1998), pp. 1900–1906
16. G. Maeda, S. Singh, D. Rye, Improving operational space control of heavy manipulators via open-loop compensation. in *Proceedings of the IEEE/RSJ International Conferencen on Intelligent Robots and Systems (IROS)*, (2011), pp. 725–731
17. J. Marshall, P. Murphy, L. Daneshmend, Toward autonomous excavation of fragmented rock: full-scale experiments. IEEE Trans. Autom. Sci. Eng. **5**(3), 562–566 (2008)
18. F. Petit, A. Albu-Schaffer, Cartesian impedance control for a variable stiffness robot arm. in *Proceedings of the IEEE International Conference on Intelligent Robots and Systems (IROS)*, (2011), pp. 4180–4186
19. A. Reece, The fundamental equation of earth-moving mechanics. Proc. Inst. Mech. Eng. **179**, 16–22 (1964)
20. W. Richardson-Little, C. Damaren, Position accommodation and compliance control for robotic excavation. J. Aerosp. Eng. **21**, 27 (2008)
21. S. Salcudean, S. Tafazoli, P. Lawrence, I. Chau, Impedance control of a teleoperated mini excavator. in *Proceedings of the 8th IEEE International Conference on Advanced Robotics*, (Citeseer, 1997)
22. X. Shi, P. Lever, F. Wang, Experimental robotic excavation with fuzzy logic and neuralnetworks. in *Proceedings of IEEE International Conference on Robotics and Automation*, vol. 1 (1996)
23. S. Singh, Learning to predict resistive forces during robotic excavation. in *Proceedings of the IEEE International Conference on Robotics and Automation*, vol. 2 (1995), pp. 2102–2107
24. S. Vahed, K. Althoefer, L. Seneviratne, X. Song, J. Dai, H. Lam, Soil estimation based on dissipation energy during autonomous excavation. in *Proceedings of the 17th International Federation of Automatic Control (IFAC)*, (World Congress, 2008)

Rock Recognition Using Stereo Vision for Large Rock Breaking Operation

Anusorn Iamrurksiri, Takashi Tsubouchi and Shigeru Sarata

Abstract At the work front in a quarry, many large rocks are generated by rock blasting. Since some of these rocks are too large to be fed into a rock crusher machine, a hydraulic breaker is used to break the oversized rocks into suitable sizes. The purpose of this study is an automation of rock breaking operation in working front of an open-pit quarry. In this paper we describe an approach using stereo vision to recognize position and shape of large rocks. For rock recognition and rock moving experiments, we set up scaled down experimental environment in laboratory and use small rocks and a robotic manipulator in experiments.

1 Introduction

This paper concerns rock recognition by means of stereo vision that will be applied for rock breaking operation of a hydraulic breaker (an excavator with a hydraulic hammer) in an open-pit quarry. The motivation of this research arises from anautomation of the operation. At the work front in a quarry, since some of the rocks generated by rock blasting are too large to be fed into a rock crusher machine breaking the large rocks to a suitable size, a hydraulic breaker is used to break the oversized rocks into small pieces. However, the working condition in a quarry is generally dangerous and dirty from flying stone fragments and rock dust; therefore an automation of the rock breaking operation is required, which is the motivation of this study.

A hydraulic breaker is shown in Fig. 1. The breaker is equipped with a chisel at the top end of the arm for breaking rocks into small pieces. Since the range suitable for breaking a rock by the chisel is limited, an operator of the breaker has

A. Iamrurksiri · T. Tsubouchi (✉) · S. Sarata
University of Tsukuba, 1-1-1 Tennoudai, Tsukuba 305-8573, Japan
e-mail: tsubo@roboken.esys.tsukuba.ac.jp

S. Sarata
e-mail: sarata@roboken.esys.tsukuba.ac.jp

K. Yoshida and S. Tadokoro (eds.), *Field and Service Robotics*,
Springer Tracts in Advanced Robotics 92, DOI: 10.1007/978-3-642-40686-7_26,

Fig. 1 A hydraulic breaker

to use the chisel to move a rock into the "breaking area" when the rock is out of the range before breaking it. To automate the rock breaking operation, an automation of (1) rock recognition, (2) rock moving process and (3) rock breaking process are required. In this paper we focus mainly on the first part, i.e. recognition of rock position and shape using stereo vision and focus partly on the rock moving process with taking the result of the recognition into account.

As a previous work of this study, Namiki et al. [1, 2] proposed an algorithm for rock recognition using stereo vision and used a small manipulator to realize an automation of rock moving operation in laboratory environment. However the proposed algorithm for rock recognition cannot separate and recognize individual rocks those are piled up or lie close together. In this paper we describe an approach to improve an accuracy of rock recognition by separating piled rocks based on a "convex–concave determination" method. As related works, Takahashi et al. [3, 4] identified position and posture of the rocks in 3D using laser beam and a CCD camera for rock breaking operation in the production plant of a quarry. And they also used force sensor attached on a chisel of a hydraulic breaker to detect the status of the breaker in rock breaking operation [5]. For the use of 3D information of the rocks, Commonwealth Scientific and Industrial Research Organisation (CSIRO) in Australia [6] developed a human–machine interface for tele-operated breaker task which several cameras are used and 3D environment information is extracted based on image processing. Tadano et al. [7] recognized a shape of a rock by means of laser scanner and realized rolling movement of the rock on flat experiment table using a small sized industrial manipulator. Matthew J. Thurley [8] measured the size distribution of limestone particles on conveyor belt using 3D data by means of laser triangulation.

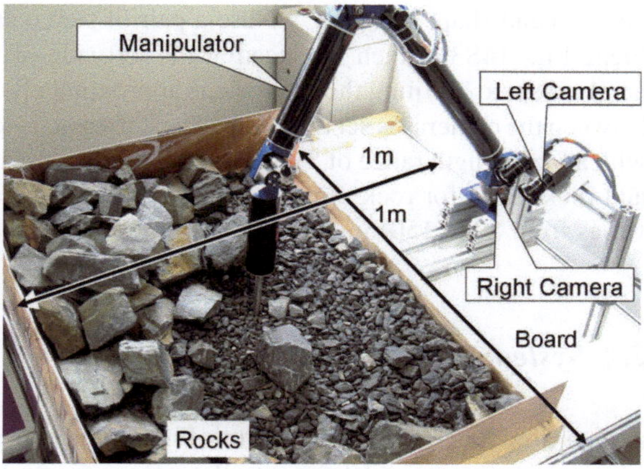

Fig. 2 An experiment table

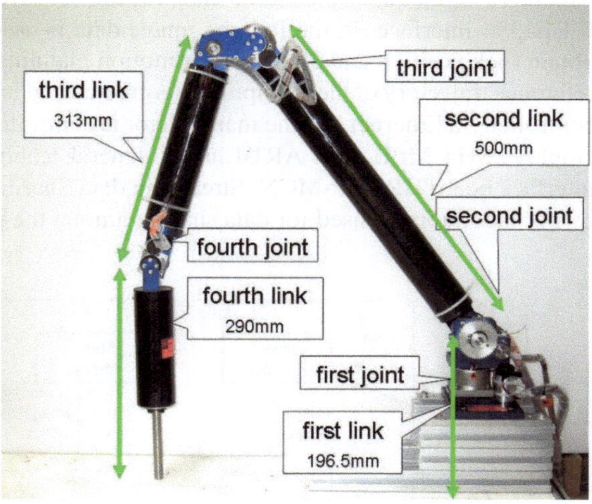

Fig. 3 Manipulator

2 Experimental Setup

2.1 Experiment Environment

Since it is difficult to realize experiments using a real hydraulic breaker, we set up a scaled down environment similar to realistic field (see Fig. 2). A specially designed manipulator (Technocraft M4FS07, see Fig. 3) is used in rock moving experiments. The manipulator has four DOF and the same ratio of the length of the links as same as the real hydraulic breaker. The scale of this manipulator and the rocks used in experiments are approximately 1/12 of the real scale in realistic environment. The manipulator is placed on a square experiment table of 1 × 1 m. To recognize the position and shape of rocks, we use a pair of stereo cameras (Point Grey Research Flea2 FL2-14S3) which is placed at the left side of the manipulator. The camera height is set to 306 mm, the tilt angle is approximately 22° and the base line distance between the cameras is set to 60 mm. The lenses of the cameras are Kowa LM5JC1M with a focal length range of 5 mm. Rubber (C-40) is scattered on the experiment table and rocks used for rock moving experiment are placed on it. The size of the rocks varies from 80 to 150 mm.

2.2 System Setup

A data flow diagram is illustrated in Fig. 4. The experimental system consists of a manipulator, a manipulator controller, stereo vision system and a PC. Image data which are captured by the stereo cameras are sent from the cameras to the PC via IEEE1394 interface. In the PC, the image data is used for stereo vision process to obtain rock position and 3D shape for motion planning of the manipulator. Then the reference trajectory of the manipulator is created and is transferred to the manipulator controller via Ethernet and the manipulator is controlled. The manipulator controller employs SH4 MPU with ART-Linux (General Robotics) HRP-3P-CN-A and I/O interface board HRP-3P-MCN. Streaming data Sharing Manager (SSM), developed by our laboratory, is used for data sharing among the processes [9].

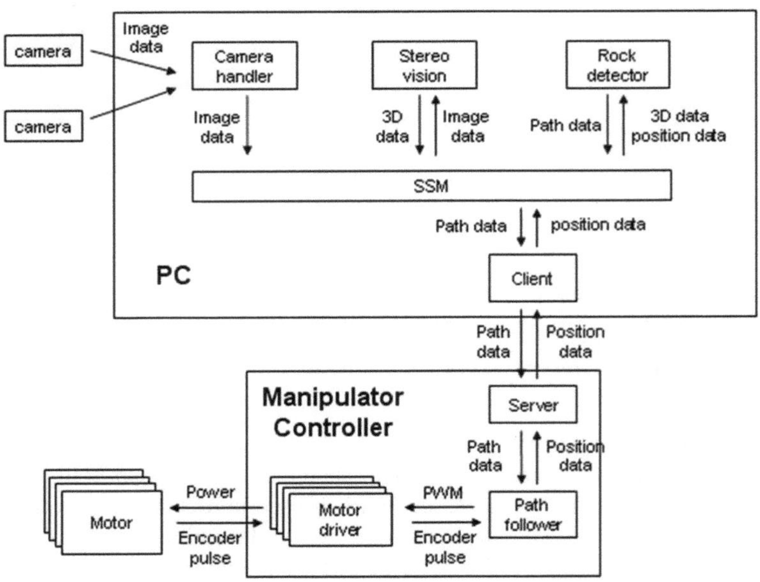

Fig. 4 Data flow diagram

3 Stereo Vision Based Recognition of Multiple Rocks

3.1 Convex–Concave Determination

Large rocks generated by rock blasting in a quarry generally are piled up or lie close together. Therefore it is difficult to obtain the position and shape of individual rocks accurately. In this chapter we propose an approach using stereo vision to separate a rock pile and to obtain position and 3D shape of individual rocks based on "convex–concave determination" method.

We can observe that most of the large rocks just after blasting have convex shape. With this hypothesis, if there is a part of the shape of the rocks obtained by stereo vision that is concave, it can be assumed that two or more rocks adjoin each other on the concave part. For example, two box-shaped rocks are illustrated in Fig. 5. We choose two faces from the rocks and determine whether they are convex or concave. If they are convex, we can hypothesize it that they are the faces of the same rock. If otherwise, the two faces belong to the different ones as hypothesis. In this chapter we describe the rock separation method using the convex–concave determination of rock faces. Computing a disparity map and 3D data from camera images using stereo vision is presented in Sect. 3.2. Extracting rock faces is described in Sect. 3.3. Rock separation using convex–concave determination is described in Sect. 3.4. Finally, an evaluation of the proposed rock recognition method is described in Sect. 3.5.

3.2 Computing Disparity Map and 3D Data Using Stereo Vision

Before stereo matching process, we first remove lens distortion and rectify the left and right images. The parameters needed for the distortion removal and the rectification are obtained by Camera Calibration Toolbox for Matlab in advance [10]. Next we compute a disparity map from the camera images using Semi-Global Block Matching algorithm (SGBM) which is available in OpenCV library (see Figs. 6, 7). Then we calculate 3D coordinates (X, Y, Z) of each pixel in the disparity map (Fig. 8). The 3D coordinates are used to create an elevation map and we extract the areas in the

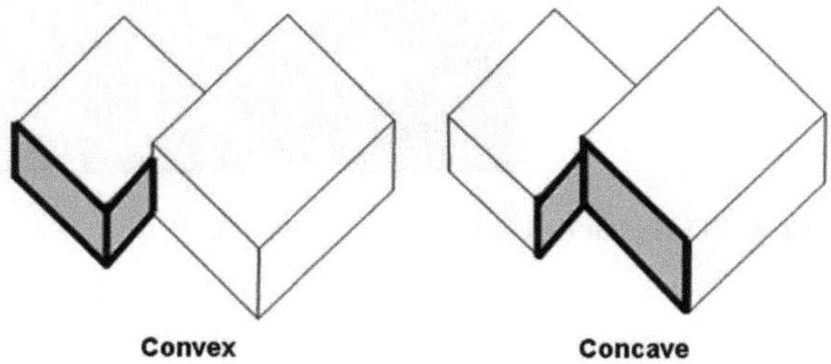

Fig. 5 Convex–concave determination

Fig. 6 Camera images (left and right)

Fig. 7 Disparity map

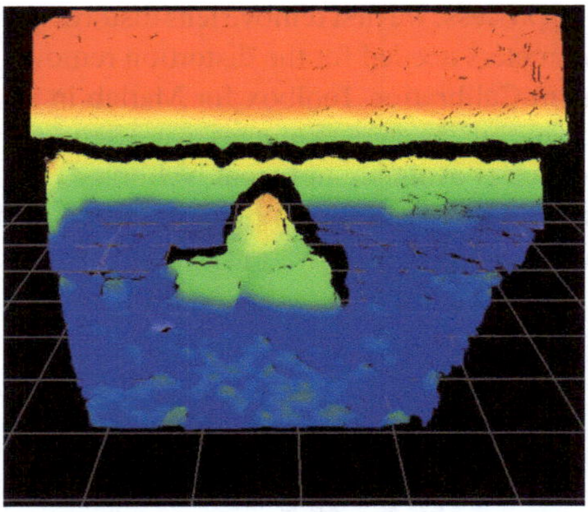

Fig. 8 3D elevation map

Fig. 9 Binarized 2D elevation map

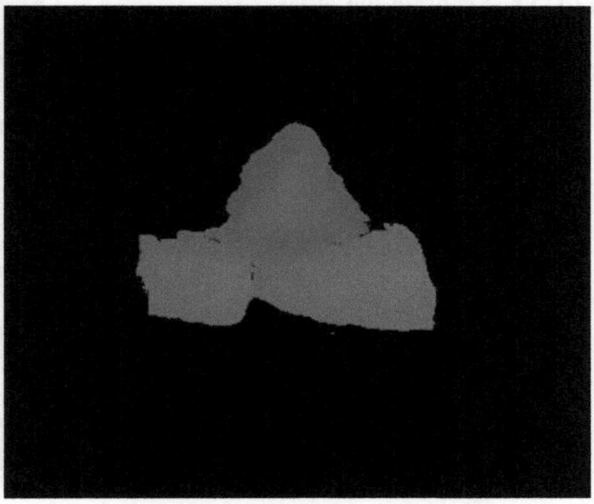

Fig. 10 Rock area in the disparity map

elevation map those are higher than a threshold value (set to 5 cm) assumed to be rock areas and choose the nearest one as a target rock area (Fig. 9). After that we extract the area in the disparity map corresponding to the target rock area in the elevation map; call it a "rock area" (Fig. 10). This rock area is used for extracting rock faces described in Sect. 3.3.

3.3 Extracting Rock Faces

To extract rock faces, first we calculate 3D normal vectors of each pixel in the rock area obtained in Sect. 3.2. Then we cluster the 3D normal vectors using K-Means algorithm. The number of cluster is fixed to 5 and the clustering process is repeated three times. To calculate the 3D normal vectors, the 3D coordinates of 9 * 9 pixels around the pixel of attention are used. We approximate the 9 * 9 neighborhood pixels to a 3D plane and calculate a normal vector of the plane and use it as a normal vector

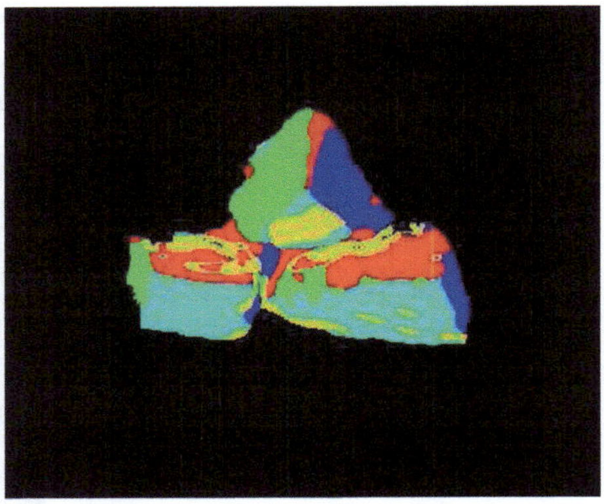

Fig. 11 Clustering result

Fig. 12 Separate the areas those are in the same cluster but do not connected each other to different clusters

of the pixel of attention. After clustering the 3D normal vectors, we delete small clusters and separate the areas those are in the same cluster but are not connected each other to different clusters (Figs. 11, 12). Finally we create 3D faces from the 3D coordinates of each cluster using least mean square method. To be easy to use in rock separation described in Sect. 3.4, the faces are approximated to rectangular shape (Fig. 13).

3.4 Identification of a Rock in the Pile of Multiple Rocks

Convex-concave determination method is used to separate a rock pile and recognize the position and shape of individual rocks. As mentioned in Sect. 3.1 with the assumption that the rocks have convex shape, if the two faces extracted from the rocks are convex, it can be assumed that they are faces of the same rock. If otherwise,

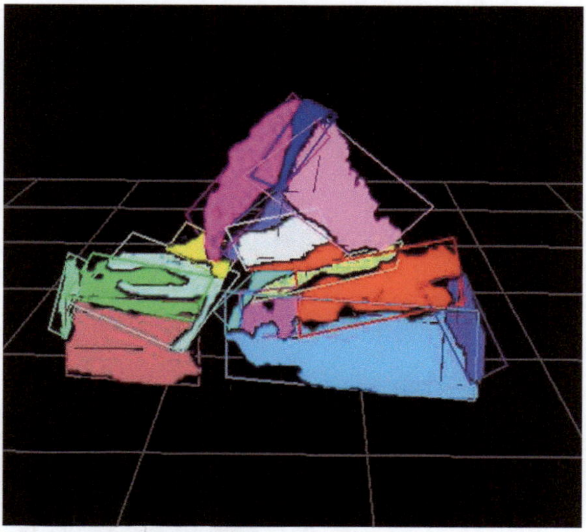

Fig. 13 3D faces approximated to *rectangular* shape

they are the faces of different rocks. A summary of the rock separation procedure is as follows:

1. Choose randomly a face that does not belong to any group; call "non-group face" and create a new group for the face.
2. Do the convex–concave determination between the face and other non-group faces around it. If they are convex, put the non-group face into the group.
3. Do step 2 for all new group members until there is no new member left.
4. Do the convex–concave determination again between group members. If there is a pair of the faces that is concave, do the convex–concave determination between the two faces and other members and exclude the face that has less convex results from the group.
5. Back to step 1 and repeat until there is no non-group face left.
6. Merge the groups those have less than two members with other groups around them.

The conditions for the convex–concave determination are as follows. Two faces, for example face A and face B shown in Fig. 14 are convex when:

1. More than half of the area of face A is on the opposite side of face B from the cameras, and
2. More than half of the area of face B is on the opposite side of face A from the cameras

The result of rock separation using the convex–concave determination is shown in Fig. 15. To be easy to use in path planning of the manipulator, we approximate the rocks to cuboids. We binarize the elevation map of the rocks created in Sect. 3.2 and then approximate the rock areas to straight lines using least mean square method and

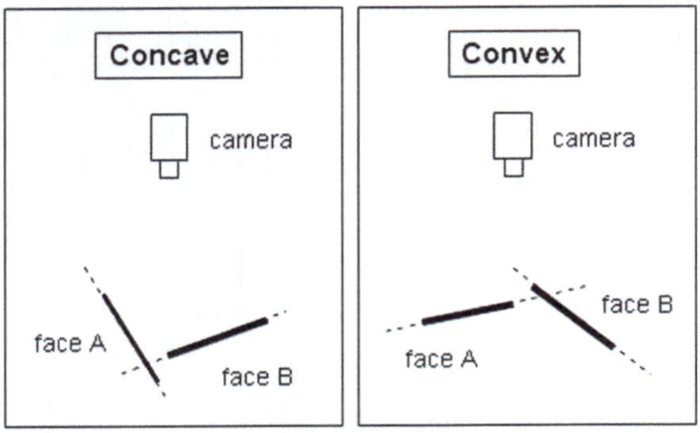

Fig. 14 Convex–concave determination

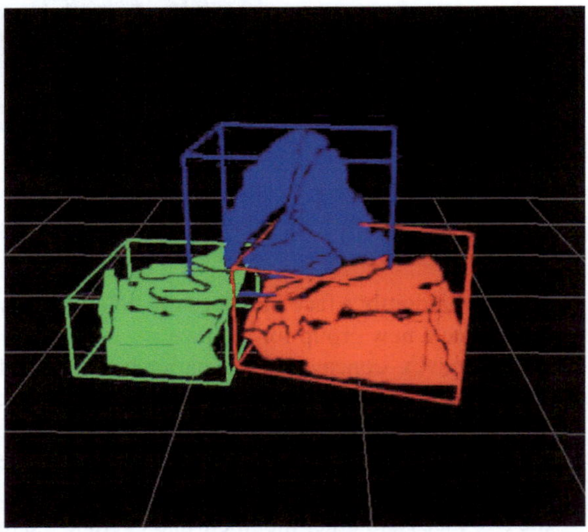

Fig. 15 A result of rock separation and cuboid-shaped approximation

wrap the rocks with cuboids parallel to the straight lines. These cuboids are used for creating a trajectory of the manipulator in rock moving operation.

3.5 Evaluation

Rock recognition experiments using convex–concave determination method are performed with 10 different images and the sample results are shown in Figs. 16, 17, 18, 19. Table 1 shows the numbers of the rocks separated by the proposed recognition method. The numbers of rocks in Table 1 do not include the rocks those are recognized as a different rock pile; for example the rock at the back right in Fig. 17. These rocks are recognized as a different rock pile when extracting rock areas from

Fig. 16 Results of rock recognition (*1*)

Fig. 17 Results of rock recognition (*2*)

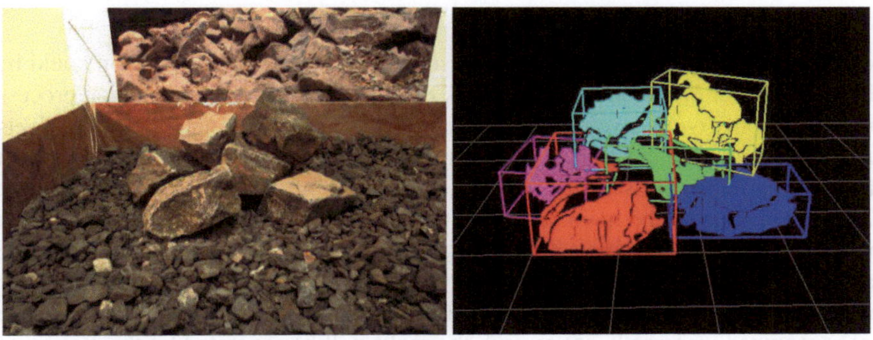

Fig. 18 Results of rock recognition (*3*)

the elevation map described in Sect. 3.2; therefore these rocks are not counted in the numbers of rocks.

From theses results, we can see that some of the rocks are not recognized correctly; two or more rocks are recognized as parts of a single rock, or one rock is separated

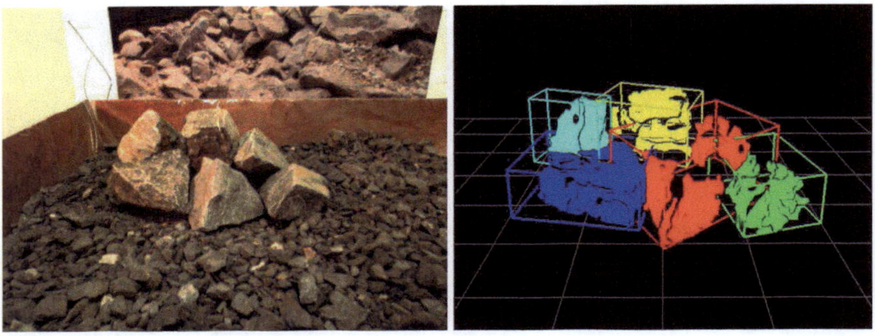

Fig. 19 Results of rock recognition (*4*)

Table 1 The result of rock recognition experiments

No.	Number of rocks recognized	Number of all rocks
1	6	7
2	4	5
3	6	6
4	5	6
5	7	7
6	6	7
7	6	6
8	6	6
9	4	6
10	5	6

to multiple rocks. According to our investigations, the wrong recognition could be due to a number of factors, such as a lack of accuracy in the face extraction process or wrong convex–concave determination. Due to using $9*9$ neighborhood pixels in normal vector calculation, the normal vectors are more or less smoothed with the neighborhood pixels which induce a false face extraction. The size of the rock faces is also one of the main causes of the wrong recognition. Due to using least mean square method to create rock faces from 3D data, the smaller the faces are, the larger the error of the position and pose become. However for our purpose, 100 or near high recognition rate is not necessarily desired. Because when the rocks those are recognized incorrectly are moved, the rocks will be separated by the motion of the manipulator so that we can recognize the rocks correctly after all. The wrong recognition problem is described again in Chap. 4.

4 Experiments

In this chapter we describe a simple rock moving experiment using the proposed rock recognition method. The goal of this experiment is moving one of the rocks those are placed approximately 60 cm in front of the manipulator to the destination. The destination is set to 30 cm in front of the manipulator and the size of the rocks used in the experiments is approximately 8–15 cm. A flow chart of the rock moving process is shown in Fig. 20. First we recognize rocks using the proposed rock recognition method described in Chap. 3 and select the nearest rock as a target to move. Then the chisel is raised up over the target rock height and is moved to the back of the target rock. After that the target rock is moved to the destination by sliding. A trajectory of the chisel for moving the rock is set with a straight line from the center of the target rock to the destination. The rock position is obtained by stereo vision throughout the moving operation to detect whether the target rock is still on the trajectory. If the rock is out of the trajectory, the chisel will be moved to the back of the rock and do the moving operation again. The rock moving process finishes when the target rock reaches the destination.

According to snapshots of a rock moving experiment shown in Fig. 21, the rightmost rock was chosen as a target rock and it was moved to the destination properly. In this experiment we used only sliding operation to move the rock along the trajectory. With sliding operation, a target rock was sometimes stuck with other rocks or ground and stopped on the way. This problem could be avoided by using other operation,

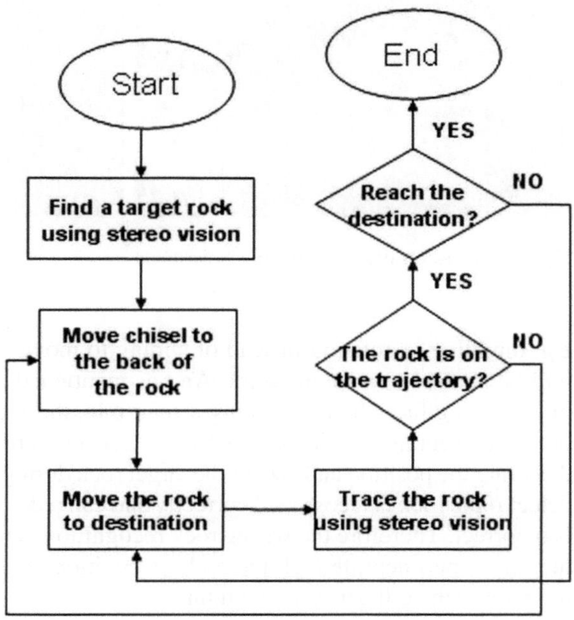

Fig. 20 A flow chart of rock moving process

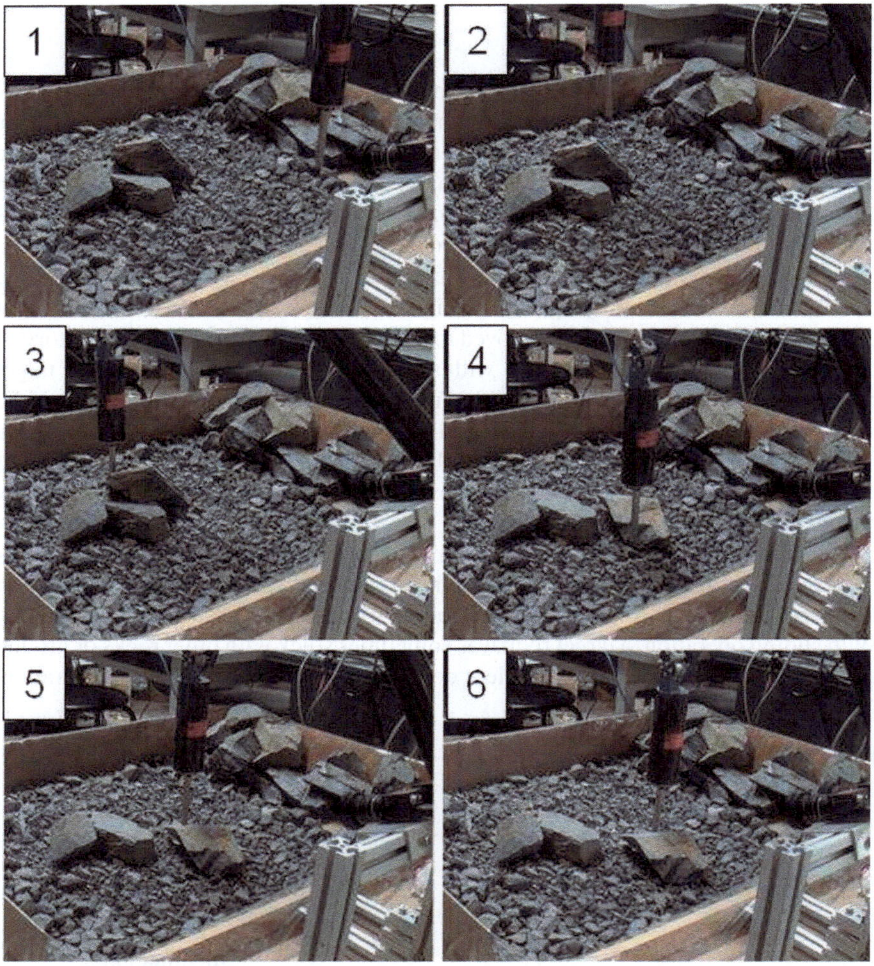

Fig. 21 Snapshots of a rock moving experiment

e.g. tumbling or rotating instead of sliding to move a rock when it is stuck, which will be studied in a future work. And as mentioned in Sect. 3.5 due to inaccuracy of extracting faces and separating a rock pile, the rock position and shape are not obtained accurately which sometimes induce a failure of the operation. However by detecting the position and size of the target rock during the moving operation, we can detect if the rock is recognized correctly and can redo the operation if the recognition is incorrect. Therefore the wrong rock recognition is not a fatal problem in this rock moving experiment, though the rock recognition method is needed to be improved to increase the efficiency in the future.

5 Conclusions and Future Works

In this paper, we presented a method for recognition of rocks using stereo vision, which will be applied for an automation of rock breaking operation in an open-pit quarry. In our approach, we used convex–concave determination to separate a rock pile and to obtain the position and shape of individual rocks. To prove effectiveness of the proposed rock recognition method, we also described a simple rock moving experiment using the proposed rock recognition method.

For a future work, we plan to improve efficiency of rock recognition, for example by using edge to separate a rock pile. Moreover we will develop path planning algorithm for a manipulator to move multiple rocks efficiently.

References

1. M. Namiki, T. Tsubouchi, A Research of Rock Moving Operation Using A Rod-Shaped Manipulator. Master's Thesis, University of Tsukuba, 2010
2. M. Namiki, T. Tsubouchi, S. Sarata and E. Koyanagi, The Analysis for Automation of Massive Rock Handling with a Breaker *In Proceedings of the 11th Symposium on Construction Robotics in Japan*, pp. 305–314, 2008
3. H. Takahashi, K. Sano, Automatic detection of large rocks by image processing. J. Min. Mater. Process. Inst. Jpn. **113**(3), 169–174 (1997)
4. H. Takahashi, K. Sano, E. Nakano, Study on automatic detection and breaking system for huge rocks by use of image processing. J. Min. Mater. Process. Inst. Jpn. **114**(8), 543–548 (1998)
5. H. Takahashi, T. Monden, K. Kaneko, Study on the automation of boulder breakage by use of force sensors and image processing. J. Min. Mater. Process. Inst. Jpn. **115**(10), 725–730 (1999)
6. Commonwealth Scientific and Industrial Research Organisation: Rock breaking Mine tool goes automatic. Earthmatters magazine, issue 19, (2009), pp. 12–13
7. K. Tadano, T. Tsubouchi, M. Kurisu and S. Sarata, Massive Rock Handling by a Breaker -Graspless Manipulation and Object Recognition*In 23rd International Symposium on Automation and Robotics in Construction*, pp. 487–492, 2006
8. J. Matthew, Thurley: automated online measurement of limestone particle size distributions using 3D range data. J. Process Control **21**(2), 254–262 (2011)
9. E. Takeuchi and T. Tsubouchi, Open Navigation Software for a Mobile Robot using Multi Sensor. JSME Conference on Robotics and Mechatronics 2008, 1P1-E22, (CD-ROM), 2008
10. Jean-Yves Bouguet. Complete Camera Calibration Toolbox for Matlab http://www.vision.caltech.edu/bouguetj/

Plowing for Rover Control on Extreme Slopes

David Kohanbash, Scott Moreland and David Wettergreen

Abstract Planetary rovers are increasingly challenged to negotiate extreme terrain. Early destinations have been benign to preclude risk, but canyons, funnels, and newly discovered holes present steep slopes that defy tractive descent. Steep craters and holes with unconsolidated material pose a particularly treacherous danger to modern rovers. This research explores robotic braking by plowing, a novel method for decreasing slip and improving mobility while driving on steep unconsolidated slopes. This technique exploits subsurface strength that is under, not on, weak soil. Starting with experimental work on Icebreaker, a tracked rover, and concluding with detailed plow testing in a wheel test-bed the plow is developed for use. This work explores using plows of different diameters and at different depths as well as the associated braking force. By plowing the Icebreaker rover can successfully move on a slope with a high degree of accuracy thereby enabling science targets on slopes and crater walls to now be considered accessible.

1 Introduction

The robotic exploration of planetary bodies requires rovers that can safely explore a variety of environments and have the ability to reach points of scientific interest. Rovers significantly expand the possible exploration areas and increase the scientific and programmatic return from a mission. The surface of the moon and planets such as Mars are covered in fine and unconsolidated regolith making mobility considerations of utmost importance [7]. There are many craters and steep slopes that are also covered in this fine and unconsolidated regolith. The ability to traverse steep slopes, funnels, and holes becomes even more compelling in the case of Mars where there is

D. Kohanbash (✉) · S. Moreland · D. Wettergreen
Carnegie Mellon University, Pittsburgh, PA 15213, USA
e-mail: dkohanba@cmu.edu

D. Wettergreen
e-mail: dsw@cmu.edu

K. Yoshida and S. Tadokoro (eds.), *Field and Service Robotics*,
Springer Tracts in Advanced Robotics 92, DOI: 10.1007/978-3-642-40686-7_27,

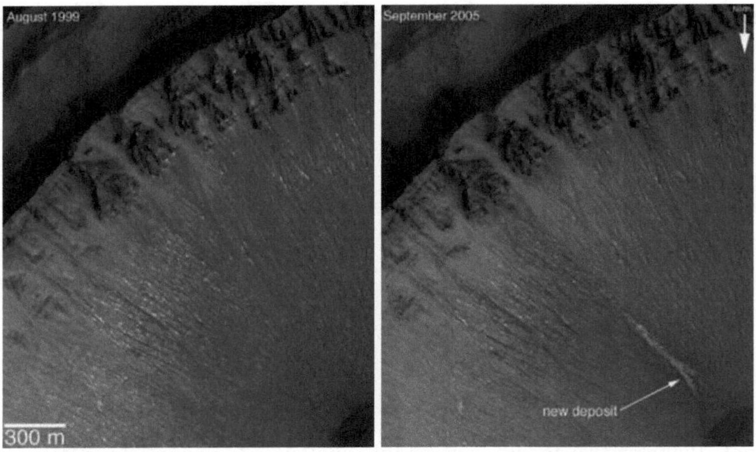

Fig. 1 New deposits found on a crater wall in Centauri Montes [10]

now evidence of new deposits forming on the side of crater walls (Fig. 1) [10] and in the case of the moon with resources in permanently shadowed polar craters. Many of these crater walls have slopes that exceed 25° [1, 5].

1.1 Background

In this work a plow mechanism is tested both on a rover named Icebreaker as well as in a wheel test-bed to throughly characterize how a plow can be used for increased control authority on a slope.

The "plow" is a dagger like rod that is driven into the ground by a robot to create a braking force by exploiting the subsurface soil strength (Fig. 2). This braking force is used to counter downhill forces in order to minimize slip (undesried/uncommanded downhill motion). The plow is a circular rod located at the center point of the robot to facilitate turning. This is a mechanical analog to a climbers use of an axe or pole to arrest or control a downhill slide. The plow on the rover is actuated using a rack and pinion and has a lexan tip to help the plow pierce the ground to minimize vertical plow forces (Fig. 3).

It is generally recognized that the interaction of the rover and the terrain is complex and difficult to model. In addition, standard models do not work in very unconsolidated material with low moisture levels [15]. Prior work with model based terrain controls have problems with accurately modeling the terrain and seek better methods to simulate the wheel-terrain interaction [2, 6]. For these reasons having a device that is purely reactive and not dependant on possibly inaccurate models is important.

Fig. 2 Icebreaker's actuated plow. The box contains the motor and pinion for driving the plow into the ground [17]

Fig. 3 Icebreaker actuated plow. Conical lexan tip is used to penetrate into the ground [17]

1.2 Prior Research in Plowing

In farming tines that are similar to the plow used in this work have been studied [8]. It is important to note that most tools that interact with the ground have been designed more on the basis of field experience than on scientific knowledge since soil-tool interaction is not well defined and quantified [12]. It has been found that soil moisture and shear strength have the largest impact on the soil forces, and at small speeds, the effect of speed can be neglected [3]. The disturbance pattern in the soil has been shown to be repeatable in a wide range of materials and soil moistures for a given tine design [14].

There are two distinct depths that need to be considered when analyzing the plow forces. Plowing above the critical depth results (Fig. 4) in brittle failure, which does not provide a significant resistance force to the plow. This failure is deemed crescent failure where the disturbed soil is pushed to the sides, in front, and upward. Plowing below the critical depth results in the surrounding soil compacting in the direction of motion, thereby providing the subsurface strength that is needed for plowing. This is called lateral failure (Fig. 5) [3]. This means that in order to have significant subsurface strength the plow depth must be deeper than this critical depth.

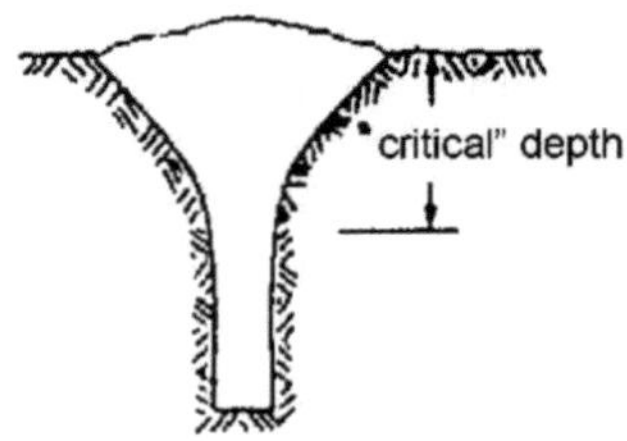

Fig. 4 Critical depth in plow track [3]

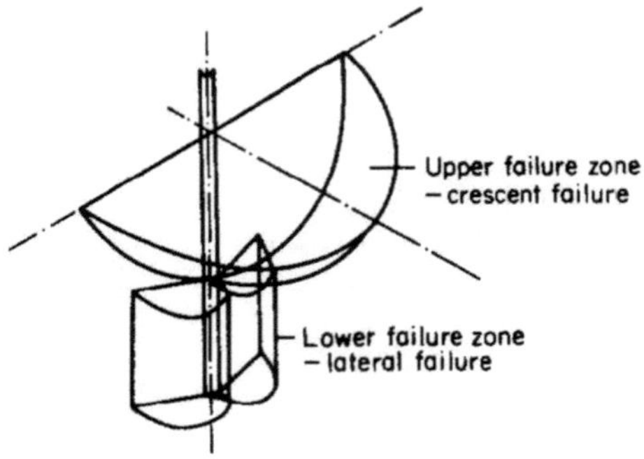

Fig. 5 Both soil failure modes of a tine are shown [4]

2 Rover Testing

Icebreaker is a prototype rover designed to traverse into steep planetary craters. The rover is 70.4 kg with 47.9 % of the weight to the left and 55.0 % to the rear. It is 1.1 m wide and 1.4 m long with a track width of 15.5 cm. The top of the rover deck is 22 cm from the ground. It is a tracked rover since tracks can offer the best solution for operating over complex terrain. This is especially important for steep and loose slopes [13]. The chassis provides a rigid frame to which the tracks are attached and the plow is mounted. The plow is mounted 11 cm from the center of the rover and is actuated to allow for easy plow adjustments (some work has been done creating a slip controller to dynamically adjust the plow). The tip of the plow has a lexan cone which is needed to help pierce into the soil and proceed with lowering the plow further into the ground. All of the electronics and batteries for the rover are contained within the track side frame. This helps the rover maintain a very low center of gravity at 13.2 cm from the ground.

2.1 Experiment Design

Testing with Icebreaker was conducted at the Simulated Lunar Operations (SLOPE) facility at the NASA Glenn Research Center in Ohio. The SLOPE facility has a large sandbox of GRC-1 lunar simulant that is capable of tilting to 30°. The soil is prepared using a standard method called T3 [16] to achieve conditions similar to the lunar terrain [11]. This process involves using a shovel to disturb and loosen the subsurface soil where the rover will be operating and then it is smoothed lightly using a rake to insure uniform conditions for the test. A Leica TCRA1103 + Total Station is used to track the rovers position with high accuracy in order to determine slip (Fig. 6). There is also an adjustable load system and a load cell for conducting drawbar pull tests (Fig. 7).

Four types of tests were conducted at the SLOPE test facility: decent, drawbar pull, cross slope, and point turns. A driving test on flat simulant was done to confirm that there is no bias between the tracks and to get the speed differential between commanded and actual drive distance. Before starting the tests it was qualitatively observed that the vehicle had minimal slip at around 17 cm plow depth. In order to bound this testing was done at 14 and 20 cm plow depth and then at 7 cm plow depth to bridge the gap from 0 to 14 cm in many of the tests. The plow used in this test is 5 cm (2 in.) in diameter and the vehicle moved at 0.02 m/s.

2.1.1 Descent

Descent tests were done with Icebreaker on a slope at 31° (Fig. 8) with plow depths of 0, 7, 14, and 20 cm where 0 cm is plow disengaged and 20 cm is the plow fully embedded into the ground. This gives a clean comparison of vehicle performance and how the plow improves mobility.

Fig. 6 Test setup with surveying instrument (on the *left*) and prism (*bright spot*) mounted on a rover at NASA GRC SLOPE facility

Fig. 7 Drawbar pull rig with adjustable load at the SLOPE facility [16]

Fig. 8 Descent test at 31°

2.1.2 Drawbar Pull

Drawbar pull tests were conducted with various loads at the four plow depths. For this test the drawbar pull rig at NASA Glenn Research Center was used. This rig allows for an adjustable load to be applied to the rover and for recording of vehicle position. A load cell was attached between the rover and the adjustable load rig. This helps us understand the braking force being applied by the plow to the rover (Fig. 9).

2.1.3 Cross Slope

Cross slope tests were done at slope angles of 10, 15, 20, 25, and 30° with no plow engaged (Fig. 10). The plow was not used in this case since the primary benifit of the plow is downslope.

Fig. 9 Draw bar pull test setup with lines going to a load cell and the variable load source

Fig. 10 Cross Slope test at 30°

2.1.4 Point Turns

Point turn tests were done at slope angles of 10, 15, 20, 25, and 30° for each plow depth of 0, 7, 14, and 20 cm. Figure 11a shows how the rover is positioned to start a test, Fig. 11b shows the rover in the middle of a test as it is pointed at about 45°, and Fig. 11c shows the position of the rover at the end of the test where it is pointed downhill.

2.2 Experimental Results

2.2.1 Descent

Descent tests exhibit an improvement as can be seen from having no plow where the rover slipped about 27 cm to when the plow is in the full 20 cm depth and acts like a brake to fully counteract the slip and even slightly impede rover motion (Fig. 12).

Fig. 11 Point turning test on 25° slope (clockwise rotation)

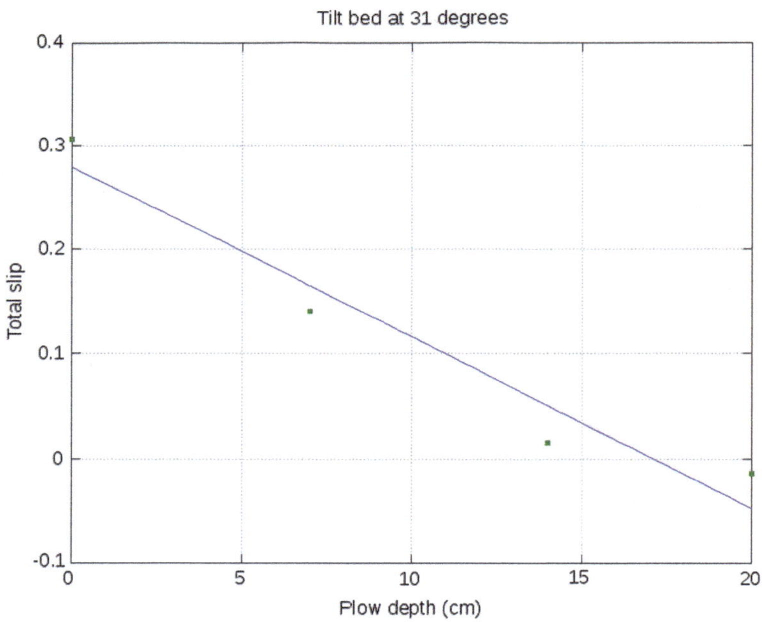

Fig. 12 Descent tests on a 31° slope at 4 plow depths

2.2.2 Drawbar Pull

In the drawbar pull tests results are consistent with what was observed in the slope tests. The deeper the plow is into the ground the greater the resistance to slip at a given force. This becomes interesting for the 20 cm plow depth where at an applied force of 100 % vehicle weight the rover still has trivial slip (Fig. 13).

2.2.3 Cross Slope

The tracked rover demonstrates excellent cross slope abilities. Downhill slip was below 4 % up to a terrain grade of 25°. At 30° the slip became 7.2 % which is still very good for such an extreme slope angle on unconsolidated GRC-1 terrain (Fig. 14).

2.2.4 Point Turns

Point turning with the plow also has a tremendous influence (Fig. 15). While with full plow depth all of the turns had less than 10 cm of slip the effects at the 30°

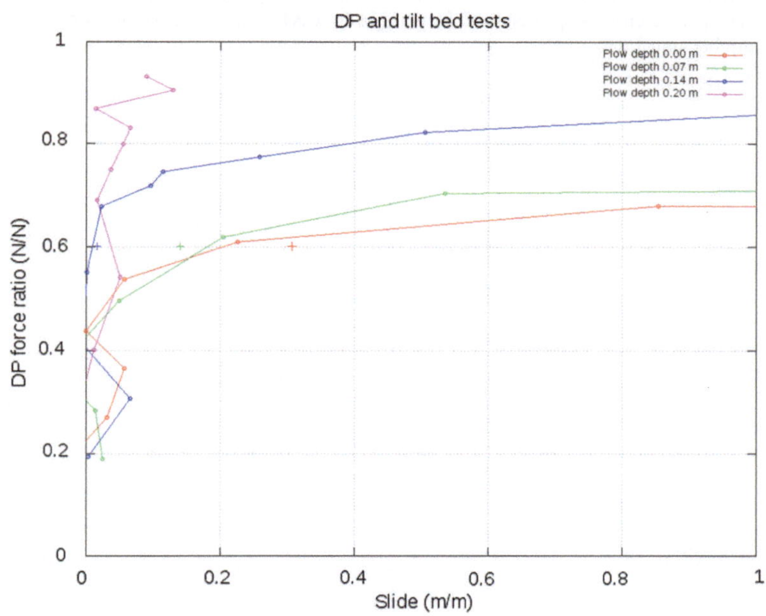

Fig. 13 Drawbar pull with the plow at 4 depths. *Plus* points represent the slip values from the descent tests (Smaller DP force ratio curves correspond to smaller plow depths)

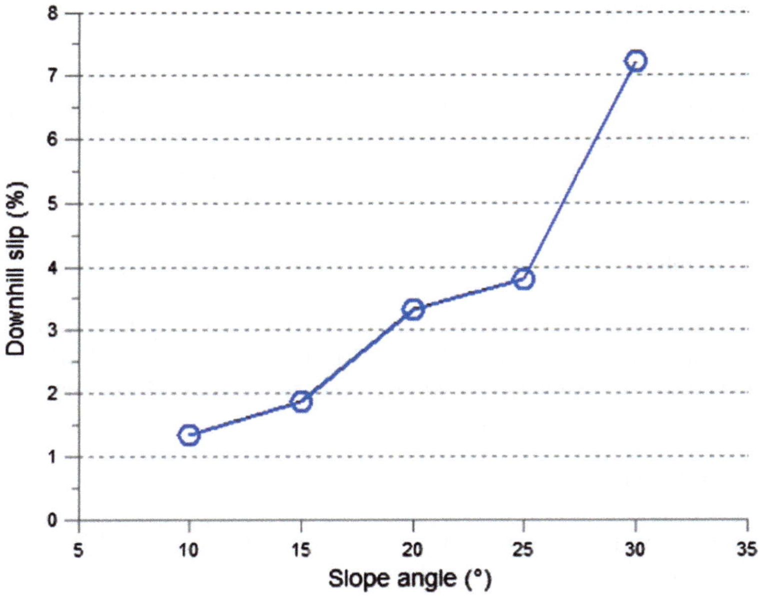

Fig. 14 Cross-slope tests results on GRC-1. Downhill slip is the percentage of downhill displacement with respect to cross-slope distance. Downhill slip was below 4 % up to a terrain grade of 25°. At 30°, slip became 7.2 %

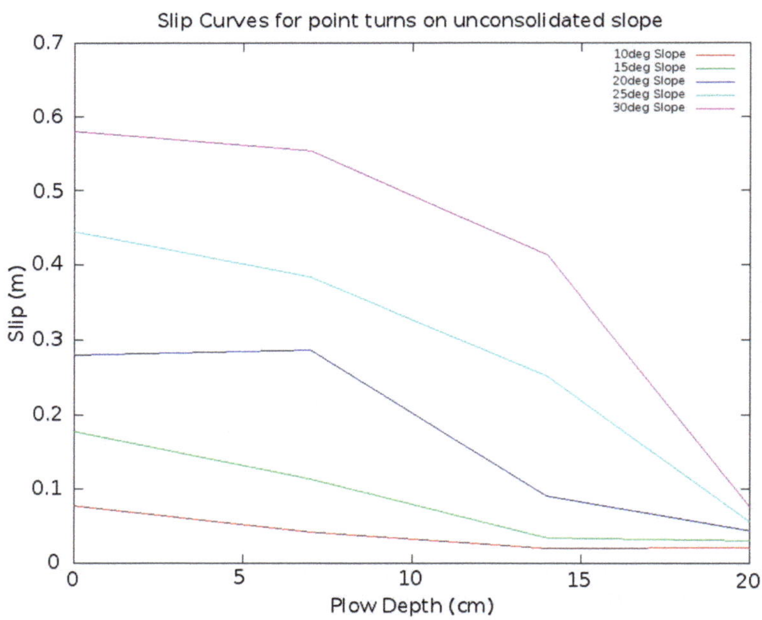

Fig. 15 Point turns slip curves (Slip curves increase in value as the slope angle increases)

maximum slope are the most telling. At 30° with no plow the vehicle slipped 0.58 m, however with the plow fully engaged the slip was only 0.077 m.

2.3 *Summary*

The addition of the plow to the rover can clearly be seen to benefit mobility, above what can be achieved with just tractive ability. Using the plow reduces slip in a variety of situations that are important when traversing on extreme slopes. This is important as since mission planners can now develop plans for operating on steep slopes in order to meet objectives and to reach areas of interest.

3 Plow Analysis

In addition to rover testing this research examines the plow as a mechanism in order to develop the tool and to create a guide that future developers can use to design and evaluate the plow. This work focuses on the different variables that affect plow design independent of the vehicle.

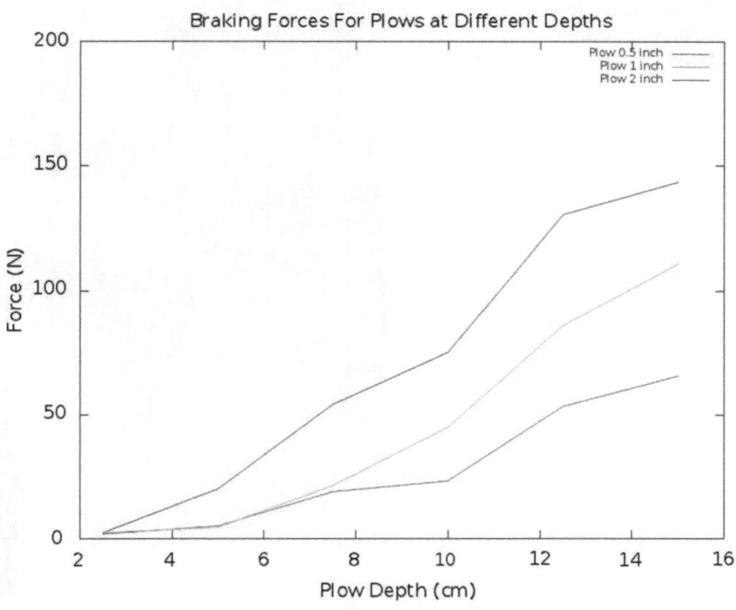

Fig. 16 Force (N) for a given plow diameter at different depths. *Note* Plow diameter is in inches (Force curves increase in value as the plow diameter increases)

3.1 Experiment Design

In order to quantify the plow performance, tests were conducted in the wheel test-bed at CMU [9]. These tests are designed to look at how the plow performance scales as the plow diameter changes and with plow depth.

The wheel test-bed has a sandbox filled with the GRC-1 lunar simulant that is prepared in a similar fashion to the large sandbox at the Glenn Research Center SLOPE facility before each test. In the sandbox a spade is used to loosen all of the subsurface sand and then it is lightly smoothed with the blade of a small dustpan. The wheel test-bed has a motor that can pull the plow at a constant velocity of 2 cm/s and record the forces on the plow in the direction of travel with an ATI Delta Transducer six axis force torque sensor. Tests were conducted with plows of 1.27 cm (0.5 in.), 2.54 cm (1 in.), and 5 cm (2 in.) diameter. A 7.62 cm (3 in.) plow was tested, however the forces were too large for the calibration range of the load cell. All of the plows

Table 1 Percent force increase from 2.5 to 15 cm plow depth for each plow diameter

Plow diameter (cm)	Percent change
1.27	33
2.54	45
5	56

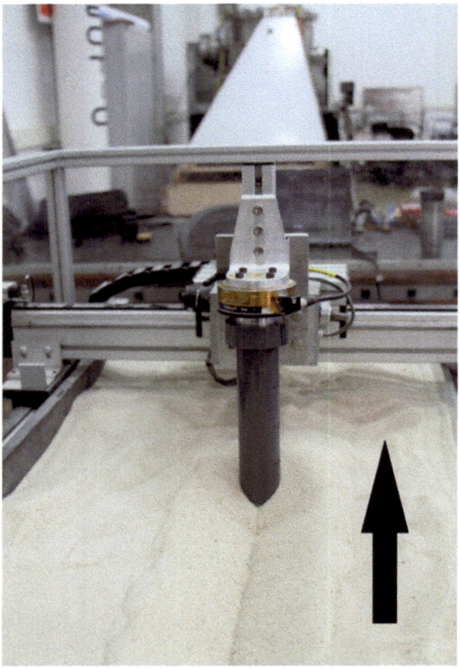

Fig. 17 Shape of path created by 5 cm (2 in.) diameter plow. *Black arrow* shows direction of plow motion

were tested at 2.5, 5, 7.5, 10, 12.5, and 15 cm depth. A depth of 20 cm was attempted, however the forces exceeded the range of the load cell.

3.2 Experimental Results

Testing done in the wheel test-bed is very informative and helps us understand how the plow works. The results can be seen in Fig. 16. As expected increasing the plow diameter and/or increasing the plow depth showed an increase in braking force. Since all of these plows had a critical depth less than 2 cm a braking force is achieved at the shallowest depth tested of 2.5 cm. For each plow the force at 15 cm depth increases 10 fold as shown in Table 1. This plow testing is important and useful for engineers developing plowing systems.

One of the basic ideas that aligns with theory is that all of the track patterns formed from the plow cutting through the sand were an identical shape as can be seen in Fig. 17. The depth of the chevron shaped mark from the plow is roughly the critical depth (Table 2). There is also a considerable amount of bulldozing at the front of the plow (Fig. 18).

Table 2 Critical depths for each plow tested determined experimentally

Plow diameter (cm)	Critical depth (cm)
1.27	2.26
2.54	3.33
5	4.5

Fig. 18 Bulldozing at end of 5 cm (2 in.) diameter plow. *Black arrow* shows direction of plow motion

4 Conclusion

This research affirms quantitatively that plowing can be used in practice for control on extreme planetary surfaces that defy tractive descent. This work has shown a 7.5 fold reduction in slip when using the plow on 30° slopes as well as a complete reduction of slip on straight downhill descents on slopes up to 31°. A new understanding of the plow teramechanics is now known based on a thorough review of farming literature and tine research. The trade offs of plow depth vs plow size has been explored. This gives the ability to design a plow and know what the braking forces will be. By using the plow rovers now have the ability to traverse into steep unconsolidated craters, canyons, funnels, and holes to explore both the slopes and the bottoms.

4.1 Future Work

There are many ways that this work can be extended and future directions for this work to follow. I think the most important one is to further develop a reactive plow controller based on on-board slip estimation. The ability to estimate slip on the fly and react is critical to steep slope navigation. Initial results with these type of controllers look very promising. This work has focused on a uni-directional plow however there is a lot of work that can be done with a directional and steerable plow (similar to a boats rudder) this might give even better performance. Adding a science instrument to the plow can make the plow even more valuable. Some instruments that could be integrated are cone penetrometers, water detecting tuned laser diodes, a drill and many others. Another area to look at it is the tradeoff between mobility methods and power. In the course of this work we started to look at this and realized a lot more work is needed.

Acknowledgments The authors would like to thank Colin Creager, Vivake Asnani, and Ryan Gilligan at NASA Glenn Research Center for all of their help and insight in preparing these tests and evaluating Icebreakers performance. The authors would also like to thank Daniel Loret De Mola Lemus for his help with testing.

References

1. B.M. French, G.H. Heiken, D.T. Vaniman (eds.), *Lunar Sourcebook: A User's Guide to the Moon* (Cambridge University Press, NY, 1991)
2. Michael Furlong, Thomas Howard, David Wettergreen, Model predictive control for mobile robots with actively reconfigurable chassis. Field Serv. Robot. **62**, 469–478 (2010)
3. R.J. Godwin, A review of the effect of implement geometry on soil failure and implement forces. Soil Tillage Res. **97**, 331–340 (2007)
4. R.J. Godwin, G. Spoor, Soil failure with narrow tines. J. Agric. Eng. Res. **22**, 213–228 (1977)
5. V.V. Gromov, W.D. Carrier, Mechanical properties of lunar soil and simulant. Enginnering, Construction, and Operations in Space, pp. 518–527, 1992

6. G. Ishigami, Terramechanics-based Analysis and Control for Lunar/Planetary Exploration Robots. Ph.D thesis, Tohoku University, March 2008
7. Genya Ishigami, Akiko Miwa, Keiji Nagatani, Kazuya Yoshida, Terramechanics-based model for steering maneuver of planetary exploration rovers on loose soil. J. Field Robot. **24**, 233–250 (2007)
8. D. Kohanbash, Plowing for rover control on extreme slopes, Carnegie Mellon University, Master's thesis, Dec 2011
9. S. Moreland, K. Skonieczny, D. Wettergreen, C. Creager, V. Asnani, Soil motion analysis system for examining wheel-soil shearing. in *International Conference of the International Society for Terrain-Vehicle Systems*, Sept 2011
10. I.A.D. Nesnas, P. Abad-Maneterola, J.A. Edlund, J.W. Burdick, Axel mobility platform for steep terrain excursions and sampling on planetary surfaces, *IEEE Aerospace Conferance*, 2008
11. H.A. Oravec, V.M. Asnani, X. Zeng, Design and charachterazation of grc-1:a soil for lunar terramechanics testing in earth-ambiant conditions. J. Teramech. **47**(6), 361–377 (2010)
12. V.M. Salokhe, B.K. Pathak, Effect of aspect ratio on soil failure pattern generated by vertical flat tines at low strain rates in dry sand. J. Agric. Eng. Res. **53**, 169–180 (1992)
13. S. Shoval, Stability of a multi tracker robot traveling over steep slopes. in *IEEE International Conferance of Robotics and Automation*, pp. 4701–4706, 2004
14. G. Spoor, R.J. Godwin, An experimental investigation into the deep loosening of soil by rigid tines. J. Agric. Eng. Res. **23**, 243–258 (1978)
15. J.Y. Wong, *Terramechanics and Off-Road Vehicle Engineering: Terrain Behavior, Off-road Vehicle Performance and Design*, 2nd edn. (Butterworth-Heinemann, Oxford, 2010)
16. A.C. Woodward, Experimental analysis of the effects of the variation of drawbar pull test parameters for exploration vehicles on grc-1 lunar soil simulant, Master's thesis, Virginia Polyttechnic Institute, June 2011
17. J. Zigler, D. Kohanbash, D. Wettergreen, W. Whittaker, Technolgies toward lunar crater exploration. Technical Report 0740, Carnegie Mellon University, 2007

Complementary Flyover and Rover Sensing for Superior Modeling of Planetary Features

Heather L. Jones, Uland Wong, Kevin M. Peterson, Jason Koenig, Aashish Sheshadri and William L. Red Whittaker

Abstract This paper presents complementary flyover and surface exploration for reconnaissance of planetary point destinations, like skylights and polar crater rims, where local 3D detail matters. Recent breakthroughs in precise, safe landing enable spacecraft to touch down within a few hundred meters of target destinations. These precision trajectories provide unprecedented access to bird's-eye views of the target site and enable a paradigm shift in terrain modeling and path planning. High-angle flyover views penetrate deep into concave features while low-angle rover perspectives provide detailed views of areas that cannot be seen in flight. By combining flyover and rover sensing in a complementary manner, coverage is improved and rover trajectory length is reduced by 40 %. Simulation results for modeling a lunar skylight are presented.

H. L. Jones (✉) · U. Wong · K. M. Peterson · A. Sheshadri · W. L. R. Whittaker
Robotics Institute, Carnegie Mellon University, 5000 Forbes Ave,
Pittsburgh, PA15213, USA
e-mail: hlj@cs.cmu.edu

U. Wong
e-mail: uyw@andrew.cmu.edu

K. M. Peterson
e-mail: kp@cs.cmu.edu

A. Sheshadri
e-mail: aashish.sheshadri@gmail.com

W. L. R. Whittaker
e-mail: red@cs.cmu.edu

J. Koenig
Computer Science Department, Carnegie Mellon University, 5000 Forbes Ave,
Pittsburgh, PA15213, USA
e-mail: jrkoenig@andrew.cmu.edu

K. Yoshida and S. Tadokoro (eds.), *Field and Service Robotics*,
Springer Tracts in Advanced Robotics 92, DOI: 10.1007/978-3-642-40686-7_28,

1 Introduction

This paper presents complementary flyover and surface exploration for reconnaissance of point destinations, like skylights and polar crater rims, where local 3D detail matters (See Fig. 1). In contrast to past missions where regional characterization was the goal, missions to point destinations will detail local terrain geometry, composition, and appearance. Characterization of this type requires high density sampling and complete coverage. Standard rover-only approaches are inefficient and cannot generate the coverage required for complete 3D modeling. Complementary flyover and surface exploration meets the requirements for modeling point features with higher efficiency than alternative approaches.

Persistent light illuminates polar locations on the Moon and Mercury. These destinations could serve as bases of operations or power stations for exploitation of polar resources, but for polar destinations, even small rocks cast long shadows, and unexpected shadows can be mission-ending for small rovers. Precise knowledge of 3D structure on the meter-scale and smaller is needed to predict where shadows will fall.

Sub-surface caverns may harbor life on Mars. They may be the best hope for human habitation on the Moon. They can provide windows into a planet's past geology, climate, and even biology. Skylights, formed by partial cave ceiling collapse, provide access to sub-surface voids. They have been conclusively shown to exist on Mars [1] and the Moon [2], and evidence supports their existence on other planetary bodies throughout the solar system [3]. Surface robots can approach and scan skylight walls, but skylight geometry prevents viewing the hole floor from a surface perspective.

Orbiters currently in service around the Moon and Mars are generating higher resolution data than ever before, but there are limits to what can be done from orbital distances. Even with a very good laser, the Lunar Reconnaissance Orbiter (LRO) sees a 5 m radius laser spot on the ground from its nominal 50 km mapping orbit [4], limiting modeling precision. LRO's camera is higher resolution, at 0.5 m per pixel for the 50 km orbit [5]. Stereo processing can be used to create a 2 m per post digital

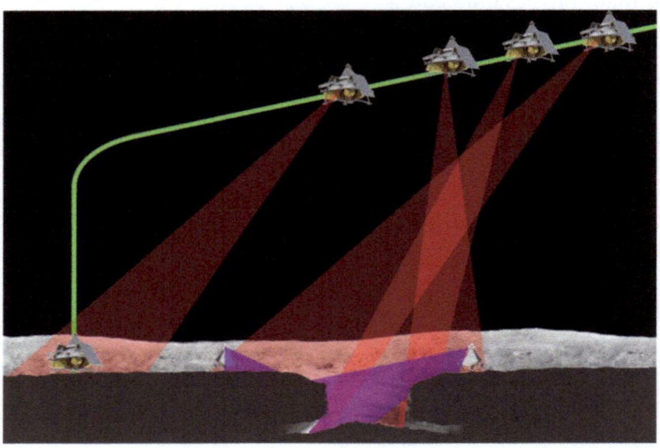

Fig. 1 Complementary flyover and surface modeling concept: a lander captures views of a terrain feature during final descent flyover. A rover carried by the lander returns to examine the feature in more detail

elevation map (DEM) from a pair of these images, but this only works for lit terrain. Skylights and polar craters contain terrain that is always in shadow. More detail, captured by flyover, is needed to see hazards on the scales that matter for robotic explorers.

New breakthroughs in terrain-relative navigation enable unprecedented precision in lander trajectory. This makes possible, for the first time, low-altitude lander flyover exploration of point targets. Precise, safeguarded landing can be achieved with real-time data from cameras and light detection and ranging (LIDAR), enabling a lander to identify a safe landing location and maneuver past hazards to safely touch down. Flyover data can further inform subsequent rover exploration for effectiveness, safety and coverage not possible in traditional missions with multi-kilometer landing ellipses. The combination of two perspectives, flyover birds-eye and rover on-the-ground, enables construction of the high-quality models needed to plan follow-on skylight exploration and science missions or develop detailed shadow prediction for crater rims. This paper presents a simulation of combined lander and rover modeling of a lunar skylight. A comparison is made between a model built with lander data only, a model built with rover data only, and a model built from combining lander and rover data, in which the rover views are chosen based on holes in the lander model.

Section 2 discusses related work in planetary exploration and "next best view" modeling. Section 3 discusses the approach to complementary flyover and surface modeling for point features where 3D detail matters. Specifics of the experiments conducted are presented in Sect. 4. Results are presented in Sect. 5. Sections 6 and 7 discuss conclusions and directions for future research.

2 Related Work

Modeling and localization are closely related: the robot location when a given frame of data was captured must be known to fit that data accurately into a model, and the most accurate localization estimate is often produced by building a model from multiple frames of data. Maps and 3D models of terrain have been created from a combination of orbiter, lander and rover imagery and used for rover localization, but not in a fully autonomous manner, and not for planetary features where 3D really matters. For the Mars Exploration Rovers (MERs), the DIMES system took three images of the landing site at about 1,000 m altitude during descent, aiming to determine the lander motion [6]. The MERs computed visual odometry onboard, although the computation was quite slow at 2 min per frame [6]. Visual odometry estimates of rover motion were more accurate than wheel odometry due to wheel slip, but position estimates still drifted over time, so bundle adjustment was performed on Earth to improve estimates of rover position. Tie points were selected automatically within a stereo image pair or panorama, and in some cases across different rover positions. DIMES imagery from the lander and HiRISE orbital imagery was used in localizing the rover and building maps, but the registration between rover and

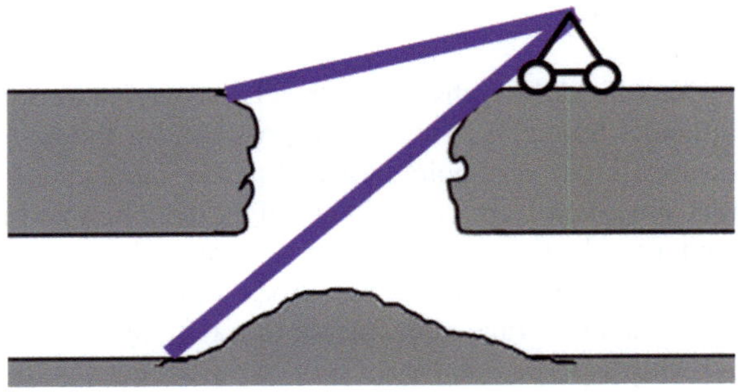

Fig. 2 Skylight geometry restricts visibility from a rover perspective. *Blue cone* shows example of visible area from a rover positioned at the skylight edge

overhead imagery was done manually [7]. While the models built by MER provide a fascinating glimpse of Martian terrain, they do not take on point features with geometries that severely restrict visibility. Victoria Crater is perhaps the closest—it has been modeled from orbit and investigated extensively by the Opportunity rover [8, 9], but at 750 m across and 75 m deep, Victoria Crater is not a point feature, and does not have visibility-restricting geometry. In contrast, the Marius Hills Hole, a lunar skylight, is estimated to be 48–57 m in diameter and approximately 45 m deep [2]. See Fig. 2 for an example of how skylight geometry prevents viewing the floor from a surface perspective.

The MER waypoints were chosen by operators on Earth, but significant work done in autonomous mapping and modeling can be leveraged to automate this part of the process. Work on laser scanning of unknown objects has used a "next best view" approach, choosing the next position from which to scan based on the amount of new information gained while maintaining overlap with existing data to facilitate model building [10]. This approach has also been applied to the robotic exploration of unknown environments [11].

Kruse, Gutsche and Wahl present a method for planning sensor views to explore a previously unknown 3D space [12]. This space is represented by a 3D grid, and each voxel in this grid is marked as either occupied, free or unknown. The value of a given view is evaluated by estimating the size of the unknown regions that become known after the measurement and determining the distance between that view and the current position in robot configuration space. The estimation of size for the unknown regions that can be seen in a given view is done using ray tracing, with a relatively small number of rays to limit computation time. This value function is re-evaluated after each view. The next view is chosen by following the gradient of the value function, starting from the current configuration. If the value function drops below a threshold, the gradient search is re-started from the best of a randomly chosen set of configurations.

Sawhney, Krishna and Srinathan use amount of unseen terrain visible and distance to determine the next best view for individuals in a multi-robot team. They find that the metric computed as (amount of unseen terrain)/distance is the most successful out of several evaluated [13].

Hollinger et al. use uncertainty to plan sensor views for a ship inspection robot [14]. They use a Gaussian process to model the surface of the ship hull. Because the shape of the ship is relatively well known before inspection, the approach assumes there will not be large changes to the model surface. This assumption would not hold in a skylight exploration case when it cannot be determined from the prior model whether a region inside the skylight is void space or collapsed ceiling.

3 Complementary Flyover and Surface Modeling Approach

3.1 Overview

This work combines lander flyover and rover exploration data to autonomously model point destinations where 3D detail matters. Lander and rover use both cameras and active sensors, such as LIDAR. Active sensing is needed to peer into shadowed regions, but active sensors are range-limited by available power and lack the high resolution of cameras.

Satellite imagery is used for terrain relative navigation, enabling landers to precisely position themselves as they fly over the features of interest. This technology enables landers to fly within 30 m of their intended trajectory within the final 500 m of descent and model regions on order of 50 m across from very low altitude. Hazard detection and avoidance technology, combined with precise navigation, enables safe and autonomous landings near features even without guaranteed-safe zones of landing-ellipse size.

Rover modeling begins at the lander location, providing a common tie-point between surface and flyover models. On-board hazard detection and avoidance ensure safety as a rover moves. Rover paths and sensor views can be autonomously chosen, using a "next best view" approach, to fill holes in a lander model.

Lander flyover captures detailed overview data, as well as perspectives that cannot be observed from a rover viewpoint. Rovers can capture close-up images of the terrain, and they can linger to capture multiple views from stationary locations, though always from low, grazing perspectives. Alternately, landers can acquire bird's-eye views but with less detail and resolution since their one-pass, always-moving trajectories are constrained by fuel limitations. Lander and rover data are combined, using lander data to localize and plan rover paths, to autonomously construct quality 3D models of point destinations.

3.2 Lander and Rover Trajectory and Sensing

For complementary flyover and surface modeling, the portion of the lander trajectory of interest is the final 500 m of descent. By this point, the lander has already canceled most of its forward velocity. It pitches over to a vertical orientation and cancels gravity to maintain a constant velocity. The lander points its sensors toward the feature of interest. After passing over the feature, the lander uses its LIDAR to detect hazards and follows a trajectory to avoid detected hazards in the landing zone. Above its target landing point, it cancels the rest of its forward velocity and descends straight down.

There is a trade-off between time to capture data and fuel used: flying slowly over a feature leaves more time to capture data but requires more fuel to maintain altitude for a low flyover; flying quickly over the feature saves fuel but may result in sparse data coverage. With this architecture, sparse data collected by the lander can be filled in by the rover, enabling the lander to move quickly and save fuel.

For complementary flyover and surface modeling, rover trajectories are described by waypoints. Desired view angles at those waypoints are specified. A rover drives from one waypoint to the next and stops to capture desired views at each waypoint.

3.3 Complementary Flyover and Surface Modeling

Following lander flyover, a point cloud model built from lander data is binned into a voxel array. Voxels which contain at least one point are marked as occupied, and all others as unseen. Ray tracing is then done from each camera perspective, and all voxels that a ray passes through before it hits an occupied voxel are marked as seen and empty. The Amanatides and Woo algorithm is used to determine ray-voxel intersections [15].[1]

The area of interest is gridded into possible rover positions. For each possible position, a discrete set of possible views are evaluated. The voxel array with seen, unseen and empty voxels is used to predict unseen areas that can be observed from each rover view. The list of previously unseen voxels visible from each view is stored, as well as the total number that can be seen in all of the views from a given position.

A metric is then computed as (number of unseen voxels visible)/ (distance from current position). The distance is computed along a straight line rover path. The position with the highest value for this metric is the next rover position. The voxels that were predicted to be visible in views from the new rover position are then marked as seen, and the metric is recomputed. This is repeated until there are no rover positions for which previously unseen voxels are visible. This approach does not always result in a particularly efficient path between waypoints. Given a set of rover positions with planned views which cover the space of visible but as-yet

[1] The code for this operation is a vectorized version of an implementation by Jesus Mena-Chalco, available on MATLAB Central: http://www.mathworks.com/matlabcentral/fileexchange/26852

unseen voxels, the order in which those positions are visited can be changed without affecting the total number of as-yet unseen voxels observed, so the positions were re-ordered using a greedy algorithm to pick the next-closest position among the set of selected rover positions.

4 Experiments

4.1 Overview

A mission to a lunar skylight is examined as an archetypical example of the complementary flyover and surface modeling approach. Lander and rover data for skylight modeling are collected in simulation. Three mission scenarios for modeling a skylight are evaluated: (1) lander alone, (2) rover alone, (3) lander and rover, where lander model informs rover path. Case 1 mimics the architecture in which there is no rover, and the lander is the sole robotic explorer. Case 2 mimics the case in which the lander does not attempt to fly over or near the feature of interest to capture overhead views. A rover travels from the landing site to the feature of interest and is the only robotic explorer to view the feature. Case 3 follows the architecture presented in Sect. 3, autonomously selecting the rover path to improve the model generated by the lander. Data for cases 1 and 2 were generated independently. For case 3, the lander-only model from case 1 was used to determine the rover path and views. The data for this alternate rover path was then generated and combined with the case 1 model.

The simulated skylight in this work is 30 m in radius and approximately 40 m deep. Modeling is restricted to a target area $100\,\text{m}^2$, centered on the skylight, although sensor views may go beyond this area.

Models built using each of the three scenarios are compared using the metric of percent coverage. In these experiments, percent coverage is defined as the percent of voxels that are filled in the ground truth model that are also filled in the generated model.

For robotic planetary landers and ground vehicles, the size of a feature considered to be a hazard is on order of 0.2–1 m [16–18]. For a terrain model to be valuable for robotic operations, the resolution should be in this range or better. In the experiment presented here, the voxelization of the lander model used in the complementary flyover and surface modeling method was 0.5 m, and the models for each scenario were also discretized to 0.5 m for comparison with ground truth. Higher resolution models can easily be built by using a smaller voxel size in the complementary flyover and surface modeling method.

Fig. 3 Bird's eye view of simulated terrain containing a skylight. Section shown is 600 m long. Skylight diameter is 60 m, depth is 40 m

4.2 Simulation

A 3D model of a lunar-like site containing a skylight was fabricated for this work (Fig. 3). A 2 m/post digital elevation map (DEM) of the Apollo 11 region, created using a stereo pair from the Lunar Reconnaissance Orbiter, was used as the base terrain. Small craters and rocks were overlaid following a statistical distribution derived from Surveyor data [19]. A procedural model was used, so that generated features did not need to be explicitly stored and only details visible in a given image were processed. A static 3D mesh of the skylight was manually modeled using Blender software [20], based on imagery and measurements of the Marius Hills skylight [3]. Terrain components were combined, detail textures were added, and simulated color and depth data were generated using Blender's built-in raytracer. Images were synthesized using a pinhole camera model with the appropriate field of view and resolution for each sensor. For LIDAR, depth channel instead of color channel information was used. Range data were transformed from the orthographic projection of the depth buffer to a perspective projection with rays from the LIDAR center.

4.3 Lander Modeling

For purposes of this study, it is assumed that the lander has a camera and a flash LIDAR mounted on a gimbal. The camera has a field of view (FOV) of 15° by 15° and a resolution of 1944×1944 pixels. The LIDAR also has a 15° FOV, but its resolution is 128×128 pixels. The optical centers of the camera and LIDAR are assumed to be co-located, which could be accomplished using a cold mirror.

The lander targets a constant glide slope of approximately 7.5 m/s along-track and 3.5 m/s down while scanning the landing zone and the skylight. After passing over the skylight it zeros its forward velocity and begins a vertical descent to the ground. The ground track follows a straight line across the center of the hole to the landing site. Due to the limited visibility that would occur when the sensors were mounted on a physical lander, it is assumed that the sensor views are restricted to look forward along the lander's direction of travel and to the sides—no backward-facing sensor views are allowed. Lander camera and LIDAR views are specified with the camera

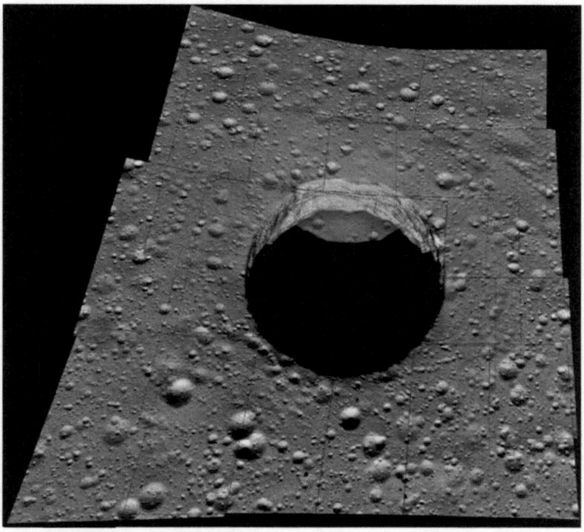

Fig. 4 Lander images stitched together using SIFT features

position and sensor view direction. Views are combined assuming that the lander position and orientation are known for each sensor view, so ground truth values are used for these parameters. The data from each view are first combined assuming that the lander position and orientation are known for each sensor view. LIDAR point clouds are stitched into a single model using ground truth lander poses, and the combined model is compared to the ground truth 3D hole model to determine coverage. In the case where the lander trajectory is not known, SIFT features [21] can be detected in each image, and the relative camera poses can be determined by matching features between images. Figure 4 shows an example of images stitched using this method. These poses could also used to stitch together the LIDAR data, although that is not explored in this work.

4.4 Rover Modeling

The rover is assumed to have a 45° FOV flash LIDAR with a resolution of 128×128 pixels. It is also assumed to have a pair of stereo cameras with 45° FOV and 1000×1000 pixel resolution. LIDAR and cameras are located on a pan-tilt head.

A naive rover path is selected to achieve full coverage of the skylight walls and surrounding terrain. The tilt angle of the rover sensors is set to achieve a desired resolution of data on the ground given the sensor resolution and FOV. The desired percentage overlap in sensor frames along the rover's path and between neighboring parallel paths is set to ensure that no gaps will be left between neighboring paths, and a resulting zig-zag pattern is generated. So long as the desired resolution can

Fig. 5 Simulated camera image from rover's-eye view, looking at the skylight edge

be achieved with the selected FOV, the FOV does not significantly affect the sensor coverage on the ground.

In order to maintain rover safety, there is a keep-out zone around the hole which the rover cannot enter. For this work, the width of the keep-out zone is 7.5 m. When the rover's coverage pattern intersects the edge of the keep-out zone for the first time, the rover makes a circle of the skylight, aiming its sensors inward to capture the ground in the keep-out zone. As it travels around the skylight rim, it stops periodically to scan the opposing skylight wall. The frequency of these stops is determined by the portion of the far wall that would be visible given the sensor FOV, assuming the hole is cylindrical. If the sensor resolution on the far wall is predicted to be lower than the desired resolution, multiple overlapping frames of data are captured to achieve the desired resolution. Once the rover completes the circle, it continues its zig-zag pattern.

Data from each view are combined assuming that the rover position and sensor pointing direction are known for each sensor view, and ground truth values are used for these parameters. See Fig. 5 for an example rover camera image.

4.5 Lander and Rover Modeling with Informed Path Selection

For this experiment, the method in Sect. 3 is applied to the lander model from Sect. 4.3. The rover is assumed to have camera and LIDAR with the FOVs and resolutions described in Sect. 4.4. In this experiment, the grid of possible rover positions had

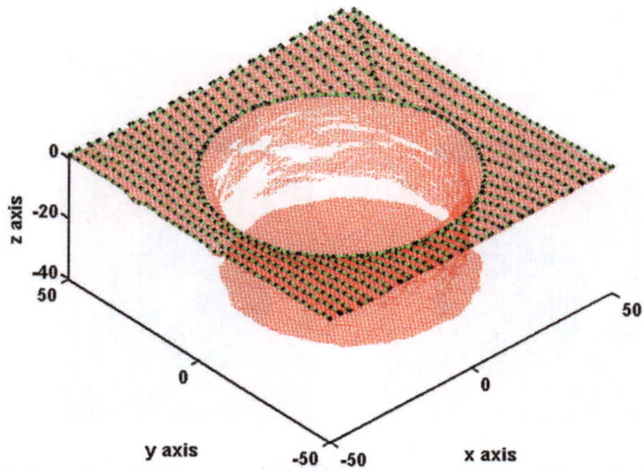

Fig. 6 Rover camera positions (*black*) and path (*green*) chosen naively to achieve full coverage, overlaid on voxel model of skylight. Axes are marked in meters

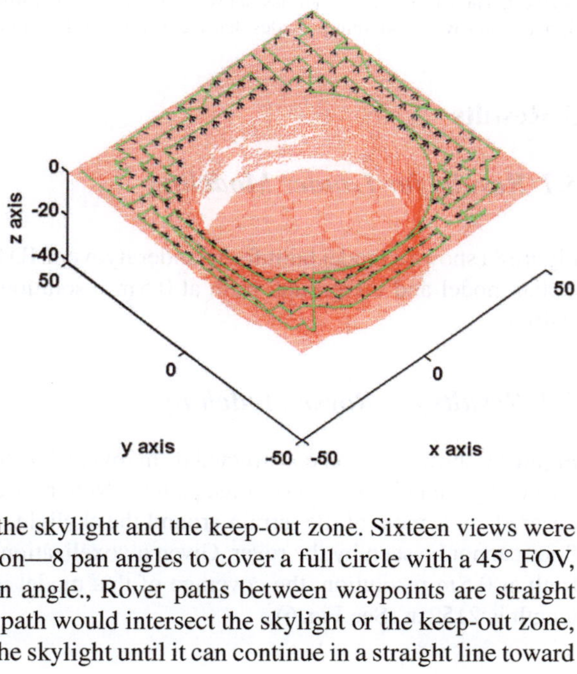

Fig. 7 Rover path planned using method from Sect. 3, overlaid on voxel model of skylight

squares of 5 m and excluded the skylight and the keep-out zone. Sixteen views were evaluated at each rover position—8 pan angles to cover a full circle with a 45° FOV, and 2 tilt angles for each pan angle., Rover paths between waypoints are straight lines, unless the straight-line path would intersect the skylight or the keep-out zone, in which case the path skirts the skylight until it can continue in a straight line toward the target waypoint.

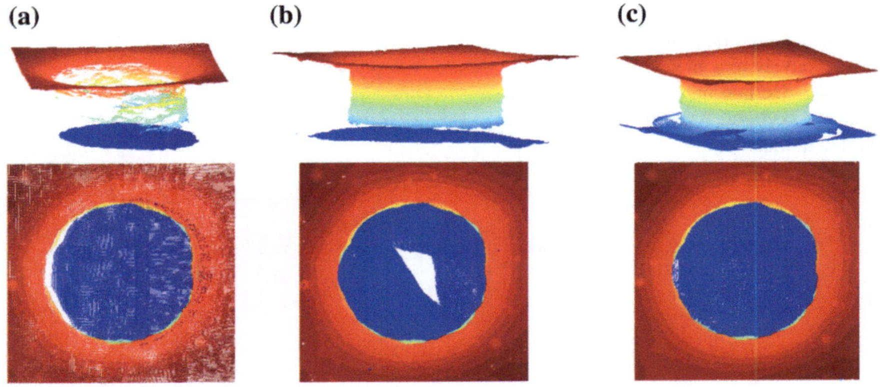

Fig. 8 Side and top views of point clouds constructed from **a** flyover LIDAR data, **b** rover LIDAR data, and **c** flyover and rover LIDAR data. The flyover-only model misses detail on the walls of the skylight. The rover-only model has dense wall coverage, but misses detail on the floor. Combined flyover and rover modeling provides dense coverage of the entire feature

5 Results

5.1 Results for Lander Modeling

Figure 8a shows a model built from lander flyover LIDAR data. Given a voxelization of the model and the ground truth at 0.5 m resolution, the coverage of this model is 46 %.

5.2 Results for Rover Modeling

Figure 8b shows a model constructed from rover LIDAR data. These data were generated using the naive rover coverage pattern. Note how the skylight walls are densely covered, as is most of the terrain around the skylight, but the central portion of the floor cannot be seen by the rover. Given a voxelization of the model and the ground truth at 0.5 m resolution, the coverage of this model is 85 %. The naive rover path length is 2152 m (See Fig. 6).

5.3 Results for Lander and Rover Modeling with Informed Path Selection

Figure 8c shows a model constructed from flyover and rover LIDAR data using the "next best view" approach for rover path planning. The lander model is used to generate the rover plan. Skylight walls and surrounding terrain are densely covered. The resulting model covers 92 % of the ground truth with a rover path length of

1281 m, a 40 % reduction in path length compared to a the naive coverage pattern. Figure 7 shows the planned rover path and views.

6 Conclusions

This paper has presented approach and analysis of a new mission concept: complementary flyover and rover sensing for planetary features. In order to explore and model planetary terrain features, rovers will be delivered to the surface by landers. In complementary flyover and surface modeling, the lander is leveraged to serve a dual purpose as a flyover explorer, capturing 3D data at a much higher resolution than possible from orbit and obtaining views of the feature not possible from a rover perspective. By planning rover sensor views from this prior lander model, complementary flyover and surface modeling achieves greater coverage with a significantly shorter rover traverse.

Experiments performed for a simulated skylight reconnaissance mission demonstrate significant improvements in coverage and quality relative to a rover-only or lander-only mission. The model built by combining lander and rover data following the complementary flyover and surface modeling method achieves 92 % terrain coverage, compared to 46 % coverage for the lander alone and 85 % coverage for a rover-only approach. The rover-only approach, using a path planned with very limited prior information about the skylight feature, has a path length of 2152 m. The complementary flyover and surface modeling approach, by planning the rover path and views using lander data, is reduced to 1281 m, a 40 % reduction in distance traveled. Although not explored in this work, high resolution data from the lander can also be used to precisely localize the rover and identify hazards, further improving the efficiency of rover operations.

7 Future Work

Lander and rover positions were assumed known for this work, but in the future, accuracy of localization and effects of localization error will be investigated. If only relative navigation techniques are used and models are built purely from these navigation estimates, position will drift over time. This means, for example, that a longer rover traverse will tend to result in a less accurate model. Model building can also contribute to localization, as was done for the MER rovers.

The effects of noise in the LIDAR data and in the camera and LIDAR commanded orientations will be investigated in future work. Noisy data would tend to reduce the accuracy of the model, and it could also reduce coverage.

Obstacle avoidance was not considered for these simulations. For a real mission, however, a rover would need to detour to avoid obstacles. These detours would change the path length and could provide obstructions to viewing certain areas of terrain that had been predicted to be visible. Additional methods of path planning for the rover views given the flyover data will also be explored.

Acknowledgments This work is partially supported by the NASA Innovative Advanced Concepts program under contract NNX11AR42G. The authors would also like to thank NVIDIA for donation of a machine that was used in the high-fidelity simulation of terrain done for this work.

References

1. G. Cushing, T. Titus, E. Maclennan, Orbital observations of martian caveentrance candidates, in *First International Planetary Caves Workshop*, Carlsbad, NM 2011
2. J.W. Ashley, M.S. Robinson, B.R. Hawke, A. Boyd, R.V. Wagner, E.J. Speyerer, H. Hiesinger, C.H. van der Bogert, Lunar caves in mare deposits imaged by the LROC narrow angle camera, in *First International Planetary Caves Workshop*, Carlsbad, NM 2011
3. J.W. Ashley, A.K. Boyd, H. Hiesinger, M.S. Robinson, T. Tran, C.H. van der Bogert, LROC science team: lunar pits: sublunarean voids and the nature of mare emplacement, in *LPSC*, The Woodlands, TX 2011
4. H. Riris, J. Cavanaugh, X. Sun, P. Liivia, M. Rodriguez, G. Neuman, The Lunar Orbiter Laser Altimeter (LOLA) on nasa's lunar reconnaissance orbiter (LRO) mission, in *International Conference on Space Optics*, Rhodes, Greece 2010
5. M. Robinson, S. Brylow, M. Tschimmel, D. Humm, S. Lawrence, P. Thomas, B. Denevi, E. Bowman-Cisneros, J. Zerr, M. Ravine, M. Caplinger, F. Ghaemi, J. Schaffner, M. Malin, P. Mahanti, A. Bartels, J. Anderson, T. Tran, E. Eliason, A. McEwen, E. Turtle, B. Jolliff, H. Hiesinger, Lunar Reconnaissance Orbiter Camera (LROC) instrument overview. Space Sci. Rev. **150**, 81–124 (2010)
6. R. Li, S.W. Squyres, R. Arvidson, B. Archinal, J. Bell, Y. Cheng, L. Crumpler, D.J. Des Marais, K. Di, T.A. Ely, M. Golombek, E. Graat, J. Grant, J. Guinn, A. Johnson, R. Greeley, R. Kirk, M. Maimone, L.H. Matthies, M. Malin, T. Parker, M. Sims, L.A. Soderblom, S. Thompson, J. Wang, P. Whelley, F. Xu, Initial results of rover localization and topographic mapping for the 2003 mars exploration rover mission. Photogram. Eng. Remote Sens. **71**(10), 1129–1142 (2005)
7. R. Li, S. He, Y. Chen, M. Tang, P. Tang, K. Di, L. Matthies, R.E. Arvidson, S.W. Squyers, L.S. Crumpler, T. Parker, M. Sims, MER Spirit rover localization: Comparison of ground image-and orbital image-based methods and science applications. JGR **116**, 1–12 (2011)
8. R. Li, K. Di, J. Hwangbo, Y. Chen, A.S. Team, Rigorous photogrammetric processing of HiRISE stereo images and topographic mapping at mars exploration rover landing sites, in *LPSC*, League City, 2008
9. S. Squyres, A. Knoll, R. Arvidson, J. Ashley, J.I. Bell, W. Calvin, P. Christensen, B. Clark, B. Cohen, P. de Souza, L. Edgar, W. Farrand, I. Fleischer, R. Gellert, M. Golombeck, J. Grant, J. Grotzinger, A. Hayes, H. Herkenhoff, J. Johnson, B. Jolliff, G. Klingelhofer, A. Knudson, R. Li, T. McCoy, S. McLennan, D. Ming, D. Mittlefehldt, R. Morris, J.J. Rice, C. Schroder, R. Sullivan, A. Yen, R. Yingst, Exploration of Victoria Crater by the mars rover opportunity. Science **324**(5930), 1058–1061 (2009)
10. R. Pito, A sensor-based solution to the gnext best viewh problem, in *Proceedings of International Conference on Pattern Recognition* 1996
11. K. Nagatani, T. Matsuzawa, K. Yoshida, Scan-point planning and 3-d map building for a 3-d laser range scanner in an outdoor environment, in *FSR*, 2009
12. E. Kruse, R. Gutsche, F.M. Wahl, Efficient, iterative, sensor based 3-d map building using rating functions in configuration space, in *ICRA*, vol. 2, (Minneapolis, Minnesota, 1996), pp. 1067–1072

13. R. Sawhney, K. Krishna, K. Srinathan, On fast exploration in 2d and 3d terrains with multiple robots, in *Proceedings of International Conference on Autonomous Agents and Multiagent Systems*, 2009
14. G.A. Hollinger, B. Englot, F. Hover, U. Mitra, G.S. Sukhatme, Uncertaintydriven view planning for underwater inspection, in *ICRA*, 2012
15. J. Amanatides, A. Woo, A fast voxel traversal algorithm for ray tracing, in *Proceedings of Eurographics*, 1987
16. M.S. Darms, P.E. Rybski, C. Baker, C. Urmson, Obstacle detection and tracking for the urban challenge. IEEE Trans. Intell. Transp. Syst. **10**(3), 475–485 (2009)
17. M. Wagner, S. Heys, D. Wettergreen, J. Teza, D. Apostolopoulos, G. Kantor, W. Whittaker, Design and control of a passively steered, dual axle vehicle. in *International Symposium on Artificial Intelligence, Robotics and Automation in Space*, 2005
18. R.D. Braun, R.M. Manning, Mars exploration entry, descent and landing challenges, in *IEEE Aerospace Conference*, 2006
19. NASA Marshall Space Flight Center: Lunar surface models. NASA space vehicle design criteria (Environment) NASA-SP-8023, NASA (1969)
20. Blender Foundation: Blender 2.59 (2011), www.blender.org
21. D. Lowe, Distinctive image features from scale-invariant keypoints. IJCV **60**(2), 91–110 (2004)

Path Planning and Navigation Framework for a Planetary Exploration Rover Using a Laser Range Finder

Genya Ishigami, Masatsugu Otsuki and Takashi Kubota

Abstract This chapter presents a path planning and navigation framework for a planetary exploration rover and its experimental tests at a Lunar/Martian analog site. The framework developed in this work employs a laser range finder (LRF) for terrain feature mapping. The path planning algorithm generates a feasible path based on a cost function consisting of terrain inclination, terrain roughness, and path length. A set of navigation commands for the rover is then computed from the generated path. The rover executes those navigation commands to reach a desired goal. In this paper, a terrain mapping technique that uses a LRF is described along with an introduction to a cylindrical coordinate digital elevation map (C^2DEM). The gird-based path planning algorithm is also presented. Field experiments regarding the path planning and navigation that evaluate the feasibility of the framework developed in this work are reported.

1 Introduction

The surface terrain of the Moon and Mars is covered with fine grained soil, or boulders, rocks, and stones. Because of this challenging terrain, the rover needs to evade mobility hazards, which includes vehicle rollover, immobilizing wheel slippage, or

G. Ishigami (✉)
Dept. of Mechanical Engineering, Keio University, Yokohama, Kanagawa, Japan
e-mail: ishigami@mech.keio.ac.jp

M. Otsuki · T. Kubota
Institute of Space and Astronautical Science, Japan Aerospace Exploration Agency, Sagamihara, Kanagawa, Japan
e-mail: otsuki.masatsugu@jaxa.jp

T. Kubota
e-mail: kubota@isas.jaxa.jp

K. Yoshida and S. Tadokoro (eds.), *Field and Service Robotics*,
Springer Tracts in Advanced Robotics 92, DOI: 10.1007/978-3-642-40686-7_29,

collision with obstacle rocks. Therefore, the planetary rover needs to navigate itself by sensing the environment as well as planning a feasible path for rough terrain traverse.

The rover uses visual information taken by an onboard stereoscopic camera to obtain terrain features and to find a path for travel [1–3]. The drawbacks of the camera-based terrain mapping are that it is generally a time-consuming task for the low-power flight CPU of the rover because the stereo images should be correlated (stereo matching), requiring a relatively long computational time [3]. Additionally, the quality of visual information may vary with the intensity of sunlights/shadows.

Another technique for terrain mapping involves the use of a laser range finder (LRF) that can determine the distance from a laser emitter to an object based on the time of flight principle. There has been an extensive research and development for the LRF technique in a geometrical feature mapping and terrain classification [4–8].The LRF can directly measure three-dimensional distances from the sensor to the objects, providing "Point cloud" data of the scene without any additional processes (c.f., the camera-based mapping needs stereo matching for the 3D mapping).

The primary focus of this paper is to implement a path planning and navigation framework as an application for the planetary exploration rover, that utilizes the LRF technique for terrain mapping. The point cloud data of the terrain features provided by the LRF are converted into a digital elevation map (DEM). A square-shaped reference grid is generally used for conventional DEM; however, in this work, a sector-shaped reference grid is applied for the DEM conversion, resulting in an elevation map with cylindrical coordinates, called C^2DEM. The path planning algorithm in the framework considers three indices: terrain inclination, terrain roughness, and path length. A cost function, composed of these indices with varied weighting factors, generates various paths on the map. The generated path is then discretized into several way points, which contain relative distances and headings from one to another. The way points are sequentially sent to a rover as navigation command. Field experiments were demonstrated at a Lunar/Martian analog terrain to evaluate the feasibility of the path planning and navigation framework. The main contribution of this work is the development and experimental validation of the LRF-based path planning and navigation framework that uses C^2DEM for the terrain representation.

This chapter is organized as follows: Sect. 2 describes the LRF-based terrain mapping method; Sect. 3 explains the path planning and navigation framework; and the experimental tests are summarized in Sect. 4.

2 Terrain Mapping Technique with a LRF

2.1 LRF Scanning System

The LRF used for terrain mapping is UXM-30 LX-E developed by Hokuyo Corp. [9]. The laser emitter/detector inside of the LRF rotates 360° in the yaw direction from an actuator, which realizes the two-dimensional plane scanning. This LRF is mounted on

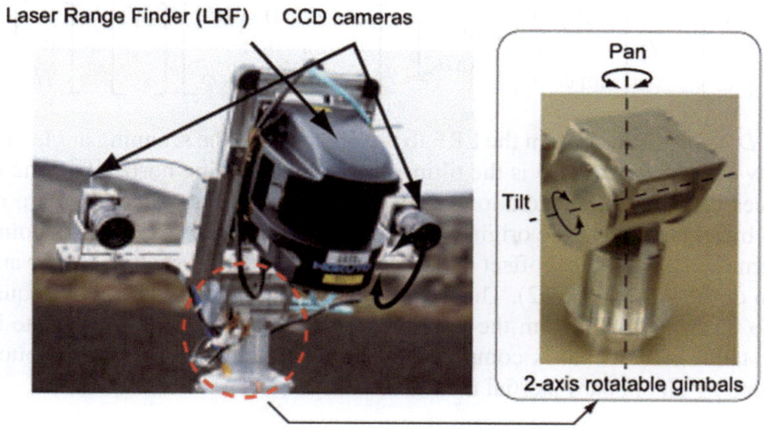

Fig. 1 Gimbaled-LRF sensory system for terrain mapping

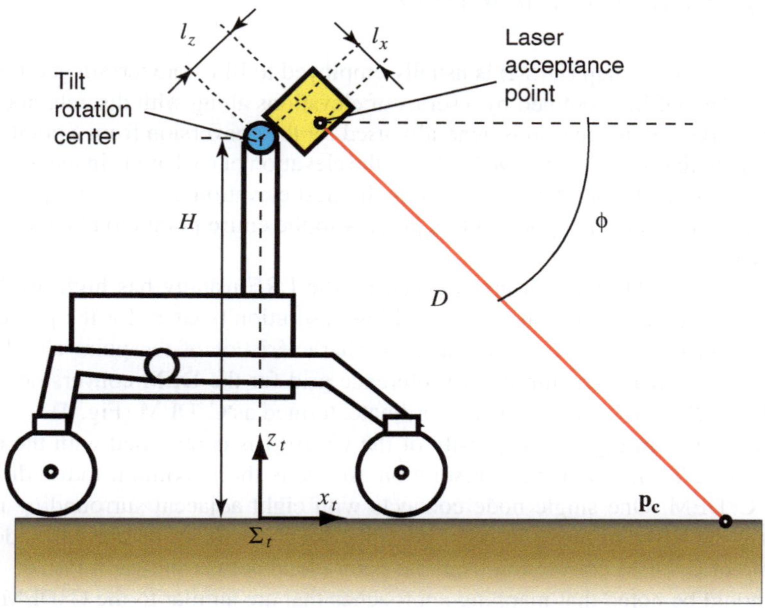

Fig. 2 Geometrical analysis of LRF based terrain mapping

the gimbals (Fig. 1). Controlling the tilting motion of the gimbals, along with the two-dimensional scanning of the LRF, three-dimensional terrain mapping is achieved.

Each scanned point of the terrain feature provided by the LRF includes the distance measured with respect to the scanning angle and the tilting angle. One single point p_c of the point cloud data is determined with the following equation (Fig. 2):

$$p_c = \begin{bmatrix} x \\ y \\ z \end{bmatrix} = D \begin{bmatrix} \cos\phi & 0 \\ 0 & 1 \\ \sin\phi & 0 \end{bmatrix} \begin{bmatrix} \cos\theta \\ \sin\theta \end{bmatrix} + \begin{bmatrix} \cos\phi & 0 & \sin\phi \\ 0 & 1 & 0 \\ -\sin\phi & 0 & \cos\phi \end{bmatrix} \begin{bmatrix} l_x \\ l_y \\ l_z \end{bmatrix} + \begin{bmatrix} 0 \\ 0 \\ H \end{bmatrix} \quad (1)$$

where D is the distance from the LRF to the objects, θ is the scanning angle (around the yaw of the LRF), and ϕ is the tilting angle with 0° being horizontal. The offset distances between the tilt rotation center and the light acceptance point are represented by l_x, l_y, and l_z. The origin of the terrain coordinate is located at a point that is determined by a vertical offset distance H between the tilt rotation center and the bottom of the wheels (Fig. 2). The point cloud data, defined by the above equation, need to be transformed from the terrain coordinate to the global coordinate based on a rotation matrix that is composed of the rover orientation (roll and pitch are measured by an onboard inertial measurement unit).

2.2 Cylindrical Coordinate DEM

A digital elevation map (DEM) is usually employed to illustrate terrain geometrical features. The DEM is defined by a series of elevations along with discrete nodes. A square-shaped reference grid is generally used for the conversion from a point cloud data point to the conventional DEM. Then, the elevation of each node in the reference grid is determined from the point with the highest elevation among multiple points inside the grid (Fig. 3). Applying this process to the entire point cloud data set, the DEM is obtained.

The point cloud data that are scanned by the LRF usually has high resolution (dense) for the points near the rover, and low resolution (sparse) for the points that are far from the rover. To represent these characteristics of the point cloud data, this work considers a sector-shaped reference grid for the DEM conversion which provides a DEM in the cylindrical coordinate, termed as C^2DEM (Fig. 4).

As observed in Fig. 5, the density of the C^2DEM is determined with the radial resolution ΔR and the angular resolution $\Delta\theta$. R is the maximum radial distance of the C^2DEM. One single node connects with eight adjacent surrounding nodes allowing the rover to move in eight possible directions from its current node. An example of C^2DEM conversion from the point cloud data is shown in Fig. 6.

It should be noted that mapping approaches that are similar to the C^2DEM have been reported. For example, a log-polar grid map representation with an ultrasonic sensor has been proposed in [10]. A comparison between a local multi-resolution representation and a log-polar representation has been presented in [11]. Additionally, a hyperbolic polar coordinate map for infinitely long distance vision and path planning has been described in [12]. Compared to these previous studies, the C^2DEM in this work is a fully three-dimensional map that can be applicable for path planning on rough terrain. Field experiments in this work validates the usefulness of the C^2DEM over the conventional DEM, as described in Sect. 4.

Fig. 3 Conventional DEM

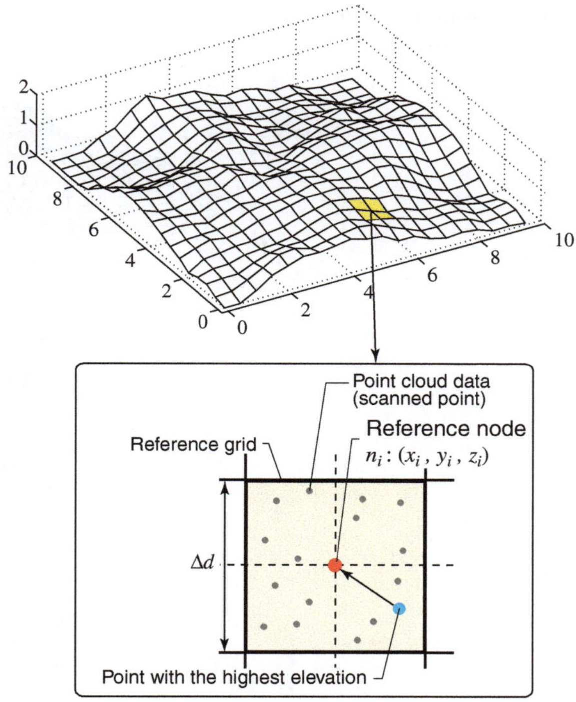

3 Path Planning and Navigation Framework

3.1 Path Planning Algorithm

There have been substantial studies dealing with path/motion planning algorithms for mobile robots, such as the $A*$ and $D*$ methods [13], the potential field approach [14], the probabilistic roadmap technique [15], and the rapidly-exploring random tree (RRT) algorithm [16].

The path planning algorithm in this work uses a classical graph search algorithm (Dijkstra's algorithm) to generate a path from a starting point to a goal point. A cost function used for the algorithm is composed of the three indices: terrain inclination, terrain roughness, and path length. Each index can be computed based on the C^2DEM representation.

3.1.1 Terrain Inclination Index

A rover experiences relatively high wheel slippage when it climbs up or traverses on a sloped terrain. This slippage is due to the traction load from the gravity, which

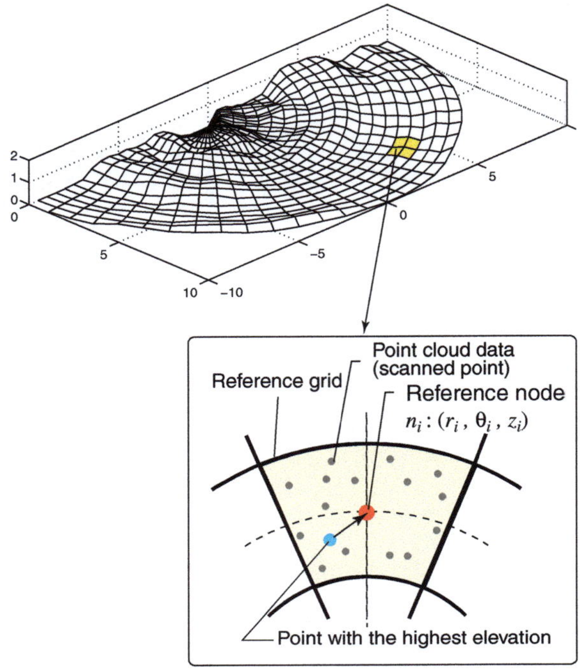

Fig. 4 Cylindrical coordinate DEM (C^{2}DEM)

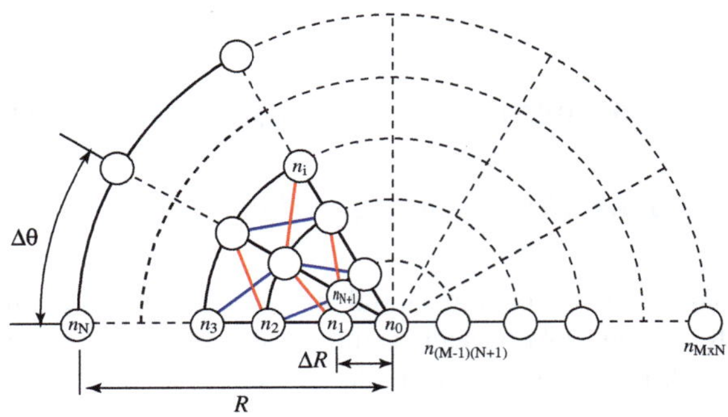

Fig. 5 Node connectivity between the adjacent nodes on the C^{2}DEM

becomes significant on the sloped terrain. Additionally, rollover of a rover traversing on a steep slope is a mobility hazard. The terrain inclination index is employed for the cost function to represent such risks on a sloped terrain.

The terrain inclination index is divided into two axes, roll and pitch of the rover (Fig. 7). Each axis is geometrically calculated as an angle between the inertial coordinate and a pseudo-plane composed of the wheel contact points at a projection region of the rover R_{ij}: Multiple terrain inclinations can be calculated from each subset of three contact points between the multiple wheels. The terrain inclination index is the largest inclination between these values.

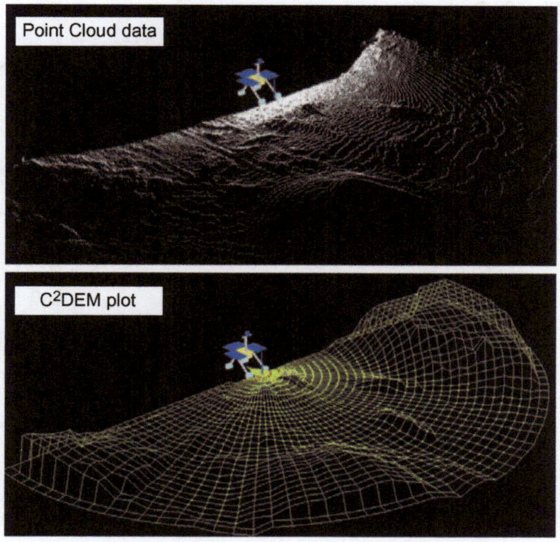

Fig. 6 An example of the conversion from point cloud data to the C^{2}DEM

$$\begin{aligned}\text{Roll} &: \Theta_{xij} = \max(\theta_x(R_{ij})) \\ \text{Pitch} &: \Theta_{y_{ij}} = \max(\theta_y(R_{ij}))\end{aligned} \tag{2}$$

The projection region R_{ij} is determined with the dimension of the rover. The heading direction is aligned along with the vector $\overrightarrow{n}_{ij}$, which is composed of the current node n_i and adjacent node n_j (Fig. 8).

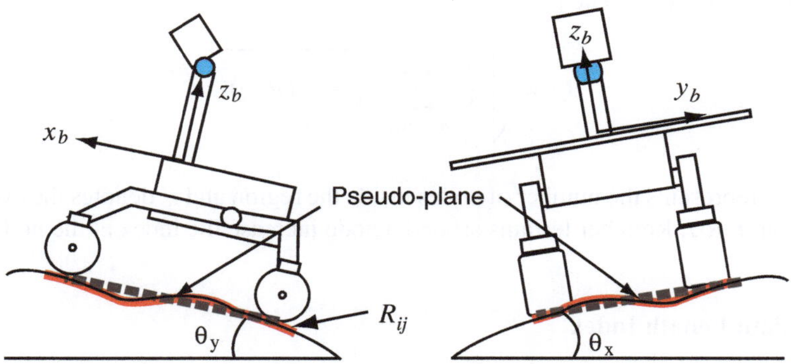

Fig. 7 Terrain inclination angles. The curve *bold line* depicts the projection region R_{ij} and the *dotted line* depicts the pseudo-plane, which is composed of the wheel contact points

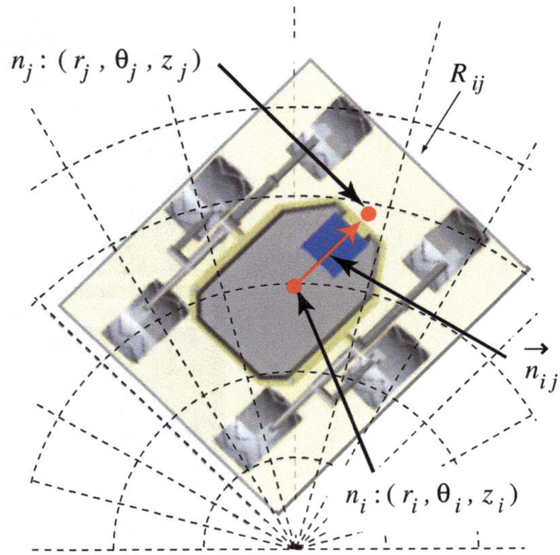

Fig. 8 Projection region *yellow-colored area* of the rover on a terrain map

3.1.2 Terrain Roughness Index

The terrain roughness is related to the traversability of a rover. To avoid uneven bumpy areas for travel path, the terrain roughness index B_{ij} is employed. First, the nodes inside the projection region R_{ij} are rotated around the x and the y with Θ_{xij} and $\Theta_{y_{ij}}$, such that the nodes are represented in the terrain coordinate system. The terrain roughness is then calculated as the standard deviation of the local elevations z' at the terrain coordinate [17]:

$$B_{ij} = \sqrt{\frac{1}{N} \sum_{n_k \in R_{ij}} \left(z'_k - \bar{z}'(R_{ij})\right)^2} \tag{3}$$

where N represents the number of nodes inside the region and $\bar{z}'$ denotes the average elevation in R_{ij}. Rougher terrains around a node increase the index at the node.

3.1.3 Path Length Index

The path length index is used to find the shortest path from a starting point to a goal point. The path length index L_{ij} is the distance from the current node n_i to adjacent nodes n_j:

$$L_{ij} = |\vec{n}_{ij}| = \sqrt{r_i^2 + r_j^2 - 2r_i r_j(\cos(\theta_i - \theta_j)) + (z_i - z_j)^2} \tag{4}$$

3.1.4 Cost Function for Path Planning

A cost function for path planning is defined by the following equation that is comprised of the indices introduced in the above subsections:

$$C(\boldsymbol{p}) = \sum_{n_i \in \boldsymbol{p}} \left(W_{\theta_x} \frac{\Theta_{xij}}{N_{\theta_x}} + W_{\theta_y} \frac{\Theta_{yij}}{N_{\theta_y}} + W_B \frac{B_{ij}}{N_B} + W_L \frac{L_{ij}}{N_L} \right) \tag{5}$$

where W_{θ_x}, W_{θ_y}, W_B, and W_L are the weighting factors, which assign specific priorities to the corresponding indices. The weighting factors for the terrain inclination indices have a value of infinity if the indices exceed predetermined threshold angles. The threshold angles are determined based on the slope traversability of the rover (i.e., maximum slope climbing angle). N_{θ_x}, N_{θ_y}, N_B, and N_L are the normalization factors. Each factor is the maximum value of the corresponding index calculated from the terrain map. The path $\boldsymbol{p}$ consists of a series of neighboring/chained nodes:

$$\boldsymbol{p} = \{n_{\text{start}}, \ \cdots, \ n_i, \ n_j, \ \cdots, \ n_{\text{goal}}\} \tag{6}$$

For the cost function in Eq. 5, smaller index values result in lower mobility hazard levels on a path. Therefore, the path planning problem is a least-cost search problem. The conventional Dijkstra's algorithm is used in this work to find the least-cost path, providing a minimum value for the cost function. The least-cost path varies in accordance with the values of the weighting factors for the individual indices.

3.2 Navigation Command

The path generated by Eq. 5 is discretized into several way points that are equivalent to the nodes in Eq. 6. A navigation command $NavCmd_i = (d_i, \ \psi_i)$, consisting of the relative distance d_i and heading ψ_i from one way point to the next, can be geometrically calculated based on the terrain map data (Fig. 9). Then, the navigation commands are sequentially sent to the rover so that it travels through each way point.

3.3 Path Planning and Navigation Framework

The path planning and navigation framework are summarized as follows (Fig. 10):

1. The rover executes the LRF-based terrain mapping and obtains the point cloud data of the terrain map (the rover does not move while scanning).
2. The point cloud data are converted to the C^{2}DEM.
3. A rover operator selects a desired goal node located inside of the C^{2}DEM.

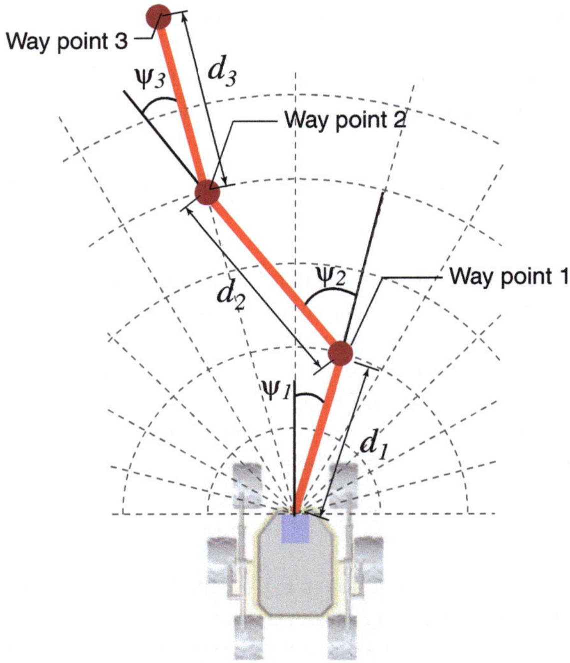

Fig. 9 Navigation command: relative distance and heading between each way point

4. The path planning algorithm generates a path from the rover position (origin of the map) to the goal node. Here, the goal node and the weighting factors may need to be changed if the cost $C(\boldsymbol{p})$ exceeds a predefined threshold. Exceeding this threshold indicates the path may include high mobility hazard areas.
5. The rover then starts traveling through the consecutive way points based on the navigation commands until the rover reaches the goal.
6. Once the rover reaches a (intermediate) goal, the rover repeats the above-mentioned tasks to advance its travel on rough terrain.

4 Field Experiments

Field experiments with a rover test bed were performed at a Lunar/Martian analog field (Izu-Oshima Island, Japan) to evaluate the usefulness of the path planning and navigation framework with the LRF-based terrain mapping. The terrain in the field is mostly covered with scoria, scattered rocks and stones. The test field also consists of sloped terrain, ditches, and vegetations.

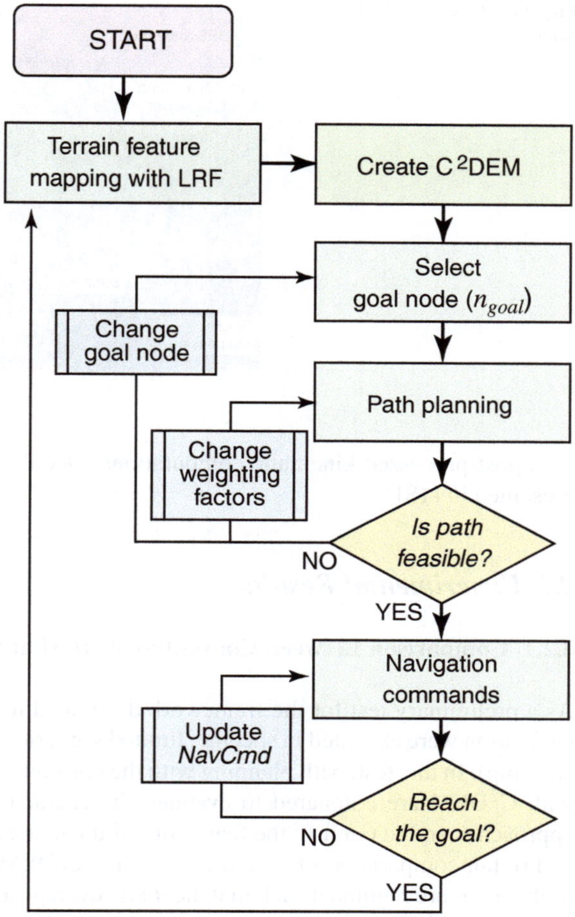

Fig. 10 Path planning and navigation framework

4.1 Rover Test Bed Overview

Micro-6 (Fig. 11) is the rover prototype developed by the authors to demonstrate the mobility and navigation on rough terrain and power management control. The rover has a four-wheeled drivable/steerable unit. The left and right wheel pairs are connected with passive rocker link suspension. The navigation subsystem of the rover has two stereo camera pairs and the LRF. The power management subsystem provides electrical power to the rover through the solar array panels with rechargeable batteries. The communication between the rover and its operators is handled via a wireless LAN with a 2.4 GHz band. The GPS antenna mounted on the rover measures the rover position as a ground truth to evaluate the rover trajectory after the experimental tests. The GPS measurement achieves an accuracy of a few centimeters

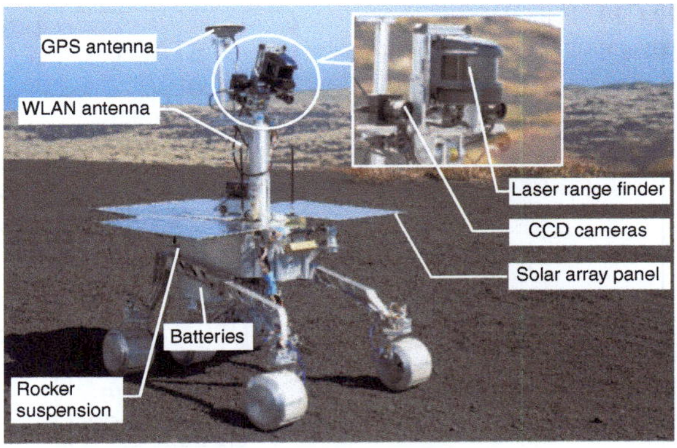

Fig. 11 Rover test bed: Micro-6

with post-processed kinematic computations. Details of the rover subsystems were presented in [18].

4.2 Experimental Results

4.2.1 Comparison Between Conventional DEM and C^{2}DEM

As a preliminary test for the framework developed in this work, path planning and navigation were executed in specific, limited scenarios (obstacle avoidance and ditch crossing). In this test, path planning with the conventional rectangular DEM and that with C^{2}DEM are compared to evaluate the characteristics of these two mapping approaches and to confirm the feasibility of the generated paths.

For fair comparison between the conventional DEM and C^{2}DEM, the dimensions of the maps are regulated such that the total coverage areas of both maps are of equal size of 157 m^2. The dimension of the conventional DEM is set as 8.86 (longitudinal) $\times$ 17.72 (lateral) m, while the maximum radial distance of the C^{2}DEM is 10.0 m and the angular range is between $\pm\pi/2$. Additionally, the resolutions of the map are set as the total number of nodes in each map are equal of 2,700. The grid interval of the conventional DEM is 0.245 m, while the C^{2}DEM has a radial interval of 0.2 m and an angular interval of 3.4 ° . The weighting factors are given as $W_B = 0.2$, $W_{\theta_x} = 0.3$, $W_{\theta_y} = 0.3$, and $W_L = 0.2$. The threshold angles for the roll and pitch axes are set as 5° so that W_{θ_x} and W_{θ_y} have a value of infinity for the node which exceeds these thresholds.

Figure 12 shows the results of terrain mapping and path planning in an obstacle avoidance scenario. Table 1 summarizes quantitative results of the generated path, which includes the total cost $C(\boldsymbol{p})$, the path length, and the maximum roll/pitch angles on the path. It can be observed that the paths generated in both maps are nearly equivalent. The total cost of the path and the path length are almost same values. The maximum pitch angle of the path on the C^{2}DEM is smaller than that on the

Table 1 Comparison between conventional DEM and C^2DEM: path planning resutls

Scenario	Map	Total cost	Path length (m)	Roll (deg)	Pitch (deg)
Obstacle	DEM	13.7	12.0	5.0	4.4
	C^2DEM	13.9	12.2	4.3	2.6
Ditch	DEM	14.2	17.5	3.9	3.9
	C^2DEM	10.2	17.2	2.7	1.9

conventional DEM. The path on the C^2DEM largely avoids the second obstacle that is located in the left front of the rover. It is deduced that the C^2DEM has dense nodes near the rover and, therefore, the path planner on the C^2DEM can more carefully select appropriate nodes in comparison with the planner on the conventional DEM.

The experiment with the ditch crossing scenario is shown in Fig. 13. For ditch crossing, the path of the conventional DEM traverses close to the edge of the ditch, which increases the total cost of the path and the maximum roll/pitch angles as summarized in Table 1. This is because the conventional DEM cannot properly represent the ditch near the rover, resulting in such challenging path. By contrast, the path of the C^2DEM evades the ditch and selects the modest area on the map, resulting in the less values for the total cost and the roll/pitch angles. Based on the experiments, the C^2DEM has a potential to provide a more feasible path than the conventional DEM.

4.2.2 Short-Range and Long-Range Navigation Tests

The experimental results for the short-range navigation are shown in Fig. 14. In this scenario, the rover reached the final goal (depicted with a red circle in the figure), after traveling through three intermediate goals (depicted with yellow circles), where the rover re-scanned the terrain features and re-generated the path to the next

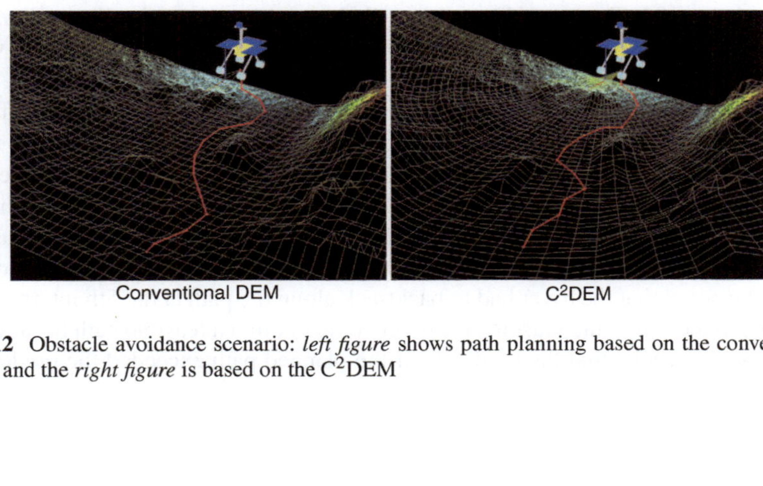

Fig. 12 Obstacle avoidance scenario: *left figure* shows path planning based on the conventional DEM and the *right figure* is based on the C^2DEM

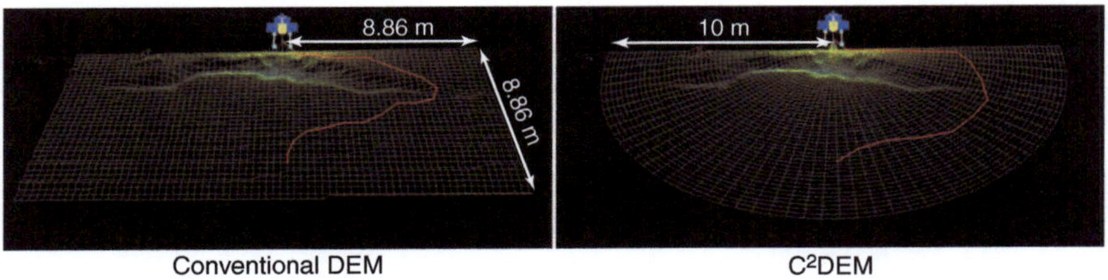

Fig. 13 Ditch crossing scenario: *left figure* shows path planning based on the conventional DEM and the *right figure* is based on the C^{2}DEM

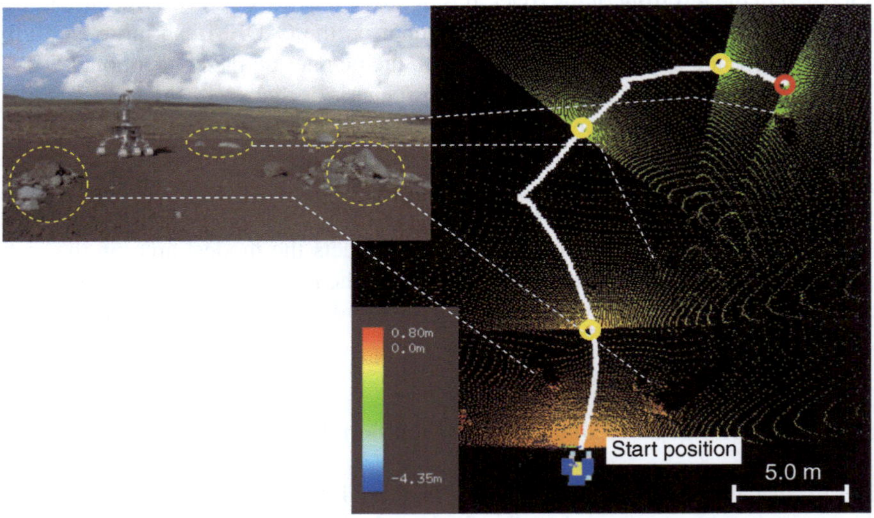

Fig. 14 Short-range navigation: The *left photo* indicates an overview of the experimental scene. The *red circle* depicts the final goal and the *yellow circles* depict the intermediate goals

intermediate goal. The total distance traveled was approximately 40 m. From the figure, the rover traversed a relatively gentle sloped terrain toward the goals while avoiding the obstacle rocks.

The long-range navigation test was also performed three times with different types of terrain. A typical result among these tests is shown in Fig. 15. The rover executed terrain mapping, path planning and navigation at multiple intermediate goals (20 in total), and successfully traveled 0.31 km in approximately three hours. The top section of Fig. 15 illustrates that the rover evaded several obstacles (bushes) throughout the terrain, and selected feasible paths toward each intermediate goal. It should be noted that the rover had to backtrack along the path at the 5th intermediate goal due to the following reasons: the rover could not find a feasible path because the terrain was so rocky that the total cost of the planned path exceeded the predefined

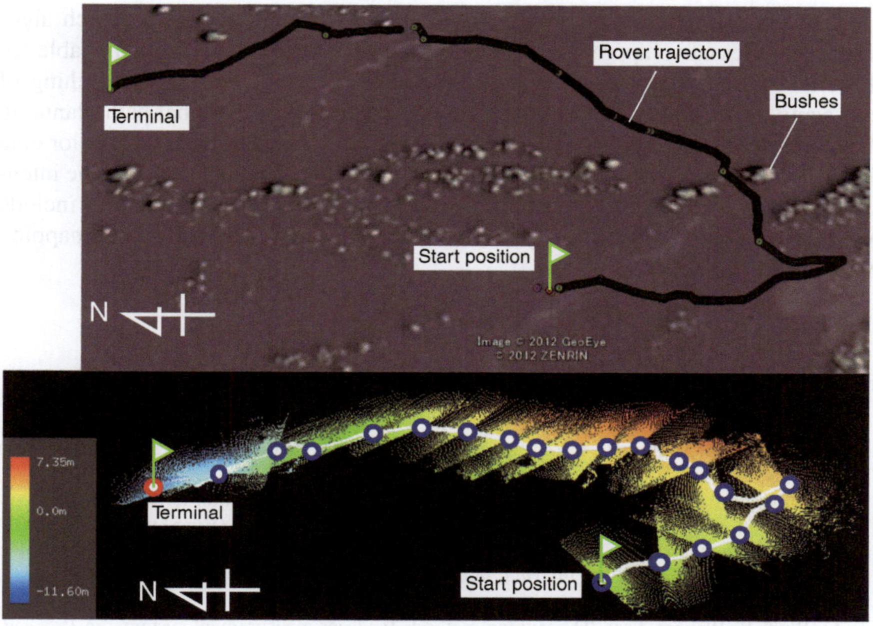

Fig. 15 Long-range navigation: The *upper figure* shows an overview of the rover trajectory, plotted on Google Earth. The *red circle* in the lower figure depicts the final goal and the *blue circles* depict the intermediate goals

threshold. Additionally, the rover needed to head north to obtain enough sunlight for the solar array panels to recharge the batteries.

5 Conclusions

This chapter has described a path planning and navigation framework for a planetary exploration rover that employs the LRF-based terrain mapping. In this work, a sector-shaped reference grid was applied to convert the point cloud data from a terrain map into the cylindrical coordinate digital elevation map, termed as C^2DEM. The path planning algorithm that is used in this framework considers three indices, including terrain inclination, terrain roughness, and path length, to generate a feasible path.

The field experiment with the rover test bed at the Lunar/Martian analog site was also presented in this paper. In the experiment, path planning with a conventional rectangular DEM and with a C^2DEM were compared. The results suggest that the C^2DEM enables the path planner to generate more feasible path than the conventional DEM. Additionally, the LRF-based path planning and navigation framework was able to execute long-range navigation on rough terrain.

In this work, the path planner is based on the conventional graph search algorithm, but alternative approaches, such as D^* or RRT would also be applicable for this framework. Additionally, the future work will include an online matching of consecutive terrain maps that will enable the rover to execute global path planning. A drawback of LRF-based mapping is that the LRF cannot provide pure color data for the terrain (grayscale or RGB) even though it can measure distance and the intensity of the light that is reflected from objects. Therefore, future research may include incorporating the LRF with a single camera that enables three-dimensional mapping with terrain texture information.

References

1. L. Matthies, M. Maimone, A. Johnson et al., Computer vision on mars. Int. J. Comput. Vision **75**(1), 67–92 (2007)
2. S. Golberg, M. Maimone, L. Matthies, Stereo vision and rover navigation software for planetary exploration. in *Proceedings of the IEEE Aerospace Conference*, vol 5 (Big Sky, Montana, 2002)
3. M. Maimone, A. Johnson, Y. Cheng, Autonomous navigation results from the mars exploration rover (MER) mission. in *9th International Symposium on Experimental Robotics* (Singapore, 2004)
4. O. Wulf, B. Wagner, Fast 3D scanning methods for laser measurement systems. in *Proceedings of the International Conference on Control Systems and Computer Science* (Bucharest, Romania, 2003), pp. 312–317
5. S. Thrun, S. Thayer, W. Whittaker et al., Autonomous exploration and mapping of abandoned mines. Robot. Autom. Mag. IEEE **11**(4), 79–91 (2004)
6. K. Nagatani, T. Matsuzawa, K. Yoshida, Scan-point planning and 3-D map building for a 3-D laser range scanner in an outdoor environment. in *Proceedings of the 7th International Conference on Field and Service Robots* (Cambridge, Massachusetts, 2009)
7. M. Buehler, K. Iagnemma, S. Singh (eds.), in *The 2005 DARPA Grand Challenge: The Great Robot Race*. Springer Tracts in Advanced Robotics (STAR) Series, vol 36 (Springer, Heidelberg, 2005)
8. M. Buehler, K. Iagnemma, S. Singh (eds.), *The DARPA Urban Challenge: Autonomous Vehicles in City Traffic*. Springer Tracts in Advanced Robotics (STAR) Series, vol 56 (Springer, Heidelberg, 2009)
9. UTM-30LX, Okuyo Automatic CO., LTD. http://www.hokuyo-aut.jp/
10. L. Longega, S. Panzieri, F. Pascucci, G. Ulivi, Indoor robot navigation using log-polar local maps. in *Prep. of 7th International IFAC Symposium on Robot, Control*, pp. 229–234 (2003)
11. M. Nieuwenhuisen, R. Steffens, S. Behnke, Local multiresolution path planning in soccer games based on projected intentions. in *Proceedings of the 15th RoboCup International Symposium*, Istanbul (2011)
12. R. Hadsell, M. Scoffier, U. Muller, Y. LeCun, Mapping and planning under uncertainty in mobile robots with long-range perception. in *Proceedings of the, IEEE/RSJ Int Conference on Intelligent Robots and Systems*, (Nice, France, 2008), pp. 2525–2530
13. A. Stentz, Optimal and efficient path planning for partially-known environments. in *Proceedings of the, IEEE International Conference on Robotics and Automation*, (San Diego, CA, 1994), pp. 3310–3317
14. J. Barraquand, B. Langlois, J. Latombe, Numerical potential field techniques for robot path planning. IEEE Trans. Syst. Man Cybern. **22**(2), 224–241 (1992)
15. L. Kavraki, P. Svestka, J. Latombe et al., Probabilistic roadmaps for path planning in high-dimensional configuration spaces. IEEE Trans. Robot. Autom. **12**(4), 566–580 (1996)

16. P. Cheng, Z. Shen, S. LaValle, RRT-based trajectory design for autonomous automobiles and spacecraft. Arch. Control Sci. **11**(3–4), 167–194 (2001)
17. K. Iagnemma, S. Dubowsky, *Mobile Robots in Rough Terrain: Estimation, Motion Planning, and Control with Application to Planetary Rovers*. Springer Tracts in Advanced Robotics (Springer, Berlin, 2004)
18. M. Otsuki, G. Ishigami, T. Shimada et al., Experimental study on mobility and navigation for exploration rover in natural rough terrain. in *Proceedings of the 28th International Symposium on Space Technology and Science*, (Okinawa, 2011), d-87 p

Motion Analysis System for Robot Traction Device Evaluation and Design

Scott J. Moreland, Krzysztof Skonieczny and David S. Wettergreen

Abstract Though much research has been conducted regarding traction of tires in soft granular terrain, little empirical data exist on the motion of soil particles beneath a tire. A novel experimentation and analysis technique has been developed to enable detailed investigation of robot interactions with granular soil. This technique, the Shear Interface Imaging Analysis method, provides visualization and analysis capability of soil shearing and flow as it is influenced by a wheel or excavation tool. The method places a half-width implement (wheel, excavation bucket, etc.) of symmetrical design in granular soil up against a transparent glass sidewall. During controlled motion of the implement, high-speed images are taken of the sub-surface soil, and are processed via optical flow software. The resulting soil displacement field is of very high fidelity and can be used for various analysis types. Identification of clusters of soil motion, shear interfaces and shearing direction/magnitude allow for analysis of the soil mechanics governing traction. The Shear Interface Imaging Analysis Tool enables analysis of robot-soil interactions in richer detail than possible before. Prior state-of-art technique relied on long-exposure images that provided only qualitative insight, while the new processing technique identifies sub-millimeter gradations in motion and can do so even for high frequency changes in motion. Results are presented for various wheel types and locomotion modes: small/large diameter, rigid/compliant rim, grouser implementation, and push-roll locomotion.

1 Introduction

The mobility systems of past and current planetary surface exploration missions are still limited by terrain encountered, demonstrating the need for continued development of traction devices for robotic vehicles. Targets, even whole regions, of scientific

S. J. Moreland (✉) · K. Skonieczny · D. S. Wettergreen
Robotics Institute, Carnegie Mellon University, Pittsburgh, PA, USA
e-mail: smoreland@cmu.edu

K. Yoshida and S. Tadokoro (eds.), *Field and Service Robotics*,
Springer Tracts in Advanced Robotics 92, DOI: 10.1007/978-3-642-40686-7_30,

interest exist in terrain that is frequently beyond the capability of all flown mobility platforms resulting in the loss of potential scientific return. Of specific challenge on the Moon and Mars are flat ground and slopes covered by loose, low strength regolith. Wheeled mobility systems can become entrenched in these terrains due to excessive slip and sinkage. With the continued exploration of Mars and other planetary bodies the study of wheel-soil behavior in loose, granular material remains imperative for achieving future scientific discoveries. Methods available utilizing common terramechanics approaches do not achieve high fidelity results, especially for non-idealized wheels with complex rim-soil surface interaction. Additionally, soil mechanics processes are not well understood for lightweight vehicles with wheels operating in loose, granular materials where significant soil flow is present.

The design of traction devices such as wheels for planetary rovers rarely involves the detailed analysis of soil shearing and failure patterns. The stress applied to the soil mass is the result of both external loading and the operating state of the wheel-soil system. The 'shear interface' are failure planes that develop in the soil below the region of interaction with the rim (Fig. 3). Shear interfaces can indicate the soil failure type, thrust/motion resistance processes and important soil flows present. Geometry of the rim, presence of grousers, wheel stiffness, contact shape and many other properties have a large effect on soil shearing during operation of a wheel. The performance of a traction device in loose, granular soil is ultimately governed by the soil properties and the shear failures that occur.

As examples, a rotating wheel may induce a forward flow leading to motion resistance and energy loss. Furthermore, a small diameter rigid wheel will cause a rearward flow, forcing particles deep into the soil and then back up again in a "*v*"-like shape terminating at the rim exit point. In contrast, a footpad utilized by vehicles with walking locomotion generates a significantly different mode of soil failure, which has a shear interface extending well beyond the wheel confinement. The distinct modes of soil failure and flow processes occurring provide insight into the development of traction of broad range of traction devices. Minute details of the wheel rim geometry or mechanics of the wheel carcass have a profound affect on the shearing processes. Geometric features such as the commonly implemented grousers are not able to be incorporated into existing terramechanic modeling techniques, thus inadequate results and misleading representations of the shearing processes are produced by most methods.

This chapter will (1) discuss the importance of the investigation of sub-surface soil shearing of traction devices in terramechanic evaluation, (2) introduce a method of analysis called Shear Interface Imaging Analysis [11] and (3) by use of examples, discuss the affect commonly considered wheel properties have on soil behavior utilizing the proposed analysis method.

In effort to directly investigate the soil shearing processes occurring due to wheel operation, an experimental approach was taken for this body of work. The study of the principles of traction was investigated by soil behavior analysis rather than bulk performance. The development of a technique we call "Shear Interface Imaging Analysis (SIIA)" was undertaken in order to measure the location and characteristics of the soil shearing in great detail. This method relies on the use of photographing soil grains through a glass-walled bin as a traction device operates. The SIIA technique has proven to produce accurate results and allowed for in-depth investigation of unconventional locomotion modes such as push-roll [11] and various wheel types for planetary surface vehicles [10].

2 Shear Interface Imaging Analysis

Prior efforts have been made to image soil effects due to travel of a wheel [4, 16] and operation of other devices such as excavation implements [8]. These researchers produced extensive results, but a limited amount of this work can provide direct insight into the application of design for planetary mobility systems. Previous works focused on validation of existing terramechanic theory, however the analytical models of interest are limited to simplified wheels and soil interaction. As such, this limited the experimentation space to simple/idealized wheels (rigid, wide, no grousers, low slip ratios) and without an investigation of soil behaviors due to a wide range of common design parameters. Existing techniques of soil imaging also produced low fidelity results, were not quantitative and could not assess time varying responses present in the soil. Results in this work show this often does not produce observable shearing processes similar to that of a rolling wheel. Previous imaging work concluded that terramechanic theory does not well represent flow of soil and that similar experimentation methods should be utilized for model development and validation of design [16]. This further motivates the work being conducted utilizing the novel Shear Interface Imaging Analysis method.

Prior methods of imaging relied on long exposure of film to indicate the shear interface. This produced low precision results that could only indicate whether soil was either moving or static and a course average of direction; no other information could be measured. The shear interface imaging analysis technique described in this work is a new method that is capable of recording many types of soil shearing information at high precision. Other techniques for lower precision visualization of soil motion have included tracking tracer particles exposed to UV light [14], observing changes to a grid pattern of different-colored particles [9], and applying white light speckle autocorrelation to an arrangement of natural and colored sand grains [7].

2.1 Description of Technique: Hardware and Software

The experimental apparatus constructed to analyze the soil shearing below a wheel consists of a glass-walled soil bin filled with regolith simulant, a traction device specimen, an actuated horizontal axis of motion (Fig. 1) and a high-speed camera. The wheel module (Fig. 2) of the imaging test bed is position or velocity controlled in coordination with the horizontal axis to create a commanded, constant slip ratio

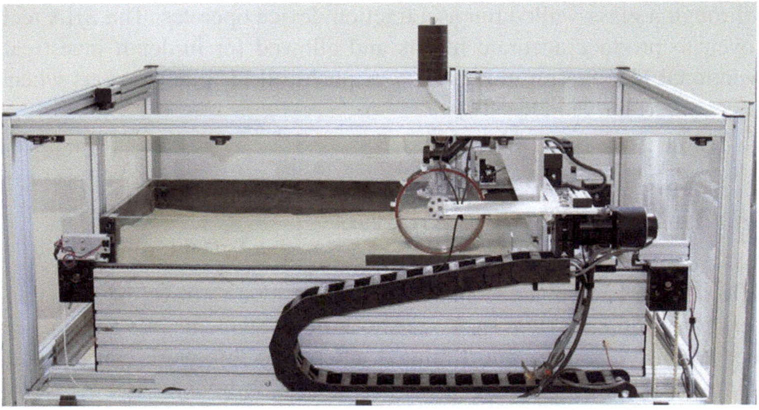

Fig. 1 Single Wheel Soil Imaging testbed. Wheel travels from left to right with controlled slip along a belt-driven linear axis

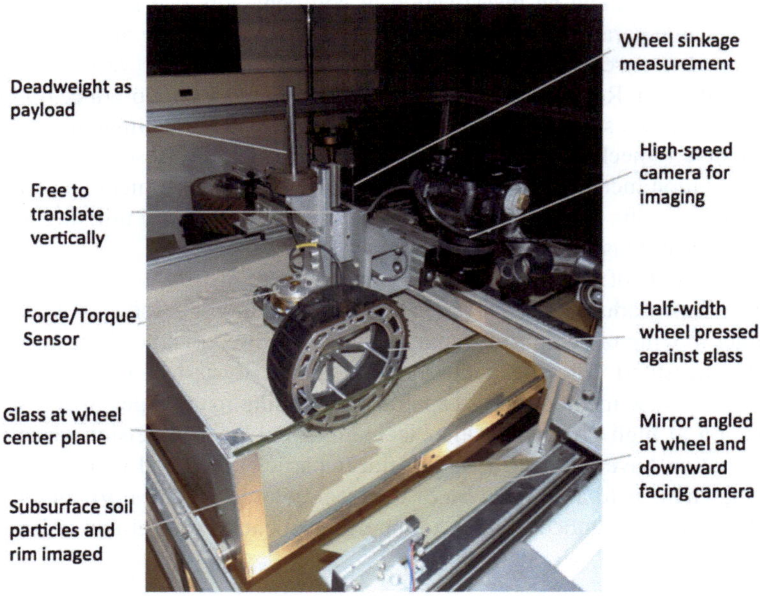

Fig. 2 Wheel module, carriage and glass-walled soil bin. A 31 cm by 22 cm cross-section of soil below the wheel is imaged with a high-speed camera

as the wheel travels forward. A linear rail allows the wheel to translate freely in the vertical direction allowing for natural sinkage to occur and payload weight to be transmitted. A 6-d.o.f. force/torque sensor is incorporated to measure the reaction loads, specifically in the travel direction as a result of traction generated. Sinkage is measured via an optical encoder affixed to the vertical free linear axis. All telemetry;

wheel angular velocity, travel velocity, slip, sinkage, load and power are logged simultaneously at 20 Hz or higher.

The wheel for all shear imaging analysis experiments is pressed against a sheet of tempered glass that extends the depth of the soil bin (27 cm depth for rig shown). Of importance is the use of a wheel of half the width of the actual specimen of interest and the application of half the payload weight. This aspect has been relied upon for over 50 years [2] and it has been experimentally validated that if shear stress between glass and soil is negligible, the glass surface acts as a plane of symmetry and the soil behaves as it would directly below an implement twice as wide due to the same boundary conditions [15, 16].

The shearing analysis requires the ability to track soil motion. A digital SLR camera with a 50 mm macro lens was used to image the wheel-soil interface, logging frames simultaneously with the rest of the telemetry. A frame rate of 8 frames-per-second was used and is sufficiently fast for the slow speeds of wheel travel applied (2 cm/s). The camera is mounted perpendicular to the soil bin glass wall and travels with the wheel in the horizontal direction as the carriage moves. For most wheel specimens (23–50 cm) a 31 cm wide by 22 cm high (soil depth) patch of soil is framed and able to capture the complete shear interface produced by the wheel in the regolith simulants utilized. External halogen flood lights at a high angle normal to the glass illuminate the soil particles.

Image processing comprises of optical flow and clustering techniques. The optical flow algorithm [5] tracks displacement of soil regions relative to a prior frame and calculates a motion vector at each pixel. Initial clustering separates each image into "soil" and "not soil" regions. Additional processing is continued only for "soil" regions. The magnitude of flow at each pixel of the soil regions is calculated from the optical flow vector fields. Soil flow is clustered into "significant" and "insignificant" magnitudes of motion. No explicit threshold is used to demarcate these clusters, but rather automatically adaptive clustering is used. The shear interface is derived from the boundary between significant and insignificant motions (Fig. 3). Soil flow direction is calculated from the optical flow vector fields, for soil regions exhibiting significant soil flow. Soil flow in any direction (360°) is visualized, and an additional boundary is identified at points where the soil transitions between forward and rear flow. Figure 3 demonstrates sample output of the process, showing soil flow magnitude, shear interface between significant and insignificant flow, soil flow direction and boundary between forward and rear flow.

2.2 Processed Results

The soil displacement field plots generated during a single wheel experiment are used to identify the effect wheel parameters and design features have on performance. Performance metrics such as net thrust produced (referred to as drawbar pull [3]) are critical for evaluation but there little information provided that aids in investigation of the underlying principles governing the measured performance. As an example,

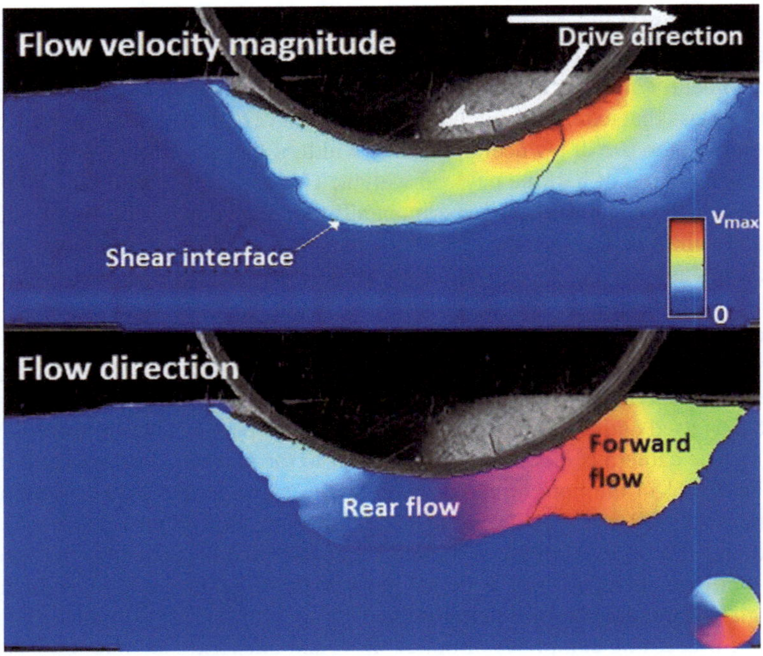

Fig. 3 Shear Interface Imaging plots. Plots show soil flow magnitude (*top*) and direction (*bottom*). Magnitude is plotted from *dark blue* (stationary soil) to *red* (soil being displaced at highest speed). Direction (within the shear interface) is plotted as shown in the circular color legend. Note a convention of all plots are wheel travel left to right and at a 20 % slip rate unless otherwise noted

parametric studies of grousered wheels are repeatedly encountered [1, 6]. To date, these investigations into the function of grousers has taken the approach of relying solely on performance measurements. Net traction, sinkage, slippage, wheel torques, power and reaction forces are typically measured during single wheel testing over a wide range of parameter changes, such as grouser spacing or height. Trends in the data are used to determine optimal parameter combination and conclusions sometimes inferred from these results. This approach is suitable for determining the response and performance of specific designs but provides limited information on the actual mechanisms and processes occurring in the soil that govern traction. Therefore, little knowledge can be gained in understanding how grousers function in general and how they should be implemented in design and vehicle operation.

Observing the soil shearing planes allows for qualitative analysis of how soil structure develops and react thrust loads or produces resistance to travel. The shear interface is indicative of the soil failure process and type. Analysis of this and the flows present within the soil can aid design of traction devices and study of terramechanic fundamentals. Figure 3 shows the processed results of a single wheel shear interface imaging analysis experiment. The Flow Velocity Magnitude and Flow Direction plots are used to analyze the wheel. These plots show processes typically present within the soil of a wheel operating in loose, granular soil.

The flow velocity field plot uses the optical flow displacement field measured between image pairs and clustering methods for classification to display the soil flow speed. These plots (Fig. 3, top) scale from dark blue (stationary soil) to red (representing the soil flowing at maximum speed, $V_{\max}$). This type of plot allows for the evaluation of the soil flow due to shearing. The shear interface is a key indicator of the means by which the wheel produces traction. This term, for purposes of this study, is defined as the region (line or band like in shape) where soil transitions from measured shear displacement (flowing) to near static (not flowing).

The Flow Direction plot (Fig. 3, bottom) displays the direction of soil particle shearing as measured by the flow velocity field. The multi-colored wheel is the legend that maps color to direction with respect to the wheel coordinate frame. 'Dark blue' indicates soil particles moving completely horizontal in the left hand direction, opposite the direction of wheel travel. The direction of shearing aids in determining what type of soil failure process occurs, design features that may contribute to the failure and the identification of multiple flows, such as resistive types at the wheel front. The separation of two flows (Fig. 3, bottom), as detected by the developed analysis software, allows for the identification of forward flows and the measurement of the location of point of maximum shear stress along the rim. This occurs at the intersection of the wheel rim and flow separation point.

2.3 General Experiment Considerations and Procedure

As an example of this process, the experimentation of the wheel shown in Fig. 4 will be described. The single wheel imaging testbed is prepared with GRC-1 lunar soil simulant [13] before each test run. The soil is loosened to a state of lowest relative density and slightly compacted by use of a drop tamper method to produce repeatable soil properties. The wheel specimen shown in Fig. 4 is rigid, 23 cm diameter by 5.72 cm wide (11.5 cm effective width), with the rim covered by course grain sandpaper. A 10 kg payload is applied in the vertical direction. All experiments are analyzed at steady-state response of the soil and reaction loads. The test run begins at static sinkage and then travels under a controlled slip rate for approximately five wheel diameters in distance. All rigid wheels (rough rim or grousers) quickly entered steady-state sinkage, reaction loading and soil shearing behavior within the first wheel revolution. The testbed controls the wheel slip while maintaining a wheel tangential rim speed of 2 cm/s by varying horizontal travel (carriage) speed. Most experiments, such as shown in Fig. 4, are evaluated at 20 % slip. Generally, wheel peak performance in loose, granular soil occurs between 10–30 % slip and as such, 20 % was chosen as a point of study for most evaluations. However, for each wheel configuration a full range of slip values (5–65 % slip, with 5 % slip intervals) are collected, but only key points such as 20 % and 50 % slip undergo soil shearing analysis. Additionally, for each experiment, at least three repeats were conducted.

The focus of this paper is to introduce the technique of improved shear interface imaging analysis (SIIA). An overview of the investigation of common wheel design features and resulting soil behavior through examples of experiments conducted will be utilized in the following section in effort to shed light on this technique.

3 Analysis Examples

There is a set of important parameters commonly decided upon in the design of a wheel for planetary surface systems. Choices between a rigid or compliant rim, tread surface parameters such as rough or with grousers and wheel geometry such as diameter. These all have significant results on the soil behavior during operation of the vehicle and the resulting traction performance. The study of the effect of some of these parameters was conducted and preliminary results are shared.

To study the development of the net thrust produced by rigid wheels, soil shearing was investigated over a range of slip ratios, and over changes in wheel diameter, tread surface, and locomotion mode. The approach was to conduct a preliminary survey of design features that have effect on performance and attempt to associate soil shearing behavior with potential gains or losses.

3.1 Rigid Wheel Soil Failure

It is evident in Fig. 11 that soil fails sharply at the shear interface and that there is significant forward flow. The shape of the shear interface, the point at which it originates at the rim (flow separation point) and magnitude of shearing indicate a forced type failure due to wheel rotation. This is further supported by the direction of the flow at the shear interface. For this wheel, the soil shearing direction is near vertical at the front of the wheel and returns to the surface behind the wheel at a near

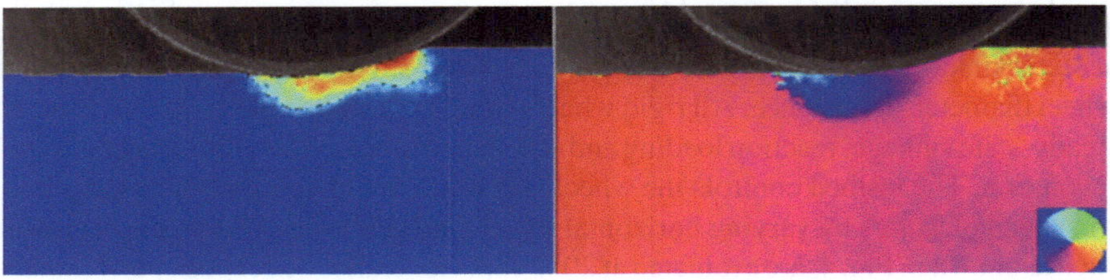

Fig. 4 Plots shown are particle velocity magnitude (*left*) and direction (*right*). Unlike Fig. 3, there is no threshold on the direction plot (*right*). The '*pink*' soil in the bulk direction plot is stationary (except directly under the wheel). Compaction of soil in forward direction (*yellow*) is visible in front of leading edge (*right*) of wheel. This is evidence of forward motion resistance

vertical direction. These types of behaviors are typical of small/medium diameter rigid wheels with mid-range performance where significant slip induced sinkage (slip-sinkage) is present. This leads to a high entrance angle. This type of observation of the soil behavior under a wheel rim is valuable in understanding terramechanic fundamentals and how a specific wheel design functions.

The degree of wheel slippage affects not only the net traction produced but also the state of soil behavior. Figure 12 shows the drawbar-slip curve for a 23 cm diameter (5.72 cm wide) rigid wheel with a sand-paper like tread carrying a 10 kg payload. The shape of the curve is typical of most wheels in loose, granular soil relevant to planetary vehicles.

There are distinct changes of the soil behavior at key points of the drawbar-slip curve as seen in Fig. 5. Three important observations can be made: (1) the shear interface size and shape does not change between 0.05 and 0.20 slip ratio although the drawbar pull produced quadrupled. (2) From 0.05 to 0.20 slip ratio, the soil shearing process within the shear interface transitions from a large gradient of shearing to near zero gradient within the shear interface (3) Lastly, above 0.20 slip ratio, the forward flow appears to diminish while the shearing zone (region within shear interface) begins to reduce in depth. There are a number of hypotheses that can be made from these observations. First, the shear interface of this wheel may be governed primarily by a mechanism other than applied thrust. This is evident since even at low slip and low load (0.05 slip and lower), the extent of the affected soil in shearing is similar to that at 0.20 slip ratio. It is possible that shearing induced by the rotation of the rim is excessive since large displacements must occur due to wheel sinkage (i.e. soil must be displaced downwards, rearwards, then back up). The second important hypothesis is that the knee in the drawbar-slip curve (about 0.20 slip in this example) may occur when the soil shearing is fully developed within the shear interface. Operating at a slip ratio above 0.20 appears to transition from a thrust generation type behavior of the wheel and soil, to an excavation behavior at increasing slip.

3.2 Walking

To investigate the effect of the rotating rim on the generation of thrust, a study comparing walking locomotion to rolling locomotion was conducted. Push-roll like locomotion (use of walking and rolling wheels) has been demonstrated to produce high drawbar pull for increased locomotion capability [11]. Utilizing shear interface imaging analysis, it was shown that the soil failure type of a walking wheel was different from a rigid rolling wheel (Fig. 6). This figure shows the "ground type failure" created by a walking wheel. This type of soil failure, due to the minute degree of shearing required, is able to produce multiple times the thrust of a rolling rigid wheel.

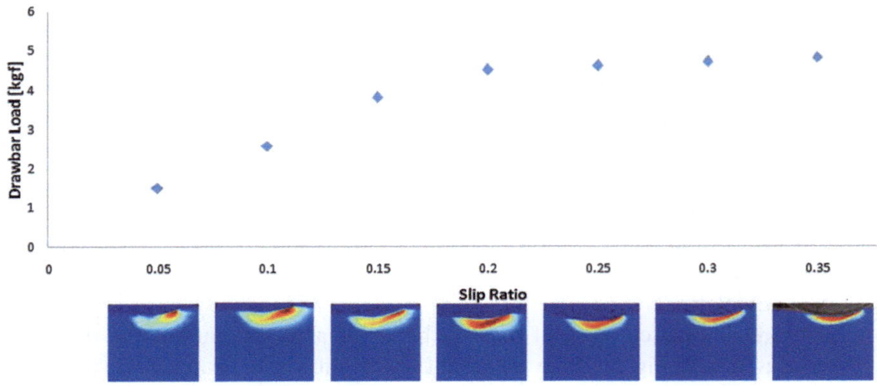

Fig. 5 Drawbar-Slip curve with corresponding soil flow magnitude plots at 5 % slip intervals (see value above *inset* plots). Distinct changes in soil shearing behavior at key slip (0.2) and load points are observed

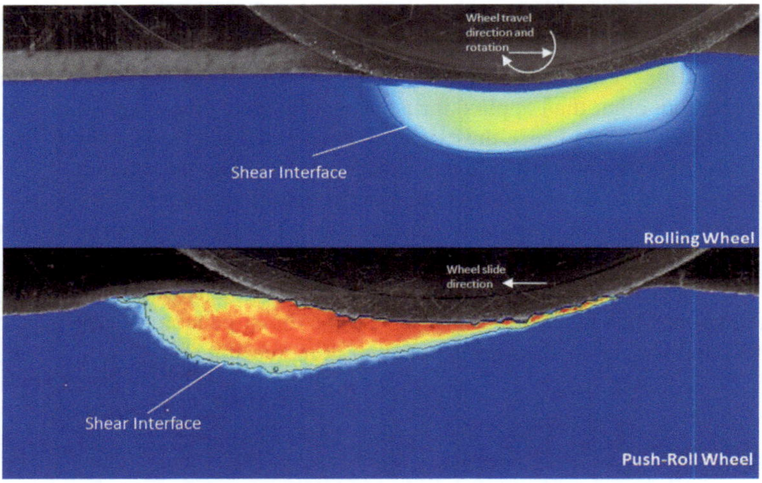

Fig. 6 Shear interface analysis comparing rolling wheel to push-roll locomotion (walking wheel). "Ground type failure" of the soil is observed for the walking wheel, identifying a source of tractive gains

3.3 Wheel Diameter

The diameter of a wheel has a profound effect on the behavior of the soil shearing and the resulting traction performance. An experiment comparing a 23 cm diameter rigid wheel to a 41 cm diameter rigid wheel of equal width, payload and slip is shown in Fig. 7. Performance using the drawbar pull metric measured a 33 % increase in traction for the larger diameter wheel. The behavior of the soil shearing is also

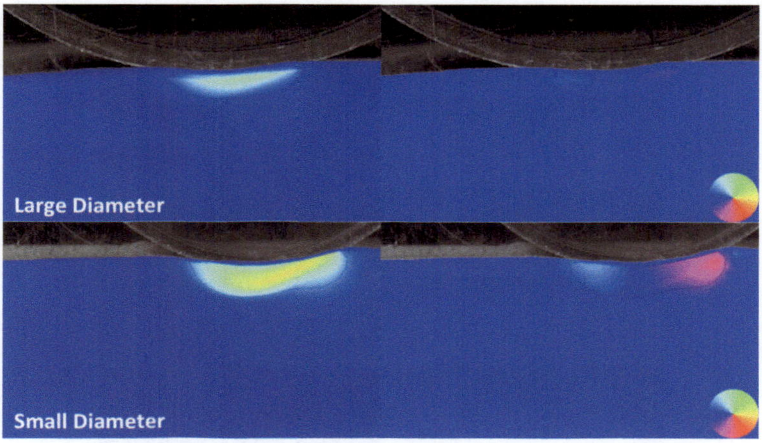

Fig. 7 Variation of shear interface with change in wheel diameter. The large wheel shows nearly horizontal flow compared to large changes in flow direction (*down* then *up*) under the small wheel

observably different. The large diameter wheel lacks an observable forward flow and the soil shearing is in a near horizontal direction. Lowered resistive flows may not be the only contributor to the increase in measured drawbar pull. Soil shearing induced by the rim rotation is in the direction of travel (horizontal) and may account for the increased drawbar pull. The soil particles need not shear as much when only acted on horizontally. This creates a larger portion of the shear interface that can operate at the peak stress of the stress-strain curve of the soil (granular, cohesionless soil shear strength peaks at low shear strain). A lower shearing magnitude is also evident for the large diameter wheel.

3.4 Grousers

Grousers are often employed in wheel designs for planetary rovers. The affect of soil shearing can be studied when analyzing these features [12]. Figure 8 shows the periodic nature of soil shearing due to individual grouser effects. It appears that the grouser at the front of the wheel entering the soil have the greatest effect. Experiments with very close spaced grousers were also conducted and showed similar results but with periodicity proportional to spacing. As the optical flow algorithm utilizes overlapping image pairs, high fidelity movies of the grouser shearing effects can be utilized to observe individual grousers interacting with soil as the rim rotates.

The traction performance of the wheel without grousers (sand paper rim) and the same wheel with 48 grousers at 13 mm height act as extremes of the configurations and drawbar pull performance amongst those tested. Studying these two cases leads to an important realization. The soil displacement behavior of the grouserless wheel

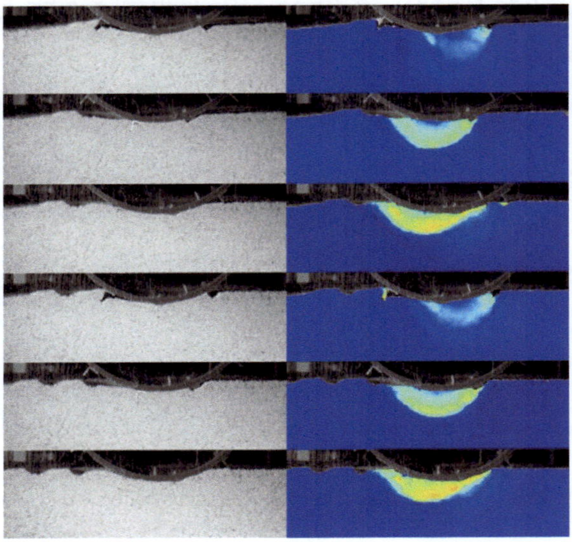

Fig. 8 Time-lapse images (*top* to *bottom*) of soil shearing by a wheel with grousers over two cycles, two grouser plunges (flow magnitude plot shown). Distinct, periodic soil motion present as each new lug rotates into the soil. This occurs for even small grouser spacing's however the magnitude of soil shearing is lowered. Wheel travel is to the right

(Fig. 4) and the 48 grouser wheel (Fig. 9) are very different at the leading edge of the wheel. Figure 4 direction plot, no grousers, shows a yellow patch of soil in front of the wheel that moves in a horizontal direction forward (low magnitude and within compaction regime). This is evidence of a significant motion resistance that would be reacted against the rim as soil is pushed forward and compacted downwards. This observed motion resistance would reduce the drawbar pull of the wheel. A wheel with high drawbar pull utilizing high performance grouser parameters (48 grousers at 13 mm height) however does not show evidence of a motion resistance. Additionally, a low resistance large diameter wheel did not have observable forward flow. It is reasonable to conclude from this observation that grousers can be implemented to significantly reduce compaction resistance and lead to higher drawbar pull. An increase in drawbar could also arise from an increase in thrust. Observations of soil shearing may also suggest whether the grousers provide additional thrust.

The rear shear interface of grousered and non-grousered (wheels are quite similar (Figs. 4 and 9 respectively). If it is assumed that in the rear flow region, the grousers are full of compacted soil, it will act like a wheel of larger diameter (only for the rear flow region, not front entrance area). With this assumption, the length, depth and shape of the shear interfaces of the two wheels are remarkably similar. If the magnitude of shearing and the soil density are similar, than the strength along the shear interface should be similar. The absolute shearing magnitude of the two wheels was measured to be similar using absolute particle velocity plots. Additionally, a reasonable assumption can be made that the soil density (as GRC-1 is not highly compactable) in the rear flow region behind the wheel center would have undergone similar compaction do to wheel payload. For these reasons, the soil strength can be assumed to be similar. As the shear interface shape/size and soil strength are similar, it is reasonable to conclude that the thrust component of the drawbar pull are of similar value for both grousered and non-grousered wheels. Differences in thrust due to drastically different shear interfaces can cause high changes in thrust, however the failure modes are of the same type for both shear interface observed. As such, it is theorized that the gain in drawbar pull due to implementation of grousers arises solely from the reduction of compaction resistance.

Single wheels test load cell measurements shows that wheels with grousers can be configured to increase drawbar pull. Compaction in the forward flow region in front of the contact area was not observed for wheels with grouser that generate high drawbar. Additionally, the shear interface of the region associated with generating thrust is similar for wheels with and without grousers. It is reasonable to conclude that increases in drawbar pull of wheels with grousers arises primarily from the decrease in compaction resistance. The increase in drawbar pull most likely is due to a decrease in motion resistance, not from an increase in thrust.

3.5 Compliant Wheels

Compliant wheels may provide large gains in performance for future surface exploration missions. Drawbar pull testing of individual wheels and full vehicles with compliant wheels has shown high tractive and energy performance. The study of the affect on soil behavior in the generation of thrust is essential. Experiments were conducted using a 23 cm diameter by 10 cm wide compliant wheel with sandpaper-like tread. The construction of the wheel did not produce uniform contact pressure as a pneumatic tire would, but a flat contact area was achieved. SIIA testing shows significant differences in soil behavior between the rigid wheel (Fig. 4) and compliant type wheel (Fig. 10). The direction of the soil displacement is completely horizontal.

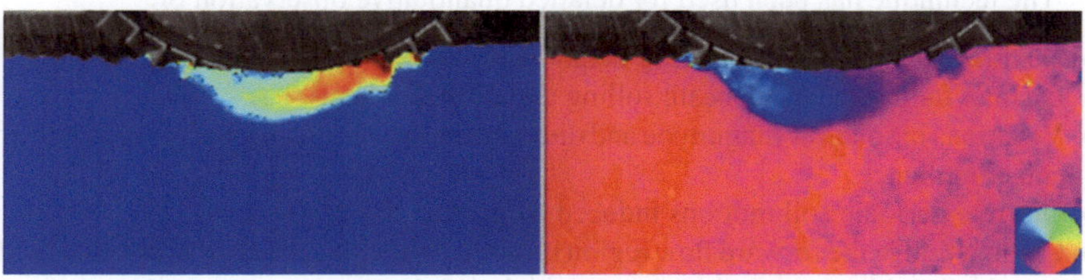

Fig. 9 Rigid, grousered wheel SIIA. For a high drawbar grousered wheel (48 lugs at 13 mm height), no observable evidence of forward compaction exists at the leading edge of the wheel. Therefore, little motion resistance should be present. Wheel motion to right

Fig. 10 Compliant wheel shearing analysis showing low sinkage, no forward flow, low shearing magnitude and completely horizontal motion of soil displacement (may all be within compaction process)

This may occur due to the extraordinary low sinkage and the flat shape of the contact along the length of the deformed rim.

It should be noted that the soil flow magnitude shown in Fig. 10 is actually an order of magnitude lower than with the rigid wheel of Fig. 4. As such, the soil shearing was so low; it appeared to be within the compaction regime of the soil simulant (as initial state was of low relative density). This limited study illustrates the importance of study the specific shear interface and soil behavior of a compliant wheel design when designing for high performance wheels. An understanding of how the observed behavior of the wheel-soil system affects performance will aid in developing more capable traction systems.

4 Conclusion

A technique for studying wheel-soil behavior and analyzing shear interfaces was developed and demonstrated. The technique can aid in design of high performance systems and increase the knowledge of terramechanics for wheels in a planetary environment. Examples investigating common wheel design parameters show a wide variation of shearing behavior that is intimately linked to traction performance.

The technique has been used for detailed quantitative observation of:

- Distinct changes in soil shearing behavior at key slip and load points.
- Distinct failure modes beneath rolling and pushed wheels.
- Variations in soil flow magnitude and direction between wheels of varying diameter and compliance.
- Discrete periodic soil motions induced by grousers.
- Compaction in the forward flow region in front of the contact area was not observed for wheels with grousers that generate high drawbar.
- Reduction of forward soil motion for grousers wheels correlated with grouser spacing/height that excavated at the leading edge enough to drastically lower the contact angle.

The state of maturity of the measurement tool (software and hardware) is high enough that it can now used by designers of mobility platforms for loose, granular soil. Additionally, the technique provides the benefit of a different perspective on terramechanics fundamentals.

References

1. R. Bauer, W. Leung, T. Barfoot, Experimental and simulation results of wheel-soil interaction for planetary rovers, *IROS* (2005)
2. M.G. Bekker, Photographic method of determining the soil action beneath footings, *Proceedings of the 2nd International Conference on Soil Mechanics and Foundation Engineering*, Rotterdam, pp. 193–194 (1948)
3. M. Bekker, *Theory of Land Locomotion* (The University of Michigan Press, Ann Arbor, 1956)
4. M.G. Bekker, *Off-the-Road Locomotion* (The University of Michigan Press, Ann Arbor, 1960)
5. M.J. Black, P. Anandan, The robust estimation of multiple motions: parametric and piecewise-smooth flow fields. Comput. Vis. Image Und. **63**(1), 75–104 (1996)
6. L. Ding, H. Gao, Z. Deng, K. Nagatani, K. Yoshida. Experimental study and analysis on driving wheels' performance for planetary exploration rovers moving in deformable soil. J. Terramech. **48**, 27–45 (2011)
7. P. Gachet, G. Klubertanz, et al., Interfacial behavior of unsaturated soil with small-scale models and use of image processing techniques. Geotech. Test. J. **26**, 12–21 (2003)
8. W.L. Harrison, Soil failure under inclined loads—II. J. Terramech. **10**(1), 11–50 (1973)
9. L.R. Khot, V.M. Salokhe, H. Jayasuriya, H. Nakashima, Experimental validation of distinct element simulation for dynamic wheel-soil interaction. J. Terrramech. **44**, 429–437 (2007)
10. S. Moreland, K. Skonieczny, D. Wettergreen, C. Creager, V. Asnani, Soil motion analysis system for examining wheel-soil shearing, *Proceedings of the 17th International Conference of the International Society for Terrain-Vehicle Systems*, Blacksburg, September 2011
11. S. Moreland, K. Skonieczny, D. Wettergreen, V. Asnani, C. Creager, H. Oravec, Inching locomotion for planetary rover mobility, *IEEE Aerospace Conference*,Big Sky, MT, March 2011
12. S. Moreland, K. Skonieczny, H. Inotsume, D. Wettergreen, Soil behavior of wheels with grousers for planetary rovers, *IEEE Aerospace*. IEEE, Big Sky, MT (2012)
13. H. Oravec, V. Asnani, X. Zeng. Design and characterization of GRC-1: a soil for lunar terramechanics testing in earth-ambient conditions. J. Terramech. **47**, 361–377 (2010)
14. S.A. Raschke, R.D. Hryciw, G.W. Donohoe, Microdeformations in sands by digital image processing and analysis. Transp. Res. Record **1548**, 31–37 (1996)
15. J.Y. Wong, A.R. Reece, Soil failure beneath rigid wheels, *Proceedings of the 2nd International Conference of the International Society for Terrain-Vehicle Systems*, Quebec City, pp. 425–445 (1966)
16. J. Wong, Behavior of soil beneath rigid wheels. J. Agric. Eng. Res. **12**(4), 257–269 (1967)

Image-Directed Sampling for Geometric Modeling of Lunar Terrain

Uland Wong, Ben Garney, Warren Whittaker and Red Whittaker

Abstract Geometric modeling from range scanners can be vastly improved by sampling the scene with a Nyquist criterion. This work presents a method to estimate frequency content *a priori* from intensity imagery using wavelet analysis and to utilize these estimates in efficient single-view sampling. The key idea is that under certain constrained and estimable image formation conditions, images are a strong predictor of surface frequency. This approach is explored in the context of lunar application to enhance robotic modeling. Experimentation on simulated data and in artificial lunar terrain at aerial and ground rover scales is documented. Results show up to 40 % improvement in MSE reconstruction error. Lastly, a class of image-directed range sensors is described and a hardware implementation of this paradigm on a structured light scanner is demonstrated.

1 Introduction

Despite proliferation of motion–coupled industrial scanners in field robotics, there remain important applications which require actuated sensors and intentional sampling. These applications must consider the question of:"Where to sample the data?" Planetary exploration is perhaps the best example of the need to plan geometric

U. Wong (✉) · W. Whittaker · R. Whittaker
Robotics Institute, Carnegie Mellon University, Pittsburgh, USA
e-mail: uyw@andrew.cmu.edu

W. Whittaker
e-mail: warrenw@andrew.cmu.edu

R. Whittaker
e-mail: red@andrew.cmu.edu

B. Garney
The Engine Company, LLC, Baltimore, USA
e-mail: ben.garney@gmail.com

K. Yoshida and S. Tadokoro (eds.), *Field and Service Robotics*,
Springer Tracts in Advanced Robotics 92, DOI: 10.1007/978-3-642-40686-7_31,

sampling. Correct photometric and geometric classification of rocks and craters could enhance automated sample collection and manipulation for drilling on science missions. Steered and optically reconfigurable flash LIDAR is of great interest in automated planetary landing for its capability in handling a spectrum of ranges and scales. Lastly, cooperative orbital and ground mapping, where the robot itself is a bore-sighted, Dirac sampler, will likewise benefit from a saliency-based approach to resolving terrain obstacles first detected in aerial sensing [1]. In these applications the importance of intelligent sampling is twofold: (1) the relative expense of range scanning places a premium on sampling well and (2) the requirements of model quality dictate the application.

The quality of 3D reconstruction from point measurements is a function of the density, distribution and order of sampling. Only grid-based or uniform angular sampling strategies – which are commonly used – inform *a priori* the total number of readings required to scan a scene from a single viewpoint. In spite of this, these are poor sampling strategies for reconstruction. Triangulation of the resultant point clouds produces glaring artifacts like polygonal slivers and incorrect connectivity. Aliasing of grid-based approaches creates high gradient planes out of depth discontinuities. Wasteful and duplicate measurement increases sensing time for little information gain. These effects are particularly problematic in applications where the 3D detail is used in object recognition or understanding. One of the compelling motivations of this work stems from the observation that sensors which are not limited by grid sampling are the same (low-throughput) sensors which would most benefit from sampling in a principled manner.

Cameras are natural complements to range sensors [2]; the authors advocate that it is possible to utilize color imagery to *direct* range sensing of a scene to avoid these problems. At the core of this idea is the observation that under certain image formation assumptions, the frequency content of color images correlates strongly with that of geometry images. By sampling the scene with Nyquist–informed density distribution – and by utilizing camera information to plan such a sampling – vast improvements can be made in both the quality and efficiency of range sensing.

This paper identifies frequency features for use with the image-directed scanning approach, compares feature performance based on reconstruction metrics and demonstrates new sensor designs and visualization techniques which utilize this paradigm. The advantages of diffuse planetary environments are discussed in tailoring domain-specific salient image features. Lastly, the aforementioned planetary applications are used to test the efficacy of the method.

2 Related Work

Image-direct sampling can be seen as a complementary approach to that detailed in the prior camera/LIDAR range fusion work of the authors [2]. Specific motivation for this work arises from the observation that the quality of geometric fusion depends as strongly on the quality of the underlying range data as it does on the image-based shape estimation. By affecting the nature of range samples, image-directed sampling can be utilized in tandem with a spectrum of traditional fusion techniques [3] to significantly enhance range models.

Research towards determining and generating optimal point samples is well documented. Work of particular note includes [4], which devised a method to reduce the post-processing time and increase the convergence probability of multi-scan alignment by storing only high–saliency points from scans. Related multi-view extensions of the sampling problem include the body of next-best-view work [5, 6]. Frequency content is also distinguished as a strong predictor of novelty in exploratory map building in [7], specifically for planetary terrains.

Wavelets and other multi-scale frequency features are widely used in data understanding and reconstruction. They include approaches to LIDAR [8] and intensity images [9, 10]. In particular image based approaches have been successfully used in interest seeking for planetary science [11]. Range and intensity fusion with wavelets has also been studied for forest region mapping [12].

While this research draws inspiration from prior work, it distinguishes itself in two important aspects. Prior approaches have enhanced unstructured range data using image content in a post-processing manner, this work tightly couples image analysis in the range *acquisition* phase. Secondly, this work is motivated by analysis of surface reflectance in environmental appearance. The authors believe that proper domain knowledge can constrain image features to physical phenomena.

3 Frequency-based Modeling

Frequency is one of the most effective predictors of information content in a signal. Continuous time or spatially-varying signals can be decomposed into their frequency components by utilizing a number of different transforms to the frequency domain. While, there is no exact definition of this domain, transforms generally have desirable properties that indicate the change-over-time of the signal.

The frequency components of range models cannot be directly measured, as the only sensors that exist for range modeling are Dirac samplers. ToF and Triangulation based methods all produce point estimates; such discrete samples are subject to aliasing. It is therefore prudent to consider the Nyquist sampling theorem for spatially varying samples [13] when reconstructing a range image. This states that perfect reconstruction is possible from severely sub-Nyquist spatial samples, if the signal has *known* compact support and shift invariance in the frequency domain.

These conditions are impossible to satisfy in practice due to sensor noise and discretization. Thus, only effective use of samples can be made by apportioning low and high information areas with some sample weight distribution and relative regard for Nyquist. This corresponds to oft-opposing objectives for modeling in a frequency sense: (1) accuracy of volumetric reconstruction and (2) localization of high spatial resolution detail to important areas. The approach taken here is that a range of sample importance priors should be explored for application-specific reasons, rather than attempting theoretical optimality.

3.1 An Image-Directed Approach to Sampling

A prior over the Nyquist frequency distribution is sought, which can guide spatially-varying sample acquisition. Two dimensional range images are common manifold representations of 3D models. The wavelet transform [14] is a simple and elegant representation of the spatial frequency content that can be used for such range images. Sum-squared detail coefficients at each pixel, the wavelet energy, are one possible indicator of the information distribution. Sample priors can be estimated using this metric on data from a complementary camera, which can instantaneously capture high-density intensity with the same perspective as a range image at a low constant time cost.

However, intensity is a generally inseparable composition of 3D geometric, material and illumination properties projected onto a 2D plane [15]. This work considers a subset of intensity images, whether raw or feature-transformed, which correlate significantly with scene geometry. Features from these images are viewed as noisy approximations of high resolution range or geometry cues.

Consider a simplified form of the rendering equation [16], which describes light transport for geometric optics at each scene point:

$$L_o = L_e + \int_{\Omega} f \times L_i(w) \times \left(w \cdot \hat{n}\right) dw \tag{1}$$

where L_o, L_e, L_i are the light intensities outgoing, emitted and incoming respectively; f is the BRDF; w is the incoming light vector and $\hat{n}$ is the surface normal for calculating incident irradiance. This equation can be reduced to the manageable form of (5) using common assumptions of planetary environments, which are *barren*, *dry*, *rocky*, and *simple illumination* (point or direct solar) [17]. Barren describes geometric smoothness, whereby surface normals can be approximated by geometric samples. Dry indicates a lack of participatory media, such as atmosphere or dust, which significantly scatters light. Rocky means scene reflectances are Lambertian and albedo change is minimal. Lastly, simple illumination assumes an additive set of discrete point sources, which in most cases reduces to a single sun source approximated by a point.

$$E = L_o = L_e + \sum\nolimits_k f L_{i,k}(w_k) \left(w_k \cdot \hat{n}\right) \tag{2}$$

$$E = \sum\nolimits_k \frac{\rho}{\pi} L_{i,k}(w_k) \left(w_{i,k} \cdot \hat{n}\right) \tag{3}$$

$$E = \sum\nolimits_k \frac{\rho}{\pi} I_k \left(w_{i,k} \cdot \hat{n}\right) \tag{4}$$

$$E \propto \left(w_{i,k} \cdot \nabla x\right) \quad (5)$$

Simplifications of the rendering equation result from applying *simple illumination* (2), *rocky* (3), *dry* (4), and *barren* (5) assumptions. Equation (5) relates image intensity (E) to a proportion of the dot product between the source and the gradient of the geometry (i.e. range image). The frequency relationship between a signal and its gradients is given by Frankot, which enables heuristic, but strongly correlated frequency sample planning for artifact reduction [18]. Note that the planetary assumptions above do not allow for cast shadowing, so only camera-centered illumination is strictly valid for real data.

The reconstruction problem presented here is viewed as selecting $x_k = \langle s_1, \ldots s_k \rangle$ from all possible samples, $s_i \in X$, such that the reconstruction error is minimized: $e = \arg\min_s \|t(x_k) - X\|$ under some interpolating function t. The objective is to determine features from intensity images (i) and learn a mapping such that $f(i, X) \rightarrow \hat{x}_k$, where $\hat{x}_k \sim x_k$.

4 Modeling for Planetary Robots

Validation of the image-directed sampling approach is conducted in the specific context of lunar robotics. Private enterprises, such as the Google Lunar X-Prize, have renewed interest in automation of landing and exploration on the moon. There is particular emphasis on creating high quality maps and models of the moon using robots as precursors to humans. Experiments were conducted with simulated aerial and ground-robot data to demonstrate scale robustness.

4.1 Lunar Aerial Mapping

Aerial-scale lunar terrain was procedurally generated in Blender^{TM}, utilizing randomized fractal crater and rock distributions taken from the Surveyor six mission [19]. The data includes independent overhead depth and RGB values for each voxel in the scene at 5 vox/m. Three different scenes were generated and are shown in Fig. 1.

The datasets simulate lunar sensing during terminal descent, at a scale 50 m above the ground and were generated for related research. An automated lander, for example, might identify geometric hazards (rocks and craters) and divert to a suitable landing spot. Image-feature based methods have been proposed to identify these hazards, including detection of shadows [20]. However, in this scenario a gimbaled altimeter or boresight flash LIDAR can also work in tandem with image-based methods to acquire a full 3D model.

Simulation of the sampling process includes selecting samples (voxels) from the highest resolution depth map, triangulating the subsamples and linearly interpolating to create a depth map at the native resolution. This depthmap is then compared against the ground truth digital elevation map to produce reconstruction error scores. Artifacts such as sensor noise are not considered in this analysis. Several scanning strategies were compared on this dataset. They include:

Uniform Grid – An $NxM = k$ element uniform grid sampling across the scene, rounded to voxel edges.

Uniform Random – Random sampling of k elements in the scene with each voxel receiving uniform weight.

Gradient Weighted – Random sampling weighted by the 2-norm of the partial image derivatives, a precursor to image-based edge detection. Calculated using the matlab command gradient:

$$w = \left\| \nabla_x I + \nabla_y I \right\|_2 \tag{6}$$

where $\nabla_x I$ and $\nabla_y I$ are the image partial derivatives in the x and y direction respectively.

Wavelet Weighted – Random sampling weighted by the sum-square of the wavelet coefficients, normalized by subregion size: The wavelet transform is generated with DB2 wavelet using the command wavedec.

$$w = \sum_{j=1}^{N} \frac{1}{2^j} \left(D_j\right)^2 \tag{7}$$

where j is the pyramid level and D_j are the detail coefficients at level j.

Entropy Weighted – Random sampling weighted entropy in a 9×9 image neighborhood centered about the query pixel, calculated with the matlab command entropyfilt. This transform is frequently used as a texture cue.

Three separate metrics were used to compare the reconstruction error between sampled depth maps and ground truth. These metrics each emphasize a different requirement of the model.

Mean Squared Reconstruction Error (MSE)–Penalizes for any large difference between the reconstruction and the ground truth.

Gradient Weighted Reconstruction Error (GRAD) – Penalizes for incorrect reconstruction at depth discontinuities.

Fig. 1 Image features in diffuse environments, such as the moon, correlate strongly with geometric features. These three simulated lunar terrains are utilized in this work

Table 1 Performance of sampling strategies on simulated aerial terrain

Scene #1	Random	Grid	Gradient	Entropy	Wavelet
MSE	1	1.02	0.88	0.87	1.11
GRAD	1	1.1	1.10	1.13	1.12
SAL	1	0.93	0.93	0.88	1.07
Scene #2	*Random*	*Grid*	*Gradient*	*Entropy*	*Wavelet*
MSE	1	1.01	0.87	0.85	0.99
GRAD	1	1.05	1.12	1.16	1.12
SAL	1	0.82	0.96	0.95	1.05
Scene #3	*Random*	*Grid*	*Gradient*	*Entropy*	*Wavelet*
MSE	1	0.91	0.83	0.80	1.21
GRAD	1	0.98	1.02	1.06	1.01
SAL	1	1.02	0.82	0.79	1.31

Saliency Weighted (SAL) – Assigns higher weight to in reconstruction salient features such as rocks and craters correctly. The saliency map is generated using using the Saliency Toolbox [21].

Random samples were generated 20 times using each strategy and the resultant scores were averaged. In the case of the grid strategy, the grid was shifted horizontally and vertically so that sampling did not always occur in the same voxels or on the edge of the scene. Optimal linear coefficients were found using fminsearch. The experiments were performed with fractional sampling density of the ground truth resolution in log spaced increments, i.e. $k = 1/2, 1/4, 1/8, \ldots, 1/2,048$.

The results of the sampling simulation are detailed in Table 1. The scores are given as the mean ratios of the error between random sampling and the strategy in question, weighted by inverse sample density.

$$\text{score} = \frac{1}{\sum k} \sum_{k \in S} \frac{T}{k} \cdot \frac{e_{\text{rand}|k}}{e_{\text{strategy}|k}} \tag{8}$$

Thus, scores represent an improvement multiplier over random sampling. Reweighting by inverse density accounts for performance across the entire logarithmic sampling densities of interest while discounting minute reconstruction noise due to lack of selection replacement at the highest density trials.

It is noted that in sparse samplings ($k < 15,000$), wavelets performed the best across all metrics (graphs of all curves are omitted for succinctness, see Fig. 3). This is particularly promising when high reductions in the amount of geometry acquired are necessary. Wavelets perform more modestly (20 % improvement) given higher numbers of points, as sample selection becomes constrained in the downsample simulation. Gradient and entropy approaches both perform relatively poorly. Gradient weighting fails because the sharpest edges occur in shadowed craters, not depth discontinuities. A similar problem occurs with entropy, which utilized a fixed scale and neighborhood. While the wavelet transform is also susceptible to overweighing shadow features, the multi-scale capability also captures the crater and rock features. The grid approach comes out ahead of other methods in deficiently sparse reductions. While grid sampling is subject to bias and aliasing, there is the possibility for large holes in random sampling. A better pseudo-random sampling approach for all these features might first compute a Delaunay triangulation within similarly valued regions to ensure adequate density.

4.2 Lunar Ground Mapping

Unlike aerial sensing, which is difficult to scale for lab experimentation, high-fidelity robot scale terrain can be constructed with readily available materials. This section describes experimental verification of the approach on lunar-like terrain constructed in a 2 × 1.3 m moonyard. Macro-scale rocks were placed on an undulating bed of garden lime with size and position determined by lunar rock distribution data. The moonyard was then dusted with CMU1, a nonhazardous, optical regolith simulant developed at Carnegie Mellon and consisting primarily of sieved coal dust and powered lime. Micro craters were created by ballistic placement of pebbles and further dusting. Figure 2 shows the ideal distribution of features generated in simulation and the actual placement of these features as-built.

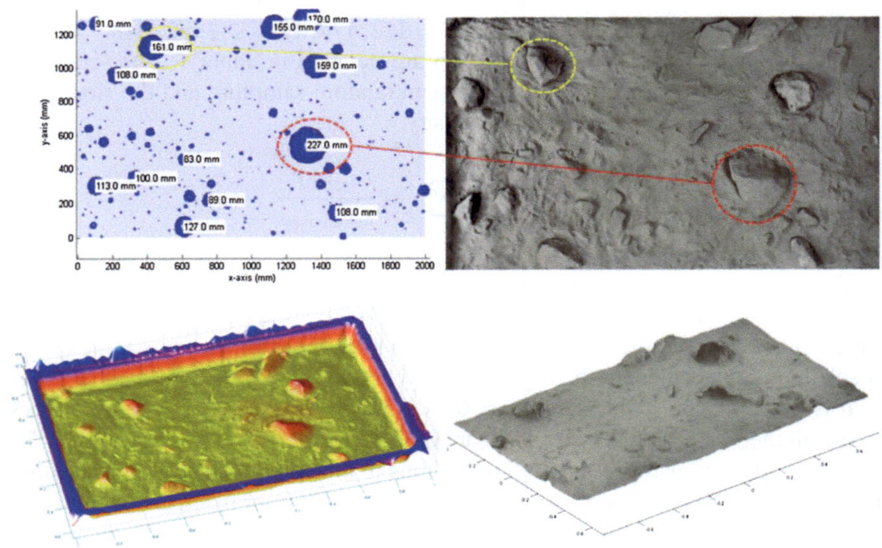

Fig. 2 A simulated lunar rock distribution utilizing Surveyor data (*top left*) and realization in the construction of an artificial moon scene (*top right*). The bottom row illustrates a z-colorized mesh model of the scene generated with LIDAR scanning (*bottom left*) and color-mapping of the terrain from DSLR images (*bottom right*)

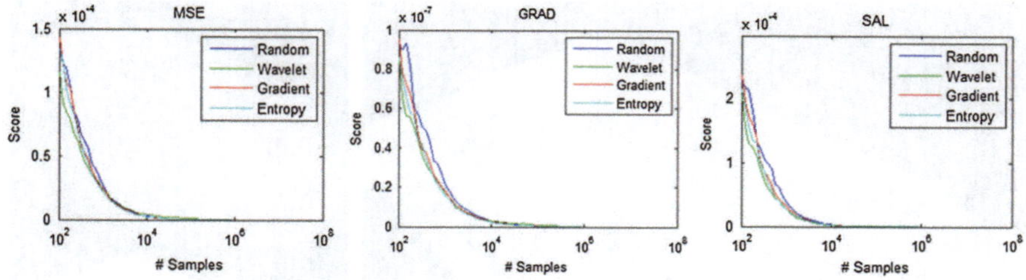

Fig. 3 Comparison of selected strategies over all subsample amounts and metrics on lunar terrain model. Visually, wavelets are the best performers (*green*)

The terrain was placed in a light simulator frame with calibrated sensor and sun-scale light source positions for data collection (an overhead sun position was used). Range data was collected with a survey-grade phase-shift LIDAR by scanning from multiple positions to ensure adequate resolution and elimination of range shadows. Likewise, HDR imagery was collected from multiple locations utilizing DSLRs. The raw LIDAR point cloud, consisting of five million points, was colored with RGB imagery, cropped to the edge of the sandbox and averaged into a 971 × 1674 voxel (1.6 M samples) orthographic digital elevation map. Figure 5 in the next section shows the data acquisition setup.

As the scanner does not have independent sampling control, it was necessary to scan at maximum resolution, voxelize into a uniform grid and sub-select samples post measurement for testing. Quantization into lower resolution voxels enables independent readings for each position without interpolation and provides resilience against the natural angular biases of the sensor.

The sampling strategies described in the previous section were tested and the same metrics described are used in this analysis without change. The results of the experiment in artificial lunar terrain are summarized in Table 2. The wavelet-based sampling is clearly the best performer in this experiment ($\sim$ 40 % improvement). Curiously, all of the strategies performed significantly better than in the simulated aerial imagery. It is believed that this effect is due to the strong cast shadows in the simulated imagery, while this experiment (though using real data) featured softer shadows and few large, negative features. While results are promising, it must be noted that this single dataset cannot be representative of the entire spectrum o lunar appearance and geometry and additional testing is required.

Table 2 Performance of sampling strategies on lunar terrain

	Grid	Random	Gradient	Entropy	Wavelet
MSE	1.1	1	1.09	1.15	1.33
GRAD	0.82	1	1.21	1.32	1.42
SAL	0.96	1	1.2	1.30	1.50

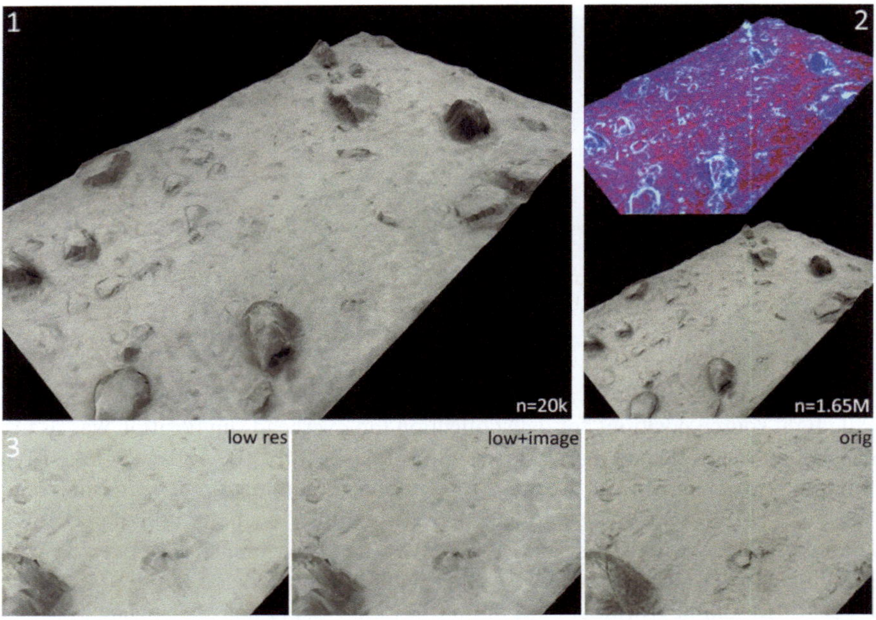

Fig. 4 Lunar terrain model generated with image-directed sampling. *1* An extremely sparse sampling rendered with roughness-modulated Oren-Nayar BRDF retains useful geometry and is visually appealing. *2* The frequency map of the terrain generated using wavelet decomposition. *3* Detail of roughness visualization demonstrates that a data deficient model can retain much of the optical appearance of the highest resolution model

Visual results of terrain reconstruction using wavelet-weighted, image-directed samples are illustrated in Fig. 4 . Figure 4 shows a rendering of the terrain with twenty thousand sample points, representing an 80x density reduction from the ground truth model of 1.6 million points (Fig. 4). As a result of well-placed samples, the macroscopic geometry is surprisingly well-formed for such a sparse mesh; there are few glaring differences when compared to the ground truth. The downsampled model is rendered with an inferred roughness from the wavelet decomposition. High frequency features estimated in the image are "painted" onto the sparse geometry by modulating the roughness term in an Oren-Nayar vertex shader proportional to the energy (Oren-Nayar is close to the true BRDF of lunar soil) [22]. A threshold can be applied to ensure major geometric features remain metrically true. This approach to visualization conveys the high resolution surface characteristics to the viewer while preserving mesh compactness with minimal computation. Figure 4 further illustrates this technique. With roughness inference turned off, the surface is an aliased collection of triangles, but rendering with roughness infuses many of the characteristics of the true surface.

5 A Class of Image Directed Scanners

In the prior sections, a discrete grid based approach is compared to random sampling approaches. Significant improvement was shown when random sampling is combined with image frequency estimates. However, it can be argued that this comparison is unfair. Few existing range sensors can acquire random, independent samples. Most of the applications that are considered in this work would use steered beam sensors if taken at the current state of sensor art. The authors believe that the approach is still beneficial when considering motion-constrained devices with trajectory planning and the possibility of utilizing region-based iterative enhancement in conjunction.

However, it is of interest to consider the class of sensors that provide pixel-independent sampling, as they are capable of implementing the approach in the purest form. Perhaps the most obvious and common of these are structured light sensors based on digital projectors. Though, foundational work has been conducted in camera-integrated smart laser scanners [23].

A structured light scanner was utilized in developing a proof-of-concept hardware implementation of image-directed scanning. Figure 5 illustrates the optical configuration of this device. A high resolution DSLR camera is optically co-located with the projector through the use of a half-silvered mirror. The system is calibrated such that each pixel in the camera corresponds to an exact outgoing ray of projected light. This configuration enables the system to sample the exact location viewed by the camera without shadowing. Finally, a second camera is optically offset from the other two devices and measures depth via triangulation. This setup is closely related to that presented in [24], which is a co-located system for augmented reality. Image-directed sampling is implemented on this scanner as follows. The projector first lights all pixels so that the scene can be imaged by the DSLR and analyzed. Afterwards, scanning proceeds in either of the two modes.

The first mode is *ambiguity enhancement*. In temporally-coded stripe scanners, the highest resolution stripes often result in ambiguities or errors (illustrated in Fig. 6). Among the sources for these errors include the albedos and complex reflectance

Fig. 5 An image-directed scanner consisting of a high resolution camera optically co-located with a DLP projector (*left*). Experimental setup utilized in this paper showing mounting positions of all the sensors (*right*)

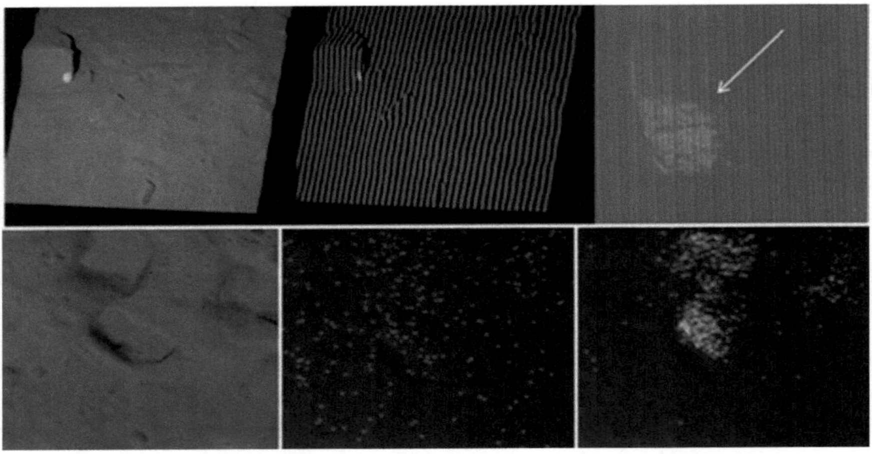

Fig. 6 Linear gray code scanning of the lunar terrain causes errors at the finest resolution due to differences in underlying scene albedos (*top*). Image cues can direct the scanner to resample salient regions such as rocks a pixel at a time (*bottom*). Samples are shown simultaneously illuminated for illustrative purposes; scanning occurs one dot at a time

of materials present in the scene to the resolving power of the camera lens. Gray codes can alleviate this problem, but result in lower effective resolution when such errors occur [25]. Image-directed sampling can be applied to identify regions which require dense, accurate samples, which can then be resampled one pixel at a time (reducing errors). The bottom row of Fig. 6 illustrates salient features (rocks) that require more samples. The locations of these samples are all illuminated by the projector for graphical clarity, but are scanned individually. Results on the moon terrain showed 25 % improvement in reconstruction MSE by re-scanning $\sim 10\,\%$ of gray code pixels using a single dot. Though the time cost of this extra scanning as implemented is large, significant reductions are possible with parallelism.

The second mode of operation explored is *data reduction*. Temporal coding in structured light scanning often precludes motion while scanning is in progress. Spatial codes, which utilize color or pseudorandom dot distributions, enable all depth information to be captured in a single frame. However, these techniques do not lend themselves to tightly-packed high resolution samples [25]. Image frequency can be utilized to plan a sampling of the scene with optimally placed dot locations. Reconstruction quality is thus improved while maintaining sparse samples. While not implemented at the time of this writing, using clique coloring in only high density regions and simple spatial pattern distortion estimation in low density regions can greatly reduce the number of colors to be discriminated.

Micro mirror arrays, like that used in the projector unit, are much higher resolution than beam sensors and even flash LIDAR. More importantly they require no actuation. The authors envision future intelligent sensors coupling low-resolution, low-throughput range receivers, micromirrors and co-located color cameras to determine optimal samples in the same vein our sensor.

6 Conclusion

An approach to image-directed range scanning utilizing wavelet analysis was described in this paper. Lunar application was targeted, where the particulars of surface appearance constrain image formation and correlates intensity with geometry. Validation of the approach was conducted for lunar aerial and ground mapping through simulation and terrestrial recreation respectively. Preliminary results are promising: reconstruction improvements of 40 % can be achieved over grid-based scanning techniques from a single viewpoint. Simple, but effective visualization for sparse meshes was demonstrated for combining image-estimated texture with the geometric portion of the lunar surface BRDF. Lastly, a hardware implementation on a pixel-independent structured light sensor was explored.

In the immediate future, we hope to analyze the effects of sensor noise on the approach. Noise in both the scanner and the camera can contribute to inaccurate prediction and wasted samples. Though this problem was mitigated in these experiments by limiting the maximum regional density that could be sampled, the sensors were low noise to begin with. The ability to handle a larger number of different albedos could also enhance the robustness of the method. Possibilities to crosscheck albedo changes (which contribute to frequency content in images) and shape changes include the use of solar-motion and probabilistic albedo estimation techniques. Similarly, while it was shown that this approach is defeated with strong shadows, the shadows can be mitigated with active illumination.

Acknowledgments The authors acknowledge Kevin Peterson, Heather Jones and Jason Koenig for use of lunar model data.

References

1. H. Jones, U. Wong, K. Peterson, Koenig et al., *Complementary flyover and rover sensing for superior modeling of planetary features.* Submitted for publication (2012)
2. U. Wong, B. Garney, W. Whittaker, R. Whittaker, Camera and LIDAR fusion for mapping of actively illuminated subterranean voids, in *Proceedings of Field and Service Robotics* (2009)
3. J. Diebel, S. Thrun, An application of markov random fields to range sensing, in *Neural Information Processing Systems* (2005)
4. J. Martinez, A. Reina, A. Mandow, Spherical laser point sampling with application to 3D scene genetic registration, in *Proceedings of ICRA* (2007)
5. K. Shahid, O. Galina, Intelligent LIDAR scanning region selection for satellite pose estimation. Comput. Vis. Image Underst. **107**(3), 203–209 (2007)
6. L.H. Mark, G. Okouneva, P. Saint-Cyr, D. Ignakov, C. English, Near optimal selection of views and surface regions for ICP pose estimation, in *Advanced in Visual Computing* (2010)
7. D. Thompson, T. Smith, D. Wettergreen, Information-Optimal selective data return for autonomous science and survey, in *Proceedings of ICRA* (2008)
8. H. Wei, M. Bartels, Unsupervised segmentation using gabor wavelets and statistical features in LIDAR data analysis, in *Proceedings of Pattern Recogntion (ICPR)* (2006)
9. E. Loupias, N. Sebe, S. Bres, J. Jolion, Wavelet-based salient points for image retrieval, in *Proceedings of 4th International Conference on Advances in Visual Information Systems* (2000)

10. R. Achanta, S. Hemami, F. Estrada, S. Susstrunk, Frequency-tuned salient region detection, in *Proceedings of CVPR* (2009)
11. H. Dunlop, D. Thompson, D. Wettergreen, Multi-scale features for detection and segmentation of rocks in mars images, in *Proceedings of CVPR* (2007)
12. Z. Wang, R. Boesch, C. Ginzler, Color and lidar data fusion: application to automatic forest boundary delineation in aerial images, in *International Archives of the Photogrammetry Remote Sensing and Spatial Information Sciences* (2007)
13. A. Aldroubi, K. Grochenig, Nonuniform sampling and reconstruction in shift-invariant spaces. SIAM Rev. **43**(4), 585–620 (2001)
14. I. Daubechies, Ten lectures on wavelets, in *Society for Industrial and Applied Mathematics* (1992)
15. B.K.P. Horn, *Robot Vision* (MIT Press, Cambridge, 1986)
16. J. Kajiya, The rendering equation, in *Proceedigns of SIGGRAPH* (1986)
17. U. Wong, *Lumehancement: Exploiting Appearance for Planetary Modeling* (Carnegie Mellon University, PhD Dissertation, 2012)
18. R. Frankot, R. Chellappa, A method for enforcing integrability in shape from shading algorithms. IEEE PAMI **10**(4), 439–451 (1988)
19. G. Heiken, D. Vanniman, B. French, *The Lunar Sourcebook* (Cambridge University Press, (Cambridge, 1991)
20. S. Hata, K. Sumioka , Detection of obstacles on the moon,in *IEEE Symposium on Industrial Electronics* (2004)
21. D. Walther, C. Koch, Modeling attention to salient proto-objects. Neural Netw. **19**, 1395–1407 (2006)
22. M. Oren, S.K. Nayar, Generalization of lambert's reflectance model, in *SIGGRAPH* (1994)
23. A. Cassinelli, A. Zerroug, M. Ishikawa, Camera-less smart laser projector, in *ACM SIGGRAPH Emerging Technologies* (2010)
24. K. Fujii, M. Grossberg, N. Shree , A projector-camera system with real-time photometric adaptation for dynamic environments, in *Proceedings of CVPR* (2009)
25. J. Salvi, J. Pages, J. Battle, Pattern codification strategies in structured light systems, in *Pattern Recognition*, pp. 827–849 (2004)

Efficient Large-Scale 3D Mobile Mapping and Surface Reconstruction of an Underground Mine

Robert Zlot and Michael Bosse

Abstract Mapping large-scale underground environments, such as mines, tunnels, and caves is typically a time consuming and challenging endeavor. In April 2011, researchers at CSIRO were contracted to map the Northparkes Mine in New South Wales, Australia. The mine operators required a locally accurate 3D surface model in order to determine whether and how some pieces of large equipment could be moved through the decline. Existing techniques utilizing 3D terrestrial scanners mounted on tripods rely on accurate surveyed sensor positions and are relatively expensive, time consuming, and inefficient. Mobile mapping solutions have the potential to map a space more efficiently and completely; however, existing commercial systems are reliant on a GPS signal and navigation- or tactical-grade inertial systems. A 3D SLAM solution developed at CSIRO, consisting of a spinning 2D lidar and industrial-grade MEMS IMU was customized for this particular application. The system was designed to be mounted on a site vehicle which continuously acquires data at typical mine driving speeds without disrupting any mine operations. The deployed system mapped over 17 km of mine tunnel in under two hours, resulting in a dense and accurate georeferenced 3D surface model that was promptly delivered to the mine operators.

1 Introduction

In April 2011, researchers at CSIRO were contracted to map several kilometers of the decline and drive at the Northparkes copper and gold mine in New South Wales, Australia. The mine operators were interested in moving large equipment

R. Zlot (✉) · M. Bosse
CSIRO ICT Centre, Autonomous Systems Laboratory, Brisbane, Australia
e-mail: robert.zlot@csiro.au

M. Bosse
e-mail: michael.bosse@csiro.au

K. Yoshida and S. Tadokoro (eds.), *Field and Service Robotics*,
Springer Tracts in Advanced Robotics 92, DOI: 10.1007/978-3-642-40686-7_32,

underground for future operations. In order to do so, a 3D model of the decline and drive was essential to determine the level of disassembly of the equipment required to manage the clearances in the tunnels. The primary requirement was that the model must be locally accurate: drift in the data acquisition trajectory could largely be tolerated, as the most critical aspects were the negotiation of curves and changes in grade.

Our research team at CSIRO had an existing relationship with Northparkes Mine based on previous work in automating an underground load-haul-dump vehicle [5]. More recently, our group has been investigating 3D SLAM in challenging environments, and in particular had developed solutions for mobile mapping in GPS-denied areas. Of particular relevance to the proposed mine mapping application, we had been working on a SLAM solution capable of estimating the six d.o.f pose of a spinning 2D lidar while the platform on which it is mounted is continuously moving [1]. In its simplest form, this solution requires no more than the raw 3D lidar returns as input; however, the use of readings from an inertial measurement unit (IMU) can aid the solution to provide increased robustness and accuracy. Variants of our system had been deployed for mapping suburban streets, industrial environments, forests, caves [8], indoor spaces, and a small section of an underground coal mine. Over the course of these experiments and deployments, our system has been improved considerably in a number of ways, some of which are detailed in this paper.

Three-dimensional mapping can play a highly critical role in underground mine development and maintenance, including planning, monitoring, safety, and vehicle localization; however, to date the existing 3D mapping and localization solutions are inefficient, labor intensive, or have not been demonstrated to be reliable, robust, and scalable. The predominant practice for lidar mapping of underground voids (such as mines, tunnels, and caves) is the use of tripod-mounted terrestrial scanners coupled with traditional survey methods to accurately estimate the scanner location [7, 9, 14, 16]. Several solutions using a mobile platform have been proposed in the robotics literature, but thus far these have predominantly been time-consuming "stop-and-scan" solutions where the platform must stop every few meters to acquire a scan [6, 10, 11, 15, 17]. These examples have further been confined to small test areas in their respective mines or test environments; none of the approaches have demonstrated applicability to large-scale deployments of more than a few hundred meters. One exception is a deployment in the San Jose silver mine in Mexico, where a small tracked robot with a terrestrial scanner covered 2.2 km of the mine using a stop-and-scan strategy; however, the data acquisition reportedly required three and a half days.[1] In contrast, above-ground mapping has been revolutionized by mobile mapping systems in recent years [12]. Such systems typically consist of high-accuracy lidars and tactical-grade inertial navigation systems mounted on the rooftop of a street vehicle. While the point clouds produced by mobile street mapping systems can be highly accurate and of large-scale, these systems heavily rely on maintaining a GPS signal (and are considerably expensive). Mobile mapping technology that can

[1] "Robot Scans Silver Mine", Photonics.com, 6 May 2008, accessed 1 June 2012. http://www.photonics.com/Article.aspx?AID=33755

reliably operate underground in the absence of GPS would be ideal for similarly transforming the underground mapping field. Our work realizes such a system, and here we demonstrate its application to a real-world large-scale underground mine mapping scenario, which required only about one hour of driving to generate a dense and accurate 3D surface model over a 17.1 km trajectory.

The contribution of this work is a SLAM solution that is able to map an underground mine in the time it takes to drive through it (up to maximum mine traffic speeds). We were able to reliably and accurately localize within and map more than 17 km of the Northparkes mine in under two hours, and effectively in real-time.[2] The hardware used to acquire the data was particularly minimal, making all of the setup, measurement, and teardown activities possible within only a few hours. The purpose of this deployment was not merely to demonstrate our technology, but to produce a 3D surface model to a particular specification for use by the mine operators for planning future operations. Our results surpassed the specified requirements and have been successfully delivered to the end customer.

The remainder of this paper is organized as follows. In Sect. 2 we first describe the hardware used, followed by the steps required to generate a dense, accurate, and georeferenced 3D surface model of the mine decline and drive in Sect. 3. At each stage, the resulting solution up to that point is illustrated and analyzed. Conclusions and lessons learned are discussed in Sect. 4.

(a) **(b)**

Fig. 1 The vehicle and sensing hardware used for data collection. The plastic sheeting over the sensor cart was used to protect the hardware from mist spray used for dust supression in some sections of the mine. **a** The vehicle and sensor cart during setup. **b** The vehicle and sensor cart inside the mine

[2] While we were not running our SLAM solution while acquiring data in the mine, we can process the data at a rate that is considerably faster than real-time.

2 Hardware and Procedure

The sensing platform used for mapping Northparkes Mine is illustrated in Fig. 1. The primary sensor utilized was a SICK LMS291 2D lidar on a rotating mount which spins the laser about the central scanning ray at one revolution every two seconds. As the LMS291 only measures in a single plane, the additional degree of rotation extends the sensor's field of view to cover a hemispherical region each second. In our previous publication [1], a 3D SLAM algorithm is introduced which is capable of estimating the sensor's 6DoF motion while the platform on which it is mounted moves continuously through an environment. For this application, we have also fixed a MicroStrain 3DM-GX2 MEMS IMU on the back of the spinning laser mount (i.e., the IMU does not spin with the laser). The IMU is utilized as an aiding sensor in the SLAM solution, providing additional robustness in rotation and globally eliminating drift with respect to the gravity vector. The lidar spin axis was pitched with respect to the horizontal in order to obtain some coverage above and in front of the vehicle. This configuration increases the reobservation of surfaces when at driving speeds, which is beneficial to the SLAM algorithm.

The spinning laser and IMU were mounted on a steel-frame cart, which was adapted from a recent experiment in which we were evaluating the SLAM solution in an underground coal mine. Given the short turnaround time expected for the Northparkes Mine deployment, we concluded that a simple solution would be to strap the existing cart to the bed in the rear of a site utility vehicle (pickup truck), thereby eliminating the need for developing a vehicle mount and minimizing the setup time prior to data collection. In addition to the sensors, the cart also included batteries and electronics, as well as a laptop which ran the sensor drivers and recorded data using the robot operating system (ROS) [13].

In addition to the spinning laser, two fixed LMS291 lidars were mounted on the cart back-to-back with their scanning planes oriented vertically, covering a 360° field of view in a vertical plane. As the vehicle drove, these sensors were trawled across the environment, covering the mine surface with a more uniform sampling density than the spinning laser (Fig. 2). The data from these sensors were used for generating mine surface models given the trajectory solution from the SLAM algorithm. The configuration of the vertical lidars is illustrated in Fig. 1. A small gap between the lidar scanning planes produces a 160 mm blind spot in the scanned surface. While this gap could be filled in with data from the spinning laser, it was not considered critical for the application and so was left empty in the final 3D models.

With the cart strapped into the bed of the site vehicle, a Northparkes employee drove the vehicle through the desired trajectory down into the mine, and returned back up to the start location just outside the entrance. The width and height of the mine tunnels vary between 5 and 6 m. The vehicle was driven according to site regulations and within the speed limit of 30 km/h (typically 20 to 30 km/h). When encountering other vehicles, the driver briefly stopped to allow them to pass in order to both avoid mapping these vehicles into the surface model and to eliminate shadows due to the associated occlusions. Any points measured from these vehicles could be easily

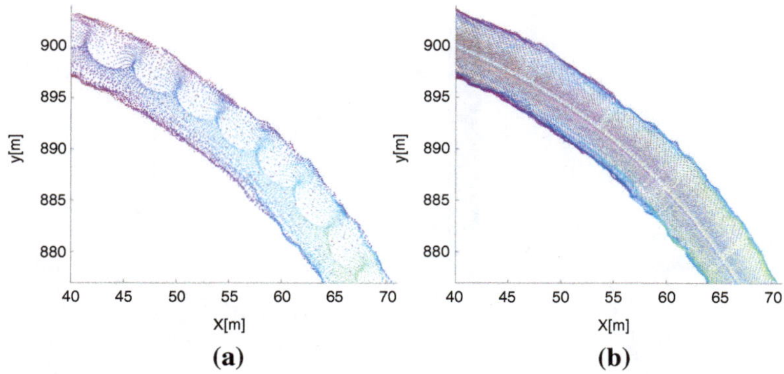

Fig. 2 Point densities of data acquired by the spinning lidar and the two vertical lidars (over approximately eight seconds), viewed from overhead. **a** The spinning laser's scan density is uneven due to the orientation of the spin axis (pitched 25° from the vertical). As the vehicle drives, the scan rate is effectively higher on the side of the vehicle where the laser spins with the direction of vehicle motion. **b** The data from the vertical lasers are more evenly sampled, with the density depending on the vehicle speed. The small longitudinal gap is due to the blind spot between the two lidar units. The sporadic radial gaps are due to scans dropped by the LMS driver (discussed further in Sect. 3.5)

removed by ignoring any measurements taken with the vertical lidars while stationary. The entire data acquisition process covered approximately 17.1 km of drive in 1 h and 53 min (including 45 min of stops for tag boards, ventilation doors, and equipment checks). Importantly, the entire mapping process was completed without any negative impact on mine operations. A mine survey shown in Figs. 3d and 4d, based on reduced-level data provided by the mine operators, illustrates the layout of most of the area that was covered and provides ground truth for some of our processing phases.

3 Data Processing

The acquired data are processed through a series of steps to produce the final surface model. First, an open-loop incremental scan-matching algorithm generates an initial estimate of the platform trajectory based on the spinning laser and IMU data. The initial trajectory is then corrected and georeferenced in two offline global registration stages. Finally a 3D surface model is reconstructed based on the resulting trajectory and scan data from the vertically-oriented lidars.

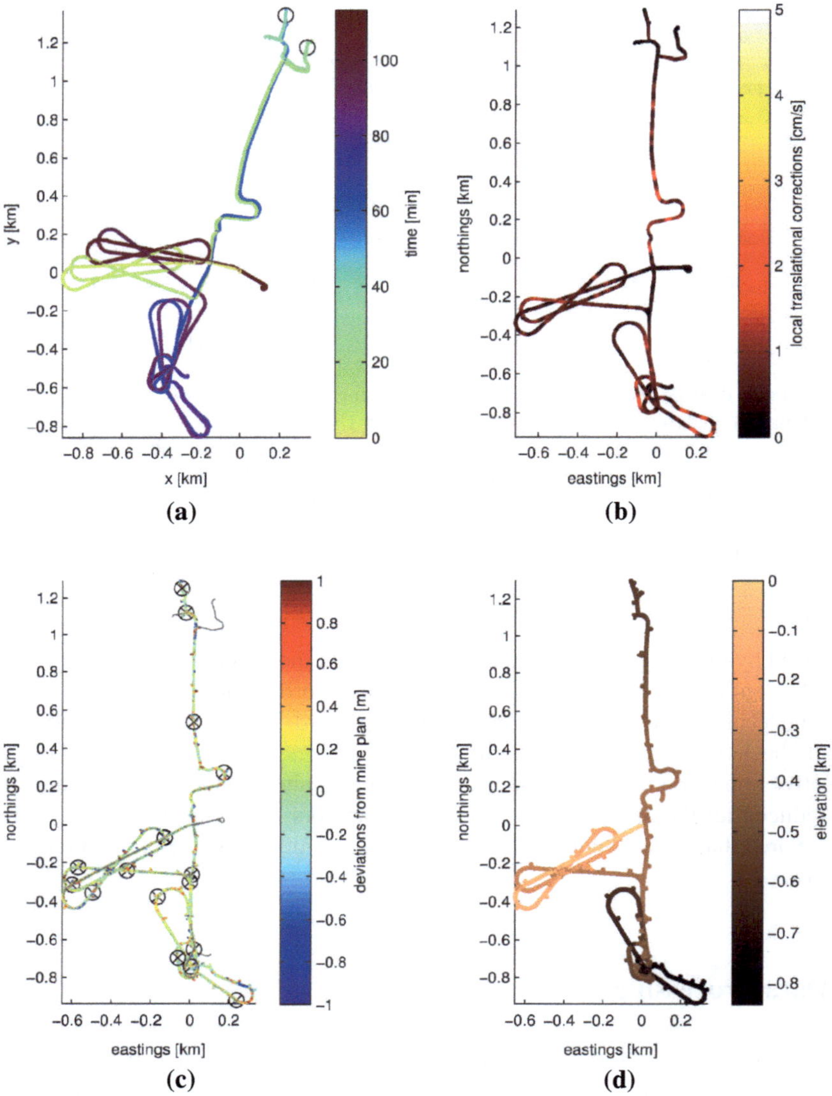

Fig. 3 Overhead views of the trajectory resulting from each of the stages of data processing. **a** The initial open-loop trajectory resulting from the online incremental scan-matching algorithm, colored according to elapsed time. Locations where sensor data briefly dropped out (see Sect. 3.1) are indicated by black circles in the upper right. **b** The trajectory after applying the first-pass global registration algorithm. The trajectory is colored according to the local translational stresses applied to correct the open-loop trajectory. **c** The trajectory resulting from the mine survey registration step. Seventeen anchor points are indicated by ⊗ symbols. Illustrated are the mine survey surface elements colored by the positional correspondence errors to the matched lidar point cloud surfels. The thin grey trajectory segments are areas we mapped that were not covered in the available mine survey data. The thin grey segment just east of the origin is outside the mine entrance. **d** The mine survey colored by elevation.(b), (c), and (d) are georeferenced in UTM coordinates, but offset here by a constant value in order to make the axis labels legible

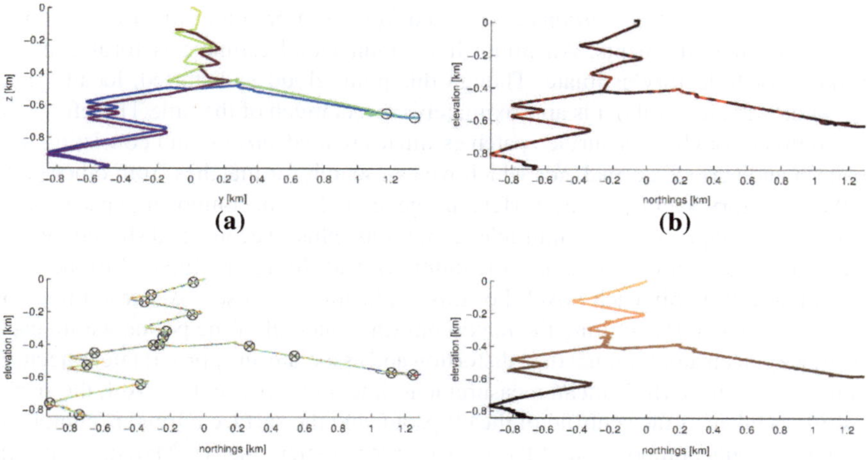

(a) **(b)** **(c)** **(d)**

Fig. 4 Side views of the trajectories from Fig. 3. All colors, symbols, and axes follow the same conventions as in the previous figure

3.1 Initial Trajectory: Open-Loop SLAM

An initial 6DoF sensor trajectory estimate was computed using an incremental scan-matching 3D SLAM solution. The solution is based on work we previously published [1], with several improvements and enhancements incorporated that have been developed since initial publication. The algorithm at each timestep is presented with data from at least two sweeps (half-revolutions) of the spinning lidar (referred to as a *segment*), and computes a smooth trajectory that best explains the lidar data. The trajectory segment window is shifted at each timestep by an amount less than or equal to the segment length. For this application, we use a trajectory segment length of 2.3 s, and shift it by 0.7 s at each timestep.

Initially, at the start of processing a segment, the true trajectory followed over the full duration is unknown. However, a rough prior trajectory can be estimated using a constant velocity model for the previously unobserved portion of the trajectory. The predicted trajectory is based on an assumed translational velocity from the end of the previous segment, and the rotational velocity measured by the IMU. The prior trajectory estimate is generally considerably inaccurate, and the associated point cloud can still be severely distorted. The goal of processing the current trajectory segment is then to solve an optimization problem to estimate a correction to the prior trajectory. The optimization is modeled as a linear system composed of a set of constraints of four types: surface match constraints, smoothness constraints, gravity constraints, and initial condition constraints. The system state solved for is a discrete set of corrections to the prior trajectory segment regularly sampled at fixed intervals.

The *surface match constraints* are derived from reobservations of surface patches in the lidar measurements. An initial lidar point cloud estimate is formed based on the prior trajectory estimate. Though this point cloud is distorted, locally (i.e., over short time intervals) it is approximately correct much of the time.Therefore, we can compute rough local surface patch estimates (called *surfels*, and consisting of a position and surface normal) for which we can search for matches from other parts of the trajectory segment. The surfels are generated by decomposing space into a grid of cube-shaped voxels at multiple resolutions (plus a second grid shifted by half of a cell in each dimension at each resolution), and fitting an ellipsoid to the lidar points that fall within each voxel. For this application, we use voxel resolutions of 0.5, 1, 2, and 4 m. By limiting the maximum time interval of the points within each voxel, we can ensure minimal data distortion and associate an approximate trajectory sample given the surfel's mean measurement time. For a particular voxel, the surfel position is taken as the centroid of the ellipsoid, and the surface normal is computed from the eigendecomposition of the second order moment matrix. Matches between the surfels are found using a *kd*-tree based approximate k-nearest neighbor search, where each vector in the search space is a weighted concatenation of the position and surface normal of a surfel. Each match constrains the relative pose between their corresponding trajectory samples.

The *smoothness constraints* ensure that the trajectory is physically realistic by modeling physical limitations on linear and rotational accelerations. The *gravity constraints* globally reduce rotational drift that would bring the 'up' direction out of alignment with gravity (as measured by the IMU accelerometers after compensating for the estimated motion of the vehicle). The *initial condition constraints* ensure continuity across multiple trajectory segment windows by fixing the start point of the current segment.

The various constraints are of the form $\boldsymbol{A}\boldsymbol{x} = \boldsymbol{b}$, where $\boldsymbol{x}$ is the state vector consisting of the corrections to the initial trajectory segment estimate at the samples. The system is solved as an iteratively reweighed least squares problem in an M-estimator framework, where the match constraints are weighted according to a Cauchy distribution to mitigate the effect of outliers. The entire optimization process is repeated until convergence or a maximum number of iterations is reached, where in each iteration the surfels are first transformed according to the previous (improved) solution, new matches and other constraints are computed, and the new linear system is solved.

To reduce the accumulation of drift errors over trajectory segment windows, the algorithm also maintains a set of surfels from past *views* that are no longer in the active window, from which additional match constraints are generated. The corresponding match constraints only contain terms for correcting the current observations (i.e., the trajectory corresponding to the fixed surfels is not further corrected). Views are occasionally saved at fixed distance intervals along the trajectory. A small constant number of fixed views is buffered (two in this application) in order to avoid unbounded growth in computation.

Key differences from the solution described in the original publication [1] include: the use of an IMU as an aiding sensor (and the accompanying reformulation of the smoothness constraints); the trajectory segment is no longer separated into two well-defined halves (matches can be taken from any two distinct times); the use of fixed views; as well as many efficiency improvements that allow the system to run in real-time.

The trajectory generated from this phase of the processing sequence is illustrated in Figs. 3a and 4a. The time required to generate this solution (on a desktop computer with a 3.2 GHz Intel Xeon CPU) is 70 min, which is approximately 62 % of the total data acquisition time. As expected, since the solution uses an incremental scan-matching technique, the resulting open-loop trajectory is affected by drift; however, given the scale of the trajectory, the accumulated drift is reasonably small. Note that there are two instances at which there are short gaps in the data (indicated in the figures), which is responsible for some of the largest apparent offsets in the visualization of the trajectory. These gaps occurred when an ethernet cable in the logging computer became disconnected. In each instance, the disconnection was noticed reasonably quickly, the driver was requested to stop, and the data logging soon recommenced (after approximately 25 s in the longest case). The open-loop trajectory simply continues from the last known location when the gaps occur; therefore, large instantaneous errors appear at these locations in the figures. These splits in the trajectory are repaired in the subsequent processing steps.

The open-loop trajectory produced at this stage is locally accurate and would be sufficient for the purposes of the application (determining clearance for transporting equipment). Nonetheless, we decided to further improve the result by globally registering the data using the techniques presented in the sections that follow.

3.2 Loop Closure

Given the open-loop trajectory, we next perform a global optimization that registers the surfaces that have been observed in multiple passes through the environment. For this procedure, we are able to apply essentially the same algorithm that is used for generating the open-loop solution; however, instead of solving for a short trajectory segment over a few-second time window, we input the entire (nearly two-hour-long) trajectory. In the same manner as above, surface elements are generated from the lidar point cloud; match, smoothness, and gravity constraints are formed; and the resulting linear system $\boldsymbol{A}\boldsymbol{x} = \boldsymbol{b}$ is solved in an iterative fashion. In addition to the standard constraints, constraints are introduced which encourage the local modifications to the original (open-loop) trajectory to be small.

The global optimization algorithm is effective only when the initial trajectory is reasonably *locally* accurate, and has difficulties when the error between reobservations from different passes is large (largely due to the use of nearest neighbor search to identify surface matches). In this type of environment, perceptual aliasing can occur where different sections of tunnel are relatively close together in the initial solution, which can pull together incorrect matches. This effect is particularly troublesome given the difference in elevation between the downward and upward traverses of the lower decline (left side of Fig. 4a), resulting in incorrect loop closures.

A simple solution to this problem is to use a small number of *anchor points*, which are rough locations manually extracted from the 2D mine survey (e.g., at particular intersections) and associated with the times that the trajectory passed near these locations. The trajectory is first rigidly transformed to minimize (in a least squares sense) the distance between the anchor points and the associated trajectory locations. Constraints encouraging the trajectory to pass within 7 m of the anchor points at the appropriate times are also incorporated into the first few iterations of the global optimization. The anchor point constraints are relaxed after several iterations due to the fact that the exact time the anchor points were passed through, and the proximity of the trajectory to the anchor points, is not known precisely. Therefore, the anchor points are primarily used to help move the initial solution into the algorithm's catchment basin, avoiding local minima. The inclusion of the anchor points also roughly georeferences the trajectory.

Figures 3b and 4b illustrate the closed-loop trajectory resulting from the global optimization process. Computing the closed-loop trajectory required approximately 45 seconds for generating the surfels, and 3.3 minutes to run the optimization. At this stage, it is clear to see the resemblance to the mine survey shown in Figs. 3d and 4d. In Figs. 3b and 4b the trajectory is colored according to the magnitude of the local translational corrections applied to the initial trajectory in order to arrive at the registered solution. The RMS of the local translational corrections is 1.8 cm/s, while the RMS of the local rotational corrections (not pictured) is 0.027°/ s.

We observe that some of the anchor points deviate from the trajectory due to the relaxation of the associated constraints in the final iterations of the optimization. Relaxing these constraints allows the surface match constraints to converge more freely, and the deviations are tolerable at this stage since they can be corrected in the next stage of processing.

3.3 Registration with Mine Survey

As a final step in correcting the trajectory, we register our 3D point cloud to the available mine survey data. By doing so, we can more accurately georeference the result, and correct some of the finer-scale drift not accounted for in previous steps. The mine survey data is a set of 17,942 points, which were previously acquired by surveyors collecting 2D scans throughout the tunnels of the decline and drive at variable local densities ranging from tens of centimeters to a few meters apart. We first preprocess the mine survey to extrapolate artificial surface elements given the 2D scan points. As the mine survey points are given as the tunnel outline at floor level, we can generate a wall surfel and floor surfel for every neighboring point pair (resulting in 35,612 mine survey surfels). The location of a surfel are taken to be the average position of the associated point pair, while the normals are constructed under the assumption that the walls are vertical and floor is aligned with the tunnel slope. Registration to the mine survey can be performed using the same global optimization algorithm described above, but now using the mine survey surfels as a fixed view (as described in Sect. 3.1).

Figures 3c and 4c shows the results after registering the closed-loop point cloud with the mine survey. This stage of the solution required 4.9 min of computation. We evaluate the quality of the match according to the positional error between the surfel center and the center of the corresponding lidar surfel measured along the surface normal (visualized by the coloration in the figures). Based on 78,945 such correspondences (we allow up to three matches per surfel), the error distribution is observed to be normally distributed with a mean of 3 cm and standard deviation of 26 cm. Based on a visual inspection of the results, the largest outliers are believed to be in places where the mine survey is inaccurate, likely due to inaccuracy or alterations to the mine since the original measurements were taken.

3.4 Surface Reconstruction

Three dimensional surface reconstruction of the mine was achieved by meshing the data acquired from the vertical lidars. The poses of the vertical lasers with respect to the spinning laser were calibrated by minimizing the error between the point clouds they produced in a small, representative section of the dataset. To avoid scan data bunching up in areas where the vehicle stopped, we filter out any scans acquired while the vehicle was moving below 20 cm/s. In addition to having a more uniform point density (Fig. 2), it is considerably more straightforward to triangulate the points from the vertical lasers according to the scan manifold order. The spinning lidar data was not used here for surface reconstruction due the fact that a similar manifold-based triangulation of these points results in a non-orientable surface (topologically a *real projective plane* for each revolution of the mount).[3] Some preprocessing was undertaken to filter out returns from water droplets and mist, which was present in some sections of the mine to suppress dust. For this purpose, a 2D (in the scan manifold) mean filter was used to replace range values that deviated from their local neighborhood by a significant amount with the median range.

Representative views of the reconstructed surface are pictured in Fig. 5. The models appear to accurately represent the environment, with the exception of some long triangles connecting objects at occlusion boundaries (e.g., cables appear connected to walls). While a simple threshold-based removal of long triangles performs somewhat adequately in removing these connections, a more complex algorithm is likely required to more elegantly clean these artifacts. The resolution of the reconstructed surface depends on the vehicle speed and surface geometry, but is roughly at 10 cm resolution at top speeds.

[3] We have developed more complex volume-based methods for surface reconstruction from spinning lidar data [8]; however, for this application we decided a simpler point-based solution would be more accurate.

Fig. 5 Two views of the reconstructed surface model of the Northparkes Mine. **a** A typical section of the decline. The gaps on the floor and ceiling are due to the blindspot between the two vertically-oriented lasers. Electrical cables are seen along the left side wall. **b** A section of the mine including open ventilation doors and ductwork. Some long surface triangles are seen along the door frame where the surface is essentially parallel to the scanning plane

3.5 Other Technical Considerations

Many other technical issues required consideration in order to produce a successful outcome. Selected issues and lessons learned from this deployment are highlighted below.

Beam Divergence. Time of flight laser range scanners return a scalar range value for each measurement, which is typically assumed to be measured along an infinitesimally narrow ray. In reality, lidar beams can have a substantial footprint diameter, and neglecting to model the conical beam shape may have significant ramifications. The long-range, high incidence angle measurements that are typical in tunnel-like environments result in underestimates in range returns if a naive ray beam model is used. At sharp incidence angles, the leading edge of the beam footprint reflects from the surface earlier than the beam center. This effect becomes more significant at longer distances, since the beam footprint diameter grows linearly with range. The overall effect is that long, straight walls appear to bow inwards. In our example, since we had the spinning lidar predominantly focused on the ceiling, this bowing resulted in a pitch bias that caused the trajectory and tunnel map to arc (due to symmetry, the biases from the side walls cancelled each other out). We therefore adjust surfel positions according to a conical model for the LMS291, which has a beam divergence angle of approximately 0.7°.

Timing. When using multiple sensors, timing is of critical importance. Any noise, latencies, or errors in the timing, directly contributes to errors in the point locations. Care needs to be taken to correct, smooth, calibrate, and synchronize timestamps to ensure that they are clean and accurate.

One unexpected challenge we encountered was in the quality of the vertical laser data timestamps. We used the LMS driver from the SICK Toolbox [4] for data acquisition,[4] and later discovered that the driver consumed more CPU than expected resulting in many dropped frames. Compounding the issue is the fact that the driver did not record any internal timing information from the sensor. Assuming a known sampling rate and that the received scans arrive in order, we were able to correct many of the timing irregularities; however, due to USB buffering overflows, there is occasionally some ambiguity in precisely which frames were lost. There are therefore some residual timing irregularities in the scan sequence that could not be eliminated (e.g., see Fig. 2).

On-Board Diagnostics. Another device we had anticipated utilizing for additional robustness is an on-board diagnostics (OBD-II) sensor, which is able to read the vehicle speed and engine RPM directly from the on-board computer. The availability of speed data could assist in developing a better motion prior than the constant velocity model, which would be especially useful in case of any tunnel segments in areas lacking in surface normals along the tunnel direction. Unfortunately, the only site vehicle available for the data acquisition was slightly too old to have a compatible OBD port.

4 Conclusions

Accurately mapping an underground mine is a challenging problem, given the lack of GPS coverage and the scale of the environment. We presented an approach which enabled us to map a traverse of over 17 km in a copper and gold mine in under 2 h, producing a 3D surface model required by the mine operators for planning an major equipment transport activity. To our knowledge, the scale and efficiency of our data collection method is orders of magnitude beyond what had been demonstrated previously in the context of 3D mapping of underground mines, and our solution offers the potential for a new methodology to be adopted for rapidly generating maps of underground spaces. We plan to provide more details on our approach and further quantitative analysis of our results in a more comprehensive forthcoming publication. A video further demonstrating our results is available at http://youtu.be/QQeJ1xd_sOU.

One limitation of the approach as presented here is the reliance on survey data to globally register and georeference the data (though this type of data would be available at operating mines). An alternative strategy for global registration is to use place recognition to identify reobserved surfaces in the environment. We have had success applying a regional point descriptor-based technique [2] to coarsely

[4] At the time we were in the process of switching our middleware from DDX [3] to ROS [13], and had not yet extensively field tested all of the drivers. We have since moved to using a version of the LMS driver from DDX (modified with ROS wrappers), which is considerably less CPU intensive and contains the necessary scan metadata required for robust timing.

register the point cloud prior to running the global registration algorithm (though some surveyed points are still required to georeference the results and remove the drift accumulated when traversing long tunnels). General applicability to other types of mines and tunnels has also not yet been established. In particular, the surfaces in underground coal mines are typically much smoother, making it more difficult to track longitudinal motion along long tunnels where there is a lack of surface normals in the tunnel direction. Additional aiding sensors may be required in such scenarios.

While the core of the 3D SLAM algorithm used for this application was originally developed more a year prior to the deployment [1], it is through fielding our systems in real conditions and applications that motivates the development of truly reliable, robust, accurate, and efficient implementations. We have applied the spinning lidar SLAM system to map a variety of environments including streets, forests, caves, industrial facilities, building interiors, and mines. The challenging scenarios presented by each new application inspire key extensions and modifications to the system. Investigating the specifics of each problem helps to develop the approach into a more general and robust solution.

Acknowledgments The authors would like to acknowledge Northparkes Mine and the CSIRO Minerals Down Under Flagship for their support and assistance. We also thank Paul Flick for hardware support and So Jung Yun for generating the surface model imagery and animations.

References

1. M. Bosse, R. Zlot. Continuous 3D scan-matching with a spinning 2D laser, in *IEEE International Conference on Robotics and Automation*, (2009)
2. M. Bosse, R. Zlot. Place recognition using regional point descriptors for 3D mapping, in *International Conference on Field and Service Robotics*, (2009)
3. P. Corke, P. Sikka, J. Roberts, E. Duff. DDX: a distributed software architecture for robotic systems, *in Australasian Conference on Robotics and Automation*, (2004)
4. J.C. Derenick, T.H. Miller, J.R. Spletzer, A. Kushleyev, T. Foote, A. Stewart, J. Bohren. The Sick LIDAR Matlab/C++ Toolbox: Software for rapidly interfacing/configuring Sick LIDARs. Technical Report LU-CSE-08-008, Lehigh University Department of Computer Science and Engineering, 2008
5. E. Duff, J. Roberts, P. Corke. Automation of an underground mining vehicle using reactive navigation and opportunistic localization, in *Proceedings of the IEEE/RSJ International Conference on Intelligent Robots and Systems*, (2003)
6. N. Fairfield, D. Wettergreen. Evidence grid-based methods for 3D map matching, in *IEEE International Conference on Robotics and Automation*, (2009)
7. S. Fekete, M. Diederichs, M. Lato, Geotechnical and operational applications for 3-dimensional laser scanning in drill and blast tunnels. Tunn. Undergr. Space Technol. **25**, 614–628 (2010)
8. C. Holenstein, R. Zlot, M. Bosse. Watertight surface reconstruction of caves from 3D lidar data, in *Proceedings of the IEEE/RSJ International Conference on Intelligent Robots and Systems*, (2011)
9. F. Lemy, S. Yong, T. Schulz. A case study of monitoring tunnel wall displacement using laser scanning technology, in *International Association for Engineering Geology*, (2006)
10. M. Magnusson, H. Andreasson, A. Nüchter, A.J. Lilienthal. Appearance-based loop detection from 3D laser data using the normal distributions transform, in *IEEE International Conference on Robotics and Automation*, (2009)

11. A. Nüchter, H. Surmann, K. Lingemann, J. Hertzberg, S. Thrun. 6D SLAM with an application in autonomous mine mapping, in textitIEEE International Conference on Robotics and Automation, (2004)
12. G. Petrie, Mobile mapping systems. An introduction to the technology. GeoInformatics **13**(1), 32–43 (2010)
13. M. Quigley, B. Gerkey, K. Conley, J. Faust, T. Foote, J. Leibs, E. Berger, R. Wheeler, A. Ng. ROS: an open-source robot operating system, in *Proceedings of the Open-Source Software Workshop at the IEEE International Conference on Robotics and Automation*, (2009)
14. H. Rüther, M. Chazan, R. Schroeder, R. Neeser, C. Held, S. Walker, A. Matmon, L. Howritz, Laser scanning for conservataion and research of African cultural heritage sites: the case study of Wonderwerk Cave. South Afrika. J. Archaeol. Sci. **36**, 1847–1856 (2009)
15. D. Silver, D. Ferguson, A. Morris, S. Thayer. Topological exploration of subterranean environments. J. Field Robot. **23**(6/7), 395–415 (2006)
16. R. van Gosliga, R. Lindenbergh, N. Pfeifer, Deformation analysis of a bored tunnel by means of terrestrial laser scanning. Image Eng. Vision Metrol. **36**, 167–172 (2006)
17. U. Wong, A. Morris, C. Lea, J. Lee, C. Whittaker, B. Garney, W. Whittaker. Comparative evaluation of range sensing technologies for underground void modeling,in *Proceedings of the IEEE/RSJ International Conference on Intelligent Robots and Systems*, (2011)

Large Scale Monocular Vision-Only Mapping from a Fixed-Wing sUAS

Michael Warren, David McKinnon, Hu He, Arren Glover, Michael Shiel and Ben Upcroft

Abstract This paper presents the application of a monocular visual SLAM on a fixed-wing small Unmanned Aerial System (sUAS) capable of simultaneous estimation of aircraft pose and scene structure. We demonstrate the robustness of unconstrained vision alone in producing reliable pose estimates of a sUAS, at altitude. It is ultimately capable of online state estimation feedback for aircraft control and next-best-view estimation for complete map coverage without the use of additional sensors. We explore some of the challenges of visual SLAM from a sUAS including dealing with planar structure, distant scenes and noisy observations. The developed techniques are applied on vision data gathered from a fast-moving fixed-wing radio control aircraft flown over a 1 × 1 km rural area at an altitude of 20–100 m. We present both raw Structure from Motion results and a SLAM solution that includes FAB-MAP based loop-closures and graph-optimised pose. Timing information is also presented to demonstrate near online capabilities. We compare the accuracy of the 6-DOF pose estimates to an off-the-shelf GPS aided INS over a 1.7 km trajectory. We also present output 3D reconstructions of the observed scene structure and texture that demonstrates future applications in autonomous monitoring and surveying.

M. Warren (✉) · D. McKinnon · H. He · A. Glover · M. Shiel · B. Upcroft
Queensland University of Technology, 2 George St, Brisbane, QLD, Australia
e-mail: michael.warren@qut.edu.au

D. McKinnon
e-mail: d.mckinnon@qut.edu.au

H. He
e-mail: h.hu2@qut.edu.au

A. Glover
e-mail: aj.glover@qut.edu.au

M. Shiel
e-mail: michael.shiel@qut.edu.au

B. Upcroft
e-mail: ben.upcroft@qut.edu.au

K. Yoshida and S. Tadokoro (eds.), *Field and Service Robotics*,
Springer Tracts in Advanced Robotics 92, DOI: 10.1007/978-3-642-40686-7_33,

1 Introduction

Low-flying small Unmanned Aerial Systems (sUAS), otherwise known as Unmanned Aerial Vehicles (UAVs), have received increasing interest in recent years as a potentially cost-effective method of mapping and monitoring large areas of terrain. In contrast to other methods of environment mapping such as high-flying aerial surveys using manned aircraft and satellite-based sensing, sUAS provide a number of unique advantages in terms of reduced size, weight, infrastructure and cost. Additionally, they are often not subject to the restrictions of full-sized aircraft, meaning they can fly closer to the ground and in areas of potential sensitivity, increasing resolution and accuracy. Our interest is in using only visual sensors on the platform to estimate pose while simultaneously generating high resolution, high accuracy maps of vast areas in a single Euclidean frame with fast turnaround and minimal human interaction, facilitating accurate reconstructions of environments for research and commercial analysis.

Vision is rapidly becoming the sensor of choice in robotic pose estimation and has the ability to produce dense, 3D point clouds of the environment. These sensors are small, lightweight and have low-power requirements. Motivated by these properties and recent advances in visual Structure from Motion (SfM), loop closure detection and pose optimisation techniques, this paper presents a high-performance monocular visual Simultaneous Localisation and Mapping (SLAM) system. The pipeline is applied on data from a low-flying sUAS that determines 6-DOF aircraft pose (up to scale) and scene structure (Fig. 1) over large (>1 km) trajectories.

While traditionally LIDAR and other laser based systems have been used in mapping from the air, including some autonomous applications [14], their bulk, cost and power requirements mean they are restricted by both platform size and flight time. In addition, many airborne mapping systems are dependent on Global Positioning System (GPS), Differential GPS (DGPS) and inertial measurement, often in a filtering framework, for accurate vehicle pose estimation. However, their deficiencies (such as multi-pathing, lock-on failure, sensor drift) and heavy dependence on external infrastructure are well known [27].

Fig. 1 A dense 3D mesh of rural farmland computed from sequential poses of visual SLAM from the air

We demonstrate the ability of vision alone to generate pose capable of rivalling and ultimately complementing other sensors (GPS, INS etc.) in the airborne scenario for use in online state estimation feedback. We achieve this by careful implementation of algorithms for feature detection, pose estimation and feature triangulation, both in terms of speed and accuracy. In addition, our algorithm detects visual loop closures using openFAB-MAP [8] and applies these constraints in the pose-graph optimiser HOG-Man [9] to generate a refined pose and scene structure estimate. We show timing results demonstrating near online operation of the system and present a pose comparison to a GPS aided INS system as ground-truth. In this paper we refer to visual pose estimation (or visual odometry) as the recovery of aircraft pose from visual SfM techniques in addition to simultaneous estimation of scene structure.

The rest of this paper is outlined as follows: Sect. 1.1 comprehensively reviews the literature on visual pose estimation and scene reconstruction on the ground and in the air. Section 2 describes our SLAM algorithm for pose and structure estimation. Section 3 describes the robotic platform and a collected dataset. Finally, Sect. 4 shows the results of the algorithm in generating pose on the gathered dataset. We compare results between raw SfM and a pose-graph optimised SLAM solution and compare both outputs to a GPS aided INS pose estimate. Additionally, an output mosaic and 3D reconstruction generated from recovered scene structure are demonstrated to indicate additional uses for the data.

1.1 Prior Work

Visual pose estimation without additional input has been demonstrated to great effect on the ground using iterative SfM techniques, both in iterative SfM based pose estimation [22, 28] and with the inclusion of loop-closure in the greater SLAM problem [17, 20]. Additionally, visual SfM has demonstrated highly accurate reconstruction of environments [23]. Such results demonstrate the suitability of vision to large scale pose estimation and mapping tasks.

In flying applications, vision has been used in a wide variety of scenarios [14]. It has received significant interest in small-scale online pose estimation tasks, particularly in quadcopter applications [2], but has often made assumptions about the environment such as texture [4] or geometry [7] to assist the estimate. Outdoors, vision has been used within a number of filtered algorithms to produce high quality pose estimates [3] and to generate both qualitative [5] and ground-truthed reconstructions of large scenes [16].

Iterative vision-only pose estimation, however, has only been used on small scale ($<$20 m) airborne tasks, on relatively slow-moving craft such as airships and helicopters, and received little quantitative analysis. It has been shown in simulation [24] and small outdoor tasks [5], but with only qualitative assessments of accuracy.

A number of solutions exist that perform large scale visual mapping from aircraft [25] and sUAS [11] but these are characterised by their batch, strictly offline methods using photogrammetry techniques for image registration. Such methods are not suited to iterative online pose estimation and from a field and services point of view cannot be used to generate 3D maps or estimates in online time, meaning that autonomous decision-making cannot be performed. As a result, tasks such as ensuring full-coverage or view point path planning (next-best-view) cannot be achieved in a single flight, causing extended operational time and costs. In contrast, this paper details an approach that could be deployed as a SLAM pipeline for real-time visual pose estimation of a sUAS.

2 Methodology

Visual SLAM from a fixed wing aircraft at altitudes greater than 20–30 m presents unique algorithmic challenges (particularly for SfM techniques), which has limited attempts at large-scale visual SLAM in this scenario. Firstly, the highly distant features impact accurate scene triangulation for small inter-camera baselines and introduce planarity issues for monocular cameras. Secondly, fast motion means feature tracks are fleeting and have only a short lifetime. Ultimately, the airborne scenario requires extreme robustness in the SfM algorithm to reliably estimate pose. This is dependent on reliable feature detection and tracking in addition to accurate triangulation and removal of noisy scene points. We have addressed these issues to demonstrate a visual SLAM pipeline for online aerial scenarios. This pipeline can be separated into:

- Pose and structure initialisation,
- An SfM approach for iteratively estimating camera pose and 3D structure of the observed scene,
- OpenFAB-MAP based loop-closure detection and,
- Pose-graph optimisation to generate a final SLAM estimate of pose and structure.

We additionally describe some algorithmic differences to the current literature. Finally, we generate 3D meshes from the optimised pose and scene structure as a demonstration of the quality of the final estimate.

2.1 Pose Initialisation

In order to set up the iterative SfM algorithm, an initial estimate of pose and scene is required. Initial pose is setup by computing the essential matrix $\mathbf{E}_{1\leftrightarrow 2}$ using 5 matched features between the first and second images [21]. From this we fix the initial camera $\mathbf{P}_1 = \mathbf{K}\,[\mathbf{I}|0]$ at the origin of the global reference frame and extract the relative pose of the second camera as $\mathbf{P}_2 = \mathbf{K}\,[\mathbf{R}|\mathbf{t}]$. The essential matrix is computed inside a MLESAC routine to eliminate poor initialisations and find the best subset of features for a good essential matrix.

Further, it is well known that a potential ambiguity exists in the pose generated from an essential matrix estimated from observing planar scenes [26]. The configuration of the airborne scenario often reflects this due to the distance of the scene and flat terrain. To avoid degenerate initialisations, we implement a test for degeneracy based on structure. As described in [5], a degenerate essential matrix will result in an unnatural spread in depth of a reconstructed scene. We use a similar algorithm as the structural degeneracy test. We first find the subset of points $\mathbf{X}$ with depth $\mathbf{Z}$ greater than their median depth $\tilde{\mathbf{Z}}$ in the coordinate frame of the origin camera:

$$\mathbf{Z_l} = \left\{ \mathbf{Z} : \mathbf{Z} > \tilde{\mathbf{Z}} \right\} \tag{1}$$

We then find the mean of the depths of this subset, $\mathbf{Z_l}$, and divide it by the original median to generate the heuristic h:

$$h = \frac{\bar{\mathbf{Z}}_\mathbf{l}}{\tilde{\mathbf{Z}}} \tag{2}$$

This heuristic is then evaluated on a strict condition, where if $h > 1.2$, the initialisation is rejected and a new essential matrix computation is performed. We find that a significant number of initialisations are degenerate when applied to airborne data, requiring up to five repeats of the initialisation step. Once an initial camera pair is accepted, observed 3D structure is triangulated directly from the pair and their matched features.

2.2 Structure from Motion

Following a correctly initialised camera pair and 3D structure, our algorithm then follows an SfM routine to iteratively generate camera pose and scene structure from incoming frames. We split the algorithm into four main components:

1. Feature detection, matching and tracking
2. Motion update
3. Structure update
4. Sliding window bundle-adjustment

Additionally, we include openFAB-MAP based place recognition as a final step in the loop. The aim of our pose estimation and scene reconstruction task is to only use visual features. We do not consider motion models, filters or any additional sensors such as an IMU or GPS to aid the solution. Ultimately, however, this pipeline would be used inside a redundant framework that includes these sensors.

2.2.1 Feature Detection, Matching and Tracking

SIFT [18] features are detected in the image according to a bucketing scheme (400 equally spaced buckets per frame) to improve the spread of features, similar to that in [20]. This avoids grouping high density features in highly salient regions to help improve the pose constraint and more reliably track features throughout the image. We use a GPU implementation of SIFT detection and matching to approach an online time pose-update step.

We place requirements on descriptor matching that is stricter than other implementations to ensure that feature matches are accurate and tracks are generated only for the most salient features. We use SIFT as this has proven the most reliable in this scenario for both inter-frame and wide baseline matching, in part due to its 128-dimension descriptors. This is in contrast to the generally faster and more widely used SURF descriptor often used in ground applications where upright descriptors (64-dimensions) are often acceptable. The dot product is used as the metric of a match between two descriptors instead of the more common Euclidean distance.

2.2.2 Motion Update

Using feature matches between the new and previous frames that have well initialised scene structure, the new camera pose is extracted using calibrated 3-point pose estimation [10], and uses a fourth point to disambiguate the 3 generated pose solutions. This is again performed inside a MLESAC estimator to generate the best possible camera location.

2.2.3 Structure Update

After a new camera pose is estimated, new scene points $\boldsymbol{X}$ that meet the minimum track length (four sequential views) are computed using a least-squares triangulation. At each update step, additional observations of a point are used to recompute a least-squares triangulation from all views.

In this algorithm, a strict upper limit is placed on the reprojection error of any scene point. A scene point with a reprojection error $e_r > 0.4$ pixels in any image is discarded from the estimate. This actively removes any scene point that is not accurately triangulated at the extreme depths indicative of this scenario, reducing the number of active tracks. As a consequence of these strict feature tracking routines we compensate by detecting and matching a high number of features per frame. We find that the culling routine actively removes more than 90 % of features in each image, and only 30–40 are actively tracked frame-to-frame.

2.2.4 Bundle Adjustment

After each motion and structure update, a bundle-adjustment nonlinear optimisation is performed on the last five camera poses and observed scene. We use the analytical derivatives in the Jacobian calculation to improve optimisation speed and accuracy.

The SfM routine is then continuously repeated in a loop such that new poses are computed, new structure initialised and the estimate optimised via bundle adjustment to provide an updated and refined estimate in a sequential manner.

2.2.5 Place Recognition

After each pose update, openFAB-MAP generates loop closure hypotheses between the current and all previous images in the trajectory. The feature codebook and Chow-Liu Tree used by the algorithm are precomputed offline from separate airborne vision data. In comparison to the SURF detector used in the original FAB-MAP [6], our algorithm uses the STAR detector (based on CenSurE [1]). This alternative detector produces more reliable loop-closure results on airborne data, where scenes have few unique features and are very self similar.

If openFAB-MAP determines a location probability for a frame greater than 99 %, features are matched (similarly to Sect. 2.2.1) between the current and other images at that location. However, a minimum difference of 1,000 frames is required to avoid naive matching against spatially close frames. If the ratio of matched feature inliers to the number of features in the current frame is greater than 15 % the match is considered a positive loop closure and recorded for use in subsequent pose optimisation (Sect. 2.4).

2.3 Frame Striding

In contrast to other methods that often use a key-framing approach [17, 24] to discard images with small inter-frame movements, we use a frame-striding technique to actively skip images in the input stream. In the airborne scenario our algorithm uses a basis stride length of three frames.

By processing only every third frame, speed is significantly improved and frames where relative motion is small are actively avoided. In situations where the pose estimate between frames fails due to frame drops or rapid rolling/pitching of the aircraft, a recursive fallback is implemented to generate the next pose. When a failure to generate a pose estimate between frames i and $i + 3$ is detected, a pose estimate between frames i and $i + 2$ is attempted and so on until a reliable pose estimate is found, then returns to a three frame stride.

2.4 Pose Optimisation

The pose estimates computed from SfM and the constraints imposed by the detected loop closures can then be represented as a pose-graph and subsequently optimised using HOG-Man. All camera poses generated by the pose estimation routine are represented as nodes, with edges applied between sequentially adjacent poses. The loop-closure hypotheses generated by openFAB-MAP are used to apply additional edge constraints in the graph.

Similar to the method described in Sect. 2.2.2, a pose at time j matched to a 'base' pose i is re-computed from the structure observed by the camera at pose i. Any false-positive matches generated by openFAB-MAP are discarded at this point as they will not meet the required geometry test when generating a new camera pose.

In the pose-graph the loop-closure edge is generated by computing the 3D homography between the base camera and recomputed camera. These nodes and edge constraints are then input to the graph optimiser and processed in a sequential method to generate the optimised camera poses. As our graph only considers poses, we need to recover scene points from the optimised poses. All scene points are re-triangulated via least-squares based on their original projections while ensuring that all meet the new epipolar constraints generated from the camera poses.

2.5 3D Scene Reconstruction

As a demonstration of the quality of the optimised solution, we generate a 3D reconstruction from the imagery and final pose estimate using a methodology described in [19]. Dense depth maps are generated to create oriented 3D points in a single consistent Euclidean space. A Poisson Surface estimation [12] is performed from this set of oriented points to generate a reconstruction of the environment and textured by projecting the coloured image data to the surface. In comparison to other reconstruction work which creates meshes from optimally selected, high resolution views, our method generates meshes from sequential, relatively low resolution images in addition to estimates of aircraft pose.

3 Experimental Setup

The flight platform is a 1/3 scale Piper Cub with a wingspan of 3.6 m and fuselage length of 2.3 m (Fig. 2). It is capable of speeds of 30–110 km/h with a maximum payload of 6 kg.

The aircraft includes an off-the-shelf mini-ITX computer system running an Intel Atom processor (1.6 GHz), with two 64 GB solid-state drives in a RAID0 configuration. The sensor payload consists of a IEEE1394B colour Point Grey Flea 2 camera. The camera is downward facing towards the terrain in the fuselage of the platform, behind the engine and logging system (Fig. 2). A 6 mm lens is used with a field of view of approximately $42° \times 32°$. The camera is calibrated before flight using a checkboard pattern and a modified version of the RADDOC Calibration toolbox [15].

Fig. 2 The experimental platform, showing the location of logging system, camera, INS and GPS antenna

3.1 Dataset

Data was collected over a 90 s portion of flight, at an altitude of 20–100 m and a speed of ∼20 m/s. Bayer encoded colour images are logged at a resolution of $1,280 \times 960$ pixels at 30 Hz. Shutter time for each frame was set at 8.5 μS to counteract motion blur. The area was rural farmland with relatively few trees, animals and buildings. Some difficulties in the dataset include rapid lighting transitions, and frame drops occur at semi regular intervals due to buffer overflows leading to difficulties in feature tracking. An XSens MTi-G INS/GPS system is used as the ground truth measurement system on the aircraft, with a manufacturer claimed positional accuracy of 2.5 m CEP. Size and weight restrictions prevent the use of more accurate DGPS systems, however, the MTi-G itself provides a reasonably accurate estimate of pose over broad scales. The MTi-G unit is rigidly attached to the onboard camera, while the GPS receiver is installed directly above the camera. GPS, unfiltered IMU data and filtered INS pose were recorded at 120 Hz from the XSens MTi-G.

4 Results

The algorithm was performed offline on the collected images to generate 879 camera poses. The dataset consisted of 2,670 frames. Some key parameters for the processing include a stride length of 3 frames, 400 feature buckets, 10 features per bucket and a sliding window bundle adjustment of 5 frames.

OpenFAB-MAP produced 91 loop closure events with $p > 0.99$, as seen in Fig. 3. Of these, 71 passed a minimum feature inlier count of 15 % and the MLESAC camera resectioning routine, successfully removing all false positive events identified by openFAB-MAP, and hence used to generate an additional edge constraint. The generated edges and poses were used by HOG-Man to produce an optimised SLAM estimate over the 879 poses.

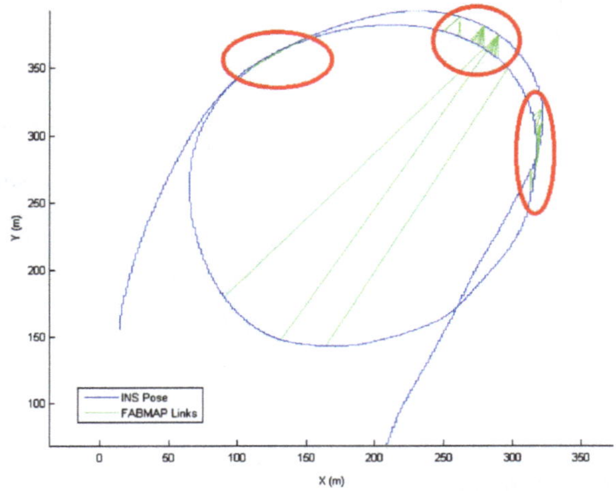

Fig. 3 Loop-closure events, highlighted in *red*, with probability $p > 0.99$ detected by openFAB-MAP overlaid on the ground truth GPS/INS pose. Some expected link locations are not observed due to differences in camera orientation at similar translational poses

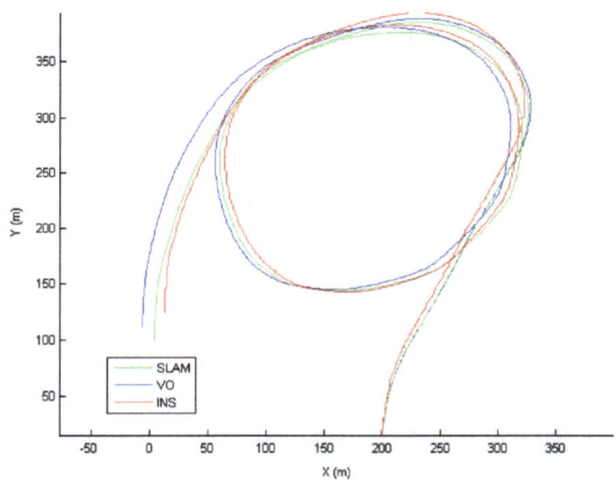

Fig. 4 Diagram in X, Y, showing SfM (VO) only path (*blue*), SLAM path (*green*) and INS path (*red*)

The monocular pose results for both the raw SfM and optimised pose estimates were then converted to a metric scale by calculating the ratio of distances between two spatially distant ground truth poses and their corresponding reconstructed poses. This scale ratio is then applied via a homography to the reconstructed poses to achieve metricity. Both the raw and optimised poses are then registered to the ground truth in all 6 degrees of freedom [13] on the first 30 camera poses.

The results of the SfM only (VO) and optimised (SLAM) pose estimates are shown in Figs. 4 and 5. The SfM only estimate clearly drifts, and has a final pose error of 40.6 m. The SLAM pose has significantly reduced error due to the optimisation, with a final pose error at the end of the trajectory of 27.2 m. The length of the entire set of poses is 1.70 km, meaning a translational drift of approximately 1.6 % over the length of trajectory. This is consistent with the accumulated error in other presented

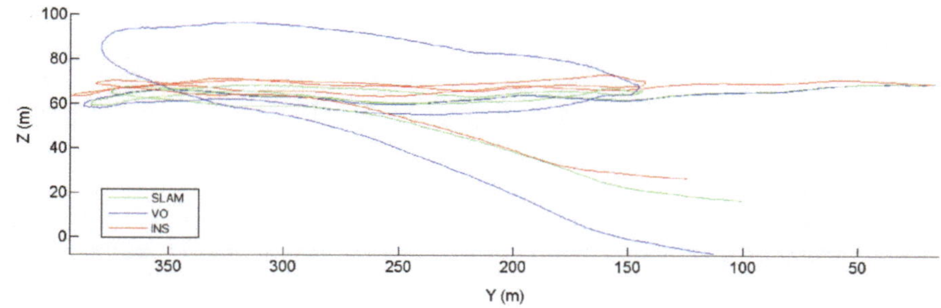

Fig. 5 Diagram in Y, Z, showing SfM (VO) only path (*blue*), SLAM path (*green*) and GPS/INS path (*red*)

works in ground scenarios [17, 28]. We speculate that some of the error is due to scale drift observable towards the end of the trajectory of both the raw and optimised estimates in Fig. 5.

We also compare the orientation produced by both estimates to ground truth, shown in Fig. 6. From this, it is clear that the algorithm is capable of accurately estimating orientation, with a maximum error of approximately 10.3° from the SfM only pose estimates, and a significantly reduced maximum error of 5.7° in the optimised estimate. The slightly positive pitch visible in Fig. 6 is a result of the slightly backward facing orientation of the camera and INS rig in the aircraft.

4.1 Timing Results

The SfM algorithm, openFAB-MAP loop-closure detection and HOG-Man pose estimation were all performed using Windows 7 64-bit on an Intel Core i5 650 Processor at 3.2 GHz with NVIDIA Quadro 600 GPU and 16 GB of system RAM. Aggressive memory management in the SfM algorithm meant that total memory consumption at the end of the sequence was 2.04 GB. In performing timing tests page-outs and disk-writes were not included in the time estimates. From Table 1 it can be seen that the SfM algorithm is capable of performing at just over 3.1 Hz if considered as a single frame stride (where every frame is processed). If we consider the 3 frame stride of this algorithm, the effective computed frame rate increases to just over 9.4 Hz.

From the computed poses and loop closure links, HOGMan produced an optimised result over the 879 poses in 2.1 s. While both the SfM algorithm and openFAB-MAP loop-closure detection were performed in a single thread, multi-threading the algorithm would lead to efficiency gains approaching online operation. We also anticipate that with strict memory management the algorithm is capable of performing similarly over much larger datasets.

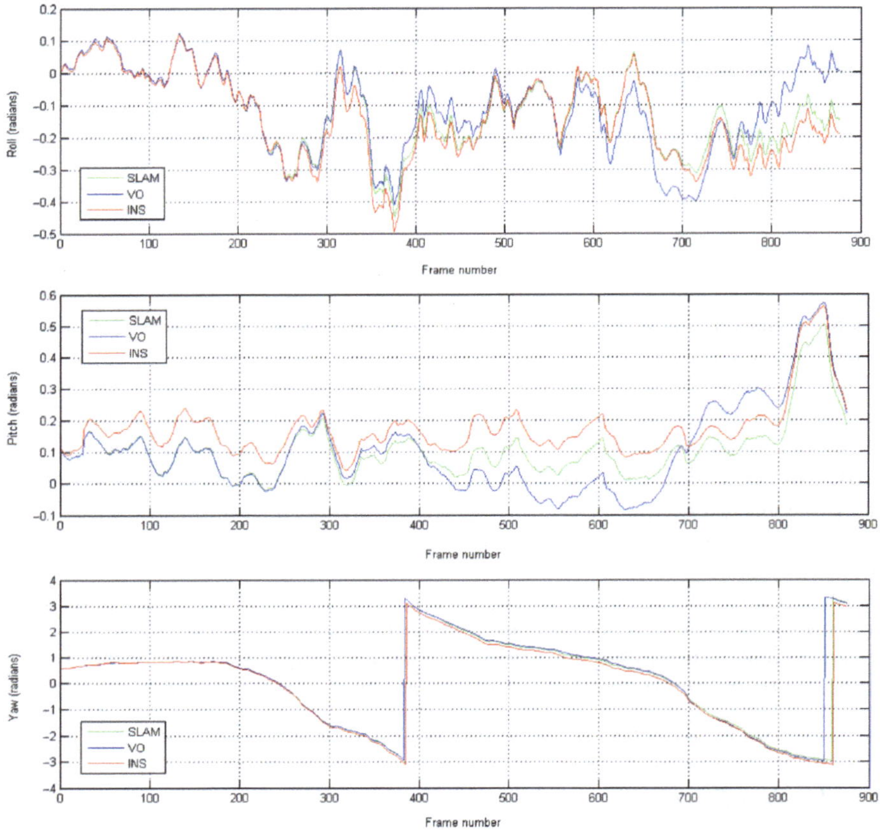

Fig. 6 Roll, pitch, yaw estimates for SfM (VO) only (*blue*), SLAM estimate (*green*) and GPS/INS (*red*), showing strong correlation

4.2 Reconstruction

From the optimised pose estimates, 3D scene points were re-triangulated using their feature projections to reconstruct the optimised scene. In Figs. 1 and 7 we present reconstruction outputs generated from this optimised estimate. Figure 7 shows a 2D mosaic of the observed images projected to a ground plane estimated from the 3D scene features. This mosaic is compared to satellite imagery of the area for qualitative analysis. It should be noted that the mosaic is only computed from pose estimates of the camera and no explicit feature matching is performed to create the 2D reconstruction.

Figure 1 shows a subsection of the final 3D reconstruction. From this reconstruction 3D structure is readily apparent, showing buildings, a parked aircraft and trees on predominantly flat terrain. These results demonstrate the viability of our airborne

Table 1 Timing results from monocular SLAM algorithm

Process	Minimum (ms)	Average (ms)	Maximum (ms)	Note
Initialisation	725	–	–	Performed only once
Feature detection	129	202	1,294	
Feature matching	65	77	196	
Pose update	2	15	195	
Structure update	1	9	62	
Bundle adjustment	0	15	102	
Total SfM time per frame	197	318	1,849	Average fps: 3.14 Hz, @ three frame stride: 9.43 Hz
openFAB-MAP	21	54	90	None
Loop closure matching	0	206	1,857	Performed only on loop closure detection

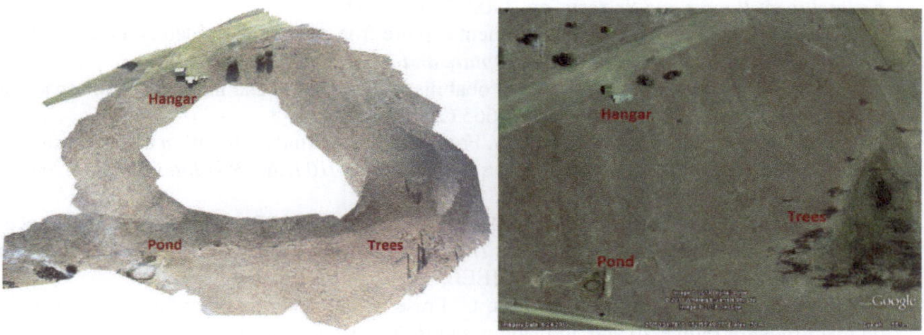

Fig. 7 *Left* A densely reconstructed ground plane using only camera pose to inform map generation over 879 frames. *Right* A comparison of the same area on Google Earth, showing qualitative accuracy of the final SLAM estimate

SLAM algorithm in producing up-to-date, 2D and 3D textured maps of environments at high resolution with rapid turnaround.

5 Conclusions

We have successfully demonstrated that visual SLAM on a fixed-wing airborne robotic platform is capable of a high degree of accuracy without additional inputs. This demonstration shows capability for use in more complicated filtered algorithms and in conjunction with additional sensors in the air. In future we intend to apply this algorithm in real-time that will facilitate online navigation and mapping for airborne robotic vehicles. Additionally, we intend to demonstrate the algorithm using multi-camera rigs to increase accuracy, remove initialisation degeneracies and remove scale issues.

References

1. M. Agrawal, K. Konolige, M. Blas, Censure: center surround extremas for realtime feature detection and matching, in *International Conference on Computer Vision, The Proceedings of the*, pp. 102–115 (2008)
2. M. Blosch, S. Weiss, D. Scaramuzza, R. Siegwart, Vision based MAV navigation in unknown and unstructured environments, in *Robotics and Automation (ICRA), 2010 IEEE International Conference on*, pp. 21–28. IEEE (2010)
3. M. Bryson, M. Johnson-Roberson, S. Sukkarieh, Airborne smoothing and mapping using vision and inertial sensors, in *2009 IEEE International Conference on Robotics and Automation*, pp. 3143–3148. IEEE (2009)
4. A. Cherian, J. Andersh, V. Morellas, N. Papanikolopoulos, B. Mettler, Autonomous altitude estimation of a UAV using a single onboard camera, in *2009 IEEE/RSJ International Conference on Intelligent Robots and Systems*, pp. 3900–3905 (2009)
5. R. Clark, M. Lin, C. Taylor, 3D environment capture from monocular video and inertial data, in *SPIE on Three-Dimensional Image Capture and Applications* (2006)
6. M. Cummins, P. Newman, FAB-MAP: probabilistic localization and mapping in the space of appearance. Int. J. Rob. Res. **27**(6), 647–665 (2008)
7. D. Eynard, P. Vasseur, C. Demonceaux, V. Frémont, UAV altitude estimation by mixed stereoscopic vision, in *Intelligent Robots and Systems (IROS), 2010 IEEE/RSJ International Conference on*, Iros, pp. 646–651. IEEE (2010)
8. A.J. Glover, W.P. Maddern, M.J. Milford, G.F. Wyeth, FAB-MAP + RatSLAM: appearance-based SLAM for multiple times of day, in *The IEEE International Conference on Robotics and Automation*, May 2010, pp. 3507–3512. IEEE, Alaska, USA (2010)
9. G. Grisetti, R. Kuemmerle, C. Stachniss, U. Frese, C. Hertzberg, Hierarchical optimization on manifolds for online 2d and 3d mapping, in *Robotics and Automation (ICRA), 2010 IEEE International Conference on*. IEEE, Anchorage, USA (2010)
10. B. Haralick, C. Lee, K. Ottenberg, M. Nölle, Review and analysis of solutions of the three point perspective pose estimation problem. Int. J. Comput. Vis. **13**(3), 331–356 (1994)
11. V. Hiep, R. Keriven, P. Labatut, J.P. Pons, Towards high-resolution large-scale multi-view stereo, in *Conference on Computer Vision and Pattern Recognition (CVPR)*, Miami, USA (2009)
12. H. Hoppe, Poisson surface reconstruction and its applications, in *Proceedings of the 2008 ACM Symposium on Solid and Physical Modeling—SPM '08*, p. 10 (2008)
13. B.K.P. Horn, Closed-form solution of absolute orientation using unit quaternions. J. Opt. Soc. Am. A **4**(4), 629 (1987)
14. T. Kanade, O. Amidi, Q. Ke, Real-time and 3D vision for autonomous small and micro air vehicles, *2004 43rd IEEE Conference on Decision and Control (CDC) (IEEE Cat. No.04CH37601)*, vol. 2, pp. 1655–1662 (2004)
15. A. Kassir, T. Peynot, Reliable automatic camera-laser calibration, in *Australasian Conference on Robotics and Automation*, Brisbane, Australia (2010)
16. J. Kim, S. Sukkarieh, Autonomous airborne navigation in unknown terrain environments. IEEE Trans. Aerosp. Electron. Syst. **40**(3), 1031–1045 (2004)
17. K. Konolige, M. Agrawal, Frameslam: from bundle adjustment to real-time visual mapping. IEEE Trans. Rob. **24**(5), 1066–1077 (2008)
18. D. Lowe, Object recognition from local scale-invariant features, in *International Conference on Computer Vision, The Proceedings of the*, vol. 2, pp. 1150–1157 (1999)

19. D. McKinnon, H. He, B. Upcroft, R.N. Smith, Towards automated and in-situ, near-real time 3-D reconstruction of coral reef environments, in *Oceans*, September 2011
20. C. Mei, G. Sibley, P. Newman, A constant-time efficient stereo SLAM system, in *Systems Engineering*, pp. 1–11 (2009)
21. D. Nistér, An efficient solution to the five-point relative pose problem. IEEE Trans. Pattern Anal. Mach. Intell. **26**, 756–777 (2004)
22. D. Nistér, O. Naroditsky, J. Bergen, Visual odometry for ground vehicle applications. J. Field Rob. **23**(1), 3–20 (2006)
23. M. Pollefeys, D. Nistér, J.M. Frahm, A. Akbarzadeh, P. Mordohai, B. Clipp, C. Engels, D. Gallup, S.J. Kim, P. Merrell, C. Salmi, S. Sinha, B. Talton, L. Wang, Q. Yang, H. Stewénius, R. Yang, G. Welch, H. Towles, Detailed real-time urban 3D reconstruction from video. Int. J. Comput. Vis. **78**(2–3), 143–167 (2007)
24. H. Strasdat, J. Montiel, A. Davison, Scale drift-aware large scale monocular SLAM, in *Proceedings of Robotics: Science and Systems (RSS)*. Citeseer, Zaragoza, Spain (2010)
25. E. Tola, C. Strecha, P. Fua, Efficient large-scale multi-view stereo for ultra high-resolution image sets. Mach. Vis. Appl. **23**(5), 903–920 (2011)
26. P. Torr, A. Fitzgibbon, A. Zisserman, The problem of degeneracy in structure and motion recovery from uncalibrated image sequences. Int. J. Comput. Vis. **32**(1), 27–44 (1999)
27. J. Volpe, Vulnerability assessment of the transportation infrastructure relying on the global positioning system, *Transportation* (2001)
28. M. Warren, D. McKinnon, H. He, B. Upcroft, Unaided stereo vision based pose estimation, in *Australasian Conference on Robotics and Automation*, December 2010, Brisbane, Australia (2010)

Super-Voxel Based Segmentation and Classification of 3D Urban Landscapes with Evaluation and Comparison

Ahmad Kamal Aijazi, Paul Checchin and Laurent Trassoudaine

Abstract Classification of urban range data into different object classes offers several challenges due to certain properties of the data such as density variation, inconsistencies due to holes and the large data size which requires heavy computation and large memory. A method to classify urban scenes based on a super-voxel segmentation of sparse 3D data obtained from Lidar sensors is presented. The 3D point cloud is first segmented into voxels which are then characterized by several attributes transforming them into super-voxels. These are joined together by using a link-chain method rather than the usual region growing algorithm to create objects. These objects are then classified using geometrical models and local descriptors. In order to evaluate the results, a new metrics is presented which combines both segmentation and classification results simultaneously. The proposed method is evaluated on standard datasets using three different evaluation metrics.

1 Introduction

Characterization and classification of cluttered urban scenes remains a challenging task for different field robots and autonomous vehicles operating in such outdoor environments. Lately, the automatic segmentation and classification of 3D urban data

A. K. Aijazi (✉) · P. Checchin · L. Trassoudaine
Clermont Université, Université Blaise Pascal, Institut Pascal, BP10448 F-63000, Clermont-Ferrand, France
e-mail: kamalaijazi@gmail.com

P. Checchin
e-mail: paul.checchin@univ-bpclermont.fr

A. K. Aijazi · P. Checchin · L. Trassoudaine
CNRS, UMR 6602, Institut Pascal, F-63171, Aubiere, France

L. Trassoudaine
e-mail: laurent.trassoudaine@univ-bpclermont.fr

K. Yoshida and S. Tadokoro (eds.), *Field and Service Robotics*,
Springer Tracts in Advanced Robotics 92, DOI: 10.1007/978-3-642-40686-7_34,

have gained widespread interest and importance in the scientific community due to the increasing demand of urban landscape analysis and cartography for different popular applications, coupled with the advances in 3D data acquisition technology. Whereas the automatic (or partially supervised) extraction of important urban scene structures such as roads, vegetation, lamp posts, and buildings from 3D data have proved essential in different autonomous perception tasks and mission planning, it has also been found to be an attractive approach to urban scene analysis because it can tremendously reduce the resources required for analyzing the data for subsequent use in 3D city modeling.

In order to fully exploit 3D point clouds, effective segmentation has proved to be a necessary and critical pre-processing step in a number of autonomous perception tasks. Earlier works including [1, 14, 18] used small sets of specialized features, such as local point density or height from the ground, to discriminate only few object categories in outdoor scenes, or to separate foreground from background. Lately, segmentation has been commonly formulated as graph clustering. Instances of such approaches are Graph-Cuts including Normalized-Cuts and Min-Cuts. Golovinskiy and Funkhouser [7] extended Graph-Cuts segmentation to 3D point clouds by using k-Nearest Neighbors (k-NN) to build a 3D graph. In this work, edge weights based on exponential decay in length were used. But the limitation of this method is that it requires prior knowledge of the location of the objects to be segmented. Another segmentation algorithm for natural images, recently introduced by Felzenszwalb and Huttenlocher (FH) [4], has gained popularity for several robotic applications due to its efficiency. Triebel et al. [26] modified the FH algorithm for range images to propose an unsupervised probabilistic segmentation technique. In this approach, the 3D data is first over-segmented during pre-processing. Schoenberg et al. [24] have applied the FH algorithm to colored 3D data obtained from a co-registered camera laser pair. The edge weights are computed as a weighted combination of Euclidean distances, pixel intensity differences and angles between surface normals estimated at each 3D point. The FH algorithm is then run on the image graph to provide the final 3D partitioning. The evaluation of the algorithm is done on road segments only. Strom et al. [25] proposed a similar approach, modifying the FH algorithm, to incorporate angle differences between surface normals in addition to the differences in color values. Our approach differs from the above mentioned methods as, instead of using the properties of each point for segmentation resulting in over segmentation, we have grouped the 3D points based on Euclidian distance into voxels and then assigned normalized properties to these voxels transforming them into super-voxels. This not only prevents over segmentation but in fact reduces the data set by many folds thus reducing post-processing time. A spanning tree approach to the segmentation of 3D point clouds was proposed in [20]. Graph nodes represent Gaussian ellipsoids as geometric primitives. These ellipsoids are then merged using a tree growing algorithm. Unlike this method, our approach uses cuboids of different sizes as geometric primitives and a link-chain method to group them together. In the literature survey we also find some segmentation methods based on surface discontinuities such as [17] who used surface convexity in a terrain mesh as a separator between objects.

In the past, research related to 3D urban scene classification and analysis had been mostly performed using either 3D data collected by airborne LiDAR for extracting building structures [28] or 3D data collected from static terrestrial laser scanners for extraction of building features such as walls and windows [21]. In [13] the authors extracted roads and objects just around the roads like road signs. They used a least square fit plane and RANSAC method to first extract a plane from the points followed by a Kalman filter to extract roads in an urban environment. A method of classification based on global features is presented in [8] in which a single global spin image for every object is used to detect cars in the scene while in [23] a Fast Point Feature Histogram (FPFH) local feature is modified into global feature for simultaneous object identification and view-point detection. Classification using local features and descriptors such as Spin Image [9], Spherical Harmonic Descriptors [10], Shape Distributions [19], 3D SURF feature [12] is also found in the literature survey. There is also a third type of Classification based on Bag Of Features (BOF) as discussed in [16]. In [15] a method of multi-scale Conditional Random Fields is proposed to classify 3D outdoor terrestrial laser scanned data by introducing regional edge potentials in addition to the local edge and node potentials in the multi-scale Conditional Random Fields. This is followed by fitting plane patches onto the labeled objects such as building terrain and floor data using the RANSAC algorithm as a post-processing step to geometrically model the scene. Douillard et al. [2] presented a method in which 3D points are projected on to the image to find regions of interest for classification. In our work, we use geometrical features and local descriptors to successfully classify different segmented objects represented by groups of voxels in the urban scene (discussed in Sect. 3). Ground is assumed to be flat and is used as an object separator.

The main contribution of our work includes: (1) a voxel based segmentation using a proposed Link-Chain method (discussed in Sect. 2); (2) introduction of a new evaluation metrics which combines both segmentation and classification results simultaneously (presented in Sect. 4); (3) evaluation of the proposed algorithm on standard data sets using three different evaluation methods (see Sect. 5).

2 Voxel Based Segmentation

The proposed voxel based segmentation method consists of three main parts which are the voxelisation of data, the transformation of voxels into super-voxels and the clustering by link-chain method.

2.1 Voxelisation of Data

When dealing with large 3D data sets, the computational cost of processing all the individual points is very high, making it impractical for real time applications. It is

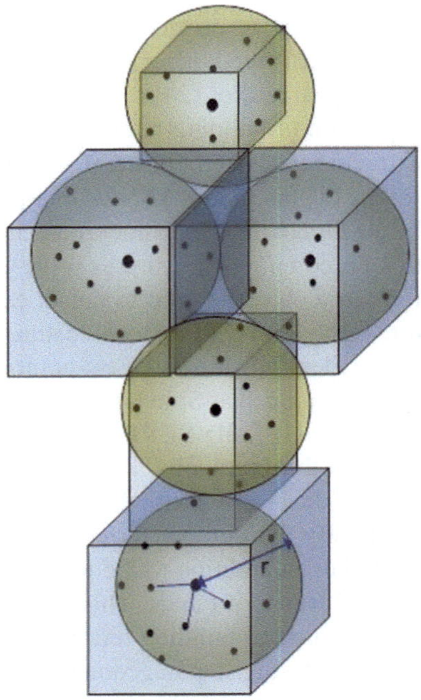

Fig. 1 A number of points is grouped together to form cubical voxels of maximum size $2r$. The actual voxel sizes vary according to the maximum and minimum values of the neighboring points found along each axis to ensure the profile of the structure

therefore sought to reduce this number of points by grouping together or removing redundant or un-useful points. Similarly, in our work the individual 3D points are clustered together to form a higher level representation or voxel as shown in Fig. 1.

For p data points, a number of s voxels ($s << p$) is computed based on r-NN, where r is the radius of ellipsoid and with $w = 1/d$ given as the weight to each neighbor (d being the Euclidian distance to the neighbor). The maximum size of the voxel $2r$ depends upon the density of the 3D point cloud. In [15] color values are also added in this step but it is observed that for relatively smaller voxel sizes, the variation in properties such as color is not profound and just increases computational cost. For these reasons, we have only used distance as a parameter in this step. The other properties are used in the next step of clustering the voxels to form objects. Also we have ensured that each 3D point which belongs to a voxel is not considered for further voxelisation. This not only prevents over segmentation but also reduces processing time.

For the voxels we use a cuboid because of its symmetry which avoids fitting problems while grouping and also minimizes the effect of voxel shape during feature extraction. Although the maximum voxel size is predefined, the actual voxel sizes vary according to the maximum and minimum values of the neighboring points found along each axis to ensure the profile of the structure.

2.2 Transformation of Voxels into Super-Voxels

A voxel is transformed into a super-voxel when properties based on its constituting points are assigned to it. These properties mainly include:

- $\mathbf{V}_{X,Y,Z}$: geometrical center of the voxel;
- $\mathbf{V}_{R,G,B}$: mean R, G, and B value of constituting 3D points;
- $\text{Var}(R, G, B)$: maximum of the variance of R, G, and B values;
- $\mathbf{V}_I$: mean laser reflectance intensity value of constituting 3D points;
- $\text{Var}(I)$: variance of laser reflectance intensity values;
- $s_{X,Y,Z}$ is the voxel size along each axis X, Y and Z;
- Surface normals: a surface normal is calculated using PCA (Principal Component Analysis). The PCA method has been proved to perform better than the area averaging method [11] to estimate the surface normal. Given a point cloud data set $\mathcal{D} = \{x_i\}_{i=1}^{n}$, the PCA surface normal approximation for a given data point $p \in \mathcal{D}$ is typically computed by first determining the k-Nearest Neighbors, $x_k \in \mathcal{D}$, of p. Given the K neighbors, the approximate surface normal is then the eigenvector associated with the smallest eigenvalue of the symmetric positive semi-definite matrix $P = \sum_{k=1}^{K}(x_k - \overline{p})^T(x_k - \overline{p})$ where $\overline{p}$ is the local data centroid $\overline{p} = \frac{1}{K}\sum_{j=1}^{K} x_j$. The estimated surface normal is ambiguous in terms of sign; to account for this ambiguity it is homogenized using the dot product. Yet for us the sign of the normal vector is not important as we are more interested in the orientation. A surface normal is estimated for all the points belonging to a voxel and is then associated with that particular voxel.

With the assignment of all these properties, a voxel is transformed into a super-voxel. All these properties would then be used in grouping these super-voxels (from now onwards referred to as s-voxels) into objects and then during the classification of these objects. Instead of using thousands of points in the data set, the advantage of this approach is that we can now use the reduced number of s-voxels to obtain similar results for classification and other algorithms. In our case, the data sets of 110, 392, 53, 676 and 27,396 points were reduced to 18,541, 6,928 and 7,924 s-voxels respectively which were then used for subsequent processing.

2.3 Clustering by Link-Chain Method

When the 3D data is converted into s-voxels, the next step is to group these s-voxels to segment into distinct objects.

Usually for such tasks a region growing algorithm [27] is used in which the properties of the whole growing region may influence the boundary or edge conditions. This may sometimes lead to erroneous segmentation. Also common in such type of methods is a node based approach [17] in which at every node, boundary conditions have to be checked in all five different possible directions. In our work, we

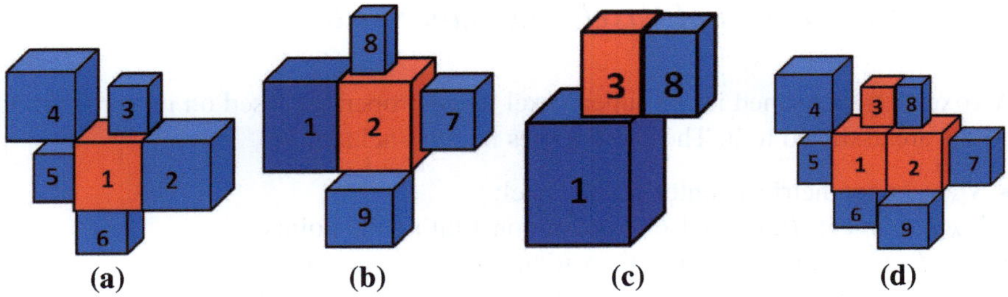

Fig. 2 Clustering of *s*-voxels using a link-chain method is demonstrated. **a** shows *s*-voxel 1 taken as principal link in red and all secondary links attached to it in blue. **b, c** show the same for *s*-voxel 2 and 3 taken as principal links. **d** shows the linking of principal links (*s*-voxels 1, 2 & 3) to form a chain removing redundant secondary links

have proposed a link-chain method instead to group these *s*-voxels together into segmented objects. In this method, each *s*-voxel is considered as a link of a chain. All secondary links attached to each of these principal links are found. In the final step, all the principal links are linked together to form a continuous chain removing redundant secondary links in the process as shown in Fig. 2.

Let $\mathbf{V}_P$ be a principal link and $\mathbf{V}_n$ be the nth secondary link. Each $\mathbf{V}_n$ is linked to $\mathbf{V}_P$ if and only if the following three conditions are fulfilled:

$$\left|\mathbf{V}_{P_{X,Y,Z}} - \mathbf{V}_{n_{X,Y,Z}}\right| \leq (w_D + c_D) \tag{1}$$

$$\left|\mathbf{V}_{P_{R,G,B}} - \mathbf{V}_{n_{R,G,B}}\right| \leq 3\sqrt{w_C} \tag{2}$$

$$\left|\mathbf{V}_{P_I} - \mathbf{V}_{n_I}\right| \leq 3\sqrt{w_I} \tag{3}$$

where, for the principal and secondary link *s*-voxels respectively:

- $\mathbf{V}_{P_{X,Y,Z}}$, $\mathbf{V}_{n_{X,Y,Z}}$ are the geometrical centers;
- $\mathbf{V}_{P_{R,G,B}}$, $\mathbf{V}_{n_{R,G,B}}$ are the mean RGB values;
- $\mathbf{V}_{P_I}$, $\mathbf{V}_{n_I}$ are the mean laser reflectance intensity values;
- w_C is the color weight equal to the maximum value of the two variances $\mathrm{Var}(R, G, B)$;
- w_I is the intensity weight equal to the maximum value of the two variances $\mathrm{Var}(I)$.

w_D is the distance weight given as $\frac{\left(\mathbf{V}_{P_{s_{X,Y,Z}}} + \mathbf{V}_{n_{s_{X,Y,Z}}}\right)}{2}$. Here $S_{X,Y,Z}$ is the voxel size along X, Y and Z axis respectively. c_D is the inter-distance constant (along the three dimensions) added depending upon the density of points and also to overcome measurement errors, holes and occlusions, etc. The value of c_D needs to be carefully selected depending upon the data. The orientation of normals is not considered in this stage to allow the segmentation of complete objects as one entity instead of just planar faces.

This segmentation method ensures that only the adjacent boundary conditions are considered for segmentation with no influence of a distant neighbor's properties. This may prove to be more adapted to sharp structural changes in the urban environment. The segmentation algorithm is summarized in Alg. 1. With this method 18,541, 6,928 and 7,924 s-voxels obtained from processing three different data sets were successfully segmented into 237, 75 and 41 distinct objects respectively.

Algorithm 1 Segmentation

1: **repeat**
2: Select a 3D point for voxelisation
3: Find all neighboring points to be included in the voxel using r-NN within the maximum voxel length specified
4: Transform voxel into s-voxel by first finding and then assigning all the properties including surface normal, found by using PCA, to it.
5: **until** all 3D points are used in a voxel
6: **repeat**
7: Specify a s-voxel as a principal link
8: Find all secondary links attached to the principal link
9: **until** all s-voxels are used
10: Link all principal links to form a chain removing redundant links in the process

3 Classification of Objects

In order to classify these objects, we assume the ground to be flat and use it as separator between objects. For this purpose we first classify and segment out the ground from the scene and then the rest of the objects. This step leaves the remaining objects as if suspended in space, i.e distinct and well separated, making them easier to be classified. The ground or roads followed by these objects are classified using geometrical and local descriptors. These mainly include:

a. **Surface Normals:** The orientation of the surface normals is essential for classification of ground and building faces. For ground object, the surface normals are along Z-axis (height axis) whereas for building faces the surface normals are parallel to the $X - Y$ axis (ground plane), see Fig. 3.
b. **Geometrical Center and Barycenter:** The height difference between the geometrical center and the barycenter along with other properties is very useful in distinguishing objects like trees and vegetation, etc., where h(barycenter − geometrical center) > 0, with h being the height function.
c. **Color and Intensity:** Intensity and color are also an important discriminating factor for several objects.

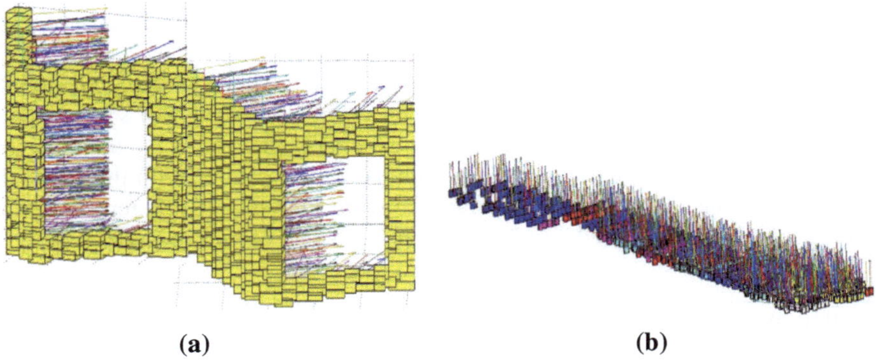

Fig. 3 **a** shows surface normals of building *s*-voxels which are parallel to the ground plane. In (**b**) it can be clearly seen that the surface normals of road surface *s*-voxels are perpendicular to the ground plane

Fig. 4 Bounding boxes for buildings, trees, cars, pedestrians and poles

d. **Geometrical Shape:** Along with the above mentioned descriptors, geometrical shape plays an important role in classifying objects. In 3D space, where pedestrians and poles are represented as long and thin with poles being longer, cars and vegetation are broad and short. Similarly, as roads represent a low flat plane, the buildings are represented as large (both in width and height) vertical blocks (as shown in Fig. 4).

Using these descriptors we successfully classify urban scenes into five different classes (mostly present in our scenes) i.e. buildings, roads, cars, poles and trees. The object types chosen for classification are so distinctly different that if they are correctly segmented out, a simple classification method like the one proposed may be sufficient.The classification results and a new evaluation metric are discussed in the following sections.

4 Evaluation Metrics

In previous works, different evaluation metrics are introduced for both segmentation results and classifiers independently. Thus in our work we present a new evaluation metric which incorporates both segmentation and classification together.

The evaluation method is based on comparing the total percentage of s-voxels successfully classified as a particular object. Let T_i, $i \in \{1, \ldots, N\}$, be the total number of s-voxels distributed into objects belonging to N number of different classes, i.e this serves as the ground truth, and let t_{j_i}, $i \in \{1, \ldots, N\}$, be the total number of s-voxels classified as a particular class of type-j and distributed into objects belonging to N different classes (for example a s-voxel classified as part of the building class may actually belong to a tree) then the ratio S_{jk} (j is the class type as well as the row number of the matrix and $k \in \{1, \ldots, N\}$) is given as $S_{jk} = \frac{t_{jk}}{T_k}$. These values of S_{jk} are calculated for each type of class and are used to fill up each element of the confusion matrix, row by row (refer to Table 2 for instance). Each row of the matrix represents a particular class. Thus, for a class of type-1 (i.e. first row of the matrix) the values of:

- True Positive rate **TP** $= S_{11}$ (i.e the diagonal of the matrix represents the **TP**s)
- False Positive rate **FP** $= \sum_{m=2}^{N} S_{1m}$
- True Negative rate **TN** = (1 - **FP**)
- False Negative rate **FN** = (1-**TP**)

The diagonal of this matrix (or **TP**s) gives the Segmentation ACCuracy **SACC**, similar to the voxel scores recently introduced by Douillard et al. [3]. The effects of unclassified s-voxels are automatically incorporated in the segmentation accuracy. Using the above values, the Classification Accuracy **CACC** is given as:

$$\mathbf{CACC} = \frac{\mathbf{TP} + \mathbf{TN}}{\mathbf{TP} + \mathbf{TN} + \mathbf{FP} + \mathbf{FN}} \tag{4}$$

This value of **CACC** is calculated for all N types of classes of objects present in the scene. Overall classification accuracy (**OCACC**) can then be calculated as

$$\mathbf{OCACC} = \frac{1}{N} \sum_{i=1}^{N} \mathbf{CACC}_i \tag{5}$$

where N is the total number of object classes present in the scene. Similarly, the Overall Segmentation ACCuracy **OSACC** can also be calculated.

The values of T_i and t_{j_i} used above are laboriously calculated by hand matching the voxelised data output and the final classified s-voxels and points.

Table 1 Results of 3D data sets of Blaise Pascal university

Data Set #	Number of 3D data points	Number of segmented s-voxels	Number of segmented objects
#1	$27,396$	$7,924$	41
#2	$53,676$	$6,928$	75
#3	$110,392$	$18,541$	237

5 Results

In order to test our algorithm, two different data sets were used: 3D data sets of Blaise Pascal University and 3D Urban Data Challenge data set [5].

5.1 3D Data Sets of Blaise Pascal University

These consist of 3D data acquired from different urban scenes on the Campus of Blaise Pascal University in Clermont-Ferrand, France, using a LEICA HDS-3000 3D laser scanner. In these data sets the 3D points are coupled with the corresponding RGB and laser reflectance intensity values. The results of three such data sets are summarized in Table 1 and the results of one such data set are presented in Fig. 5. The evaluation results using the new evaluation metrics for two of the data sets are presented in Tables 2 and 3. These results are evaluated using a value of maximum voxel size equal to 0.3 m and $c_D = 0.25$ m. The choice of an optimal value of maximum voxel size and c_D is important as it is observed that with the reduction in voxel size the segmentation and classification results improve but at the cost of processing time. These values can simply be determined by plotting the **OCACC** and the corresponding processing time for a range of values of maximum voxel size and c_D for any one of the scenes from the respective data sets. The optimal values depending upon the requirements can be chosen using these two plots.

5.2 3D Urban Data Challenge Data Set

The algorithm was further tested on the static data set of the recently concluded 3D Urban Data Challenge 2011, acquired and used by the authors of [5]. This standard data set contains a rich collection of 3D urban scenes of New York city mainly focusing on building facades and structures. These 3D points are coupled with corresponding RGB and reflectance intensity values. A value of maximum voxel size equal to 0.5 m and $c_D = 0.15$ m were used for this data set. The evaluation results

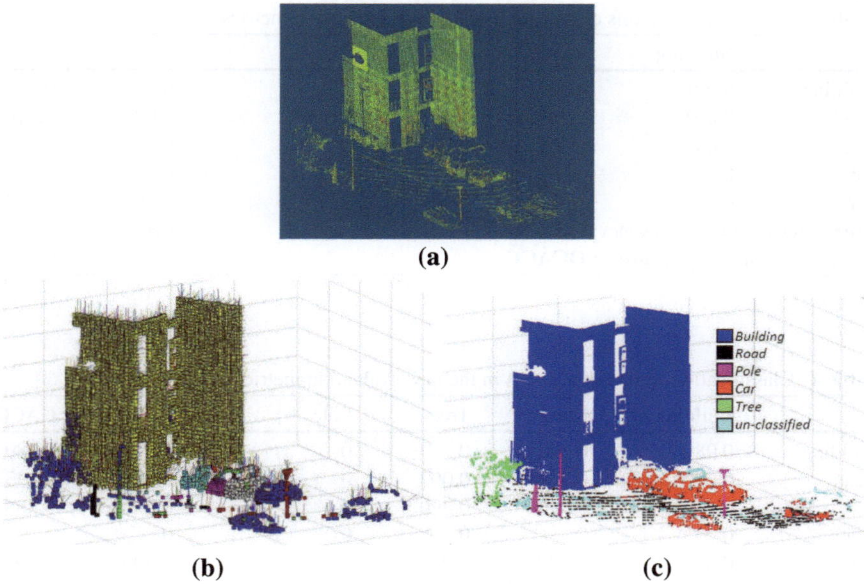

Fig. 5 **a** shows 3D data points of data set 2. **b** shows *s*-voxel segmentation of 3D points. **c** shows classification results (labeled 3D points)

Table 2 Classification results of data set 2 in the new evaluation metrics

	Building	Road	Tree	Pole	Car	CACC
Building	0.996	0.007	0	0	0	0.995
Road	0	0.906	0.028	0.023	0.012	0.921
Tree	0	0.045	0.922	0	0	0.938
Pole	0	0.012	0	0.964	0	0.976
Car	0	0.012	0	0	0.907	0.947
Overall segmentation accuracy: **OSACC**					0.939	
Overall classification accuracy: **OCACC**						0.955

using the new evaluation metrics for two such scenes from this data set are presented in Tables 4 and 5. Results of one such scene from this data set are shown in Fig. 6.

5.3 Comparison of Results with Existing Evaluation Methods

The classification results were also evaluated using already existing methods along with the proposed evaluation metrics for comparison purpose. Firstly, one of the more

Table 3 Classification results of data set 3 in the new evaluation metrics

	Building	Road	Tree	Pole	Car	CACC
Building	0.901	0.005	0.148	0	0	0.874
Road	0.003	0.887	0.011	0.016	0.026	0.916
Tree	0.042	0.005	0.780	0	0	0.867
Pole	0	0.002	0	0.966	0	0.982
Car	0	0.016	0.12	0	0.862	0.863
Overall segmentation accuracy: **OSACC**					0.879	
Overall classification accuracy: **OCACC**						0.901

Table 4 Classification results of scene-*A* in the new evaluation metrics

	Building	Road	Tree	Pole	Car	CACC
Building	0.980	0.002	0	0	0	0.989
Road	0.002	0.950	0.002	0	0.080	0.933
Tree	0	0.040	0.890	0	0.080	0.885
Pole	0	0	0	0	0	-
Car	0.040	0.020	0.030	0	0.900	0.905
Overall segmentation accuracy: **OSACC**					0.930	
Overall classification accuracy: **OCACC**						0.928

Table 5 Classification results of scene-*B* in the new evaluation metrics

	Building	Road	Tree	Pole	Car	**CACC**
Building	0.985	0.002	0	0	0	0.991
Road	0.002	0.950	0.002	0	0.080	0.933
Tree	0	0.012	0.680	0.080	0	0.794
Pole	0	0.006	0	0.860	0.016	0.919
Car	0.060	0.050	0.020	0.050	0.970	0.895
Overall segmentation accuracy: **OSACC**					0.889	
Overall classification accuracy: **OCACC**						0.906

frequently used metrics: F-measure which is based on the calculation of Recall and Precision as described in [6] is used. Secondly, a conditional entropy based metrics: V-measure which is based on the calculation of Homogeneity and Completeness as presented in [22] is used. The later method overcomes the problem of matching suffered by the former and evaluates a solution independent of the algorithm, size of the data set, number of classes and number of clusters as explained in [22]. Another advantage of using these two metrics is that just like the proposed metrics they have the same bounded score. For all three metrics, the score varies from 0 to 1 and higher score signifies better classification results and vice versa. The results are summarized in Table 6.

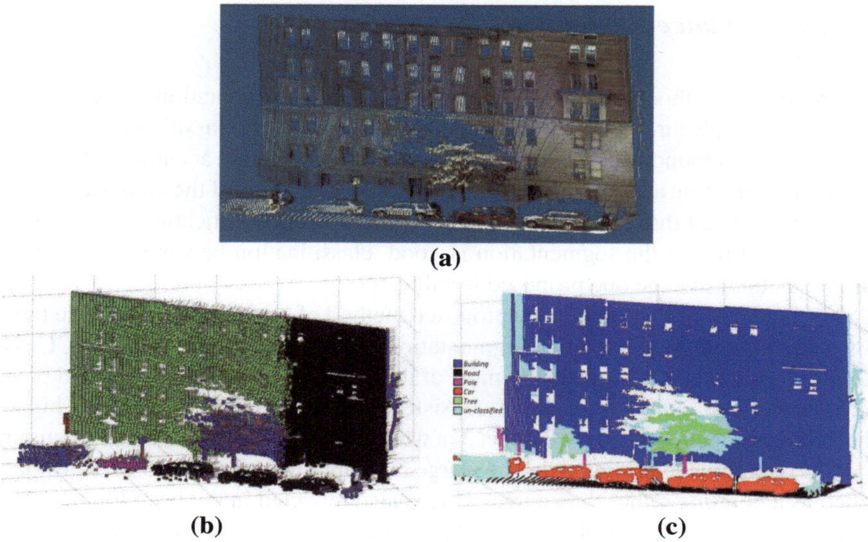

Fig. 6 Segmentation and classification results for a particular scene-C of static scenes from 3D Urban Data Challenge 2011, image # `ParkAvenue_SW14_piece00` [5]. **a** shows 3D data points of data set 1. **b** shows *s*-voxel segmentation of 3D points. **c** shows classification results

Table 6 Classification results evaluated using three different metrics. For the calculation of *V*-measure the value $\beta = 1$ is used

Data Set #	OCACC	*F*-measure	*V*-measure
#1	0.943	0.922	0.745
#2	0.955	0.942	0.826
#3	0.901	0.831	0.733
#A	0.928	0.917	0.741
#B	0.906	0.860	0.734

From Table 6, it can be seen that the results evaluated by all the three evaluation metrics are consistent with data set 2 receiving the highest scores and data set 3 the lowest. The results not only validate the proposed metrics but also indicate that it can be used as an alternative evaluation method. The results evaluated using these standard existing evaluation methods also permits to compare the performance of the proposed algorithm with other published techniques evaluated using them.

5.4 Performance Evaluation and Discussion

The proposed method gives good (in terms of scores) segmentation and classification results in all three evaluation methods. In general, the classification accuracy (**OCACC**) was found to be slightly better than the segmentation accuracy (**OSACC**). Not taking anything away from the segmentation method, one of the main reasons for this being the fact that the 5 types of objects chosen for classification are distinctly different and that if the segmentation is good, classification becomes easier and a simple method like the one proposed is sufficient.

As compared to V-measure, the proposed method of evaluation can provide more information regarding individual segmentation and classification results (**SACC** and **CACC**). These results show that in most of the cases, the buildings, roads and poles have been classified the best with a consistent scores of **SACC** and **CACC** higher than 90 % except in the case of data set 3 in which the building classification accuracy **CACC** is slightly deteriorated due to a large overlapping tree which is wrongly classified as a building rather than a tree. This is also reflected in the low Homogeneity value of 0.670 obtained when calculating V-measure for this data set. The classification of cars is generally good and the results are consistent but they are slightly hampered due to occlusions in some scenes (data set 3: **CACC** 86.3 %, scene-B: **CACC** 89.5 %). In case of trees, the **SACC** and **CACC** are found to vary the most. This is mainly due to the fact that the proposed classification method is based on local descriptors and geometrical features which in the case of trees are very difficult to define (due to large variation of shapes, sizes and types). Thus, where the proposed algorithm succeeded in classifying smaller trees of more classical shapes with a higher **SACC** and **CACC** scores, it produced a low **SACC** and **CACC** score of 68 % and 79.4 % respectively for scene-B. The Recall and Precision scores obtained during the calculation of F-measure for the tree class of this scene were similarly found low as well (0.682 and 0.614 respectively).

6 Conclusion

In this work we have presented a super-voxel based segmentation and classification method for 3D urban scenes. For segmentation a link-chain method is proposed. It is followed by the classification of objects using local descriptors and geometrical models. In order to evaluate our work we have introduced a new evaluation metric which incorporates both segmentation and classification results. The results show an OSACC of 87 % and an OCACC of about 90 %. The results indicate that with good segmentation, a simplified classification method like the one proposed is sufficient. The comparison of the results using three different metrics not only validates the proposed metrics but also permits to compare the performance of the proposed algorithm with other published techniques evaluated using them. The proposed method can also be used as an add-on boost for other classification algorithms.

Acknowledgments This work is supported by the Agence Nationale de la Recherche (ANR - the French national research agency) (ANR CONTINT iSpace & Time – ANR-10-CONT-23) and by "le Conseil Général de l'Allier". The authors would like to thank Pierre Bonnet and all the other members of Institut Pascal who contributed to this project.

References

1. D. Anguelov, B. Taskar, V. Chatalbashev, D. Koller, D. Gupta, G. Heitz, A. Ng, Discriminative learning of Markov random fields for segmentation of 3D scan data, in *IEEE Computer Society Conference on Computer Vision and Pattern Recognition*, vol. 2 (IEEE Computer Society, Los Alamitos, CA, USA, 2005), pp. 169–176
2. B. Douillard, A. Brooks, F. Ramos, A 3D laser and vision based classifier, in *5th International Conference on Intelligent Sensors, Sensor Networks and Information Processing (ISSNIP)*, Melbourne, Australia (2009), p. 6
3. B Douillard, J. Underwood, N. Kuntz, V. Vlaskine, A. Quadros, P. Morton, A. Frenkel, On the segmentation of 3D LIDAR point clouds, in *IEEE International Conferene on Robotics and Automation (ICRA)*, Shanghai, China (2011), p. 8
4. P. Felzenszwalb, D. Huttenlocher, Efficient graph-based image segmentation. Int. J. Comput. Vision **59**, 167–181 (2004)
5. S. Friedman, I. Stamos, Real time detection of repeated Structures in point clouds of urban scenes, in textitThe First Joint 3DIM/3DPVT (3DIMPVT) Conference, Hangzhou, China (2011), p. 8
6. B.C.M. Fung, K. Wang, M. Ester, Hierarchical document clustering using frequent itemsets, chap. I, in *Proceedings of the Third SIAM International Conference on Data Mining. SIAM*, San Francisco, CA (2003), pp. 59–70
7. A. Golovinskiy, T Funkhouser, Min-cut based segmentation of Point clouds, in *IEEE Workshop on Search in 3D and Video (S3DV) at ICCV*, (2009), pp. 39–46
8. A. Halma, F. ter Haar, E. Bovenkamp, P. Eendebak, A. van Eekeren, Single spin image-ICP matching for efficient 3D object recognition, in *Proceedings of the ACM workshop on 3D object retrieval, 3DOR '10*, NY, USA, New York (2010), pp. 21–26
9. A. Johnson, Spin-images: a representation for 3-D surface matching. Ph.D. thesis, Robotics Institute, Carnegie Mellon University, Pittsburgh, PA, 1997
10. M. Kazhdan, T. Funkhouser, S. Rusinkiewicz, Rotation invariant spherical harmonic representation of 3D shape descriptors, in *Proceedings of the 2003 Eurographics/ACM SIGGRAPH symposium on Geometry processing, SGP '03* (Eurographics Association, Aire-la-Ville, Switzerland, Switzerland, 2003), pp. 156–164
11. K. Klasing, D. Althoff, D. Wollherr, M. Buss, Comparison of surface normal estimation methods for range sensing applications, in *IEEE International Conference on Robotics and Automation*, Kobe, Japan (2009), pp. 3206–3211
12. J. Knopp, M. Prasad, L.V. Gool, Orientation invariant 3D object classification using hough transform based methods, in *Proceedings of the ACM workshop on 3D object retrieval, 3DOR '10* (ACM, New York, NY, USA, 2010), pp. 15–20
13. J. Lam, K. Kusevic, P. Mrstik, R. Harrap, M. Greenspan, Urban scene extraction from mobile ground based LiDAR data, in *International Symposium on 3D Data Processing Visualization and Transmission*, Paris, France (2010), p. 8
14. E. Lim, D. Suter, Conditional random field for 3D point clouds with adaptive data reduction, in *International Conference on Cyberworlds*, Hannover (2007), pp. 404–408
15. E.H. Lim, D. Suter, Multi-scale conditional random fields for over-segmented irregular 3D point clouds classification, in *Computer Vision and Pattern Recognition Workshop* (IEEE Computer Society, Anchorage, AK, USA, 2008), pp. 1–7

16. Y. Liu, H. Zha, H. Qin, Shape topics—a compact representation and new algorithms for 3D partial shape retrieval, in *IEEE Conference on Computer Vision and Pattern Recognition*, vol. 2 (IEEE Computer Society, New York, NY, USA (2006), pp. 2025–2032
17. F. Moosmann, O. Pink, C. Stiller, Segmentation of 3D lidar data in non-flat urban environments using a local convexity criterion, in Proceedings of the IEEE Intelligent Vehicles Symposium (IV) (Nashville, Tennessee, USA, 2009), pp. 215–220
18. D. Munoz, N. Vandapel, M. Hebert, Onboard contextual classification of 3-D point clouds with learned high-order Markov random fields, in *IEEE International Conference on Robotics and Automation (ICRA)*, Kobe, Japan (2009), pp. 2009–2016
19. R. Osada, T. Funkhouser, B. Chazelle, D. Dobkin, Shape distributions. ACM Trans. Graph. **21**, 807–832 (2002)
20. F. Pauling, M. Bosse, R. Zlot, Automatic segmentation of 3D laser point clouds by ellipsoidal region growing, in *Australasian Conference on Robotics and Automation*, Sydney, Australia (2009), p. 10
21. S. Pu, G. Vosselman, Building facade reconstruction by fusing terrestrial laser points and images. Sensors **9**(6), 4525–4542 (2009)
22. A. Rosenberg, J. Hirschberg, V-measure: a conditional entropy-based external cluster evaluation measure, in *Proceedings of the 2007 Joint Conference on Empirical Methods in Natural Language Processing and Computational Natural Language Learning (EMNLP-CoNLL)* (2007), pp. 410–420
23. R. Rusu, G. Bradski, R. Thibaux, J. Hsu, Fast 3D recognition and pose using the viewpoint feature histogram, in *IEEE/RSJ International Conference on Intelligence Robots and Systems (IROS)*, Taipei, Taiwan (2010), pp. 2155–2162
24. J. Schoenberg, A. Nathan, M. Campbell, Segmentation of dense range information in complex urban scenes, in *IEEE/RSJ International Conference on Intelligence Robots and Systems (IROS)*, Taipei, Taiwan (2010), pp. 2033–2038
25. J. Strom, A. Richardson, E. Olson, Graph-based segmentation for colored 3D laser point clouds, in *Proceedings of the IEEE/RSJ International Conference on Intelligence Robots and Systems (IROS)* (2010), pp. 2131–2136
26. R. Triebel, J. Shin, R. Siegwart, Segmentation and unsupervised part-based discovery of repetitive objects, in *Proceedings of Robotics: Science and Systems*, Zaragoza, Spain (2010), p. 8
27. M. Vieira, K. Shimada, Surface mesh segmentation and smooth surface extraction through region growing. Comput. Aided Geom. Des. **22**(8), 771–792 (2005)
28. G. Vosselman, P. Kessels, B. Gorte, The utilisation of airborne laser scanning for mapping. Int. J. Appl. Earth Obs. Geoinf. **6**(3–4), 177–186 (2005)

Classification of 3-D Point Cloud Data that Includes Line and Frame Objects on the Basis of Geometrical Features and the Pass Rate of Laser Rays

Kazunori Ohno, Takahiro Suzuki, Kazuyuki Higashi, Masanobu Tsubota, Eijiro Takeuchi and Satoshi Tadokoro

Abstract The authors aim at classification of 3-D point cloud data at disaster environment. In this paper, we proposed a method of classification for 3-D point cloud data using geometrical features and the pass rate of laser rays. Line and frame objects often trap robots, which causes the damages of sensors, motors, mechanical parts etc. at remote operation. Using our proposed method, the line and frame objects can be classified from the 3-D point cloud data. Key-point is use of the pass rate of laser rays. It is confirm that recognition rate of line and frame objects can be increased using the pass rate of laser rays. In addition, it is confirm that the proposed classification method works in the real scene. A training facility of Japan fireman department is used for the evaluation test because it is similar to the real disaster scene comparing the laboratory's test field.

K. Ohno (✉)
JST/Tohoku University, Sendai, Japan
e-mail: kazunori@rm.is.tohoku.ac.jp

T. Suzuki · K. Higashi · M. Tsubota · E. Takeuchi · S. Tadokoro
Tohoku University, Sendai, Japan
e-mail: suzutaka@rm.is.tohoku.ac.jp

K. Higashi
e-mail: higashi@rm.is.tohoku.ac.jp

M. Tsubota
e-mail: tsubota@rm.is.tohoku.ac.jp

E. Takeuchi
e-mail: takeuchi@rm.is.tohoku.ac.jp

S. Tadokoro
e-mail: tadokoro@rm.is.tohoku.ac.jp

K. Yoshida and S. Tadokoro (eds.), *Field and Service Robotics*, Springer Tracts in Advanced Robotics 92, DOI: 10.1007/978-3-642-40686-7_35,

1 Introduction

The authors have studied 3-D map construction of disaster areas, and its application [1, 2]. We constructed 3-D maps using our tracked vehicle equipped with a 3-D laser scanner at subway stations in Japan, at training facilities of fire department in JAPAN, and at that of Federal Emergency Management Agency (FEMA) in Disaster City, USA. Figure 1 illustrates the tracked vehicle equipped with 3-D laser scanner [3, 4], and 3-D point cloud data measured in Disaster City. 3-D maps were made from 3-D point cloud data because raw 3-D point cloud data could show the detailed 3-D shapes. These 3-D maps were highly evaluated by Japan rescue members and FEMA. However, 3-D point cloud data are hard to understand at a glance, without view angle change.

The aim of this study is to detect important geometric information for the robot's operation or human environment recognition, and to display them with different colors according to their geometric types. One of problems at remote control is the difficulty of finding line and frame objects (e.g., thin poles, lines, pipe, mesh or lattice wall, and frame) from dark camera images and 3-D maps. A tracked vehicle is often trapped by these objects, which results in damages of equipped sensors and the tracked vehicle. Therefore, we develop a method of classifying 3-D point cloud data and displaying 3-D shapes with different colors according to their geometrical types.

Fig. 1 Tracked vehicle equipped with a 3-D laser scanner and 3-D point cloud data measured in Disaster City, TX, USA

In the field of mobile robots, one of typical methods of classifying 3-D point cloud data is the method using geometrical features studied by Vandapel et al. [5]. The geometrical features are represented by eigenvalues of 3-D point cloud in each voxel. We also use the geometrical features in our classification method because they are much useful for classifying 3-D point cloud data. However, line and frame objects cannot be distinguished from noise and flat walls by geometrical features. Hence, we distinguish line and frame objects from the noise and the flat walls using the pass rate of laser rays. We also distinguish rugged floor with rubbles from the flat floor because a tracked vehicle with sub-tracks can run on rubble environments by changing sub-track motions. The rugged floor is judged according to the roughness estimated from 3-D point cloud data.

The contribution of this paper is improving the detection of line and frame objects using the pass rate of laser rays. Using the number of passing laser rays, we can confirm that the observation of 3-D scanner is sufficient. Geometrical features are calculated when the number of point cloud data in voxel is low. Using the pass rate of laser rays allows us to distinguish line and frame objects from noise and flat wall.

In this paper, we explain related works in Sect. 2. In Sect. 3, we propose the method of classifying 3-D point cloud data using geometric features and the pass rate of laser rays, and explain the flowchart of our method and the definition of each feature. Then, we verify the efficiency of detecting pole and frame objects in our method in Sect. 4. We show the result of classifying 3-D points cloud data collected in a training environment of Japan rescue members in Sect. 5. We conclude in Sect. 6.

2 Related Works

A typical method of environmental recognition based on geometrical features is classifying 3-D points cloud data according to eigenvalues [5, 6]. Vandapel et al. proposed a method of classifying 3-D point cloud data of forest into grass, tree, and the ground. 3-D point cloud data were classified by comparing three eigenvalues derived from the covariance matrix of the point cloud in each voxel. "Scatter" was judged as grass, and "Surface" was judged as the ground or large tree trunk. "Curve" was judged as telegraph lines or tree branches, but the judgment sometimes fails because such data are considered as noise if the number of points in the voxel is low. It is necessary to confirm the point cloud data in each voxel can be used for the judgment. We thought that the number of passing laser is useful to confirm it. In addition, point cloud data of frame objects are similar to those of flat walls. It is too hard to classify them using only the eigenvalues. We considered that the difference between the wall and the frame is the pass rate of laser rays. Hence, we used the number of passing laser and the pass rate of laser rays to detect line and frame objects.

A ray tracing score has been used for making occupancy grid map in Voxel space [7]. A ray tracing score is equivalent to the number of passing laser. Matthew et al. [7] succeeded in detecting ground in forest by using the ray tracing score. It enables to distinguish the ground from others if we can assume that the ground never exists in the space where laser rays pass. The ray tracing score has been also used for classification of grasses. If an object is grass, laser reflection and transmission are observed simultaneously. Alberto et al. [8] succeeded in detecting wide grass by using the laser reflection and the transparencies. In our research, the number of passing laser is used to confirm that the observation of 3-D laser scanner is sufficient. In addition, the pass rate of laser rays is used to find line and frame objects from 3-D point cloud data. A major difference between our study and previous studies is to use both the pass rate and the geometric features for detecting line and frame objects. The safety of remote operation can be increased by using our method.

Saitou [9] proposed a method of classifying road surface by using the reflection strength of laser. The accuracy of our classification method may be improved by incorporating it.

3 Classification of 3-D Point Cloud Data Based on Geometrical Features and the Pass Rate of Laser Rays

3.1 Target Objects for Classification

At first, we define our target objects to be recognized from 3-D point cloud data by using our classifying method. We consider that target objects for robot's remote operation and space recognition are

1. the ground and rough terrain (where the robot or human can move),
2. walls (used for landmarks), and
3. line and frame objects (used for risk avoidance). Figure 2 shows a part of the target objects.

Plane whose slope is similar to the robot pose is defined as the ground. Tracked vehicles with sub-tracks need to change the sub-tracks motion according to the ground shape. So, it is necessary to distinguish rough terrain from the ground. The threshold of roughness is defined as 150 mm because our tracked vehicle can climb gaps whose height is less than 150 mm with no sub-tracks motion. Plane whose normal vector is perpendicular to the normal vector of the ground is defined as wall. Thin pole objects whose width is between 10 and 80 mm are defined as line and frame objects. Frame objects includes metal frame structures, pipe frame structure, and lattice objects. The width value is decided on the basis of Japanese standard of electrical and gas line (ex. Width of electrical cable is between 6 and 11 mm. So, 10 mm width is selected as minimum width. Width of pipes inside walls and ceilings is between 19 and 76 mm. Width of grab rail at the side of corridors and stairs is between 30 and 40 mm.). Since voxel size must be larger than these thresholds, voxel size is defined as 200 mm width.

Fig. 2 Target objects for segmentation

3.2 Classification of 3-D Point Cloud Data that Includes Line and Frame Objects

3-D point cloud data that includes line and frame objects are segmented using geometrical features and the pass rate of laser rays. Figure 3 illustrates the process flow of our proposed classifying method. 3-D point cloud data $S_i \cdots S_t$ are measured by a 3-D laser scanner named HD scanner. 256,000 point data are obtained every 2.5 s. The point cloud is projected inside a voxel space $\boldsymbol{V}$ whose voxel size is 200 mm. The number of passing laser rays N_{pass} and the number of reflected points N_{point} are counted in each voxel $V_{i,j,k}$. If enough number of laser rays passed inside the voxel, geometrical features $\boldsymbol{\lambda}_{i,j,k} = \{\lambda_0, \lambda_1, \lambda_2\}$, the pass rate of laser rays $\gamma_{i,j,k}$, and roughness $\beta_{i,j,k}$ are calculated from the point cloud data inside the voxel. Using $\boldsymbol{\lambda}_{i,j,k}, \gamma_{i,j,k}, \beta_{i,j,k}$, each voxel $V_{i,j,k}$ is labeled class label C_n ($n = 1$:line and frame objects, $n = 2$:walls, $n = 3$:the ground, $n = 4$:rough terrain, and $n = 5$:others).

1. 3-D point cloud data are obtained: $S_1 \cdots S_t$,
2. Stored 3-D point cloud data into voxel space: $V_{i,j,k}$
   - No. of passing laser rays: N_{pass}
   - No. of points: N_{point}
   - No. of total points: $N_{total} = N_{point} + N_{pass}$
3. Segmentation of each voxel: $V_{i,j,k}$

   if ($N_{total} > N_{thresh1}$ or $N_{pass} > N_{thresh1}$)
   if($N_{pass} > N_{thresh2}$)
   Eigenvalues: $\lambda_{i,j,k} = \{\lambda_0, \lambda_1, \lambda_2\}$
   Pass rate of laser rays: $\gamma_{i,j,k}$
   Roughness value: $\beta_{i,j,k}$
   $\rightarrow$ Class: C_n
4. Iterate above process during 3-D mapping.

Fig. 3 Process flow of 3-D point cloud segmentation.

Figure 4 illustrates three patterns of point cloud distribution and its geometrical features (λ_0, λ_1, λ_2). λ_0, λ_1, λ_2 are eigenvalues calculated from the covariance matrix of the 3-D point cloud in $V_{i,j,k}$. The subscript number of λ shows the order of the eigenvalues: 0 is the largest eigenvalue. The distribution of point cloud in the voxel can be classified into plane, pole, and ellipsoidal object by using three eigenvalues [5]. In our research, plane and pole like objects are found using the eigenvalues λ_0, λ_1, λ_2. Ellipsoidal shape is judged as other objects.

However, it is hard to identify thin line and frame objects from only the eigenvalues λ_0, λ_1, λ_2 and the number of reflected points N_{point}. Figure 5 illustrates three cases

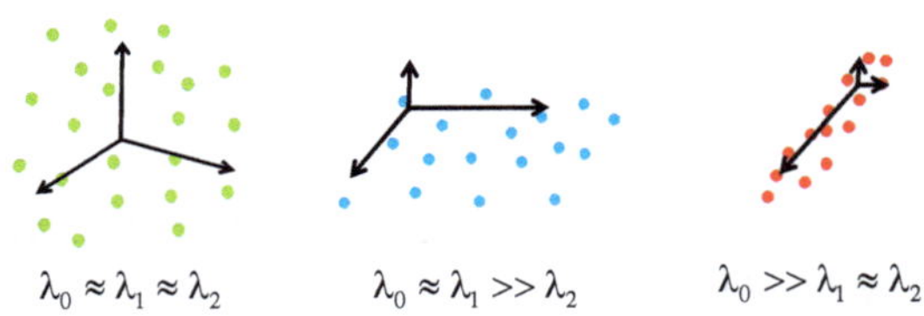

Fig. 4 Segmentation of 3D point clouds by Eigenvalues

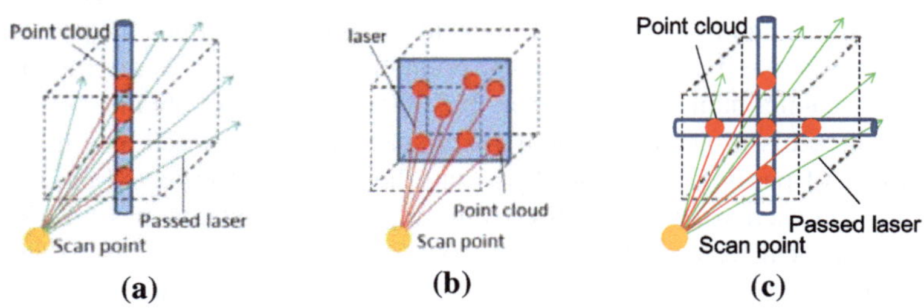

Fig. 5 Passing Laser Rays inside the Voxel which Contains Pole, Plane, and Lattice Objects **a** Pole **b** Plane **c** Lattice

where a thin line, a plane, and a lattice object are measured respectively. When the number of reflected points is small in a voxel, these points are considered as noise. Therefore, the eigenvalues of these points are not calculated even when these points represent the thin object (Fig. 5a). We would like to improve the judgment accuracy when the number of points is small. We notice that many laser rays pass inside the voxel in the case of Fig. 5a. Therefore, we use the number of passing laser rays N_{pass} to confirm that the observation is sufficient.

Frame object is also segmented by using the eigenvalues. We explain a lattice object which is one of the frame objects. The eigenvalues of the lattice object is similar to the eigenvalues of a plane (Fig. 5b, c). We cannot judge whether the object is a lattice object using only the eigenvalues. In the case of the lattice object (Fig. 5c), many laser rays also pass inside the voxel (Fig. 5c). So, we use the pass rate of laser rays $\gamma_{i,j,k}$ to judge whether the object is the lattice object. The details are described in Sect. 3.3.

Walls are distinguished from the ground using a geometrical constraint that the wall is perpendicular to the ground. Uneven terrains are distinguished from the ground using the roughness α, β. The details are described in Sect. 3.4.

3.3 Line and Frame Objects Recognition Using the Pass Rate of Laser Rays

Figure 6 illustrates a model of laser rays in the voxel space. All voxels between the sensor and the reflected point are labeled "pass". The voxels where the laser ray passes are judged using Bresenham algorithm [11]. The number of passing point N_{pass} in each voxel is counted when the reflected point is observed.

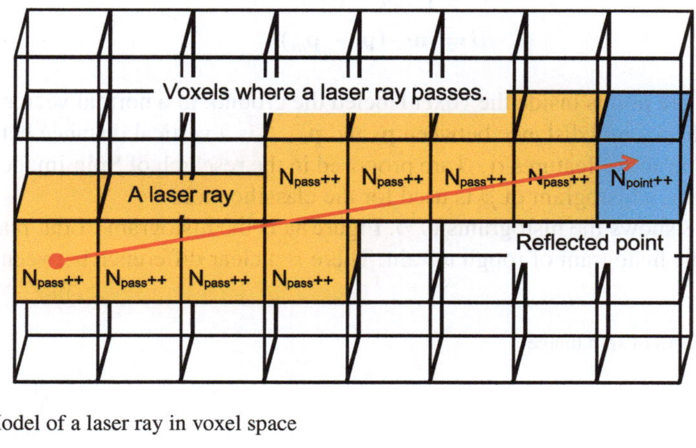

Fig. 6 Model of a laser ray in voxel space

The pass rate of laser rays $\gamma_{i,j,k}$ enables to judge whether the number of the reflected points is small due to the shortage of observation data. The pass rate of laser rays $\gamma_{i,j,k}$ proposed by the authors is represented as follows:

$$\gamma_{i,j,j} = 1 - \frac{N_{\text{point}}}{N_{\text{point}} + N_{\text{pass}}} \tag{1}$$

where N_{pass} and N_{point} are the number of the passing laser rays and the number of the reflected points in Voxel $V_{i,j,k}$, respectively. The pass rate of laser rays is in the range of $0.0 \leq \gamma_{i,j,k} \leq 1.0$, and it becomes closer to 1.0 as N_{pass} increases.

3-D point cloud data in a voxel is classified when $N_{\text{total}} = N_{\text{pass}} + N_{\text{point}}$ in the voxel is above a threshold. The object is classified as "line" when the pass rate of laser rays is close to 1.0 and the geometrical feature is pole (Fig. 5a). The object is classified as "lattice wall" when the pass rate of laser rays is above a threshold and the geometrical feature is plane (Fig. 5b). Although these thresholds are empirically defined, this method can be applied to various observation data because the pass rate of laser rays is clearly different between thin objects and others.

3.4 Classification of Rough Terrain Using the Roughness

Segments that belong to the ground are obtained using the eigenvalues. These segments compose of the flat ground and rough terrain. It is necessary to distinguish the rough terrain from the segments using the roughness. For calculating the roughness, we used parameters α, β (Eq. 4) that show the relative distance between points inside a voxel (Fig. 7).

$$\alpha = \sqrt{\|\mathbf{p}_l - \mathbf{p}_m\|^2 - \beta^2} \tag{2}$$

$$\beta = |\mathbf{n}_l \cdot (\mathbf{p}_l - \mathbf{p}_m)| \tag{3}$$

$\mathbf{p}_l$ and $\mathbf{p}_m$ are points inside the voxel labeled the ground. $\mathbf{n}_l$ a normal vector of point $\mathbf{p}_l$. α is a horizontal distance between $\mathbf{p}_l$ and $\mathbf{p}_m$. β is a vertical distance between $\mathbf{p}_l$ and $\mathbf{p}_m$. The above features α, β are proposed in the research of Spin Image [10]. In our research, a histogram of β is used for the classification.

Figure 8 shows the histograms of β. Figure 8a is the histogram of flat plane. Figure 8b is the histogram of rough terrain. There is a clear difference between the two

Fig. 7 Features of spin image

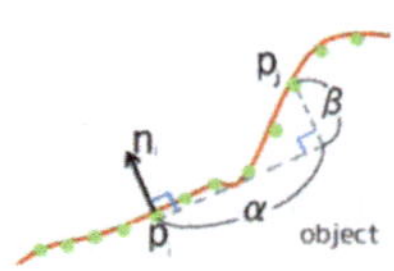

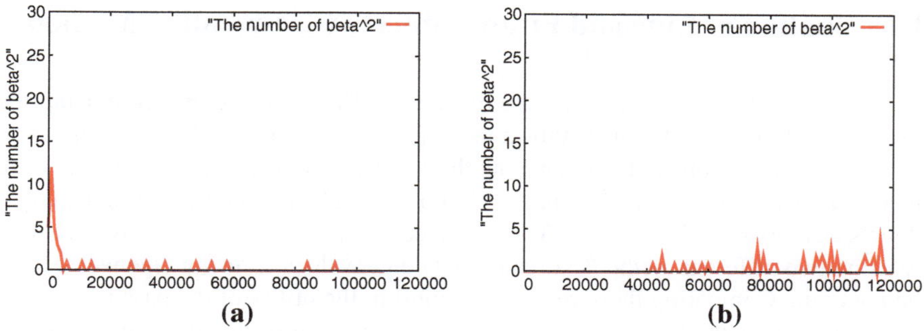

Fig. 8 Histogram of β^2: **a** Flat ground and **b** Rough Terrain

histograms. The histogram of the plane has one peak, the histogram of the rough terrain has no peaks and wide distribution. For the classification, we used the following standard deviation σ_β.

$$\sigma_\beta^2 = \frac{1}{N}\Sigma_i(\beta_i - \bar{\beta}_i)^2 \tag{4}$$

$\bar{\beta}$ is the average of β. If the σ_β is small, the segment is judged as the ground (flat ground). If σ_β is large, the segment is judged as the rough terrain.

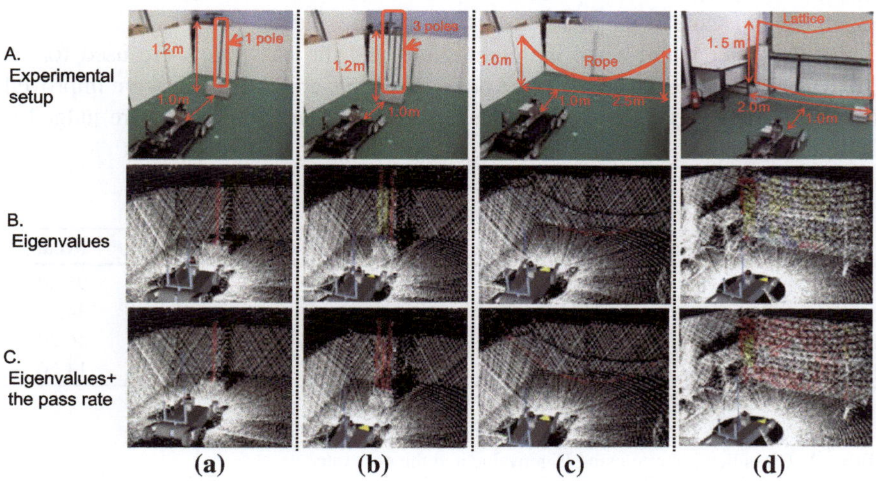

Fig. 9 Experimental setup and classification results for each experiment

4 Evaluation of Line and Frame Objects Classification Method

We confirm that the classification rate of line and frame objects increase using the pass rate of laser rays. Figure 9 shows the experimental setup and the classification results. 3-D point cloud data of one pole, three poles, a rope, and lattice objects were measured using the 3-D laser scanner. 3-D point cloud data that belong to the target objects were extracted by human. The extracted 3-D point cloud data were classified by using two different methods: 1. using only eigenvalues, 2. using eigenvalues and the pass rate. Comparing these results, we confirm the utility of the pass rate.

Middle row data in Fig. 9 illustrates the classification results using eigenvalues. Table 1 shows the classification results and the accuracy (recall). Bottom row data in Fig. 9 illustrates the classification results using both eigenvalues and the pass rate. Table 2 shows the classification results and the accuracy. Color of each point in Fig. 9 shows the classification result (red: line and fame objects, yellow: walls, sky blue: the ground, blue: rough terrain, white: others). In the evaluation, N_{total} was used because the number of points was small for thin pole objects, a lot of voxel could not be used for the classification.

When the eigenvalues were used for the classification, the recall of three poles, rope and lattice were less than 46.2 % except for recall of one pole, it is quite small. Sixty percentage point cloud data of the three poles was judged as wall because the distribution was similar to wall's one. Twenty-five percentage point cloud data of the rope was judged as rough terrain. Thirty percentage point cloud data of the lattice was judged as wall. Fifteen percentage point cloud data of the lattice was judged as rough terrain. Most of the point cloud data were misjudged.

On the other hand, when the eigenvalues and the pass rate were used for the classification, the recall of the three poles, the rope, and the lattice were improved. No points were judged as wall in the three poles case. No points were judged as

Table 1 Classification rate using only Eigenvalues

	Line & frame	Wall	The ground	Rough terrain	Others	Recall[%]
1 pole	217	7	0	0	3	95.59
3 poles	215	352	0	42	19	34.24
Rope	146	3	0	92	75	46.20
Lattice	489	1,236	129	512	970	14.66

Table 2 Classification rate using Eigenvalue and the pass rate

	Line & frame	Wall	The ground	Rough terrain	Others	Recall[%]
1 pole	217	7	0	0	3	95.59
3 poles	609	0	0	0	19	96.97
Rope	241	0	0	0	75	76.27
Lattice	2,025	327	14	0	970	60.7

rough terrain in the rope case. No points were judged as terrain in the lattice case. The number of points judged as wall in the lattice decreased to 327. It was confirmed that the recall of line and frame objects classification increased using the pass rate of laser rays. The number of points judged as others in the rope and the lattice objects is large because the observation is not sufficient. We need more observation to decrease the number of others objects. In the real situation, the robot can measure 3-D point cloud data during the exploration. So, this problem can be solved as the observation time increases.

Fig. 10 Experimental field and correct answer data made by hand

5 Classification of 3-D Point Cloud Data Measured at a Training Facility of Japan Fireman Department

We evaluated the validity of our proposed method using 3-D point cloud data measured in a training facility of Japan fireman department. We would like to confirm that the proposed classification method works in 3-D point cloud data of disaster area. Figure 10 shows the experimental field. There were a frame structure, walls, the ground, rough terrain. The correct answer data was made by hand (Fig. 10). Eigenvalues, the pass rate of laser rays, and roughness were used for the classification. Noise filter was used to decrease the misjudgment of the classification.

Figure 11 shows the classification result, and Table 3 shows the classification rate. We calculated two rates (recall, precision) for the evaluation. Recall of lien and frame objects was up to about 93 %. Using the pass rate of laser rays can help us to find the line and frame objects. Recall of the ground was up to 70 %. However, recall of walls and rough terrain was about 55 %.

Twenty-five percentage point cloud data in the wall case was judged as other because the walls were located around edge of scan and the observation was not

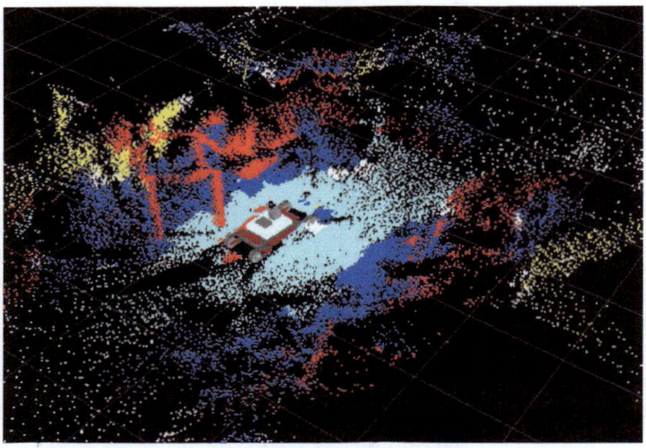

Fig. 11 Classification result of rough terrain using geometrical features, the pass rate of laser rays, and noise filter

Table 3 Classification rate using geometrical features and the pass rate of laser rays

	Line & frame	Wall	The ground	Rough terrain	Others	Recall[%]
Line & frame	1,448	26	7	75	0	93.06
Wall	336	3,480	32	638	1057	54.59
The ground	1,421	38	18,336	4,636	1,529	70.63
Rough terrain	3,914	193	2,011	10,023	1,478	56.89
Precision[%]	20.34	90.61	89.94	65.20	0.0	

sufficient. Same problem can be found in the walls and uneven cases. These problem can be solved as the observation time increases.

However, precision of line and frame objects was low. The misjudgment decrease the recall of ground and rough terrain cases. This problem is caused by the pass rate of laser rays at the edge of the scan data. We found a solution to improve the problem. Unfortunately, the evaluation is not enough. So, we will report the solution in next presentation.

6 Conclusion and Future Works

The authors aim at classification of 3-D point cloud data at disaster environment. In this paper, we proposed a method of classification using geometrical features and the pass rate of laser rays. Eigenvalues are used as geometrical feature. The pass rate of laser rays is calculated in voxel space. It was confirmed that recognition rate of line and lattice objects can be increased using the pass rate of laser rays. In addition, it was confirmed that the proposed classification method works in the case of a training facility of Japan fireman department. It is similar to the real disaster scene comparing the laboratory's test field. From the evaluation, recalls of line and frame objects and the ground was up to 90 and 70 % respectively. However, the recall of wall and rough terrain was about 50 %. We need to improve the recall of them. The improvement is on going works.

In the current classification method, some thresholds are used for the decision. As the feature work, we would like to decrease the number of the threshold using machine learning techniques. In studies of machine learning, it is well known that the use of good feature promises the good segmentation result if simple classification algorithms are used (ex. K-NN, SVM). Combination of geometrical feature and the pass rate can provide you a good segmentation result.

Acknowledgments This research has been partially supported by the NEDO Project for Strategic Development of Advanced Robotics Elemental Technologies, High-Speed Search Robot System in Confined Space, the PRESTO JST: "Environment Recognition based on Visual and Tactile Innovation's for Mobile Robot", and JST.

References

1. K. Ohno, S. Tadokoro, K. Nagatani, E. Koyanagi, T. Yoshida, 3-D Mapping of an Underground Mall Using a Tracked Vehicle with Four Sub-tracks. in *Proceedings of 2009 IEEE International Workshop on Safety, Security, and Rescue, Robotics*, 2009
2. K. Ohno, S. Tadokoro, K. Nagatani, E. Koyanagi, T. Yoshida, Trials of 3-D Map Construction Using the Tele-operated Tracked Vehicle Kenaf at Disaster City. in *Proceedings of IEEE International Conference on Robotics and Automation*, (2010), pp. 2864–2870
3. T. Yoshida, E. Koyanagi, S. Tadokoro, K. Yoshida, K. Nagatani, K. Ohno, T. Tsubouchi, S. Maeyama, I. Noda, O. Takizawa, Y. Hada, A High Mobility 6-Crawler Mobile Robot 'Kenaf'.

in *Proceedings of 4th International Workshop on Synthetic Simulation and Robotics to Mitigate Earthquake Disaster (SRMED2007)*, (2007), pp. 38

4. K. Ohno, T. Kawahara, S. Tadokoro, Development of 3D Laser Scanner for Measuring Uniform and Dense 3D Shapes of Static Objects in Dynamic Environment. in *Proceedings of ROBIO*, 2008
5. N. Vandapel, D. Huber, A. Kapuria, M. Hebert, Natural terrain classification using 3-d ladar data. in *IEEE International Conference on Robotics and Automation*, vol. 5 (2004), pp. 5117–5122
6. D. Wolf, G. Sukhatme, D. Fox, W. Burgard. Autonomous Terrain Mapping and Classification Using Hidden Markov Models. in *IEEE International Conference on Robotics and Automation*, (2005) pp. 2026–2031
7. W. Matthew, T. Nishitani, C. A. Brooks, K. Iagnemma, Ground Plane Identification Using LIDAR in Forested Environments. in *IEEE International Conference on Robotics and Automation*, (2010) pp. 3831–3836
8. A. Lacaze, K. Murphy, M. DelGiorno, Autonomous Mobility for the Demo III Experimental Unmanned Vehicles. in *Proceedings of AUVSI*, 2002
9. T. Saitoh, Y. Kuroda, Online Road Surface Analysis using Laser Remission in Structured Outdoor Environments. in *41th International Symposium on, Robotics (ISR2010)*, 2010
10. A. Johnson, M. Hebert, Usibg spin images for efficient object recognition in cluttered 3Dscenes. IEEE Trans. Pattern Anal. Mach. Intell. **21**(5), 433–449 (1999)
11. J.E. Bresenham, Ambiguities in incremental line rastering. Comput. Graphics Appl. **7**(5), 31–43 (1987)

Solid Model Reconstruction of Large-Scale Outdoor Scenes from 3D Lidar Data

Ciril Baselgia, Michael Bosse, Robert Zlot and Claude Holenstein

Abstract Globally consistent 3D maps are commonly used for robot mission planning, navigation, and teleoperation in unstructured and uncontrolled environments. These maps are typically represented as 3D point clouds; however other representations, such as surface or solid models, are often required for humans to perform scientific analyses, infrastructure planning, or for general visualization purposes. Robust large-scale solid model reconstruction from point clouds of outdoor scenes can be challenging due to the presence of dynamic objects, the ambiguitiy between non-returns and sky-points, and scalability requirements. Volume-based methods are able to remove spurious points arising from moving objects in the scene by considering the entire ray of each measurement, rather than simply the end point. Scalability can be addressed by decomposing the overall space into multiple tiles, from which the resulting surfaces can later be merged. We propose an approach that applies a weighted signed distance function along each measurement ray, where the weight indicates the confidence of the calculated distance. Due to the unenclosed nature of outdoor environments, we introduce a technique to automatically generate a thickened structure in order to model surfaces seen from only one side. The final solid models are thus suitable to be physically printed by a rapid prototyping machine.The approach is evaluated on 3D laser point cloud data collected from a mobile lidar in unstructured and uncontrolled environments, including outdoors and inside caves. The accuracy of the solid model reconstruction is compared to a previously developed binary voxel carving method. The results show that the weighted signed

C. Baselgia (✉) · C. Holenstein
Autonomous Systems Laboratory, CSIRO ICT Centre, Brisbane, Australia
e-mail: Ciril.Baselgia@csiro.au

M. Bosse
CLA E13 Tannenstrasse 3, 8049 Zürich, Switzerland
e-mail: michael.bosse@gmail.com

R. Zlot
Robotic Perception Autonomous Systems Laboratory Computational Informatics CSIRO, Brisbane, Australia

K. Yoshida and S. Tadokoro (eds.), *Field and Service Robotics*, 541
Springer Tracts in Advanced Robotics 92, DOI: 10.1007/978-3-642-40686-7_36,

distance approach produces a more accurate reconstruction of the surface, and since higher accuracy models can be produced at lower resolutions, this additionally results in significant improvements in processing time.

1 Introduction

In recent years, 3D scanning has become increasingly widespread both in the research community (robotic perception, remote sensing), and in the commercial space (mobile mapping systems, consumer-grade depth sensors). Models generated by these systems are typically represented as 3D point clouds; however, in some applications more physically relevant models are required. An application we have been investigating recently in collaboration with scientists from ANSTO is the generation of 3D models of natural cave systems. The ANSTO scientists require watertight 3D surface models of the cave interiors to understand airflows within the cave system. It became apparent that the 3D models produced would also be useful for tourism and cultural heritage purposes. Though we could produce interactive 3D visulations or fly through animations using the surface models or point clouds, tangible 3D printed models are often more accessible to the public, and could be useful for managing or studying the caves.While our previous work [1] assumed a fully enclosed environment, there are several outdoor connections between the caves where this assumption does not hold. Therefore, to produce a complete model of the cave system and surrounds that is printable, we must be able to generate 3D solid models of outdoor scenes. As an important factor in this application is the large scale of the environments (tens of kilometers), the approach should be automated requiring minimal manual assistance. We present a method for generating 3D solid models of outdoor scenes from point cloud data. In addition, aspects of the approach can be applied to improve on our previous work in 3D surface reconstruction of cave interiors.

There are several methods for surface reconstruction, which can be roughly categorized into three classes: model-based, point-based, or volume-based methods. The first group tries to fit geometric primitives such as planes to the points [2–5]. Even though this method works reasonably well for artificial objects, they are not useful for the high structural variance in natural scenes. Point-based algorithms create a surface from the points using triangulation and interpolation to generate a mesh [6, 7]. Typically, these techniques require additional pre- or post-processing steps to ensure watertight surface reconstruction, to deal with noise, or to simplify the mesh structure [8, 9]. A disadvantage of both model- and point-based methods, is that while they may cope with a reasonable amount of noise, they are sensitive to outliers, such as those originating from dynamic objects and misregistrations. Volume-based techniques rasterize the space into a regular voxel grid [10–12] or an octree [13, 14]. Each measurement is rendered into the volumetric grid as a ray or beam of free space eminating from the sensor to the measured point, thereby erasing non-static points as rays from other viewpoints pass through the voxels that were occupied in previous scans. A surface model can then be extracted from the volume by calculating an isosurface at an appropriate isovalue.

Solid model reconstruction is typically focussed on modeling objects [15–17], rather than environments. Compared to a laboratory setting, environment modeling is relatively challenging as neither complete coverage nor adequate sampling density can be guaranteed. Existing work in environment modeling from 3D point clouds has focused on surface reconstruction [2, 18, 19], rather than solid models.

In September 2010, we mapped the Jenolan Cave System in New South Wales, Australia with a mobile 3D lidar system consisting of a spinning 2D SICK lidar mounted on a two-wheeled furniture dolly (Fig. 1a). Data was recorded while maneouvering the dolly inside the caves and outdoors between and around the cave entrances. At a walking pace, we collected data along a trajectory of more than 9 km mainly using pathways and stairways constructed throughout the cave system. Due to the confined nature of the environment, it was impossible to ensure that the mapping team remained outside of the sensor's wide field of view. To handle unavoidable lidar returns from dynamic objects, we adopt a volume-based space-carving approach using a signed distance function [10]. A triangulated surface can be extracted from the volumetric model by computing an isosurface at the zero level set. In closed environments (such as cave interiors), one can print the air void as a solid model; however, in open spaces this approach would not produce desirable results. Instead we model the volumes of objects, rather than empty space; however, to ensure sufficient thickness for producing a stable printed model, we must extend the regions behind the measured surfaces (Fig. 1b). Simply extruding surfaces would introduce solid artifacts into areas of known free space.

The contribution of this work is the development of an algorithm which is able to robustly reconstruct solid models of large-scale, unstructured, open environments. In addition, the application of the proposed signed distance function improves the

(a) The *Hannibal* mapping platform **(b)** 3D printed solid model

Fig. 1 **a** Mapping an outdoor area at the Jenolan Caves with a spinning laser. **b** Photograph of a solid model of a single tile printed on a 3D printer at 1:250 scale. The distance along the longest dimension is 60 m. A building, some trees, and a hillside (seen from behind) are visible accuracy and robustness of our previous framework for watertight 3D surface reconstruction in enclosed environments.

The remainder of the paper is organized as follows. Section 2 gives a brief overview of the signed distance function followed by the surface thickening technique. Experimental setup and its evaluation is provided in Sect. 3. Finally, Sect. 4 concludes the paper and provides outlook for future work.

2 Approach

2.1 Data Acquisition

A spinning 2D SICK LMS291 mounted on a two-wheeled hand trolley is used to acquire range measurements within the caves. The scanner updates at a frequency of 75 Hz, capturing 180 points with one degree angular resolution per scan, with a maximum scanning range of approximately 80 m (surface dependent). The mount rotates the laser about the central scan ray at a rate of 0.5 Hz which results in a hemispherical field of view containing 13,500 points each second (with scans spaced at 2.4° intervals). The data is transformed into a consistent 3D point cloud using a incremental scan-matching-based SLAM algorithm [20] followed by a global registration procedure based on the same algorithm.

2.2 3D Reconstruction

In order to reconstruct a solid model or surface from a point cloud, we use a volume-based approach, where continuous space is split into discrete voxels with a fixed uniform resolution. Each voxel stores a value related to the distance from the nearest surface and a weight indicating the confidence of this value. Distance and weight values for a voxel are updated whenever a new measurement ray passes through it. After processing all measurements, the resulting solid or surface can then be derived from the volumetric model. Volume-based methods have the advantage that they automatically erase points which originate from dynamic objects as rays from multiple viewpoints pass through the associated voxel. Our approach processes the data in scan order, and follows four steps for each scan. Steps 1, 2, and 4 are the same as discussed in our previous paper [1], but for the sake of completeness we briefly review them here as background.

Triangulation of Scans. Space-carving involves tracing each measurement ray from the sensor location to the reflecting surface and marking each voxel along that path. When the surfaces are sufficiently distant from the sensor, a simple carving of each measurement ray will result in noticeable gaps in the voxel grid. This effect produces a spiky structure which does not accurately reflect the underlying volume.

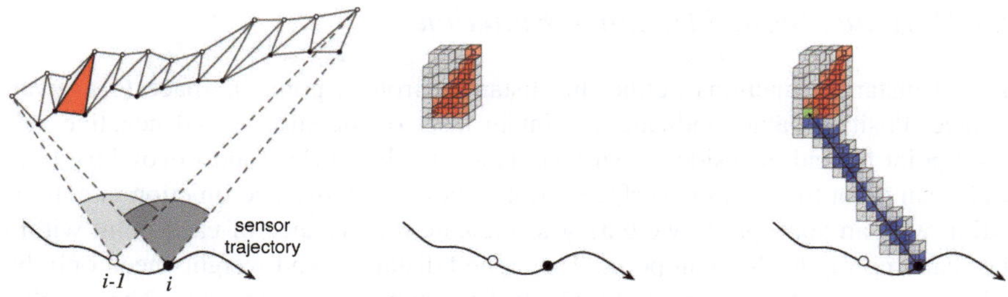

(a) Traingulation **(b)** Scrface Rasterization **(c)** Ray casting

Fig. 2 Steps in the rasterization of a laser scan. **a** Scan points from scan *i* (*small black circles*) are triangulated with the points from the previous scan (*small white circles*). The two *larger circles* at the bottom of the figure represent the sensor position at the time of each scan measurement). A representative surface *triangle* is highlighted in *red*. **b** The surface *triangle* is rasterized in a global 3D voxel grid, indicated as *red-shaded voxels* (the *unshaded voxels* are untouched in this step). Only a small local subset of the voxels are illustrated for clarity. **c** A ray is cast from the associated sensor position to each surface voxel, and all sampled voxels along the ray are marked as free space. An example surface voxel is shaded in *green*, and the marked voxels along the ray are colored in *blue*. This step is repeated for all surface voxels resulting from the original triangle, and the process is applied to all other triangles in the scan triangulation. (Reused from previous publication [1])

To alleviate this issue, we triangulate between neighboring points in the scan manifold in order to interpolate across the gaps [1]. Due to the spinning lidar configuration, naive triangulation along the scan manifold would result in non-orientable surfaces, from which we would not be able to directly construct valid solid models or closed surfaces. Instead, the manifold triangles provide an intermediate surface which is sampled to produce rays at a sufficient density to fill the gaps (Fig. 2a).

Voxelization of Surface Triangle. Sampling of the intermediate surface is achieved by rasterizing the triangles into the voxel grid (Fig. 2b). Rasterization is performed by recursively splitting the longest edge of a triangle to form two new triangles. This procedure is continued until the longest edge of a triangle is smaller than half of the voxel resolution.[1] The voxel containing the centroid that small triangle is then marked as belonging to the surface for the subsequent steps.

Voxel Carving. The third step consists of carving the entire volume between the rasterized intermediate surface and the sensor. The carving is performed by tracing a ray through the voxels from the surface patch to the sensor using a 3D version of the well-known Bresenham algorithm [21]. While tracing the rays, the algorithm updates the corresponding signed distance and weighting values for each voxel traversed. More details about the weighted signed distance function is given in Sect. 2.3.

Isosurface extraction. The last step of the surface reconstruction algorithm is the extraction of the surface itself. We use MATLAB's *isosurface* function to extract a zero level set surface from the voxel grid volume.

[1] We use the Polygon2Voxel algorithm by Dirk-Jan Kroon, available on the Mathworks MATLAB Central website.

2.3 Weighted Signed Distance Function

Signed distance functions define the distances from a point in space to a given surface. Positive values indicate a point in front of the surface and negative values a point behind or inside the surface, respectively. Curless and Levoy [10] treat each scan point to be on the surface and use the signed distance function in combination with an appropriate weighting to indicate the distance of each point within the scanning ray to the scan point. The signed distances and weights need only be calculated for each voxel $\boldsymbol{x}_j$ with which the scanning ray intersects. The weights are an indicator for the uncertainty of the measured distance. The overall signed distance function $D(\boldsymbol{x}_j)$ at a given voxel $\boldsymbol{x}_j$ is recursively updated by combining the single signed distance functions $d_1(\boldsymbol{x}_j), d_2(\boldsymbol{x}_j), \ldots, d_n(\boldsymbol{x}_j)$ and corresponding weights $w_1(\boldsymbol{x}_j), w_2(\boldsymbol{x}_j), \ldots, w_n(\boldsymbol{x}_j)$ obtained from several range measurements $r_1, \ldots, r_n$ [10]. Finally, the underlying surface can be extracted by computing the isosurface at $D(\boldsymbol{x}) = 0$ (*i.e.*, where the points have zero distance to the surface). It can be shown that, under a certain set of assumptions, the extracted surface is optimal in the least squares sense [10].

The cumulative weighted signed distance $D_I(\boldsymbol{x}_j)$ with the corresponding weights $W_I(\boldsymbol{x}_j)$ of a voxel $\boldsymbol{x}_j$ to the surface can be written as

$$D_I(\boldsymbol{x}_j) = \frac{\sum w_i(\boldsymbol{x}_j) d_i(\boldsymbol{x}_j)}{\sum w_i}, \qquad W_I(\boldsymbol{x}_j) = \sum w_i(\boldsymbol{x}_j), \tag{1}$$

where $d_i(\boldsymbol{x}_j)$ is the signed distance and $w_i(\boldsymbol{x}_j)$ is the corresponding weight for a measurement i. Note that these quantities can also be computed recursively, so the voxel states can be updated iteratively.

In defining the distance function and corresponding weights, various effects are accounted for by decomposing the ray into multiple regions (Fig. 3). Starting from the sensor, the regions are: free space between the sensor and surface, where we are highly confident there are no objects; the front surface transition, an interval which contains the measurement and should contain the physical surface; the interior region; and the back surface transition, which may not correspond to any physical surface but thickens the surface to create a solid model. The distance and weight functions are determined by the incidence angle of the ray with the surface and two user-specified parameters. The parameter δ defines the width of the transition area and is dependent on the sensor noise and voxel resolution. The parameter Δ determines the maximum thickness of the solid, and should be larger than 2δ.

$$d(x_j) = \begin{cases} \delta & \delta < n^T x_j \\ n^T x_j & -\delta < n^T x_j < \delta \\ -\delta & \delta - \Delta < n^T x_j < -\delta \\ -n^T x_j - \Delta & -\Delta - \delta < n^T x_j < \delta - \Delta \\ \delta & n^T x_j < -\Delta - \delta, \end{cases} \tag{2}$$

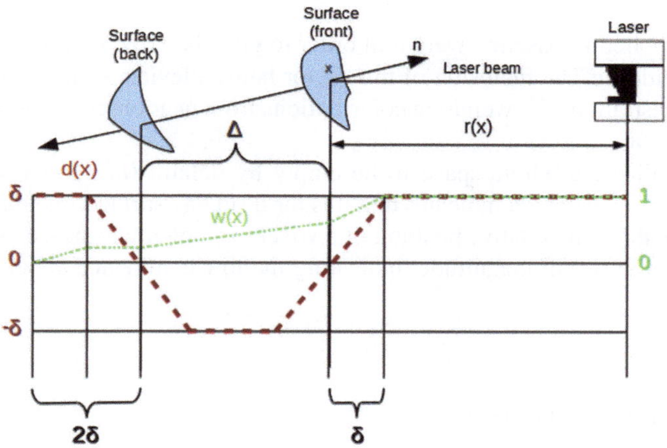

Fig. 3 Signed distance function and the corresponding weight for a single ray. The *red dashed line* is the signed distance function which is assigned to the voxel where the laser beam traverses. The *green dotted line* denotes the value of the corresponding weight. For visualization purposes, the weight is not scaled with $\frac{1}{r^2}$ in this figure. The traced ray is reoriented to the surface normal behind the surface to ensure an equal thickness throughout the whole structure (see Sect. 2.4)

where $\boldsymbol{n}$ is the surface normal and $\boldsymbol{x}_j$ the coordinates of the voxel. Note that although the distance function values are assigned to voxels along the ray, the values are calculated according to the distance from the surface in the normal direction. We truncate the distance function at a maximum value of δ away from the surface in order to prevent outliers from dominating the voxel state.

The weighting function indicates the uncertainty of the distance values assigned to the voxel and is defined as

$$w(x_j) = \frac{1}{\|r^2\|}\begin{cases} 1 & \delta \leq n^T x_j \\ (1 - w_\alpha)\frac{1}{\delta}n^T x_j + w_\alpha & 0 \leq n^T x_j < \delta \\ \frac{w_\alpha}{2\Delta}\frac{1}{\delta}n^T x_j + w_\alpha & -\Delta \leq n^T x_j < 0 \\ \frac{w_\alpha}{2} & -\Delta - \delta \leq n^T x_j < -\Delta \\ \frac{w_\alpha}{2\delta}\frac{1}{\delta}n^T x_j + \frac{w_\alpha}{2\delta}(\Delta + \delta) & -\Delta - 2\delta \leq n^T x_j < -\Delta - \delta \\ 0 & \text{otherwise.} \end{cases} \quad (3)$$

Due to errors associated with high incidence angled reflections, in a manner similar to that of Curless and Levoy [10], we define one part of the weight w_α to be the cosine of the angle α between the incoming ray and the surface normal. That effect is only considered valid close to the surface, so the baseline weight linearly increases to a value of one at a distance δ from the surface. Behind the surface, the weight linearly drops to $w_\alpha/2$ at the back surface transition. The weight value remains constant for a distance of δ behind the back surface, before dropping to zero within another distance of δ further back. This choice ensures there is sufficient support for isosurface near the back transition. An overall factor of $1/r^2$, where r is the distance from the sensor, is applied to baseline weight in order to give the voxels closer to the sensor higher confidence.The inclusion of this factor helps alleviate some artifacts due to scan misregistrations, by weighing contributions from near observations higher than those from afar.

We initialize the whole space to be empty by default (*i.e.*, every voxel has a distance $d(x) = 1$) which indicates that it is far from any surface. As this is a strong assumption about the relative position of a voxel to a potential surface, we initialize the weights to a small magnitude, indicating the low confidence about the voxel's actual state.

2.4 Surface Thickening

Surface thickening is required to ensure that a solid model can be generated from surface measurements even when objects are not observed from all sides. Having a model with a thickness is especially useful for outdoor scenes as these environments are not closed as is the case for the fully enclosed cave interiors. Without adding some thickness, it would not be possible to render a model of an outdoor environment on a 3D printer due to the thin surface. Thickening of the model is achieved by "back-carving" each ray a given distance into the surface while assigning a negative distance value to each voxel behind the surface. Due to the initialization of the full voxel set as empty (*i.e.*, the space is set to a positive value), the back-carving results in another zero transition some distance behind the surface. Thus isosurface generation produces an additional surface at the back of the structure, resulting in a set of closed solid models (Fig. 4). Ray carving into the surface is performed in the direction of the surface normal. By doing so, we can achieve an approximately equal thickness throughout the structure (Fig. 4b), which would not be the case were we to simply carve the ray into the surface along its incidence direction (Fig. 4a). Carving along the ray would result in an irregular background because the distance a ray carves into a surface is dependent on the angle of incidence.

In Eq. 2 we define negative distances $d(x_j)$ to increase again for negative values less than $\delta - \Delta$ (Fig. 3). Defining the distances in this manner avoids a large step in the zero transition at the back surface which would arise were we to simply assign increasingly negative values while carving deeper into the surface before terminating the the back-carving. Due to the well-defined zero transition at the back surface (see Fig. 7a), the final isosurface extraction produces a smoother surface than would be the case for a step transition.

2.5 Coarse Spatial Decomposition

Given the scale of the environments and size of the available memory, the algorithm can work with only a limited volume at any given time. Therefore the working

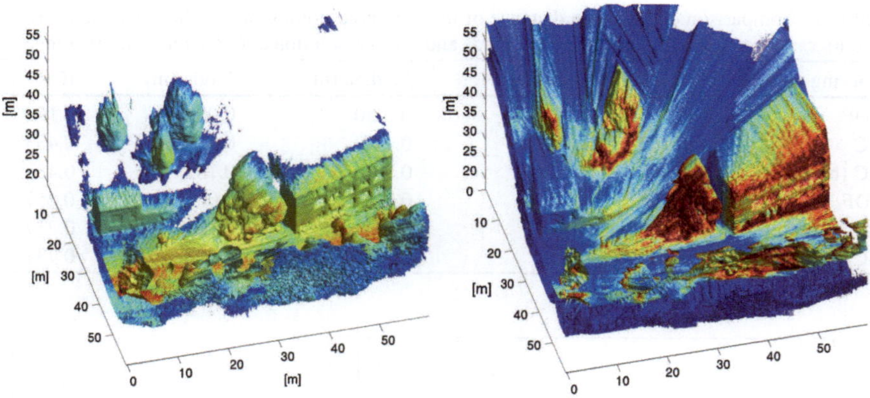

(**a**) Carving along the normal direction (**b**) Carving along the ray direction

Fig. 4 Comparison of a carving in the normal **a** and along the ray **b** direction. It can be observed that the simple carving along the ray produces a very noisy back surface. The surface coloration indicates the weights assigned to the voxels (*blue* low, *red* high). Note that the resulting structure with the surface in the back and the front is closed. The open structure appearing on the right of both images is due to a tile boundary (see Sect. 2.5). The structure continues into the neighboring tile and forms a *closed solid*

space is decomposed into coarse sub-volumes we call *tiles* [1]. Each of the tiles is processed individually in order to circumvent memory limitations. Using this technique generally reduces computation efficiency as single scans which intersect with multiple tiles have to be processed several times. After processing the full dataset, the solid or surface models generated from all of the tiles are merged together.

As we process each tile individually, we need to initially create a lookup table in a pre-processing step in order to determine the set of beams that intersect each tile. Due to the back-carving of the surface, we also need to take those beams into account which do actually not intersect with a tile, but their back-carving does.

An alternative and more memory efficient approach would be to process the measurements into a single octree-based volumetric data structure [22], which we are investigating as future work. Using an octree would allow for the model resolution to better match the data density, thus regions of empty space and sparse sampling could be represented by lower-resolution voxels.

3 Experiments

While scanning the Jenolan Caves, we also acquired a considerable amount of data from areas outside the caves, including around and between the cave entrances. As the area was not closed off to the public, there are many measurements from pedestrians and vehicles passing through the field of view of our scanner.

Table 1 Comparison of the mean distance of the measured point cloud to the calculated surface and its variances for different carving methods and voxel resolutions. IQR = interquartile range

Carving method	Voxel resolution (m)	Median (m)	Mode (m)	IQR (m)
SDF	0.2	-0.030	0.00038	0.190
BC	0.2	0.270	0.057	0.483
BC (biased)	0.2	0.308	0.182	0.483
SDF	0.5	0.0018	0.00015	0.257
BC	0.5	0.512	0.108	0.717
BC (biased)	0.5	0.600	0.103	0.731

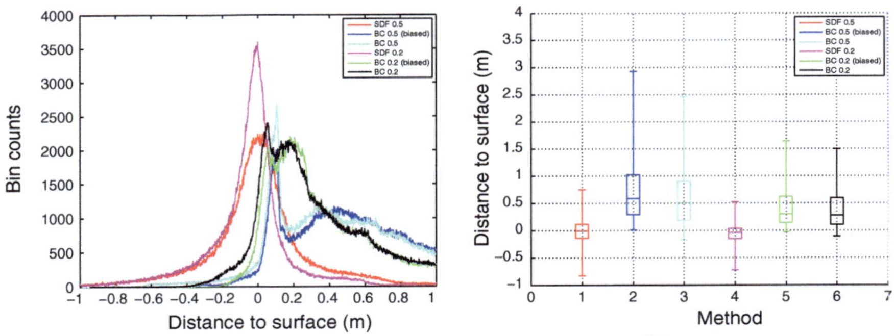

(a) Error histograms **(b)** Box plots

Fig. 5 **a** Distributions of point-to-surface distances. **b** Boxplots for the different methods and resolutions. The numerical values are given in Table 1. *SDF* signed distance function; *BC* binary carving. 0.5 and 0.2 are the voxel resolutions in meters. Results labeled as "biased" are based on a comparison the binary carving surfaces to the unmodified point cloud

We show a comparison for our method (SDF) developed in Sect. 2.3 and the binary carving (BC) method developed in our previous paper [1]. In the BC method, we simply initialize all space to occupied, then mark any voxels through which a ray passes as free space. The evaluation of the two methods is performed by comparing the average distance of the generated surface to the originally recorded point cloud for different voxel resolutions. To do so, we compute the normal vector for each surface triangle as the cross product of the vertices spanning the triangle. The distance of a point in the point cloud is then simply the distance from the point along the normal to the plane spanned by the closest triangle in the mesh.

Figure 5 shows the distributions of the point-to-surface distances for the two carving approaches at two different resolutions applied to an outdoor scene containing both natural (trees, hills) and artificial (buildings, roads) structure (Fig. 6). Table 1 presents some of the relevant statistics from these distributions. We observe that the point-to-surface distances in the BC method are biased towards the associated voxel resolution. This phenomenon occurs as the BC algorithm sets a voxel to 'free space' when a single ray intersects with it. Thus, even when a ray enters only a few

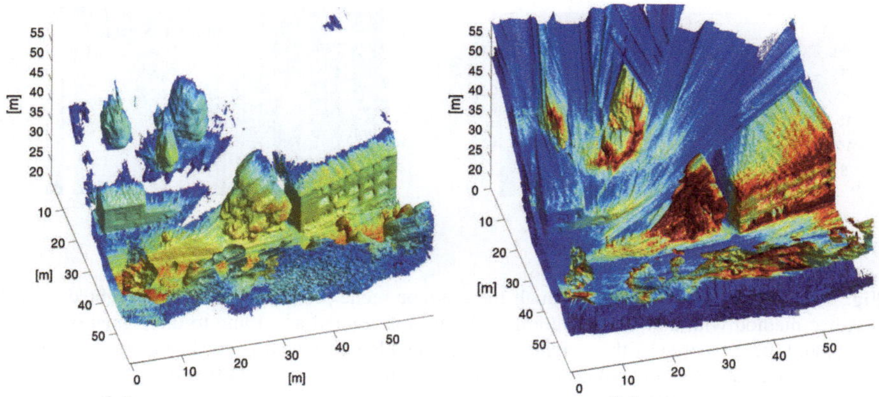

(a) Signed distance carving   (b) Binary carving

Fig. 6 An outdoor scene processed with the (**a**) signed distance method and the (**b**) binary carving method. The colors in (**a**) visualize the logarithm of the weights which are assigned to the voxels where the surface passes through. For (**b**), the colors indicate a count of the number of times the corresponding voxel was back-carved during processing. A 3D printed model of (**a**) can be seen in Fig. 1b

millimeters into the voxel, the voxel is marked as being free space. To reduce this bias, we can shrink the point cloud towards the corresponding sensor position by the voxel resolution. Results both with and without this modification are presented. The results from the SDF method do not display the bias observed in the binary carving case: the distributions appear more symmetric and centered close to zero, regardless of the resolution. The ability of the SDF approach to fuse multiple measurements, rather than aggressively carve away all intersected space, results in smoother and more accurate surfaces. More accurate results can be obtained as the resolution is increased, but as there is a cubic dependency between the resolution and both the computation time and memory usage, there is a clear tradeoff that must be considered. In the remaining experiments, results are reported for the unbiased version of the BC method (*i.e.*, with the modified point cloud).

Figure 6 compares the solid models generated with the SDF and BC methods for a single tile from an outdoor scene. As expected, the BC method is considerably more aggressive especially for unconstrained, unstructured objects. This effect is best observed when looking at the trees which are less voluminous with the BC method compared to SDF carving. The fuller trees from the SDF approach result from the distance and weight pairing of the algorithm. SDF can reverse its estimate about the occupancy of the space if there is enough evidence for a voxel to be assigned as full, even though it was assigned to be empty from previous measurements. Hence, it also takes those scans into account which are reflected at the outermost leaves of the trees and does not necessarily delete the information provided by this scan if there is another scan which hits the tree in an inner layer. In contrast, BC is a one-way algorithm: if a voxel is assigned to be empty space, it will remain empty

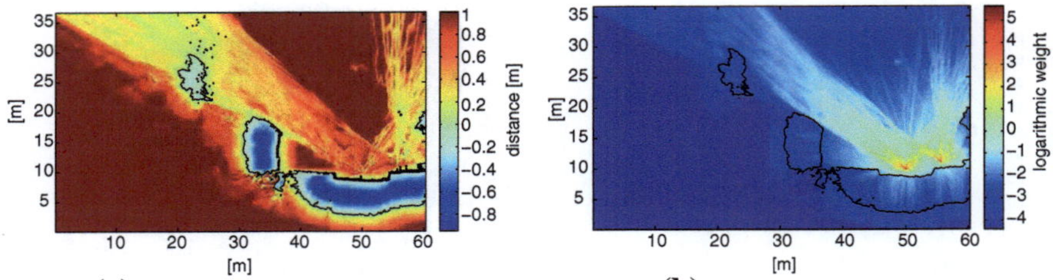

(a) Colored by signed distance

(b) Colored by weight

Fig. 7 A vertical cross section through the outdoor scene processed with the weighted signed distance method with a voxel resolution of 0.2 m. **a** is colored according to the signed distances $D(x_j)$ and (**b**) is colored by the weights $W(x_j)$, scaled logarithmically. The black points in (**a**) are the raw input data (*i.e.*, the point cloud at the position of the slice) and the *black line* shows the extracted isosurface at the zero transition

forever. Thus, the BC method cuts the leaves/branches away when it processes scan points which hit the tree in an inner layer or even produces holes when a single outlier beam passes through the entire tree or structure. In our previous paper [1], we treated all rays with maximum range as outliers and neglected them.This assumption is reasonable when measurements are made in an enclosed area where it is very unlikely to observe scans with maximum range, but it does not hold in outdoor environments. In fact, rays which have maximum range (such as non-reflected rays shooting into the sky) provide relevant information regarding free space in outdoor environments. Maximum range measurements resulting from outlier measurements produce holes when using the BC approach; however, valid measurements are likely to overwrite the effect of isolated outliers in the SDF approach.

Figure 7 shows a vertical cross section through the voxels of the same area as shown in Fig. 4. Note that the distance in the back of the structure increases slowly towards a positive value and thus produces a well defined zero transition as discussed in Sect. 2.4.

Figure 8 shows a surface model of two connected caves in the Jenolan Cave system, generated using the signed distance function and 20 cm voxels. The distance between the lower left and upper right extremes of the visualized model is approximately 360 m.

4 Conclusions

We developed a new approach to reconstruct solid models and surfaces for unstructured large-scale outdoor environments. Our approach is based on the space carving using a weighted signed distance function in a 3D uniform grid. The algorithm for surface reconstruction is demonstrated in an unstructured outdoor scene including natural (*e.g.*, trees, bushes) and artificial (*e.g.*, buildings, streets) objects. We have observed a significant improvement over the binary carving in terms of accuracy of

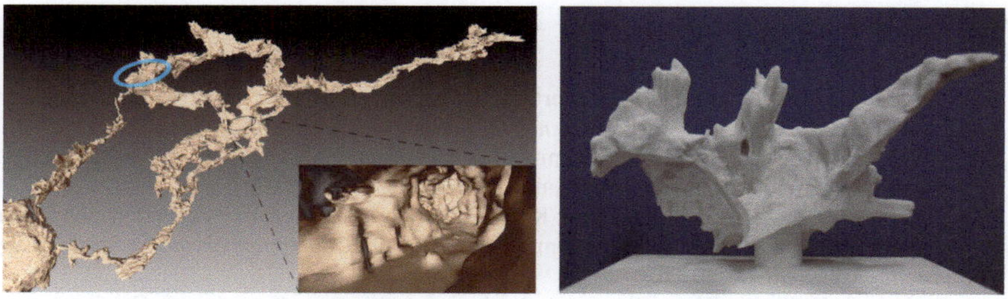

(a) Surface model: Chiefly and Imperial Caves (b) 3D printed model

Fig. 8 **a** Reconstructed surface for Chifley and Imperial caves, produced using the signed distance function with 20 cm voxel resolution. The inset in the lower left illustrates a view from the interior. **b** A 3D printed model of a 50 m section of Chifley Cave, printed at 1:250 scale. The printed area is roughly indicated by the *blue* ellipse in the upper left area of **a**

the reconstructed surfaces. As the new approach for surface reconstruction is more precise even for lower voxel resolutions, we can achieve computational and memory improvements as the costs increase with $O(1/resolution^3)$.

However, there are inherently a few limitations to our approach. Although outliers are better handled by the weighted signed distance function method, outliers can still affect the overall result when they occur in concentrated bunches as is often the case with windows, surfaces with high specular reflectance, or bodies of water. Furthermore, thin structures which are seen from only one side will appear much thicker in the reconstruction as a result of the thickening process. Alternatively, thin structures can disappear if they are seen from both sides and their thickness is less than a voxel resolution. Another limitation is that the algorithm does not necessarily produce connected surfaces over the whole model. Disconnections often occur between structures where too few or no scans are present. In order to produce a connected surface, one would either have to make sure to densely sample the whole environment which can become rather tedious (or even impossible) or a method must be developed which fills up the missing structure in an intelligent way. In some cases, it may be sufficient to add simple supporting structure to each connected component in the model.

Future plans also include the combination of the outdoor and enclosed surface reconstruction, such that the full Jenolan Cave system dataset can be processed at once. Such a model requires integrating variants for both open and enclosed environments as the back carving in indoor scenes is not desirable.

Acknowledgments The authors would like to thank Paul Flick for leading the hardware development of the sensor cart system. We would also like to thank Chris Waring and the Australian Nuclear Science and Technology Organisation (ANSTO), as well as the Jenolan Caves Reserve Trust for providing the opportunity to map the caves and for assisting with data collection.

References

1. C. Holenstein, R. Zlot, M. Bosse, Watertight surface reconstruction of caves from 3D laser data, in *IEEE/RSJ International Conference on Intelligent Robots and Systems*, 2011
2. D. Hähnel, W. Burgard, S. Thrun, Learning compact 3D models of indoor and outdoor environments with a mobile robot. Robot. Auton Syst. **44**(1), 15–27 (2003)
3. T. Liu, M. Carlberg, G. Chen, J. Chen, J. Kua, A. Zakhor, Indoor localization and visualization using a human-operated backpack system, in *International Conference on Indoor Positioning and Indoor Navigation*, 2010
4. I. Mahon, S. Williams, Three-dimensional robotic mapping, in *Australasian Conference on Robotics and Automation*, 2003
5. J. Weingarten, R. Siegwart, EKF-based 3D SLAM for structured environment reconstruction, in *IEEE/RSJ International Conference on Intelligent Robots and Systems*, 2005
6. N. Amenta, S. Choi, R.K. Kolluri, The power crust, in *Proceedings of the Sixth ACM Symposium on Solid Modeling and Applications*, 2001
7. M. Gopi, S. Krishnan, A fast and efficient projection-based approach for surface reconstruction, in*Brazilian Symposium on Computer Graphics and Image Processing*, 2002
8. T.K. Dey, Provable surface reconstruction from noisy samples, in *Proceedings of the Twentieth Annual Symposium on Computational Geometry*, 2004
9. T.K. Dey, S. Goswami, Tight Cocone: a water-tight surface reconstructor, in *ACM Symposium on Solid Modeling Applications*, 2003
10. B. Curless, M. Levoy, A volumetric method for building complex models from range images, in *ACM SIGGRAPH*, 1996
11. H. Moravec, A. Elfes, High resolution maps from wide angle sonar, in *Proceedings of the 1985 IEEE International Conference on Robotics and Automation*, vol. 2 (1985), pp 116–121
12. R.A. Newcombe, S. Izadi, O. Hilliges, D. Molyneaux, D. Kim, A. Davison, P. Kohli, J. Shotton, S. Hodges, A. Fitzgibbon, KinectFusion: real-time dense surface mapping and tracking, in *IEEE International Symposium on Mixed and and Augmented Reality*, 2011
13. K. Pulli, T. Duchamp, H. Hoppe, J. McDonald, L. Shapiro, W. Stuetzle, Robust meshes from multiple range maps, in *International Conference on Recent Advances in 3-D Digital Imaging and Modeling*, 1997
14. K. Zhou, M. Gong, X. Huang, B. Guo, Data-parallel octrees for surface reconstruction. IEEE Trans. Vis. Comput. Graph. **17**(5), 669–681 (2011)
15. J. Carr, R. Beatson, H. Cherrie, T. Mitchel, W. Fright, B. McCallum, T. Evans, Reconstruction and representation of 3D objects with radial basis functions, in *ACM SIGGRAPH*, 2001
16. M. Kazhdan, Reconstruction of solid models from oriented point sets, in *Proceedings of the Eurographics Symposium on Geometry Processing*, 2005
17. G. Turk, J. O'Brien, Modelling with implicit surfaces that interpolate. ACM Trans. Graph. **21**(4), 855–873 (October 2002)
18. S. Gemme, J. Bakambu, I. Rekleitis, 3D reconstruction of environments for planetary exploration. In *Canadian Conference on Computer and Robot Vision*, 2005
19. Z.C. Marton, R.B. Rusu, M. Beetz, On fast surface reconstruction methods for large and noisy point clouds, in *IEEE International Conference on Robotics and Automation*, 2009
20. M. Bosse, R. Zlot, Continuous 3D scan-matching with a spinning 2D laser, in *IEEE International Conference on Robotics and Automation*, 2009
21. J.E. Bresenham, Algorithm for computer control of a digital plotter. IBM Syst. J. **4**(1), 25–30 (1965)
22. M. Bolitho, M. Kazhdan, R. Burns, H. Hoppe, Multilevel streaming for out-of-core surface reconstruction, in *Eurographics Symposium on Geometry Processing*, 2007

Lightweight Laser Scan Registration in Underground Mines with Band-based Downsampling Method

James Lee, David Wettergreen and George Kantor

Abstract Robots operating in underground mines must accurately track their location and create maps. The rough, undulating floors typical of mine environments preclude the 2D representation of scene integral to many existing real-time mobile robot simultaneous localization and mapping systems. On the other hand, a full 3D solution is made unrealistic by the computational expense of aligning large point clouds. This paper presents an approach that extracts high-density, horizontal bands of laser scans and uses them to represent the scene with detail sufficient to capture the moderate non-planar motion typical of mining robots. Our approach is able to operate in real-time, building maps and localizing in pace with range scanning and is fast enough to allow continuous vehicle motion. We present details of the approach which has been validated in an underground mine. Trials runs have shown a significant decrease in computation time without an appreciable decrease in accuracy over a full 3D strategy.

1 Introduction

Mining is a classic example of the types of dull, dangerous and dirty jobs that humans are not particularly well suited for. In present-day manned mining operations, active measures must be taken to bring down heat and pump in air thereby ensuring the safety of miners. As mines venture further underground it becomes increasingly more difficult to exploit mineral resources both safely and cost-effectively [1].

J. Lee (✉) · D. Wettergreen · G. Kantor
Robotics Institute, Carnegie Mellon University, Pittsburgh 15213, USA
e-mail: jslee@cmu.edu

D. Wettergreen
e-mail: dsw@cmu.edu

G. Kantor
e-mail: kantor@cmu.edu

K. Yoshida and S. Tadokoro (eds.), *Field and Service Robotics*, 555
Springer Tracts in Advanced Robotics 92, DOI: 10.1007/978-3-642-40686-7_37,

Autonomous robotic mining is increasingly being turned to as a solution to these problems. Moreover, mine accidents, such as one at Quecreek, Pennsylvania, where the breach of a poorly documented mine caused a flood that trapped nine miners underground for three days [2], underscore the need for highly accurate mapping of mines. The tens of thousands of these undocumented mines estimated to exist under the United States [3] pose a threat to miner safety until they can be reliably documented. Accurate maps generated by subterranean robots can alleviate these safety concerns.

Conditions in mining environments dictate sensing modality. In mines where darkness blinds visual sensors, scanners that sense geometric properties thrive [4]. Light detection and ranging (LIDAR) devices are one means of sensing structure and scan matching range images is an integral component of several mobile robot simultaneous localization and mapping systems. Furthermore, these systems do not require that a robot carry its own light source and function in both active and abandoned mines. Many current algorithms represent the world 2-dimensionally in a flat plane, either with a single scan or by flattening horizontal scans. This 2D approach is unsuitable for mapping in situations where non-planar movements of a vehicle confound scan matching. Mine floors, with dips, rises, railways, and debris, are an example of such an environment. As robot rolls and pitches while travelling over uneven terrain, strictly 2D scan matching fails because scans no longer overlap. The full 3D solution is typically made unrealistic by the computational expense of aligning and matching large numbers of range measurements. Robots that excavate rock and manipulate their environments cannot rely solely on static, pre-calculated maps and must be able to create maps in real time to plan as well as to inspect rock face and ceiling surfaces for damage.

2 Related Research

The backbone of robotic mapping is simultaneous localization and mapping (SLAM). In order to produce accurate maps a robot must solve the chicken and egg problem of inferring pose and maps from one another. In recent years much progress has been made in extending SLAM to large-scale environments but many of these approaches assume either an in-plane motion rare in mining environments or operate on features that can suffer from ambiguous data association in the subsurface context [5–8]. A 3D approach presented by Fairfield et al., utilizing a highly compact octree structure in conjunction with an Rao-Blackwellized Particle Filter, has demonstrated real-time 3D mapping of subterranean voids [9]. This method is able to track the full 6DoF (degree of freedom) movement of a robot. However, it stores scene structure as occupancy probabilities in an coarse octree producing ordered, grid-like worlds that, while suitable for planning, are unfavorable for survey quality maps. Other 3D approaches uniformly or randomly downsample scans [10] or use computationally expensive probabilistic methods to drive downsampling [11, 12]. These approaches seek to reduce the number of points in each scan because the computational complexity of aligning scans using ICP (iterative closest point) grows exponentially with the number of points [15]. The many lightweight approaches preserve the structure of the entire scene by lowering overall point density. However, naively downsampling scans by either selecting random points, every nth point or points to preserve a constant density retains points from geometrically uninteresting regions that are unlikely to assist in registration [14]. Alternatively, Cole et al. have shown that matching with smaller, high-density regions increases both matching speed and accuracy [14]. They further note that finding these smaller high-density regions is often difficult and time consuming and that poor selection criteria can produce regions that do not overlap across multiple scans or miss important surface features that constrain the transformation [13].

This paper proposes and evaluates a simple method of selecting suitable high-density regions to be used in scan matching. In particular we propose to reduce the size of a 3D point cloud by extracting a band, which is defined to be the set of points that is swept through by a n-degree rotation of a plane that passes through and rotates about midpoints on the front and back of a robot. The resulting approach is able to operate in real-time, building maps and localizing in pace with range scanning. It combines the benefits of having high-density, geometrically interesting regions and is fast enough to allow continuous vehicle motion. We present details of the approach which has been validated in an underground mine and show that match stability and geometric saliency are persevered. Trial runs have shown a significant decrease in computation time without an appreciable decrease in accuracy over a full 3D strategy.

3 Method

As a robot traverses through a mine it collects thousands of 3D scans. In our case, these scans were collected from a SICK LMS200 line-scanning laser spun about its forward axis to sweep a hemisphere of range measurements. A high-precision inertial measurement unit, in conjunction with wheel odometry information, is used to perform the initial alignment creating 3D scans. Band regions are extracted from a pre-determined point in each scan. A band, in this system, can be simplified to a point cloud created by stitching n consecutive scans from the spinning laser. We determine

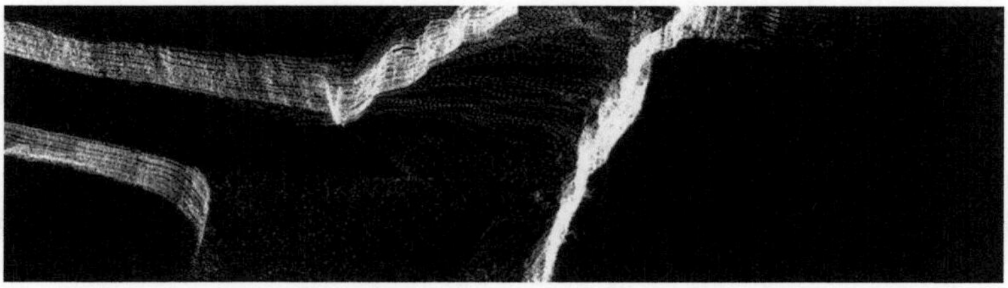

Fig. 1 Perspective view of reconstructed map mine showing sampled bands

the size and placement of the band region that we extract through stability analysis (described in Sect. 3.1) and to supplement other functions of the robot. These band regions are then registered into a global frame using a standard point- to-plane ICP (Fig. 1) by matching it to a map that combines previous bands. The new band as well as the transform between it and the last band is stored in a pose graph structure similar to the one described in [17]. Loop closure adds additional edges to the graph which, upon optimization, generate accurate estimations of pose. Loops are detected by looking for intersections within laser scans. When two intersections are found that are within a threshold distance from one another, we generate two map segments by combining scans taken around the two intersection scans and align the two segments to determine the transform used to close the loop.

3.1 Band Stability

Gelfand et al. defines a stable point sampling in the context of point-to-plane ICP as a sample set that constrains all degrees of freedom of the rigid body transform [16]. Moreover, they show that ICP registration with a good sampling of stable points not only improves the stability of the ICP match but also converges to the correct solution faster [16]. Samplings that only constrain some of these degrees of freedom introduce slip. For instance, points that describe a flat plane are free to slide about in two axes and rotate about one. Likewise, points lying on a cylinder have one free axis about which they can translate and rotate. Point stability for a set of points P with k number of points p is defined in [16] as the number of eigenvalues above a threshold of a 66 matrix C where pn is the nth point in set P and nn is the normal of the point in a overlapping set Q that correspondes to the nth point in set P. C is defined as

$$C = \begin{bmatrix} p_1 \times n_1 & \cdots & p_k \times n_k \\ n_1 & \cdots & n_k \end{bmatrix} \begin{bmatrix} p_1 \times n_1 & \cdots & p_k \times n_k \\ n_1 & \cdots & n_k \end{bmatrix}^T \qquad (1)$$

4 Experiments

This band matching method was validated with field experiments conducted at a research coal mine near Pittsburgh, Pennsylvania. Two approximately 300 m long traverses around large loops were collected in geometrically distinct sections of the mine to demonstrate suitability in a variety of mining environments.

Data was collected on the robot CaveCrawler (Fig. 2) equipped with a spinning LMS200 SICK scanner capable of scanning a 180° hemisphere at a rate of 3 s per scan.

Fig. 2 Cave Crawler platform with spinning SICK in coal mine. Featureless floors, such as the one in this picture, are prevalent throughout this mine and provide little additional information to aid scan matching. Picture by Uland Wongg

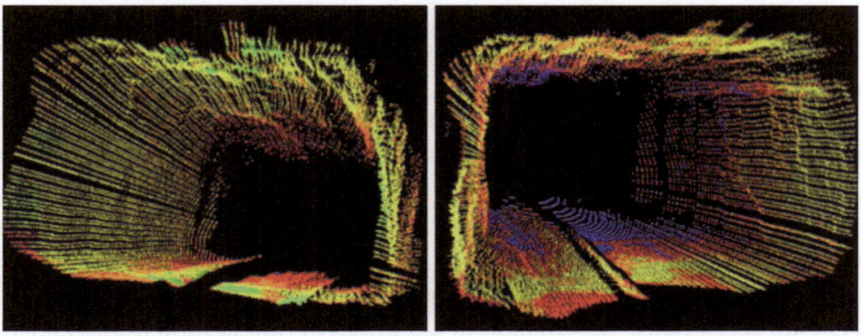

Fig. 3 Two views of mine corridors showing point stability. *Green* points represent the most stable points where all 6 DOFs are constrained. *Yellow*, *orange*, *red* and *blue* are progressively less stable points that constrain 5, 4, 3, and 2 or less degrees of freedom

4.1 Band Stability

The stability metric from Sect. 3.1 was used to analyse 3D point clouds col- lected at a coal mine. Figure 3 shows two 3D scans from within a coal mine with points colorized to show stability. Points stable in all 6 DoFs were colored green; table in 5, yellow and 4, orange. Points only stable in 3 DoF or 2 or fewer were colored red and blue, respectively. The plots indicate that mine walls contain more stable points than do floors and ceilings. Figure 4 plots the ratio of 4 or more DoF stable points to total band points of a 10° band as a function of the center angle. The most stable band in this

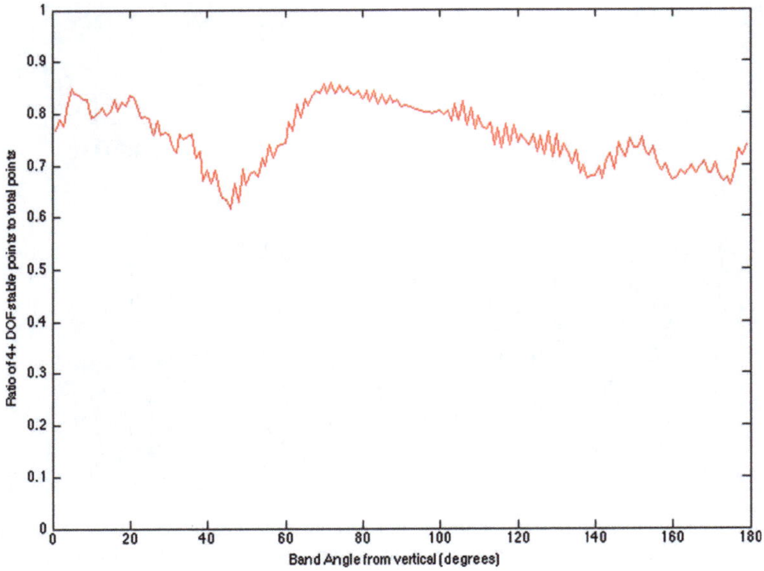

Fig. 4 Band stability versus band angle

scan is centered near the horizontal at 73°. We found that using either scan matching using the full 3D scan or recalculating stability at each timestep to aid downsampling took more time than it took for the robot to traverse its path. The robot would return scans at a rate of 1 scan every 3 s, whereas the full 3D and stability assisted methods took amount 8 and 11 s, respectively, to process each scan. Furthermore, the exponential increase in the time required to detect and close loops was for the full 3D method [15] and made it intractable for real-time operation. Although calculating this per-point analysis of ICP matching stability is computationally expensive and also unrealistic for real-time applications, it can be used to develop constant time segmentations for real-time use. In the mining case, we observe that stability lies around the horizontal plane and choose to segment a band that captures this stability. An additional benefit of this method is that feature detection becomes very repeatable. In mining environments we can guarantee that two downsampled scans will retain overlap provided the thickness of the band selected is large enough to account for the uneven floors. Other methods are sometimes unable to sample the same regions over large perspective or point density changes.

4.2 Band-based Mapping

We estimated poses and created maps using band-based (described in Sect. 3.1), as well as 2D, 3D, uniform downsampled, and constant density down- sampled methods and compared the relative runtimes and root mean square errors of each. Uniform downsampling reduces the number of points by only retaining every nth point in a scan; constant density downsampling achieves this by using an octree structure to prune out points that are close together. It does this by selecting a leaf-size, binning all the points based on which octree node they reside in, and storing the mean x, y, z coordinate of all the points in each bin.

The band parameters were selected to meet a pre-determined maximum time allotment for the initial laser scan match. This upper bound was found to be about 1 s per each scan, which was determined by timing the length of time the robot took to complete a traverse and subtracting the time required for loop detection and closure. A similar approach was taken to determine the downsample factor for uniform downsampling as well as the desired point density for the constant density approach. Care was taken to ensure that no algorithm was given an unfair advantage. The distance between each 3D scan was determined by the speed of the robot and the angular rate of rotation of the laser spinner. In our tests this distance was fixed at approximately 1 m for all of the compared methods.

We found that a 10° horizontal band was sufficient to both meet the time requirement and for tracking 3D motion in this environment.

We compared accumulated drift from ICP matching these downsampled scans by intentionally not closing loops under the assumption that the method that performs

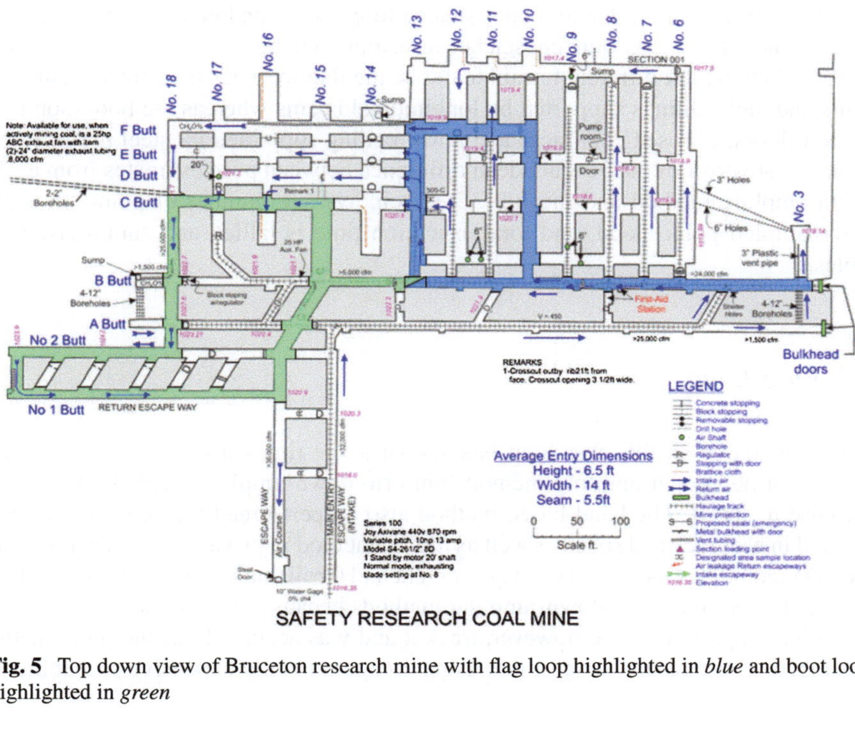

Fig. 5 Top down view of Bruceton research mine with flag loop highlighted in *blue* and boot loop highlighted in *green*

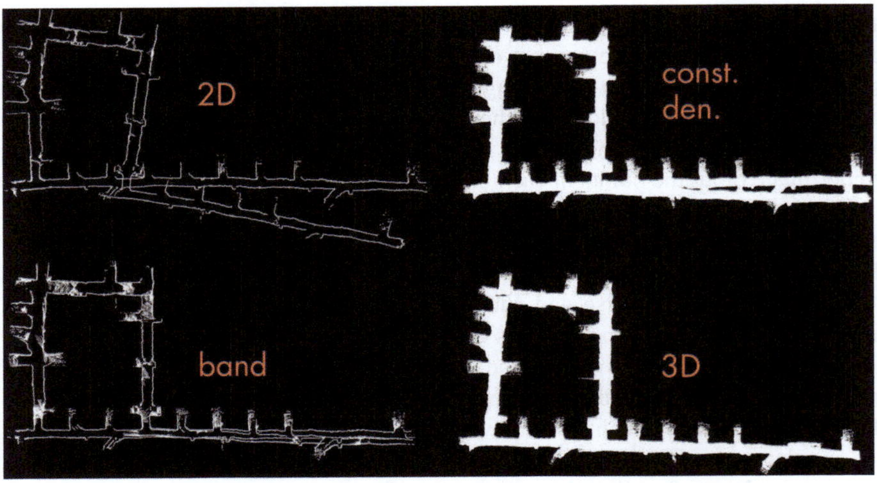

Fig. 6 Top-down view of flag loop maps generated from 4 different methods

most like the 3D method is the most accurate method. This is a reasonable assumption to make because the 3D match utilizes all of the data points from the scan and is often used to calculate 6DoF transforms in many existing 3D SLAM algorithms. Two trials were run on flag and boot shaped loops (blue and green, respectively in Fig. 5). These two loops were chosen because, although they are from the same mine facility, they exhibit different terrain features: the flag loop has train tracks, gunnite walls and high ceilings supported by longitudinal beams whereas the boot loop has natural floors, exposed coal faces and a low ceiling supported by steel mesh. Root mean square error of the 3D euclidean difference between pose estimates from each downsampling method and the full 3D match, average number of points in each downsampled point cloud, and total execution time is tallied and summarized in tables.

4.3 Flag Loop

Experimentation in this flag loop has shown a 1.9 times decrease in error over an uninformed downsampling method (uniform downsampling) with a 1.69 times increase in speed. The band-based method also outperformed the constant density method in accuracy and speed as well as the 2D method in just accuracy. We find that the 2D method that assumes 0 z-displacement and 0 roll and pitch angles does better than both other real-time downsampling methods because of the relative flatness of this mine loop. It does not, however, track x and y as accurately as the other methods Fig. 6. These results suggest that using computationally expensive probabilistic

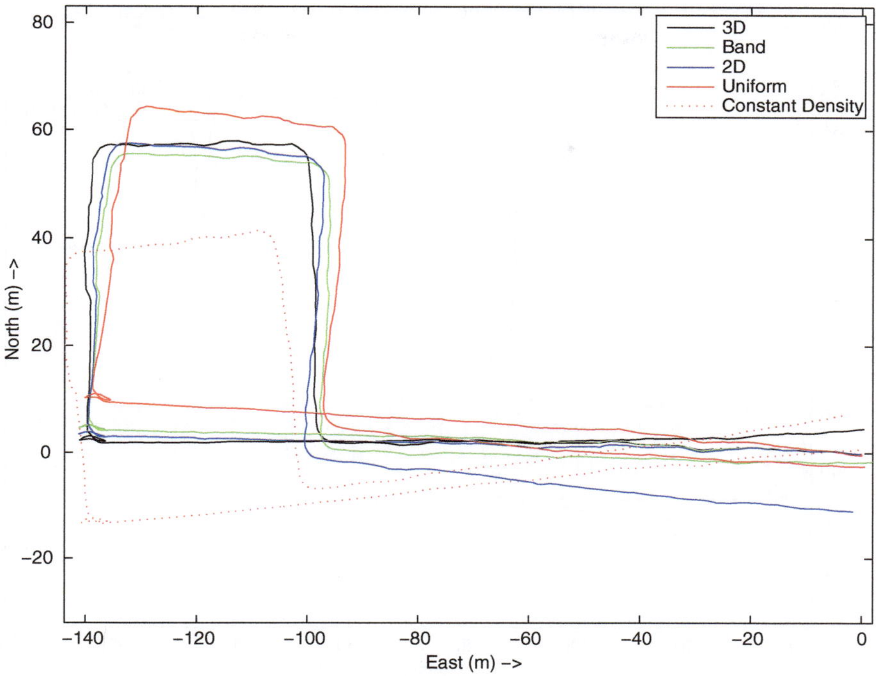

Fig. 7 Top-down view of estimated robot pose from the 5 different methods on flag shaped loop (*Black* 3D, *Green* Band, *Blue* 2D, *Solid Red* Uniform, *Dotted Red*: Constant Density)

	3D	Band	2D	Uniform	Constant Density
RMS Error (m)	N/A	2.220	3.813	4.150	9.380
Average Number of Points	29444.102	1975.154	361.195	2587.325	2593.517
Computation Time (s)	1518.458	117.045	59.600	198.580	230.908

Fig. 8 Table comparing RMS error, average number of points, and runtime on flag loop

methods to make informed choices for segmentation can aid ICP matching accuracy and speed of convergence (Figs. 7, 8).

4.4 Boot Loop

We further applied the same band-based segmentation to another mining dataset while retaining the same horizontal band. This simulates the effect of a change in environment midrun that may alter the location of stable bands.

Experimentation in the boot shaped loop has shown that band-based matching produces comparable results to those from the band downsampled run of the flag

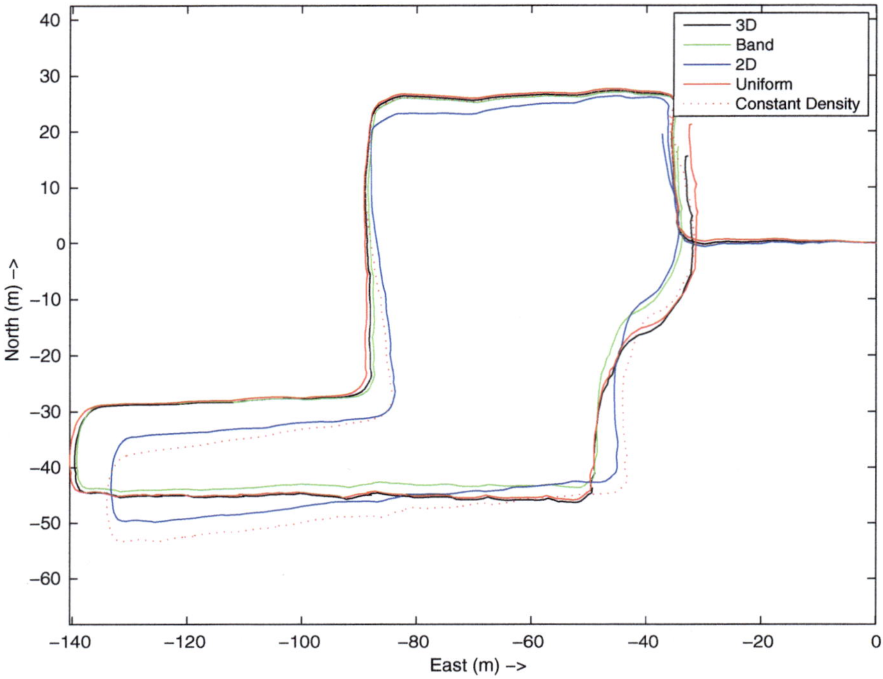

Fig. 9 Top-down view of estimated robot pose from the 5 different methods on boot shaped loop (*Black* 3D, *Green* Band, *Blue* 2D, *Solid Red* Uniform, *Dotted Red* Constant Density)

	3D	Band	2D	Uniform	Constant Density
RMS Error (m)	N/A	1.0940	3.5768	1.0295	4.320
Average Number of Points	29722.391	3588.055	360.401	2972.274	2949.371
Computation Time (s)	1926.296	343.099	59.128	359.883	284.459

Fig. 10 Table comparing RMS error, average number of points, and runtime on boot loop

shaped loop (Figs. 9, 10). While these two loops exhibit different sets of features the band down-sampled point clouds still produce accurate ICP alignments. This demonstrates that band based downsampling can be applied to varying mine environments successfully. In addition, the uniform downsampling method does remarkably better in this dataset than it did on the flag loop data. While not significantly better than comparable lightweight alternatives, the band based approach is at least a good the alternatives tested.

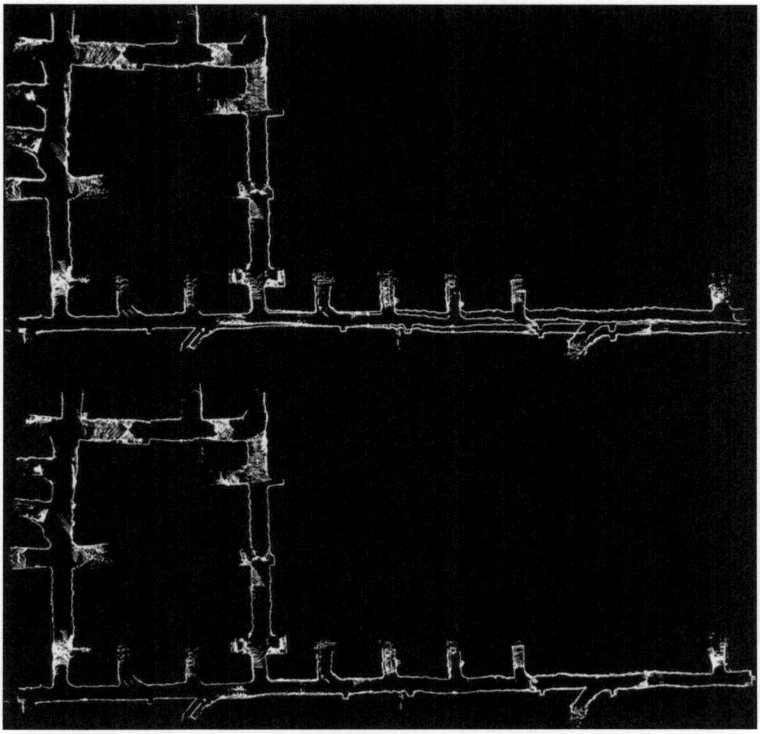

Fig. 11 Flag loop without loop closure (*top*) and with closure (*bottom*)

4.5 Loop Closure

The matched band segments are then used to predict and close loops. Reducing the size of each scan by nearly 15-fold reduces the time it takes to perform loop closures exponentially [15]. The benefit that this method has over methods that further reduce the number of points by recalculating stable or salient regions is that this is both faster and also guarantees that the same regions will be preserved regardless of orientation or scale changes. This speedy loop closure, in conjunction with efficient band based ICP alignment, is demonstrated on the two loops above. Figures 11 and 12 show both flag and boot loops after the loop closure step.

5 Conclusion

As robots continue to explore vast and unfamiliar worlds, smart decimation algorithms will be required for a robot to create on-the-fly maps that aid navigation and autonomy. This approach proves that accurate, constant time decimation can be

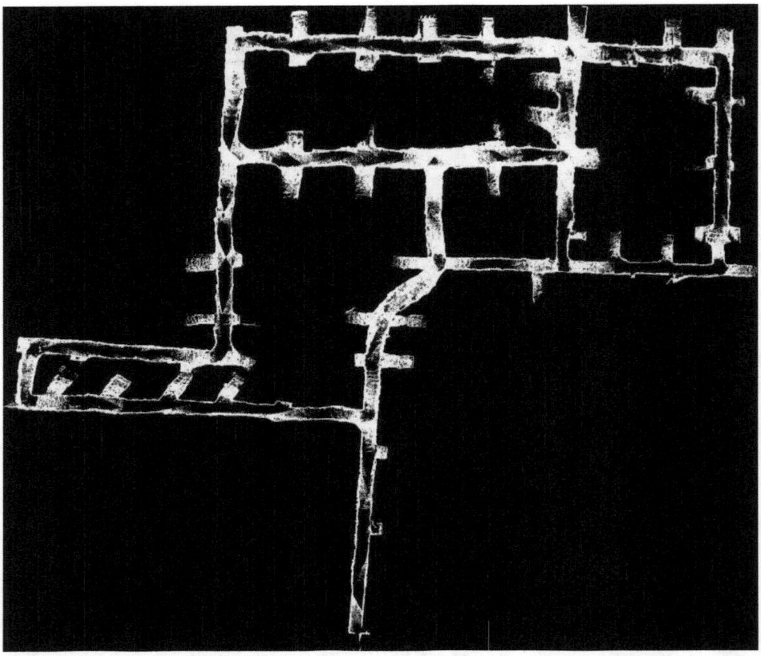

Fig. 12 Boot loop and partial flag loop with closure

achieved by using computational expensive stability analysis to learn good segmentations ahead of time. Trials have shown that in subterranean environments a band based matching approach is able to match with an accuracy approaching that of a full 3D solve but with runtimes similar to that of a 2D one.

References

1. MRA, in *Safety in Deep Mining conference* http://www.spintelligent-events.com/deep-mining2008/en/index.php
2. Bureau of Deep Mine Safety, Department of Environmental Protection, Commonwealth of Pennsylvania. Report of Investigation, Black Wolf Coal Company, Inc., Quecreek No. 1 Mine (2003)
3. J. Belwood, R. Waugh, Bats and mines: abandoned does not always mean empty. Bats **9**(3), 13–16 (1991)
4. A. Morris, D. Ferguson, Z. Omohundro, D. Bradley, D. Silver, C. Baker, S. Thayer, C. Whittaker, W. Whittaker, Recent developments in subterranean robotics. J. Field Robot. **23**(1), 35–57 (2006)
5. M. Bosse, P. Newman, J. Leonard, S. Teller, Simultaneous localization and map building in large-scale cyclic environments using the atlas framework. Int. J. Robot. Res. **23**(12), 1113–1139 (2004)

6. M. Kaess, H. Johannsson, R. Roberts, V. Ila, J. Leonard, F. Dellaert, iSAM2: incremental smoothing and mapping with fluid relinearization and incremental variable reordering. ICRA **24**(6), 1365–1378 (2011)
7. M. Montemerlo, S. Thrun, D. Koller, B. Wegbreit, FastSLAM 2.0: An improved particle ltering algorithm for simultaneous localization and mapping that provably converges. Proc. Int. Jt. Conf. Artif. Intell. **18**, 1151–1156 (2003)
8. S. Rusinkiewicz, M. Levoy. Effcient variants of the ICP algorithm. 3D Digital Imaging Model. **2001**, 145–152 (2001)
9. N. Fairfield, G. Kantor, D. Wettergreen, Segmented SLAM in three-dimensional environments. J. Field Robot. **27**(1), 85–103 (2010)
10. M. Pauly, M. Gross, L. Kobbelt, Efficient simplification of point-sampled surfaces. IEEE Visual. **2**(4), 163–170 (2002)
11. J. Bowers, R. Wang, L. Wei, D. Maletz. Parallel Poisson disk sampling with spectrum analysis on surfaces. ACM Trans. Graph. **29** 166:1–166:10 (2010)
12. A. Torsello, E. Rodola, A. Albarelli. Sampling, Relevant Points for Surface Regis- tration. International Conference on 3D Imaging. Model. Process. Visual. Transm. **2011**, 90–295 (2011)
13. J. Yao, M. Ruggeri, P. Taddei, V. Sequeira, Robust surface registration using N-points approximate congruent set. EURASIP J. Adv. Signal Process. **72**, 1–22 (2011)
14. D. Cole, A. Harrison, P. Newman. Using naturally salient regions for SLAM with 3D laser data, in *International Conference on Robotics and Automation SLAM, Workshop* (2005)
15. A. Nuchter, H. Surmann, K. Lingemann, J. Hertzberg, S. Thrun, 6D SLAM with an application in autonomous mine mapping. ICRA **2**, 1998–2003 (2004)
16. N. Gelfand, L. Ikemoto, S. Rusinkiewicz, M. Levoy. Geometrically stable sampling for the ICP algorithm. 3D Digital Imaging Model. **2003**, 260–267 (2003)
17. E. Olson, J. Leonard, S. Teller, Fast iterative optimization of pose graphs with poor initial estimates. ICRA **2009**, 2262–2269 (2006)

Featureless Visual Processing for SLAM in Changing Outdoor Environments

Michael Milford and Ashley George

Abstract Vision-based SLAM is mostly a solved problem providing clear, sharp images can be obtained. However, in outdoor environments a number of factors such as rough terrain, high speeds and hardware limitations can result in these conditions not being met. High speed transit on rough terrain can lead to image blur and under/over exposure, problems that cannot easily be dealt with using low cost hardware. Furthermore, recently there has been a growth in interest in lifelong autonomy for robots, which brings with it the challenge in outdoor environments of dealing with a moving sun and lack of constant artificial lighting. In this paper, we present a lightweight approach to visual localization and visual odometry that addresses the challenges posed by perceptual change and low cost cameras. The approach combines low resolution imagery with the SLAM algorithm, RatSLAM. We test the system using a cheap consumer camera mounted on a small vehicle in a mixed urban and vegetated environment, at times ranging from dawn to dusk and in conditions ranging from sunny weather to rain. We first show that the system is able to provide reliable mapping and recall over the course of the day and incrementally incorporate new visual scenes from different times into an existing map. We then restrict the system to only learning visual scenes at one time of day, and show that the system is still able to localize and map at other times of day. The results demonstrate the viability of the approach in situations where image quality is poor and environmental or hardware factors preclude the use of visual features.

M. Milford (✉) · A. George
School of Electrical Engineering and Computer Science, Queensland University of Technology, Brisbane, Australia
e-mail: michael.milford@qut.edu.au

K. Yoshida and S. Tadokoro (eds.), *Field and Service Robotics*,
Springer Tracts in Advanced Robotics 92, DOI: 10.1007/978-3-642-40686-7_38,

1 Introduction

Visual mapping and navigation on robots has advanced rapidly in the last decade. There are now many vision-based techniques including FAB-MAP [1], MonoSLAM [2], FrameSLAM [3], V-GPS [4], Mini-SLAM [5] and others [6–10] that are competitive with or superior to range sensor-based algorithms, with routes as long as 1,000 km being mapped [1]. The majority of these systems have been developed and demonstrated largely under certain conditions: high quality imaging sensors have been used, on relatively stable vehicle platforms and in bright illumination conditions, minimizing problems such as motion blur and changes in appearance. However, these are restrictive constraints, especially as robots are expected to operate over longer periods of time and with lower hardware costs. Many growing fields such as environmental monitoring could benefit greatly from the availability of a small, low cost robot platform with an all-day, all weather mapping and navigation capability that is not reliant on GPS or environmental beacons. Towards that aim, in this paper we seek to address two of the major challenges facing visual mapping systems:

1. The difficulty of obtaining high quality images required by feature-based techniques, when using low cost hardware at speed on off-road terrain and in poor lighting.
2. The problem of achieving reliable place recognition in an outdoor environment over the course of a day and during all types of weather.

Figure 1 illustrates these two challenges using camera images. Large changes in illumination (compare panels of Fig. 1a, d) or changes in the weather (see rain drops on lens in Fig. 1c) can radically alter the types of features detectable by a state of the art algorithm such as Scale-Invariant Feature Transforms (SIFT) [11] and Speeded Up Robust Features (SURF) [12]. Furthermore, in poor lighting with low cost hardware and on off-road terrain, image blur is hard to avoid (Fig. 1c, also Figs. 4–8). Motion blur affects both the place recognition and odometry components of a mapping system, while change in appearance over the course of a day primarily affects place recognition.

To some degree these problems can be reduced by using more capable sensing equipment and implementing techniques such as high dynamic range [13]. However,

Fig. 1 Visual change in an environment over the course of a day and in varying weather—**a** dawn, **b** morning, **c** rain and **d** dusk. As well as changing illumination other challenges are present such as motion blur from the jerky motion of the platform when travelling off-road

high dynamic range techniques degrade in viability as the speed of the platform increases. Without active illumination of an environment, even long exposure images can look very different to an image obtained in sunlight during the day. Motion estimation from motion blurred images can be achieved by tracking edges, but is more difficult to incorporate into a mapping process [14]. More capable sensors and lenses are expensive, usually bulkier and heavier to accommodate larger imaging sensors and lenses, and require more power. While this approach is viable on large expensive platforms where the sensor cost is relatively small, there is an increasing interest in cheap robot platforms for large scale operations such as ecology monitoring. On these platforms size and cost considerations make such an approach unfeasible. Ultimately, even with sophisticated hardware, there are physical limits to optics which are unlikely to be solved in the near future.

In this paper, we describe research towards enabling any-time vision-based SLAM for outdoor robots in changing environments equipped with cheap consumer-grade cameras. The focus is on scenarios where, due to the combination of cost limitations, illumination changes and challenging terrain, the ability to reliably recognize traditional visual features is limited. We present a lightweight visual recognition algorithm based on patch normalization techniques that provides a high degree of invariance to changes in environment conditions such as lighting. A patch tracking algorithm provides visual odometry information, while the pose and visual filtering is provided by the biologically inspired RatSLAM system. We demonstrate the system working at real-time speed in a mixed off-road and urban environment at four different times of day with different environmental conditions—at dawn, during the morning, during a rain shower, and in fading light at dusk. The visual recognition algorithm is able to consistently recognize familiar places despite the changes in conditions. The visual odometry system is able to provide "good enough" motion information to perform reliable mapping and localization over all the datasets when combined with the visual loop closures. We also demonstrate the system is able to map and localize off all the datasets even when restricted to learning visual templates only at one time of day, showing that a single exposure to the environment is enough to enable navigation at the other times of day.

The work presented here builds on previous research including mapping of a suburban road network at different times of day [15, 16] and sequence-based localization on road networks [17, 18]. Unlike the highly constrained nature of a road network, this system is applied in a mixed urban and vegetated environment with off-road areas. The degree of perceptual change encountered in the datasets presented here is qualitatively larger than in [15]. We present a featureless approach to visual matching, rather than the feature and intensity profile-based techniques used in [15]. In contrast to [17, 18], which were localization only studies, we implement a full SLAM solution that calculates and uses motion information to build a map and localize within that map.

2 Approach

In this section we describe the visual recognition and visual odometry algorithms, and give a brief overview of the RatSLAM system.

2.1 RatSLAM System

Processing of the data output by the visual recognition and visual odometry algorithms is performed by the RatSLAM system. RatSLAM is a robot SLAM system based on models of the navigation processes thought to occur in the rodent brain, specifically the rodent hippocampus [19]. RatSLAM has been demonstrated mapping a large road network in static conditions [20] and a smaller road network with moderately varying illumination [15].

The RatSLAM system consists of three modules, shown in Fig. 2. The local view cells encode visual scenes in the environment, with cells incrementally recruited to represent new distinct visual scenes as they are encountered. The pose cells are a network of highly interconnected neural units connected by both excitatory (positive or reinforcing) and inhibitory (negative) connections. They encode an internal representation of the robot's pose state, and filter both the place recognition and self-motion information provided by the visual recognition and visual odometry processes. Finally, the experience map is a graphical map made up of nodes called experiences that encode distinct places in the environment, and connected by transitions that encode odometry information. A graph relaxation algorithm [20] is run

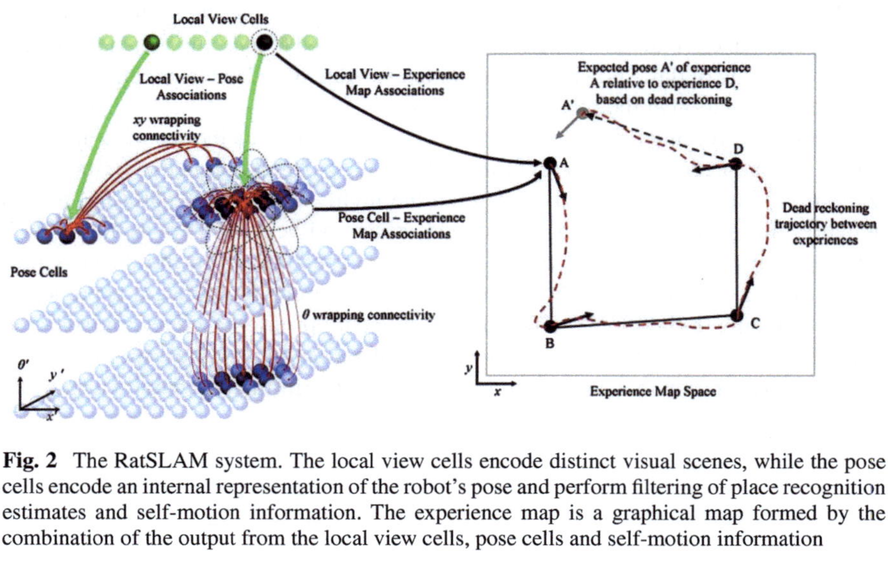

Fig. 2 The RatSLAM system. The local view cells encode distinct visual scenes, while the pose cells encode an internal representation of the robot's pose and perform filtering of place recognition estimates and self-motion information. The experience map is a graphical map formed by the combination of the output from the local view cells, pose cells and self-motion information

continuously on the experience map, resulting in the continuous map evolution seen in the video accompanying the paper and also shown in Figs. 9 and 11. Further information on the RatSLAM system can be found in [20, 21].

2.2 Patch-Based Visual Odometry

The visual odometry system is a modified version of the system deployed on a quad rotor in [22]. The system tracks movement of two image patches to calculate translational speed and yaw of the platform, as shown in Fig. 3a. The primary assumptions are that of a non-holonomic platform at a consistent height above the ground surface. Frame to frame motion of the top patch provides the yaw information and bottom patch motion provides the translational speed. The odometry gain was calibrated by running the car along a known length of ground and calculating the required gain constant, given in Table 2. Patch comparisons were performed by calculating the mean of the intensity difference between each pixel in the patch compared to the corresponding pixel in the previous image. Further implementation details are provided in [22].

2.3 Patch-Normalized Visual Template Learning and Recognition

The visual place recognition process is illustrated in Fig. 3b. Camera images are captured and the bottom half removed. While the ground is useful for patch-based visual odometry, its proximity means that its appearance, when using a "whole of image" based recognition process, is sensitive to slight changes in vehicle pose when closing the loop, which tends to make place recognition brittle.

Once cropped, the image is resolution reduced to 48×24 pixels. Patch normalization is applied to the image in discrete square patches (rather than continuously over the image). Patch normalized pixel intensities, I', are given by:

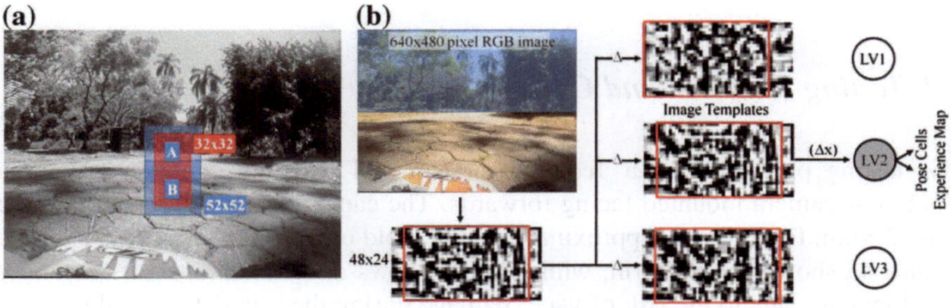

Fig. 3 **a** Patch-based visual odometry and **b** patch-normalized template matching

$$I'_{xy} = \frac{I_{xy} - \mu_{xy}}{\sigma_{xy}} \tag{1}$$

where μ_{xy} and σ_{xy} are the mean and standard deviation of pixel values in the patch of size P_{size} that (x, y) is located within. Mean image differences between the current visual scene and all the learnt visual templates are calculated using a normalized sum of intensity differences, performed over a range of horizontal offsets:

$$D_j = \min_{\Delta x \in [-\sigma, \sigma]} g(\Delta x, i, j) \tag{2}$$

where σ is the template offset range, and $g()$ is given by:

$$g(\Delta x, i, j) = \frac{1}{s} \sum_{x=0} \sum_{y=0} \left(p^i_{x+\Delta x, y} - p^j_{x,y} \right) \tag{3}$$

where s is the area in pixels of the template sub frame. If the minimum difference across all existing templates and relative offsets is larger than a threshold D_t, a new template is learned. Otherwise an existing template is matched, leading to activation of pose cells associated with that visual scene and a possible loop closure event. The range of horizontal offsets provides (assuming the majority of objects in the image are relatively distal) some invariance to camera pose. This invariance enables loop closure even when routes are repeated at slightly different lateral offsets or at different orientations. This capability is important for off-road motion (in contrast to movement along a road network) where repeated paths vary due to environmental change or variation in the path executed by the human or autonomous navigation system.

3 Experimental Setup

This section describes the testing platform, camera, environment and datasets used for this work.

3.1 Testing Platform and Camera

The testing platform was a Team Losi Mini-LST2 remote control car with a Contour + camera mounted facing forwards. The camera has a fisheye wide-angle lens (2.8 mm focal length, approximately 170° field of view) and logged GPS data. Figure 4a shows the platform, while Fig. 4b shows an autonomous version under development. Due to the risk of water damage during the rain dataset and extreme nature of some of the off-road terrain (small logs, deep leaf litter) the non-autonomous platform was used. The video feed and GPS coordinates were logged onboard and processed offline. To reduce the effect of vibration and jerkiness due to the rough terrain and small size of the vehicle, videos were run through a stabilizing filter (VirtualDub Deshaker filter, available at [23], default values used). The use of a stabilizer introduces a one frame lag between image capture and the image being available to the localization and odometry routines, equivalent to 33 ms at real-time speed.

3.2 *Testing Environment and Datasets*

Experiments were run over a 1-week period in an area including the Queensland University of Technology campus and the City Botanic Gardens in Brisbane, Australia (Fig. 4c). The testing area measures approximately 200×200 m and contains a mixture of open grass, pathways, gravel baths, shrubbery, garden beds and buildings. The car was remotely driven by an operator following the vehicle.

A set of four datasets were gathered under a range of environmental conditions and at different times of the day (Table 1). Each dataset repeated the same route, although minor deviations were inevitable due to pedestrian traffic, construction work and the difficulty of the terrain in sections. A single traverse of the route was

Fig. 4 **a** Testing platform, a small but capable off-road enthusiast hobby car with mounted consumer camera, and **b** an autonomous version under development. **c** The vehicle path, with order indicated by the letter sequence. The numbers show the sample frame match locations from Fig. 8. Aerial photos from google maps

Table 1 Dataset descriptions. Times in Australian Eastern Standard Time (AEST)

Dataset name	Time and comments
Dawn	5:45 am. Sun just above local horizon, most areas in shade, excessive sun flare in sections.
Morning	10:00 am. Sun high up in sky, large ground areas in bright sunlight.
Rain	10:30 am. Rain drops on lens, wet ground, overcast and dark.
Dusk	6:45 pm. Sun setting, extremely dark in heavily vegetated areas, significant motion blur and lack of ground texture.

Table 2 Parameters

Parameter	Value	Description
r	32 pixels	Odometry patch size
ς	0.375°/pixel	Yaw gain constant
ν	0.0125 m/pixel	Translational speed constant
ρ	10 pixels	Odometry patch offset range
s	48 × 24 pixels	Template sub frame size
D_t	0.06 pixel	Template learning threshold
σ	4 pixels	Template offset range

approximately 1,310 m in length (calculated by tracing the route on an aerial map) and took an average of approximately 15 min to complete. The car was jammed twice by branches and leaf litter and was stopped temporarily to remove the obstructing objects. These sections of video were cut, resulting in several discontinuous jumps in the footage. Frames were logged at 30 frames per second, with every frame processed by the visual odometry system but only every 5th frame processed by the visual template system, due to the high degree of overlap between neighboring frames. The four datasets are available online.[1]

An attempt was made to use GPS tracking (CEP 10 m) as a ground truth measure. However, due to the heavily vegetated and urban canyon nature of much of the environment, the quality of the GPS tracking was too poor to be useful (far worse than specifications), as shown in Fig. 5.

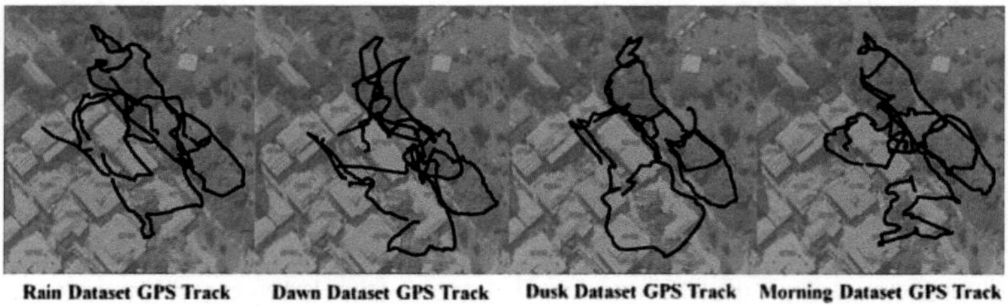

Fig. 5 GPS was unreliable especially under tree cover and around buildings

[1] https://wiki.qut.edu.au/display/cyphy/Michael+Milford+Datasets+and+Downloads

4 Results

In this section we present the visual odometry, place recognition and mapping results as well as computational statistics.

4.1 Visual Odometry

Figure 6 shows the trajectory output by the patch-based visual odometry system for all four datasets, for the common starting pose of $(x, y, \theta) = (0\,m, 0\,m, 0\,°)$. Although the trajectories clearly do not match on a global scale, subsections of the route are similar for all four datasets, such as the small loop (sequence *EFGEF*) in Fig. 4. The differences in the odometry-only trajectories were primarily caused by underestimation of yaw angles and translational speeds in the rain dataset, probably due to reflections in the water lying on the ground, and underestimation of the translational speed in the dusk dataset, due to the poor illumination and consequent lack of ground textures. The differences in translational speed calculations are most easily seen by looking at the length of the first section of each trajectory starting at (0, 0) leading up to the first right turn.

4.2 Visual Place Recognition

Figure 7 displays a graph of the active (recognized or just learnt) visual template versus frame number over all four datasets in the order they were processed, starting with the dawn dataset. The area of the graph below the dashed line is the area in

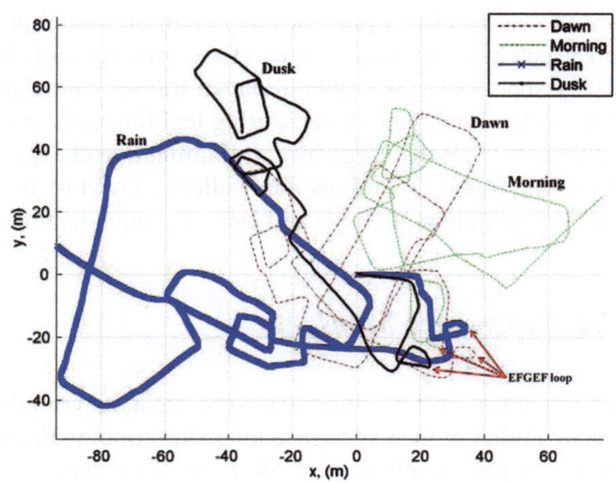

Fig. 6 Vehicle trajectories calculated by the patch-based visual odometry system for the four datasets

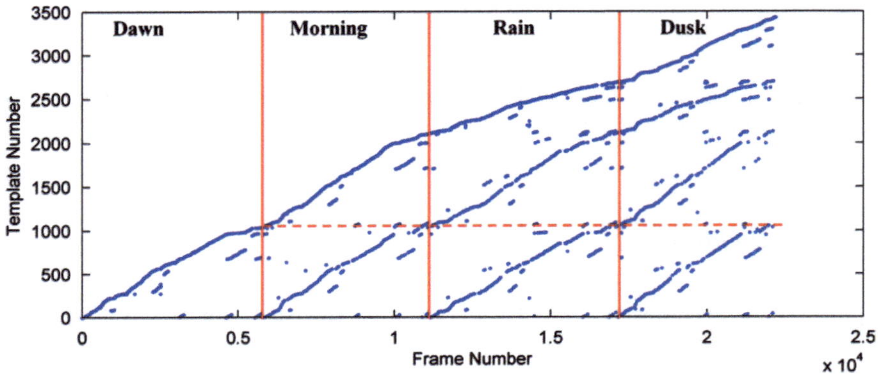

Fig. 7 Visual template learning and recognition over the four datasets

which visual templates learned during the first dawn traverse of the environment were recognized during the subsequent datasets. The system was able to recognize places from the dawn dataset at regular intervals throughout the other three datasets. However, the graph also shows additional templates representative of the subsequent datasets being learnt in parallel and bound to those locations in the map. Learning of new templates was due to the zigzag nature of much of the robot's movement through the environment, resulting in different image sequences each time a section was traversed.

4.3 Matched Frames

Figure 8 shows a selection of ten pairs of frames that were matched by the visual template system for locations throughout the entire route. The original video frames are shown for clarity purposes, although the actual processed images were 48×24 pixel patch-normalized images. The corresponding locations are shown in Fig. 4. The visual system was able to match frames with significantly varying appearance due to (1, 3) sun flare, (2) obscuring leaf litter, (4) motion blur, (5–7) major shadow change, (3, 6, 9–10) large overall illumination change and (10) water on the camera lens. The frames also show the challenge faced by the visual odometry system due to jerky vehicle motion (4) and lack of ground texture in low light (1, 3, 6–10).

4.4 Experience Maps

The final test of the system was to create a map of all four datasets. Figure 9 shows the evolution of the experience map after running through each dataset in order. The map is topologically correct after the dawn and morning datasets, although globally it is warped. The map shrinks slightly, primarily due to the underreporting of

Fig. 8 Matched visual templates over the four datasets. Corresponding locations are shown in Fig. 4

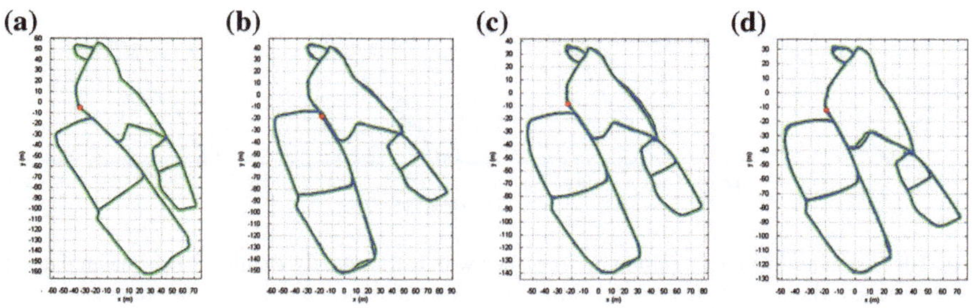

Fig. 9 Experience map evolution over time. Experience maps are from after the **a** dawn, **b** morning, **c** rain and **d** dusk datasets

translational velocity in the dusk dataset and to a lesser extent the rain dataset. However, the constant loop closure within and across datasets ensures the map topology remains correct. The final map layout, although not metrically precise, has the correct topology. A video of the experience map and frame matching processes is available online. [2]

---

[2] https://wiki.qut.edu.au/display/cyphy/Michael+Milford+Datasets+and+Downloads

4.5 SLAM with Only Visual Templates from a Single Time

To test the ability of the system to map and localize with only the visual templates learned at one particular time of day, we conducted an additional experiment where template learning was disabled after the first dawn dataset. From that point onwards the visual template system either recognized a familiar template or reported no match, but did not learn any additional templates (Fig. 10). Figure 11 shows the evolution of the experience map under these conditions. There are three locations where place recognition failed briefly, all at places where the vehicle was turning corners and actual physical paths varied significantly. Although successful loop closures were achieved surrounding these points, the variation in visual odometry meant that the graph relaxation process was not able to draw these trajectories together to correctly overlap. The local topology in these areas is incomplete but correct, meaning navigation could still be achieved but might be suboptimal.

4.6 Compute and Storage

To demonstrate the feasibility of real-time performance on low cost hardware, we present some pertinent computational statistics. The primary storage requirements come from the visual template library. Over all four datasets, a total of 3,353 templates were learned, taking up 5.8 MB of storage. Compute wise, the system performs

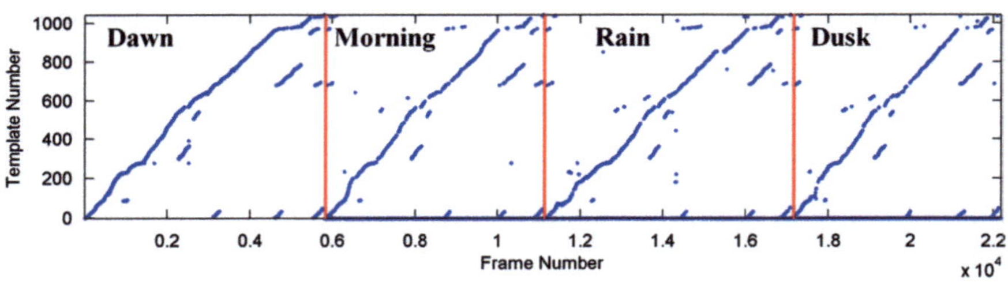

Fig. 10 Visual template recognition performance with learning only enabled for the dawn dataset. Non-matches where a template would normally be learned appear as number zero templates

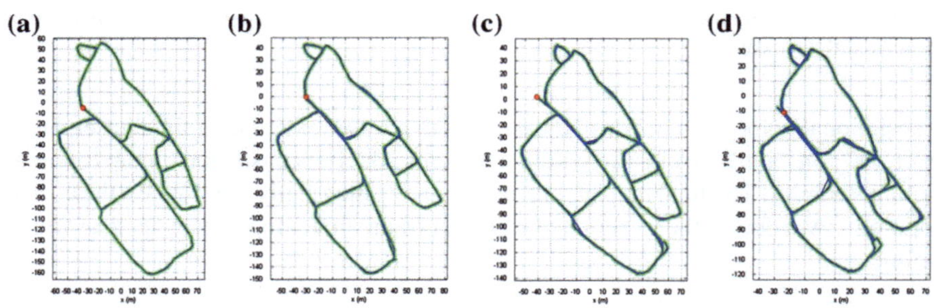

Fig. 11 Experience map evolution with template learning disabled after the first dataset. Map shown after the **a** dawn, **b** morning, **c** rain and **d** dusk datasets

all computation on a fixed time basis, except for visual template comparison and experience map graph relaxation which are both order O(N) (experience map graph relaxation approximates to order O(N) in a typical sparsely interconnected map). Each of these two processes was run on a separate CPU on a standard desktop PC. At the end of the dusk dataset when the system was at maximum load the visual template system was performing 104 million pixel to pixel comparisons per second of data, which ran at real-time speed in unoptimized Matlab code. Experience map graph relaxation is performed with leftover compute cycles. At the end of the experiment, an average of 156 global graph relaxation iterations were performed per second of real-time. This figure can be compared with the 8 iterations per second performed at the end of a previous indoor mapping experiment [24], which was still sufficient to maintain a map that was used for robot navigation. A low power onboard CPU (such as the 1 GHz processor on the robot shown in Fig. 4b) should be capable of running the entire system in real-time for an environment of this size. The RatSLAM system used as the mapping backend has had lightweight versions implemented on a Lego Mindstorms NXT [25] and a small mobile robot called the *iRat* [26], demonstrating the feasibility of running the system on a cheap platform.

5 Discussion

This paper presents a study into the feasibility of using a lightweight, "whole of image" approach to vision-based SLAM on small, low cost vehicles expected to operate in a wide range of environments and in highly varied conditions. The visual processing techniques require no prior training[3], and are demonstrated to enable topological mapping in a varied vegetated and urban environment. Furthermore, the results demonstrate the viability of the approach in a wide range of conditions such as varying time of day and weather. Lastly, the techniques are able to create and consistently localize within a topological map even when it is not possible to obtain high quality visual odometry, such as during the rain and dusk datasets, and when traditional visual features are not available in blurred or very dark images. Here we discuss the limitations of the presented approach and areas for future work.

We used a forward facing camera only, and hence had no ability to close the loop when retracing a route in the opposite direction. However, past work has demonstrated that such a forward facing system can be adapted to utilize omnidirectional imagery [24, 27]. The ability of the system to function with low resolution imagery would also be likely to enable the combination of cheap and compact panoramic imaging rigs with a low cost camera (the mirror could be mass produced with loose specifications). In contrast, much current robot research makes use of high end panoramic imaging setups such as the Point Grey Ladybug 2 ($\sim$ 10, 000 USD). Alternatively, two perspective cameras mounted in opposite directions along the primary vehicle axis would provide forward-backwards recognition capability.

[3] No training is required to generate a topological map. To obtain a map with absolute scale, a short calibration of the translational gain constant is required.

The visual template system is not suited to open-field operation in large open environments where movement is unrestricted and paths are not necessarily repeated. However, this restriction is also present in many vision-based SLAM systems developed to date. One common approach to overcoming this limitation is to combine a SLAM system with absolute positioning information provided by GPS, when available. It is interesting to note that GPS availability and visual SLAM viability tend to be complementary, at least in the system presented in this paper. In the mixed urban and vegetated environment, when GPS was unavailable the vehicle was usually travelling along urban canyons or off-road paths where paths are constrained, situations in which the presented approach works well.

Future work will pursue a number of research directions beyond those mentioned above. The first will be to pursue optimization of the template matching algorithm, which is predicted to be the computationally limiting factor as environments get larger. Secondly, we will investigate how best to add a feature-based mapping technique such as FAB-MAP; FAB-MAP will provide a higher degree of pose invariance when features are detectable, while the visual template method will bind together map locations where features are not reliably detected. Lastly, the quality of the maps exceeds that of those used successfully for robot navigation previously [24], suggesting navigation using these maps is feasible. We will investigate combining state of the art local obstacle avoidance techniques with RatSLAM navigation algorithms [24] in order to enable navigation under challenging and changing environmental conditions.

Acknowledgments This work was supported by an Australian Research Council Discovery Early Career Researcher Award DE120100995 to MM.

References

1. M. Cummins and P. Newman, Highly scalable appearance-only SLAM - FAB-MAP 2.0, in *Robotics: Science and Systems*, University of Washington, Seattle 2009
2. A.J. Davison, I.D. Reid, N.D. Molton, O. Stasse et al., MonoSLAM: real-time single camera SLAM. IEEE Trans. Patt. Anal. Mach. Intell. **29**, 1052–1067 (June 2007)
3. K. Konolige, M. Agrawal, FrameSLAM: from bundle adjustment to real-time visual mapping. IEEE Trans. Robot. **24**, 1066–1077 (2008)
4. D. Burschka and G. D. Hager, V-GPS (SLAM): vision-based inertial system for mobile robots, vol. 1, p. 409–415, 2004
5. H. Andreasson, T. Duckett, A. Lilienthal, Mini-SLAM: minimalistic visual SLAM in large-scale environments based on a new interpretation of image similarity, in *on Robotics and Automation*, ed by International Conference (Rome, Italy, 2007), p. 4096–4101
6. H. Andreasson, T. Duckett, A. Lilienthal, A minimalistic approach to appearance-based visual SLAM. IEEE Trans. Robot. **24**, 1–11 (2008)
7. L.M. Paz, P. Pinies, J.D. Tardos, J. Neira, Large-scale 6-DOF SLAM with stereo-in-hand. IEEE Trans. Robot. **24**, 946–957 (2008)
8. E. Royer, J. Bom, M. Dhome, B. Thuilot, M. Lhuillier, F. Marmoiton, Outdoor autonomous navigation using monocular vision, in *IEEE International Conference on Intelligent Robots and Systems*, 1253–1258 (2005)

9. A.M. Zhang, L. Kleeman, Robust appearance based visual route following for navigation in large-scale outdoor environments. Int. J. Robot. Res. **28**, 331–356 (March 1, 2009)
10. K. Konolige, M. Agrawal, R. Bolles, C. Cowan, M. Fischler, B. Gerkey, Outdoor Mapp. Navig. Using Stereo Vision, **39**, 179–190 (2008)
11. D.G. Lowe, *Object recognition from local scale-invariant features, in International Conference on Computer Vision* (Kerkyra, Greence, 1999)
12. H. Bay, T. Tuytelaars, L. Van Gool, SURF: speeded up robust features, in *Computer Vision - ECCV 2006*, p. 404–417, (ed 2006)
13. I. Kiyoshi, Y. Tomoaki, T. Masahiro, A high dynamic range vision approach to outdoor localization, in *International Conference on Robotics and Automation* (Shanghai, China, 2011)
14. G. Klein and D. Murray, Improving the agility of keyframe-based SLAM, in*European Conference on Computer Vision*, p. 802–815, 2008
15. A.J. Glover, W.P. Maddern, M.J. Milford, G.F. Wyeth, FAB-MAP + RatSLAM: appearance-based SLAM for multiple times of day, in **International Conference on Robotics and Automation** (Anchorage, United States, 2010)
16. M. Milford, G. Wyeth, Improving recall in appearance-based visual SLAM using visual expectation, in *Australasian Conference on Robotics and Automation* (Brisbane, Australia, 2010)
17. M. Milford, G. Wyeth, SeqSLAM: visual route-based navigation for sunny summer days and stormy winter nights, in *IEEE International Conference on Robotics and Automation* (St Paul, United States, 2012)
18. M. Milford, Towards condition-invariant sequence-based route recognition, in *Australasian Conference on Robotics and Automation* (Melbourne, Australia, 2011)
19. M.J. Milford, *Robot Navigation from Nature: Simultaneous Localisation, Mapping, and Path Planning Based on Hippocampal Models* (Springer, Berlin-Heidelberg, 2008)
20. M. Milford, G. Wyeth, Mapping a suburb with a single camera using a biologically inspired SLAM system. IEEE Trans. Robot. **24**, 1038–1053 (2008)
21. M. Milford and G. Wyeth, Persistent navigation and mapping using a biologically inspired SLAM system, Int. J. Robot. Res. 2009
22. M. Milford, F. Schill, P. Corke, R. Mahony, G. Wyeth, Aerial SLAM with a single camera using visual expectation, in *International Conference on Robotics and Automation* (Shanghai, China, 2011)
23. G. Thalin, Deshaker - video stabilizer, 2.5 ed, 2010
24. M. Milford, G. Wyeth, Persistent navigation and mapping using a biologically inspired SLAM system. Int. J. Robot. Res. **29**, 1131–1153 (2010)
25. RatSLAM NXT - http://code.google.com/p/rsnxt08/, ed by The University of Queensland, 2008
26. D. Ball, S. Heath, M. Milford, G. Wyeth, J. Wiles, A navigating rat animat, ed by MIT Press, p. 804–811, 2010
27. D. Prasser, M. Milford, G. Wyeth, Outdoor simultaneous localisation and mapping using RatSLAM, in *International Conference on Field and Service Robotics* (Port Douglas, Australia, 2005)

Gold-Fish SLAM: An Application of SLAM to Localize AGVs

Henrik Andreasson, Abdelbaki Bouguerra, Björn Åstrand and Thorsteinn Rögnvaldsson

Abstract The main focus of this paper is to present a case study of a SLAM solution for Automated Guided Vehicles (AGVs) operating in real-world industrial environments. The studied solution, called Gold-fish SLAM, was implemented to provide localization estimates in dynamic industrial environments, where there are static landmarks that are only rarely perceived by the AGVs. The main idea of Gold-fish SLAM is to consider the goods that enter and leave the environment as temporary landmarks that can be used in combination with the rarely seen static landmarks to compute online estimates of AGV poses. The solution is tested and verified in a factory of paper using an eight ton diesel-truck retrofitted with an AGV control system running at speeds up to 3 m/s. The paper includes also a general discussion on how SLAM can be used in industrial applications with AGVs.

1 Introduction

Simultaneous localization and mapping (SLAM) have been a main research topic in mobile robotics [4]. SLAM algorithms can be run either online or offline. An online SLAM algorithm computes a robot pose estimate at runtime while at the same time

H. Andreasson (✉) · A. Bouguerra
AASS, Dept. of Technology, Örebro University, Örebro, Sweden
e-mail: Henrik.Andreasson@oru.se

A. Bouguerra
e-mail: Abdelbaki.Bouguerra@oru.se

B. Åstrand · T. Rögnvaldsson
CAISR, Dept. of Information Science, Computer and Electrical Engineering, Halmstad University, Halmstad, Sweden
e-mail: Bjorn.Astrand@hh.se

T. Rögnvaldsson
e-mail: Thorsteinn.Rognvaldsson@hh.se

K. Yoshida and S. Tadokoro (eds.), *Field and Service Robotics*, Springer Tracts in Advanced Robotics 92, DOI: 10.1007/978-3-642-40686-7_39,

computing a map of the environment. Offline algorithm on the other hand operate on previously recorded sensor data.

Despite the abundance of various SLAM approaches, the number of reported real world applications for online SLAM methods is small [5]. Offline SLAM, also denoted surveying, is on the other hand used in many applications, for example, for creating a map of reflectors in a factory environment [1]. Another more recent example is reported in [8] where autonomous driving in a parking garage was made possible by first creating a 3D map.

In this chapter we address a different setup where there exists a predefined map with static landmarks that is used at run-time in combination with an online dynamic SLAM approach to localize vehicles in real industrial environments. The main contribution lies in using dynamic features of the environment to compute reliable estimates even when the static features are not seen for longer periods of times. The dynamic features that are utilized in this work are basically goods manipulated by the vehicle. The approach has been implemented and tested on an eight tons diesel forklift providing pose estimates used by the on board AGV controller to smoothly and reliably follow predefined paths at speeds of up to 3 m/s.

1.1 Applications of Online SLAM

The main purpose with online SLAM is to determine a localization estimate within a incrementally generated map. This estimate (and map) could either be provided to a human operator or to an autonomous agent.

One application area that requires online SLAM is exploration of unknown environments such as in urban search and rescue or surveillance tasks [2]. For a human operator an online and up to date map is a valuable source of information to teleoperate a vehicle in a safe manner [16].

From an industrial perspective, exploration needs to be safe which means different (possibly expensive) sensors would be needed to detect different type of obstacles in the environment. The major problem for industrial vehicles is the increased cost for perception. For most AGVs the perception typically consists of a safety classified 2D laser scanner facing the direction of travel mounted at a height less than 10 cm about ground level to be able detect a person lying flat on the ground to follow the required safety regulations (EN 1525 - Safety of industrial trucks - Driverless trucks and their systems). The map will also require further manual intervention and fitted with outer properties, which typically include, predefined paths and loading/unloading areas.

2 Related Work

There is a large amount of different approaches to perform SLAM and the area is simply too wide to cover in this Sect. 3. The work presented here is a classical SLAM approach, where landmark observations and ego-motion estimates are combined to create a map. Landmarks are also used in extended Kalman Filter (EKF) based SLAM approaches [13]. The main drawback, however, of an EKF based approach is the computational complexity $O(n^2)$, where n is the number of landmarks, which makes the method only applicable online for smaller sized maps. Another drawback with an EKF approach is errors due to linearization.

Other landmark based methods are, for example, FastSLAM [12], a particle filter approach where each particle consists of a robot trajectory and where each landmark is treated independently. SEIF based SLAM [15] utilizes the sparseness of the information matrix instead of the full correlation matrix used in EKF. The Treemap algorithm by Frese [6], is an efficient approach $O(log(n))$, which uses a hierarchical treelike structure to subdivide the map into different regions which allows for an efficient update. Graph-SLAM approaches addresses the full-SLAM problem and are typically only used for offline mapping, however, there are some works (see, for example, Kretzschmar [7]) which enable Graph-SLAM approaches to be used over long time periods.

The work reported in Meyer et al. [10], is related to our work where they use a particle filter based SLAM approach that combines a predefined static map of buildings close to a parking lot and online dynamic map containing parked cars to facilitate the task of localization.

3 Gold-Fish SLAM

This paper presents an approach of how SLAM can be used in a scenario where semi-dynamic landmarks (stacks of paper reels) are used in combination with static landmarks estimated (concrete pillars) to localize an eight ton diesel truck retrofitted with an AGV system. The environment used for evaluation is a warehouse for paper reels with an area of approximately 8,000 m^2.

The key idea is that the vehicle position is initially known with respect to the static map and observations which can be related to the static map are used to correct the vehicle pose estimate and the position estimate of the semi-dynamic landmarks.

Besides improving odometry between areas with sparse static landmarks, SLAM is used to limit the pose uncertainty if the robot operates in the same area without seeing any static landmarks. In addition, landmarks might consists of objects that the vehicle manipulates, in our specific context: paper reels, therefore it is indeed useful to estimate their position and to have them in a map.

The vehicle does not have to keep track of all seen dynamic landmarks. It only needs to keep a limited set of them, which is reminiscent of the popular belief of a gold-fish memory.

3.1 Requirements from the AGV Systems

AGVs require very good absolute localization estimates, which is used by the on board controller to follow paths. If an obstacle occurs, the AGV will stop and wait until the path is free. Therefore there is no requirement of path re-planning or obstacle avoidance behavior. Accuracy in providing localization estimate is very important for safe navigation, since industrial environments are cluttered with goods and other vehicles. It is also important for productivity because bad localization estimates can lead to unnecessary stops, for example, if a fleet of AGVs are used, other AGVs could be detected as obstacle and together cause 'dead locks'. On the other hand, on research platforms the inaccuracy of the absolute position localization estimates can be taken care of by the obstacle avoidance module, for example, vector field histogram (VFH) [17] or nearest diagram (ND) [11]. A popular localization approach used in research is Monte Carlo based localization [3], which often has a rather low update rate and where the output is typically not smooth enough to be directly fed into a control system. The output is instead used to determine if the goal state of the obstacle avoidance has been achieved where as the actual control are performed by the obstacle avoidance mechanism.

Just taking any SLAM approach and applying it directly in an industrial environment would simply not work. Firstly, the pose estimates has to be given within a specific coordinate system and bounds. This makes it clear that we also have to "anchor" the global localization into the SLAM representation and also to bound the uncertainty independently on the distance traveled.

Secondly, smooth pose estimates needs to be provided all the time, however, this is indeed related to loop closure. Unless the localization estimate are 'on the spot', jumps in the pose estimates will occur. Depending on the environment, the sensors, and the performance of the data association this jump could be large. This all comes down to the same conclusion as above; that the uncertainty in the pose estimate has to be bounded.

The uncertainty in the pose estimate is directly affected if the robot sees statical landmarks. In case the robot pose uncertainty grows beyond a predefined application specific threshold the vehicle will stop.

3.2 Overview and Difference Compared to Online SLAM

An overview of the proposed method is depicted in Fig. 1 where input (static landmark maps, initial pose estimate and sensory data) are shown in rounded shaped boxes. The output fed into the control system is the estimated pose X_{pose} and the output of the "Safety stop" box. In the following we give more details about the functionalities of the main boxes.

- Incremental Pose Estimator: Integrates the odometry data to estimate the relative pose and its covariance.
- Prediction: Progresses the current pose estimate according to the kinematic model of the vehicle.
- Static Submap Selection: Based on the current pose estimate selects a subset of static landmarks that can be potentially observed by the system.

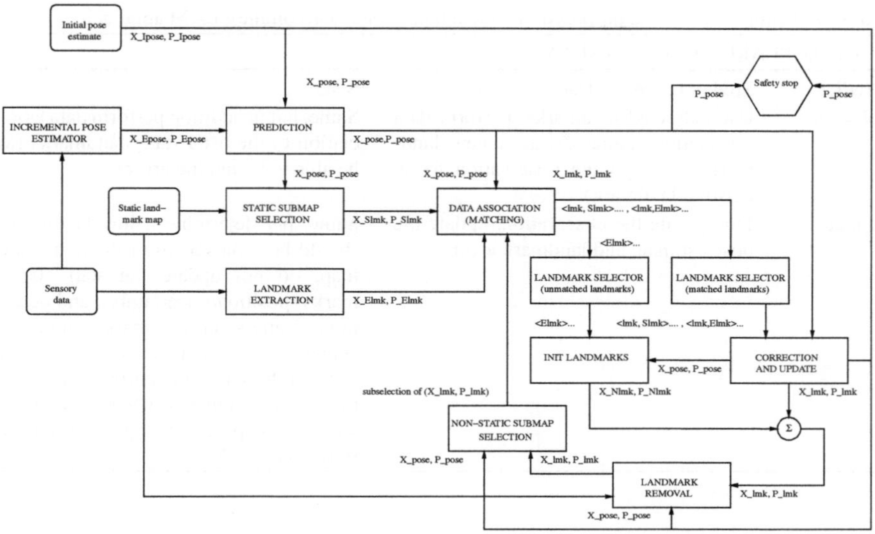

Fig. 1 Schematic overview of the system. The *rounded shaped boxes* indicates inputs to the system. The *arrows* indicates the flow of the state variables X and the corresponding parts in the correlation matrix P. The notation $pose$, $Ipose$, $Epose$ corresponds to the current predicted/estimated robot pose, initial pose, incremental pose estimate (from odometry) respectively; lmk, $Elmk$, $Slmk$, $Nlmk$ corresponds to the current predicted/estimated landmark position, the extracted landmarks, static landmarks and newly initiated landmarks

- Data Association: find the correspondence between landmarks (static and dynamic) and their observations.
- Init Landmarks: creates new dynamic landmarks for observations that were not matched in the data association step.
- Correction and Update: Corrects the estimates of both the vehicle pose and landmarks positions based on the observations.
- Landmark removals: Removes landmarks from the dynamic map based on age, uncertainty, size of the map and the max number of landmarks that can be kept.
- Non-static submap selection: Based on the current pose estimate selects a subset of dynamic landmarks that can be potentially observed by the system.

To avoid that the size of the static map influences the computational complexity of the system, static landmarks can be added/replaced/removed from the SLAM backend by looking at the current estimated pose. Static landmarks that are simply invisible or out of range for the sensors can safely be removed. Whether a static landmark should be added can be determined from the data-association step. The central advantage here is that the size of the sum of dynamic and static landmarks will be bounded independently on the size of the environment. One key requirement, however, is that the global pose of the vehicle is approximately known when the

Table 1 Comparison of the proposed method against a standard online SLAM approach

Online SLAM vs Gold-Fish SLAM		
Predict	Predict based on dead reckoning	Same
Observe	Use extracted landmarks, perform data association (either create a new landmark or assign the observation to an existing landmark)	Same, but in addition perform data association to the used static landmarks (to be able to bound the error)
Update	Incorporate the observation, update the pose estimate and landmark estimates	Same, but determine if any landmarks should be removed from the non-static map, do not update the static landmarks (position/uncertainty), and determine if any static landmarks should be added/replaced/removed in the SLAM backend based on the current pose estimate - this simply to keep the computational complexity independent on the static map size)

proposed localization system starts. In Table 1 the proposed idea is compared with an online SLAM approach to better illustrate the differences.

3.3 Landmark Extraction

Two types of landmarks are extracted from the 2D laser scanner data: paper reels and corners. Paper reels are the dynamic landmarks, while the pillars, created from a set of corners forms the static landmarks, see Fig. 2. Another major problem utilizing reflectors apart from the need to install additional infrastructure etc. in this type of environment (see Fig. 4) is they will be occluded by stacks of paper reels, therefore using reflectors alone is simply not feasible, but could be used as a static landmark in the proposed method.

3.3.1 Paper Reel Extraction

Paper reels appear as sets of point of a circular arc, note that paper reels are in our settings always standing upright. To determine the position as well as the radius of each reel, a method based on Taubin's work for fitting a circle to data points [14] is used.

To validate the extracted reels, all segments with too few points are ignored. Moreover, if the estimated reel radius is not within a predefined interval or the mean square fit (MSE) is above a threshold, the reel is rejected. To avoid to fit reels into a concave area, the estimated reel position has to be located further away than the

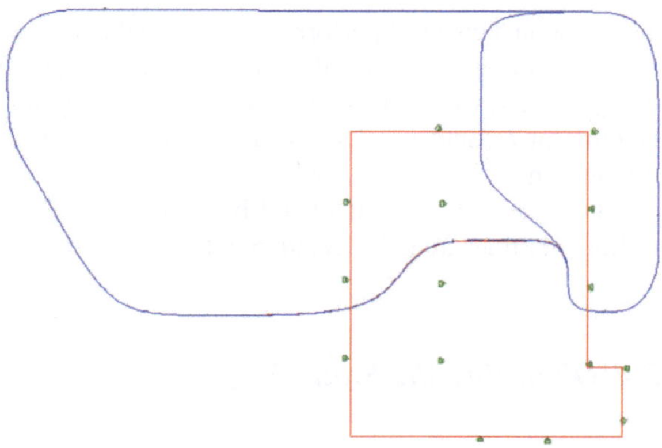

Fig. 2 The predefined paths used in the autonomous localization experiment (*blue*). The (*green*) dots represents the location of the installed reflector and the enclosed (*red*) *area* is where the reflector based localization potentially be used depending on how many reflectors are visible due to occlusions of paper reel stacks

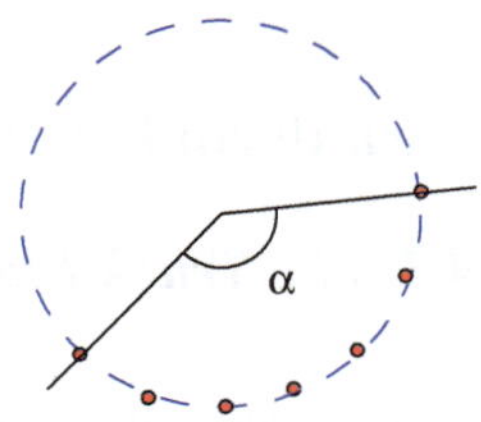

Fig. 3 The utilized *arc angle* α of the fitted *circle*, which could be interpreted how much of the actual *circle* was seen in the laser data. This measure is also useful to reject wrong estimates. A small value indicates that the data typically do not come from a reel (or a heavily occluded reel). A large value indicates that a *corner* like object are fitted instead

mean position of the segment. The utilized arc angle on the circle, see Fig. 3, has to be located in a specified interval. Basically, a small value indicates that this is not likely to be a reel or that it is heavily occluded. A high value instead indicates that this is a corner like object.

An uncertainty estimate is calculated based on the distance between the laser points and the fitted circle along the radial direction.

3.3.2 Pillar Extraction

Segments that were not detected as reels are further processed to check if they form corners. We do not explicitly extract pillars as landmarks but instead use corners directly. A corner is defined as an intersection of two walls (or lines in 2D) at an intersection of 90° . The first step is to check that the size of the segment corresponds to a pillar and are within a specified interval. The segment is split into to two sub parts S_a and S_b at the location of the highest curvature. For both segments (S_a and S_b) a line is obtained using least square fit to each segment resulting in two lines L_a and L_b. If the MSE of the line fit is less than a predefined threshold for both segments an orthogonal check is performed to assure that the two lines are approximately 90 ° apart $\pm$ another predefined threshold. The corner position is set to be the intersection of L_a and L_b, whereas the orientation is set using the normalized mean angle of the heading of L_a and L_b. The extracted corner consists of both a 2D position and an orientation.

The uncertainty is calculated based on the distance between the data and two orthogonal lines at $\pm$ 45° relative to the orientation of the corner.

3.4 Obtaining the Static Map

The static map was created by manually driving the truck around in the warehouse while collecting odometry, laser data from the safety sensor and also from the reflector based localization laser. The latter was used to more easily align the predefined map used in the AGV controller consisting of reflector poses and predefined paths. After the map was built, only the corners were saved to the static map.

4 Platform Description and Environment

4.1 The Truck Platform

The platform is based on a modified Linde H 50 D diesel forklift truck that has a load capacity of 5,000 kg (see Fig. 4). The standard version of the truck was modified by shortening the mast and replacing the forks with a clamp. The truck was retrofitted with an off-the-shelf AGV control system developed by Kollmorgen. The AGV control system comprises a set of hardware and software components (PC, IO modules, field bus controller, rotating laser ranger, etc.). The control system interfaces the actuators and sensors of the truck through the already built-in local

Fig. 4 *Left* The industrial truck used in the MALTA project. The truck is retrofitted with an AGV controller. The *bottom* lasers are used as safety sensors and in this work also for landmark detection. *Right* Stacked paper reels waiting to be loaded and one of the concrete pillars

CAN network. To detect paper reels, other landmarks and obstacles, three extra SICK laser rangers were incorporated into the truck (see Fig. 4), in the following experiments only the low mounted front and rear lasers were used.

The main functionality of the AGV controller is to navigate the truck from an initial location to a goal location. To do so, an operator defines and uploads a layout of drivable paths specified as collections of line segments and B-splines, see also Fig. 2. The controller achieves navigation tasks by following an appropriate path. The position of the truck can be tracked using a spinning laser (installed on the top of the truck) and reflective markers installed in the environment or by specifying external pose estimates, which in these experiments were provided with the SLAM based localization system. Here it was found to be very critical to give a smooth and fast update rate (10Hz) to the AGV controller.

5 Experiments

The environment used in the experiments are paper reel warehouses, see Fig. 4.

Due to the lack of ground truth data, the comparison is done towards an autonomous run where the platform was controlled using localization estimates from the proposed system.

For the experimental evaluation two approaches were used. Treemap [6] and a standard EKF based method. Treemap operates on a very efficient tree-structure and have good computational properties. Treemap is a generic backend for any least square problems, however, in this approach the Treemap method is used in a similar way as a standard EKF method would be used by marginalizing out all robot poses except the last one. Input to the method are: odometry estimates with uncertainty and relative observation to landmarks with uncertainty, where as the output consists of an estimate of the robot pose including estimates of all landmarks. Due to the dependency between already added landmarks it is not straight forward to remove landmarks in Treemap therefore an EKF based approach was used in some evaluations.

The classical drawbacks of an EKF based approach are partly avoided in the proposed approach. The linearization error will not be problematic since the bounding of the error in the estimate and the complexity is somewhat covered by having a bounded map size.

5.1 Relating SLAM Estimates to a Predefined Map

Even though the estimates and the map are consistently built, the problem is to express the current estimates in the reference frame of the predefined map. Typically any reasonable offset in x and y directions do not cause any problem, however, only a very minor offset in orientation θ will give tremendous problems even if the distance

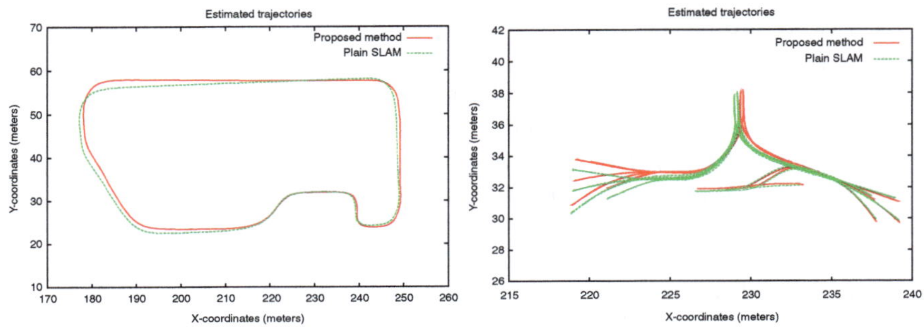

Fig. 5 Trajectories of using a 'plain' SLAM method without any pre existing landmarks and the proposed method. *Left* A complete loop. *Right* Using a pick'n'place scenario

Table 2 Difference in positions and heading in the proposed method while changing the amount of used dynamic landmarks

Distances/max nb landmarks	0	10	20	30	40	60	80	100
avg. abs. position error	0.20	0.19	0.19	0.17	0.15	0.16	0.15	0.16
avg. abs. angular error	0.95	0.71	0.69	0.56	0.42	0.40	0.38	0.38

is rather short, which can be seen in Fig. 5 where the truck was autonomously driven around in a predefined path consisting of a loop with a path length of 204 m and the position difference is up to 2.5 m.

In a pick'n'place scenario (see Fig. 5) paper reels were moved from a loading zone to an unloading zone using relative measurements to determine the path to drive to each reel and where each reel should be placed. Here one could argue that, given an initial estimate of the pose a plain-SLAM method would work, however, here instead the globally defined loading/unloading zones needs to be transferred into the plain-SLAM coordinate frame.

5.2 Changing the Amount of Dynamic Landmarks

The proposed method 'Gold-fish SLAM' implies that we need to constantly forget parts of the dynamic landmark map. To evaluate the impact of the number of dynamic landmarks used, one of the autonomous runs was evaluated offline. The system was rerun using recorded raw sensor data approximately in real-time for each evaluation. The results are presented in Table. 2.

In the evaluated data set the number of seen static landmarks is enough to create a consistent map, however due to the smoothness criteria this approach would cause difficulties as seen in Fig. 6 where the avg. absolute position and angular error are shown using 0, 40 and 100 dynamic landmarks. From this figure it is clear that

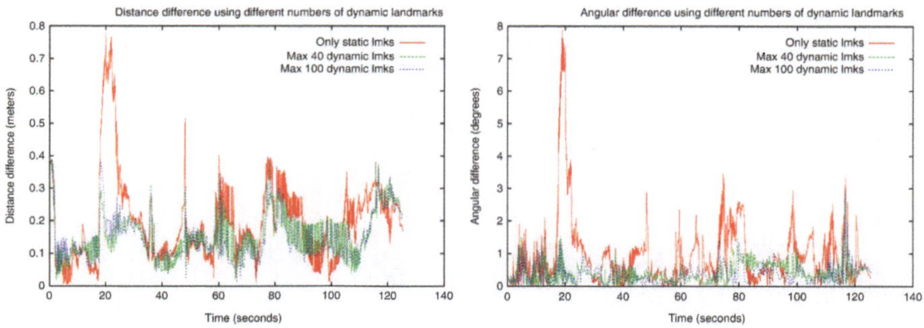

Fig. 6 Absolute difference between the estimated pose in an autonomous run with different number of dynamic landmarks. *Left* distance difference. *Right* angular difference

despite the rather accurate average error, there is significant difference especially in the angular estimate, which was an important control input in our platform.

5.3 Using Sparser Static Landmark Maps

In the previous section, the complete static landmark map was used in the evaluation. Here, the number of static landmarks used are instead decreased to check the system dependency on the amount of static landmarks. To simplify the presentation, the max number of dynamic landmark used was always 40. The accuracy of the system depends on the amount of observed static landmarks and their location rather than the number of landmarks in the static map. Therefore the evaluation also contains the ratio r of observation containing at least one static landmark (showing the number of static observation we have) and the number of unique static landmarks observed (indicating the map size actually used). In the evaluation 12 submaps were randomly created by subsampling the full static map. In Fig. 7, the angular and position difference are shown toghether with the number of static landmarks observed at each time for two different static sub-maps. The key difference is that in the left figure static landmarks are not seen for a longer period despite roughly the same map size.

Another result is depicted in Fig. 8, where the ratio r, the angular and position difference are plotted with the number of used static landmarks. One interesting aspect is that the angular difference seems not to depend on the number of observed static landmarks. This indicates that even using few static landmarks the orientation of the vehicle can be determined, whereas to correctly determine the position is more sensitive to the density of static landmarks and preferable that the multiple landmarks are visible at the same time.

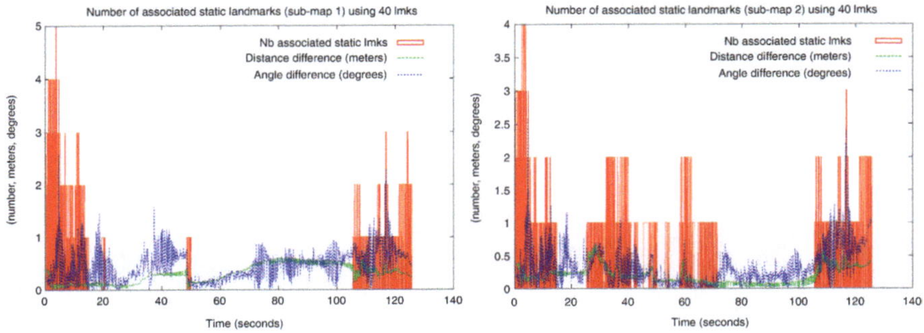

Fig. 7 Absolute difference between the estimated pose compared to an autonomous run using 40 dynamic landmarks with sub-sampled static landmarks maps. *Left* Sub-map 1 with 12 used landmarks (position/angular difference : 0.3 m/0.46). *Right* Sub-map 2 with 14 used landmarks (position/angular difference : 0.20 m/0.33°)

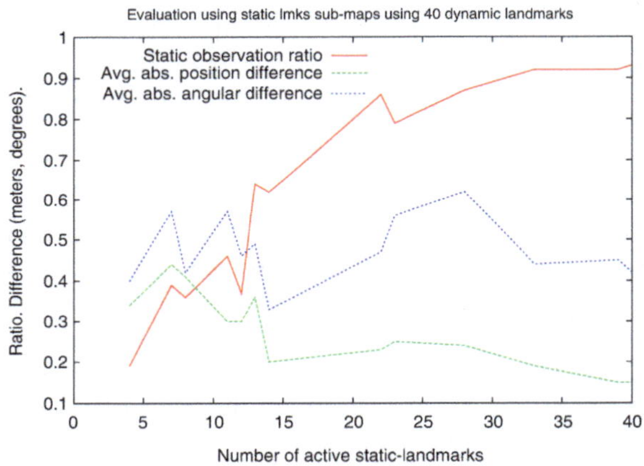

Fig. 8 Absolute difference between the estimated pose compared to an autonomous run using 40 dynamic landmarks with sub-sampled static landmarks maps. *Left* Sub-map 1 with 12 used landmarks. *Right* Sub-map 2 with 14 used landmarks

5.4 An Autonomous Localization Loop

One critical aspect was to get the complete system to run in real-time on the platform traveling up to 3 m/s. A movie of one of the autonomous localization runs is located at the projects web page [9]. This utilized Treemap as a backend without limiting the number of dynamic landmarks.

6 Conclusions and Future Work

This paper presents an online SLAM approach to localize an AGV in a warehouse or factory with static and dynamic landmarks. The results are so far preliminary and rather show a proof of concept. The evaluation should be extended with data covering larger areas, longer operational time and with reasonable ground truth. Future work includes investigation of methods regarding selection of relevant dynamic landmarks. An interesting work would be to evaluate the approach on other types of environments.

Acknowledgments The Authors would like to thank the teams from Kollmorgen, Linde Material Handling, and Stora Enso for their contribution to the MALTA project. The Authors would also like to acknowledge the support of the Swedish KK foundation.

References

1. K. Åström, Where am I and what am I seeing? Algorithms for a laser guided vehicle, Master's thesis, Centre for Mathematical Sciences LTH, Lund University, Sweden, 1991
2. M. Batalin, G.S. Sukhatme, Coverage, exploration and deployment by a mobile robot and communication, network. Telecommun. Syst. J. **26**(2), (2004)
3. F. Dellaert, D. Fox, W. Burgard, S. Thrun, Monte Carlo Localization for Mobile Robots, in *Proceedings of the IEEE International Conference on Robotics & Automation (ICRA), IEEE* (1999)
4. H. Durrant-Whyte, T. Bailey, Simultaneous localisation and mapping (slam): Part i the essential algorithms. IEEE Robot. Autom. Mag. **2**, 2006 (2006)
5. U. Frese, R. Wagner, T. Röfer, A slam overview from a user's perspective. KI-Zeitschrift **24**(3), (2010), http://www.springerlink.com/content/p052055h1377gp85/
6. U. Frese, Treemap: An $O(log n)$ algorithm for indoor simultaneous localization and mapping. Auton. Robot. **21**(2), 103–122 (2006)
7. G.G. Henrik Kretzschmar, C. Stachniss, Lifelong map learning for graph-based slam in static environments. KI-Zeitschrift **24**(3), (2010)
8. R. Kummerle, D. Hahnel, D. Dolgov, S. Thrun, W. Burgard, Autonomous driving in a multi-level parking structure, in *Proceedings of the 2009 IEEE International Conference on Robotics and Automation (ICRA-09)*, Kobe, Japan (2009) (To Appear0.
9. MALTA Project: http://aass.oru.se/Research/Learning/malta/index.html
10. D. Meyer-Delius, J. Hess, G. Grisetti, W. Burgard, Temporary maps for robust localization in semi-static environments, in *Proceedings of the IEEE/RSJ International Conference on Intelligent Robots and Systems (IROS)* Taipei, Taiwan (2010).
11. J. Minguez, A. Member, L. Montano, Nearness diagram (nd) navigation: collision avoidance in troublesome scenarios. IEEE Trans. Robot. Autom. **20**, 2004 (2004)
12. M. Montemerlo, S. Thrun, D. Koller, B. Wegbreit, Fastslam: A factored solution to the simultaneous localization and mapping problem, in *Proceedings of the AAAI National Conference on Artificial Intelligence*, pp. 593–598, AAAI (2002)
13. R. Smith, M. Self, P. Cheeseman, Estimating uncertain spatial relationships in Robot. pp. 167–193 (1990)
14. G. Taubin, Estimation of planar curves, surfaces, and nonplanar space curves defined by implicit equations with applications to edge and range image segmentation. IEE Trans. Pattern Anal. Mach. Intell.13(11), 1115–1138 (1991). DOI: 10.1109/34.103273.
15. S. Thrun, W. Bugard, D. Fox, *Probabilistic robotics* (MIT Press, Cambridge, 2005)

16. T. Tsubouchi, A. Tanaka, A. Ishioka, M. Tomono, S. Yuta, A slam based teleoperation and interface system for indoor environment reconnaissance in rescue activities, in *Proceedings of the IEEE/RSJ International Conferrence on Intelligent Robots and Systems (IROS)* (2004)
17. I. Ulrich, J. Borenstein, Vfh+: Reliable obstacle avoidance for fast mobile robots (1998)

Design, Development, and Mobility Test of an Omnidirectional Mobile Robot for Rough Terrain

Genya Ishigami, Elvine Pineda, Jim Overholt, Greg Hudas and Karl Iagnemma

Abstract Omnidirectional vehicles have been widely applied in several areas, but most of them are designed for the case of motion on flat, smooth terrain, and are not feasible for outdoor usage. This paper presents an omnidirectional mobile robot that possesses high mobility in rough terrain. The omnidirectional robot employs four sets of mobility modules, called active split offset caster (ASOC). The ASOC module has two independently-driven wheels that produce arbitrary planar translational velocity, enabling the robot to achieve its omnidirectional mobility. Each module is connected to the main body of the robot via a parallel link with shock absorbers. In this paper, a design and development of the ASOC-driven omnidirectional mobile robot for rough terrain are described. Also, a control scheme that considers a kinematics of the omnidirectional mobile robot is presented. The omnidirectional mobility of the robot regardless of ifs heading direction is experimentally evaluated based on a metric called omnidirectional mobility index.

1 Introduction

An omnidirectional vehicle is capable of moving in any arbitrary direction and performing complex maneuvers that cannot be achieved by typical Ackermann steered wheeled vehicles. Omnidirectional vehicles have been investigated and

G. Ishigami (✉)
Dept. of Mechanical Engineering, Keio University, Yokohama, Kanagawa, JAPAN
e-mail: ishigami@mech.keio.ac.jp

E. Pineda · K. Iagnemma
Massachusetts Institute of Technology, Cambridge, MA, USA
e-mail: e_pineda@mit.edu

K. Iagnemma
e-mail: kdi@mit.edu

J. Overholt · G. Hudas
U.S. Army Tank Automotive Research, Development and Engineering Center,
Warren, MI, USA

K. Yoshida and S. Tadokoro (eds.), *Field and Service Robotics*,
Springer Tracts in Advanced Robotics 92, DOI: 10.1007/978-3-642-40686-7_40,

widely applied in many practical areas, such as mobile robotic bases for research, materials handling vehicles for logistics, and wheelchairs [1–3]. Most omnidirectional vehicles to date have employed specialized wheel designs, including roller, Mecanum, or spherical wheels [4–10] (Fig. 1). These wheels, however, are not practical for the use in rough terrain because of the following reasons: small rollers may be easily clogged with debris or dirt. They also have a constraint on wheel diameter and width relative to the small slender rollers, and thus they cannot tolerate with large loads or cannot produce enough thrust power.

This paper introduces an omnidirectional mobile robot in rough terrain driven by active split offset caster (ASOC), which was initially proposed in [11] for indoor use. The ASOC module employs two independently drivable wheels. The ASOC modules are integrated in the robot main body with suspension mechanism, enabling an agile traversability on uneven terrain (Fig. 2). In this paper, design and development of the ASOC-driven omnidirectional mobile robot for rough terrain are described. A kinematic control scheme that can coordinate each ASOC module (i.e., wheel velocities) to achieve a desired maneuver of the omnidirectional robot is also presented. Experimental tests with different configurations of the robot were performed to evaluate the omnidirectional mobility of the robot regardless of ifs heading direction.

Fig. 1 Examples of roller wheels [9] and Mecanumm wheels [10]

Fig. 2 ASOC-driven omnidirectional mobile robot

This paper is organized as follows: Sect. 2 introduces the design analysis of the ASOC module and the omnidirectional mobile robot; Sect. 3 describes the kinematic control of the robot; the experimental test for the mobility evaluation of the robot is presented in Sect. 4

2 ASOC-driven Omnidirectional Mobile Robot

2.1 Active Split Offset Caster: ASOC

Figure 3 shows the ASOC module developed in this work. The module has independently-driven dual wheels that are attached to the chassis of the module. The chassis is connected with the battery housing via a freely rotatable joint (around the roll axis of the assembly). The roll axis works to let the wheel pair maintain its contact on sloped or bumpy terrain surfaces. Another freely rotatable axis, which is located at the top of the battery housing, enables the module to rotate 360° around the pivot (yaw) axis of the module assembly. The angle of rotation of the roll and pivot axes can be measured by the potentiometers.

The ASOC module can produce arbitrary (planar) translational velocities at a point along its pivot axis, by independently controlling each wheel's velocity. Two or more ASOC modules attached to a rigid robot body can thus produce arbitrary translational and rotational robot velocities. A control scheme is introduced in Sect. 3.

2.1.1 Kinematic Isotropy for ASOC Design

All omnidirectional mobile robots are able to travel in any planar direction instantaneously. However, while some omnidirectional mobile robots exhibit preferred

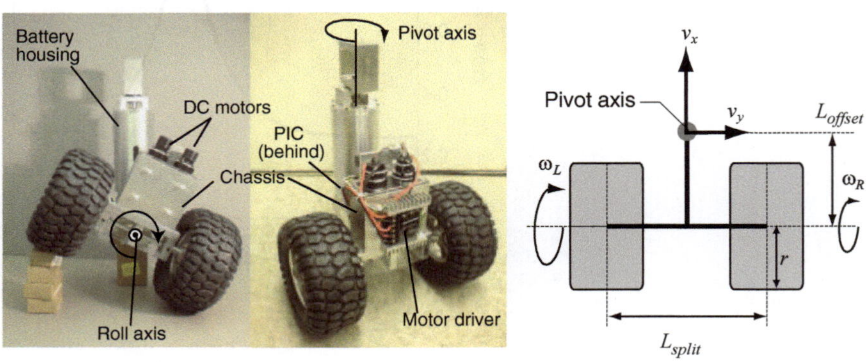

Fig. 3 Assembly of ASOC (*left*) and its schematic illustration (*right*)

directions of travel, others exhibit equal mobility characteristics in all directions, i.e., "isotropic mobility". Hence, a kinematics isotropy metric is used to quantify the system's omnidirectional mobility.

Kinematic isotropy is defined as the condition in which a robot possesses a constant input velocity/output velocity ratio for all possible output velocity directions [12]. An isotropy metric is a measure of how close a robot is to the isotropy condition, and increases from 0 for a singular configuration (i.e., purely anisotropic, or non-omnidirectional) to 1 for kinematic isotropy. Ideally, an omnidirectional robot should possess a metric value of 1 for any configurations (i.e., wheel allocation with respect to the robot body), and thus not have a preferred direction of travel.

To evaluate the effect of the ASOC module parameters on the isotropy, the kinematic parameters of the module such as L_{offset}, L_{split}, and wheel radius r were selected. As illustrated in Fig. 3, the Jacobian between the wheel angular velocities (ω_L and ω_R) and the translational velocities of the ASOC module is given as follows:

$$\begin{bmatrix} v_x \\ v_y \end{bmatrix} = r \begin{bmatrix} 1/2 & 1/2 \\ -L_{offset}/L_{split} & L_{offset}/L_{split} \end{bmatrix} \begin{bmatrix} \omega_L \\ \omega_R \end{bmatrix} \quad (1)$$

where v_x and v_y are the longitudinal and lateral translational velocities of the ASOC at the pivot axis, respectively. From the above equation, the wheel radius can be neglected for calculating the ratio of the eigen values, and therefore, the module isotropy is independent of the wheel radius.

Figure 4 shows an isotropy values over a range of L_{split}/L_{offset}. It can be seen that the isotropy depends on both L_{offset} and L_{split} and has a maximum value when L_{split}/L_{offset} ratio is equal to 2.0. For the practical design of the ASOC module, the ratio of L_{split}/L_{offset} becomes 2.07 ($L_{split} = 0.228$ m, and $L_{offset} = 0.110$ m).

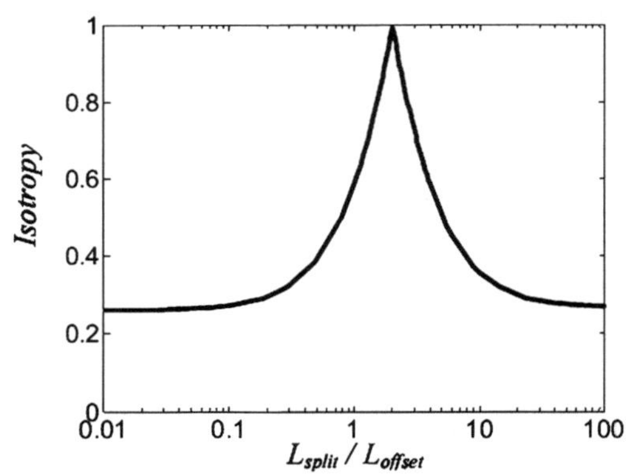

Fig. 4 Average isotropy for a robot driven by ASOC modules as a function of L_{split}/L_{offset} (from [13])

2.1.2 ASOC Components

The ASOC is a self-sustained "robotic" module, comprised of a power supply, two actuators, a microcontroller, a wireless device (Xbee), and a motor driver. The total weight of the module is 5.6 kg.

An off-the-shelf rubber coated wheel is used; the wheel radius is 0.085 m and width is 0.086 m. Two actuators (MAXOS RE35) with planetary gear heads are located vertically between the wheels (see Fig. 3). The motor torque is transferred to each wheel via bevel gear. The maximum speed of the ASOC module is approximately 2.2 m/s (8.0 km/h).

A microcontroller, PIC18F4431 (40 MHz), executes a local feedback control of wheel angular velocity. The wheel angular velocity is measured by a tachometer mounted on the tip of actuator. The angle of rotations around the pivot and roll are measured by potentiometers on their axes. These sensory data are handled by the microcontroller and sent to the onboard computer mounted on the robot body via Xbee wireless communication device.

A serially connected 6 cells of Li-Ion batteries (3.2 Ah) allocated inside the battery housing activates each module up to approximately two hours for continuous driving. The supplied voltages are conditioned by DCDC converters.

2.2 System Overview of Omnidirectional Mobile Robot

The omnidirectional mobile robot developed in this work consists of a main body and four ASOC modules. The ASOC modules, that are evenly spaced with 90° intervals one another, are connected to the main body via parallel links with shock absorbers (Fig. 5). The suspension mechanism allows the robot to operate on rough terrain with moderate terrain adaptability. The total weight is 35 kg.

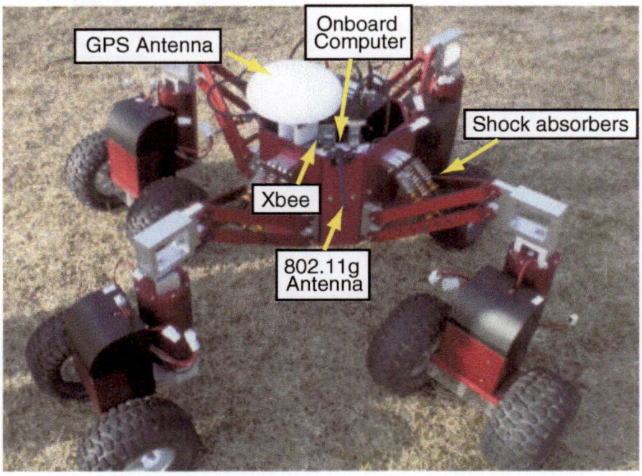

Fig. 5 The ASOC-driven omnidirectional mobile robot

The onboard computer, Gumstix Overo Earth (600 MHz) running Linux, supervises all ASOC modules via Xbee wireless links such that the onboard computer kinematically coordinates ASOC modules to achieve the omnidirectional maneuver. The GPS data is collected as ground truth data for outdoor experiments. These data are sent to an operator via IEEE 802.11 g, along with ASOC motion data.

To avoid the interference between neighboring ASOCs, the distance D between the centroid of the main body and the ASOC pivot axis must be more than the square root of two times of the ASOC module workspace, $r_{workspace}$ (Fig. 6). This workspace radius is the length from the pivot axis to the most distal point of the wheel edge. Therefore, the geometric constrains of the robot configuration is as follows:

$$D > \sqrt{2}r_{workspace} = \sqrt{2(L_{offset} + r)^2 + (L_{split} + b)^2/2} \tag{2}$$

where b is the wheel width. The distance D can be regulated by the length of the parallel links and stiffness of the shock absorbers. The geometric constraint in the above equation determined the dimension of the robot: $1.05 \times 0.96 \times 0.39$ m in its traveling configuration and $0.81 \times 0.81 \times 0.39$ m in its stowed configuration (Fig. 7).

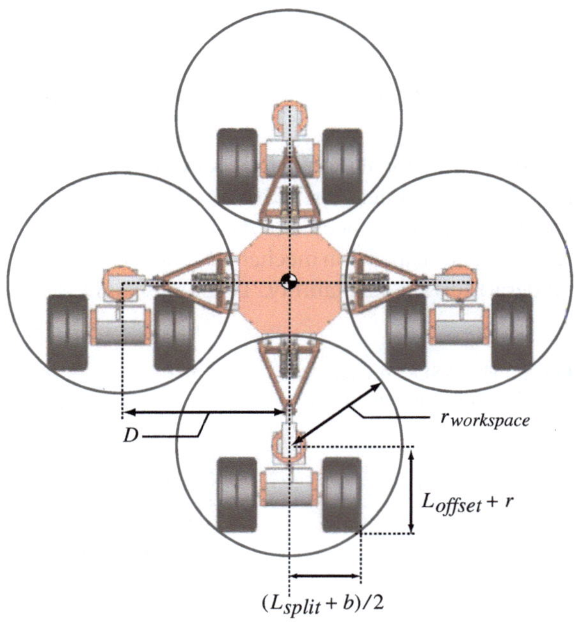

Fig. 6 ASOC module workspace. The circles represent the boundaries of the workspace

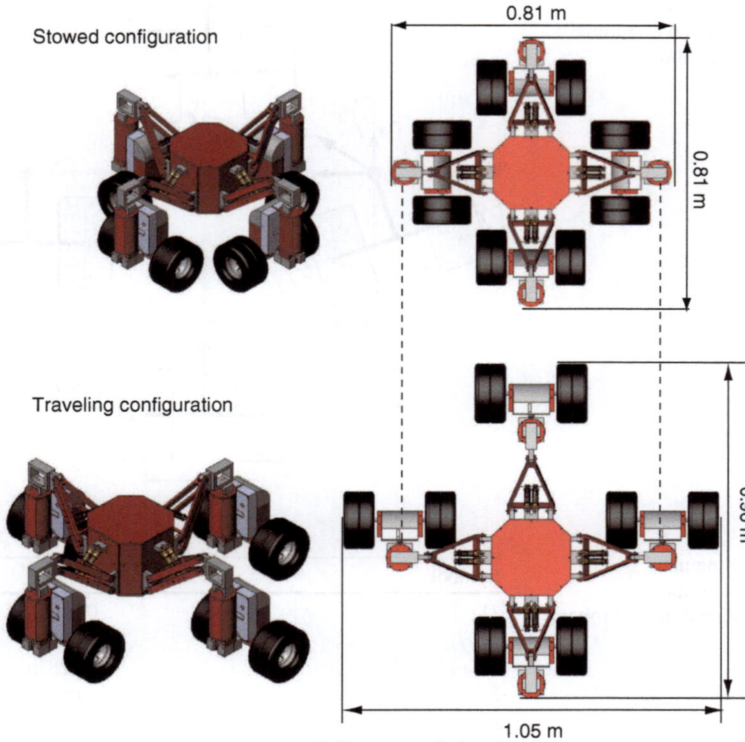

Fig. 7 CAD model of the ASOC-driven omnidirectional mobile robot. Top figures show the stowed configuration, and bottom ones show the traveling configuration

3 Kinematic Control of the Omnidirectional Mobile robot

Figure 8 illustrates a kinematic model of the ASOC-driven omnidirectional mobile robot. The coordinate frame for the main body Σ_b is fixed on the centroid of the body and defined as the right-hand frame, depicting the longitudinal direction as x. The coordinate frame for each ASOC module Σ_i $(i = 1 \cdots 4)$ is defined such that the z axis is aligned to the pivot shaft and fixed at a point along its pivot axis (Σ_i does not rotate along with the ASOC rotation around its pivot axis). D and ξ_i locate each ASOC module with regard to the main body. Table 1 summarizes kinematic parameters that are used in the experiments described later.

The kinematic control explained here calculates all wheel angular velocities that satisfy desired body translational and rotational velocities, which are defined in an inertial coordinate frame. First, the relationship between the planer velocity vector at the i-th ASOC coordinate frame $\dot{\boldsymbol{x}}_i$ and that at the body frame $\dot{\boldsymbol{x}}_b = (\dot{x}_b,\ \dot{y}_b)$ can be represented as:

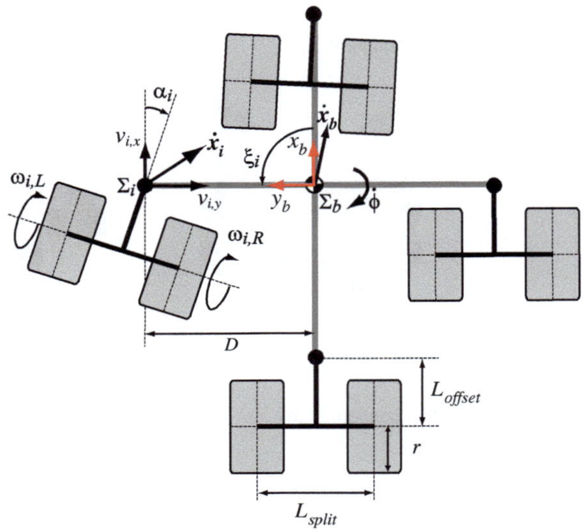

Fig. 8 Kinematic model of the ASOC-driven omnidirectional mobile robot

Table 1 Kinematic parameters for the omnidirectional mobile robot

Symbol	Value
D	0.353 m
L_{offset}	0.110 m
L_{split}	0.228 m
r	0.085 m
ξ	$[0, 0.5\pi, \pi, 1.5\pi]$

$$\dot{\boldsymbol{x}}_i = \begin{bmatrix} v_{i,x} \\ v_{i,y} \end{bmatrix} = \dot{\boldsymbol{x}}_b + \dot{\phi} D \begin{bmatrix} \cos \xi_i \\ \sin \xi_i \end{bmatrix} \qquad (i = 1 \cdots 4) \tag{3}$$

where $\dot{\phi}$ is the yaw rate of the main body. The distance D is assumed as constant value in the experiment because the shock absorbers installed are relatively stiff enough so that the displacement of D is negligible. Here, $\dot{\boldsymbol{x}}_i$ is also formulated based on the kinematic relationship with the wheel angular velocities:

$$\begin{aligned} \dot{\boldsymbol{x}}_i = \begin{bmatrix} v_{i,x} \\ v_{i,y} \end{bmatrix} &= r \begin{bmatrix} \cos \alpha_i & -\sin \alpha_i \\ \sin \alpha_i & \cos \alpha_i \end{bmatrix} \begin{bmatrix} 1/2 & 1/2 \\ -L & L \end{bmatrix} \begin{bmatrix} \omega_{i,L} \\ \omega_{i,R} \end{bmatrix} \\ &= \frac{r}{2} \begin{bmatrix} \cos \alpha_i + 2L \sin \alpha_i & \cos \alpha_i - 2L \sin \alpha_i \\ \sin \alpha_i - 2L \cos \alpha_i & \sin \alpha_i + 2L \cos \alpha_i \end{bmatrix} \begin{bmatrix} \omega_{i,L} \\ \omega_{i,R} \end{bmatrix} \\ &= \boldsymbol{C}_i \cdot r \cdot \left[\omega_{i,L} \; \omega_{i,R}\right]^T \qquad (i = 1 \cdots 4) \end{aligned} \tag{4}$$

where α_i is the angle of the pivot axis measured by the potentiometer, and $L = L_{offset}/L_{split}$. The desired wheel angular velocities of each ASOC module that satisfies the given body velocity and the yaw rate are then obtained by substituting

Eqs. (3)-(4):

$$\begin{bmatrix} \omega_{i,L} \\ \omega_{i,R} \end{bmatrix} = \frac{C_i^{-1}}{r} \begin{bmatrix} 1 & 0 & D\cos\xi_i \\ 0 & 1 & D\sin\xi_i \end{bmatrix} \begin{bmatrix} \dot{x}_b \\ \dot{y}_b \\ \dot{\phi} \end{bmatrix} = \frac{1}{r} \cdot C_i^{-1} \cdot \boldsymbol{J} \cdot \boldsymbol{u} \tag{5}$$

For the kinematic control, $\boldsymbol{u}$ is the control input variables given by an operator. The control method described above aligns the velocity vector of each ASOC in the desired direction of travel, minimizing energy loss due to internal forces.

4 Experimental Tests for Omnidirectional Mobility

In this section, first, an evaluation metric for an omnidirectional mobility of the robot is introduced. The experimental results are also presented along with the mobility evaluation based on the metric termed as the omnidirectional mobility index.

4.1 Omnidirectional Mobility Index

Several metrics have been studied for the mobility analysis of mobile vehicles in rough terrain. For example, a mobility index of an off-road vehicle was considered as a function of contact pressure, vehicle weight, and wheel load [14]. In addition, a tractive efficiency (ratio of input and output powers) of vehicle has been employed for the comparison of off-road vehicle performance. These indices basically consider a dynamic interaction between vehicle and terrain. Another metric related to a body motion (i.e., velocity, acceleration, or jerk) is commonly used to evaluate the mobility of conventional passenger vehicles or mobile robots.

Focusing on the mobility of an omnidirectional mobile robot, a particular requirement for its mobility is to achieve a high agility for a near-arbitrary omnidirectional maneuver. In the case of the ASOC-driven omnidirectional mobile robot, each ASOC needs to be kinematically coordinated to follow a given maneuver. Therefore, in this work, a metric related to the ASOC motion is employed for the mobility evaluation. The metric, termed as an omnidirectional mobility index, is defined as a root mean square error between the desired profile of the ASOC pivot angle and its actual profile measured by the potentiometer on its axis (Fig. 9). The index has a unit of degrees. Higher agility of the omnidirectional robot decreases the magnitude of the index. In the experiment, a gross omnidirectional mobility index of the robot is calculated as a mean value between the indices obtained from four ASOCs.

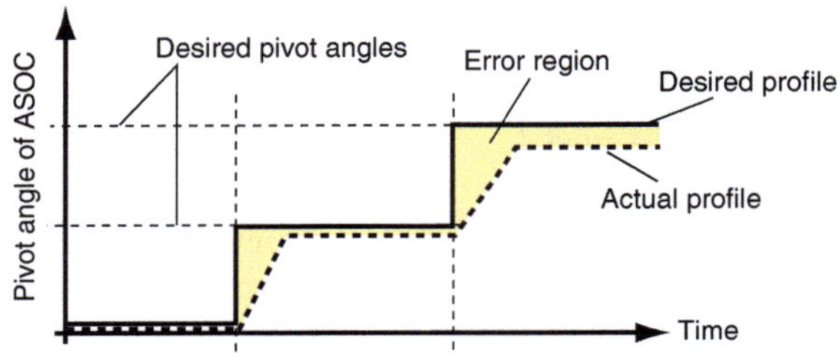

Fig. 9 Schematic graph of a time history of pivot angle of an ASOC. The *solid line* is the desired profile (given maneuver) and the *dotted line* is the measured profile (actual maneuver). The region surrounded by these two lines indicates an error between desired and measured pivot angles

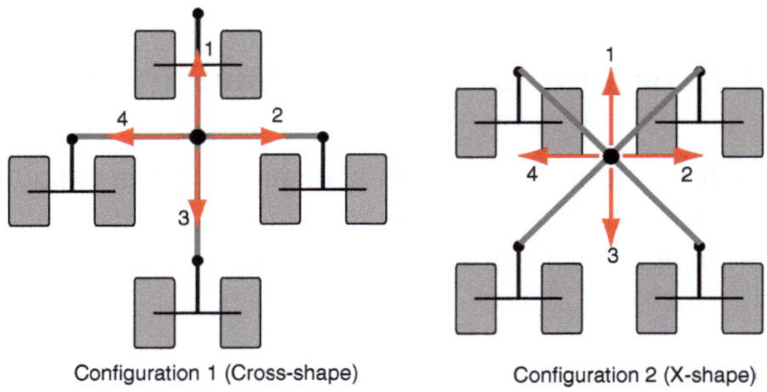

Fig. 10 Two configurations tested in the experiment. The arrows indicate the given velocity vectors. The direction of these vectors changes by 90° in every 5 s for square motion

4.2 Experimental Description

Two different configurations of the robot have been experimentally tested as shown in Fig. 10: in Configuration 1, the velocity vector of the robot is always aligned with the ASOC module orientation (i.e., a cross-shape configuration); in Configuration 2, the velocity vector is diagonal (45°) with regard to the ASOC module orientation (i.e., X-shape configuration).

In the experimental test, the robot changes its velocity vector by 90° in every 5 s, drawing a square motion path. The kinematic control method presented in Sect. 3 calculates the wheel angular velocities that are required for the maneuver. The traveling body velocity of the robot is controlled to maintain a constant value of 0.36 m/s. During each run, the pivot angle and wheel velocities at each ASOC are measured and sent to the onboard computer for the mobility evaluation of the robot.

4.3 Results and Discussion

Figures 11 and 12 show time histories of the ASOC pivot angle in different configurations with and without compliant suspension, respectively. Table 2 summarizes the omnidirectional mobility index for each configuration.

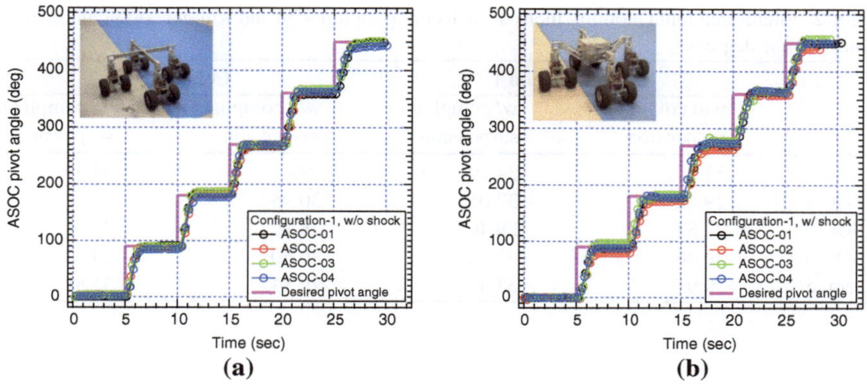

Fig. 11 Time history of the pivot angle for mobility evaluation (Configuration 1)

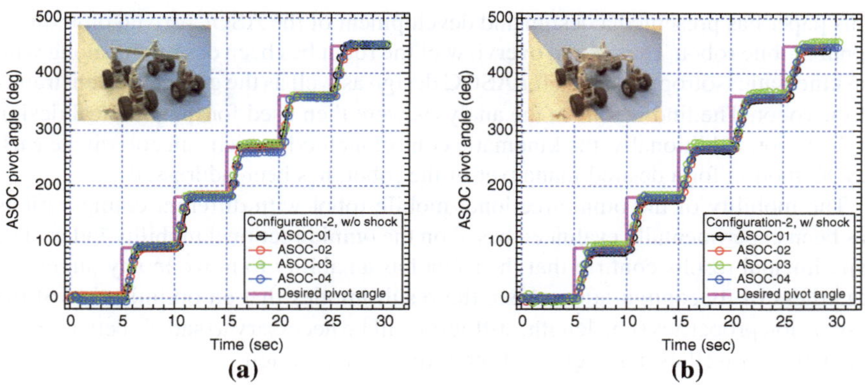

Fig. 12 Time history of the pivot angle for mobility evaluation (Configuration 2)

From the figures and table, the omnidirectional mobility indices between the two configurations are negligible: the difference between them is less than 0.1° in the case of the robot without the compliant suspension, and 1.6° with. This indicates that the ASOC-driven omnidirectional robot possesses relatively high agility that is independent of its configuration.

In addition, it can be seen that the omnidirectional mobility of the robot with rigid links (without compliant suspension) is better than the robot with compliant suspension. This is due to the fact that the shock absorbers for the compliant suspension mitigate sudden velocity change. The shocks absorbers also reduce thrust energy generated at the wheel contact patch while turning, resulting in a less agile turning maneuver. On the other hand, the robot with the rigid links can efficiently coordinate each ASOC with less energy loss, enabling more agile maneuver. This result implies that a trade-off between high terrain adaptability with compliant suspension and high omnidirectional mobility with rigid links is necessary.

Table 2 Omnidirectional mobility index at each configuration with and without compliant suspensions (Unit is degrees)

	Configuration 1		Configuration 2	
	w/o compliant suspension	w/ compliant suspension	w/o compliant suspension	w/ compliant suspension
ASOC 1	31.67	34.19	29.61	34.39
ASOC 2	29.68	37.05	30.95	35.77
ASOC 3	27.83	29.46	28.19	27.51
ASOC 4	30.46	33.02	30.60	29.62
Average	29.91	33.43	29.84	31.82

5 Conclusions

This paper has presented a design and development of the ASOC-driven omnidirectional mobile robot. The system overview of the robot has been described, along with the kinematic isotropy analysis for ASOC design as well as the geometric constraints of the robot. The findings from the analyses were then used for the practical design of the robot. Additionally, the kinematic control method, which can coordinate each ASOC motion for a desired maneuver of the robot, has been addressed.

The mobility of the omnidirectional mobile robot with different configurations has been experimentally evaluated based on the omnidirectional mobility index. The experimental results confirm that the robot has an ability to move in any directions regardless of its configuration. Also, the result implies that an optimization of the suspension properties (i.e., length, stiffness) will be necessary to satisfy better terrain adaptation as well as high agility of omnidirectional motion.

References

1. P. Muir, C. Neuman, *Kinematic Modeling for Feedback Control of an Omnidirectional Wheeled Mobile Robot*, in Proceeding of the IEEE international conference on robotics and automation, Raleigh, NC, USA 1987. pp. 1772–1778
2. S. Fujisawa, K. Ohkubo, T. Yoshida et al., *Improved Moving Properties of an Omnidirectional Vehicle Using Stepping Motor*, in Proceeding of the 36th conference on decision and control. San Diego, CA, 1997, pp. 3654–3657
3. R. Williams, B. Carter, P. Gallina, G. Rosati, *Wheeled Omnidirectional Robot Dynamics Including Slip*, in Proceeding of the 2002 ASME design engineering technical conference, Montreal, Canada, pp. 201–207
4. J. Blumrich, Omnidirectional wheel. US Patent 3,789,947 (1974)
5. R. Smith, Omnidirectional vehicle base. US Patent 4,715,460 (1987)
6. L. Ferriere, B. Raucent, *ROLLMOBS, A New Universal Wheel Concept*, in Proceeding of the, international Conference on robotics and automation, Leuven, Belgium, 1998, pp. 1877–1882
7. M. West, H. Asada, Design of ball wheel mechanisms for omnidirectional vehicles with full mobility and invariant kinematics. ASME J. Mech. Des. **117**(2), 153–161 (1997)

8. K. Tadakuma, R. Tadakuma, J. Berengeres, *Development of Holonomic Omnidirectional Vehicle with "Omni-Ball": Spherical Wheels* in Proceeding of the, IEEE/RSJ international conference on intelligent robotics and systems, San Diego, CA, 2007, pp. 33–39
9. Kornylak Corporation, http://www.kornylak.com/wheels/wheels.html. Accessed Oct 2012
10. Vehicle Technologies Inc., http://www.vetexinc.com/vehicles/sidewinder.html. Accessed Oct 2012
11. H. Yu, S. Dubowsky, A. Skwersky, Omnidirectional mobility using active split offset castors. ASME J. Mech. Des. **126**(5), 822–829 (2004)
12. T. Park, J. Lee, B. Yi, W. Kim, B. You, *Optimal Design and Actuator Sizing of Redundantly Actuated Omnidirectional Mobile Robots*, in Proceeding of the, IEEE international conference on robotics and automation, Washington DC, 2002, pp. 732–737
13. M. Udengaard, K. Iagnemma, *Design of an Omnidirectional Mobile Robot for Rough Terrain*, in Proceeding of the, IEEE international conference on robotics and automation, pp. 1666–1671
14. J. Wong, *Theory of Ground Vehicles*, 4th edn. (Wiley, New York, 2008). 4th edn.

A Vector Algebra Formulation of Mobile Robot Velocity Kinematics

Alonzo Kelly and Neal Seegmiller

Abstract Typical formulations of the forward and inverse velocity kinematics of wheeled mobile robots assume flat terrain, consistent constraints, and no slip at the wheels. Such assumptions can sometimes permit the wheel constraints to be substituted into the differential equation to produce a compact, apparently unconstrained result. However, in the general case, the terrain is not flat, the wheel constraints cannot be eliminated in this way, and they are typically inconsistent if derived from sensed information. In reality, the motion of a wheeled mobile robot (WMR) is restricted to a manifold which more-or-less satisfies the wheel slip constraints while both following the terrain and responding to the inputs. To address these more realistic cases, we have developed a formulation of WMR velocity kinematics as a differential-algebraic system—a constrained differential equation of first order. This paper presents the modeling part of the formulation. The *Transport Theorem* is used to derive a generic 3D model of the motion at the wheels which is implied by the motion of an arbitrarily articulated body. This *wheel equation* is the basis for forward and inverse velocity kinematics and for the expression of explicit constraints of wheel slip and terrain following. The result is a mathematically correct method for predicting motion over non-flat terrain for arbitrary wheeled vehicles on arbitrary terrain subject to arbitrary constraints. We validate our formulation by applying it to a Mars rover prototype with a passive suspension in a context where ground truth measurement is easy to obtain. Our approach can constitute a key component of more informed state estimation, motion control, and motion planning algorithms for wheeled mobile robots.

A. Kelly (✉) · N. Seegmiller
Robotics Institute, Carnegie Mellon University, Pittsburg, PA 15213, USA
e-mail: alonzo@cmu.edu

N. Seegmiller
e-mail: nseegmiller@cmu.edu

K. Yoshida and S. Tadokoro (eds.), *Field and Service Robotics*,
Springer Tracts in Advanced Robotics 92, DOI: 10.1007/978-3-642-40686-7_41,

1 Introduction

Wheeled mobile robots (WMRs) are perhaps the most common configuration of terrestrial mobile robot, and although decades of research are behind us, little has been revealed about how to model them effectively in anything other than flat floor environments. The motion model of the robot is nonetheless central to pose estimation, control, and motion planning.

Unlike for their predecessors, manipulators, modeling the articulations of the mechanisms involved is not the fundamental issue. WMRs need to know how they move over the terrain and such models are intrinsically differential equations. For WMRs, these equations are also constrained, the constraints are nonholonomic, the system is almost always overconstrained to some degree, and even if it was not, the constraints are typically violated in ways that are only partially predictable. In this light, it is perhaps less surprising that so little has been written on this problem. While its importance is clear, its solution is less clear.

Our own historical approaches to the problem [6] have avoided the issues by formulating inputs in state space, where constraints (and constraint consistency) are not an issue. Terrain following was treated after the fact by integrating the unconstrained dynamics and then forcing the constraints to be satisfied in a separate optimization process. While this was adequate, it was hardly principled.

While service robots may operate exclusively in flat floor environments, almost any useful field robot will have to operate competently on uneven, sloped, and slippery terrain for extended periods of time. The first step toward competent autonomy in these conditions is the incorporation of faster-than-real-time models that predict the consequences of candidate actions well. Fast and accurate WMR models are therefore a fundamental problem and we propose a general approach to designing such models in this paper.

1.1 Prior Work

Muir and Newman published one of the earliest general approaches to kinematic modeling of wheeled mobile robots [10]. Following Sheth-Uicker conventions they assign coordinate systems and derive a graph of homogenous transforms relating wheel and robot positions. By differentiating cascades of transforms, Jacobian matrices are computed for each wheel (relating wheel and robot velocities) which are combined to form the "composite robot equation". They provide a "sensed forward" solution (in which the robot velocity is determined from sensed steer angles and wheel velocities) as well as an "actuated inverse" solution.

Several researchers extended this *transformation* approach to WMR kinematics modeling. Alexander and Maddocks proposed an alternative forward solution when rolling without slipping is impossible, derived from Coulomb's Law of friction [1]. Rajagopalan handled the case of inclined steering columns [11]. Campion et al.

classified WMR configurations into five mobility types based on degrees of mobility and steerability, which they define [2]. Yet others proposed *geometric* approaches to WMR kinematics modeling [5, 7].

However, these earlier approaches and analyses are limited to *planar* motion. More recently in 2005, Tarokh and McDermott published a general approach to modeling full 6-DOF kinematics for articulated rovers driving on uneven terrain [13]. Their approach resembles Muir and Newman in requiring the derivation of homogenous transform graphs and the differentiation of transforms to compute wheel Jacobians. Others have derived and simulated full-3D WMR kinematics on rough terrain with specific objectives, such as mechanisms that enable rolling without slipping [3, 4], precise localization [8], and control of passively-steered rovers [12].

In contrast to prior transformation and geometric approaches, we derive the kinematics and constraint equations for WMR using vector algebra. This new approach is intuitive and, unlike [13], does not require differentiation. Our method for propagating velocities forward through a kinematic chain is a classical one that has also been used in robot manipulation [9].

2 Kinematics of Wheeled Mobile Robots

In the general case, a wheeled mobile robot may be articulated in various ways and it may roll over arbitrary terrain with any particular wheel lying either on or above the nominal terrain surface. Assuming terrain contact is assured by geometry or a suspension, there are two principal difficulties associated with wheeled mobile robot (WMR) kinematic modeling: nonlinearity and overconstraint. Nonlinearity occurs in steering control because trigonometric functions of the steer angles appear in the mapping between body and wheel velocities. Overconstraint can occur in estimation contexts where the set of $m > n$ measurements of velocities and/or steer angles lead to an inconsistent solution for the n degrees of velocity freedom available in the vehicle state vector. This section develops solutions for both the control and estimation problems using a vector algebraic formulation.

We will first develop the basic kinematic relationships between a) the linear and angular velocity of a distinguished coordinate frame on the body of the mobile robot and b) the linear velocity of an arbitrarily positioned point corresponding to a wheel. In contrast to all prior work, we will formulate the transformation using vector algebra, leading to a very straightforward expression for even the general case.

2.1 Transport Theorem

The key element of the technique is a basic theorem of physics, commonly used in dynamics and inertial navigation theory. Known either as the *Coriolis Equation* or the *Transport Theorem*, it concerns the dependence of measurements in physics on the state of motion of the observer. The notation $\vec{u}_a^{\,b}$ will mean the vector quantity u of frame a with respect to frame b. Let the letter f refer to a frame of reference associated with a fixed observer, whereas m will refer to one associated with a moving observer. Due to their relative, instantaneous angular velocity $\vec{\omega}_m^{\,f}$, our observers would compute (or measure) different time derivatives of the *same vector* $\vec{v}$ that are related as follows:

$$\left.\frac{d\vec{v}}{dt}\right|_f = \left.\frac{d\vec{v}}{dt}\right|_m + \vec{\omega}_m^{\,f} \times \vec{v} \tag{1}$$

2.2 Velocity Transformation

Now, let these two frames have an instantaneous relative position of $\vec{r}_m^{\,f}$. Suppose that the moving observer measures the position $\vec{r}_o^{\,m}$ and velocity $\vec{v}_o^{\,m}$ of an object o, and we wish to know what the fixed observer would measure for the motion of the same object. The position vectors can be derived from vector addition thus:

$$\vec{r}_o^{\,f} = \vec{r}_o^{\,m} + \vec{r}_m^{\,f} \tag{2}$$

The time derivative of this position vector, computed in the fixed frame is:

$$\left.\frac{d}{dt}\right|_f (\vec{r}_o^{\,f}) = \left.\frac{d}{dt}\right|_f (\vec{r}_o^{\,m} + \vec{r}_m^{\,f}) = \left.\frac{d}{dt}\right|_f (\vec{r}_o^{\,m}) + \left.\frac{d}{dt}\right|_f (\vec{r}_m^{\,f}) \tag{3}$$

Now we can apply the Coriolis equation to the first term on the right to produce the general result for the transformation of apparent velocities of the object o between two frames of reference undergoing arbitrary relative motion:

$$\vec{v}_o^{\,f} = \vec{v}_o^{\,m} + \vec{v}_m^{\,f} + \vec{\omega}_m^{\,f} \times \vec{r}_o^{\,m} \tag{4}$$

We have used the fact that, for any frames a and b, $\left.\frac{d}{dt}\right|_b (\vec{r}_a^{\,b}) = \vec{v}_a^{\,b}$.

2.3 Wheel Equation

We define a wheel frame (w) and a vehicle frame (v). We allow the point around which a wheel may steer (frame s) to be offset from the contact point (frame c) between the wheel and the ground (Fig. 1). In such a case, we can write the position vector for the wheel contact point as follows:

$$\vec{r}_c^{\,w} = \vec{r}_v^{\,w} + \vec{r}_s^{\,v} + \vec{r}_c^{\,s} \tag{5}$$

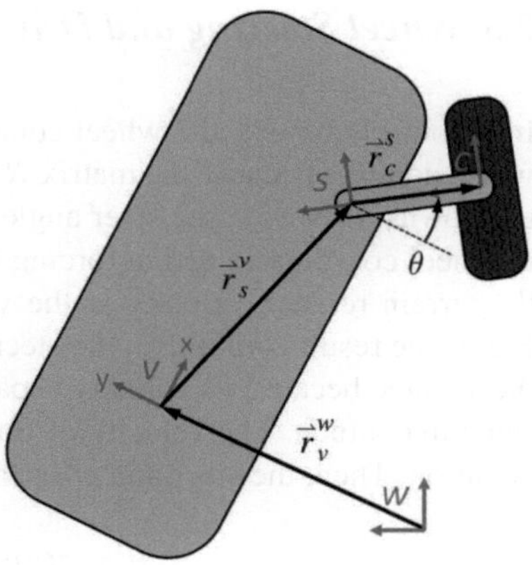

Fig. 1 Frames for WMR Kinematics. The four frames necessary for the relation of wheel rotation rates and to vehicle speed and angular velocity

Next, we associate any ground-fixed frame with the fixed observer and the body-fixed frame with the moving observer and we can use the above velocity transformation to write a kinematic equation for each wheel. Differentiating the position vector in the world frame, substituting the Coriolis equation, and using $\vec{v}_s^{\,v} = 0$ yields:

$$\vec{v}_c^{\,w} = \vec{v}_v^{\,w} + \vec{\omega}_v^{\,w} \times \vec{r}_s^{\,v} + \vec{\omega}_v^{\,w} \times \vec{r}_c^{\,s} + \vec{\omega}_c^{\,v} \times \vec{r}_c^{\,s} \tag{6}$$

This is important enough to give it a name: the *wheel equation*. In the case of no offset, the last two terms vanish and the steer velocity ($\vec{\omega}_s^{\,v}$ or $\vec{\omega}_c^{\,v}$) no longer matters. The formula is valid in 3D and it also applies to cases with arbitrary articulations between the v and s frames because only the vector $\vec{r}_s^{\,v}$ is relevant. In other words, this is the general case.

2.4 Inverse Velocity Kinematics: Body to Wheels

Let the term *inverse* kinematics refer to the problem, relevant to control, of computing the wheel velocities from the body velocity. Given the above, the problem is solved by writing a wheel equation for each wheel. To do so, the physical vectors $\vec{u}$ must be expressed in a particular coordinate system. Let ${}^{c}\underline{u}_a^b$ denote the vector quantity u of frame a with respect to frame b, expressed in the coordinates of frame c (and let $\underline{u}_a^b$ imply ${}^{b}\underline{u}_a^b$). Then, if R_s^v is the rotation matrix that converts coordinates from the steer frame to the vehicle frame, it becomes possible to express the wheel equation for any wheel in the vehicle frame where many of the vectors are typically known:

$$^{v}\underline{v}_c^w = {}^{v}\underline{v}_v^w + {}^{v}\underline{\omega}_v^w \times \underline{r}_s^v + {}^{v}\underline{\omega}_v^w \times R_s^v \underline{r}_c^s + \underline{\omega}_c^v \times R_s^v \underline{r}_c^s \tag{7}$$

2.5 Wheel Steering and Drive: Control and Estimation

In a control context, the wheel equation cannot be used directly as written to find wheel controls because the matrix R_s^v depends on the steer angle, which is one of the unknowns. However, the steer angle can be found by expressing the wheel velocity in wheel coordinates and enforcing the constraint that the lateral (y) component of the terrain relative velocity in the wheel frame must vanish. For the geometry in Fig. 1, the result is intuitive, the steer angle can be determined from the direction of the s frame because its velocity is parallel to that of c, though not necessarily of the same magnitude. The velocity of frame s is simply the first two terms of the wheel equation. Then, the steer angle for the wheel is:

$$\theta = atan2[\ ({}^{v}\underline{v}_s^w)_x\ ,\ ({}^{v}\underline{v}_s^w)_y\ ] \tag{8}$$

Once the steer angle is known, the wheel velocity along the forward (x) axis of the wheel frame can be determined from the x component the wheel equation in wheel coordinates. Then the drive velocity (around the axle) can be computed using the wheel radius.

For the opposite problem of wheel sensing, measurements of wheel rotation rate provide the wheel velocities along the x axis of the wheel frame. Then a measurement of steer angle provides the rotation matrix needed to convert to a vector expressed in the vehicle frame.

2.6 Forward Velocity Kinematics: Wheels to Body

Let the term *forward* kinematics refer to the problem, relevant to estimation, of computing the body velocity from the wheel velocities. The wheel equation can be written in matrix form by using skew symmetric matrices to represent the cross products as a matrix products (specifically, $\underline{a} \times \underline{b} = -\underline{b} \times \underline{a} = -[\underline{b}]_\times \underline{a} = [\underline{b}]_\times^T \underline{a}$):

$${}^{v}\underline{v}_c^w = {}^{v}\underline{v}_v^w + [\underline{r}_s^v]_\times^T ({}^{v}\underline{\omega}_v^w) + [{}^{v}\underline{r}_c^s]_\times^T ({}^{v}\underline{\omega}_v^w) + [{}^{v}\underline{r}_c^s]_\times^T (\underline{\omega}_c^v) \tag{9}$$

For multiple wheels, stacking all the equations and grouping the first three terms together produces a matrix equation of the form:

$$\underline{v}_c = H_v(\underline{\theta}) \begin{bmatrix} {}^{v}\underline{v}_v^w \\ {}^{v}\underline{\omega}_v^w \end{bmatrix} + H_\theta(\underline{\theta})\underline{\omega}_c^v = H_v(\underline{\theta})\underline{V} + H_\theta(\underline{\theta})\dot{\underline{\theta}} \tag{10}$$

where $\underline{v}_c$ represents wheel velocities, and $\underline{V}$ represents the linear and angular velocity of the vehicle with respect to the ground. Both $\underline{v}_c$ and $\underline{V}$ are in body coordinates. $\underline{\theta}$ is the steer angles and it can include other articulations if desired. The last term in (10) is the increment to wheel velocity due to the steering rates.

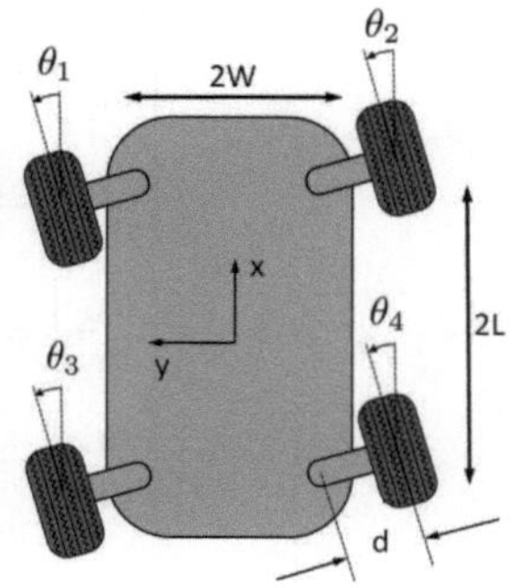

Fig. 2 A four wheel steer vehicle. Arbitrary motions are possible

2.7 Example: Four Wheel Steer

While the equations are linear in velocity, there can easily be more of them than there are degrees of freedom, making the problem overconstrained. Control will typically try to steer the wheels to be consistent with a single instantaneous center of rotation but errors can never be completely eliminated. A straightforward way to estimate the vehicle linear and angular velocity is to use the pseudoinverse—after removing the effect of steering rates. Such an approach minimizes the squared residual of wheel velocities and weights them equally:

$$\underline{V} = H_v(\underline{\theta})^+(\underline{v}_c - H_\theta(\underline{\theta})\dot{\underline{\theta}}) \tag{11}$$

This case (Fig. 2) presents a particularly difficult example of a vehicle with four wheels which are both driven and steered (from an offset position). The equations were implemented and tested on such a vehicle. Let the velocities of the body frame in body coordinates be denoted $\underline{V} = [V_x\ V_y\ \omega]^T$ and the steer angles $\underline{\theta} = [\theta_1\ \theta_2\ \theta_3\ \theta_4]^T$. Unlike a car, this vehicle is not constrained to move in the direction it is pointed. Indeed, it can drive with any linear and angular velocity that is consistent with the wheel speed and steering limits. The steer frame centers are positioned relative to the body frame as follows:

$$\underline{r}^v_{s1} = [L\ W]^T,\ \ \underline{r}^v_{s2} = [L\ -W]^T,\ \ \underline{r}^v_{s3} = [-L\ W]^T,\ \ \underline{r}^v_{s4} = [-L\ -W]^T \tag{12}$$

The contact point offsets in the body frame depend on the steer angles. They are:

$${}^v\underline{r}^{s1}_{c1} = d[-s_1\ c_1]^T,\ {}^v\underline{r}^{s2}_{c2} = d[s_2\ -c_2]^T,\ {}^v\underline{r}^{s3}_{c3} = d[-s_3\ c_3]^T,\ {}^v\underline{r}^{s4}_{c4} = d[s_4\ -c_4]^T \tag{13}$$

where (s_1 in the vector denotes $sin(\theta_1)$ etc.). If we denote the elements of these position vectors as $\underline{r}^v_s = [x\ y]^T$ and ${}^v\underline{r}^s_c = [a\ b]^T$, the set of wheel equations is as follows:

$$\begin{bmatrix} v_{1x} \\ v_{1y} \end{bmatrix} = \begin{bmatrix} 1 & 0 & -(y_1 + b_1) \\ 0 & 1 & (x_1 + a_1) \end{bmatrix} \underline{V} + \begin{bmatrix} -b_1 & 0 & 0 & 0 \\ a_1 & 0 & 0 & 0 \end{bmatrix} \underline{\dot{\theta}}$$
$$\begin{bmatrix} v_{2x} \\ v_{2y} \end{bmatrix} = \begin{bmatrix} 1 & 0 & -(y_2 + b_2) \\ 0 & 1 & (x_2 + a_2) \end{bmatrix} \underline{V} + \begin{bmatrix} 0 & -b_2 & 0 & 0 \\ 0 & a_2 & 0 & 0 \end{bmatrix} \underline{\dot{\theta}} \quad (14)$$
$$\begin{bmatrix} v_{3x} \\ v_{3y} \end{bmatrix} = \begin{bmatrix} 1 & 0 & -(y_3 + b_3) \\ 0 & 1 & (x_3 + a_3) \end{bmatrix} \underline{V} + \begin{bmatrix} 0 & 0 & -b_3 & 0 \\ 0 & 0 & a_3 & 0 \end{bmatrix} \underline{\dot{\theta}}$$
$$\begin{bmatrix} v_{4x} \\ v_{4y} \end{bmatrix} = \begin{bmatrix} 1 & 0 & -(y_4 + b_4) \\ 0 & 1 & (x_4 + a_4) \end{bmatrix} \underline{V} + \begin{bmatrix} 0 & 0 & 0 & -b_4 \\ 0 & 0 & 0 & a_4 \end{bmatrix} \underline{\dot{\theta}}$$

3 Wheel Constraints

So far, we have proposed control and estimation mechanisms that satisfy wheel slip constraints in both the forward and inverse kinematics describing the motion of the vehicle in the instantaneous terrain tangent plane. Steering and propulsion are actively controlled in a vehicle, so some measures can be taken to try to satisfy wheel slip constraints. Doing so enhances controllability and avoids the energy loss that would be associated with doing (sliding) work on the terrain.

On non-flat terrain, another constraint of interest is terrain following. Assuming an adequate suspension, wheels should neither penetrate nor rise above the terrain. Such constraints determine altitude (z), and attitude (pitch and roll). These constraints are satisfied passively by the suspensions of most vehicles, so the inverse kinematic problem of active suspension occurs less often. We will now present methods to incorporate both types of constraints in the context of *motion prediction*: the problem of estimating or predicting position and attitude by integrating the system differential equation.

3.1 Constrained Dynamics

We will find it convenient to formulate the WMR motion prediction problem as the integration of a differential-algebraic equation (DAE) where the constraints remain explicit. We will use a nonstandard formulation of the form:

$$\begin{aligned} \underline{\dot{x}} &= \underline{f}(\underline{x}, \underline{u}) \\ \underline{c}(\underline{x}) &= \underline{0} \\ \underline{d}(\underline{x})^T \underline{\dot{x}} &= \underline{0} \end{aligned} \quad (15)$$

The m constraint equations in $\underline{c}$ and $\underline{d}$ are understood to be active at all times. Each element of $\underline{d}$ is a particular form of *nonholonomic* constraint known as a *Pfaffian* velocity constraint. Each specifies a disallowed direction restricting the admissible values of the state derivative. The equations in $\underline{c}$ are *holonomic* constraints that restrict the admissible values of the state $\underline{x}$ and therefore, through the differential equation, they ultimately restrict the state derivative as well.

Both forms of constraints are ultimately treated identically because, as is commonly performed in DAE theory, the gradient of $\underline{c}$ produces the associated disallowed directions of the holonomic constraints. It will turn out that terrain following will be expressible as holonomic constraints and wheel slip will be nonholonomic.

3.2 Wheel Slip Constraints

In the case of rolling without *lateral* slipping, the disallowed direction for the wheel is clearly aligned with the y axis of the contact point c frame. However, to use the constraint in a DAE, it must be converted to an equivalent disallowed direction in state derivative space. The simplest way to do so is to write (10) in wheel coordinates thus (assuming $R_v^s = R_v^c$):

$$^{c}\underline{v}_c^w = R_v^s H_v(\underline{\theta})\underline{V} + R_v^s H_\theta(\underline{\theta})\underline{\dot{\theta}} \tag{16}$$

Note that $\underline{V}$ is exactly the relevant components of the state derivative, so the first row of $R_v^s H_v(\underline{\theta})$ is both the gradient of the lateral wheel velocity with respect to the state derivative, and the associated disallowed direction. Note that in full-3D, a transformation from Euler angle rates to angular velocity may be required (see Sect. 4.1). As long as the steer angles $\underline{\theta}$ are not in the state vector, the second term is irrelevant, but if they are, the first row of the gradient can be extracted for these as well. If there were any other articulations in the kinematic chain from the body frame to the wheel contact point frame, they can be treated similarly.

3.3 Terrain Following Constraints

It is tempting to extract the z component of the wheel velocity in an analogous manner to produce a terrain following constraint, but the problem is slightly more complicated. It is a basic assumption that the location of the wheel contact point is known. This point is on the bottom of the wheel on flat terrain and it must be computed for uneven terrain. In any case, the axes of the c frame are aligned with the wheel by assumption.

A terrain following constraint can be generated by noting that the terrain normal at the contact point is the other disallowed direction for wheel motion. Indeed, to be precise, the wheel y axis should ideally be projected onto the terrain tangent plane for lateral slip constraints as well. We can enforce terrain following by requiring the dot product of the terrain normal and contact point velocity vectors to equal zero:

Fig. 3 Zoë's axles are free to rotate in both the steer and roll angles

$\hat{n} \cdot \vec{v}_c^{\,w} = 0$. Accordingly, the gradient of out-of-terrain wheel motion with respect to the state derivative $\underline{V}$ is:

$$\underline{d}(\underline{x})^T = {}^v\hat{n}^T H_v(\underline{\theta}) \tag{17}$$

where ${}^v\hat{n}$ is the terrain normal expressed in vehicle coordinates.

The more common approach (proposed by [16]) is to differentiate the holonomic constraints $\underline{c}(\underline{x})$ with respect to the state to obtain the gradient $\underline{c}_{\underline{x}}$. The holonomic constraints are then enforced to first order by requiring that $\underline{c}_{\underline{x}}\, \dot{\underline{x}} = \underline{0}$. Here we computed the disallowed gradient $\underline{c}_{\underline{x}}$ using vector algebra and avoided the differentiation.

4 Results

We present results on the Zoë rover, which previously surveyed the distribution of microscopic life in Chile's Atacama desert [15]. Zoë has four independently driven wheels on two passively articulated axles. The axles are free to rotate in both the steer (θ) and roll (ϕ) angles, as seen in Fig. 3. A roll averaging mechanism constrains the front and rear axle roll angles to be symmetric ($\phi_f = -\phi_r$).

One of the authors previously developed a transform-based 3D kinematic model for the Zoë rover and applied it to control [12]. Here we re-derive the kinematic and constraint equations using the new, vector algebra formulation and apply them to estimation and simulation.

4.1 Zoë Forward Velocity Kinematics

Here we present Zoë's forward velocity kinematics. We begin by defining variables in the wheel equation, arranged according to (10):

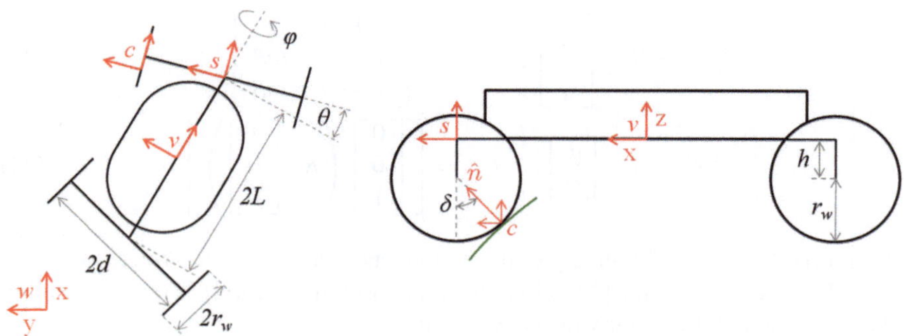

Fig. 4 A diagram of Zoë's coordinate systems, dimensions, and steering/suspension angles

$$ {}^{v}\underline{v}_{c}^{w} = \left[I \; [\underline{r}_{c}^{v}]_{\times}^{T} \right] \begin{bmatrix} {}^{v}\underline{v}_{v}^{w} \\ {}^{v}\underline{\omega}_{v}^{w} \end{bmatrix} [{}^{v}\underline{r}_{c}^{s}]_{\times}^{T} \underline{\omega}_{c}^{v} \tag{18} $$

The vector $\underline{r}_c^v$ is the position of the wheel contact point with respect to the vehicle, expressed in vehicle coordinates:

$$ \begin{aligned} \underline{r}_c^v &= \underline{r}_s^v + {}^{v}\underline{r}_c^s && (19) \\ &= \underline{r}_s^v + R_s^v \underline{r}_c^s && (20) \\ &= \begin{bmatrix} \pm L \\ 0 \\ 0 \end{bmatrix} + Rot_x(\phi) Rot_z(\theta) \begin{bmatrix} -r_w \sin(\delta) \\ \pm d \\ -h - r_w \cos(\delta) \end{bmatrix} && (21) \end{aligned} $$

Refer to Fig. 4 for the meaning of dimensions L, r_w, d, and h. The dimension L is positive for front axle wheels and negative for rear wheels. d is positive for left wheels (1, 3) and negative for right wheels (2, 4). The contact angle, δ, specifies where along circumference of the wheel contact with the terrain is made.

To simulate the constrained dynamics according to (15) we must define the state vector:

$$ \underline{x} = \left[x \; y \; z \; \gamma \; \beta \; \alpha \; \theta_f \; \phi_f \; \theta_r \; \phi_r \right]^T \tag{22} $$

The first three states are the position of the vehicle in world coordinates ($\underline{r}_v^w$). The second three are Euler angles (roll, pitch, and yaw), which specify the orientation of the vehicle with respect to the world frame. Let $\underline{\Omega}$ denote the vector of Euler angles: $\underline{\Omega} = [\gamma \; \beta \; \alpha]^T$. The last four states are the steer (θ) and roll (ϕ) angles for the front and rear axle joints.

We can compute $\dot{\underline{x}}$ from (18), but we must first transform the angular velocities to Euler and axle angle rates as follows:

$$
{}^{v}\underline{\omega}_{v}^{w} = T_{\omega_v}\begin{bmatrix}\dot{\gamma}\\ \dot{\beta}\\ \dot{\alpha}\end{bmatrix}, \quad T_{\omega_v} = \begin{bmatrix}1 & 0 & -s\beta\\ 0 & c\gamma & s\gamma c\beta\\ 0 & -s\gamma & c\gamma c\beta\end{bmatrix} \tag{23}
$$

$$
{}^{s}\underline{\omega}_{s}^{v} = T_{\omega_s}\begin{bmatrix}\dot{\theta}\\ \dot{\phi}\end{bmatrix}, \quad T_{\omega_s} = \left[\begin{bmatrix}0\\0\\1\end{bmatrix}\left(R_v^s\begin{bmatrix}1\\0\\0\end{bmatrix}\right)\right] \tag{24}
$$

The matrix T_{ω_v} (for the Euler angle convention where $R_v^w = Rot_z(\alpha)Rot_y(\beta)Rot_x(\gamma)$) is widely used in navigation [14]. Given the transforms in (23) and (24) and combining wheel equations for all four wheels, we obtain:

$$
\underline{v}_c = H_v(\underline{\theta})\underline{V}' + H_\theta(\underline{\theta})\underline{\dot{\theta}} \tag{25}
$$

$$
\begin{bmatrix}{}^{v}\underline{v}_{c1}^{w}\\ \vdots\\ {}^{v}\underline{v}_{c4}^{w}\end{bmatrix} = \begin{bmatrix}R_w^v[\underline{r}_{c1}^{v}]_\times^T T_{\omega_v}\\ \vdots \quad\quad \vdots\\ R_w^v[\underline{r}_{c4}^{v}]_\times^T T_{\omega_v}\end{bmatrix}\begin{bmatrix}\underline{v}_v^w\\ \underline{\dot{\Omega}}\end{bmatrix} + \begin{bmatrix}[{}^{v}\underline{r}_{c1}^{s}]_\times^T R_s^v T_{\omega_s} & 0_{3\times 2}\\ [{}^{v}\underline{r}_{c2}^{s}]_\times^T R_s^v T_{\omega_s} & 0_{3\times 2}\\ 0_{3\times 2} & [{}^{v}\underline{r}_{c3}^{s}]_\times^T R_s^v T_{\omega_s}\\ 0_{3\times 2} & [{}^{v}\underline{r}_{c4}^{s}]_\times^T R_s^v T_{\omega_s}\end{bmatrix}\begin{bmatrix}\dot{\theta}_f\\ \dot{\phi}_f\\ \dot{\theta}_r\\ \dot{\phi}_r\end{bmatrix} \tag{26}
$$

Each wheel corresponds to three rows of (26). Note that $\underline{V}'$ differs from $\underline{V}$ as defined in (10) because it contains linear velocities in *world* coordinates and Euler angle rates $\underline{\dot{\Omega}}$. Note also that variables containing s in the superscript or subscript are different for the front and rear axles, i.e. for wheels (1, 2) and (3, 4).

Because Zoë's steering and suspension joints are passive, it is necessary in simulation (or prediction) contexts to solve for the joint angle rates $\underline{\dot{\theta}}$ simultaneously with the vehicle velocity $\underline{V}'$:

$$
\underline{v}_c = \begin{bmatrix}H_v(\underline{\theta}) & H_\theta(\underline{\theta})\end{bmatrix}\begin{bmatrix}\underline{V}'\\ \underline{\dot{\theta}}\end{bmatrix} = H(\underline{\theta})\underline{\dot{x}}. \tag{27}
$$

The system is overdetermined and can be solved for $\underline{\dot{x}}$ using the pseudoinverse.

4.2 Zoë Constraint Equations

As formulated, there are nine total constraints on Zoë's forward kinematics. As explained in Sect. 3.2 the nonholonomic, no-lateral-slip constraints are enforced by disallowing wheel velocity along the y axis of the c frame for each wheel. To compute the constraint for a single wheel, we extract the corresponding 3 rows of (27) and left multiply by $R_v^s (= R_v^c)$ to convert to wheel coordinates:

$$
{}^{c}\underline{v}_{c}^{w} = R_v^s H_c(\underline{\theta})\underline{\dot{x}} \tag{28}
$$

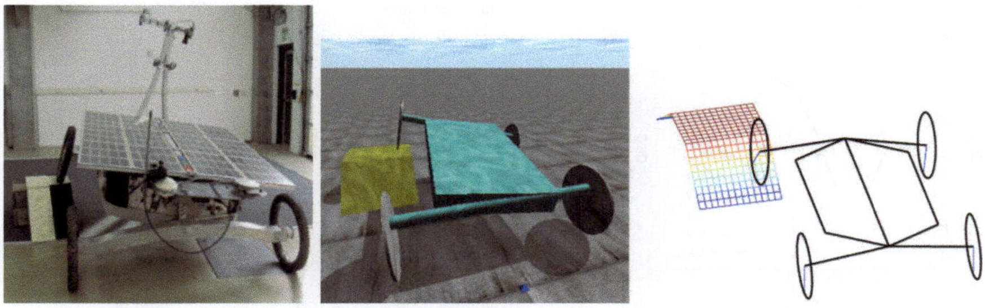

Fig. 5 Photographs and screenshots of the Zoë rover captured during (from *left* to *right*) a physical experiment, Lagrangian dynamics simulation, and kinematic simulation

where H_c denotes the three rows of H corresponding to the chosen wheel. The disallowed direction in state space $\underline{d}(\underline{x})^T$ is simply the second row of $R_v^s H_c(\underline{\theta})$. Because, in this case, left and right wheels on the same axle generate identical no-lateral-slip constraints, one redundant constraint may be eliminated per axle.

As explained in Sect. 3.3, the four holonomic terrain following constraints are enforced, to first order, by disallowing wheel velocity in the terrain-normal direction for each wheel. Given that the dot product ${}^v\hat{n} \cdot {}^v\underline{v}_c^w$ must be zero, where ${}^v\hat{n}$ is the terrain normal vector expressed in vehicle (or body) coordinates, the disallowed direction in state space is ${}^v\hat{n}^T H_c$.

The roll-averaging mechanism generates one additional holonomic constraint that $\phi_f + \phi_r = 0$. This is enforced to first-order by constraining $\frac{d\phi_f}{dt} + \frac{d\phi_r}{dt} = 0$.

4.3 Terrain Following Experiment

Here we present the results of a terrain-following experiment. Zoë is commanded to drive straight at 0.15 m/s while its left wheels traverse a ramp obstacle (1.71 m length × 0.41 m height, 36° slope) that causes the body to roll. A simple proportional controller drives the front and rear steer angles to zero degrees (Fig. 5).

Results are shown (from left to right in Fig. 6) for a physical experiment, a Lagrangian dynamics simulation (implemented using Open Dynamics Engine), and a kinematic simulation using the vector algebra formulation presented here. Note that all three plots match closely; the same changes in suspension and steer angles are observed as the front and rear wheels encounter the obstacle, and both simulations correctly predict a terminal heading error of approximately 2.5°. The kinematic simulation, however, is computationally much cheaper than the full dynamics simulation. Surface contact parameters were tuned to minimize wheel slip in the dynamic simulation. Minor disagreements between the physical experiment and simulations are due to unmodeled peculiarities in Zoë's construction, such as hysteresis in the roll-averaging mechanism.

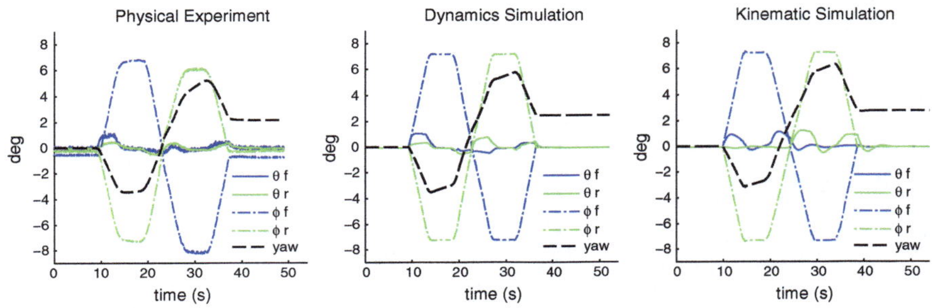

Fig. 6 Plots of steering (θ), suspension (ϕ), and vehicle yaw angles versus time recorded during (from *left* to *right*) a physical experiment, Lagrangian dynamics simulation, and kinematic simulation using our presented vector algebra model

5 Conclusion

This paper has shown how the Transport Theorem provides the basis for modeling the kinematic relationships between the body velocities and the wheel velocities of a wheeled mobile robot. Our formulation produces the solution for the general three dimensional case for arbitrary robot articulations and arbitrary terrain. In contrast to all prior work in WMR kinematics, we use an intrinsic velocity transformation in coordinate system independent form. This approach has the key advantage of eliminating the need to differentiate the relevant pose transforms in order to produce a mechanism Jacobian for each wheel. Furthermore, the resultant equation for wheel motion also provides a very convenient basis for imposing constraints for wheel slip and terrain following in a differential-algebraic system. Although we have not justified it here, this turns out to enable efficient predictor-corrector integration.

We have also shown advantages relative to a full second order dynamics model. Our formulation needs to be integrated only once but it satisfies the same constraints and those constraints ultimately determine the trajectory followed. Furthermore, because we express our constraints explicitly, our formulation permits them to be arbitrary. We have shown how to model an example of passive articulations here, but we can also accommodate models of how wheels are actually slipping both longitudinally and laterally. In short, ease of derivation, faster-than-real-time computation, and highly realistic motion make our formulation a natural choice for the modeling of any wheeled mobile robot in arbitrary terrain.

Acknowledgments This research was made with U.S. Government support under and awarded by the DoD, Air Force Office of Scientific Research, National Defense Science and Engineering Graduate (NDSEG) Fellowship, 32 CFR 168a. The NASA funded Life in the Atacama project (NAG5-12890) supported development of the Zoë rover.

References

1. J.C. Alexander, J.H. Maddocks, On the kinematics of wheeled mobile robots. Int. J. Rob. Res. **8**(5), 15–27 (1989)
2. G. Campion, G. Bastin, B. D'Andrea-Novel, Structural properties and classification of kinematic and dynamic models of wheeled mobile robots. IEEE Trans. Robot. **12**(1), 47–62 (1996)
3. N. Chakraborty, A. Ghosal, Kinematics of wheeled mobile robots on uneven terrain. Mech. Mach. Theory **39**(12), 1273–1287 (2004)
4. B.J. Choi, S.V. Sreenivasan, Gross motion characteristics of articulated mobile robots with pure rolling capability on smooth uneven surfaces. IEEE Trans. Robot. **15**(2), 340–343 (1999)
5. I.J. Cox, G.T. Wilfong, *Autonomous Robot Vehicles* (Springer, New York, 1990)
6. T. Howard, A. Kelly, Optimal rough terrain trajectory generation for wheeled mobile robots. Int. J. Robot. Res. **26**(2), 141–166 (2007)
7. D.S. Kim, H.C. Lee, W.H. Kwon, Geometric kinematics modeling of omni-directional autonomous mobile robot and its applications. IEEE Int. Conf. Robot. Autom. **3**, 2033–2038 (2000)
8. P. Lamon, R. Siegwart, 3D position tracking in challenging terrain. Int. J. Robot. Res. **26**(2), 167–186 (2007)
9. J.Y.S. Luh, M.W. Walker, R.P.C. Paul, On-line computational scheme for mechanical manipulators. J. Dyn. Sys. Meas. Control **102**(2), 69–76 (1980)
10. P.F. Muir, C.P. Neuman, *Kinematic Modeling of Wheeled Mobile Robots* (CMU Robotics Institute Technical Report, 1986)
11. R. Rajagopalan, A generic kinematic formulation for wheeled mobile robots. J. Robot. Syst. **14**(2), 77–91 (1997)
12. N. Seegmiller, D. Wettergreen, Control of a passively steered rover using 3-D kinematics. in *IEEE/RSJ International Conference on Intelligent Robots and Systems* (2011)
13. M. Tarokh, G. McDermott, Kinematics modeling and analyses of articulated rovers. IEEE Trans. Robot. **21**(4), 539–553 (2005)
14. D. Titterton, J. Weston, *Strapdown Inertial Navigation Technology*, 2nd edn. AAAI (2004)
15. D. Wettergreen et al., Second experiments in the robotic investigation of life in the Atacama desert of Chile. in *Proceedings of 8th International Symposium on Artificial Intelligence, Robotics and Automation in Space* (2005)
16. X. Yun, N. Sarkar, Unified formulation of robotic systems with holonomic and nonholonomic constraints. IEEE Trans. Robot. **14**(4), 640–650 (1998)

A Self-Learning Ground Classifier Using Radar Features

Giulio Reina, Annalisa Milella and James Underwood

Abstract Autonomous off-road ground vehicles require advanced perception systems in order to sense and understand the surrounding environment, while ensuring robustness under compromised visibility conditions. In this paper, the use of millimeter wave radar is proposed as a possible solution for all-weather off-road perception. A self-learning ground classifier is developed that segments radar data for scene understanding and autonomous navigation tasks. The proposed system comprises two main stages: an adaptive training stage and a classification stage. During the training stage, the system automatically learns to associate appearance of radar data with class labels. Then, it makes predictions based on past observations. The training set is continuously updated online using the latest radar readings, thus making it feasible to use the system for long range and long duration navigation, over changing environments. Experimental results, obtained with an unmanned ground vehicle operating in a rural environment, are presented to validate this approach. Conclusions are drawn on the utility of millimeter-wave radar as a robotic sensor for persistent and accurate perception in natural scenarios.

G. Reina (✉)
Department of Engineering for Innovation, University of Salento, 73100 Lecce, Italy
e-mail: giulio.reina@unisalento.it

A. Milella
Institute of Intelligent Systems for Automation (ISSIA), National Research Council (CNR), via G. Amendola 122/D, 70126 Bari, Italy
e-mail: milella@ba.issia.cnr.it

J. Underwood
Australian Centre for Field Robotics, School of Aerospace, Mechanical and Mechatronic Engineering, The University of Sydney, Sydney, NSW 2006, Australia
e-mail: j.underwood@acfr.usyd.edu.au

K. Yoshida and S. Tadokoro (eds.), *Field and Service Robotics*,
Springer Tracts in Advanced Robotics 92, DOI: 10.1007/978-3-642-40686-7_42,

1 Introduction

Research in mobile robotics aims to develop technologies that allow vehicles to travel longer distances with limited human supervision in unstructured environments. One of the main challenges toward this objective is accurate and robust scene understanding to perform important tasks including environment segmentation and classification, mapping and identification of terrain regions that can be safely traversed [1, 2]. In field scenarios, these tasks are even more difficult due to variable visibility conditions. Day/night cycles change illumination. Weather phenomena such as fog, rain, snow and hail impede visual perception. Dust clouds rise in excavation sites and agricultural fields, and they are expected during planetary exploration. Smoke also compromises visibility in fire emergencies and disaster sites. Laser and vision are common imaging technologies, but they are known to have difficulties under compromised visibility [3]. Sonar is a sensor that is not affected by visibility restrictions. However, it is considered of limited utility for field robots due to high atmospheric attenuation, noise, and reflections by specular surfaces.

As an alternative or complementary sensor to conventional range devices, radar can be employed to develop persistent and accurate perception systems in field scenarios. Radar operates at a wavelength that penetrates dust and other visual obscurants. Furthermore, it can provide information of distributed and multiple targets that appear in a single observation, whereas other range sensors are generally limited to one target return per emission, although last pulse-based lasers solve this problem to some extent, and are becoming more common.

The ability of radar to perceive the environment in low visibility conditions was demonstrated in numerous papers, for example in [4] and [5]. This aspect is not explicitly considered in the context of this paper. Nevertheless, radar has shortcomings as well, such as a large footprint, specularity effects, and limited range resolution, all of which may result in poor environment survey or difficulty in interpretation. Relatively limited research has been devoted to investigate explicitly millimeter-wave radar for short-range perception and scene segmentation, and three-dimensional terrain mapping. For example, previous work presented the implementation of radar-based obstacle avoidance on large mining trucks [6]. In other work, a millimeter-wave radar-based navigation system detected and matched artificial beacons for localization in a two-dimensional scan [7]. Millimeter-wave radar has been used on a large AGV for cargo handling [8]; the radar is scanned horizontally, and measures range and bearing to a set of trihedral aluminum reflectors. The reflectors may be covered by a polarising grating to enable discrimination from other objects. Radar capability was demonstrated in an Arctic environment [9] and for mining applications [10]. Mullane et al. [11] used a millimeter-wave (MMW) radar for occupancy mapping within a probabilistic framework.

In this research, we propose an adaptive self-learning ground segmentation approach using data acquired by a MMW radar. Self-supervised systems have been recently proposed by many to reduce or eliminate the need for hand-labeled training data, thus gaining flexibility in unknown environments. Typically, with self-supervision, a reliable classification module (such as a laser-based classifier) provides labels to train another classifier (such as a visual classifier). An example can be found in Milella et al. [12, 13], where a visual classifier was trained by radar-driven labels. Brooks et al. [14] proposed a self-supervised framework that predicts the mechanical properties of distant terrain based on a previously-learned association with visual appearance. Self-supervised learning helped to win the 2005 DARPA Grand Challenge: the winning approach used a probabilistic model to identify road surface based on color information extracted immediately ahead of the vehicle as it drives [15]. Stavens and Thrun [16] used self-supervision to train a terrain roughness predictor.

In the context of this paper, "self-learning classification" refers to the automatic training of a ground classifier using features extracted from radar images of the ground. First, the system automatically learns to associate the appearance of radar data with class labels during a training stage. Then it makes predictions based on past observations, classifying new data in two broad categories, namely ground and non-ground. The ground class corresponds to returns from the terrain, whereas the non-ground class corresponds to all other returns, including reflections from above ground objects (i.e., obstacles), occluded areas, or sensor misreadings. Since the characteristics of the ground may change geographically and over time, the system is continuously retrained in every scan: new automatically labeled data are added to the ground model replacing the oldest labels in order to incorporate changes in the ground appearance.

In this investigation, a mechanically scanned MMW radar, designed for perception and navigation in low visibility conditions, is employed. Although the sensor is custom-built at the Australian Centre for Field Robotics (ACFR) [10], it is similar in performance to other commercially available systems.[1] It is a 95-GHz frequency-modulated continuous wave (FMCW) MMW radar that reports the amplitude of echoes at ranges between 1 and 120 m. The wavelength is $\lambda = 3$ mm, and the 3 dB beamwidth is about 3.0° in elevation and azimuth. The antenna scans across the angular range of 360° at a scan frequency of about 3 Hz. The raw range resolution is about 0.32 m at 20 m. The radar is integrated with the CAS Outdoor Research Demonstrator (CORD): an eight-wheel, skid-steering all terrain unmanned ground vehicle (UGV) (see Fig. 1), which has been employed for the testing and the field validation of the system. The robot's sensor suite is completed by four 2D SICK laser range scanners, a mono-charge-coupled device (CCD) color camera, a thermal infrared camera, and a real-time kinematic/differential global positioning system/ inertial navigation system (RTK DGPS/INS) unit that provides accurate pose estimation of the vehicle.

The rest of the paper is organized as follows. Section 2 illustrates the radar perception system used for this research. Section 3 describes the proposed radar-based classifier, and experimental results are presented in Sect. 4. Conclusions are drawn in Sect. 5.

[1] http://www.nav-tech.com/Industrial%20Sensors2.htm

Fig. 1 The CORD UGV employed in this research. The sensor suite can be seen, attached to a rigid frame

2 Radar Perception System

In the proposed configuration, the radar is directed at the front of the vehicle with a fixed nodding angle to produce a grazing angle γ of about 11°, so that the center of the beam intersects the ground at a look-ahead distance approximately of 11.4 m in front of the vehicle, as shown in the explanatory scheme of Fig. 2. The origin of the beam at the center of the antenna is O. The proximal and distal borders of the footprint area illuminated by the divergence beam are denoted with A and B, respectively. The height of the beam origin with respect to the ground plane is h, the

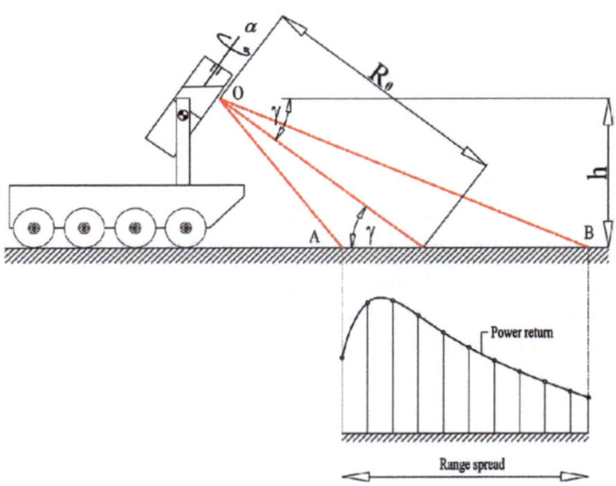

Fig. 2 Scheme of the radar configuration used for environment survey

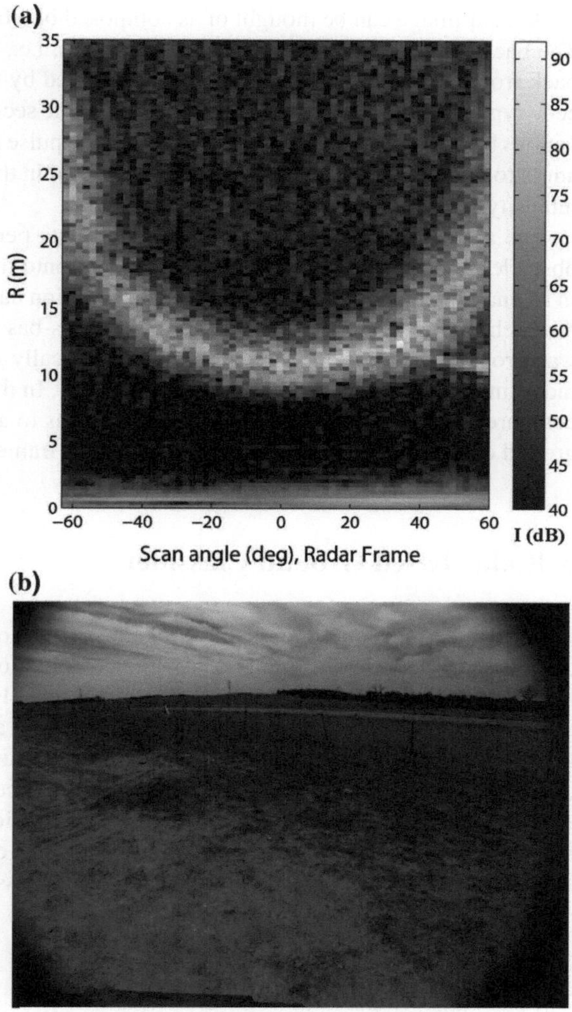

Fig. 3 A sample radar image acquired from a large, flat area: **a** scan angle-range image. **b** Camera image approximately colocated with the radar. Note the rich information content of the radar map due to its ability to sample reflectivity at multiple ranges for a single scan angle

slant range of the radar bore sight is R_0, and the scan angle is α. With a single sweep of 360°, the sensor produces an intensity map or radar image of the environment in front of the robot. As an example, a radar image, acquired from a large, relatively flat area is shown in Fig. 3a, along with its approximately colocated visual image shown in Fig. 3b. The abscissas in Fig. 3a represent the scan angle with an effective horizontal field of view (HFOV) of about 120°. The ordinates represent the range measured by the sensor. Amplitude values above the noise level suggest the presence of objects with significant reflectivity. Amplitude close or below the noise level generally corresponds to the absence of objects.

A radar image can be thought of as composed of a foreground and a background. The background is produced by the ground echo, i.e., the intensity return scattered back from the portion of terrain that is illuminated by the sensor beam. The ground echo typically appears as a high-intensity parabolic sector (see Fig. 3a). Radar observations belonging to the background show a wide pulse produced by the high incident angle to the surface. Conversely, obstacles present in the foreground appear as high-intensity narrow pulses.

The ability to automatically identify radar data pertaining to the ground and to obstacles present in the scene and project them onto the vehicle body frame results in an enabling technology for all visibility-condition navigation systems. In previous research by the authors [5], a theoretical physics-based model of the ground echo was proposed along with a method to automatically extract the background from radar images based on an edge-detection strategy. In this work, the ground model is compared against single background observations to asses their membership to the ground class within a self-learning classification framework.

3 Radar-based Ground Classifier

The first step of the classifier is the extraction of a set of features from radar images. The features provide a training set for the construction of a multivariate Gaussian model of the ground. The training set is automatically initialized at the beginning of the robot's operation via a bootstrapping approach and progressively updated, so that in every newly acquired radar scan, the latest training set is used to retrain the classifier. The latter consists of a Mahalanobis distance-based one-class classifier. It aims to generalize from training data to unseen situations in order to identify single new observations as ground or non-ground. In the rest of this section, first the feature extraction stage is discussed, then the radar-based classification system is presented.

3.1 Extraction of Radar Features

Radar features are extracted based on the ground echo model. We recall here the important properties of the model and refer the reader to [5] for more details. The power spectrum of the ground return can be expressed for a given scan angle as a function of the range R

$$P_r(R) = k\frac{G(R, R_0)^2}{\cos\gamma} \tag{1}$$

where k is a constant quantity, R_0 is the slant range, G is the antenna gain (usually modeled as Gaussian) and γ is the grazing angle, as explained in Fig. 2. Figure 4 shows a simulated wide pulse of the ground return obtained using (1).

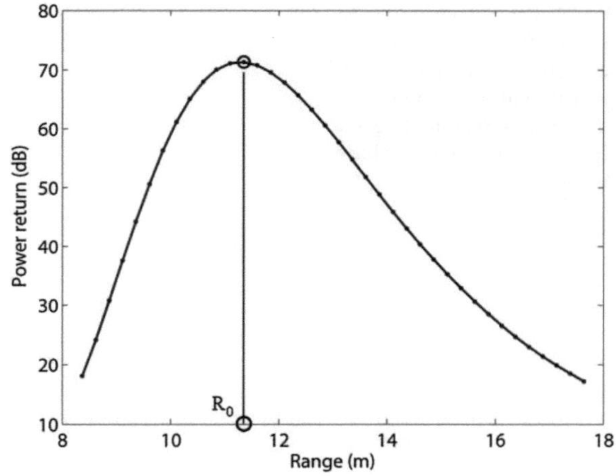

Fig. 4 Simulated power return of the ground echo for a given scan angle: the following parameters were adopted in the simulation: $k = 70\,\text{dB}$, $R_0 = 11.3\,\text{m}$, $h = 2.2\,\text{m}$, $\gamma = 11°$

The appearance of the ground is constructed upon a set of intensity and shape features that are obtained by fitting the ground model (1) to data belonging to the radar background. The underlying hypothesis is that a good match between the parametric model and the experimental data attests to a high likelihood of ground. Conversely, a poor goodness of fit suggests low likelihood due, for example, to the presence of an obstacle or to irregular or occluded terrain.

From (1), it can be seen that $Pr(R)$ is a function defined by the parameters k and R_0. Both parameters can be estimated by data fitting for the given scan angle.

The parameter k can be interpreted as the power return at the slant range R_0 and it is chosen as the first feature defining the ground appearance. A non-linear least squares approach using the Gauss-Newton-Marquardt method is adopted for data fitting. Output from the fitting process are the updated parameters R_0 and k as well as an estimate of the goodness of fit. The coefficient of efficiency was found to be well suited for this application and it is chosen as the second feature for our model:

$$E = 1 - \frac{\sum(t - y)^2}{\sum(t - \bar{t})^2} \tag{2}$$

t being the data point, $\bar{t}$ the mean of the observations, and y the output from the regression model. E ranges from $-\infty$ to 1, as the best possible value. Typical results are shown in Fig. 5. Specifically, in Fig. 5a, the model matches the experimental data very well, with a high coefficient of efficiency $E = 0.98$ and $k = 74.89\,\text{dB}$, thus attesting to high confidence of ground. Conversely, Fig. 5b shows an example where the goodness of fit is poor ($E < 0$); for this case a low confidence in ground is expected. However, relying on the first two features may be misleading in some cases. Figure 5c shows an example where a radar observation appears as a ground label due to the high coefficient of efficiency ($E = 0.91$), when there is actually no ground return. In order to solve this issue, a shape factor can be defined as

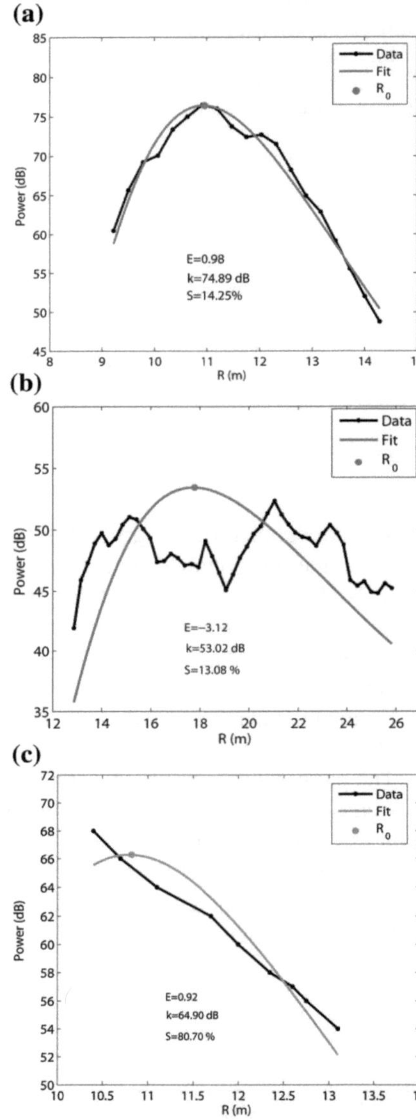

Fig. 5 Feature generation by model fitting: **a** good fit with high confidence in ground, **b** poor fit with low confidence in ground, **c** seemingly good fit with a high value of shape factor

$$S = \left| \frac{I_0 - I_{end}}{I_0} \right| \tag{3}$$

where I_0 and I_{end} are the initial and final intensity value of the ground echo. Our hypothesis is that a normal ground echo should have similar initial and final intensities due to the physical interaction between the radar emission and the ground. High values of S (for example $S = 80.7\,\%$ in Fig. 5c) indicate a discrepancy and suggest low confidence that the signal is a ground echo.

In summary, three main features define the ground model: the intensity associated with the slant range, k, the goodness of fit, E, and the shape factor, S. This set of features expresses our physical understanding of the problem and is used within a self-learning classification framework, as follows.

3.2 Algorithm Description

The ground model problem is formulated as a one-class classifier [17]. One-class classification techniques are particularly useful for two-class classification problems where one class, referred to as the target class is well-sampled, while the other class, named the outlier class is under-sampled. This is the case for our application where most of the radar background readings are reflected by ground with sparse instances of non-ground. Typically, the objective of a one class-classifier is that of constructing a decision boundary that separates the instances of the target class from all other possible objects. In our case, ground samples constitute the target class, while non-ground samples (i.e., obstacles) are regarded as the outlier class. Nevertheless, in open rural environments non-ground samples are typically sparse; in addition, the variation of all possible non-ground classes is unlimited. That makes it difficult to model the non-ground class, whereas, although it changes geographically and over time, the ground class is generally less variable in appearance than random objects. Furthermore, our objective is that of building a model of the ground. Therefore, it is reasonable to formulate the problem as one of distribution modeling, where the distribution to estimate is the ground class. Specifically, we adopt a multivariate Gaussian distribution to model positive ground samples. A Mahalanobis distance-based classifier [18] is then applied to estimate the membership likelihood of a given observation to the ground class following an outlier detection strategy. Looking at the Mahalanobis distance (MhD) and its distribution, it is possible to predict if this radar reading has an extremely low probability of belonging to ground and may be suspected to be an outlier.

Let X_t be an $n \times m$ data table representing a sample of x_i vectors with $i = 1, 2, \ldots n$, each characterized by m traits (scalar features): $X_t = \{x_1, \ldots, x_n\}$. These vectors constitute the training set at a given time t. If we compute the sample mean μ_t and the sample covariance Σ_t of the data in X_t, we can denote the ground model at this time as $M_t(\mu_t, \Sigma_t)$, where μ_t describes the location, and Σ_t the scale (shape) of the distribution. Then, in the next radar scan acquired at time $t + 1$, the single new observation z can be classified by estimating its squared Mahalanobis distance from the ground model:

$$d^2 = (z - \mu_t)\Sigma_t^{-1}(z - \mu_t)^t \tag{4}$$

Assuming that the vectors x_i are independent and have Gaussian distribution, it can be proved that the squared Mahalanobis distance is asymptotically distributed as the

m degrees of freedom chi-square distribution χ^2_m. Then, we can use the quantile β of the m degrees of freedom chi-square distribution as the delimiter (cutoff) for outlying observations. Let β denote a constant probability level: $0 < \beta < 1$. Let $\chi^2_{m;\beta}$ denote the appropriate quantile of the distribution. Then, it holds that

$$p(d^2 \geq \chi^2_{m;\beta}) = 1 - \beta \tag{5}$$

which means that values of d^2 greater than (or equal to) the value $\chi^2_{m;\beta}$ appear with a probability equal to $1 - \beta$. Now we define the cutoff for the Mahalonobis distance as

$$L_\beta = \sqrt{\chi^2_{m;\beta}} \tag{6}$$

Any radar observation with Mahalanobis distance d satisfying the inequality $d \geq L_\beta$ may be suspected to be an outlier. In other words, the pattern is an outlier, i.e. it is defined as a non-ground observation, if d^2 is greater than a threshold, which is computed as the β quantile $\chi^2_{m;\beta}$ of the m degrees of freedom chi-square distribution.

At the beginning of the robot's operation, the training set is initialized under the assumption that the vehicle starts from an area free of obstacles, so that the radar "looks" at ground only. Successively, the ground model is continuously updated during operation: the new ground feature vectors labeled in the most recent radar scans are incorporated, replacing an equal number of the oldest ground instances.

4 Experimental Results

In this section, experimental results are presented to validate our approach for ground segmentation using radar data. The system was integrated with the CORD UGV (see Fig. 1) and tested in a rural environment at the University of Sydney's test facility near Marulan, NSW, Australia. The test field features three wooden buildings and two stationary cars. The whole area is bounded by a fence and is characterized by relatively flat ground with sparse low grass. During the experiment, the CORD vehicle was remotely driven with an average speed of about 0.5 m/s and a maximum speed of 1.5 m/s. Variable yaw rates were achieved with a maximum of 1.12 rad/s (i.e., 64°/s) and roll and pitch angles of up to 4°. In this experiment, the RTK DGPS/INS unit and a high-precision 2D SICK laser range scanner provided the ground truth. The full data set is public and available online [19]. The path resulted in a total distance of 170 m traveled in about 5 min.

Figure 6 shows a typical result obtained during the experiment at the instant T = 47 s, when the vehicle moves toward the eastern area of the path, delimited by two lateral buildings with a car in front of the robot. Figure 6a shows the radar intensity image overlaid with the results obtained from the radar-based classifier. Ground labels are denoted by black dots, a black cross marks non-ground, and obstacles detected in the foreground as high-intensity narrow pulses are denoted by a black triangle.

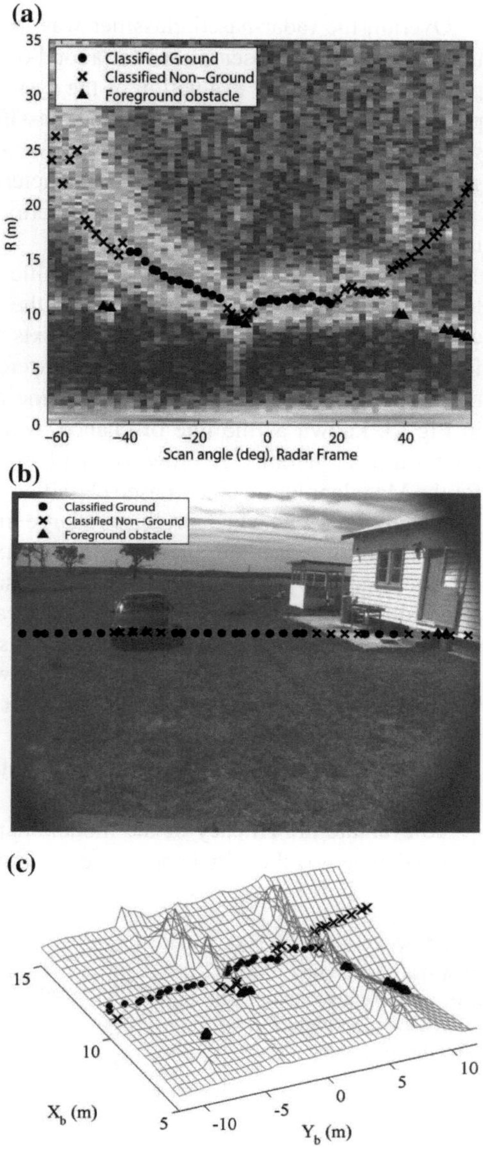

Fig. 6 Results obtained from the radar-based classifier for a scenario delimited by two side buildings and a frontal car: **a** output of the system, **b** results overlaid on the camera image, **c** and on the laser-generated ground-truth map

In Fig. 6b, the results are projected over the image plane of the camera for visualization purposes. Note that only the observations common to the field of view of both sensors are shown. Finally, a comparison with laser-generated ground truth is provided in Fig. 6c, showing the effectiveness of the proposed approach for ground segmentation.

Overall, the radar-based classifier was tested over 778 radar images, each containing 63 azimuth observations for a total of 48,960 classifications [20]. In order to provide a quantitative evaluation of the system performance, we measured the true positive and false positive rates of the classifier for the radar observations that fall within the camera field of view and that we can hand-label by visual inspection (i.e., ground truth). A receiver operating characteristic (ROC) curve was constructed by adjusting the cutoff value for terrain detection (i.e., by varying the β-quantile value in the range [0–1]).

The ROC curve for the proposed classifier is shown in Fig. 7 denoted by a black line. It plots the true positive rate (i.e., the fraction of ground patches that were correctly classified as ground) on the y-axis against the false positive rate (i.e., the fraction of non-ground samples, which were erroneously classified as ground by the system) on the x-axis. A diagonal line at 45° (shown by a black dotted line in Fig. 7), known as the line of chance, would result from a test which allocated observations randomly. In order to establish the optimal threshold value or cutpoint for the Mahalanobis distance-based classifier, we can use the ROC curve. In general, a good cutpoint is one which produces a large true positive rate and a low false positive rate. An intuitive method, therefore, is to maximize their difference. This approach assumes that true positive rate and a false positive rate are equally important. This is referred to as the Youden index [21], which can be interpreted as choosing the point on the ROC curve with the largest vertical distance from the line of chance. For our system, the point of maximum difference between the true positive rate and the false positive rate is reached at $\beta = 0.992$, and it corresponds to a true positive rate of 88.0 % and a false positive rate of 11.9 %. For this point, the overall accuracy, i.e. the fraction of correct detections with respect to the total number of classifications is of 87.8 %.

To evaluate the efficacy of the model adaptability, a static model was also constructed, whereby the initial appearance model built during the bootstrapping process

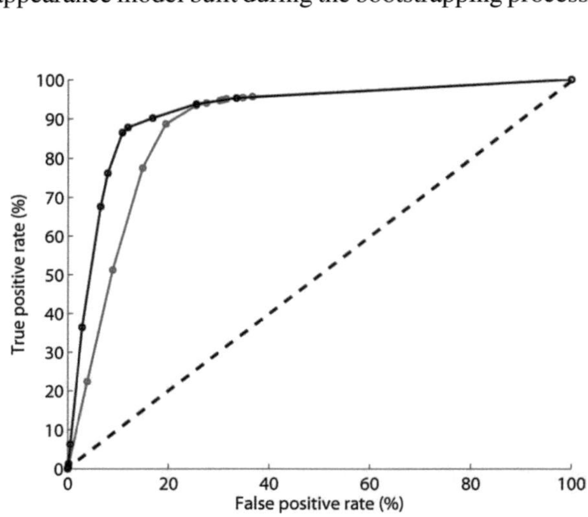

Fig. 7 ROC curve for the radar-based classifier. *Black line*: adaptive algorithm. *Grey line*: static algorithm. *Dotted line*: line of chance

was fixed and not further modified. The advantage of using an adaptive learning approach with respect to a batch training system can be shown by obtaining the ROC curve of the "static", where the ground model was constructed at the beginning of the sequence, and it was never updated. The ROC curve of the "static" classifier is denoted by the grey line in Fig. 7. The comparison between the two implementations can be performed in terms of the total area under the ROC curve (AUC). The larger the AUC, the better the overall performance of the test to correctly identify ground and non-ground observations. Therefore, the adaptive algorithm outperforms the static one, and the curve is closer to the perfect discrimination.

5 Conclusions

In this paper, a self-learning classifier was described for ground segmentation by an autonomous vehicle using MMW radar. Experimental results obtained using an UGV in field scenarios showed that the proposed radar-based classifier leads to good classification performance with the following main advantages: (a) enabling technology for all visibility-condition navigation systems, (b) self-learning training of the system, where the radar allows the vehicle to automatically acquire a set of ground samples, eliminating the need for time-consuming manual labeling, (c) continuous updating of the system during vehicle operation, thus making it adaptive and feasible for long range and long duration navigation applications.

Acknowledgments The authors are thankful to the Australian Department of Education, Employment and Workplace Relations for supporting the project through the 2010 Endeavour Research Fellowship 1745_2010. The authors would like also to thank the National Research Council, Italy, for supporting this work under the CNR 2010 Short Term Mobility program. This research was undertaken through the Centre for Intelligent Mobile Systems (CIMS), and was funded by BAE Systems as part of an ongoing partnership with the University of Sydney. The financial support of the ERA-NET ICT-AGRI through the grant Ambient Awareness for Autonomous Agricultural Vehicles (QUAD-AV) is also gratefully acknowledged.

References

1. A. Milella, G. Reina, R. Siegwart, Computer vision methods for improved mobile robot state estimation in challenging terrains. J. Multimedia. **1**(7), 49–61 (2006)
2. G. Reina, G. Ishigami, K. Nagatani, K. Yoshida, Odometry correction using visual slip-angle estimation for planetary exploration rovers. Adv. Robot. **24**(3), 359–385 (2010)
3. N. Vandapel, S. Moorehead, W. Whittaker, R. Chatila, R. Murrieta-Cid, Preliminary results on the use of stereo, color cameras and laser sensors in Antarctica, in *International Symposium on Experimental Robotics*, Sydney, Australia, 1999
4. T. Peynot, J. Underwood, S. Scheding, Towards reliable perception for unmanned ground vehicles in challenging conditions, in *IEEE/RSJ International Conference on Intelligent Robots and Systems*, St Louis, MO, USA, 2009

5. G. Reina, J. Underwood, G. Brooker, H. Durrant-Whyte, Radar-based perception for autonomous outdoor vehicles. J. Field Robot. **28**(6), 894–913 (2011)
6. R. League, N. Lay, System and method for tracking objects using a detection system. U.S. patent no. 5.587.929, 1996
7. S. Clark, H. Durrant-Whyte, The design of a high performance MMW radar system for autonomous land vehicle navigation, in *International Conference Field and Service Robotics*, Sydney, Australia, 1997
8. H. Durrant-Whyte, An autonomous guided vehicle for cargo handling applications. Int. J. Robot. Res. **15**(5), 407–441 (2002)
9. A. Foessel-Bunting, S. Chheda, D. Apostolopoulos, Short-range millimeter-wave radar perception in a polar environment, in *International Conference Field and Service Robotics*, Leuven, Belgium, 1999
10. G. Brooker, R. Hennesy, C. Lobsey, M. Bishop, E. Widzyk-Capehart, Seeing through dust and water vapor: Millimeter wave radar sensors for mining applications. J. Field Robot. **24**, 527–557 (2007)
11. J. Mullane, D.M. Adams, W.S. Wijesoma, Robotic mapping using measurement likelihood filtering. Int. J. Robot. Res. **28**(2), 172–190 (2009)
12. A. Milella, G. Reina, J. Underwood, B. Douillard, Combining radar and vision for self-supervised ground segmentation in outdoor environments, in *IEEE/RSJ International Conference on Intelligent Robots and Systems*, San Francisco, CA, USA, 2011, pp. 255–260
13. A. Milella, G. Reina, J. Underwood, B. Douillard, Visual ground segmentation by radar supervision, in *Robotics and Autonomous Systems* (2013, in press)
14. C. Brooks, K. Iagnemma, Self-supervised terrain classification for planetary rovers, in *NASA Science Technology Conference*, 2007
15. S. Thrun et al., Stanley: The robot that won the DARPA grand challenge. J. Field Robot. **23**(9), 661–692 (2006)
16. D. Stavens, S. Thrun, A self-supervised terrain roughness estimator for offroad autonomous driving, in *Conference on Uncertainty in AI*, 2006
17. D.M.J. Tax, One-class classification, concept learning in the absence of counter examples, PhD Thesis, Delft University of Technology, Delft, Netherlands, 2001
18. E.O. Duda, P.E. Hart, D.G. Stork, *Pattern Classification*, 2nd edn. (Wiley, New York, 2001)
19. T. Peynot, S. Scheding, S. Terho, The Marulan data sets: Multi-sensor perception in a natural environment with challenging conditions. Int. J. Robot. Res. **29**(13), 1602–1607 (2010)
20. G. Reina, A. Milella, J. Underwood, Self-learning classification of radar features for scene understanding. Robot. Auton. Syst. **60**(11), 1377–1388 (2012)
21. J.A. Hanley, B.J. McNeil, The meaning and use of the area under a receiver operating characteristic (ROC) curve. Radiology **143**, 29–36 (1982)

Development of a Low Cost Multi-Robot Autonomous Marine Surface Platform

A. Valada, P. Velagapudi, B. Kannan, C. Tomaszewski, G. Kantor and P. Scerri

Abstract In this paper, we outline a low cost multi-robot autonomous platform for a broad set of applications including water quality monitoring, flood disaster mitigation and depth buoy verification. By working cooperatively, fleets of vessels can cover large areas that would otherwise be impractical, time consuming and prohibitively expensive to traverse by a single vessel. We describe the hardware design, control infrastructure, and software architecture of the system, while additionally presenting experimental results from several field trials. Further, we discuss our initial efforts towards developing our system for water quality monitoring, in which a team of watercraft equipped with specialized sensors autonomously samples the physical quantity being measured and provides online situational awareness to the operator regarding water quality in the observed area. From canals in New York to volcanic lakes in the Philippines, our vessels have been tested in diverse marine environments and the results obtained from initial experiments in these domains are also discussed.

A. Valada (✉) · P. Velagapudi · B. Kannan · C. Tomaszewski · G. Kantor · P. Scerri
The Robotics Institute, Carnegie Mellon University, 5000 Forbes Ave, Pittsburgh PA 15213, USA
e-mail: avalada@cmu.edu

P. Velagapudi
e-mail: pkv@cmu.edu

B. Kannan
e-mail: bkannan@cmu.edu

C. Tomaszewski
e-mail: ckt@cmu.edu

G. Kantor
e-mail: kantor@cmu.edu

P. Scerri
e-mail: pscerri@cmu.edu

K. Yoshida and S. Tadokoro (eds.), *Field and Service Robotics*, Springer Tracts in Advanced Robotics 92, DOI: 10.1007/978-3-642-40686-7_43,

1 Introduction

Understanding and monitoring complex natural systems is critical to our ability to use the resources provided by those systems while ensuring the systems stay healthy over long periods of time. Ecologically sensitive marine areas are large and remote, making it difficult to see impacts from threats such as climate change and pollution. Scientists and governments are interested in tracking contaminants in the water, assessing environmental change, monitoring sea levels, surveying the coastline and understanding biodiversity. Unfortunately, monitoring and understanding large spatially distributed systems is difficult due in part to partial observability, complex dynamics, high cost, and the diversity of data that must be collected. Radically new approaches to sampling and monitoring that enable collection of collecting large amounts of data over large areas and extended periods are required if we are to balance economic development with conservation for future generations [6].

Since the 1990 s Unmanned Surface Vehicles (USV) or Autonomous Surface Craft (ASC) have been demonstrated to work in a broad set of applications such as bathymetric mapping, environmental sampling, defense and robotics research, yet there has been very little commercialization or adoption in these domains. Unlike ground or aerial vehicles, the extremely dynamic and frangible environment in which ASC's are deployed make the task of developing these machines far more complex. For example, in applications such as monitoring pollution around coral reefs, the system has to operate in a manner that does not damage the delicate ecosystem. For long range deployments, communication range and data transmission bandwidth are also important considerations. Although a few commercial ASC's have been developed in the past few years, the high cost of these systems has prevented many agencies from adopting them. Most of these vehicles are fitted with expensive navigation systems and sensors, thereby bringing the cost of the entire vehicle to several thousand dollars. These are only some of the challenges that the ASC community is facing today (Fig. 1).

Fig. 1 Team of CRW sampling at Taal lake in Philippines

In this paper we present our work on developing a multi-robot ASC platform which addresses some of these challenges. We present the Cooperative Robotic Watercraft (CRW) platform, a commercial smartphone based solution that leverages the built-in inertial and visual sensors of modern mobile devices for navigation and localization. In recent years, smartphones have redefined the notion of mobile computing platforms. Unrelenting advances in mobile processors, low power sensors, and battery technology have made smartphones more powerful, affordable, and portable, opening up a number of interesting novel applications. By using a combination of WiFi and 3G, we improve communication reliability, extend the range of system deployments beyond the limited range offered by standard 802.11 networks and take advantage of existing infrastructure while maintaining a low development cost for each craft—below $500. The reduced vehicle cost allows for development of a fleet of boats, which can cooperate to perform tasks and achieve goals more efficiently and expediently than a single craft. We envision very large fleets of CRW, perhaps even numbering in the hundreds, autonomously exploring large bodies of water under the supervision of a small number of operators. Previous work has detailed the challenges involved in such coordination from a multi-agent perspective, including challenges in task allocation, information sharing, and adjustable autonomy [3].

The overarching goal of our work is to develop a low-cost multi-robot marine surface platform that is easy to deploy and versatile enough to be used for a wide range of applications. This results in several unique design constraints that have not previously been explored. Section 3 outlines these considerations and describes the specific design choices that we made. We then discuss our validation experiments in cooperative adaptive water sampling, where a suite of sampling algorithms ranging from random to highest uncertainty tracking were developed and the system performance in different operating environments (lakes, rivers, and floods) as well as for varying team sizes were compared. The results from several field trials and subsequent analysis is presented in Sect. 4. Finally, Sect. 5 provides a brief conclusion along with a description of our plans for future research.

2 Related Work

Although there has been considerably less work done on ASC's compared to AUV's, there are still some notable platforms that have made significant contributions to this field. One of the first ASC's that was developed was the ARTEMIS platform, which was was used to perform bathymetry and test control algorithms [10]. After several iterations of the ARTEMIS platform, a new platform known as AutoCat [5] was developed to improve upon the shortcomings of its predecessors. Tele-supervised Adaptive Ocean Sensor Fleet [1] is a deep sea multi-robot science exploration system that combines a group of robotic boats to enable in situ study of phenomena in the ocean-atmosphere interface as well as on the ocean surface and sub-surface. The NAMOS project [9] developed an airboat which works in conjunction with static nodes that are to be deployed prior to sampling. The static nodes provide temporal coverage while the airboat samples a good spatial resolution of data. The OASIS platform is a long-duration solar-powered autonomous surface vehicle, designed for autonomous global open-ocean operations [7]. Recently there has also been some work on developing a unique ASV which runs on the energy harnessed from waves. This ASV known as the Wave Glider [2] is also equipped with solar panels and has been demonstrated to work well for long distance as well as extended deployments. There has also been some exciting work in developing multi-robot systems with a wide variety of capabilities [4], but under tightly controlled conditions.

Although these platforms are extremely capable and engineered specific to the requirements of the operating domain, the large associated cost with these platforms make them infeasible for large scale deployment.

3 Cooperative Robotic Watercraft

The primary objective of our work is to design, develop and test an application independent low cost, multi-robot platform that can be used for collecting scientific data and to monitor dynamic spatial phenomena over large distributed areas with minimal human supervision. In this chapter we discuss five challenging areas that impact the process of developing autonomous watercraft teams. These include: *hardware design*, *sensing and autonomy*, *user interface*, *communication and coordination*, and *exploration and coverage*.

As the system is equipped with a smartphone, the CRW platform has the capability to use either WiFi or 3G for communications. Application specific algorithms decide where the vessels need to go and determine which is best for each required location. In the future, we anticipate making this is a distributed process, but it has not been required until now as good wireless connectivity has been available. The exploration and coordination aspects are handled according to the application specific tasks. The following sections describe the other key areas.

3.1 Hardware Design

For the design of individual craft we chose that of a traditional airboat, which is ideally suited for navigating shallow or ecologically sensitive areas such as reefs and estuaries without damaging the marine ecosystem due to its flat bottom and absence of operating components below the waterline. Another advantage of this design is that it can be used in urban floods where debris filled water often hinders rescue operations. Throughout the development process, several design iterations of various sizes ranging from 40 to 70 cm long were fabricated and tested to achieve our desired performance characteristics.

The main design criteria was to develop a system that is low-cost as well as easy to manufacture and repair. Unlike other expensive autonomous marine vehicles, the

CRW platform was designed to work cooperatively in large teams to accomplish a task that would otherwise be impractical by a single vessel, which placed a few unusual constraints on the physical design of the vehicle. Simplicity and ease of fabrication were critical to keeping the cost of the craft as low as possible, however overall system robustness and modularity was also highly desired. Since even a relatively low failure rate for a large team would result in a cost-prohibitive amount of time for repairs, the individual airboats were designed with modular, interchangeable components, which can easily be swapped in the field.

The hull of the vessel is fabricated from multiple sheets of extruded polystyrene foam to be resilient and inexpensive. The sheets are bonded together into a single volume and carved using a heated nichrome wire setup. The hull is then sanded, painted, and coated with sealants for weatherproofing. An important feature of the platform is the self containment of the system components: the propulsion and steering system is bolted to a single deck plate which is mounted on top the vessel's hull, the electronics box slots into a dedicated compartment in the hull and is secured to the deck plate, and the smartphone is mounted to the deck plate at the bow of the craft. Any of these components can be easily detached and replaced within a couple of minutes.

3.1.1 Propulsion and Steering

The propulsion and steering system is one of the most important components of the watercraft. Our initial design was based on that of a conventional airboat, in which a shrouded propeller powered by a brushless motor is mounted at the rear end of the vessel and a pair of rudders is used to direct thrust in order to steer the vehicle. Despite the tried and true nature of the traditional design, the production and maintenance of the rudder mechanism was cumbersome and its complexity did not support our vision of a modular system. After experimenting with several initial prototypes, we designed a rudder-less steering system where the entire propulsion assembly is actuated to steer the vehicle. In our design, the motor is mounted within a close-fitting PVC duct, which serves to decrease thrust losses and noise due to the tip vortices of the propeller. The duct is mounted to a circular acrylic base, which is free to rotate on bearings within a larger acrylic retaining ring. A servo is top-mounted to the circular base, with its horn protruding below the assembly where it can engage a matching cutout on the deck plate. Using the servo, the entire duct can pivot allowing for more efficient thrust vectoring than the traditional rudder system, particularly at rudder angles further from the neutral position. At full charge the propulsion system is capable of producing up to 8.2 Newtons of thrust, which can be entirely directed along any vector to the rear of the vessel, facilitating a small turning radius. An exploded model of the propulsion assembly is shown in Fig. 2a.

The primary material used in the propulsion and steering system is weather and UV resistant acrylic, which is cut from sheets using a laser cutter. The acrylic components are then assembled and bonded together with a special adhesive before installing the PVC ducting, motor, propeller, and protective mesh. Once templates were created

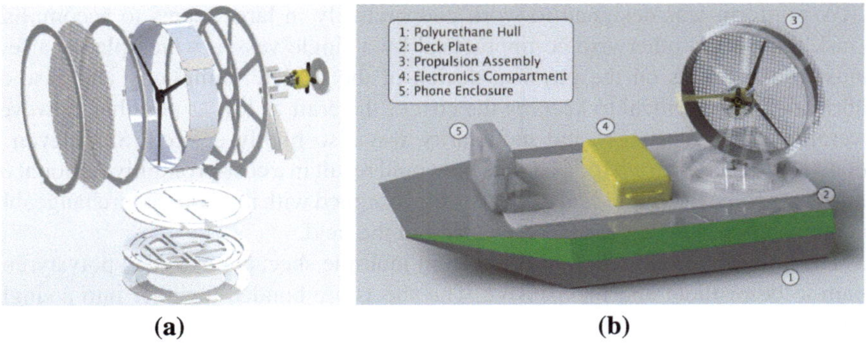

Fig. 2 CAD models of the CRW platform **a** Exploded model of the propulsion and steering system. **b** Cooperative robotic watercraft

for these acrylic components, manufacturing was seamless and many propulsion assemblies could be produced in a matter of hours. This new design proved not only to be more robust and efficient but also improved the overall maneuverability of the vehicle and could easily be swapped out in case of failure.

3.1.2 Electronics

The onboard electronics are decomposed into two major components: a phone assembly which houses the smartphone and orients it to best align its sensors with the vessel, and an electronics assembly that contains the main power source and actuation drivers for the vehicle as well as embedded electronics to interface with additional external sensors. The heavier electronics assembly is placed close to the center of mass of the vessel and is partially recessed into the deck. The assembly is powered by a 10 Ah NiMh battery, which provides about 150 min of continuous operation. An Arduino Mega provides a fast and flexible array of digital and analog I/O for controlling the propulsion system and external sensor modules. Sensors such as fluorometers and sonars are directly plugged into the Arduino through analog, digital, or serial channels. Bluetooth is used as an interface between the smartphone and the electronic components, enabling us to keep the two components decoupled physically and architecturally. Figure 3 shows the system architecture diagram.

3.2 Sensing and Autonomy

At the core of the CRW platform design is an Android smartphone which provides computing and inertial sensors in an integrated package. This design decision allows us to avoid the cost-prohibitive process of integrating sensory, computing, and

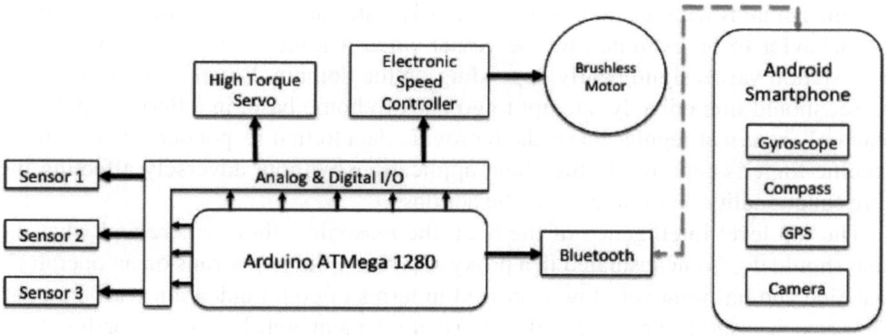

Fig. 3 System architecture diagram

communication components in a comparable form factor. Modern smartphones also provide access to multiple modes of communication such as 3G, WiFi and bluetooth. The phone is enclosed in a watertight case that is mounted upright on the deck of the vessel to best utilize the camera's field of view. The Google Nexus S makes an ideal candidate due to its powerful open source Android development environment.

Computationally intensive tasks and high-level controls are performed on the phone, while low-level sensor interfacing and high-speed control loops execute on the Arduino. The primary issue to overcome pertains to the quality of obtained data from the embedded sensors required for control, specifically the GPS, gyroscope, and compass. Layers of filters are required to smooth the data to extract sufficiently clean information to effectively control the boat.

At times throughout sampling it becomes desirable to collect a physical sample to verify sensor readings or for more detailed analysis than the onboard sensors can provide. A water sampling mechanism provides this capability, allowing for up to six unique samples to be collected before its storage must be emptied. The sampler is loaded with evacuated glass vials, which are punctured by needles affixed to individual rubber intake hoses leading beneath the boat. The vacuum in the vials is held by pinching the hoses against rounded pins from a tumbler lock, which ride against a central cam. A single actuator is then used to rotate the cam, leveraging the elasticity of the hoses to displace the tumbler lock pins and selectively draw samples into each vial. By using individually sealed vials and a unique hose for each sample, cross contamination between samples is minimized without requiring any additional actuators.

As all the computation is performed on the Android smartphone, local intelligence for each boat resides on the phone. Layers of functionality separate general modules from application specific modules. An end-user interface provides a single operator with an overview of the state of the boats and provides high and low level commands for interacting with them. The boat executes the core functionality via the boat server. Client applications and additional modules running on the phone provide the domain specific functionality. This design allows us to make subtle changes for specific domains without modifying previously tested and reliable code. For example, the behavior to be exhibited by the vessel when it loses communication with the base station varies significantly depending on the domain. When doing testing, the vessel should immediately attempt to go back to home base; in a flood response, it may only return at regular intervals to provide data to first responders. This domain specific logic is captured in the client applications without adversely affecting the core functionality that implements the actions.

The top level intelligence of the boat, the reasoning about where and what the boat should do, is encapsulated in a proxy. Currently the proxy runs on an operator's machine and has relatively low-overhead in terms of communication with the boat. The proxy is responsible for path planning to implement high-level operator directives about areas to visit or search.

3.3 Operator Interface

A centralized operator interface provides an enhanced situational awareness about the multi-robot teams and the operating domain. The interface displays information about the locations of all the vessels, overlaid on a map of the environment. Using the interface the operator can specify high-level objectives either as waypoints, paths or areas to search, or low-level direction commands to the boats. The watercraft also transmit images captured from the on-board camera at approximately 1 Hz. An image queue on the operator's side receives and reorders the images, allowing the operator to observe, discard, or save images for later use. The operator interface emphasizes simplicity and reliability over complex functionality. A snapshot of the main panel of the interface is shown in Fig. 4. The interface is also composed of subpanels that display application specific information. For example, the water sensing subpanel displays the sampling grid cells with the current mean of observations within each cell as well as buttons to modify the number cells and sensor channels. The debug subpanel displays additional vital information about the vehicle such as the yaw rate, coordinates of the current position and waypoints positions, controller gains and tuning sliders.

4 Experimental Results

The CRW platform is designed as an application independent fleet of ASV's. In order to evaluate the designed system, we performed experiments in different domains such as water quality monitoring, depth buoy verification, flood disaster mitigation and bathymetry. In this chapter we will focus on water quality monitoring as this application has attracted much attention in the recent years.

Fig. 4 Main panel of the operator interface

4.1 Water Quality Monitoring

Investigating surface pollution is a difficult and complex task that can be challenging even for an autonomous watercraft. Pollutants such as petroleum hydrocarbons, heavy metals, polycyclic aromatic hydrocarbons, pesticides, and some toxic organic compounds are the common contaminants of concern. Even though there are several protocols for safely disposing such compounds, the lack of efficient monitoring techniques has lead to illegal dumping in water bodies. Using cooperative ASC's we can drastically improve the management of water resources by using intelligent techniques such as adaptive sampling to analyze the trend of contamination and identify the sources of pollution. Analysis of surface water also gives us information on the interaction between the atmosphere and hydrosphere, which is of scientific interest. While performing sampling with an autonomous watercraft, our main objective is to collect measurements of interest in a way that best explains the hydrological processes. Different sampling applications require either spatial or temporal resolution of data; while sampling to identify the sources of pollution, it is desirable to sample at locations having greater variance, whereas while sampling to map a hydrological process, the main objective is to obtain the highest resolution possible. In either of these applications the vehicles paths have to be planned and replanned according to the measurements obtained. Sampling of the entire field has to be performed before the field varies significantly, hence the use of a single vehicle to monitor large areas is impractical.

During our initial sampling efforts using the CRW system, we implemented a highest uncertainty tracking cooperative sampling algorithm and compared the results to those obtained using random search and lawnmower patterns. We chose temperature and specific conductivity as the quantities to be sampled, as conductivity gives us a measure of the total dissolved salts (TDS) in the water and temperature is an important factor that influences biological activity in the water. The sample space is then discretized into grid cells where the number of cells are configurable. Determining the number of cells to use is a critical task as there will be an optimal number of cells for the selected sampling space for which an algorithm will work most efficiently, and choosing more or less cells than necessary might lead to over or under-sampling.

$$v_{i,j} = \sigma^2 \times \alpha^c$$
$$\alpha < 1.0 \qquad (1)$$

In random search sampling, the vehicle is given random cells to sample one after the other until the sampling time expires, while in lawnmower sampling, the vehicle sweeps the area in a lawnmower fashion. These two algorithms give us a good reference for comparison. As we use multiple vehicles for sampling, it was necessary to divide the sample space into sub regions and assign each vehicle to a specific subregion for the lawnmower sampling approach. For the random search approach, dividing the sampling region was not required since each vehicle is assigned a random cell to sample. In the maximum uncertainty tracking algorithm, each cell is assigned a value based on the variance and the number of times that cell has been visited. The standard deviation of all the observations in each cell is calculated and the new value of the cell is given by Eq. 1, where c is the number of times that cell has been visited, σ^2 is the variance and α is a constant. A path is then planned to the cell which has the greatest v. Initially when the count of all the cells are zero, v is assigned a high sensor value so that the vehicle visits all the cells at least once, this gives an opportunity to the vehicle to explore the area before sampling for the maximum uncertainty.

For this algorithm to work efficiently the standard deviation at each cell should not vary very significantly from another. The time taken for the algorithm to converge directly correlates with the difference in the standard deviation between the cells. Other factors such as number of cells also influences the performance of the algorithm. After sampling the entire field, the surface is reconstructed using interpolation. The fidelity of the reconstruction is then determined using the estimation error obtained at different periods of time.

4.1.1 Experimental Domain I: Taal Lake

The Taal lake in Philippines is famous for many reasons. It is the largest fish farm in the Philippines and rather uniquely, the world's largest lake on an island on a lake on an island. In May 2011 however, it became infamous for another reason: it was the site of one of the largest fish kills in the world. About 956 metric tons of fish appeared dead on the surface of the lake, costing the fish farming industry nearly 70 million PHP (1.6 million USD) in lost revenue. One of the major reasons cited for

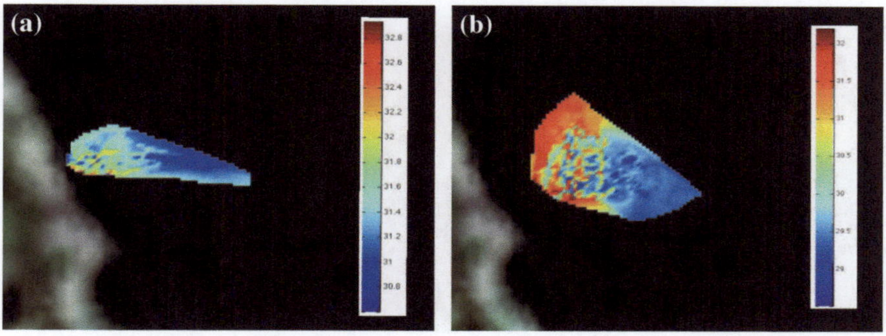

Fig. 5 Surface temperature variation before and after rainfall in Taal lake **a** Before rainfall. **b** After rainfall

this disaster is the sudden rise in temperature of the lake followed by rainfall that prevented the adequate exchange of oxygen on the surface. A team consisting of five CRW was deployed in Taal lake to observe the changes in the surface temperature before and after a tropical rainfall. Plots shown in Fig. 5 displays an overlay of the surface temperatures obtained by the CRW team and an image captured during deployment is shown in Fig. 1. It was observed that the average surface temperature varied from 31.81 °C before the rainfall to 29.96 °C after the rainfall.

4.1.2 Experimental Domain II: Gowanus Canal

The Gowanus Canal is one of the most polluted bodies of water in the United States and was recently named a superfund site by the Environmental Protection Agency. Effluents from chemical and manufacturing plants along the canal are the major source of contamination of the waterbody. A team of CRW was used to autonomously map the specific conductivity distribution in the area to analyze the spatial variation in the field. Significant spatial variation suggests manual sampling is not viable as several hundred points would have to be sampled to get a low estimation error. Figure 6c shows a plot of the sample space discretized into 10×10 cells with the standard deviation of each cell and the number measurements obtained in them indicated by red circular markers. It can be seen that even though a large number of samples are taken in most of the cells, there is a high variation in the measured value. The mean of the observations in the cells varied from 7.7050 to 7.5607 dS/m across the sensing field and the variance was calculated to be 0.0023. The spatial distribution of the specific conductivity of a section of the Gowanus Canal is shown in Fig. 6a. The average measured specific conductivity was 7.6387 dS/m, which is well above safe limits.

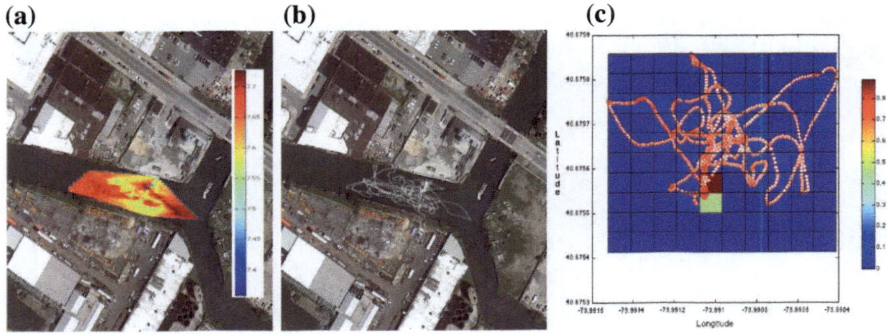

Fig. 6 Specific conductivity variation in a section of the Gowanus canal **a** Specific conductivity overlay. **b** Path traversed by CRW during sampling. **c** Plot of the standard deviation of each cell with the number of observations shown using *red markers*

4.1.3 Experimental Domain III: Allegheny River

A more rigorous experimental validation of the random sampling, lawnmower pattern, and the highest uncertainty algorithms were performed on a branch of the Allegheny river shielded from the main flow, located by Washington's Landing in Pittsburgh, PA. All three sampling algorithms were evaluated using a team of CRW consisting of three vessels equipped with sonars as well as conductivity and temperature sensors. The sampling area was about 6510 m^2, discretized into 10×10 cells and the sampling time for each algorithm was 30 min. To quantify the performance of the highest uncertainty tracking adaptive sampling algorithm, we calculated the normalized error and compared it with that of the random sampling and lawnmower patterns. The error was calculated by subtracting the value of each cell at the end of 30 min to that at each sampling interval. This process also gives us the rate of convergence to the final measurements. It is assumed that the field does not vary significantly for the period of sampling and the values of the cells at the end of the sampling time are an approximate measure of the ground truth. To analyze the final field estimation error, the sum of errors of all the cells at discrete sampling intervals was plotted. The graph in Fig. 7a shows the trend obtained from testing using one

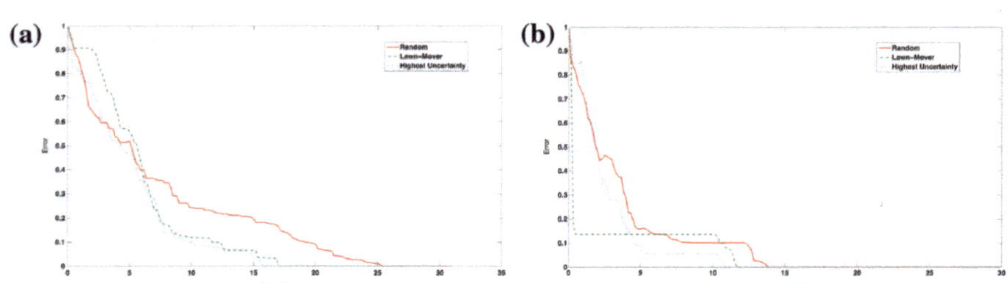

Fig. 7 Error comparison between random, lawnmower pattern and highest uncertainty sampling **a** Team consisting of one CRW. **b** Team consisting of two CRW

CRW. With the random sampling, more measurements end up being taken in the cells that are in the center of the grid, as the vessel traverses to random waypoints from one end of the sample space to another. In the case of the lawnmower pattern, a more uniform trajectory is followed, hence exploration is sequential and slow. With the highest uncertainty algorithm, priority is first given to exploration of the sample space and then to tracking the cells with highest variance. It can be seen in Fig. 7a that this algorithm has more of a uniform downward trend and the final value converges faster than both random and lawnmower patterns.

The time taken to converge to the final value can be improved by cooperatively sampling the area, thereby reducing the time required for exploration and maximizing coverage. To demonstrate the performance of the algorithms using multiple vessels, experiments on the algorithms were performed using teams of two and three CRW. In Fig. 7b, the lawnmower pattern sampling has the least initial error, this can be attributed to the fact that each vessel samples only the subregion assigned to it, hence the initial explorative sampling is fast and even, but the convergence to the final value slows down as the vessels follow a fixed pattern. With the highest uncertainty algorithm, fast convergence is achieved as the observations from both the vessels are used to calculate the uncertainty and plan the next waypoint accordingly. With a single CRW, convergence was achieved in 15.6 min, whereas with two CRW, convergence was achieved in 10.4 min. A comparison of the error for the lawnmower pattern with varying team sizes is shown in Fig. 8. With a team consisting of three vessels, convergence to the final value was achieved within 10.5 min, which is still higher than the result achieved using two vessels with the highest uncertainty algorithm.

Bathymetric mapping is one of the most successful applications for ASV's and AUV's as bathymetric maps are an important navigational aid for all marine vehicles. A team of CRW equipped with sonars were used to map the river bed near Washington's Landing. To create a three dimensional surface model, a grid is first

Fig. 8 Error comparison for lawnmower patten sampling with CRW teams of varying sizes

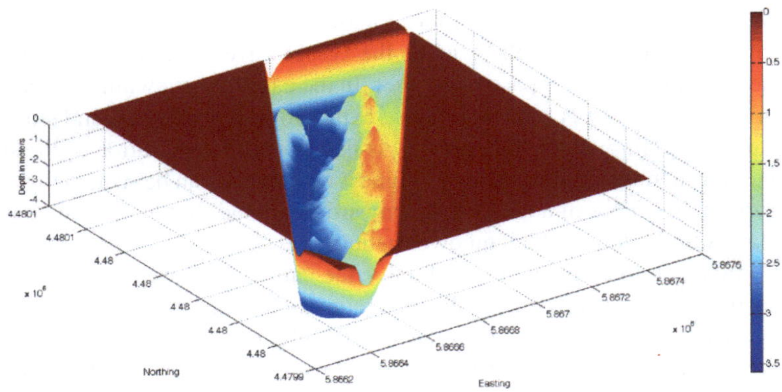

Fig. 9 Terrain model of a section of the Allegheny riverbed

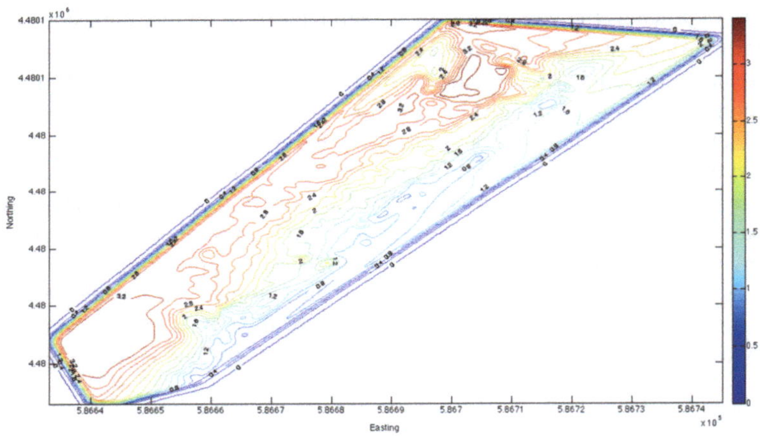

Fig. 10 Countour map of a section of the Allegheny river bed

Table 1 Deployment statistics

Experimental domain	Team size	Time in water (Hours)
Taal lake	5	95
Gowanus canal	2	15
Allegheny river	3	70

constructed using the range of the obtained values and interpolation is performed on the measured data. Gaussian smoothing is applied to smooth any noise in the measured data and a surface is fit on it. A map developed from sampling portraying the isobaths and the terrain model for the same area is shown in Fig. 9a, b.

5 Conclusion and Future Work

In this paper we described the design and architecture of a team of CRW and analyzed its performance in the water quality monitoring application. The integrated smartphone solves many inherent problems associated with ASV's relating to long range communication, situational awareness, system integration, and cost effectiveness. The custom designed propulsion assembly not only increased the efficiency of steering of CRW but also remarkably improves the fault tolerance of the entire system while compared to the traditional design. Additionally, by incorporating multiple layers of communication and a centralized coordination protocol, we ensure efficient and robust operation of the autonomous system. We compared the estimation error of the highest uncertainty tracking algorithm with random sampling and lawnmower patterns for teams of varying sizes. The highest uncertainty algorithm not only has lower average error but also converges faster than the random sampling and lawnmower patterns. Our experiments totaling to more than 180 boat hours in different operational and environmental conditions and with with teams of up to five CRW have confirmed the versatility and the serviceability of our platform. A table detailing the approximate testing time in water at each location along with the team size is shown in Table 1.

Ongoing work is focused on system modeling and development of a nonlinear controller with an integrated planner for this underactuated system. Unlike conventional airboats, the actuated propulsion assembly enables the CRW to perform extremely tight turns, thereby allowing for very accurate tracking by planning paths using the system dynamics. Future work includes developing more intelligent multi-robot adaptive sampling algorithms that reduce the estimation error and improve the spatial resolution of surface reconstruction. Another promising application for our work is depth buoy verification, where a team of CRW is used to visually identify the depth buoys which are placed in river channels to mark the safe passage of ships and barges, a process that is currently performed manually at a considerable expense (Fig. 10).

References

1. A. Elfes, G.W. Podnar, J.M. Dolan, S. Stancliff, E. Lin, J.C. Hosler, T.J. Ames, J. Higinbotham, J.R. Moisan, T.A. Moisan, et al., The telesupervised adaptive ocean sensor fleet architecture: coordination of multiple oceanic robot boats, in *Proceedings of the IEEE Aerospace Conference*, 2008
2. R. Hine, P. McGillivary, Wave powered autonomous surface vessels as components of ocean observing systems, in *Proceedings of PACON 2007*, Honolulu, HI, June 2007
3. J.C. Leedekerken, M.F. Fallon, J.J. Leonard, Mapping complex marine environments with autonomous surface craft, in *12th International Symposium on Experimental Robotics 2010*, New Delhi & Agra, India, December 2010
4. K.H. Low, G. Podnar, S. Stancliff, J.M. Dolan, A. Elfes, Robot boats as a mobile aquatic sensor network, in *Proceeding Workshop on Sensor Networks for Earth and Space Science*

Applications at the International Conference on Information Processing in Sensor Networks, 2009
5. J. Manley, A. Marsh, W. Cornforth, C. Wisema, Evolution of the autonomous surface craft autocat, in *Proceedings of Oceans 2000*, MTS/IEEE Providence, RI, October 2000
6. G. Niemi, D. Wardrop, R. Brooks, S. Anderson, V. Brady et al., Rationale for a new generation of indicators for coastal waters. Environ. Health Perspect **112**(9), (2004). doi: 10.1289/ehp.6903
7. G.W. Podnar, J.M. Dolan, A. Elfes, S. Stancliff, E. Lin, J.C Hosier, T.J. Ames, J. Moisan, T.A. Moisan, J. Higinbotham, et al., Operation of robotic science boats using the telesupervised adaptive ocean sensor fleet system, in *Proceedings of the IEEE International Conference on Robotics and Automation (ICRA '07)*, 2007
8. P. Scerri, B. Kannan, P. Velagapudi, K. Macarthur, P. Stone, M.E. Taylor, J. Dolan, A. Farinelli, A. Chapman, B. Dias, et al., Flood disaster mitigation: a real-world challenge problem for multi-agent unmanned surface vehicles, in *AAMAS '11 Workshop on Autonomous Robots and Multi-robot Systems*, 2011
9. G.S. Sukhatme, A. Dhariwal, B. Zhang, C. Oberg, B. Stauffer, D.A. Caron, The design and development of a wireless robotic networked aquatic microbial observing system. Environ. Eng. Sci. **24**(2), 205–215, Mar (2007)
10. T. Vaneck, J. Manley, C. Rodriguez, M. Schmid, Automated bathymetry using an autonomous surface craft. Navigation, J. Inst. Navig. **43**(4), Winter (1996–1997)

MIX
Papier aus verantwortungsvollen Quellen
Paper from responsible sources
FSC® C105338

If you have any concerns about our products,
you can contact us on
ProductSafety@springernature.com

In case Publisher is established outside the EU,
the EU authorized representative is:
Springer Nature Customer Service Center GmbH
Europaplatz 3, 69115 Heidelberg, Germany

Printed by Libri Plureos GmbH
in Hamburg, Germany